Energy and Water Conference 2013

Integrated Solutions for Advancing Technology and Management

Nashville, Tennessee, USA
6 – 9 May 2013

ISBN: 978-1-5108-7148-9

Energy and Water Conference 2013

Integrated Solutions for Advancing Technology and Management

Nashville, Tennessee, USA
6 – 9 May 2013

Water Environment Federation
the water quality people™

Printed from e-media with permission by:

Curran Associates, Inc.
57 Morehouse Lane
Red Hook, NY 12571

Some format issues inherent in the e-media version may also appear in this print version.

Printed by Curran Associates, Inc. (2018)

For permission requests, please contact Water Environment Federation (WEF) at the address below.

Water Environment Federation (WEF)
601 Wythe Street
Alexandria, Virginia 22314
USA

Phone: 1-800-666-0206
Fax: 1-703-684-1545

csc@wef.org

Additional copies of this publication are available from:

Curran Associates, Inc.
57 Morehouse Lane
Red Hook, NY 12571 USA
Phone: 845-758-0400
Fax: 845-758-2633
Email: curran@proceedings.com
Web: www.proceedings.com

Water Environment Federation
the water quality people™

Energy and Water 2013:
Integrated Solutions for Advancing Technology and Management

This conference is organized by the Water Environment Federation in cooperation with the Kentucky-Tennessee Water Environment Association, Water Environment Research Foundation, Alliance to Save Energy, American Council for an Energy-Efficient Economy, Imagine H2O, and the Alliance for Water Efficiency.

WORKSHOP A

Workshop A: Energy Services Performance Contracting: Innovative Delivery that GUARANTEES Energy and Cost Savings!

Monday, May 6, 2013
8:30am – 12:00pm

Room: 103/104
0.3CEUs

8:30am	**Opening Remarks** Peter Thomson, Black & Veatch
8:40am	**Introduction to Performance Contracting** Peter Thomson, Black & Veatch
9:00am	**First Breakout Session – Identify Challenges to Implementing Performance Contracting**
9:20am	**Report out from Breakout**
9:40am	**Helping a Colorado Town Save 35% on Well Pumping Costs** Justin Rundle, Honeywell Building Solutions
10:00am	**Break**
10:30am	**How a Texas Community Leveraged Guaranteed Utility Savings to Pay for Water and Wastewater Infrastructure Improvements**
10:50am	**Challenging "Business As Usual" – FWSA Looks to Energy Savings Performance Contracting** Dennis Clough, Energy Systems Group
11:10am	**Second Breakout Session – Identify Solutions to Challenges from Breakout Session 1** Jordan Lerner, Energy Solutions, Schneider Electric
11:30am	**Report out from Breakout**
11:50am	**Closing Remarks** Peter Thomson, Black & Veatch
12:00pm	**Adjourn**

WORKSHOP B

Workshop B: "Drivers Ed" for the Energy Roadmap

Monday, May 6, 2013
8:30am – 12:00pm

Room: 108/109
0.3 CEUs

8:30am	**Welcome, Workshop Format, Background Roadmap** Alicia Chakrabarti, EBMUD
8:50am	**Strategic Management / Organizational Culture** Ed McCormick, EBMUD
9:10am	**Innovating for the Future** Bill Toffey, Effluential Synergies
9:25am	**Demand Side Reduction / Energy Generation** Dave Reardon, HDR
10:00am	**Break**
10:25am	**Communications and Outreach** Jason Turgeon, USEPA
10:40am	**Instructions for Teaching Case and Break Out** Barry Liner, Water Environment Federation
10:45am	**Group Address Case Study Questions** All Speakers and Facilitators
11:40am	**Report Out on Findings** All Speakers
12:00pm	**Adjourn**

OPENING GENERAL SESSION

Monday, May 6, 2013
1:30pm – 5:00pm

Room: 108/109
3.0 PDHs

1:30 pm	**Welcome and Opening Remarks** Cindy Wallis-Lage, Black & Veatch, *Conference Chair*
1:35pm	**WEF and KY/TN WEA Introductions** Rick Warner, Washoe County, NV Wes Sydnor, Louisville MSD, KY
1:45 pm	**Robert Balzar** Vice President, Demand Side Management, Tennessee Valley Authority
2:00 pm	**Holmes Hummel** Senior Advisor, Office of the Undersecretary for Energy
2:30 pm	**Cynthia Truelove** Director and Visiting Scholar, Stanford University
3:00 pm	**Networking Break**
3:45 pm	**Panel Discussion: Energy Management Challenges and Opportunities for Wastewater Utilities** Steve Rogowski, Metro Water Reclamation District, CO Rob Taylor, Washington Suburban Sanitary Commission George Martin, Greenwood Metropolitan District, SC Moderator: Charles Bott, Hampton Roads Sanitation District
5:00 pm	**Adjourn**

Networking Reception – Center/East Exhibit Hall
Immediately following the conclusion of the Opening General Session, please join us for the Networking Reception held in Center/East Exhibit Hall from 5:00pm – 6:30pm

TECHNICAL PROGRAM

Session EW 1: Treatment

Tuesday, May 7 **Room: 103/104**
8:30am – 11:45am **2.5 PDHs**

Moderator: Jeanette Brown, Manhattan College
Peter Cavagnaro, Johnson Controls

EW1A 8:30am **Energy Evaluation of Combination Processes for Secondary Effluent Treatment** 1
Hong Y. Wang, Dongbum Im, Aki Ito, Naoya Wada, Norihide Nakada, Research Center for Environmental Quality Management, Kyoto University; Michiko Aoki, Yasuhiro Kato, Metawater Co., Ltd.; Hiroaki Tanaka, Research Center for Environmental Quality Management, Kyoto University

EW1B 9:00am **Nexus of Energy and Virus Reduction of UF Membrane Filtration Process with Coagulation/UV Irradiation for Wastewater Reclamation** 5
Suntae Lee, Naoyuki Yamashita, Masaru Ihara, Hiroaki Tanaka, Kyoto University; Kentaro Kobayashi, Hiroaki Tanaka, Hiroo Takabatake, Yuji Tanaka, Toray Industries

EW1C 9:30am **Overview and Challenges of Anaerobic Wastewater Treatment: Is there a Place for It in Our Future?** 14
Jose A. Jimenez, Brown And Caldwell; Charles Bott, HRSD

10:00am **Networking Break**

EW1D 10:45am **Achieving Positive Net Energy While Maximizing Use of Existing Assets: The Ejby Molle WWTP Story** 30
Julian Sandino, CH2M HILL; Per H. Nielsen, VandCenter Syd (VCS) Denmark; Dru Whitlock, Tom Johnson, CH2M HILL; Mads Leth, VandCenter Syd (VCS) Denmark; Thomas Jensen, Michael Nielsen, Ramboll Denmark

EW1E 11:15am **NEW (Nutrient, Energy, Water) Sustainability Principles Drive Technology Selection in Advanced Water Reclamation Application** 41
Randall Boe, Srinivas Jalla, Chuck Crandall, Julian Sandino, Jeremy Kraemer, CH2M HILL; Todd Cleaver, Tyler Richards, Holly Elmendorf, Gwinnett County, GA Department of Water Resources

11:45am **Adjourn**

TECHNICAL PROGRAM

Session EW 2: Alternative Energy

Tuesday, May 7
8:30am – 11:45am

Room: 108/109
2.5 PDHs

Moderators: Priscilla Bloomfield, CH2M HILL
Robert Lonergan, Tetra Tech

EW2A 8:30am **High Efficiency Bio-Diesel Production from a Municipal FOG Facility** 66
Michael D. Curtis, RPM Sustainable Technologies; J. Lanzoni, Siemens; R. Parnas, RPM Sustainable Technologies

EW2B 9:00am **Sustainable Energy Management Systems for Wastewater Treatment Facilities** 73
James C. McCaughey, Narragansett Bay Commission

EW2C 9:30am **eVortex: A Novel Energy Harvesting Mechanism for Waste and Storm Water Conveyance Systems** 89
Mungtra Chuslip, Arun Prakash Jaganathan, Erez Allouche, Louisiana Tech University

10:00am **Networking Break**

EW2D 10:45am **Use of Alternative Water Sources for a Trigeneration System: A Case Study** 103
Priska Grace, Dwayne Myers, Enrique Lopez Calva, Shirley Zhang, Priyanka Mehta, CDM Smith

EW2E 11:15am **Demonstrating the Business Case for Biogas-fueled Stationary Fuel Cell Power Systems at Wastewater Treatment Facilities** 105
Gregory D. Moreland, SRA International; Peter Devlin, U.S. Department of Energy

11:45am **Adjourn**

TECHNICAL PROGRAM

Session EW 3: Greenhouse Gases

Tuesday, May 7
1:30pm – 4:45pm

Room: 103/104
2.5 PDHs

Moderator: Robert Lonergan, Tetra Tech
Priscilla Bloomfield, CH2M HILL

EW3A 1:30pm **Carbon Free Water - A Successful Case Study in Sonoma County, California** 118
Cordel Stillman, Amy Bolten, Sonoma County Water Agency

EW3B 2:00pm **Estimating Carbon Emissions in Water Supply Portfolio Evaluations Including Recycled Water and Greywater** 125
Enrique J. Lopez Calva, Shirley Zhang, Dwayne R. Myers, Priska Grace, P. Mehta, CDM Smith

EW3C 2:30pm **Climate Ready Water Utilities: Helping the Water Sector Prepare for and Adapt to a Climate Change** 127
Laura Dubin, U.S. Environmental Protection Agency

3:00pm **Networking Break**

EW3D 3:45pm **Quantifying Fugitive CH_4 Emissions from Facultative Sludge Lagoons** 131
Anup Shah, Brown and Caldwell; Lauren Fillmore, Water Environment Research Foundation; A. Detweiler, NASA-AMES Research Center; John Willis, Brown and Caldwell; Brad Bebout, Adrianne Frisbee, NASA-AMES Research Center; Cheryl Kelley, University of Missouri; Jeff Chanton, Florida State University; Leslie Bebout, NASA-AMES Research Center

EW3E 4:15pm **Peak Amplification of Energy Demand, Energy Cost, and Carbon Emission in Water Reclamation and Purification Processes During Diurnal Cycles** 143
Reza Sobhani, University of California, Irvine; Lory Larson, Southern California Edison; Diego Rosso, University of California, Irvine

4:45pm **Adjourn**

TECHNICAL PROGRAM

Session EW 4: Lessons Learned

Tuesday, May 7 **Room: 108/109**
1:30pm – 4:45pm **2.5 PDHs**

Moderator: John Willis, Brown and Caldwell
Peter Cavagnaro, Johnson Controls

EW4A 1:30pm **Energy Consumption Trends in BNR Mixing Technologies** 153
Lamya Badr, Joe Rohrbacher, Patricia Stiegel, Hazen and Sawyer; Don Howard, Lori Cooper, City of Greensboro

EW4B 2:00pm **Making Energy Efficiency Happen - Experience and Lessons Learned in the EPA Region 4 - TN Department of Environment Conservation Energy Management Initiative** 167
Robert J. Freeman, US EPA Region 4 Atlanta; Jennifer Dodd, Tennessee Department of Environment & Conservation

EW4C 2:30pm **Energy Efficient, Sustainable Improvements and Technologies Implemented at Wastewater Treatment Plants** 180
Brian Sibiga, Wendel; Frank Nerone, Niagara County Sewer District No. 1

3:00pm **Networking Break**

EW4D 3:45pm **Minimizing Energy Consumption in Decentralized Wastewater Collection and Treatment Systems** 194
Kartiki S. Naik, Michael K. Stenstrom, University of California, Los Angeles

EW4E 4:15pm **Ten Years and Counting in the Wastewater Process Energy Efficiency Industry: Northern California's CalPOP Program** 207
Edward C. Myers, Lescure Engineers; John Bidwell, Quantum Energy Services and Technologies (QuEST) Inc.

4:45pm **Adjourn**

TECHNICAL PROGRAM

Session EW 5: Water Quantity

Wednesday, May 8
8:30am - 10:00am

Room: 103/104
1.5 PDHs

Moderator: Jeanette Brown, Manhattan College

EW5A 8:30am **The Water Implications of Generating Electricity: Water Consumption Across the United States Based on Different Electricity Pathways through 2050** 221
Jordan Macknick, National Renewable Energy Laboratory; Sandra Sattler, Union of Concerned Scientists; Kristen Averyt, Cooperative Institute for Research in Environmental Sciences; Steve Clemmer, John Rogers, Union of Concerned Scientsts

EW5B 9:00am **Water and Energy Utilities: Improving Collaboration** 233
Cindy Dyballa, Dyballa Consulting

EW5C 9:30am **The Importance of Water for Power Generation** 247
Michael C. Preston, Black & Veatch

10:00am **Adjourn**

Power Day
Note: This session is part of the Power Day Track geared specifically toward electric and gas utilities and their staff.

TECHNICAL PROGRAM

Session EW 6: Control Strategies

Wednesday, May 8
8:30am - 11:45am

Room: 108/109
2.5 PDHs

Moderator: Louann Unger, USEPA

EW6A 8:30am **Watergy in the U.S.: Energy Management, Policy Mechanisms and Technologies to Optimize Energy Use in Pumping and Aeration Systems** 264
Bruce Lung, Alliance to Save Energy; Lisa Tryson, John Masters, Danfoss; Ryan Graf, Policy Navigation Group;

EW6B 9:00am **Energy Savings from Advanced Process Control: Two Case Studies** 273
Steven M. Kestel, Matthew Gray, Weixin Shi, BioChem Technology; Junquan Yan, Jieyuan Draining Co.

EW6C 9:30am **Dynamic Energy Footprint Monitoring and Reduction in Energy Demand from Wastewater Aeration Systems via Full-Scale Deployment of Real-Time Off-Gas Analyzers** 291
Matthew K. Jeung, University of California, Irvine; Lory E. Larson, Southern California Edison; Michael K. Stenstrom, UCLA; Diego Rosso, University of California, Irvine

10:00am **Networking Break**

EW6D 10:45am **Advanced Controls Maximizes Energy Efficiency of a Large Effluent Pump Station** 300
Curt Wendt, CDM Smith Inc

EW6E 11:15am **Advanced Water Reclamation Minimizes the Energy Footprint: The Riverfront Water Reclamation Facility** 307
Brooks W. Newbry, Water Technology Group; Kenneth Knickerbocker, Wood, Patel and Assoc. Consulting Engineers; Dan Lueder, City of Cottonwood, Arizona

11:45am **Adjourn**

TECHNICAL PROGRAM

Session EW 7: Water Quality

Wednesday, May 8
10:45am - 11:45am

Room: 103/104
1 PDH

Moderator: Cindy Wallis-Lage, Black & Veatch

EW7A 10:45am **Reclaimed Water Quality Requirements for Electric Utilities** 322
Don Vandertulip, CDM Smith

EW7B 11:15am **Maintaining NPDES Permit Compliance at Power Plants While Using Municipal Treatment Plant Effluent with Variable Ammonia and Nitrate Levels.** 335
Christopher Varnon, Randi Kim, Jospeh Viciere, Nathan Schmaus, Tina Petersen, CDM Smith; Jaron Bergin, Calpine Corp

11:45am **Adjourn**

Power Day
Note: This session is part of the Power Day Track geared specifically toward electric and gas utilities and their staff.

TECHNICAL PROGRAM

Session EW 8: Energy Savings

Wednesday, May 8
1:30pm - 3:00pm

Room: 103/104
1.5 PDHs

Moderators: Tom Schwartz, Woodard and Curran
Priscilla Bloomfield, CH2M HILL

EW8A 1:30pm	**Get It Done! - Using Performance Contracting to Implement Energy-Related Improvements** Richard Atoulikian, HDR; Clark Wiedetz, Siemens; Thomas Kochaba, Kim Kennedy, Robert Huffmyer, HDR	363
EW8B 2:00pm	**Demand Management Strategies Resulting in Zero and Low Cost Energy Saving Opportunities for Water and Wastewater Treatment Facilities** Bryan R. Lisk, Hazen and Sawyer	377
EW8C 2:30pm	**Aeration Efficiency in Oxidation Ditches** John Lindam, Xylem Water Solutions AB	387
3:00am	**Adjourn**	

Power Day
Note: This session is part of the Power Day Track geared specifically toward electric and gas utilities and their staff.

TECHNICAL PROGRAM

Session EW 9: Biogas

Wednesday, May 8 | **Room: 108/109**
1:30pm - 3:00pm | **1.5 PDHs**

Moderators: Jeanette Brown, Manhattan College
Louann Unger, USEPA

EW9A 1:30pm **Guide to Bioaugmentation for Systems Experiencing Chemical Inhibition using an Operating Envelope** 401
Robin Schroeder, Martin Maxwell, and Dong Hyun, Environmental Business Specialists, LLC

EW9B 2:00pm **CHP System Implementation and Operations Optimization at F. Wayne Hill Water Resources Center** 410
Richard Schoeck, Gwinnett County Department of Water Resources; Srinivas Jalla, CH2M HILL; Richard Porter, Gwinnett County Department of Water Resources

EW9C 2:30pm **Energy-Recovery from Solids in St. Petersburg, FL: Biogas to Vehicle Fuel and CHP** N/A
John Willis, Brown and Caldwell; Steve Leavitt, Steven Marshall, St. Petersburg; Todd Bosso, Brown and Caldwell

3:00pm **Adjourn**

TECHNICAL PROGRAM

Session EW 10: Global Perspectives

Wednesday, May 8
3:45pm - 5:15pm

Room: 103/104
1.5 PDHs

Moderator: Bob Forbes, CH2M HILL

EW10A **3:45pm** **Life Cycle Water use for Electricity Generation: Implications of the Distribution of Collected Estimates** 425
James Meldrum, University of Colorado; Jordan Macknick, Garvin Heath, Syndi Nettles-Anderson, National Renewable Energy Laboratory

EW10B **4:15pm** **Water and Energy Efficiency through Steam Systems** 434
Angela Ganley, Lesley Smith, Audra Liubinas, Sophea Eng, City West Water

EW10C **4:45pm** **Creating New Revenue with Existing Processes: Evaluating Grid Balance with Demand-Side Loads at Wastewater Treatment Plants Using Dynamic Modeling** 446
Daryl Letto, ENBALA Power Networks; Spencer Snowling, Hydromantis Environmental Software Solutions; Raoul George, Christopher Morrissey, ENBALA Power Networks; Hank Andres, Hydromantis Environmental Software Solutions; Conrad Fox, ENBALA Power Networks

5:15pm **Adjourn**

Power Day
Note: This session is part of the Power Day Track geared specifically toward electric and gas utilities and their staff.

Session EW 11: Exemplary Programs That Save Both Energy and Water

Wednesday, May 8
3:45pm – 5:15pm

Room: 108/109
1.5 PDHs

Moderator: Sara Hayes, American Council for an Energy-Efficient Economy

This session addresses a topic at the heart of the conference: how to design programs that achieve both water and energy savings. The case studies to be presented and the speakers to be featured will provide information on best practices in designing and implementing these programs as well as lessons learned through their experiences.

3:45pm **Exemplary Programs that Save Both Energy and Water**
Mary Ann Dickinson, Alliance for Water Efficiency

4:00pm **Massachusetts Water Resource Authority (MWRA) Long Term Sustainability Program**
Stephen Estes-Smargiassi, Massachusetts Water Resources Authority (MWRA)

4:10pm **Southern California Electric Leak Pilot Detection Program**
Gene Rodrigues, Southern California Electric

4:20pm **City of Boulder Energy Performance Contracting Program**
Joe Castro, City of Boulder, Colorado

4:30pm **United Technologies Corporation (UTC) 2015 Sustainability Goals Program**
Sean West, United Technologies Corporation (UTC)

4:40pm **Questions and Discussion**

5:15pm **Adjourn**

TECHNICAL PROGRAM

Session EW 12: Research

Thursday, May 9
8:30am - 10:00am

Room: 103/104
1.5 PDHs

Moderator: Stephen Tarallo, Black & Veatch

EW12A **8:30am** **Looking at the Future: How Metropolitan Sewer District of Greater Cincinnati (MSDGC) is Turning Waste into Resources and Energy** 469
Ting Lu, Biju George, MSD of Greater Cincinnati; Joo-Youp Lee, Mingming Lu, University of Cincinnati

EW12B **9:00am** **Energy Consumption of MBR for Municipal Wastewater Treatment Current Situation and Potential** 479
Pierre Cote, COTE Membrane Separation Ltd; Thor Young, Sebastian Smoot, GHD; Jeff Peeters, GE Water & Process Technologies

EW12C **9:30am** **Environment Enhancing Energy Innovative Adsorptive-Biological Process to Recovery/Recycle Carbon and Nitrogen from Aqueous Product of Hydrothermal Liquefaction to Maximize Bioenergy Production from Wastewater** 493
Yan Zhou, Lance Schideman, Yuanhui Zhang, X. Zhao, University of Illinois

10:00am **Adjourn**

TECHNICAL PROGRAM

Session EW 13: Water MSW & Integrated Planning

Thursday, May 9 **8:30am - 11:45am**	**Room: 108/109** **3.0 PDHs**

Moderator: Jason Turgeon, USEPA

EW13A	**8:30am**	**Sustainable Solutions for the 21st Century - Integration of Water Treatment Systems with Energy Derived from Municipal Wastes** Paul Hauck, Diane Kemp, Valerie Going, CDM Smith	501
EW13B	**9:00am**	**Integrating Major Components of a New Landfill Gas System - Coordination of Multiple Contracts for Gas Treatment, Pipeline Transmission and Turbine Energy Generation** Ralph Schroedel, Brown & Caldwell; Michael Martin, Cary Solberg, William Krill, Milwaukee Metropolitan Sewerage District	519
EW13C	**9:30am**	**Integrated Net Zero for Energy, Water and Waste at Army Installations: Establishing Priorities in a Fiscally Constrained Context** Elizabeth Keysar, Concurrent Technologies Corporation (CTC)	534
	10:00am	**Networking Break**	
EW13D	**10:15am**	**Santa Rosa Case Study - Water, Electric, and Gas Utility Collaboration** Adam Ross, Brown and Caldwell; David Guhin, M. Prinz, City of Santa Rosa;	536
EW13E	**10:45am**	**Integrating Liquid and Solid Waste Planning to Make Green Infrastructure Pay** Stephen J. Salter, Farallon Consultants Limited	544
EW13F	**11:15am**	**Embodied Energy of Lost Water: Evaluating the Energy Efficiency of Infrastructure Investments** Craig P. Aubuchon, Analysis Group, Inc.; Alan Roberson, AWWA	573
	11:45am	**Adjourn**	

TECHNICAL PROGRAM

Session EW 14: Net Zero

Thursday, May 9
10:15am - 11:45am

Room: 103/104
1.5 PDHs

Moderator: Alicia Chakrabarti, East Bay Municipal Utility District
Peter Cavagnaro, Johnson Controls

EW14A **10:15am** **The Most Energy Efficient Pump is the One That's Not Running: A Case Study of Energy Savings Resulting from an I/I Reduction Program** 585
Kevin Colvett, CH2M Hill; Brent Freeman, Metro Water Services; Todd Spangler, City of Brentwood, TN

EW14B **10:45am** **Small Wastwater Treatment Facilities can be Energy Producers** 590
Joseph C. Cantwell, Science Applications Intl Corp

EW14C **11:15am** **The Treatment Plant of the Future Integrated CO_2 Capture for Enhanced Plant Growth** 599
James W. Schettler, Nancy E. Andrews, Brown and Caldwell

11:45am **Adjourn**

ADVANCED VIRTUAL CONFERENCE

The Advanced Virtual Conferences were held on January 23rd and March 20th as two recoreded webcasts that were used as an example of the content that will be presented onsite at the Energy and Water 2013 Conference.

Panelists from the Virtual Conferences included:

Low Energy Process Control - January 23, 2013

- Leiv Rieger, inCTRL Solutions Inc.
- Jose Jimenez, Brown and Caldwell
- Andy Shaw, Black & Veatch (Moderator)

Energy-Water Nexus Highlights - March 20, 2013

- Radenka Maric, University of Connecticut
- Alex West, BioteQ Environmental Technologies
- Jeanette Brown, University of Connecticut (Moderator)

To hear these live recored presentations, please visit www1.gotomeeting.com/register/547792017 and www1.gotomeeting.com/register/452252785. To listen to other live recored webcasts, please visit www.wef.org/webcasts/.

ENERGY AND WATER SESSIONS AT-A-GLANCE

Session #	Title	Credits	Room
MONDAY			
Workshop A	Energy Services Performance Contracting: Innovative Delivery that Guarantees Energy and Cost Savings!	0.3 CEUs	103/104
Workshop B	"Drivers Ed" for the Energy Roadmap	0.3 CEUs	108/109
	Opening General Session	3.0 PDHs	108/109
TUESDAY			
1	Treatment	2.5 PDHs	103/104
2	Alternative Energy	2.5 PDHs	108/109
3	Greenhouse Gases	2.5 PDHs	103/104
4	Lessons Learned	2.5 PDHs	108/109
WEDNESDAY			
5	Water Quantity *	1.5 PDHs	103/104
6	Control Strategies	2.5 PDHs	108/109
7	Water Quality*	1 PDH	103/104
8	Energy Savings*	1.5 PDHs	103/104
9	Biogas	1.5 PDHs	108/109
10	Global Perspectives*	1.5 PDHs	103/104
11	Exemplary Programs That Save Both Energy and Water	1.5 PDHs	108/109
THURSDAY			
12	Research	1.5 PDHs	103/104
13	Water MSW & Integrated Planning	3.0 PDHs	108/109
14	Net Zero	1.5 PDHs	103/104

** Power Day Track*

EXHIBITION GENERAL INFORMATION

Located in the East and Central Hall of the Nashville Convention Center, the Exhibition hall hosts morning and afternoon refreshment breaks, luncheons, and a networking reception.

The exhibition offers networking opportunities with companies displaying the latest product and service solutions directly related to the conference's technical program.

Exhibit Hall Schedule

Monday, May 6 — **5:00 pm – 6:30 pm**
5:00 pm – 6:30 pm...Networking Reception

Tuesday, May 7 — **10:00 am – 4:45 pm**
10:00 am –10:45 am ....Coffee Break
12:00 pm – 1:30 pm ....Luncheon
3:00 pm – 3:45 pm ....Refreshment Break

Wednesday, May 8 — **10:00 am – 3:45 pm**
10:30 am –10:45 am ....Coffee Break
12:00 pm – 1:30 pm ....Luncheon
3:00 pm – 3:45 pm ....Refreshment Break

Children under 18 entering the Exhibition must be accompanied by a parent or guardian at all times. The parent or guardian assumes all risk and responsibility for the child's safety.

Photographing or video recording in the Exhibition is only allowed with exhibitor's permission.

EXHIBITOR DIRECTORY

As of April 18, 2013

Alfa Laval / Ashbrook Simon-Hartley .... Booth 303
Phone: (800) 362-9041
11600 E Hardy Road
Houston, TX 77093

Expanded solutions for a cleaner environment. Ashbrook Simon-Hartley is now a part of Alfa Laval. Together, we deliver a market-leading portfolio of products and solutions for municipal and industrial water and wastewater applications.

Alka Tech, Inc. .... Booth 522
www.alka-tech.com
Phone: (515) 961-7407
1700 N. 14th Street
Indianola, IA 50125

EXHIBITOR DIRECTORY

Alpine Technology, Inc. .... **Booth 707**
www.AlpineTechnology.com
Phone: (512) 328-5829
P.O. Box 160878
Austin, TX 78716

Dewatering and Bioenergy automation. Energy Recovery. Emission and Mercury Control. ALSTOM Boilers and Heat Exchangers. CPPE Mercury Fixed Bed Systems. EnviroCare Scrubbers and WESPs. Elliott Steam Turbine Generators. Metso on line cake sensors.

Anaergia Inc. .... **Booth 302**
www.anaergia.com
Phone: (905) 766-3333
5338 John Lucas Drive
Burlington, ON L7L 6A5
Canada

Andritz Separation .... **Booth 418**
www.andritz.com/separation.htm
Phone: (817) 465-5611
1010 Commercial Blvd S
Arlington, TX 76001

ANDRITZ Separation, a global market leader, specializes in the manufacturer of solids/liquid separation technologies such as screening, dewatering, thickening, filtering and drying equipment.

BASF - The Chemical Company .... **Booth 602**
Phone: (757) 538-5296
100 Park Avenue
Florham Park, NJ 07932

BASF provides technical expertise and products to help our customers in the industrial and municipal water sectors meet their commitments and ensure reliable process operations.

BCR Environmental, Inc. .... **Booth 402**
www.bcrenv.com
Phone: (904) 819-9170
3740 St. Johns Bluff Road S
Suite 21
Jacksonville, FL 32224

BCR Environmental is a U.S. based clean technology company focused on wastewater treatment processes that convert organic waste streams into safe, valuable, and marketable end products.

EXHIBITOR DIRECTORY

BDP INDUSTRIES
Sales: 518-527-5417 Factory: 518-695-6851
Email: Kelly@BDPIndustries.com

BDP Industries, Inc. .... **Booth 600**
www.BDPIndustries.com
Phone: (518) 695-6851
354 Route 29
Greenwich, NY 84014

Manufacturer of filtration equipment: belt press, gravity belt thickener, rotary drum concentrator, screw press, pressure filter, mobile dewatering trailer, polymer make-up systems, lime stabilization systems, and equipment restorations.

Berlie Technologies Inc .... **Booth 110**
Phone: (450) 444-0566
1245 Industrielle Street
La Prairie, Quebec J5R 2E4
Canada

Berlie Technologies: expertise in water and sludge treatment (biosolids/organic residuals; specialized in turnkey projects; provides services in process and mechanical engineering in agitation, transport, thermal exchange for fluids and solids.

Bio-Nomic Services, Inc. .... **Booth 700**
www.bio-nomic.com
Phone: (800) 782-6798
516 Roundtree Road
Charlotte, NC 28217

BioChem Technology Inc. .... **Booth 619**
Phone: (610) 768-9360 x228
3620 Horizon Drive
Suite 200
King of Prussia, PA 19406

Manufacturer of ammonia/nitrate analyzers, aeration control and automatic process control of secondary wastewater treatment for optimal nutrient removal, maximum energy savings and process stabilization.

Boeger, LLC .... **Booth 404**
www.boerger.com
Phone: (612) 435-7300
2860 Water Tower Place
Chanhassen, MN 55317

Boerger specializes in reliable and cost effective Rotary Lobe Pumps and Macerating Technology for the conveyance of low to high viscous and abrasive materials.

EXHIBITOR DIRECTORY

Brown Bear Corporation .... **Booth 502**
Phone: (641) 322-4220
2248 Avenue of Industries
Corning IA 50841

Manufacturer of mobile equipment used for dewatering and composting of sludge. Also, off road vehicles for land application of sludge.

CAMBI Inc. .... **Booth 611**
Phone: (205) 275-5201
1903 Corona Avenue
Jasper, AL 35501

Cambi's patented Thermal Hydrolysis Process treats municipal and industrial sludge and biowaste prior to anaerobic digestion. It has been applied in 21 plants worldwide.

CB&I .... **Booth 113**
www.cbi.com
Phone: (815) 439-6000
14105 S Route 59
Plainfield, IL 60544

CB&I furnishes anaerobic digestion systems including Egg Shaped Digester, 2 Stage Class A digestion systems and digester gas storage, all on a constructed basis.

Centrisys Corporation .... **Booth 401**
www.centrisys.us
Phone: (262) 654-6006
9586 58th Place
Kenosha, WI 53144

USA Manufacturer of dewatering and thickening centrifuges for municipal/industrial wastewater. Focused centrifuge innovation includes 2/3 phase technologies; global service/repair cabilities; new THK thickening technology

Crowder Construction Company.. **Booth 123**
www.crowdercc.com
Phone: (919) 367-2000
1111 Burma Drive
Apex, NC 27539

Custom Conveyor Corporation .... **Booth 618**
Phone: (703) 367-6200
20550 Commerce Boulevard
Rogers, MN 55374

Design and manufacture of material handling and storage equipment for municipal wastewater treatment facilities.

EXHIBITOR DIRECTORY

Danfoss .... **Booth 620**
www.danfossdrives.com
Phone: (414) 355-8800
8800 West Bradley Road
Milwaukee, WI 53224

DRYCAKE .... **Booth 120**
www.drycake.com
Phone: (819) 470-8629
225 Boulevard Charest Est.
Suite 110
Quebec, QC G1K3G9
Canada

EMA Resources, Inc. .... **Booth 304**
www.emaresourcesinc.com
Phone: (336) 751-1441
755 Yadkinville Road
Mocksville, NC 27028

Residuals management company providing turnkey services including land application, mobile dewatering, dredging, lagoon closure, digester cleaning, alkaline stabilization, permitting, sampling & analysis, monitoring & reporting, DBO facilities.

Environmental Dynamics International .... **Booth 111**
www.wastewater.com
Phone: (573) 474-9456
5601 Paris Road
Columbia, MO 65202

FKC Co., Ltd. .... **Booth 701**
www.fkcscrewpress.com
Phone: (360) 452-9472
2708 West 18th Street
Port Angeles, WA 98363

Flottweg Separation Technology, Inc. .... **Booth 601**
www.flottweg.com
Phone: (859) 448-2310
10700 Toebben Drive
Independence, KY 41051

Flottweg is an ISO 9001 and EU manufacturer of high quality thickening and dewatering centrifuges sold worldwide. Designs include energy efficient independent scroll and bowl VFD drives with the latest advancements in decanter centrifuge design.

EXHIBITOR DIRECTORY

Fluid Dynamics (Pump Solutions Group) .... **Booth 205**
www.psgdover.com
Phone: (215) 699-8700
295 DeKalb Pike
North Wales, PA 19454

Fluid Dynamics refined and patented the dynaBLEND™ polymer blending technology. Fluid Dynamics stands as one of the premier polymer equipment companies in the world and is staffed with some of the leading polymer equipment engineers in the US. Our technical staff is specifically dedicated to the development and application of polymer equipment.

Fournier Industries Inc. .... **Booth 705**
www.fournierindustries.com
Phone: (418) 423-4341
3787 Frontenac Boulevard West
Thetford Mines, QC G6H 2B5
Canada

Garney Construction .... **Booth 703**
www.garney.com
Phone: (615) 350-7975
200 Crutchfield Avenue
Nashville, TN 37210

GEA Westfalia Separator .... **Booth 218**
www.wsus.com
Phone: (201) 767-3900
100 Fairway Court
Northvale, NJ 07647

GEA Westfalia Separator offers centrifugal technology for the treatment of water and wastewater, including decanters for dewatering and thickening sludge volumes, and treatment of drinking water.

GeoEnvironment Technologies LLC .... **Booth 505**
www.geoenvironment-technologies.com
Phone: (626) 305-0400
103 E Lemon Avenue
Suite 212
Monrovia, CA 91016

GORE® Cover System .... **Booth 203**
www.gorecover.com
Phone: (610) 733-4078
105 Vieve's Way
Elkton, MD 21921

EXHIBITOR DIRECTORY

Hitachi Zosen Inova U.S.A. LLC .... Booth 322
Phone: (678) 987-2506
3740 Davinci Court
Suite 250
Norcross, GA 30092

HZI specializes in the design and supply of equipment for the thermal treatment of biosolids, wastes, and other hard to handle fuels; additionally, HZI provides all the associated air pollution control technologies to meet new EPA standards.

HML .... Booth 222
www.hml.com
Phone: (765) 288-1124
912 W. McGalliard Road
Muncie, IN 47303-1702

HML, established 1976, provides full service microbiological and chemical analysis of biosolids, pathogen reduction studies and rapid bacterial identification by DNA (PCR) or immunoassay.

Huber Technology, Inc. .... Booth 411
www.huberforum.net/residuals
Phone: (704) 990-2055
9735 North Cross Center Court
Suite A
Huntersville, NC 28079

Experts in liquid-solid separation technologies with focus on screening, thickening, dewatering, and drying technologies. We offer the complete chain of sludge & biosolids handling processes.

Hydro International .... Booth 219
www.hydro-int.com
Phone: (866) 615-8130
2925 NW Aloclek Drive
Suite 140
Hillsboro, OR 97124

Hydro International supplies environmentally sustainable products and innovative solutions that control and treat stormwater, wastewater and combined sewer overflows, using advanced vortex and complementary technologies.

EXHIBITOR DIRECTORY

Infilco Degremont .... **Booth 311**
Phone: (804) 756-8423
8007 Discovery Drive
Richmond, VA 23229

Infilco Degremont (SUEZ Environnement) provides high performance technologies to the municipal and industrial markets with a broad suite of trusted and reliable treatment solutions for drinking water, wastewater and sludge management.

JDV Equipment Corporation .... **Booth 520**
www.JDVEquipment.com
Phone: (973) 366-6556
1 Princeton Avenue
Dover NJ 07801

JDV Equipment Corporation is a leading manufacturer and provider of safe, environmentally friendly processing equipment and services for water treatment, wastewater treatment, industrial and agricultural applications.

Jim Myers & Sons, Inc. .... **Booth 301**
www.myersequipment.com
Phone: (704) 554-8397
P.O. Box 240038
Charlotte, NC 28224

Celebrating 50 years. Design, engineering and manufacturing of municipal wastewater and water treatment. Material handling and conveyance equipment, live bottom hoppers, slide gates and more.

Komline-Sanderson .... **Booth 504**
www.komline.com
Phone: (908) 234-1000
12 Holland Avenue
Peapack, NJ 07977

Pump. Thicken. Dewater. Dry. Since 1946 K-S has provided environmental solutions. Products include biosolids drying systems, belt filter presses, gravity belt thickeners, plunger pumps.

EXHIBITOR DIRECTORY

KRÜGER

Kruger Inc. .... **Booth 210**
www.krugerusa.com
Phone: (919) 677-8310
4001 Weston Parkway
Cary, NC 27513

Kruger supplies solutions and technologies for wastewater and drinking water treatment. Through years of extensive research and development, Kruger provides integrated solutions that optimize energy and operating costs.

Kusters Water .... **Booth 101**
www.kusterswater.com
Phone: (205) 987-8976
101 Zima Park Drive
Spartanburg, SC 29301

Kusters Water offers an extensive Biosolids product line, including: Hiller decanter centrifuges, ACAT screw presses, Hi-Tech DAF thickeners, WasteTech rotary drum thickeners, Fuchs ATAD, Watropur dryers and sludge screens from FSM.

LobePro Rotary Pumps .... **Booth 711**
ww.lobepro.com
Phone: (912) 466-0304
2610 Sidney Lanier Drive
Brunswick, GA 31525

LobePro manufactures 52 displacement rotary lobe pumps models to handle abrasives, self-prime, and pump thick viscous materials, sludge, mud and slurries.

LSC Environmental Products, LLC .... **Booth 221**
www.posishell.com
Phone: (800) 800-7671
2183 Pennsylvania Avenue
Apalachin, NY 13732

Lystek International Inc. .... **Booth 323**
Phone: (226) 444-0186
1425 Bishop Street North
Unit 16
Cambridge, ON N1R 6J9
Canada

MaxWest Environmental Systems, Inc. .... **Booth 718**
www.maxwestenergy.com
Phone: (407) 328-4992
114 W. 1st Street
Suite 220
Sanford, FL 32771

MaxWest Environmental Systems provides an onsite biosolids disposal solution, leading the wastewater treatment industry with a biogasification system that is economical, environmentally-friendly, and safe.

Merrell Bros., Inc. .... **Booth 610**
www.merrellbros.com
Phone: (574) 699-7782
8811 West 500 North
Kokomo, IN 46901

Merrell Bros., Inc. is one of the leading Biosolids Management Companies in the United States providing a complete range of residuals services. We stand on our reputation of professional, ethical and dependable service you can trust.

Metso .... **Booth 709**
www.metso.com
Phone: (678) 772-9584
2900 Courtyards Drive
Norcross, GA 30071

Manufacturer of solids measurement solutions from centrate to cake solids – if you can pump it we can measure it!

Mintek Resources, Inc. .... **Booth 720**
www.mintekresources.com
Phone: (937) 431-0218
2440 Dayton-Xenia Road
Suite D
Beavercreek, OH 45434

Providing Calciment® along with other innovative and cost effective materials solutions for production Class A and B biosolids, dewatering, pH adjustment and more.

NEFCO
Recycle, Renew, Restore

NEFCO .... **Booth 519**
www.nefcobiosolids.com
Phone: (617) 773-3131
500 Victory Road
North Quincy, MA 02356

A full-service biosolids management company with experience to design a complete biosolids process facility, the ability to tailor treatment process to suit your needs, and the financial backing necessary to fully privatize your project, if desired.

EXHIBITOR DIRECTORY

NETZSCH Pumps North America, LLC .... **Booth 419**
www.netzschusa.com
Phone: (610) 363-8010
119 Pickering Way
Exton, PA 19341

NETZSCH, the world market leader of PC Pumps, offers PC Pumps, Rotary Lobe Pumps and accessories, providing customized, sophisticated solutions for various applications worldwide.

Noxon North America Inc .... **Booth 200**
www.noxon.com
Phone: (416) 843-6500
4-470 North Rivermede
Concord, ON L4K 3R8
Canada

Noxon manufactures decanter centrifuges in up to 30 different configurations; the Lamella screw option is especially suitable for dilute sludges like MBR sludge.

OpenCEL .... **Booth 318**
www.opencel.com
Phone: (847) 835-7418
900 Circle 75 Parkway
Suite 1330
Atlanta, GA 30339

Opencel uses Focused Pulse Technology to achieve cell lysis reliably and affordably. In anaerobic digestion, it increases biogas 30-60% and reduces biosolids by 20-30%. In BNR, the treated sludge can be used as a carbon source for denitrification.

Ovivo .... **Booth 400**
www.ovivowater.com
Phone: (512) 834-6047
2404 Rutland Drive
Austin, TX 78758

Aerobic Digestion, Dewatering, Biological, MBR Systems, Anaerobic Digestion.

Parkson Corporation .... **Booth 202**
www.parkson.com
Phone: 1 (888) PARKSON
1401 W Cypress Creek Road
Suite 100
Fort Lauderdale, FL 33309

With 25,000 installations in its over 50 years of existence, Parkson Corporation is a leading provider of advanced solutions in water recycling and treatment and is committed to providing clean water for the world.

EXHIBITOR DIRECTORY

Penn Valley Pump Co., Inc. .... **Booth 319**
www.pennvalleypump.com
Phone: (215) 343-8750
998 Easton Road
Warrington, PA 18976

The Double Disc Pump effectively handles higher solids sludge material and feeds dewatering equipment at the Lowest Life Cycle Cost of any PD pump on the market.

PHG Energy .... **Booth 100**
www.PHGEnergy.com
Phone: (615) 471-9299
3048 Owen Drive
Antioch, TN 37013

Prime Solution, Inc. .... **Booth 715**
www.psirotary.com
Phone: (269) 694-6666
610 S. Platt Street
Otsego, MI 49078

Rotary Fan Press for sludge dewatering. Slow rotation, less than 1 RPM.

Process Wastewater Technologies, LLC .... **Booth 300**
www.pwtech.us
Phone: (410) 238-7977
9003 Yellow Brick Road
Suite S
Baltimore, MD 21237

Provider of the world's leading range of screening technologies for CSO/SSO treatment, unique biological wastewater treatment processes, and the Volute Dewatering Press sludge dewatering system.

proMtec .... **Booth 118**
www.promtec.de
Phone: (510) 420-1716
5927 Taft Avenue
Oakland, CA 94618

RDP Technologies .... **Booth 104**
www.rdptech.com
Phone: (610) 650-9900
2495 Boulevard Of The Generals
Norristown, PA 19403

RDP provides cost effective Class A & B biosolids stabilization systems producing products that can be used within the community for a holistic and sustainable biosolids management program.

EXHIBITOR DIRECTORY

Schwing Bioset, Inc. .... **Booth 604**
www.schwingbioset.com
Phone: (715) 247-3433
350 SMC Drive
Somerset, WI 54025

Screw press dewatering, class A bioset process and fluid bed drying, sliding frames, live bottoms, piston pumps, operations and Class A end use marketing.

seepex.com
all things flow

seepex Inc. .... **Booth 511**
www.seepex.com
Phone: (937) 864-7150
511 Speedway Drive
Enon, OH 45323

Progressive cavity pumps, run-dry protector, bridge-breaker cake pumps, 10,000 hour U-joint, grinders, Tri-Cam® geometry, MAP hose pump, MDI metering pump and new BTHE high solids pump.

SH+E Group .... **Booth 421**
www.she-group.com
Phone: (262) 764-3645
9535 58th Place
Kenosha, WI 53144

The SH+E GROUP ranks amongst the world leading suppliers in the field of municipal waste water treatment, industrial water treatment, waste to energy concepts, packaging and process technology. Based on its global presence, the SH+E GROUP has completed projects in more than 160 countries.

Siemens Industry .... **Booth 702**
www.siemens.com/water
Phone: (724) 772-0044
10 Technology Drive
Warrendale, PA 15086

Siemens combines the industry's broadest portfolio of technologies with technical expertise and proven installation experience to help consulting engineers develop best treatment options for municipalities.

Southeastern Tank, Inc. .... **Booth 423**
www.southeasterntank.com
Phone: (615) 466-5220
60 Vesta Road
Lebanon, TN 37090

SPIRAC USA INC .... **Booth 717**
www.spirac.com
Phone: (770) 683-0147
75 Jackson Street
Suite 300
Newnan, GA 30263

SPIRAC - World leader in biosolids and sludge handling systems. Receiving, conveying, storing and outloading.

Synagro .... **Booth 201**
www.synagro.com
Phone: (713) 369-1757
1800 Bering Drive
Houston, TX 77057

Founded in 1986, Synagro is the country's largest provider of comprehensive waste and wastewater treatment and conversion services. We lead the field in solutions that transform waste challenges into state-of-the-art, sustainable solutions.

Tech 3 Solutions, Inc. .... **Booth 713**
www.tech3solutions.net
Phone: (305) 666-1910
7301 SW 57th Court
Suite 400
South Miami, FL 33143

Terra Renewal .... **Booth 119**
Phone: (972) 996-7549
1100 E Campbell Road
Suite 220
Richardson, TX 75081

Terra Renewal serves municipal and industrial wastewater/water systems across the US. We provide beneficial use and turn-key management solutions for biosolids and organic residuals.

Therma-Flite Inc. .... **Booth 722**
www.Therma-Flite.com
Phone: (707) 747-5949
849 Jackson Street
Benicia, CA 94510

EXHIBITOR DIRECTORY

Thermal Process Systems .... **Booth 403**
www.thermalprocess.com
Phone: (219) 663-1034
627 East 110th Avenue
Crown Point, IN 46307

Thermal Process System's ThermAer(TM) Process provides aerobic digestion for excellent TS and VS reduction while producing "Class A" biosolids without the need for an external heat source.

Unison Solutions, Inc. .... **Booth 500**
www.unisonsolutions.com
Phone: (563) 585-0967
5451 Chavenelle Road
Dubuque, IA 52002

Unison Solutions, Inc. is an industry leader in biogas conditioning, BioCNG™ vehicle fuel systems and distributed generation.

US Composting Council .... **Booth 102**
Phone: (301) 897-2715
5400 Grosvenor Lane
Bethesda, MD 20814

Uzelac Industries Inc .... **Booth 623**
www.uzelacind.com
Phone: (414) 529-0240
6901 Industrial Loop
Greendale, WI 53129

Uzelac Industries is a design/build company providing Drying solutions, expertise and equipment to the Biosolids Industry.

Vaughan Company, Inc. .... **Booth 211**
www.chopperpumps.com
Phone: (360) 249-4042
364 Monte Elma Road
Montesano, WA 98563

Since 1960, Vaughan has been issued over 35 patents and has sold more than 25,000 pumps worldwide including the Rotamix Process Mixing System.

Velocity Dynamics, Inc. .... **Booth 621**
www.polymersolution.com
Phone: (303) 530-3298
543 South Pierce Avenue
Louisville, CO 80027

EXHIBITOR DIRECTORY

VitAg Corporation .... **Booth 605**
www.vitagcorp.com
Phone: (239) 398-6127
117 Allwin Lane
Beech Island, SC 29842

VitAg employs beneficial-use technology that economically converts biosolids into a Class A controlled-release high-nitrogen granular fertilizer that it sells into the specialty and commodity fertilizer marketplace.

Vogelsang .... **Booth 105**
Phone: (330) 296-3820 x232
P.O. Box 751
Ravenna, OH 44266

Vogelsang positive displacement, rotary lobe pumps and in-line grinders provide quality, technologically advanced, economical solutions to all your wastewater pumping needs.

Wallace & Tiernan ChemFeed .... **Booth 116**
www.wtchemfeed.com
Phone: (908) 305-0057
1901 West Garden Road
Vineland, NJ 08360

WeCare Organics, LLC .... **Booth 719**
www.wecarecorganics.com
Phone: (315) 689-1937
9293 Bonta Bridge Road
Jordan, NY 13080

Specializes in the beneficial reuse of recyclable organic residuals, project development, processing operations and final product marketing of biosolids-based soil amendments for commercial use.

WesTech Engineering .... **Booth 501**
www.westech-inc.com
Phone: (801) 265-1000
3665 S. West Temple
Salt Lake City, UT 84103

WesTech Engineering provides Digester Covers, both beam and truss type, DuoSphere™ GasHolders for energy production and Extreme Duty™ Sludge Mixers to optimize Digester tank performance.

EXHIBITOR DIRECTORY

World Water Works Inc. .... **Booth 518**
http://www.worldwaterworks.com
Phone: (405) 943-9000
P.O. Box 892050
Oklahoma City, OK

A global leader in the design and manufacturing of innovative water and wastewater treatment technologies including: DEMON®, MBBR/IFAS, DAF, Nikuni, Screens, and Polypropylene Tanks.

Wyse Inc .... **Booth 503**
http://www.wysedryers.com
Phone: (201) 809-9846
1470 Bergen Boulevard
Fort Lee, NJ 07024

Super sludge drying: odorless operation and odorless by-products (for most applications), safe dust-free operation, uniformity drying temperature - elimination of overheating and baking of product, and dried product is granular in nature.

INDEX OF EXHIBITORS BY PRODUCT CATEGORY

Advanced Wastewater Treatment Processes
Boerger, LLC.................. 404
Parkson Corporation 202
SH+E U.S. CORP. 421
WesTech Engineering..... 501

Aerobic Treatment
SH+E U.S. CORP. 421
WesTech Engineering..... 501

Anaerobic Treatment
Boerger, LLC.................. 404
WesTech Engineering..... 501

Biosolids Conditioning & Sterilization
Andritz Separation, Inc.... 418
CB&I.............................. 113
RDP Technologies 104
Schwing Bioset Inc.......... 604
SH+E U.S. CORP. 421
Wyse Inc 503

Biosolids/Land Application Equipment
Boerger, LLC.................. 404
Schwing Bioset Inc.......... 604

Centrifuges
Andritz Separation, Inc.... 418
Centrisys Corporation 401
Kusters Water 101
Noxon North America Inc.200
SH+E U.S. CORP. 421

Chemical Feed Equipment
Boerger, LLC.................. 404
Noxon North America Inc.200

Clarifier Equipment
WesTech Engineering..... 501

Conveyors
Andritz Separation, Inc.... 418
Jim Myers & Sons, Inc. ... 301
Kusters Water 101
Noxon North America Inc.200
Schwing Bioset Inc.......... 604

Dewatering Equipment
Andritz Separation, Inc.... 418
Boerger, LLC.................. 404
Centrisys Corporation 401
Kusters Water 101
Noxon North America Inc.200
Parkson Corporation 202
Uzelac Industries, Inc...... 623
SH+E U.S. CORP. 421

Dewatering Services
Centrisys Corporation 401
SH+E U.S. CORP. 421

Digesters
Boerger, LLC.................. 404
CB&I.............................. 113
Kusters Water 101
SH+E U.S. CORP. 421
WesTech Engineering..... 501

Dissolved Air Flotation Systems
Kusters Water................ 101
WesTech Engineering..... 501

INDEX OF EXHIBITORS BY PRODUCT CATEGORY

Engineering & Consulting Services
SH+E U.S. CORP. 421

Filters, Biological
WesTech Engineering..... 501

Filters, Gravity
Andritz Separation, Inc.... 418

Filter Presses & Belt Filter Presses
Andritz Separation, Inc.... 418
Parkson Corporation 202
SH+E U.S. CORP. 421

Flocculation Equipment
Jim Myers & Sons, Inc. ... 301
Noxon North America Inc.200

Gas Collection & Treatment
SH+E U.S. CORP. 421

Gates & Gate Controls
Jim Myers & Sons, Inc. ... 301
Schwing Bioset Inc.......... 604

Grinders/Shredders & Comminutors
Boerger, LLC.................. 404

Grit Handling Equipment
Jim Myers & Sons, Inc. ... 301
WesTech Engineering..... 501

Heat Exchangers
SH+E U.S. CORP. 421

Incinerators
Uzelac Industries, Inc...... 623

Lime Slakers
RDP Technologies 104

Mixers
Schwing Bioset Inc.......... 604
WesTech Engineering..... 501

Odor Control Equipment & Systems
Parkson Corporation 202

Oil/Water Separators
Boerger, LLC.................. 404
Centrisys Corporation 401
Parkson Corporation 202
SH+E U.S. CORP. 421

Operations & Maintenance Services
Schwing Bioset Inc.......... 604

Oxygen Activated Sludge Systems
WesTech Engineering..... 501

Pipe Fittings
Schwing Bioset Inc.......... 604

Pipe, Clay
Schwing Bioset Inc.......... 604

Pipe, Steel
Schwing Bioset Inc.......... 604

Polymers
Boerger, LLC.................. 404
Fluid Dynamics (Pump Solutions Group)............ 205

Plate Settlers
Jim Myers & Sons, Inc. ... 301
Parkson Corporation 202
WesTech Engineering..... 501

Pump Controls
Schwing Bioset Inc.......... 604

Pumps, Chemical Feed
Boerger, LLC.................. 404

Pumps, Grinder
Boerger, LLC.................. 404

Pumps, Lift Station
Boerger, LLC.................. 404
Fluid Dynamics (Pump Solutions Group)............ 205

Pumps, Reciprocating
Schwing Bioset Inc.......... 604

Pumps, Sludge
Boerger, LLC.................. 404
Fluid Dynamics (Pump Solutions Group)............ 205
Noxon North America Inc.200
Schwing Bioset Inc.......... 604

Pumps, Submersible
Boerger, LLC.................. 404

Rental Equipment
Andritz Separation, Inc.... 418
Centrisys Corporation 401

Reverse Osmosis Equipment
SH+E U.S. CORP. 421

Screens, Fine
Andritz Separation, Inc.... 418
WesTech Engineering..... 501

Sedimentation Equipment
Kusters Water 101
WesTech Engineering..... 501

Sludge Collectors
Kusters Water 101

Sludge Drying Equipment
Andritz Separation, Inc.... 418
Kusters Water 101
Parkson Corporation 202
RDP Technologies 104
Schwing Bioset Inc.......... 604
SH+E U.S. CORP. 421

INDEX OF EXHIBITORS BY PRODUCT CATEGORY

Sludge Drying Equipment (cont.)
Uzelac Industries, Inc...... 623
Wyse Inc 503

Sludge Hauling
Schwing Bioset Inc.......... 604

Sludge Mixers
WesTech Engineering..... 501

Sludge Thickeners
Andritz Separation, Inc.... 418
Centrisys Corporation 401
Kusters Water 101
Noxon North America Inc.200
SH+E U.S. CORP. 421
WesTech Engineering..... 501

Tanks/Wastewater
CB&I................................ 113
Jim Myers & Sons, Inc. ... 301

Valves, Rotary
Boerger, LLC.................. 404

Valves, Telescoping
Jim Myers & Sons, Inc. ... 301

Water Recycling/Reuse
Parkson Corporation 202
SH+E U.S. CORP. 421

PRESENTER AND MODERATOR DIRECTORY

Samantha Abraham, *UCLA, School of Public Health*, **RB 6B**

Mohammad Abu-Orf, *AECOM*, **RB Session 15 Moderator**

Charlie Alix, *Stantec*, **RB Session 11 Moderator**

Vincent Apa, *CDM Smith*, **RB 15E**

Richard Atoulikian, *HDR Engineering*, **EW 8A**

Craig Aubuchon, *Analysis Group, Inc.*, **EW 13F**

Eric Auerbach, *Malcolm Pirnie Inc*, **RB Session 10 Alternate**

Lamya Badr, *Hazen and Sawyer*, **EW 4A**

Amber Barritt, *CDM Smith*, **RB 8D**

Robert Bates, *Louisville Metropolitan Sewer District*, **RB 2D, RB Session 16, RB Session 7 Moderator**

Ned Beecher, *North East Biosolids & Residuals Association*, **RB Workshop A**, **RB Session 14**

Matthew Berg, *CH2M HILL*, **RB 2B**

Priscilla Bloomfield, *CH2M HILL*, **EW Session 2, 3, 8 Moderator**

Randall Boe, *CH2M HILL*, **EW 1E**

Amy Bolten, *Sonoma County Water Agency*, **EW 3A**

Kelly Brenner, *Material Matters, Inc*, **RB 4A**

Jeanette Brown, *Manhattan College*, **EW Session 1, 5, 9 Moderator**

Michael Bullard, *Hazen and Sawyer*, **RB Session 2 Moderator**

Gillian Burger, *University of Waterloo*, **RB 9A**

Peter Burrowes, *CH2M HILL Canada Ltd*, **RB 7B**

Jennifer Callahan, *Material Matters, Inc.*, **RB 3A**

Alice Cannella, *City of Chattanooga, TN*, **EW Session 16**

Joseph Cantwell, *Science Applications Intl Corp*, **EW 14B**

Scott Carr, *CDM Smith*, **RB Session 7 Moderator**

Joe Castro, *City of Boulder, Colorado*, **EW Session 11**

Pete Cavagnaro, *Johnson Controls*, **EW Session 1, 4, 14 Moderator**

Alicia Chakrabarti, *East Bay Municipal Utility District*, **EW Workshop B, EW Session 14 Moderator**

Dennis Clough, *Energy Systems Group*, **EW Workshop A**

Kevin Colvett, *CH2M Hill*, **RB Session 8 Moderator, EW 14A**

Heather Crabtree, *Metro Water Services*, **RB Session 18 Moderator**

Dustin Craig, *CDM Smith*, **RB Session 10 Moderator**

Thomas Crawford, *Retired*, **RB Workshop A**

Michael Curtis, *Simplified Energy Solutions*, **EW 2A**

Roy Denney, *Metro Water Services*, **RB Session 18**

Mary Ann Dickinson, *Alliance for Water Efficiency*, **EW Session 11**

Richard Dimassimo, *Kruger, Inc.*, **RB 9E**

Jennifer Dodd, *Tennessee Department of Environment & Conservation*, **EW 4B**

PRESENTER AND MODERATOR DIRECTORY

John Donovan, *CDM Smith*, **RB 2E, RB 3B, RB 14,**

Laura Dubin, *USEPA*, **EW 3C**

David Duest, *Massachusetts Water Resources Authority*, **RB 2E**

Cindy Dyballa, *Dyballa Consulting*, **EW 5B**

Maria Elektorowicz, *Concordia University*, **RB 6A**

Mike Elenbaas, *Black and Veatch*, **RB Workshop B**

Ralph Eschborn, *AECOM*, **RB 12B**

Cigdem Eskicioglu, *University of British Columbia-Okanagan*, **RB 7C, RB 13A**

Stephen Estes-Smargiassi, *Massachusetts Water Resources Authority*, **EW Session 11**

Greg Evanylo, *Virginia Tech*, **RB Workshop A**

Lauren Fillmore, *WERF*, **RB Workshop B**

Bob Forbes, *CH2M HILL*, **RB Session 2 Moderator, EW Session 10 Moderator**

Cynthia Fort, *American Structurepoint Inc*, **RB 8C**

Robert Freeman, *US EPA Region 4 Atlanta*, **EW 4B**

Dale Gabel, *CH2M HILL*, **RB 13E**

Michael Gardner, *Bowling Green Municipal Utilities*, **RB Session 16**

Diane Gilbert-Jones, *City of Los Angeles*, **RB 3C**

Robert Gillette, *Carollo Engineers*, **RB Session 4 Moderator**

Charles Goss, *AECOM*, **RB 15B**

Priska Grace, *CDM Smith*, **EW 2D**

Steve Hall, *Terra Renewal*, **RB Workshop B**

Michelle Hatcher, *CDM Smith*, **RB Session 13 Moderator**

Paul Hauck, *CDM Smith*, **EW 13A**

Roger Haug, *Loyola Marymount University*, **RB 10D**

Sara Hayes, *ACEEE*, **EW Session 11 Moderator**

Arun Prakash Jaganathan, *Louisiana Tech University*, **EW 2C**

Jose Jimenez, *Brown and Caldwell*, **EW 1C**

James Johnson, *Metro Wastewater Reclamation District*, **RB 13B**

Trudy Johnson, *Material Matters*, **RB Session 3 Moderator, RB 13D**

John Kabouris, *CH2M HILL*, **RB 15D**

Richard Kelly, *Brown & Caldwell*, **RB 4B**

Steven Kestel, *BioChem Technology*, **EW 6B**

Greg Kester, *California Association of Sanitation Agencies*, **RB Workshop A, RB Session 1 Moderator**

Elizabeth Keysar, *Concurrent Technologies Corporation*, **EW 13C**

Wendall Khunjar, *Hazen and Sawyer*, **RB Session 14**

Richard Kuchenrither, *University of Colorado, Boulder*, **RB 5B**

Shannon Lambert, *Barge Waggoner Sumner and Cannon, Inc.*, **RB Session 18**

PRESENTER AND MODERATOR DIRECTORY

Ron Latimer, *Hazen and Sawyer*, **RB Session 14**

Jeffrey J. LeBlanc, *WeCare Organics, LLC*, **RB Session 13 Moderator**

Suntae Lee, *Kyoto University*, **EW 1B**

Peggy Leonard, *King County Washington*, **RB Workshop A**

Jordan Lerner, *Schneider Electric*, **EW Workshop A**

Daryl Letto, *ENBALA Power Networks*, **EW 10C**

F Michael Lewis, *F Michael Lewis Inc*, **RB 10E**

John Lindam, *Xylem Water Solutions AB*, **EW 8C**

Barry Liner, *WEF*, **EW Workshop B**

Bryan Lisk, *Hazen and Sawyer*, **EW 8B**

Robert Lonergan, *Tetra Tech*, **EW Session 2, 3 Moderator**

James Long, *Hazen and Sawyer*, **RB 10B**

Enrique Lopez Calva, *CDM Smith*, **EW 3B**

Ting Lu, *MSD of Greater Cincinnati*, **EW 12A**

Bruce Lung, *Alliance to Save Energy*, **EW 6A**

Jing Luo, *Pima County*, **RB 15C**

Jordan Macknick, *National Renewable Energy Laboratory*, **EW 5A, EW 10A**

Daryl McCartney, *University of Alberta*, **RB Session 6 Alternate**

James McCaughey, *Narragansett Bay Commission*, **EW 2B**

Ed McCormick, *EBMUD*, **EW Workshop B**

Michael McFarland, *Utah State University*, **RB Workshop A**

Kelley Merritt, *Metro Wastewater Reclamation District*, **RB 5A**

Alexandre Miot, *San Francisco Public Utilities Commission*, **RB 10C**

Bryan Mitchell, *Metro Wastewater Reclamation District*, **RB 5A**

Gregory Moreland, *SRA International*, **EW 2E**

Lynne Moss, *CDM Smith*, **RB Session 14**

Christopher Muller, *Brown & Caldwell*, **RB Session 15 Alternate**

Fred Mussari, *BCR Environmental*, **RB Session 1 Alternate**

Edward Myers, *Lescure Engineers*, **EW 4E**

Kartiki Naik, *University of California, Los Angeles*, **EW 4D**

Brooks Newbry, *Water Technology Group*, **EW 6E**

Robert O'Dette, *Tennessee Department of Environment and Conservation*, **RB Workshop B**

David Oerke, *CH2M HILL*, **RB Session 8 Moderator, RB 9D**

Chul Park, *UMass Amherst*, **RB 1B**

David Parry, *CDM Smith*, **RB Session 8 Alternate, RB 09B, RB Session 9 Moderator**

Jeff Peeters, *GE Water & Process Technologies*, **EW 12B**

Chris Peot, *DC Water*, **RB Session 16 Moderator**

Christine Polo, *Black and Veatch*, **RB Session 14**

Scott Potter, *Metro Water Services*, **RB Session 18**

Michael Preston, *Black & Veatch*, **EW 5C**

PRESENTER AND MODERATOR DIRECTORY

Yinan Qi, *Black & Veatch*, **RB Session 6 Moderator, RB 12A**

Tazio Qubeck, *Brown and Caldwell*, **RB Session 3, 16 Moderator**

Caroline Quinn, *Delta Diablo Sanitation Dist.*, **RB 12C**

Dave Reardon, *HDR*, **EW Workshop B**

Gene Rodrigues, *Southern California Electric*, **EW Session 11**

Adam Ross, *Brown and Caldwell*, **EW 13D**

Robert Rubin, *North Carolina State University*, **RB Session 17**

Justin Rundle, *Honeywell*, **EW Workshop A**

Stephen Salter, *Farallon Consultants Limited*, **EW 13E**

Julian Sandino, *CH2M HILL*, **EW 1D**

Dilek Sanin, *Middle East Technical University*, **RB 1E**

Devendra Saroj, *University of Surrey*, **RB 9C**

Patricia Scanlan, *Black & Veatch*, **RB 8E, RB Session 12 Moderator**

Rebecca Schaefer, *Greeley and Hansen*, **RB 1D**

Perry Schafer, *Brown and Caldwell*, **RB 4C**

James Schettler, *Brown & Caldwell*, **EW 14C**

Kristi Schnell, *Gresham Smith and Partners*, **RB 7A**

Richard Schoeck, *Gwinnett County Department of Water Resources*, **EW 9B**

Ralph Schroedel, Jr., *Brown & Caldwell*, **EW 13B**

Robin Schroeder, *Environmental Business Specialists, LLC*, **EW 9A**

Tom Schwartz, *Woodard & Curran*, **EW Session 8 Moderator**

Anup Shah, *Brown And Caldwell*, **EW 3D**

EJ Shalaby, *West County Wastewater Dist.*, **RB 12C**

Todd Shearin, *Terra Renewal*, **RB Workshop B**

Brian Sibiga, *Niagara County Sewer District No. 1*, **EW 4C**

Natalie Sierra, *RMC Water and Environment*, **RB Workshop A**

James Slaughter, *Beveridge & Diamond PC*, **RB 11A**

Roy Smith, *Guthrie Sales & Service, Inc*, **RB Session 1 Moderator**

Lesley Smith, *City West Water*, **EW 10B**

Jim Smith Jr., *Independent Environmental Consultant*, **RB 11B, RB Session 4 Moderator**

Matthew L. Snyder, *City of Chattanooga*, **TN, RB Session 16**

Reza Sobhani, *University of California, Irvine*, **EW 3E**

Michael K. Stenstrom, *Univeristy of California, Los Angeles*, **EW 6C**

Aaron Stephens, *Material Matters, Inc*, **RB Session 10 Moderator, RB 13C**

Jay Surti, *CH2M HILL*, **RB Session 7 Alternate**

Stephen Tarallo, *Black &Veatch*, **EW Session 12 Moderator**

PRESENTER AND MODERATOR DIRECTORY

Ron Taylor, *Metro Water Services*, **RB Session 18**

Peter Thomson, *Black and Veatch*, **EW Workshop A**

Bill Toffey, *Effluential Synergies,* **EW Workshop B**

Yuanming Tsai, *Infilco Degremont, Inc.*, **RB 10A**

Richard Tsang, *CDM Smith*, **RB Session 15 Moderator**

Patricia Turman, *USDA-NRCS*, **RB Session 17**

David Tucker, *Metro Water Services*, **RB Session 18**

Jason Turgeon, US *EPA*, **EW Workshop B, EW Session 13 Moderator**

Louann Unger, *US EPA*, **EW Session 6, 9 Moderator**

Matthew Van Horne, *Hazen and Sawyer*, **RB 2C**

Don Vandertulip, *CDM Smith*, **EW 7A**

Christopher Varnon, *CDM Smith*, **EW 7B**

Richard von Langen, *RVL Associates, Inc.*, **RB 15A**

Forbes Walker, *University of Tennessee*, **RB Session 17**

Cindy Wallis-Lage, *Black and Veatch*, **EW Session 7 Moderator**

Hong Y. Wang, *Research Center for Environmental Quality Management*, **EW 1A**

Curt Wendt, *CDM Smith*, **EW 6D**

Sean West, *United Technologies Corporation*, **EW Session 11**

Drury Whitlock, *CH2M HILL*, **RB 1A**

Todd Williams, *CH2M HILL*, **RB Session 6 Moderator, RB 8B**

John Willis, *Brown & Caldwell*, **RB Workshop B, RB 2A, RB Session 9 Moderator, EW Session 4 Moderator, EW 9C**

Christopher Wilson, *Greeley and Hansen*, **RB 01C**

Steve Wilson, *Brown and Caldwell*, **RB Session 5 Moderator**

Bob Wimmer, *Black & Veatch*, **RB Session 18**

Greg Woodward, *Stantec*, **RB Session 1 Moderator**

Doug Yarosz, *Brown and Caldwell*, **RB Session 12, 18 Moderator**

Yan Zhou, *University of Illinois*, **EW 12C**

Francesco Zurzolo, *University of Manitoba*, **RB Session 13 Alternate**

UPCOMING CONFERENCES

2013 WEF Forum Activated Sludge on its 100th Birthday: Challenges and Opportunities
June 9 - 11, 2013
Stonehill College
Easton, Massachusetts
www.wef.org/ActivatedSludge

Collection Systems 2013
June 9 – 12, 2013
Sacramento Convention Center
Sacramento, California
www.wef.org/CollectionSystems

WEF/IWA Nutrient Removal and Recovery 2013
July 28 – 31, 2013
Sheraton Vancouver Wall Centre Hotel
Vancouver, British Columbia, Canada
www.wef.org/nutrients

WEFTEC 2013
October 5 – 9, 2013
McCormeck Place Convention Center
Chicago, Illinois
www.weftec.org

WEF/AWWA Utility Management Conference 2014
February 25 – 28, 2014
Hyatt Regency Savannah
Savannah, Georgia
www.wef.org/UtilityManagement2014

Residuals and Biosolids 2014
May 18 - 21, 2014
Austin Convention Center
Austin Texas
www.wef.org/ResidualsBiosolids

Odors and Air Pollutants 2014
May 31 – June 3, 2014
Hyatt Regency Miami
Miami, Florida
www.wef.org/OdorsAir

Energy Evaluation of Combination Processes for Secondary Effluent Treatment

Hongyang Wang*, Dongbum Im*, Aki Ito*, Naoya Wada*, Norihide Nakada*, Michiko Aoki**, Yasuhiro Kato**, Hiroaki Tanaka*

* Research Center for Environmental Quality Management, Kyoto University, 1-2 Yumihama, Otsu, Shiga 520-0811, Japan (E-mail: wanghy06@gmail.com; im@biwa.eqc.kyoto-u.ac.jp; aki.blue.peter.1@gmail.com; wada.naoya.58r@st.kyoto-u.ac.jp; nakada.norihide.8w@ kyoto-u.ac.jp; htanaka@biwa.eqc.kyoto-u.ac.jp)
* * Metawater Co., Ltd., Yahatakaigan St. 7, Ichihara, Chiba, 290-8511, Japan (E-mail: aoki-michiko@metawater.co.jp; kato-yasuhiro@metawater.co.jp)

ABSTRACT

Water reclamation projects are established all over the world to lighten the water stress or scarcity problem. For choosing water reclamation processes not only water qualities but also energy consumption should be considered. This study evaluated the potential of ozonation and ceramic membrane filtration (CMF) combination processes application for water reuse mainly in terms of energy consumption. Energy consumption was calculated based on running experiences of one pilot plant operated by METAWATER CO., with capacity of 7000 m^3/day. Considering both PPCPs and MS2 removal target, ozonation, ozonation+CMF and ozonation +coagulaiton+CMF under 4, 6 mg O_3/L dose were potential processes. In order to meet the goal, energy consumption for ozonation was in the range from 0.05 to 0.10 $kgCO_2/m^3$. Ozonation +CMF and ozonation+coagulation+CMF consumed higher energy with higher product water quality. It was 0.14-0.19 $kgCO_2/m^3$. Considering different requirements of various reuses, which process was better should be evaluated case by case. This study showed significance through presenting basic knowledge to choose suitable technology to gain co-benefit of product water quality and energy consumption in water reclamation field.

KEYWORDS: Ceramic membrane filtration; Ozonation; Coagulation, Energy

INTRODUCTION

Many countries and regions of the world are facing water scarcity and deterioration of surface water and groundwater quality. To lighten these problems, a wider use of recycled water could be undertaken. Among various water reclamation treatments, low-pressure membrane technologies has been extensively studied, due to high removal of turbidity and bacteria, low energy consumption, small footprint. However, small molecular size contaminants could not be removed, such as pharmaceuticals and personal care products (PPCPs) residuals. Meanwhile, ozonation was reported as effective process for such contaminants controlling (Kim et al., 2009). In order to combine ozonation with membrane filtration, ceramic membrane was chosen for mechanical superior and chemical resistance. Such combination processes were found showing well performance for removal of both PPCPs residuals and MS2 phage (Wang et al., 2012). While, for choosing water reclamation processes not only water qualities but also energy consumption should be considered. Thus the objective of this work was to evaluate the potential of ozonation and ceramic membrane filtration (CMF) combination processes application for water reuse mainly in terms of energy consumption.

METHODOLOGY

Wastewater and experimental set-up

Biologically treated municipal wastewater (secondary effluent) was used as the raw water. Five processes were conducted (Table 1.1). Ozonaiton and CMF were done using continuous equipments, same as the ones used in previous work (Wang et al., 2011). The ozonation reaction time was 5 minutes. The flux for CM was 4 m/d, filtration period was 30 min. The CM, provided by METAWATER CO., Ltd., was monolithic type, with 0.042 m^2 effective membrane area. Pore size is 0.1μm. Coagulation, with polyaluminium chloride (PAC) as coagulant, was done in 8L tank continuously during CM filtration. The GT value was controlled around 17000. These equipments were combined to achieve target processes.

Table 1.1 Main parameters of each process

Process	O_3 dose (mg/L)	PAC dose (mg/L)	Flux during CMF (m/d)
Ozonation	1,2,3,4,6	-	-
CM	-	-	4.0
Ozonation +CM	2	-	4.0
	3	-	
	4	-	
	6	-	
PAC+CM	-	15	4.0
	-	25	
	-	35	
	-	50	
Ozonation +PAC+CM	2	25	4.0
	2	35	
	4	25	
	6	15	
	6	25	
	6	35	

Energy consumption method

Energy consumption was calculated based on running experiences of one pilot plant operated by METAWATER CO., with capacity of 7000m^3/day. Operation energy consumption was 0.0169 $kgCO_2$ m^3/gO_3 and 0.0666 $kgCO_2$ /m^3 for O_3 and CM, respectively. For CM filtration, chemical enhanced backwashing (CEB) should be also considered. The CEB intervals were deduced from transmembrane pressure (TMP) changing trend during 5 days running. H_2SO_4 was assumed as the CEB chemical. Before discharging wastewater during CEB, NaOH was used for neutralization. The total energy consumption for CEB procedure was 0.0033 $kgCO_2$

/m^3. For coagulant PAC, 0.4050 $kgCO_2$/kgPAC energy was needed. Using these basic information, energy could be calculated.

RESULTS

The change of total PPCPs amount removal rate with energy consumption was discussed firstly. CMF and PAC + CMF showed little effect on PPCPs removal. While, ozonation could reduce total PPCPs amount obviously. Over 90% removal could be achieved with 0.05 kCO_2/m^3 energy consumption. Adding ceramic membrane with/without coagulation after ozonation, the power required for removing 90% total PPCPs amount was increased to over 0.14 kCO_2/m^3. It could be explained by that PPCPs was mainly removed by ozonation, and coagulation and ceramic membrane showed little effect.

Then the relationship of MS2 removal rate with energy consumption was examined. Ozonation could meet over 5 logs MS2 removal rate with over 0.05 kCO_2/m^3 energy consumption. The removal rate of MS2 by CMF was only 0.45 logs. It could be enhanced obviously by incorporated with ozonation and/or coagulation. The removal rate increased to over 5.0 logs with coagulation or ozonation pretreatment under over 0.11 kCO_2/m^3 energy consumption. It was further increased to over 8 logs by adding ozonation before PAC+CMF under similar energy consumption.

DISCUSSION

The achievement of 90% removal of PPCPs detected and 5 logs MS2 removal were decided as the goal. For PPCPs, the major removal happened in the ozonation process. Product water from ozonation, ozonation +CM, ozonation +PAC+CM under 4, 6 mgO_3/L dose could meet the requirement. For MS2, both ozonation and coagulation presented high removal rate. Product water from ozonation and ozonation +CM under 4, 6 mgO_3/L dose, PAC+CM, ozonation +PAC+CM could meet the requirement. Considering both PPCPs and MS2 removal target, ozonation, ozonation +CM and ozonation +PAC+CM under 4, 6 mgO_3/L dose were potential processes. In order to meet the goal, energy consumption for ozonation was in the range from 0.05 to 0.10 $kgCO_2/m^3$. Ozonation +CM and ozonation +PAC+CM consumed higher energy. It was 0.14-0.19 $kgCO_2/m^3$. And product water quality from the combination processes was higher than ozonation.

In addition, results of TOC, UV_{254}, bacteria, turbidity and total phosphorus showed the advantage of ozonation on organic matter degradation, bacteria disinfection, merit of coagulation on total phosphorus removal, and that of the membrane on bacteria, turbidity rejection. Considering different requirements of various reuses, which process was better should be evaluated case by case.

This study showed significance through presenting basic knowledge to choose suitable technology to gain co-benefit of product water quality and energy consumption in water reclamation field.

ACKNOWLEDGEMENTS

This study has been conducted as part of Core Research for Evolutionary Science and Technology (CREST) entitled "Development and Evaluation of Water Reuse

Technologies for the Establishment of 21st century type Water Circulation System" headed by Prof. Hiroaki Tanaka, supported by the Japan Science and Technology Agency.

REFERENCES

Kim, I.H., et al 2009. Discussion on the application of UV/H_2O_2, O_3 and O_3/UV processes as technologies for sewage reuse considering the removal of pharmaceuticals and personal care products. Water Science and Technology 59(5), 945-955.

Wang, H.Y., et al (2012). Performance of ceramic membrane filtration with coagulation pretreatment for pharmaceuticals and personal care products and MS2 phage removal from secondary effluent IWA 3rd Regional Conference in Membrane Technology, Buenos Aires, Argentina.

Wang, H.Y., et al (2011). Removal of pharmaceuticals and personal care products by ozonation pretreatment and ceramic membrane process. The 4th IWA-ASPIRE Conference & Exhibition, Tokyo, Japan.

Nexus of Energy and Virus Reduction of UF Membrane Filtration Process with Coagulation/UV Irradiation for Wastewater Reclamation

Suntae Lee, RCEQM, Kyoto University
Naoyuki Yamashita, RCEQM, Kyoto University
Masaru Ihara, RCEQM, Kyoto University
Hiroaki Tanaka, RCEQM, Kyoto University
Kentaro Kobayashi, Toray Industries, Inc.
Hiroaki Tanaka, Toray Industries, Inc.
Hiroo Takabatake, Toray Industries, Inc.
Yuji Tanaka, Toray Industries, Inc.
Research Center for Environmental Quality Management(RCEQM), Kyoto University
1-2 Yumihama
Otsu, Shiga 520-0811, Japan
(077) 527-6223

ABSTRACT

In this study, focuses on the nexus of energy consumption and virus reduction of the three water reclamation processes using UF membrane. In virus removal test, the coagulation under pH control followed by UF filtration process and UF filtration without coagulation followed by UV irradiation process showed more than 5.2-log removal/inactivation of MS2. In nexus of virus removal and energy consumption, UF filtration followed by UV irradiation process can remove virus most effectively in terms of energy consumption. Because coagulant dosage can be reduced, which is resulted in the reduction of energy consumption of "Chemicals" and "Disposal" portions for coagulants. UF + UV process can be considered to be the most effective processes in this study, but more comprehensive evaluation will be necessary in application of these technologies to secondary effluent of other STPs.

KEYWORDS

Wastewater reclamation, UF membrane, UV, coagulation, energy evaluation, virus reduction.

INTRODUCTION

Wastewater reclamation and reuse has been attracting attention as a new water resource due to the global water shortage in recent years. However, hygienic safety is an important concern in using the reclaimed water. There is a need to control pathogens properly to ensure hygienic safety of the reclaimed water. For instance, "Title 22", criteria in California which is one of the most stringent regulations for water reuse in the world, demands water quality as shown in Figure 1 for unrestricted agricultural use (CDHS, 1978). On the other hand, in recent year water reclamation process needs to reduce energy consumption (CO_2 emissions) because global warming issues and the sustainability are also important. Therefore, energy reduction of water reclamation process is necessary with ensuring the hygienic safety.

Conventional full treatment, contact filtration and direct filtration are commonly applied for many unrestricted water application complying with California "Title22". On the other hand, ultrafiltration (UF) membrane is emerging technology with low pressure but high reliability on small foot print. However, to meet the requirements of the virus removal rate as high as 5.2-log complying with "Title 22" Criteria, treatment by UF membrane alone will be difficult because the viral size is smaller than the pore size distribution of the UF membrane (Lee et al, 2012). Therefore, the combination of processes of UF with coagulation and/or UV disinfection is necessary to achieve the required virus removal rate for unrestricted water reuse application.

UF membrane filtration following coagulation process can achieve the "Title22" criteria (Kobayashi et al, 2011, Lee et al, 2012). However, little studies were carried out from the viewpoint of the nexus of energy consumption and virus reduction in these processes for water reclamation. Therefore, this study focuses on the nexus of energy consumption and virus reduction of the three water reclamation processes using UF membrane.

In this study, Energy consumption was estimated by Life Cycle CO_2 (LC-CO_2) method. LC-CO_2 method can calculate energy consumption of these elements as well as electric consumption. Therefore, it is effective to use LC-CO_2 method in order to evaluate the amount of energy consumption in whole treatment process.

Parameters	Quality Criteria
Turbidity	0.2 NTU
Total Coliform	2.2 MPN / 100mL
Virus Removal Rate	5.2 - log

Figure 1 Water Quality Requirements in "Title 22"

MATERIALS AND METHODS

Raw water, Coagulant, Membrane, UF Membrane Treatment Processes

This study used the following three water reclamation processes: coagulation followed by UF filtration; coagulation under pH control followed by UF filtration; UF filtration without coagulation followed by UV irradiation as shown in Figure 2. The experiments were carried out with secondary effluent from A sewage treatment plant (STP; Okinawa, Japan), which applies conventional activated process without nitrification. Water quality of secondary effluent in STP A is shown in Figure 3.

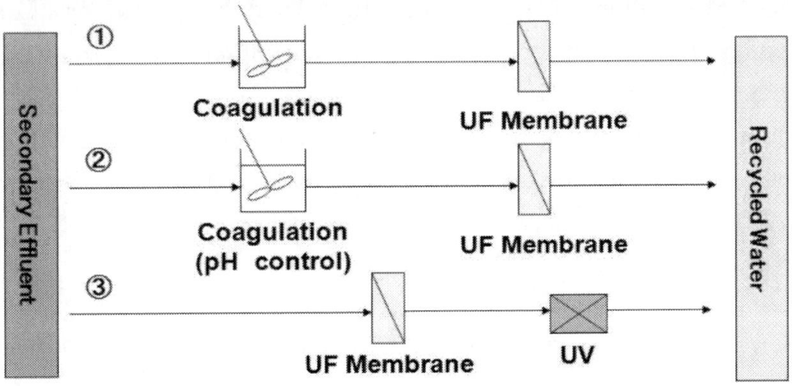

Figure 2 UF Membrane Treatment Processes Evaluated in This Research

Parameters	STP A
pH	6.9 ~ 7.4
Turbidity (NTU)	1.0 ~ 3.0
SS (mg/L)	1.0 ~ 5.0
TOC (mg/L)	7.0 ~ 11.0
E260 (/cm)	0.090 ~ 0.120

Figure 3 Water Quality of Secondary Effluent

The hollow fiber membrane HFU-2008 (Toray Industries Inc., Japan) made of polyvinylidene fluoride with 0.01 μm-pore and 150 kDa molecular weight cut-off was selected as a UF membrane that can produce 12-24 m^3/day of reclaimed water. Flux was set at 1m/d or 2m/d and initial TMP was set at 30kPa and was operated by upper limit of 200kPa in terms of increasing TMP. Polyaluminum chloride (PAC) and sulfuric acid (H_2SO_4) were selected as the coagulant and pH control reagent. As for a UV disinfection facility, Low-pressure mercury lamp UVL-15DH (SEN LIGHTS Co., Japan) was used in a cylindrical column by flowing through the UF filtrate. UV dose was changed by changing the flow rate and was measured by a bio-actinometry using the bacteriophage MS2.

Evaluation of Virus Removal Rate

The bacteriophage MS2 (NBRC 102619) was used as a model virus. MS2 was obtained from the NITE Biological Research Center (NBRC, Japan). MS2 is the prototype member of the genus *Levivirus* in the *Leviviridae*. The genomes of MS2 contain a single molecule of linear positive-sense, single-stranded RNA, which is encapsulated in an icosahedral protein capsid with a diameter of 24–26nm. MS2 was propagated for 22–24 h at 37 °C in *Escherichia coli* (NBRC 13965). MS2 culture solution was centrifuged and then passed through a 0.45-µm pore size membrane filter to remove *E.coli*. The concentration of MS2 was measured by the double layer agar method with *E.coli* as host bacterium. The numbers of MS2 were counted as plaque-forming units (PFU) in 1mL of sample (Otaki). For the sample with low concentration of MS2, 50mL of sample was poured into five Petri dishes and covered by the double layer LB base agar. Each plaque was totalized and represented as PFU/50mL (Grabow, 1986).

MS2 was spiked to secondary effluent at 10^5-10^7 PFU/mL and the mix solution was agitated slowly for 3 min. Then, UF membrane filtration was started and MS2-spiked secondary effluent, feed water and filtrated water were sampled 3 min later of the start operation for measurement of the MS2 concentration.

Estimation of Energy Consumption and Footprint

Energy consumption was calculated in terms of CO_2 emissions from electrical consumption during operation such as pumps and compressors, chemical dose that is estimated as energy for its production and disposal as sludge etc. Published value (Okinawa, 2013) and literature value (LCA, 1998) were used for basic unit of CO_2 emissions. On the basis of the results of pilot plant tests, CO_2 emissions of disposal, electricity and chemicals was calculated on the assumption of water reclamation treatment equipment and sludge treatment equipment on the scale of 50,000m^3/d as described in Figure 4. Footprint was also calculated based on the assumption of water reclamation treatment equipment, which included coagulation, sedimentation, membrane filtration and UV treatment.

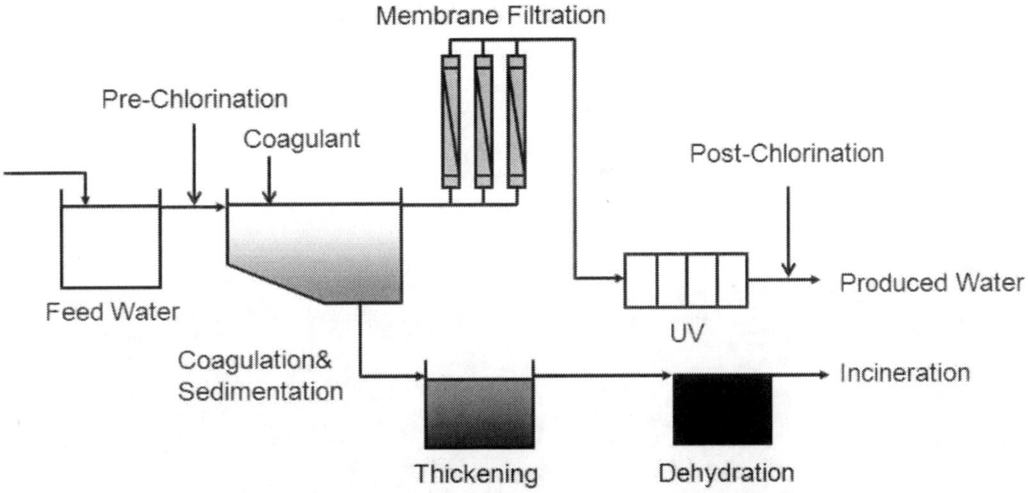

Figure 4 Treatment Flow of Water Reclamation Treatment System

RESULTS AND DISCUSSION

Virus Removal Test in UF Pilot Plant

The results of virus spiked in the UF pilot plant test at pH values of raw water pH (pH 7.2) and lowering pH (pH5.5) and same PAC dosage(50 mg/L) were shown in Figure 5. Raw water pH condition indicated 4.63-log virus removal. This log virus removal could not achieve the Title22 criteria (5.2-log removal/inactivation of viruses). However, the result at pH5.5 condition indicated more than 7.75-log virus removal after coagulation following by UF filtration, where virus concentration in the filtrated water was below measurable limits. This result complied with the Title22 criteria.

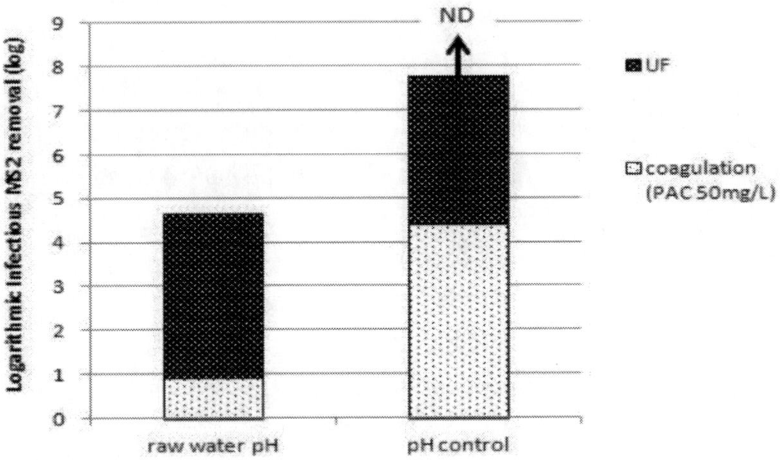

Figure 5 MS2 Log Removals of Pilot Plant Tests at pH

On the other hand, virus removal rate in the membrane plus UV process was shown in Figure 6. UF membrane process followed by UV irradiation resulted that the membrane process removed MS2 at 1.5-log in average. Increasing inactivation rate by UV irradiation followed by the UF process was observed proportionally with UV fluence increase, which showed that approximately 80 mJ/cm2 of UV irradiation can achieve 5.2-log virus removal in total.

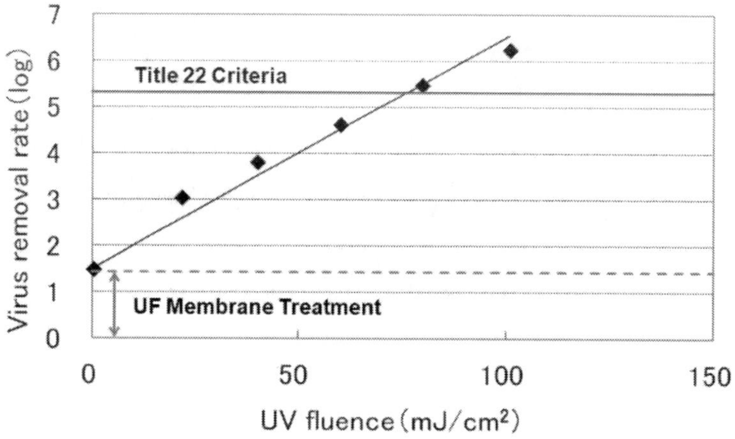

Figure 6 Virus Removal Rate by UF Membrane with UV and Effect of UV Fluence

Nexus of Virus Removal and Energy Consumption

Figure 7 shows the nexus of virus removal and energy consumption of each process. Each slope in the Figure 7 shows the reduction of virus per energy consumption of each process. The steeper slope means more efficient viral reduction in terms of the energy consumption.

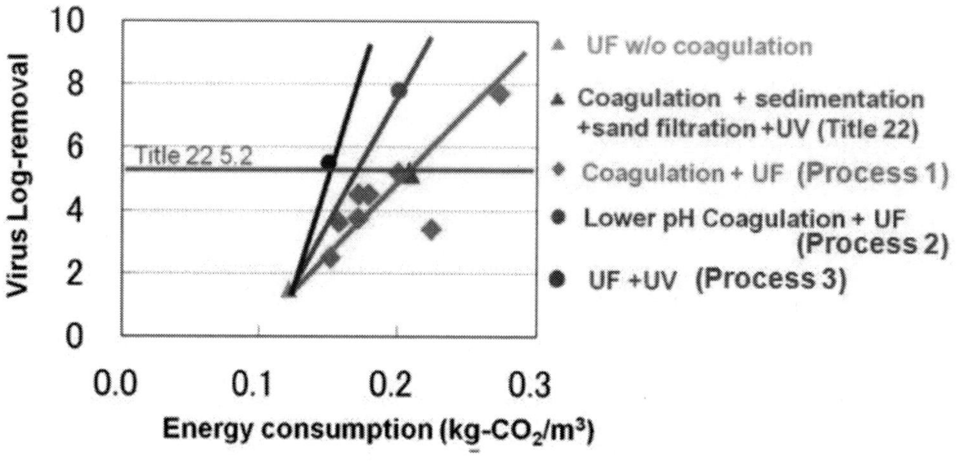

Figure 7 Relationship Between Virus Removal Rate and Energy Consumption

Figure 7 indicates that Process 3 (UF + UV) > Process 2(Coagulation (lower pH control) + UF) > Process 1(Coagulation + UF). The required energy consumption for increasing 1-log virus removal rate was estimated as E1 (Process 1 slope) = 0.016 kg-CO_2/log viral reduction, E2 (Process 2 slope) = 0.013 kg-CO_2/log viral reduction (lower pH control), E3 (Process 3 slope) = 0.007 kg-CO_2/log viral reduction for Processes 1, 2, and 3, respectively. Therefore, as for secondary effluent from A sewage treatment plant Process 3 can remove virus most effectively in terms of energy consumption.

In case, the viral removal is set at 5.2-log reduction for reclamation process that can comply with Title 22 criteria, the energy consumptions of the above three processes are estimated as shown in Figure 3. It can be concluded that energy consumption in terms of CO_2 emission of Process 3 is lowest among the three processes. While usage of electricity increases by UV irradiation, coagulant dosage can be reduced, which is resulted in the reduction of energy consumption of "Chemicals" and "Disposal" portions for coagulants.

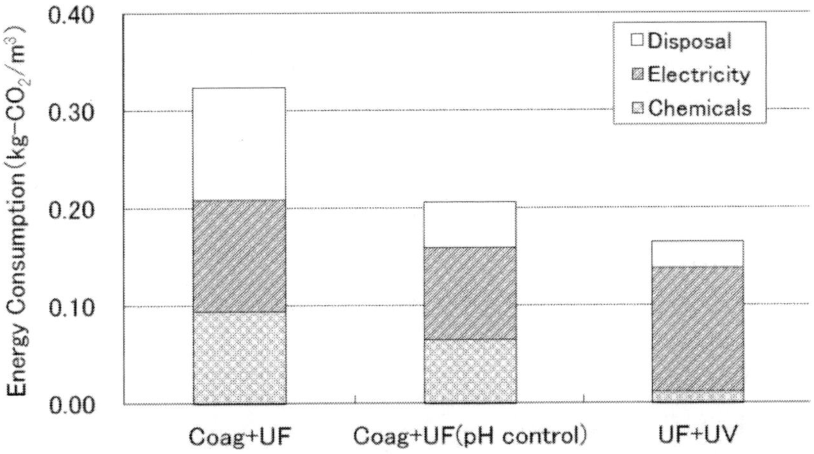

Figure 8 Energy Consumption for UF Membrane Treatment Processes

As described above, UF + UV process can be realized as most effective processes in this study for wastewater effluent reclamation, but more comprehensive evaluation will be necessary in application of these technologies to other treated wastewater. Further study will be successful in custom-made design for optimizing reduction of energy consumption and improvement of the safety in water reclamation and reuse.

CONCLUSIONS

This study focuses on the nexus of energy consumption and virus reduction of the three water reclamation processes: coagulation followed by UF filtration; coagulation under pH control followed by UF filtration; UF filtration without coagulation followed by UV irradiation.

In virus removal test, the coagulation under pH control followed by UF filtration process showed enough removal of MS2, while coagulation followed by UF filtration process at same PAC dosage showed poor removal of MS2. In UF filtration without coagulation followed by UV irradiation process, increasing inactivation rate of UV irradiation followed by the UV process was observed proportionally with UV fluence increase, which showed that approximately 80 mJ/cm2 of UV irradiation can achieve 5.2-log virus removal in total.

In nexus of virus removal and energy consumption, Process 3 (UF + UV) is more efficiently viral reduction in terms of the energy consumption than Process 2(Coagulation (lower pH control) + UF) and Process 1(Coagulation + UF). While usage of electricity increases by UV irradiation, reduction of coagulant dosage is possible which resulted in the reduction of energy consumption in "Chemicals" and "Disposal" portions of coagulants.

Further study will be successful in custom-made design for optimizing reduction of energy consumption and improvement of the safety in water reclamation and reuse.

ACKNOWLEDGEMENT

This research was performed with the financial support of the Core Research for Evolutional Science and Technology (CREST) entitled "Development and Evaluation of Water Reuse Technologies for the Establishment of 21st century type Water Circulation System" provided by the Japan Science and Technology Agency (JST).

REFERENCES

California Department of Health Services (CDHS) (1978) *California Code of Regulations. In Wastewater Reclamation Criteria; Title 22;* Sanitary Engineering Section, Berkeley, USA.

Grabow W. O. K; Coubrough P (1986) Practical direct plaque-assay for coliphages in 100-ml samples if drinking-water. *Applied and Environmental Microbiology*, 52(3), 430-433.

Kobayashi, K; Suzuki, H; Takabatake, H; Tanaka, Y; Nishida, Y; Lee, S.T; Yamashita, N; Tanaka, H (2011) *Virus Removal for Energy-Saving Sewage Reclamation System with UF Membrane;* The 4th IWA-ASPIRE Conference & Exhibition, Tokyo, Japan.

Lee, S.T; Yamashita, N; Tanaka, H; Kobayashi, K; Tanaka, H; Takabatake, H (2012) *Virus removal by ultrafiltration with coagulation process considering the pH for creation of wastewater reuse system;* IWA 3rd Regional Conference in Membrane Technology, Buenos Aires, Argentina.

LCA introductory affairs editorial committee in Japanese (1998) LCA introductory affairs

Otaki M. Method for Measurement of Coliphage, [online], available: http://www.eng.ocha.ac.jp/enveng/theme/coliphage.html [accessed 16 January 2013]

Okinawa Electric Power Co., Inc. [online], basic unit of CO_2 emissions in 2009: http://www.okiden.co.jp/environment/report2011/sec7/sec71.html [accessed 5 February 2013]

OVERVIEW AND CHALLENGES OF ANAEROBIC WASTEWATER TREATMENT: IS THERE A PLACE FOR IT IN OUR FUTURE?

Jose A. Jimenez[1] and Charles B. Bott[2]

[1] Brown and Caldwell, Maitland, FL
[2] Hampton Road Sanitation District, Virginia Beach, VA
Corresponding Author: JJimenez@BrwnCald.com

ABSTRACT

To improve energy recovery from municipal wastewater treatment and to implement the new generation of energy neutral or energy positive treatment facilities, mainstream anaerobic processes are being considered as a supplement of conventional aerobic treatment systems with anaerobic sludge digestion. Anaerobic treatment is an attractive option for the biological treatment of municipal wastewater since it has the potential to eliminate the need for aeration energy required to remove the dissolved organic fraction, significantly increase the generation of methane (CH_4) gas, and to generate substantially less sludge to be disposed. However, despite of major progress made for the last 30 years, there are still unresolved issues inhibiting the widespread application of this technology including: (i) COD removal and CH_4 production rates under psychrophilic temperatures and low influent organic concentrations, (ii) recovery of dissolved CH_4 in the anaerobic reactor effluent, and (iii) improving the removal efficiencies of the organic matter, nutrients and pathogens in the final effluent. The successful application of high-rate anaerobic systems represent a paradigm shift for the industry, offering the opportunity for sustainable wastewater treatment, energy neutral or even energy positive facilities and dramatic reductions in treatment costs, which has widespread environmental, economic and societal benefits.

KEYWORDS: anaerobic treatment, municipal wastewater, biogas, energy

INTRODUCTION

With the growing concerns over the impact of fossil fuel usage on climate change, as well as increasing energy costs, the development of clean energy and the reuse of renewable resources have drawn great attention. This has led to the reconsideration of using wastewater as a source of renewable energy and to explore how wastewater treatment can be accomplished under energy neutral or even at energy positive conditions.

Tchobanoglous and Burton (1991) indicated that municipal wastewater contains more energy than the energy needed for treatment, and that the most direct and useful energy source in wastewater is the organic fraction (measured as chemical oxygen demand, COD). The use of domestic wastewater as a source of energy has a long history, especially through the anaerobic conversion of wastewater organic content into methane (CH_4) gas, a useful biofuel. In municipal wastewater treatment, suspended solids are often concentrated in a primary settling tank and diverted as primary sludge to anaerobic digestion resulting in the conversion of the organic fraction to CH_4, but CH_4 results only from the particulate biodegradable fraction. The dissolved organic fraction in the wastewater cannot be concentrated easily and so is subjected to aerobic treatment or used for denitrification for its removal to meet effluent standards effectively. Activated sludge is an energy-intensive biological treatment process where the dissolved organic fraction in the municipal wastewater is not recovered, but is converted to carbon dioxide by applying energy for aeration (Reardon, 1995). The energy needs for a typical wastewater treatment plant employing aerobic activated sludge treatment and anaerobic sludge digestion is 0.6 kWh/m^3 of wastewater treated (Curtis, 2010), from which about half is for electrical energy to supply air for the aeration basins (EPA, 2006). Therefore, through the conventional practice of aerobic treatment combined with anaerobic sludge digestion, only a portion of the energy potential of wastewater is captured (McCarty et al., 2011), and thus constitutes a lost resource. As a result, the energy requirements for aerobic treatment are typically much greater that the energy recoverable via anaerobic sludge digestion (Foley eta el., 2010). Therefore, our efforts should be focused on not only capturing more of the energy potential in wastewater but to develop and implement treatment systems that require less energy; and only then wastewater treatment could become a net energy producer rather than a consumer.

To improve energy recovery from municipal wastewater treatment and to implement the new generation of energy neutral or energy positive treatment facilities, mainstream anaerobic processes are being considered as a replacement for conventional aerobic treatment systems with anaerobic sludge digestion (van Haandel et al., 2006). In this respect, the use of anaerobic rather than aerobic treatment processes has the potential to eliminate the need for aeration energy required to remove the dissolved organic fraction, significantly increase the generation of CH_4 gas, and to generate substantially less sludge to be disposed (Chong et al., 2012). The objective of this article is to present how anaerobic treatment can be used to maximize the recovery of the wastewater's energy content.

ADVANCES IN ANAEROBIC TREATMENT

Anaerobic wastewater treatment is well over a century old (McCarty et al., 1982). However, over the past 50 years more efficient anaerobic processes have been developed leading to suggestions that they be applied to treat domestic wastewaters (Lettinga et al., 1983; Jewell, 1987). Major progress was made when researchers began to understand the importance of designing anaerobic reactors which could maintain high solids residence times (with a buildup of high active biomass concentrations) for good process efficiency and stability, and with low hydraulic retention times for system economy (von Sperling and Chernicharo, 2005). This resulted in the development of the high-rate anaerobic reactors. The common high-rate anaerobic processes are: upflow anaerobic sludge blanket, expanded granular sludge bed, anaerobic filter, anaerobic baffled and fluidized or expanded bed reactors (Chong et al., 2012), among others.

Of particular significance in the application and implementation of high-rate anaerobic systems was the development of the upflow anaerobic sludge blanket (UASB) reactor by Lettinga et al. (1983) which has been used to the greatest extent for domestic wastewater treatment. Since its development, there have been a number of applications of full-scale mainstream anaerobic treatment of domestic wastewater, with over 1,000 UASB reactors installed worldwide (Tiwari et al., 2006) particularly in developing countries with warmer climates such as Brazil, Colombia, Mexico, Egypt, and India (van Haandel et al., 2006). A general schematic diagram of a UASB reactor is shown in Figure 1. The UASB process employs a dense granular sludge bed to maintain the biomass and to filter solids at the bottom of the reactor. A gas-solids separator is used at the top to capture digester gas while preventing solids from leaving the reactor (Lettinga et al., 1986 and 1991; Souza, 1986).

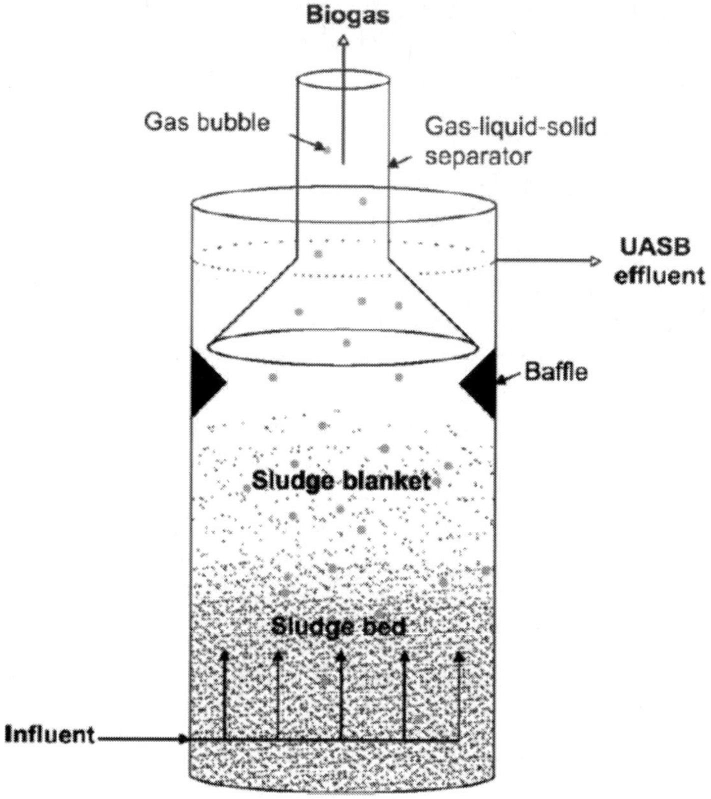

Figure 1. Schematic of UASB reactor (*adapted from Chong et al., 2012*)

Table 1 presents a compendium of advantages and disadvantages of high-rate anaerobic treatment over aerobic processes. Despite of many of the advantages of high-rate anaerobic treatment, there are still disadvantages for this technology that need further attention. One of the major drawbacks cited is the effect of low temperature and low influent organic concentrations on mainstream anaerobic treatment of domestic wastewaters. Specifically, COD removal rates and CH_4 production rates are low owing to lower microbial activity under psychrophilic (< 20°C) conditions. However, many laboratory studies and full-scale applications have shown good performance at temperatures as low as 5°C and with hydraulic retention times (HRT) of only a few hours (Tseng and Ling, 1994; Switzenbaum and Jewell, 1980). Despite this, greater insights into the relationship between process performance and microbial characteristics in the anaerobic treatment of municipal wastewaters at low temperature are required to achieve stable COD removal and biogas production. Other major drawbacks have been reported as long SRT requirements, long start-up period, impure biogas and incomplete or insufficient removal of organics, pathogens and nutrients in the final effluent.

Table 1. Advantages and disadvantages of high-rate anaerobic treatment systems over aerobic processes

Advantages	Disadvantages
• Low energy consumption • Low sludge production • Low space requirement • It can easily be applied on either a very large or a very small scale. • High energy generation potential • High efficiency • It can handle organic shock loads effectively • Low nutrient and chemical requirement	• Necessity of post-treatment to meet secondary standard • Proper temperature control (15-35°C) required for colder climates. • Low pathogen and nutrient removal • Long start-up • Odor generation due to hydrogen sulfide • Quality of biogas

Seghezzo et al., 1998; von Sperling and Chernicharo, 2005; Switzenbaum, 2007; and McCarty et al., 2011

It is important to consider dissolved CH_4 gas in the anaerobic treatment of municipal wastewaters, especially at low temperature conditions. The loss of dissolved CH_4 is enhanced at lower temperatures because of the increase in CH_4 solubility at low temperatures (Bandara et al., 2011). Discharge of residual dissolved CH_4 in the reactor effluent is usually not recovered which results in greenhouse gas emission and lost of energy (Hartley and Lant, 2006; Hatamoto et al., 2010; Matsuura et al., 2010). Few studies have investigated the removal of dissolved CH_4 in anaerobic wastewater treatment processes by biological oxidation, physical gasification based on gas-liquid equilibrium, mixing with gas or a paddle and degassing membranes (Hartley and Lant, 2006; Hatamoto et al., 2010; Matsuura et al., 2010; Pauss et al., 1990; Wasala et al., 2011). However, the recovery efficiency in most cases was low and/or the recovered CH_4 gas composition was low in these processes. Hence, further research is required on this area.

An extensive literature reviewed was conducted to document the performance of different high-rate anaerobic processes, especially at psychrophilic temperatures (less than 20°C), and the results are summarized in Table 2. The highest removal efficiency reported and presented in Table 2 was 86% at 30°C by Cherinchao and Machado (1998). At psychrophilic temperatures, Orozco (1993) reported average COD removal efficiencies averaging 82% at 15°C from an anaerobic baffled reactor. Lettinga (1980) treated wastewater (140-1100 mg/L COD) at ambient temperatures of 8-20°C using a UASB. Removal efficiency of 65-90% was achieved for influent COD greater than 400 mg/L and an efficiency of 50-65% was obtained for COD concentrations less than 300 mg/L. At temperatures of 15-30°C, a one-year pilot-scale UASB system was able to consistently remove 60-70% COD, signifying the satisfactory independence of system performance from climate conditions (La Motta et al., 2007). These two studies indicated temperature had limited effect on removal efficiency of the systems. However, pilot plant results reported by Agrawal et al. (1997) indicate significant decrease of COD removal from a pilot-plant UASB at 10°C. To date, the role of dissolved CH_4 in effluent of anaerobic systems as a contributor of COD has been overlooked and should be further researched since this could provide insights at understanding the true impact of temperature on COD removal.

In summary, COD removals expected with high-rate anaerobic reactors range from 60 to 80%, which is not quite sufficient to meet stringent Clean Water Act regulatory standards for secondary treatment. Because of this and other experiences, it has been commonly suggested that effluent polishing or a post-treatment steps are necessary to meet effluent standards (Langenhoff et al., 2000; Chong et al., 2012).

The necessity for effluent polishing or post-treatment to meet stringent effluent standards has led to the development of two-step processes, including activated sludge systems, biofilters, stabilizing ponds and wetlands to mention a few (Chong et al., 2012). Figure 2 shows an UASB-activated sludge (AS) system which consists of a UASB reactor, a continuous-flow aeration tank and a settler. Aerobic post-treatment has been used after high-rate anaerobic processes to meet lower effluent requirements (Jordao et al., 2004), particularly in industrial applications with high strength waste streams. The uniqueness of this system is that the UASB reactor replaces the primary sedimentation tank as in the conventional activated sludge system (Chernicharo, 2006). La Motta et al. (2007) demonstrated that a UASB-AS system was able to produce an effluent which meets the secondary effluent water quality requirements (with COD and TSS concentrations of 46 and 8 mg/L in the final effluent respectively). This two-step process has been proven to reduce the energy requirements for carbon removal by 45 to 55% of the conventional system according to Jordao et al (2004).

Table 2. Summary of performance of high-rate anaerobic systems

Source	Process	Temperature (°C)	COD Removal (%)
Orozco (1993)	ABR	15	82
Agrawal et al. (1997)	UASB	10	25
Cherinchao and Machado (1998)	UASB + AF	15 - 35	86
von Sperling et al. (2001)	UASB	30	68 – 85
Elmitwalli et al. (2001)	AF + AH	13	71
Gao et al. (2001)	UAFB	15 - 35	49
Lew et al. (2003)	UASB + AF	10 - 28	63
Noyala (2004)	UASB	25	70 – 80
Noyala (2004)	UASB	25	69 – 76
Mahmoud et al. (2004)	UASB-Digester	15	66
Jordao et al. (2004)	UASB	20	62
Jordao et al. (2004)	UASB	20	65
Giraldo et al. (2007)	UASB	20	75
Giraldo et al. (2007)	UASB	20	70
La Motta et al. (2007)	UASB	15-30	60-70
Mahmoud (2008)	UASB	20	80
Alvarez et al. (2008)	HUSB + UASB	14	64
Tawfik et al. (2008)	UASB	20	70
Cao and Ang (2009)	UASB	30	60-70
Oliviera and von Sperling (2009)	UASB	20	80
Mungray and Patel (2011)	UASB	20-30	40-60

UASB: upflow anaerobic sludge blanket; AF: anaerobic filter; AH: anaerobic hybrid; UAFB: upflow anaerobic fixed bed; HUSB: hydrolytic upflow sludge blanket; ABR: anaerobic baffled reactor

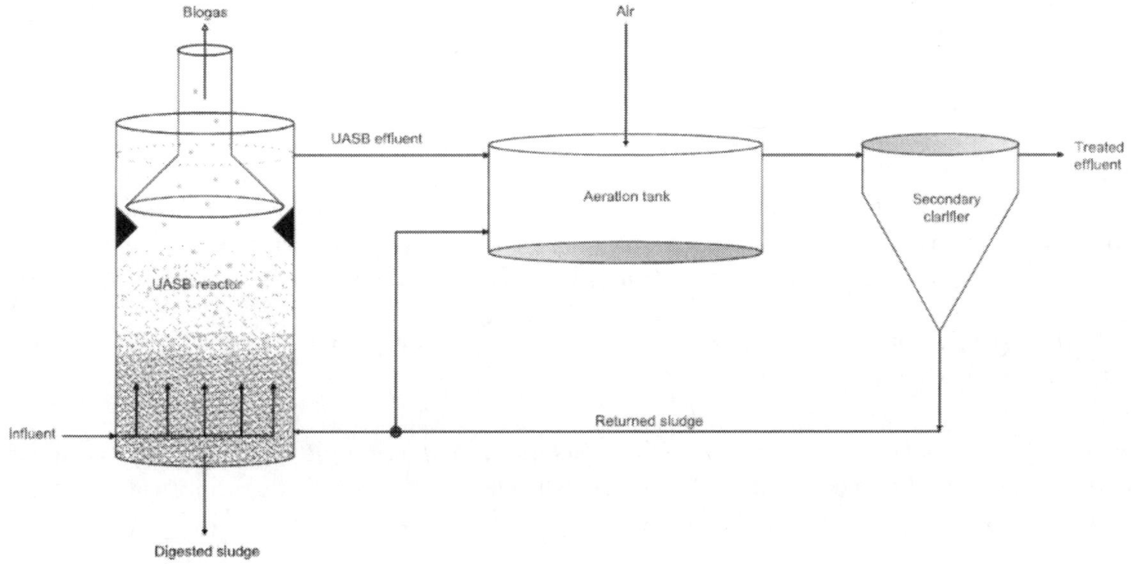

Figure 2. Typical configuration for a UASB-activated sludge system (*adapted from La Motta, 2007*)

Membrane bioreactors are widely used today for aerobic wastewater treatment, as they are capable of producing a high quality effluent with low suspended solids concentration and small footprint relative to traditional aerobic treatment systems. Therefore, recent studies with high-rate anaerobic systems have suggested the use of anaerobic membrane bioreactors (AnMBR) to provide good quality effluent with low suspended solids and COD concentrations in lieu of an aerobic polishing step (Hu and Stuckey, 2006; Kim et al., 2011; and Smith et al., 2012). Several studies have suggested that one of the main drawbacks from using AnMBR system is membrane bio-fouling. Several approaches have been applied to counteract membrane biofouling in AnMBRs, such as back-flushing (Chu et al., 2005; Ho and Sung, 2010) and biogas sparging (Dagnew et al., 2011; Gimenez et al., 2011; Huang et al., 2011; Martinez-Sosa et al., 2011). Using biogas sparging and backflushing concurrently was observed by Smith et al. (2012) to be more effective than either control method alone in a pilot-scale AnMBR.

A number of studies have been published assessing AnMBR performance for the treatment of municipal wastewater as reviewed recently by Smith et al. (2012). However, only a few studies have evaluated AnMBR performance at psychrophilic temperatures of 15°C and below. Specifically, Chu et al.(2005), Ho and Sung(2010) and Smith et al. (2012) observed average COD removals of 70 to 85% at 15°C, and Wen et al.(1999) reported an average COD removal of 88% at 12°C. Smith et al. (2012) reported final effluent quality for COD and 5-day biochemical oxygen demand (BOD_5) from an AnMBR treating municipal wastewater at 15°C, averaging 76±10 mg/L and 24±3 mg/L, respectively, indicating compliance with the U.S. EPA's standard for secondary effluent (30 mg/L BOD_5).

It is useful to compare and contrast the possible benefits of the new AnMBR process with its predecessor, the UASB. It is obvious that the main advantage of the AnMBR is with respect to effluent quality and its ability to produce low solids effluent, and pathogen removal. However,

AnMBR systems can result in approximately at least 0.65 kWh/m^3 of energy generation (assuming typical wastewater characteristics), of which 90% needs to be used internally for filtration, aeration, and pumping. One of the largest consumers of energy is the anaerobic membrane process itself, which requires methane gas sparging, which could use up to 40% of the energy produced to maintain the membranes. The AnMBR is also a complex process, requiring gas recirculation, liquid effluent pumps, and membrane cleaning chemicals and infrastructure. This means that maintenance is relatively complex and expensive. In the case of the UASB reactor, the energy requirement for feed pumping is about 20% of that for an AnMBR since the solids liquid separation is achieved by biological filtration and gravity settling. In addition, the UASB does not have the complex infrastructure or chemical demands of a membrane process.

COMPARISON OF ANAEROBIC TREATMENT AND CONVENTIONAL ACTIVATED SLUDGE TREATMENT

An evaluation was made of the potential benefits of mainstream high-rate anaerobic wastewater treatment compared to a conventional activated sludge system with sludge digestion, assuming the wastewater composition listed in Table 3.

Table 3. Summary of wastewater composition used for comparison of mainstream high-rate anaerobic wastewater treatment compared to a conventional activated sludge

Constituent	Municipal wastewater concentrations (mg/L)
COD (Organics)	
Total	500
Dissolved	190
Solids	
TSS	220
VSS	200
Nitrogen	
Ammonia	25
TKN	35
Phosphorus	
Total P	8
Ortho-phosphate	4

Figure 3 through Figure 5 show a comparison of CH_4 generation, net energy production and total sludge produced from three different process configurations as follows: 1. aerobic activated sludge with anaerobic sludge digestion; 2. aerobic activated sludge with primary sedimentation and anaerobic sludge digestion; and 3. mainstream high-rate anaerobic system. Based on these results, the mainstream high-rate anaerobic system can generate significantly more CH_4 than an aerobic activated sludge process with and without primary sedimentation and anaerobic sludge digesters. For the purpose of this analysis, it was assumed that all the CH_4 was recovered from the effluent of the high-rate anaerobic system at psychrophilic conditions. Therefore, with mainstream anaerobic treatment, the CH_4 production can be almost doubled over conventional aerobic treatment, and energy production could greatly exceed the energy needed for plant operation.

Perhaps the most readily adaptable and usable approach to reduce external energy requirements with existing treatment plants is to make full use of the CH_4 produced through use of combined heat and power (CHP) systems (cogeneration). For the purpose of this analysis, it was assumed that electricity would be generated using CH_4-driven engines to turn generators to produce electricity. Therefore, only about 30-40% of the CH_4 energy is converted into electricity (EPA, 2007), the remainder is given off as heat, which may or may not be useful. Figure 4 shows that high-rate anaerobic systems has the potential producing energy to greatly exceed the energy needs for plant operation.

Another significant advantage is that the quantity of digested sludge resulting from mainstream high-rate anaerobic treatment is much less than with aerobic treatment (Figure 5), another highly significant cost as well as energy benefit. Figure 5 shows that the anaerobic system can reduce the total sludge produced by approximately 50 to 60 percent compared to conventional aerobic system with and without primary clarifiers and anaerobic digesters.

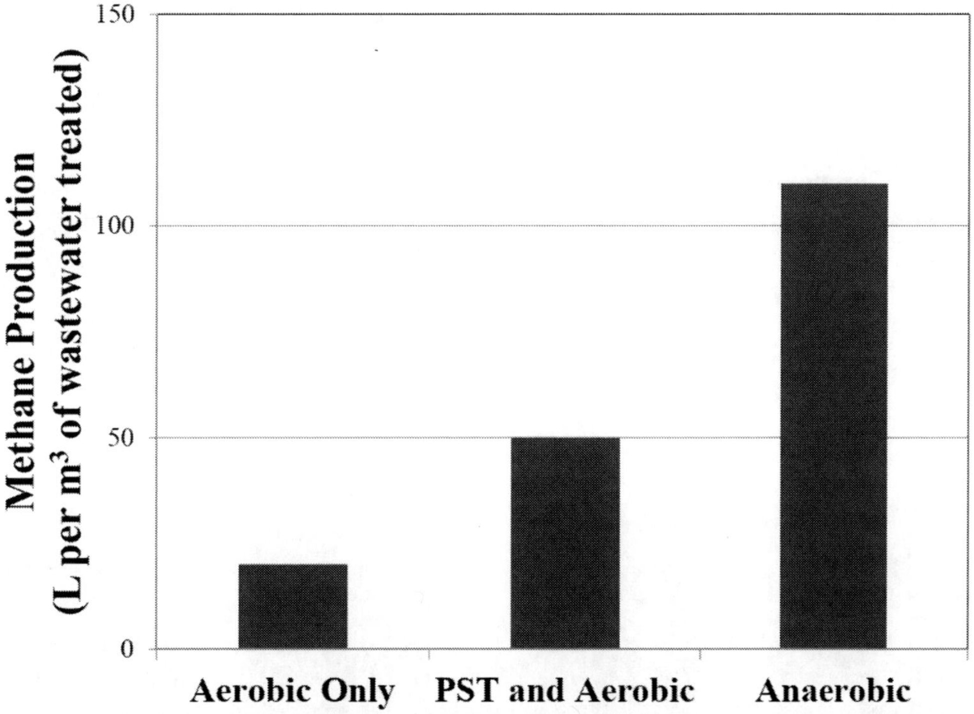

Figure 3. Comparison of methane generation from different process configurations

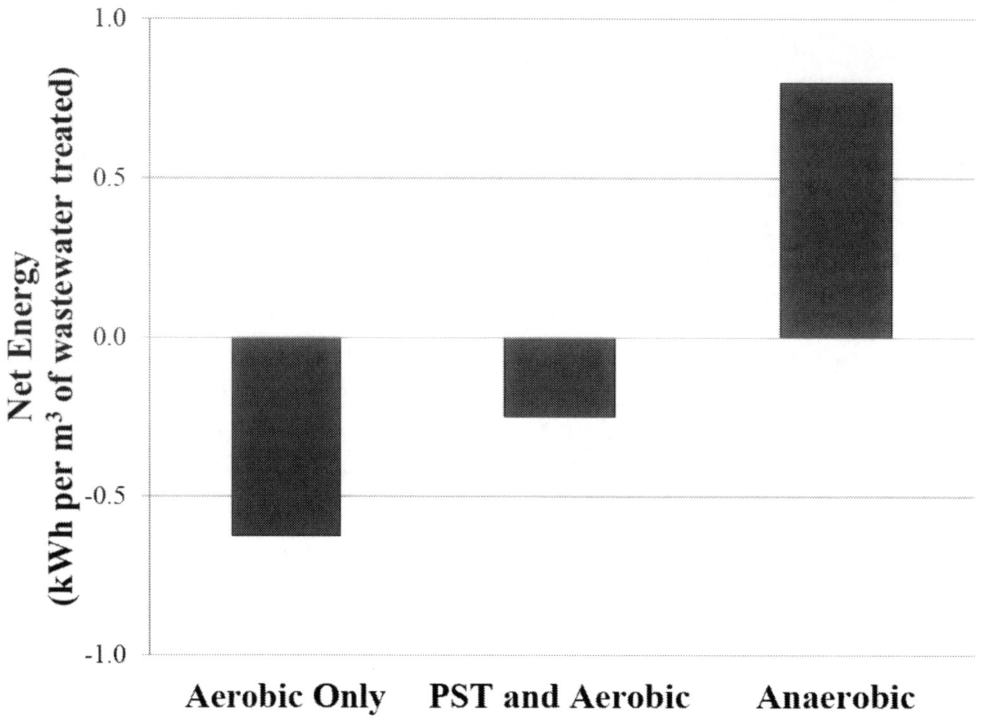

Figure 4. Comparison of net energy generation from different process configurations
[Energy generated using CH_4-driven engines with 35% efficiency, 65% of energy is given off as heat]

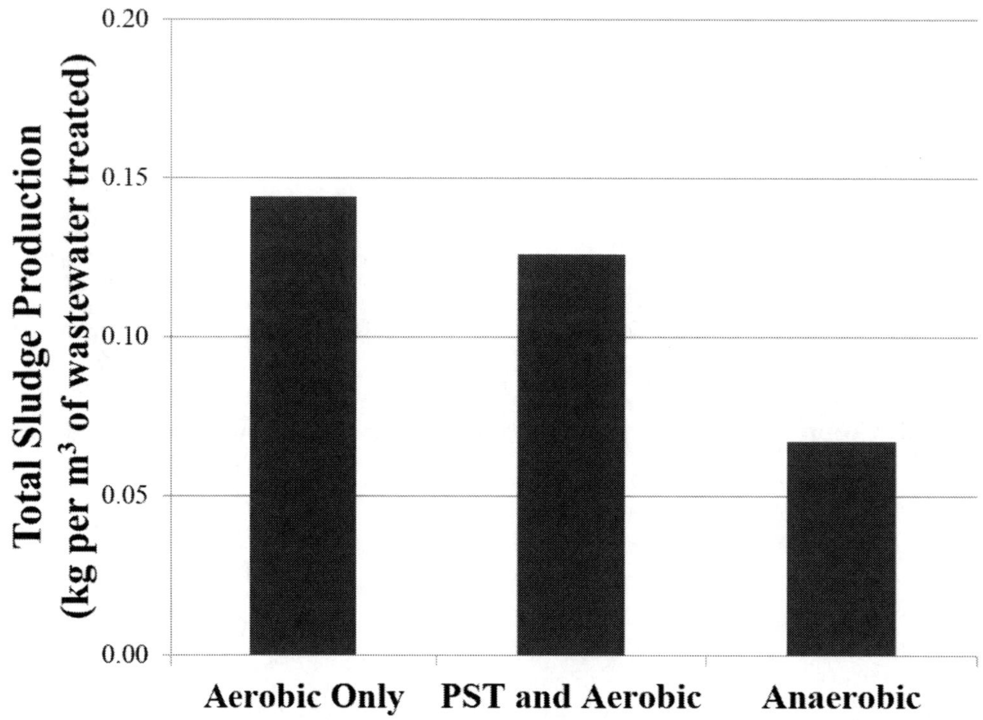

Figure 5. Comparison of total sludge production from different process configurations

FURTHER RESEARCH NEEDED FOR ANAEROBIC SYSTEMS

While complete anaerobic domestic wastewater treatment has potential energy and cost savings compared to conventional aerobic processes, there are still important issues that need to be addressed. Although the biogas produced in high-rate anaerobic reactors treating domestic wastewater usually contains high concentrations of CH_4 (70–80%), significant amounts of the dissolved CH4 escapes with the effluent (van Haandel and Lettinga 1994; Noyola et al. 2006). Therefore, for climate change concerns, CH_4 must not be allowed to escape to the atmosphere but should be collected and used. This becomes more important at systems operating at psychrophilic temperatures due to the increase in CH_4 solubility at low temperatures.

Another, associated problem that also requires more attention is sulfate (SO_4^{2-}) reduction to sulfide (S^{2-}), which competes with CH_4 production and produces hydrogen sulfide (H_2S), which is a toxic and corrosive gas (Speece, 2008; Noyola et al., 2006) and can produce effluent odor and toxicity concerns. Removal of sulfide from the biogas is required for its utilization for energy generation.

Removal of pharmaceuticals and hormones through anaerobic reactors also requires additional attention. Kujawa-Roeleveld et al. (2006) investigated the biodegradation potential of a number of selected pharmaceuticals in batch tests under different redox conditions: acetylsalicylic acid, diclofenac, ibuprofen, carbamazepine, metropolol, clofibric acid, bezafibrate and fenofibrate. Under anaerobic conditions, only three out of the eight compounds namely acetylsalicylic acid, ibuprofen and feno-fibrate were removed after a long batch digestion time of 30 days. Under aerobic conditions, the biodegradation of these three compounds was shown to be much higher. Therefore, mainstream implementation of high-rate anaerobic systems may be coupled with tertiary processes for the removal of pharmaceuticals from wastewater.

The high-rate anaerobic reactors may replace the primary sedimentation, the anaerobic sludge digester, the aerobic step (e.g. activated sludge), and the secondary settling tanks of a conventional aerobic treatment plant. However, the effluent from these anaerobic reactors usually needs further treatment. There are many approaches here that can be used with anaerobic systems and one example is presented in Figure 6. For the next generation of nitrogen removal plants, it is envisioned that separation of the carbon removal step from the nitrogen removal step is necessary to limit substrate competition. Normally, this has been accomplished with the adsorption/bio-oxidation (A/B) process (Boehnke et al., 1997). However, in this case the carbonaceous removal stage (A-stage) might be replaced with a high-rate anaerobic process instead to maximize the energy generation potential of these systems.

In this case, the removal of most of the organics (soluble and particulate) takes place in a high-rate anaerobic reactor. CH_4 gas produced during the anaerobic process would be collected, clean and use for energy generation in a CHP system, but dissolved CH_4 is difficult to recover, especially at lower temperatures. Since the release of the dissolved methane into the environment contributes to the greenhouse effect is undesirable. However, methane is an energy-rich compound that can be used as an electron donor to circumvent the need for expensive electron donors (i.e. methanol) in a downstream nitrogen removal process. Luesken et al. (2011) reported about the recent discovery and enrichment of bacteria that couple anaerobic methane oxidation to nitrite reduction (n-damo). The nitrite-dependent anaerobic oxidation of methane can be coupled with anaerobic ammonium oxidation (anammox) for the removal of nitrogen. The combination of these processes would be able to remove ammonium, dissolved methane, and nitrite simultaneously without the need for extra aeration and the addition of electron donors.

Another possibility for nitrogen removal in a downstream process would be to use the hydrogen sulfide in the biogas as an electron donor in an autotrophic post-denitrification step (Kleerebezem and Mendez, 2002). In this case, a nitrification-denitrification step would be required but instead of using external electron donor for denitrification, hydrogen sulfide would be used. Denitrification with sulfide as electron donor is an autotrophic process, suggesting significantly lower biomass yields when compared to heterotrophic denitrification.

Phosphorus removal can be accomplished via chemical precipitation and solids liquid separation in the down-stream process.

Waste activated sludge from the downstream reactor can be sent to the head of the high-rate anaerobic reactor for digestion and stabilization before final disposal, similar to the process configuration developed by La Motta et al. (2007).

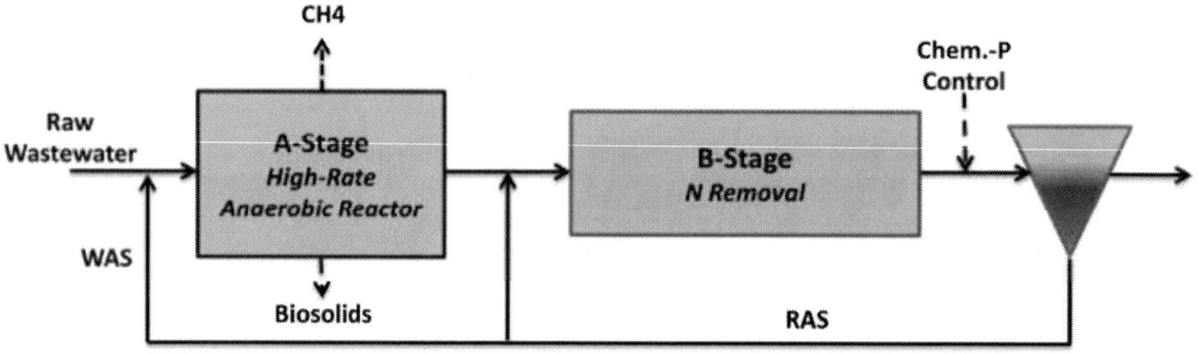

Figure 6. Possible integration of high-rate anaerobic treatment and nutrient removal

CONCLUSIONS

The carbon emission and therefore carbon footprint of water utilities is an important issue. In this respect, we should consider the opportunities to reduce carbon footprint from wastewater treatment plants. The use of anaerobic rather than aerobic treatment processes would achieve this aim because no aeration is required and the generation of methane can be used within the plant. Therefore, the successful application of high-rate anaerobic systems represent a paradigm shift for the industry, offering the opportunity for sustainable wastewater treatment, energy neutral or even energy positive facilities and dramatic reductions in treatment costs, which has widespread environmental, economic and societal benefits.

A great deal of progress has been made to understand high-rate anaerobic wastewater treatment processes and their application to the treatment of domestic wastewater. The following conclusions can be offered:

- Anaerobic treatment can be considered an appropriate technology for the removal of organics from municipal wastewater at psychrophilic temperatures.
- Effluent polishing or post-treatment to meet stringent effluent standards is necessary in most cases.
- Recovery of residual dissolved CH_4 in the reactor effluent needs to be addressed because of greenhouse gas emission and lost of energy implications.
- Hydrogen sulfide control needs to be implemented due to odor and toxicity concerns.
- Additional processes are required for nutrient and pharmaceutical removals.
- CH_4 and hydrogen sulfide have the potential to serve as electron donors for denitrification in downstream nitrogen removal processes.

REFERENCES

Agrawal, L., Harada, H., and Okui, H. (1997) Treatment of dilute wastewater in a UASB reactor at a moderate temperature: performance aspects. *Journal of Fermentation and Bioengineering*, 83(2), 179-184.

Alvarez, J., Armstrong, E., Gomez, M., Soto, M. (2008) Anaerobic treatment of low-strength municipal wastewater by a two-stage pilot plant under psychrophilic conditions. *Bioresource Technology*, 99(15), 7051-7062.

Bandara, W., Satoh, H., Sasakawa, M., Nakahara, Y., Takahashi, M., Okabe, S. (2011) Removal of residual dissolved methane gas in an upflow anaerobic sludge blanket reactor treating low strength wastewater at low temperature with degassing membrane. *Water Research*, 45(11), 3533-3540.

Boehnke, B., Diering, B., and Zuckut, S. (1997) AB process removes organic and nutrients. *Water Environment and Technology*, 9(3), 23-27.

Cao, Y., Ang, C.M. (2009) Coupled UASB-activated sludge process for COD and nitrogen removals in municipal sewage treatment in warm climate. *Water Science and Technology,* 60(11), 2829-2839.

Chernicharo, C., Machado, R. (1998) Feasibility of the UASB/AF system for domestic sewage treatment in developing countries. *Water Science and Technology,* 38 (8-9), 325-332.

Chernicharo, C. (2006) Post-treatment options for the anaerobic treatment of domestic wastewater. Reviews in *Environmental Science and Biotechnology* 5 (1), 73-92.

Chong, S., Sen, T. K., Kayaalp, A., Ang, H. M. (2012) The performance enhancements of upflow anaerobic sludge blanket (UASB) reactors for domestic sludge treatment: A State-of-the-art review. *Water Research,* 46, 3434-3470.

Chu, L. B., Yang, F. L., Zhang, X. W. (2005) Anaerobic treatment of domestic wastewater in a membrane-coupled expended granular sludge bed (EGSB) reactor under moderate to low temperature. *Process Biochemistry,* 40(3-4), 1063-1070.

Curtis, T. P. (2010) Low-energy wastewater treatment: strategies and technologies. Environmental Microbiology, 2nd edition. Wiley-Blackwell: Hoboken, NJ.

Elmitwalli, T., Zeeman, G., Lettinga, G. (2001) Anaerobic treatment of domestic sewage at low temperature. *Water Science and Technology,* 44 (4), 33-40.

Dagnew, M., Parker, W., Seto, P., Waldner, K., Hong, Y., Bayly, R., Cumin, J. (2011) Pilot testing of an AnMBR for municipal wastewater treatment. Proceeding of the 84th Annual Water Environment Federation Technical Exhibition and Conference, Los Angeles, CA.

Environmental Protection Agency (EPA) Office of Water (2006) Wastewater Management Fact Sheet, Energy Conservation, EPA 832-F-06-024; U.S. Environmental Protection Agency: Washington DC.

Environmental Protection Agency (EPA) Office of Water (2007) Opportunities for and Benefits of Combined Heat and Power at Wastewater Treatment Facilities, EPA-430-R-07 003; Environmental Protection Agency: Washington DC, 2007

Foley, J., de Haas, D., Hartley, K., and Lant, P. (2010) Comprehensive life cycle inventories of wastewater treatment systems. *Water Research*, 44(5), 111-124.

Gao, D., Tao, Y., An, R., Fu, Y., Ren, N. (2011) Fate of organic carbon in UAFB treating raw sewage: Impact of moderate to low temperature. *Bioresource Technology,* 102(3), 2248-2254.

Gimenez, J. B., Robles, A., Carretero, L., Duran, F., Ruano, M. V., Gatti, M. N., Ribes, J., Ferrer, J., Seco, A. (2011) Experimental study of the anaerobic urban wastewater treatment in a submerged hollow-fiber membrane bioreactor at pilot scale. Bioresource Technology 102(19), 8799-8806.

Giraldo, E., Pena, M., Cherincharo, C., Sandino, J., and Noyala, A. (2007) Anaerobic sewage treatment technology in Latin-American: A selection of 20 years of experience. Proceedings of WEFTEC. Water Environment Federation, Alexandria, VA.

Hartley, K., Lant, P. (2006) Eliminating non-renewable CO2 emissions from sewage treatment: an anaerobic migrating bed reactor pilot plant study. *Biotechnology and Bioengineering,* 95(3), 384-398.

Hatamoto, M., Yamamoto, H., Kindaichi, T., Ozaki, N., Ohashi, A. (2010) Biological oxidation of dissolved methane in effluents from anaerobic reactors using a down-flow hanging sponge reactor. *Water Research,* 44(5), 1409-1418.

Ho, J., Sung, S. (2010) Methanogenic activities in anaerobic membrane bioreactors (AnMBR) treating synthetic municipal wastewater. *Bioresource Technology,* 101 (7), 2191-2196.

Hu, A.Y.; Stuckey, D.C. (2006) Treatment of dilute wastewaters using a novel submerged anaerobic membrane bioreactor. *Journal of Environmental Engineering*, 132 (2), 190–198.

Huang, Z., Ong, S. L., Ng, H. Y. (2011) Submerged anaerobic membrane bioreactor for low-strength wastewater treatment: effect of HRT and SRT on treatment performance and membrane fouling. *Water Research*, 45, 705-713.

Kim, J.; Kim, K.; Ye, H.; Lee, E.; Shin, C.; McCarty, P. L.; Bae, J. (2011) Anaerobic fluidized bed membrane bioreactor for wastewater treatment. *Environmental Science and Technology*, 45, 576–581.

Kleerebezem, R., and Mendez, R. (2002) Autotrophic denitrification for combined hydrogen sulfide removal from biogas and post-denitrification. *Water Science and Technology*, 45(10), 349–356.

La Motta, E.J., Silva, E., Bustillos, A., Padron, H., Luque, J. (2007) Combined anaerobic/aerobic secondary municipal wastewater treatment: pilot-plant demonstration of the UASB/ Aerobic solids contact system. *Journal of Environmental Engineering,* 133 (4), 397-403.

Langenhoff, A. A. M.; Stuckey, D. C. (2000) Treatment of dilute wastewater using an anaerobic baffled reactor: Effect of low temperature. *Water Research*, 34 (15), 3867–3875.

Lettinga, G., van-Velsen, F., Hobma, S., De-Zeeuw, W., and Klapwijk, A. (1980) Use of the Upflow Sludge Blanket (USB) reactor concept for biological wastewater treatment. *Biotechnology and Bioengineering*, 22, 699-734.

Lettinga, G.; Roersma, R.; Grin, P. (1983) Anaerobic treatment of raw domestic sewage at ambient-temperatures using a granular-bed UASB reactor. *Biotechnology and Bioengineering*, 25 (7), 1701–1723.

Lettinga, G., Hulshoff Pol, L.W., (1991) UASB-process design for various types of wastewaters. *Water Science and Technology*, 24 (8), 87-107.

Lew, B., Belavski, M., Admon, S., Tarre, S., Green, M. (2003) Temperature effect on UASB reactor operation for domestic wastewater treatment in temperate climate regions. *Water Science and Technology*, 48 (3), 25-30.

Luesken, F. A., Sanchez, J., van Alen, T., Sanabria, J, den Camp, H., Jetten, M., Kartal, B. (2011) Simultaneous Nitrite-Dependent Anaerobic Methane and Ammonium Oxidation Processes. *Applied and Environmental Microbiology*, 77(19), 6802–6807.

Jewell, W. J. (1987) Anaerobic sewage treatment. *Environ. Sci. Technol.* 21 (1), 14–21.

Jordao, E. P., Volschan, I. (2004) Cost-effective solutions for sewage treatment in developing countries: The case of Brazil. *Water Sci. Technol.* 50 (7), 237–242.

Mahmoud, N., Zeeman, G., Gijzen, H., Lettinga, G. (2004) Anaerobic sewage treatment in a one-stage UASB reactor and a combined UASB-Digester system. *Water Research*, 38 (9), 2348-2358.

Mahmoud, N. (2008) High strength sewage treatment in a UASB reactor and an integrated UASB-digester system. *Bioresource Technology*, 99 (16), 7531-7538.

Martinez-Sosa, D., Helmreich, B., Netter, T., Paris, S., Bischof, F., Horn, H. (2011) Anaerobic submerged membrane bioreactor (AnSMBR) for municipal wastewater treatment under mesophilic and psychrophilic temperature conditions. *Bioresource Technology*, 102 (22), 10377-10385.

Matsuura, N., Hatamoto, M., Sumino, H., Syutsubo, K.,Yamaguchi, T., Ohashi, A. (2010) Closed DHS system to prevent dissolved methane emissions as greenhouse gas in anaerobic wastewater treatment by its recovery and biological oxidation. *Water Science and Technology*, 61 (9), 2407-2415.

McCarty, P.L. (1982) One hundred years of anaerobic treatment. In: Anaerobic Digestion, 1981, D.E. Hughes et al. (Editors), Elsevier Biomedical.

McCarty, P. L., Bae, J., and Kim, J. (2011) Dometict Wastewater Treatment as a Net Energy Producer – Can This be Achieved? *Environ. Sci. Techol.*, 45, 7100-7106.

Mungray, A. K., Patel, K. (2011) Coliforms removal in two UASB + ASP based systems. *International Biodeterioration and Biodegradation*, 65 (1), 23-28.

Noyola, A.; Morgan-Sagastume, J. M.; Lopez-Hernandez, J. E. (2006) Treatment of biogas produced in anaerobic reactors for domestic wastewater: odor control and energy/resource recovery. *Rev. Environ. Sci. Bio/Technol.*, 5, 93–114.

Pauss, A., Andre, G., Perrier, M., Guiot, S.R., 1990. Liquid-to-gas mass transfer in anaerobic processes: inevitable transfer limitations of methane and hydrogen in the bio-methanation process. *Applied and Environmental Microbiology*, 56 (6), 1636-1644.

Oliveira, S. C.; Von Sperling, M. (2008) Reliability analysis of wastewater treatment plants. *Water Res.*, 42 (45), 1182–1194.

Oliveira, S., von Sperling, M. (2009) Performance evaluation of UASB reactor systems with and without post-treatment. *Water Science and Technology*, 59 (7), 1299-1306.

Reardon, D. J. (1995) Turning Down the Power. *Civil Engineering*,. **65**(8) 54-56.

Seghezzo, L., Zeeman, G., van Lier, J.B., Hamelers, H.V.M., Lettinga, G., (1998) A review: the anaerobic treatment of sewage in UASB and EGSB reactors. *Bioresource Technology*, 65 (3), 175-190.

Shizas, I. and Bagley, D. (2004) Experimental Determination of Energy Content of Unknown Organics in Municipal Wastewater Streams. *J. Energy Engineering*, 130/2, 45-53.

Speece, R. E. (2008) Anaerobic Biotechnology and Odor/Corrosion Control; Archae Press: Nashville, TN, p 586.

Souza, M.E. (1986) Criteria for the Utilization, Design and Operation of UASB Reactors. *Wat. Sci. Tech.,* 18, (12) 55-69.

Smith, A., Skerlos, S., and Raskin, L. (2012), Psychrophilic anaerobic membrane bioreactor treatment of domestic wastewater. *Water Research* – In Press.

Switzenbaum, M. S.; Jewell, W. J. (1980) Anaerobic attached - film expanded bed reactor treatment. *J. Water Pollut. Control Fed.*, 52 (7), 1953–1965.

Tawfik, A., Sobhey, M., Badawy, M. (2008) Treatment of a combined dairy and domestic wastewater in an up-flow anaerobic sludge blanket (UASB) reactor followed by activated sludge (AS system). *Desalination*, 227 (1-3), 167-177.

Tchobanoglous, G.; Burton, F. L. (1991) Wastewater Engineering: Treatment, Disposal, Reuse, 3rd ed.; McGraw-Hill, Inc.: New York; p 1334.

Tiwari, M., Guha, S., Harendranath, C., Tripathi, S. (2006) Influence of extrinsic factors on granulation in UASB reactor. *Applied Microbiology and Biotechnology*, 71 (2), 145e154.

Tseng, S. K.; Lin, M. R. (1994) Treatment of organic wastewater by anaerobic biological fluidized bed reactor. *Water Sci. Technol.*, 29 (12), 157–166.

van Haandel, A. C., and Lettinga, G. (1994) Anaerobic Sewage Treatment, A Practical Guide for Regions with a Hot Climate. John Wiley & Sons Ltd.,West Sussex, p. 226.

van Haandel, A.; Kato, M. T.; Cavalcanti, P. F; Florencio, L. (2006) Anaerobic reactor design concepts for the treatment of domestic wastewater. *Rev. Environ. Sci. Bio-Technol.*, 5, 21–38.

von Sperling, M., Freire, V., Chernicharo, C. (2001) Performance evaluation of a UASB Activated sludge system treating municipal wastewater. *Water Science and Technology*, 43 (11), 323-328.

von Sperling, M., Chernicharo, C. A. L. (2005) Biological Wastewater Treatment in Warm Climate Regions. IWA Publishing.

Wen, C., Huang, X., Qian, Y. (1999) Domestic wastewater treatment using an anaerobic bioreactor coupled with membrane filtration. *Process Biochemistry*, 35 (3-4), 335-340.

Achieving Positive Net Energy While Maximizing Use of Existing Assets: The Ejby Mølle WWTP Story

Julian Sandino, CH2M HILL
Per Henrik Nielsen, VandCenter Syd (VCS) Denmark
Dru Whitlock, CH2M HILL
Tom Johnson, CH2M HILL
Mads Leth, VandCenter Syd (VCS) Denmark
Thomas Jensen, Ramboll Denmark
Michael Nielsen, Ramboll Denmark
CH2M HILL
2300 Main Street, Suite 180
Kansas City, MO 64114
+1 (816) 728 8369

ABSTRACT

A series of energy optimization options (EOOs) were identified for the Ejby Molle WWTP through the analysis of historical operational data, the use of an advanced mass/energy balance simulation tool, and the adoption of a collaborative workshop-based approach in identifying and evaluating alternatives. This biological nutrient removal facility is located in Odense, Denmark, and has a treatment capacity suitable for 385.000 population equivalents. Implementation by plant staff of some of the recommended EOOs (primarily those only requiring modifications in operational strategies of existing unit processes) has succeeded in making this plant essentially energy self-sufficient. Implementation of facility improvement related to providing side stream treatment and improving oxygen transfer efficiencies in its bioreactors will be able to take the plant into a positive net energy condition.

KEYWORDS

Positive net energy, energy optimization, nutrient removal, wastewater treatment

INTRODUCTION AND BACKGROUND

The Ejby Mølle Wastewater Treatment Plant (EMWWTP) is the largest treatment facility in Odense, Denmark, and has a treatment capacity of 385.000 population equivalents (PE). VandCenter Syd (VCS) Denmark, is the provider of water and wastewater services for Odense, and operates the EMWWTP. The load to the plant is mainly of domestic origin, and it discharges to a small river with very stringent effluent requirements. The plant is currently producing on an annual average basis, effluent concentrations of 1.5 mg/L of BOD_5, 4 mg/L of total nitrogen as N, and 0.1 mg/L of total phosphorus as P.

Figure 1 presents a simplified process schematic for the WMWWTP. The plant's liquid treatment train is comprised of pretreatment (grit removal and screening), primary clarification (capable of chemical assistance on demand), biological nutrient removal (BNR) activated sludge process (using oxidation ditches), a two-stage trickling filter process currently used for treating wet weather related excess flows (adapted from an earlier plant configuration aimed at achieving just nitrification), and tertiary sand filters. Four separate pumping stages are used in moving water through this configuration. Sludge processing consists of raw sludge thickening, conventional anaerobic digestion with a combined heat power (CHP) biogas cogeneration system, and digested sludge dewatering. Dewatered cake is hauled offsite for composting.

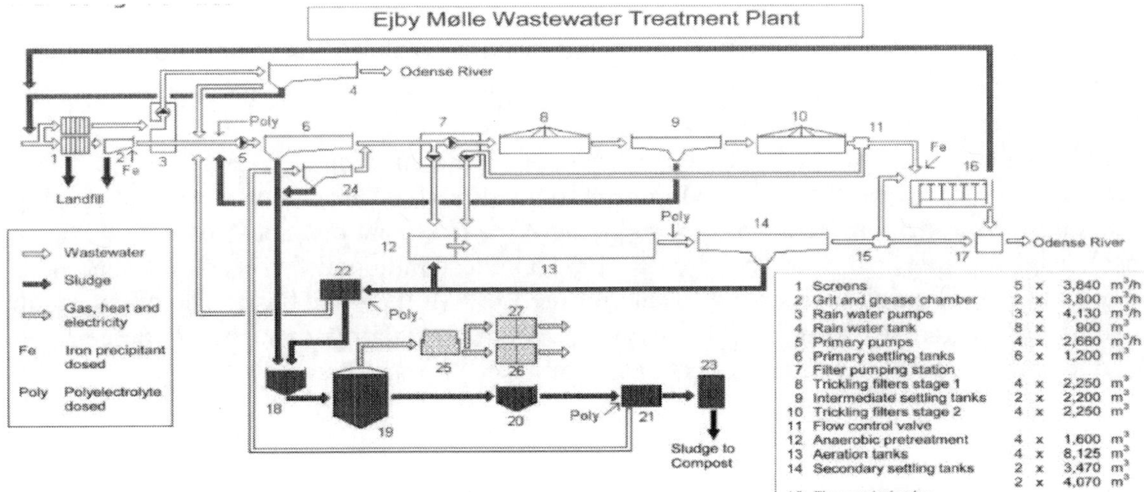

	Capacity	Load Influent	Load Effluent		Discharge limits		values
BOD_5	385,000 PE	276,700	-	PE			1.92
BOD_5		14,774	111	kg/d	8	mg/l	19.8
COD		35,030	1,142	kg/d	50	mg/l	3.96
Total Nitrogen		2,360	229	kg/d	8	mg/l	0.2
Ammonium N		1,270	12	kg/d	3	mg/l	0.12
Total Phosphorus		329	7	kg/d	0.5	mg/l	1.55
Suspended Solids		20,487	89	kg/d	10	mg/l	
Flow average		57,710		m^3/d			
maximum	6,600 m^3/h						
Wet weather maximum	19.000 m^3/h						
Sludge Production			45	tons/d			

1	Screens	5	x	3,840	m^3/h
2	Grit and grease chamber	2	x	3,800	m^3/h
3	Rain water pumps	3	x	4,130	m^3/h
4	Rain water tank	8	x	900	m^3
5	Primary pumps	4	x	2,660	m^3/h
6	Primary settling tanks	6	x	1,200	m^3
7	Filter pumping station				
8	Trickling filters stage 1	4	x	2,250	m^3
9	Intermediate settling tanks	2	x	2,200	m^3
10	Trickling filters stage 2	4	x	2,250	m^3
11	Flow control valve				
12	Anaerobic pretreatment	4	x	1,600	m^3
13	Aeration tanks	4	x	8,125	m^3
14	Secondary settling tanks	2	x	3,470	m^3
		2	x	4,070	m^3
15	Flow control valve				
16	Contact filter	8	x	100	m^3
17	Aeration unit			200	m^3
18	Sludge thickeners	2	x	140	m^3
19	Anaerobic digesters	4	x	2,800	m^3
20	Digested sludge storage			1,900	m^3
21	Centrifuges	2	x	50	m^3/h
22	Waste sludge decanter	1	x	30	m^3/h
23	Sludge skips			120	m^3
24	Supernatant buffer tank			1,200	m^3
25	Gasometer			800	m^3
26	Gas motor generators	1	x	840	kW
		1	x	306	kW
27	Furnaces	3	x	668	kW

Figure 1 - Ejby Mølle WWTP Process Flow Diagram and Summary of Key Facilities

VCS is a very progressive utility, striving to become a model for the incorporation of sustainability principles in discharging its responsibilities. To this end, it has adopted a goal of achieving CO_2 neutrality firm-wide by year 2014. Through its own efforts, VCS has been able to implement changes in the facilities (primarily through the incorporation of the CHP cogeneration scheme) and in how it operates the EMWWTP, achieving in 2011 a very impressive level of 77% energy self-sufficiency. In early 2012, VCS selected through an international tender the team of CH2M HILL and Ramboll Denmark (two global engineering firms) to execute a project aimed at identifying additional energy optimization opportunities (EOOs) at the EMWWTP that would further facilitate the achievement of their declared corporate goal.

GENERAL APPROACH

The approach adopted by the combined utility/consultant team basically aimed at first identifying EOOs that relied primarily on relatively quick to implement process modifications (considering both facility re-configuration as well as modified operational strategies), which would significantly reduce electrical power consumption and/or increase power generation from cogeneration. The adopted approach also identified (but not fully develop) somewhat more complex to implement improvements that could be considered for future implementation by

VCS. The key tool in facilitating the identification, prioritization and eventual development of the selected EOOs was a whole-plant integrated model that can simulate unit process performance and operational requirements through pollutant mass balances, while also predicting the corresponding carbon and energy (electrical, thermal, and calorific) balances.

RESULTS

Data Collection and Analysis. Historical influent, effluent, and operating data was obtained from VCS. The data was obtained primarily through supervisory control and data acquisition (SCADA) collection along with a handful of sampling reports for primary sludge, mixed-liquor suspended solids (MLSS), and dewatered solids. An additional sampling program was conducted in order to provide additional information required for the evaluation of the existing processing capabilities as well as for the modeling of the treatment plant. The data was consolidated and analyzed statistically to develop an understanding of facility flows, loadings, and operations. Electricity consumption data for the Ejby Mølle WWTP was also obtained, and it's summarized graphically in the figure below. Overall, the facility is roughly 77 percent self-sufficient with its existing combined heat and power generation system.

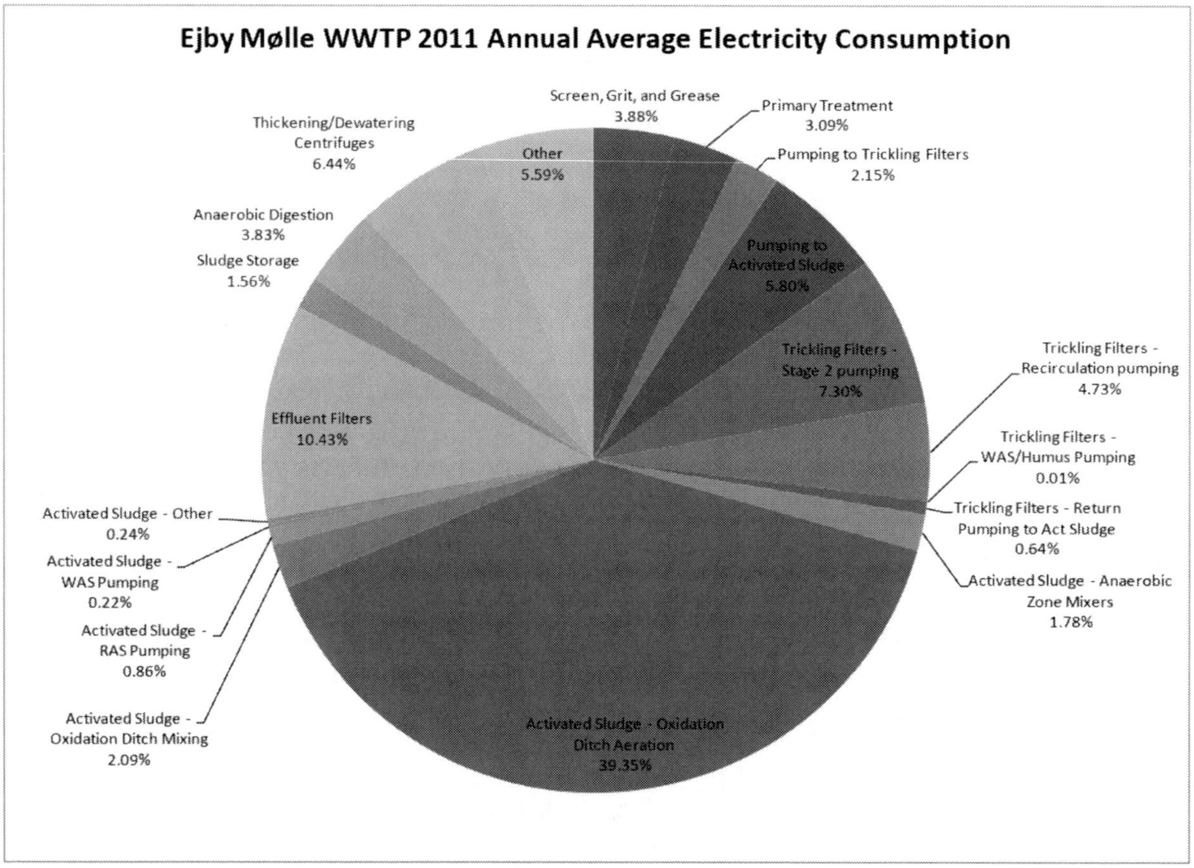

Figure 2 - 2011 EMWWTP Energy Consumption

Modeling of Existing Facility. A process model using CH2M HILL's proprietary Professional Process Design or Pro2D™was developed for the Ejby Mølle WWTP. Pro2D™ is a static, steady-state, whole-plant simulator that has the capability to integrate process and design information to evaluate existing systems or size new ones. Whole plant means that the model is not limited to just activated sludge processes, but can also evaluate and/or design primary treatment units as well as secondary treatment units including clarification systems and some tertiary systems. The model is based upon a modified version of the International Water Association (IWA) Activated Sludge Model (ASM) version 2d. ASM is the industry standard for modeling, designing, and evaluating activated sludge facilities. Inputs to the process model and used for its calibration were developed from the historical flows, loads, and operations data (including electrical energy consumption and production). The figure below shows the resulting model process flow diagram.

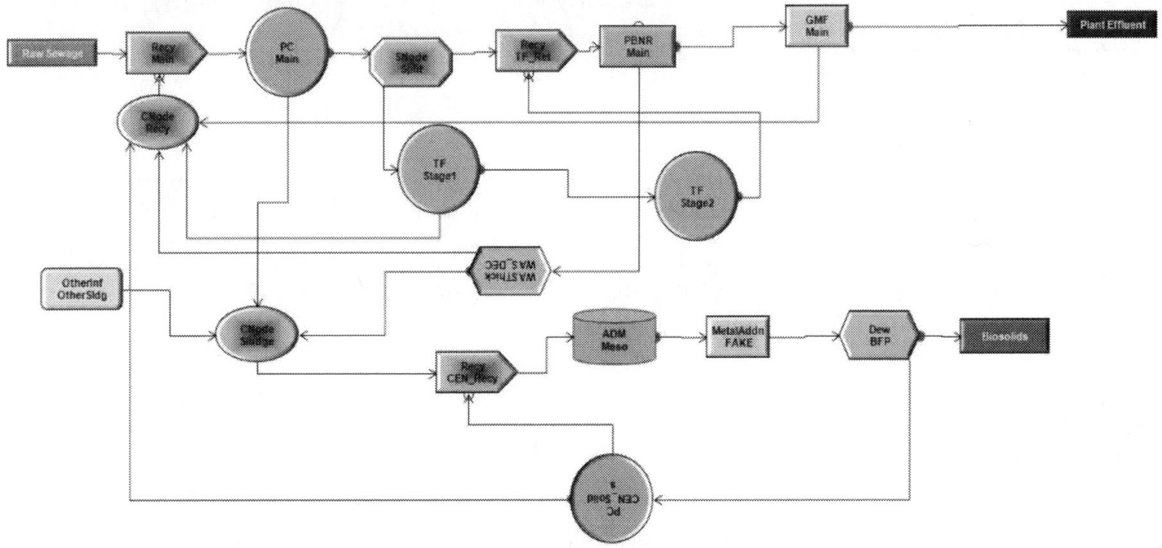

Figure 3 - Pro2D Process Flow Diagram for the EMWWTP

Identification and screening of Energy Optimization Opportunities (EOOs). A meeting was conducted at the Ejby Mølle WWTP on August 22-23, 2012 (Workshop1), and was attended by management and operations staff from VCS, as well as by representatives from the Consulting team. The main objectives of this workshop were to charter the joint VCS/Consultant team, to review the data gathering and analysis activities, to develop a "long list" of energy optimization opportunities (EOOs), to confirm the evaluation methodology, and to establish a "short list" of viable alternatives to further evaluate in greater detail. The following table includes the "long list" of identified EOOs grouped into three categories: Treatment Process Modifications; Equipment Improvements; and Energy Generation Improvements.

Table 1 – Identification of "Long List" of Energy Optimization Opportunities for the EMWWTP

ENERGY OPTIMIZATION OPPORTUNITIES	
TREATMENT PROCESS MODIFICATIONS	
Reroute discharge of centrate tank directly to TFs	Reroute discharge of supernatant buffer tank directly to TFs for the nitrification of plant sidestreams. Nitrified TF effluent would be routed to the biological nutrient removal (BNR) process for denitrification. Preliminary analysis showed 10 percent reduction in aeration basin power demand.
Implement chemically enhanced primary treatment (CEPT)	Implement CEPT at the primary settling tanks to significantly enhance removal performance and reduce organic load to aeration basins. Preliminary analysis indicated that increasing total TSS removal to 80 percent will reduce aeration basin power demand by approximately 20 percent. Additionally, CEPT will result in higher sludge quantities to digestion thus higher biogas generation potential.
ANAMMOX sidestream treatment of centrate	Implement anaerobic ammonia oxidation sidestream treatment (such as DEMON, ANNAMOX, etc.) to reduce total nitrogen load from centrate.
Operate at shorter SRTs	Reduce power consumption in BNR bioreactors by reducing oxygen (O_2) demand and mixing requirements; increase biological sludge to digester, which could result in increased biogas production.
Modify the plant hydraulic profile to reduce pumping energy	Examine opportunities to modify overall plant hydraulics to reduce energy demand. Target reducing the extent of pumping throughout the plant.
A/B (adsorption bio-oxidation) process conversion: Add small bioreactor before primaries to enhance bioflocculation	Aimed at producing high levels of combined biological and primary sludges in the "A" stage, which would then produce more biogas for cogeneration from the anaerobic digesters.
Anaerobic treatment of waste	Implement mainstream anaerobic treatment processes, such as anaerobic MBR, UASB, and anaerobic fluidized bed reactor, to reduce aeration energy demand and reuse biogas produced.
Divide discrete processes into small modules	Construct multiple small unit processes rather than single larger ones to allow process requirements to match flow and load requirements by taking units out of service. For example, constructing several parallel bioreactors allows basins to be removed from service and reduces biomass, which reduces the oxygen requirements and associated aeration energy cost.
Retain heat in bioreactors	Cover treatment basins to reduce the loss of heat energy from biological reactors allowing a higher temperature to be maintained in the biological process. This could allow reducing the minimum SRT required for nitrification.

Operational flexibility	Provide design and operational flexibility to allow operators to conserve energy. For example, provide variable-speed drives on appropriate equipment to match duty to performance requirements and install reliable and appropriate instrumentation to provide operators with rapid, direct feedback on operational requirements such as dissolved oxygen and nutrient probes.
Evaluate option of no brewery waste in influent	Brewery waste makes up a significant portion of organic loading on Ejby Mølle WWTP. By removing this loading, it will decrease energy demand. Conversely, this could reduce carbon availability for BNR process, negatively affecting effluent quality.
Micro-screening after primary clarifiers	Install 200 micron filters downstream of the primary clarifiers to reduce the loading on the secondary treatment system. Screenings could be added directly to the anaerobic digester to enhance biogas production.
Removal of tertiary filters	Bypassing the tertiary filters would decrease the pumping energy to this system and thereby decrease the energy demand. This could be attempted at Ejby Mølle WWTP to determine if the impact on secondary effluent is significant.
Replace effluent filters with lower energy alternative	Replace effluent filters with lower head / lower energy alternative process such as cloth filters.
Liquid stream ANAMMOX	Liquid stream ANAMMOX is intended to remove nitrogen from the liquid stream – similar to the side stream process. Removing TN anaerobically will significantly reduce energy demand.
Replace TF media	Replace TF rock media with plastic media thus increasing the efficiency of operation.
Eliminate Second Stage TF	Eliminating the second stage of the TF system would decrease pumping energy demand.
Run all TFs in parallel rather than in series	By running the TFs in parallel, the pumping demand would be decreased by eliminating one pumping stage therefore decreasing the energy demand.
EQUIPMENT IMPROVEMENTS	
Retrofit bioreactors with diffused aeration	Fine-bubble diffusion delivers oxygen to suspended growth systems more efficiently than mechanical aerators, hence, decreasing the energy demand. This option could replace all of the mechanical aerators or provide a combination of existing aerators and fine bubble diffusers.
Replace existing motors with smaller, higher efficiency units with variable frequency drives (VFDs)	Examine possibilities of decreasing energy demand with the addition of VFDs or higher efficiency equipment.
Retrofit digesters with linear motion mixers	Examine lower energy mixing regimes or mixing equipment to lower the energy demand.
Replace dewatering centrifuges with belt filter presses or similar lower power systems	Belt filter presses and screw presses have much lower energy demand than centrifuges.

Lighting retrofits, fixture upgrades, occupancy sensor installation	Human occupied space can draw a significant amount of energy demand. Lighting and HVAC systems should be examined for EOOs.
Develop energy use dashboard for the plant	Provide real-time feedback for Ejby Mølle WWTP personnel so that energy demand can be honed and modified on a daily basis.
Optimize operation of Grit Chamber	Examine the possibility of operating one grit chamber instead of two to decrease energy demand.
Optimize TF pumping	TF pumping makes up a large portion of the Ejby Mølle WWTP demand. Examine approaches to optimize this process. Also, as part of the new pump station modifications already scheduled, ensure that this project is congruent with other modifications validated via this project.
ENERGY GENERATION IMPROVEMENTS	
Co-digestion with high-strength industrial waste	Adding highly concentrated, readily biodegradable supplemental waste to the anaerobic digestion system is a proven method of increasing biogas production.
Increase degradability of WAS through conditioning (thermal hydrolysis, electrical pulse, sonication, etc.)	Pre-conditioning technologies such as thermal hydrolysis have shown that they can increase the rate and extent of digestion, thereby increasing biogas production. Also, these processes have shown improvement to sludge dewaterability, thereby decreasing energy demand from hauling biosolids.
Modify anaerobic digestion to increase biogas generation (Thermophilic, Temperature-phased, Enhanced Hydrolysis, Phase separation, etc.)	Advanced anaerobic processes have shown they can increase the extent of digestion, increase biogas production, and increase energy production.
Biosolids to energy (incineration, gasification, carbonization, biodiesel, etc.)	Biosolids to energy processes can be coupled with anaerobic digestion system energy recovery to enhance energy production and decrease energy demand of hauling sludge offsite.
Effluent thermal recovery	Using heat pumps to create additional heat using plant effluent as a heat sink can be an efficient means of increasing heat production.
Effluent hydropower recovery	Capturing energy via hydropower at the outfall can increase energy production.
Use a fuel cell for more power production in cogeneration system. Increase overall conversion efficiency	Fuel cells have very high electrical power conversion efficiencies. Using fuel cells in lieu of reciprocating engines would increase the power production with the same amount of biogas.
Add hydrogen to the anaerobic digesters to increase methane production	Examine the possibility of adding hydrogen to the anaerobic digestion system to enhance methane production.
Increase solids concentration to enhance digestion	Increasing solids concentration to the anaerobic digestion system will increase the SRT and therefore increase the volatile solids reduction and biogas generation. Increasing solids concentration can be accomplished via better thickening of influent solids or by recuperative thickening.

The following screening criteria were used to produce a "short list" during the workshop:

- Readily implementable
- Primarily process modifications
- Significant impact on energy profile
- Proven elsewhere

Based on the results of preliminary mass/energy balance simulations and using the above screening criteria, the joint team selected the following EOOs for further consideration:

- CEPT: Implement at the primary settling tanks (using existing ferric and polymer feed systems) to significantly enhance removal performance and reduce organic load to aeration basins. Additionally, CEPT will result in higher sludge quantities to digestion and thus have a higher biogas generation potential.
- Nitrify centrate in TFs: Reroute discharge of supernatant buffer tank directly to TFs for the nitrification of plant sidestreams. TF effluent would be denitrified in BNR bioreactors.
- Decommission TF and convert TF clarifiers to CEPT for wet weather treatment: treat 100 percent of settled primary effluent with the BNR process; blend BNR and TF clarifier effluents prior to tertiary filtration.
- Operate at shorter BNR system SRT: this should reduce bioreactor O2 demand as well as mixing requirements; increase biological sludge to digester, which could result in increased biogas production.
- Reduce effluent filtration operation to 12 hours per day: bypass all or a portion of settled secondary effluent around the effluent filtration system when it has the quality to meet regulatory requirements without additional treatment.

Two additional EOOs were identified in Workshop 1 as potentially of interest, but it was also recognized they would require additional effort and time to fully consider and implement. These two were further evaluated at a conceptual level as part of the project:

- Co-digestion with high-strength industrial waste: Adding highly concentrated, readily biodegradable supplemental waste to the anaerobic digestion system as a proven method of increasing biogas production.
- Retrofit bioreactors with diffused aeration: Fine-bubble diffusion delivers oxygen to suspended growth systems more efficiently than mechanical aerators therefore decreasing the associated energy demand. This option would replace all of the mechanical aerators or provide a combination of existing aerators and fine-bubble diffusers.

Evaluation of Short-listed EOOs. Subsequent to Workshop 1, supplemental sampling was conducted which enabled a better estimation of the primary clarifier efficiency. Also, it was determined that a key industrial contributor to the influent loads at the plant (the Dalum Papir paper mill) would be discontinuing its operation at the end of 2012, and thus it was removed from the influent flow and loadings analysis. The plant-wide Pro2D model was recalibrated on the basis of this updated information, which allowed for the more detailed evaluation of the short-listed EOOS identified. The following table presents a summary of the results of the evaluation of the individual viable alternatives described above. Results are presented in terms of savings in electrical power consumption, increase in cogeneration electrical output, net increase (as a percentage) as it relates to the existing total electrical power consumption, and net change in GHG emissions associated with the change in electrical power use. For this purpose, a GHG emission factor of 384 kilograms of carbon dioxide equivalent per megawatt hour was used, as supplied by VCS.

Table 2 - Summary of Operational EOO Results

EOO	Energy Saved kWh/d	Additional Energy Generation kWh/d	Net Energy Change kWh/d	Energy Self-Sufficiency %	Increase in Sufficiency %	Net GHG Change Tonnes/yr $CO_{2,e}$
Remove Paper mill	N/A	N/A	N/A	76.4%	N/A	N/A
CEPT	475	1,725	- 2,199	86.2%	9.9%	-67
TF Centrate	-6,950	413	+ 6,537	59.3%	-17.1%	+ 974
Turn off TFs	164	152	- 316	77.7%	1.3%	- 23
Lower SRT	268	283	- 551	78.7%	2.3%	- 38
Effluent Filter Usage 12 hrs/d	1,174	0	- 1,174	80.8%	4.4%	-165

Notes:
$CO_{2,e}$ = Carbon Dioxide Equivalent
GHG = Greenhouse Gas
Tonnes/yr = Tonnes per year

Furthermore, the next table presents the results of a series of potential "groupings" of these individual options into comprehensive energy optimization scenarios for the plant.

Table 3 - Summary of Operational EOO Combinations

EOO	Energy Saved kWh/d	Additional Energy Generation kWh/d	Net Energy Change kWh/d	Energy Self-Sufficiency %	Increase in Sufficiency %	Net GHG Change Tonnes/yr $CO_{2,e}$
CEPT & Lower SRT	548	1,890	- 2,438	87.3%	10.9%	- 77
Turn off TFs & 50% Filter Usage	1,338	152	-1,490	82.2%	5.8%	- 188
Lower SRT, Turn off TFs & 50% Filter Usage	1,720	416	-2,136	85.1%	8.7%	- 241
CEPT, Turn off TFs & 50% Filter Usage	2,088	1,915	- 4,003	94.3%	17.9%	- 293
Lower SRT, CEPT, Turn off TFs & 50% Filter Usage	2,154	2,135	- 4,289	95.8%	19.4%	- 302

As can be seen from these results, all of the EOOs have the potential to significantly improve the energy profile of the EMWWTP and reduce GHG emissions associated with electrical power consumption. Many of these options have already began to be readily implemented by plant staff in a "gradual" approach to ascertain the full extent of their benefit and to confirm/identify potential associated risks and thus establish mitigation actions, if warranted. These results also indicate that the objective of transforming the plant into a near net-zero electrical power facility is achievable in the short term, especially by combining some of the EOOs as described and summarized in the table above.

The following table presents similar estimates for the two facility-improvement EOOs also identified as part of the project: the incorporation of a side stream treatment component to reduce the nitrogen related recycle load coming into the bioreactor as a result of the dewatering of digested sludge, and the retrofit of the existing mechanical surface aerators from the bioreactors with a fine bubble diffused aeration system.

Table 4 - Summary of Facility Improvement EOO Results

EOO	Energy Saved kWh/d	Additional Energy Generation kWh/d	Net Energy Change kWh/d	Energy Self-Sufficiency %	Increase in Sufficiency %	Net GHG Change Tonnes/yr $CO_{2,e}$
Side stream treatment	548	1,890	- 2,438	87.3%	10.9%	- 77
Diffused aeration retrofit	1,338	152	-1,490	82.2%	5.8%	- 188

These results would indicate that by combining both operational and facility improvement EOOs, it is possible to make the EMWWTP a net positive treatment facility, as shown in the following figure:

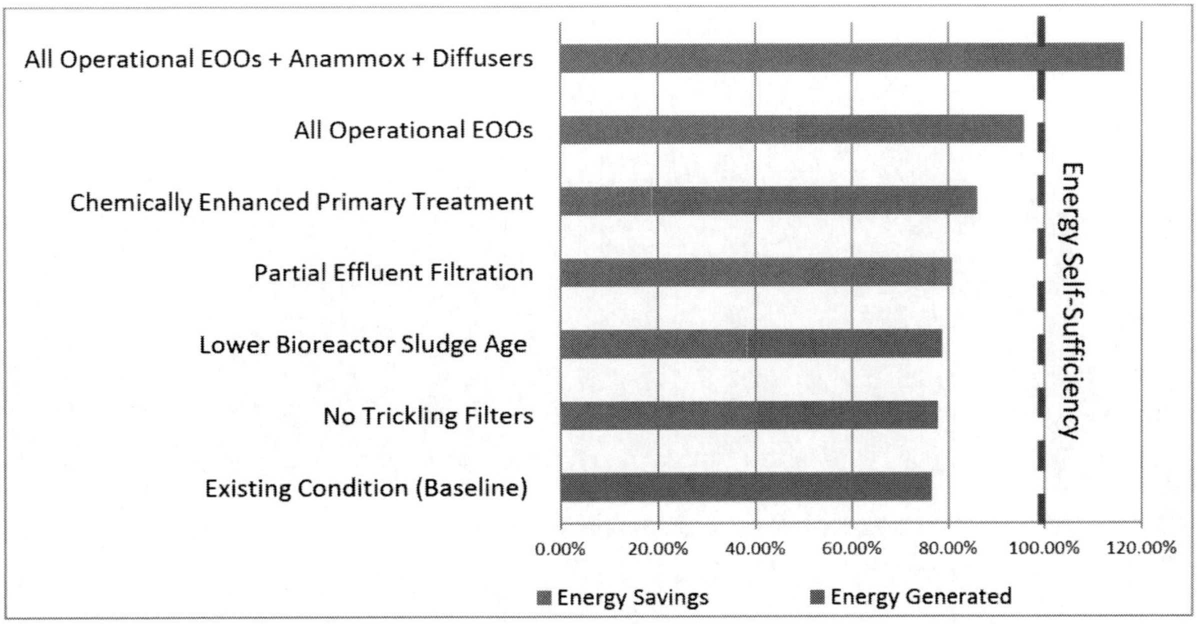

Figure 4 – Summary of Results of Operational and Facility Improvement EOOs – EMWWTP

SUMMARY AND CONCLUSIONS

Through the analysis of historical operational data, the use of an advanced mass/energy balance simulation tool, and the adoption of a collaborative workshop-based approach in identifying and evaluating alternatives, the joint VCS/Consulting Team was able to identify a series of readily implementable operational EOOs at the EMWWTP that will significantly contribute toward the Utility's adopted goal of achieving CO_2 neutrality firm-wide by year 2014. Implementation during the 2012-2013 Winter season of some of these EOOs has already resulted in several months of a slight net-positive energy balance operation, while maintaining compliance with the plant's regulated effluent requirements thus demonstrating the feasibility and real potential of the recommended optimization options. Results of this study also indicate that the implementation of side stream treatment (preferably through an anammox-based process) and the retrofit of the bioreactors with diffused aeration would clearly move this plant into a net-positive energy profile, a great contribution towards the realization of VCS's aggressive corporate environmental goal.

NEW (Nutrient, Energy, Water) Sustainability Principles Drive Technology Selection in Advanced Water Reclamation Application

Randall Boe, P.E.[1,2], Srinivas Jalla, P.E.[1], Chuck Crandall, P.E.[1], Julian Sandino, Ph.D., P.E.[1], Jeremy Kraemer, Ph.D., P.E.[1], Todd Cleaver[3], Tyler Richards, P.E.[3], Holly Elmendorf, P.E.[3]

[1]CH2M HILL, [2]randy.boe@ch2m.com, [3]Gwinnett County Department of Water Resources

ABSTRACT

The Crooked Creek Water Reclamation Facility (WRF) is the oldest of three Gwinnett County, Georgia water reclamation facilities. The original facility was constructed in 1972. Major capacity upgrades were completed in 1985 and in 1990. There has also been an almost continuous program of process upgrades since that time through the present. Current permitted maximum monthly average day flow (MMADF) capacity is 60,000 m^3/d (16 mgd). The WRF is currently rehabilitating and upgrading the activated sludge process and influent wastewater is being diverted to another facility to limit flow to approximately 30,000 to 37,500 m^3/d (8 to 10 mgd) to facilitate construction activities.

In 2012, a facility plan was prepared to identify longer-term upgrades to the liquid, solids, and odorous air treatment systems. A major project goal was that these upgrades enhance facility sustainability and meet more stringent effluent limits to allow possible future discharge to Lake Lanier, the drinking water supply for the Atlanta metropolitan area.

A comprehensive planning framework was developed to facilitate appropriate balancing of technical and cost factors applicable to the Crooked Creek WRF in a long-term capital improvements plan. This included assessments of competing process and technology selections for the liquid, solids and odorous air streams.

Following development of the project goals and vision by the project stakeholders, a long list of alternate process and treatment technologies were brainstormed and analyzed. The outcome of the screening level analysis was a short-list of the most promising solutions. The short-listed technologies and process options were then investigated in more depth using whole-plant process modeling to quantify the relative energy and carbon benefits of the alternate plant configurations aimed at producing low levels of nutrients in its effluent.

The evaluation demonstrated that sustainability principles can be incorporated into selection of a technology/approach that appropriately balances the competing aspects of nutrient removal, energy efficiency, and water reclamation needs while considering financial and operational constraints.

Key Words: sustainability, facility planning, chemically enhanced primary treatment, biologically enhanced primary treatment, energy efficiency

INTRODUCTION

The Crooked Creek Water Reclamation Facility (WRF) is the oldest of three Gwinnett County, Georgia water reclamation facilities. The original facility was constructed in 1972. Major capacity upgrades were completed in 1985 and in1990. There has also been an almost continuous program of process upgrades since that time through the present. Current permitted maximum monthly average day flow (MMADF) is 60,000 m^3/d (16 mgd). The WRF is currently rehabilitating and upgrading the activated sludge process and influent wastewater is being diverted to another facility to limit flow to approximately 30,000 to 37,500 m^3/d (8 to 10 mgd) to facilitate construction activities. The WRF meets current effluent limits. However, some of the treatment processes may underperform at higher flows. Also, existing process mechanical, electrical, control systems, and possibly structures, are nearing the end of their useful life and must be repaired or replaced.

In 2012, a Facility Plan was prepared to identify longer-term upgrades to the Crooked Creek WRF liquid, solids, and odorous air treatment systems. A major project goal was that these upgrades enhance facility sustainability and meet more stringent effluent limits to allow possible future discharge to Lake Lanier, the drinking water supply for the Atlanta metropolitan area. The facility planning approach, technology selection process, and evaluation results are discussed, illustrating how nutrient, energy, and water (NEW) sustainability principles influenced the technology selection.

FACILITY PLANNING APPROACH

The facility planning approach included the following steps:

1. Definition of goals and boundary conditions
2. Definition of regulatory requirements
3. Analysis of historical data
4. Definition of design flows and loads
5. Screening of treatment technologies
6. Detailed evaluation of short listed viable technologies
7. Selection of technology
8. Sizing and layout of facilities
9. Development of hydraulic profile
10. Preparation of cost estimates and phased implementation plan
11. Documentation of the facility plan

The heart of the facility plan was definition of project goals to set the project vision; screening of possible technologies; evaluation of those that were viable for meeting the goals and requirements of the project; and selection of the technology from among the viable options. The goal setting, screening, evaluation, and technology selection are described in more detail in the following sections.

PROJECT GOALS AND VISION

The first step in the facility planning process was development of the project goals and boundary conditions. This step was critical because it defined the vision for what the project could achieve as well as what was out of bounds or beyond the scope of the project. To set the vision, a workshop was conducted with all project stakeholders to solicit what their vision/hopes were for the project as well as what their greatest fears were regarding the project. These items were solicited from the stakeholders anonymously to ensure that everyone was free to express their concerns freely and openly.

Responses were collected and consolidated by grouping common items. The consolidated responses were discussed as a group to identify consensus aspirations, issues, and concerns. Following discussion, the project goals and vision were formalized as follows:

- The effluent quality should meet the requirements for both Chattahoochee River and Lake Lanier discharge.
- The treatment plant should demonstrate environmental stewardship through use of sustainable technologies.
- The treatment process should be effective, capable of meeting needs for the next 20 years, and be simple to operate and maintain.
- The facility should be reliable and improvements implemented on a pay-as-you-go basis (no bond issue).

Some of the greatest fears about the project included being unaffordable and having high operation and maintenance (O&M) costs.

The project boundaries were delineated as follows:

- Design flow: 60,000 m^3/d (16 mgd) maximum month average day flow (MMADF) with provisions for a possible future expansion to 95,000 m^3/d (25 mgd) MMADF, but operate efficiently at a flow of 26,500 to 30,000 m^3/d (7 to 8 mgd)
- Discharge location: initially to the Chattahoochee River with ability for future discharge to Lake Lanier
- Target effluent limits: Lake standards because these are the most stringent (see Table 1 below)
- Biosolids management: on-site biosolids management
- Cost: phase project to match available cash flow
- Improvements must fit on the existing site

Table 1. Effluent Quality Target Treatment Limits

Parameter	Target Limit
Maximum Month Average Day Flow	60,000 m^3/d (16.0 mgd)
5-day Carbonaceous Biochemical Oxygen Demand ($cBOD_5$)	2.5 mg/L
Total Suspended Solids (TSS)	3 mg/L
Turbidity	0.5 NTU
Total Phosphorus (TP)	0.08 mg/L
Total Nitrogen (TN)	10 mg/L
Dissolved Oxygen (DO)	7 mg/L
Fecal Coliform	23 colonies/100 ml

ALTERNATIVE TECHNOLOGY SCREENING

A facilitated workshop was held to brainstorm and develop a long list of technologies. These were then screened down to generate a short list of technologies to be evaluated more extensively. A relatively simple set of criteria were established to screen the wide range of potential technologies. The screening criteria included:

- Sustainability (environmental, social, financial)
- Reliability
- Performance
- Feasibility

The long list of technologies was organized by unit process. Each technology was then scored as follows: 0 = fatal flaw, 1 = below average, 2 = average, 3 = above average. Although there were few overall criteria, each encompassed many considerations as described in Table 2.

Table 2. Screening Level Technology Evaluation Criteria

Criteria	Characteristics
Sustainability	Low power consumption Low chemical/materials consumption Reduced transportation traffic Reduced sludge/residuals production Recovers resources for beneficial reuse Enhances energy production Low ownership cost

Table 2. Screening Level Technology Evaluation Criteria

Criteria	Characteristics
Reliability	Technology is robust and proven with successful implementation at multiple installations of similar size
	Not susceptible to upset
	Does not require an excessive amount of maintenance
	Consistently meets treatment criteria
	Low operational complexity – generally lower complexity is more reliable
	System-wide standardization
Performance	Able to meet required effluent limits
	Flexibility to adapt to future requirements (regulatory/disposal)
	Provides operational flexibility (multiple ways to address operational/process challenges)
	Does not require a lot of operator attention (ability to automate operation to a large extent)
Feasibility	Fits on site
	Regulatory acceptance
	Able to implement without pilot testing (not an embryonic technology)
	Low noise, odor and visual impact (requiring additional technologies for mitigation)
	Likely to be within budget, affordable
	Constructible while keeping the existing plant in operation

Although the technologies were organized and screened by unit process, they were considered in the context of an overall treatment flow sheet. Each unit process is affected by others and only when considered holistically can the sustainability aspects of each be quantified. Nowhere is this more evident than when considering the interaction between primary and secondary treatment. The short-listed technologies by unit process were:

- Liquid Treatment:
  - Primary Treatment – conventional primary treatment, chemically enhanced primary treatment (CEPT), and biologically enhanced primary treatment (BEPT)
  - Secondary/Tertiary Treatment – membrane bioreactor (MBR) with enhanced biological phosphorous removal (EBPR)
  - Disinfection - UV disinfection was selected for the facility based on previous evaluations of disinfection alternatives by DWR. Space on the site will be allocated for a future facility for ozonation or advanced oxidation.

  - Storage – equalization of influent is a utility standard and was therefore included by default
- Biosolids Management:
  - Pretreatment - Pretreatment options were rejected in general as not meeting reliability criteria which reflected the evaluation and conclusions recently completed for other DWR facilities.
  - Thickening – rotary drum thickeners (RDTs) were selected for the facility based on recently completed evaluations for other DWR facilities
  - Digestion - Conventional mesophilic anaerobic digestion (MAD) was selected for the facility based on recently completed evaluations for other DWR facilities.
  - Dewatering - Three options: belt filter press, centrifuge, and screw press were selected for further evaluation

The short-listed technologies and process options were then investigated through the use of whole-plant process modeling to examine the relative energy and carbon benefits of the different plant configurations aimed at producing low levels of nutrients in its effluent. Several different MBR process configuration options were evaluated for their merits relative to the project goals but are not discussed here. The remainder of this paper focuses on the evaluation and comparison of the short-listed liquid treatment alternatives (primary and secondary treatment) and the effect that various combinations have on sustainability considerations.

DETAILED EVALUATIONS

The evaluation of short-listed technologies focused on addressing and balancing competing aspects of nutrient removal, net energy usage, and the high quality effluent water goals while considering financial and operational constraints. As previously mentioned, water reclamation and nutrient removal goals led to the selection of membrane bioreactor and ultraviolet disinfection technology for the liquid stream.

The solids stream evaluations focused on anaerobic digestion and biogas production for combined power and heat production. This led to the evaluation of several primary treatment approaches including: conventional primary clarifiers, CEPT, and BEPT through the "A/B Process." The A/B process utilizes a highly loaded "A-stage" aeration tank and clarifier (performing the biologically enhanced primary treatment function) followed by a more conventionally loaded "B-stage" aeration tank and clarifier. In this case an MBR would be utilized in the "B-stage" process. The A/B process has been employed at many European facilities since the early 1980's and has gained notoriety as contributing to energy neutrality at the WWTP in Strass, Austria.

The facility plan also provided a holistic investigation of deammonification and phosphorus recovery from the solids stream, but is not discussed here as it was considered common for all alternatives. The investigation of the sidestream treatment alternatives found that the majority of the operational cost savings could be attributed to phosphorus removal rather than nitrogen removal.

Figures 1 through 3 illustrate the primary treatment strategies selected for further evaluation.

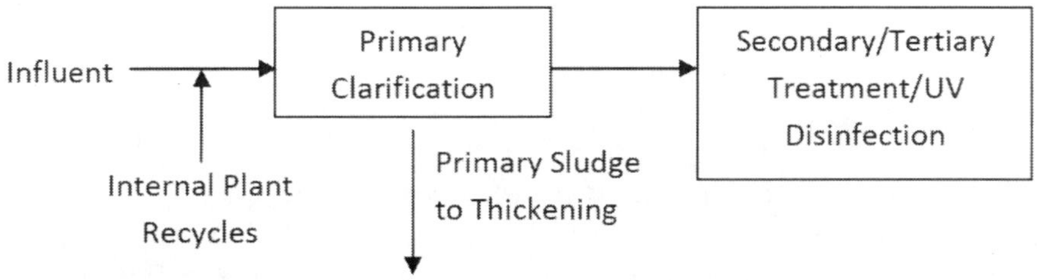

Figure 1. Conventional Primary Treatment

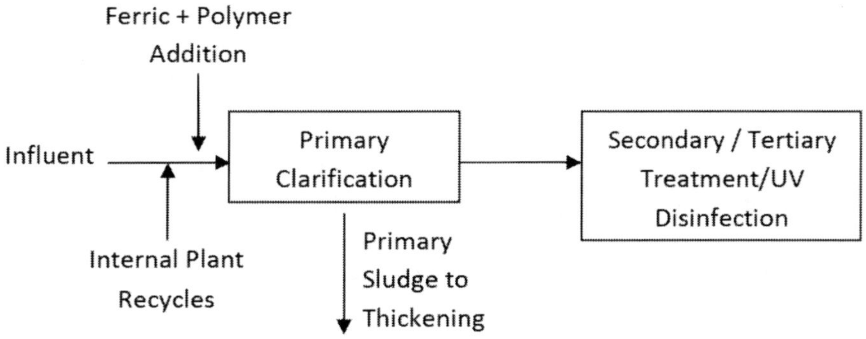

Figure 2. Chemically Enhanced Primary Treatment (CEPT)

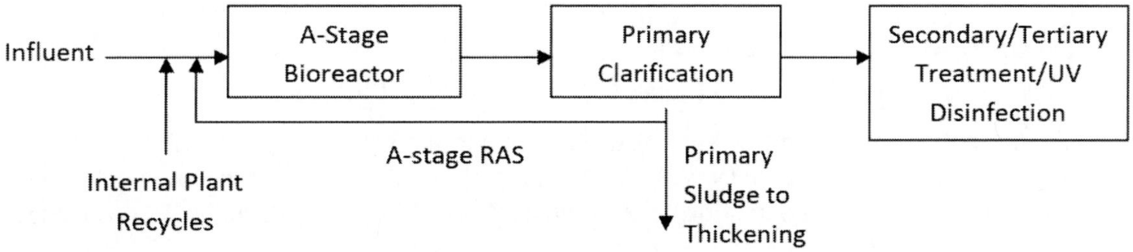

Figure 3. Biologically Enhanced Primary Treatment (BEPT)

Evaluation Scenarios

The three primary treatment strategies were combined into four scenarios that were modeled using BioWin (Version 3.1.0.838). Due to the comparative nature of the evaluation, the process models were set-up using annual average flow, loads, and temperature. Table 3 presents the details of the four scenarios.

Table 3. Summary of Four Process Models

Scenario Name	Secondary Process Sizing	WRF Operating Mode
Conventional	Conventional	Conventional
Operational CEPT	Conventional	CEPT
Full CEPT	CEPT	CEPT
BEPT	BEPT	BEPT

The "Conventional" scenario reflects the use of conventional primary clarifiers (2,000 gpd/ft^2) followed by a MBR. Conventional primaries were assumed to provide approximately 60 percent TSS removal and the MBR was sized for approximately 8,000 mg/L mixed-liquor suspended solids (MLSS) at the average conditions modeled.

The "Operational CEPT" scenario reflects the same physical plant sizing as the Conventional scenario, but CEPT is practiced in order to increase primary treatment efficiency and in turn decrease the organic loading to the downstream MBR thereby decreasing the aeration requirements and operating MLSS concentration. CEPT operation assumed the use of both a high ferric dose (approximately 37 mg/L as $FeCl_3$) as well as the use of polymer (0.75 mg/L).

The "Full CEPT" scenario reflects decreased size of primary clarifiers due to the ability to operate CEPT primaries at higher surface overflow rate (3,000 gpd/ft^2 versus 2,000 gpd/ft^2). In addition, the bioreactors of the MBR are decreased in size due to the lower organic loadings. The bioreactor volume was decreased to 50% of Conventional to achieve a MLSS of approximately 8,000 mg/L, the same as the Conventional scenario.

The "BEPT" scenario reflects the use of biological enhancement of the primary treatment process utilizing the A/B process. This process includes a highly-loaded aeration tank ahead of the A-Stage clarifiers with a surface overflow rate (SOR) the same as conventional clarifiers (2,000 gpd/ft^2). This process provides comparable overall TSS removal efficiency as the Conventional scenario, but provides higher (approximately 60 percent) $cBOD_5$ removal efficiency. Hence there is a modestly lower organic loading to the MBR, and the bioreactors for the BEPT option were decreased by 20% relative to Conventional to achieve a MLSS concentration of 8,000 mg/L comparable with the other options.

The models were configured such that the effluent quality was the same for all four scenarios. The predicted effluent quality was: $cBOD_5$ approximately 1 mg/L (less than detection), TSS less than 1 mg/L (less than detection), TP between 0.04 and 0.05 mg/L and a TN concentration of approximately 10 mg/L.

All of the process models assumed a total solids retention time of about 19 days (4-5 days each for anaerobic and anoxic, 9 days aerobic). The target mixed liquor suspended solids concentration in the bioreactors was less than 8,000 mg/L. The return activated sludge flow was assumed to be four times the plant's influent flow rate for the Conventional, CEPT and BEPT models. A lower return activated sludge flow was used for the Operational CEPT model due to the lower MLSS concentration while still achieving less than 10,000 mg/L in the membrane tank.

The same three-stage BNR process was used in the MBR for all scenarios; with 400% nitrified recycle (NRCY) and 200% anaerobic recycle (ARCY) (Figure 4).

The BEPT A-Stage aeration tank was sized to provide a 30 minute hydraulic retention time, RAS was assumed to be 50% of the plant influent flow, dissolved oxygen set point was 1 mg/L and the target mixed liquor suspended solids concentration was 3,000 mg/L.

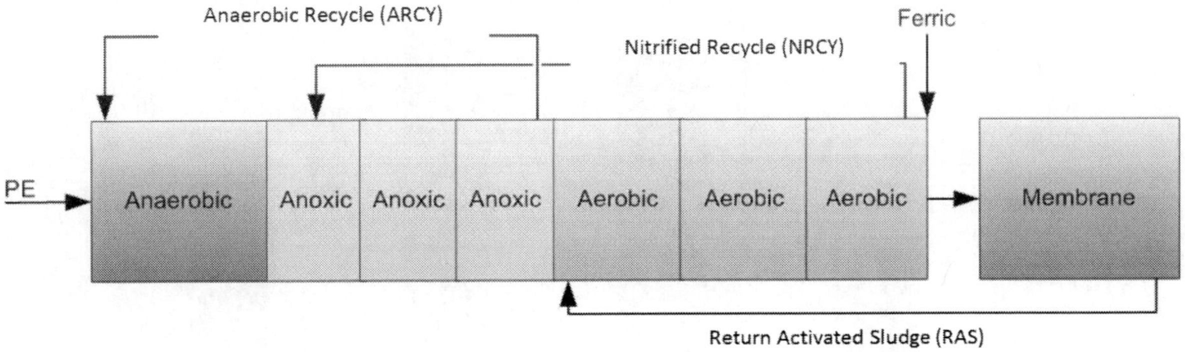

Figure 4. Process Flow Diagram Illustrating Three-Stage BNR Configuration

Oxygen requirements predicted by BioWin were converted to volumetric air rates using the following parameters: DO = 2.0 mg/L, alpha was calculated using the equation of Krampe and Krauth (2003) correlating MLSS to alpha (alpha = exp{-0.08788*MLSS/1000} where MLSS is in mg/L), F = 0.9, water temperature = 19 °C (66 F), beta = 0.95, altitude = 1000 ft, theta = 1.024, SOTE = 35%, air temperature = 15 °C (59 F), and blower efficiency = 70%.

Assumed polymer requirements were 9 lb/ton DS for thickening and 23 lb/ton DS for dewatering. Utilization of biogas in a reciprocating engine co-generator was assumed to provide 38% electrical efficiency and 42% thermal efficiency. RAS pumping was assumed using 30 ft total dynamic head (TDH), 75% pump efficiency and 90% motor efficiency.

Predicted Pollutant Removal

Table 4 outlines the predicted pollutant removal performance of the primary clarification alternatives.

Table 4. Predicted Primary Clarification Performance

Parameter	Conventional	CEPT Full (or Operational)	BEPT
TSS Removal	60 percent	80 percent	60 percent
$cBOD_5$ Removal	30 percent	60 percent	60 percent
TP Removal	15 percent	65 percent	25 percent
Primary Sludge Concentration	1 percent	1 percent	0.8 percent

Figure 5 illustrates the predicted primary sludge mass for the alternatives. The two CEPT alternatives were predicted to generate 65 percent greater primary sludge mass than conventional primary treatment. BEPT was predicted to generate about 30 percent additional mass. This demonstrates the benefits of CEPT and BEPT in terms of capturing influent carbon for transfer to anaerobic digestion and ultimate biogas production and energy recovery.

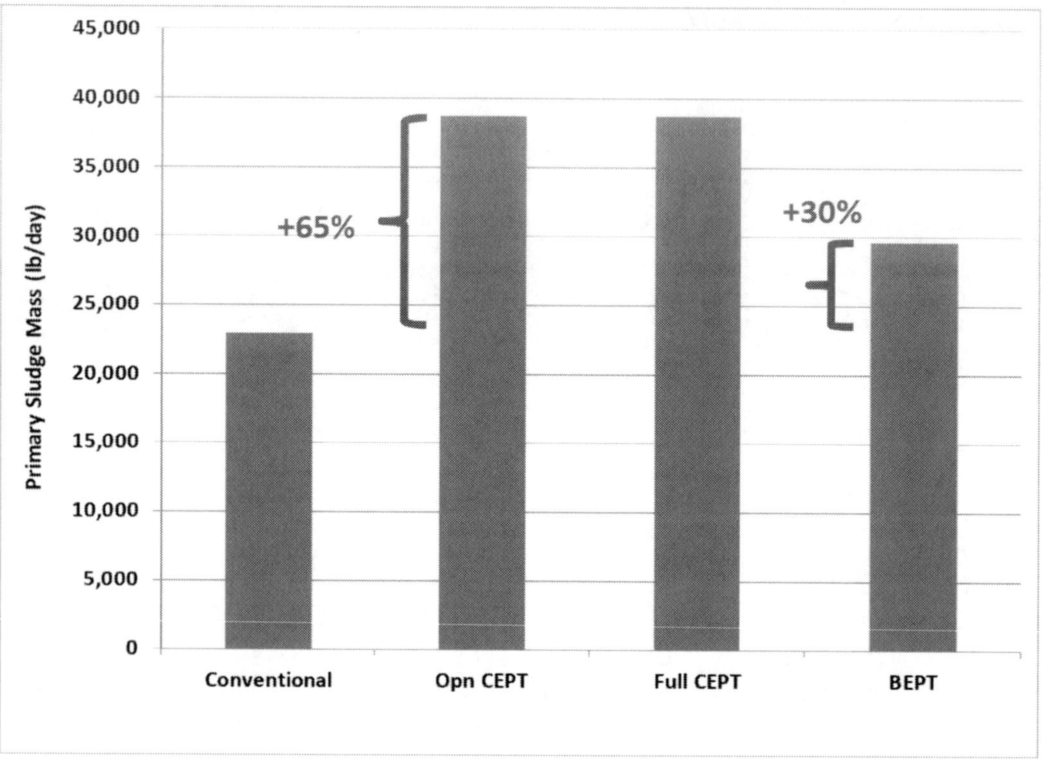

Figure 5. Predicted Primary Sludge Production

Figure 6 illustrates the differences in primary effluent COD and its associated distribution among various particulate and soluble fractions. The options employing CEPT and BEPT have a lower effluent COD than conventional due to the greater capture of carbon as primary sludge. CEPT acts to improve removal of particulate COD as well as converts colloidal COD to particulate form thereby allowing its removal. CEPT does not affect soluble COD. On the other hand, BEPT acts on both soluble and colloidal constituents; the adsorption of colloidal COD is improved, and the removal of about 50% of readily biodegradable COD and nearly all of the VFA COD is also accomplished. However, as noted later, the removal of readily biodegradable COD and VFA COD is detrimental to downstream biological phosphorus removal processes.

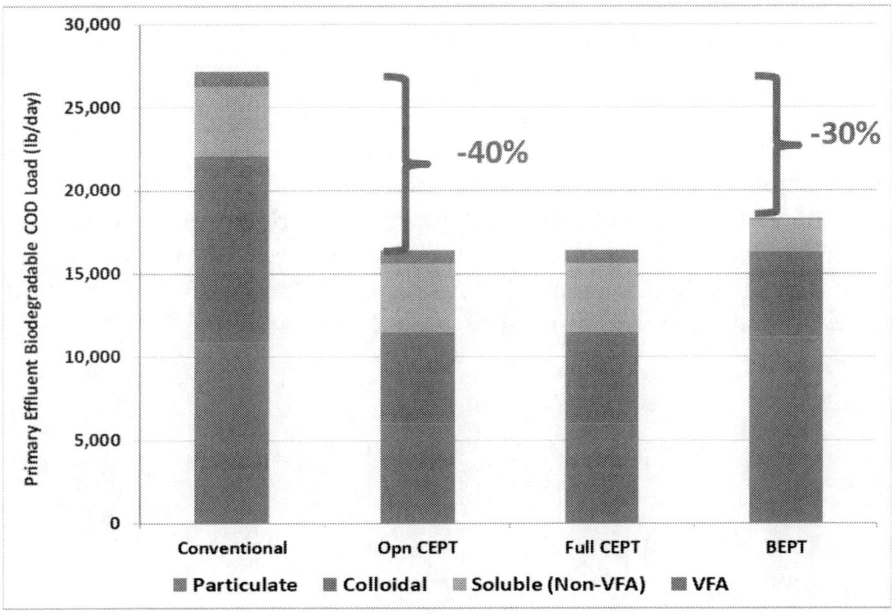

Figure 6. Predicted Primary Effluent Biodegradable Chemical Oxygen Demand

Figure 7 illustrates predicted primary effluent phosphorus. CEPT significantly decreases the phosphorus load due to conversion of soluble phosphorus to particulate form in primary treatment with coagulation and flocculation. The 15 percent decrease in soluble phosphorus in the BEPT option is a result of uptake due to growth of the A-stage biomass.

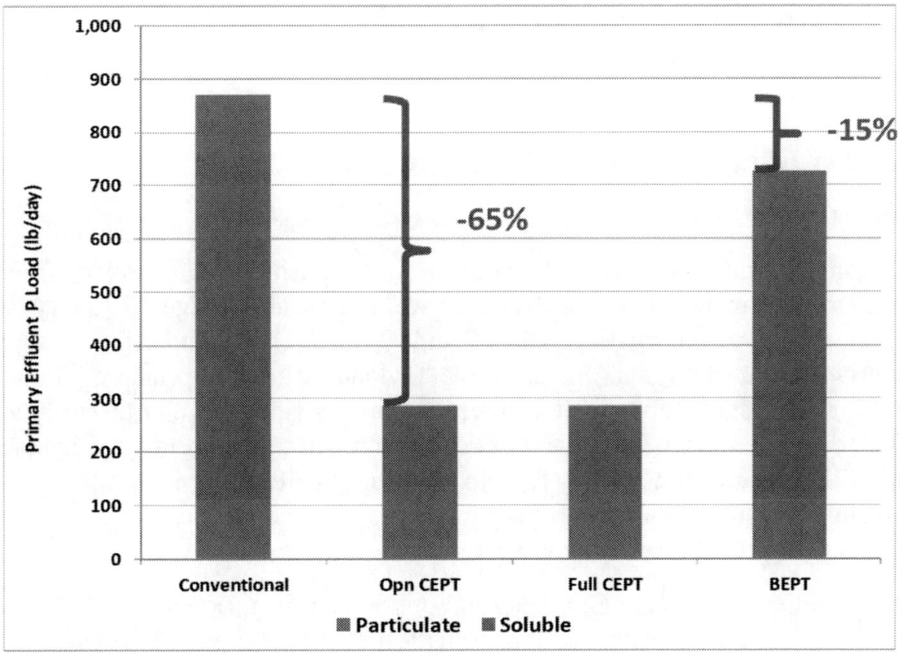

Figure 7. Predicted Primary Effluent Phosphorus

$cBOD_5$ and VFA to Total Phosphorus Ratios

Figure 8 illustrates the ratios of $cBOD_5$ to TP and VFA to TP. The conventional primary treatment option results in an approximately ideal ratio of $cBOD_5$ to TP, although the ratio of VFA to TP is below ideal, which indicates there should be sufficient carbon to ferment readily biodegradable material into VFAs within the anaerobic zone of the bioreactor. CEPT results in a high ratio of $cBOD_5$ to TP, due to the significant removal of phosphorus through primary treatment, although the ratio of VFA to TP is still below the idea range. However, as will be seen in Figure 9 there is no enhanced biological phosphorus removal (EBPR) activity for the CEPT option. As indicated in Figure 8, BEPT is detrimental for EBPR due to the removal of a portion of the readily biodegradable COD and removal of most of the influent VFA.

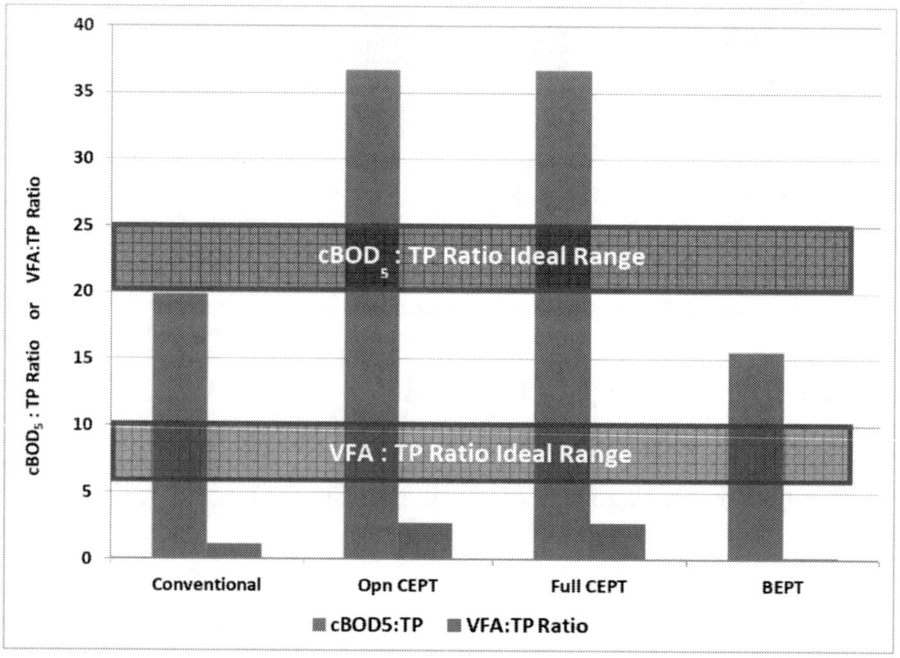

Figure 8. Predicted Primary Effluent Carbon to Phosphorus Ratio

Total Phosphorus Removal

Figure 9 illustrates a phosphorus mass balance on the liquid train treatment system, considering the TP influent to the primary treatment process and TP removal in primary sludge ("Primary Organics" and "CEPT Ferric"), WAS ("Growth", "Bio-P" and "Trim Ferric"), and effluent ("Effluent"). The cumulative height of the bars indicates the TP load influent to primary treatment considering both raw wastewater as well as internal plant recycles. As all options have the same influent TP load, the difference between the options is a result of different recycle loads within the plant: options with better EBPR activity tend to result in greater phosphate release through anaerobic digestion which then results in higher internal recycle loads i.e. Conventional has the highest recycle load versus CEPT with the lowest internal recycle load.

EBPR is able to successfully take place in the Conventional option as this option has sufficient carbon entering the anoxic and anaerobic zones for the formation of the required VFAs (in addition to any VFA in the influent). Phosphorus removal in CEPT is primarily a result of ferric addition, with a large portion removed in primary treatment compared to the other two options. Decreased EBPR activity with BEPT is a result of insufficient VFAs because of failure of inline fermentation due to deficiency of readily biodegradable carbon.

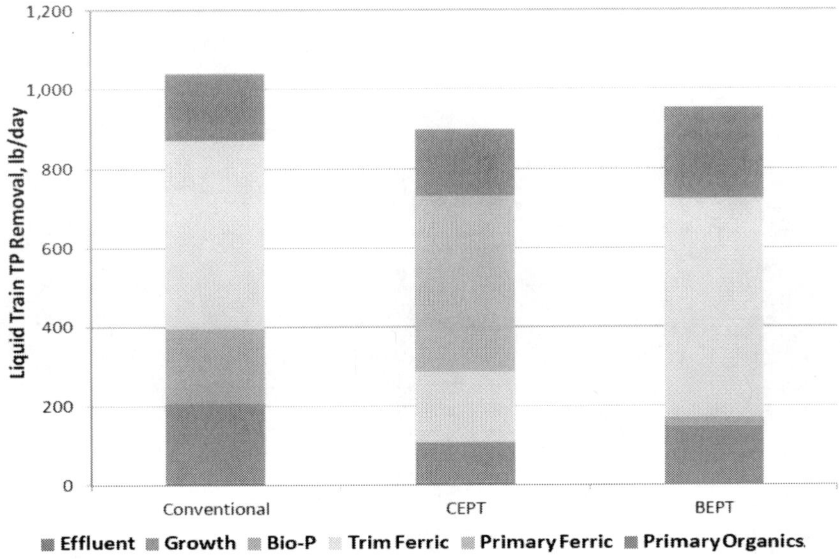

Figure 9. Predicted Phosphorus Removal

Metal Salt Usage

Figure 10 illustrates the predicted ferric usage of the four scenarios. The BRB (trim) ferric dosages were selected based on a desired effluent total phosphorus concentration between 0.04 and 0.05 mg/L. The Conventional option resulted in the lowest ferric requirement due to the best EBPR, CEPT resulted in the highest ferric usage and BEPT was in the middle. It should be noted that because the modeling assumed a CEPT dose on the low end of typical dosages that the actual ferric use in reality could be higher for the CEPT options.

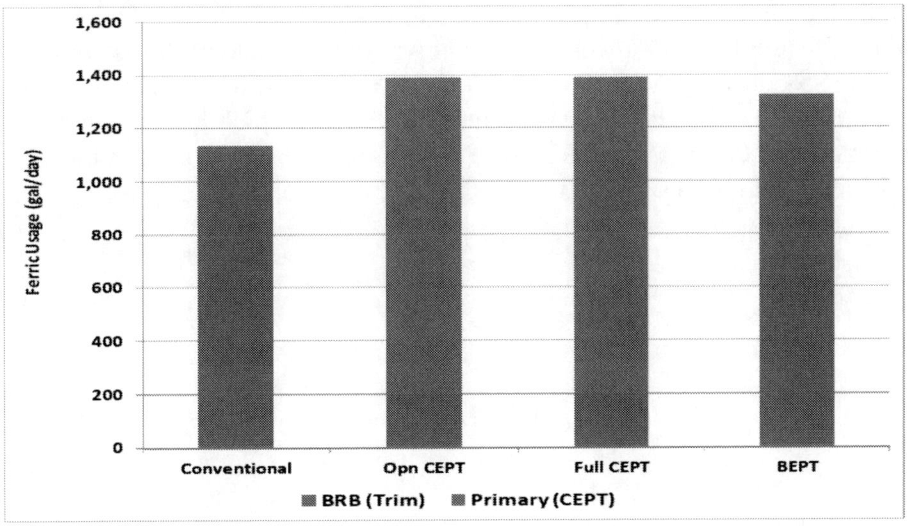

Figure 10. Predicted Ferric Chloride Requirements

Bioreactor Volume

Figure 11 illustrates the total bioreactor volume, relative to Conventional, for the four scenarios. The total bioreactor volumes were sized based on having the same mixed liquor suspended solids concentration across the various the primary treatment options, except Operational CEPT where the MLSS was lower. The use of CEPT allowed for a bioreactor volume which was half of that needed for the Conventional option, and BEPT resulted in a 20 percent smaller bioreactor.

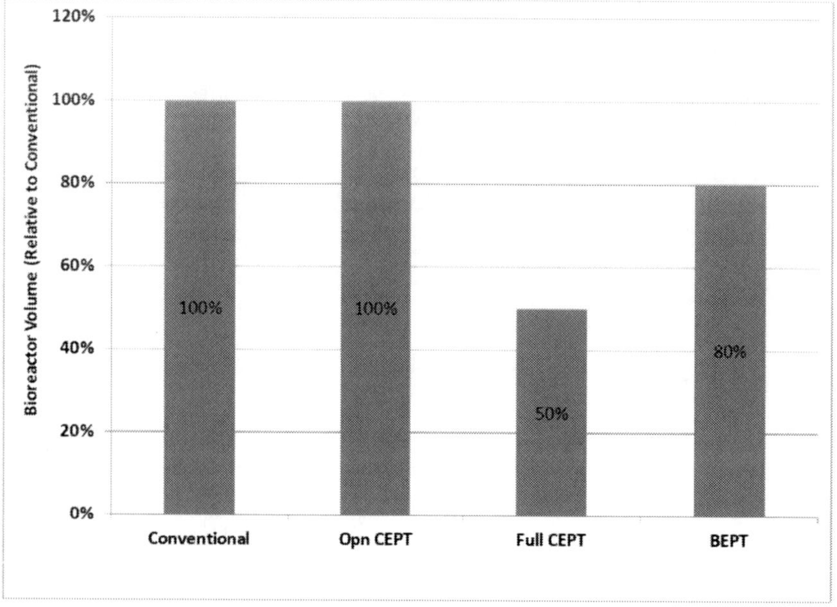

Figure 11. Required Bioreactor Volume

Process Oxygen Requirements

Figure 12 illustrates the actual oxygen requirement (AOR) and the corresponding aeration operating horsepower to supply the required air for the four scenarios. The CEPT options have lower oxygen requirements and aeration power demands because a portion of the carbon is captured in the primary sludge which results in less oxygen demand in secondary treatment. Operational CEPT results in a significantly lower aeration power demand due to the increased efficiency of aeration at MLSS concentrations around 4,000 to 5,000 mg/L which increases the alpha factor, compared Full CEPT (as well as Conventional and BEPT) which has decreased alpha factors due to operation at 8,000 mg/L MLSS. BEPT was predicted to result in a model aeration savings of 10 percent relative to Conventional.

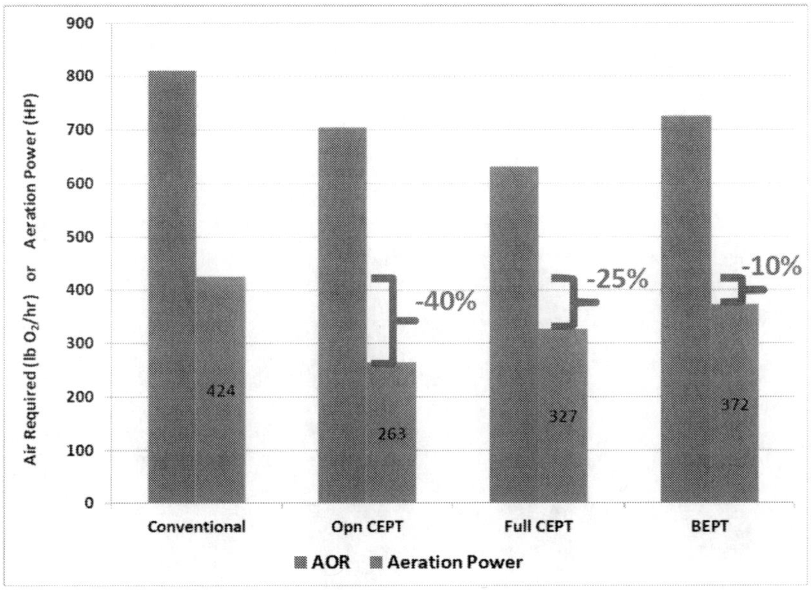

Figure 12. Predicted Oxygen Requirement and Aeration Running Horsepower

Return Activated Sludge Pumping

Figure 13 illustrates the return activated sludge operating flow, expressed as a percentage of the influent plant flow. The decreased RAS rate for Operational CEPT is available due to the lower MLSS concentration of 4,000 mg/L (compared to 8,000 mg/L for the other scenarios); decreasing to a 200% RAS rate still allows meeting the MLSS limits in the membrane basins but decreases energy consumption.

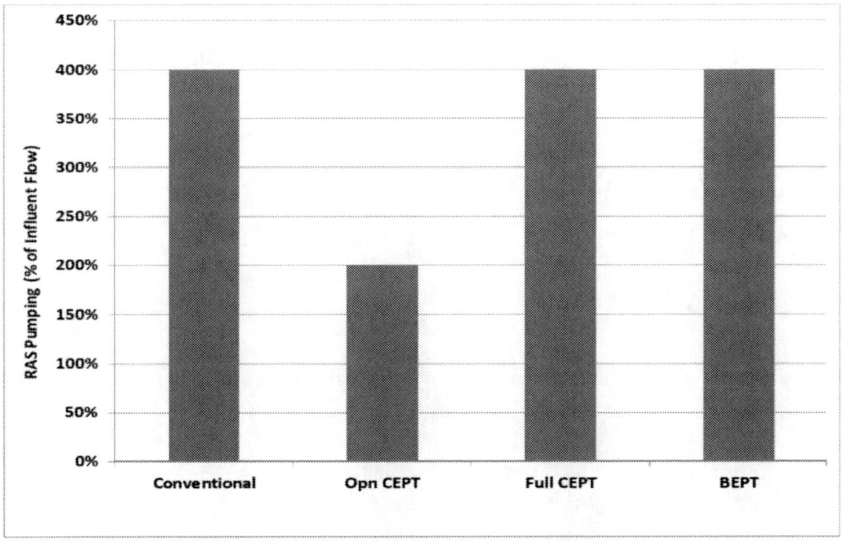

Figure 13. Predicted Return Activated Sludge Pumping

Secondary Treatment Energy Usage

Figure 14 illustrates the energy requirement of secondary treatment for four scenarios. The electricity consumption of the Operational CEPT scenario is lower due to the lower aeration energy and RAS pumping. CEPT and BEPT have modestly lower electricity consumption due to lower aeration power.

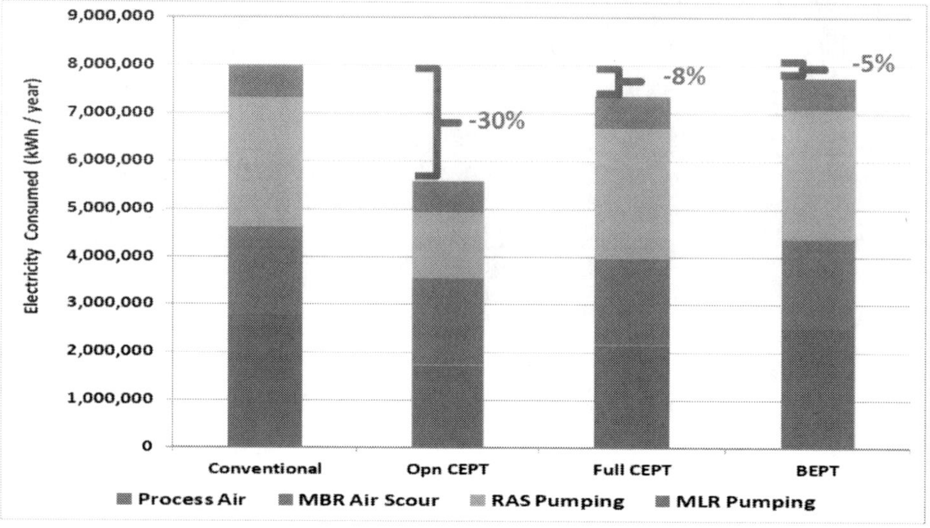

Figure 14. Predicted Secondary Treatment Energy

Waste Activated Sludge (WAS) Production

Figure 15 illustrates the predicted waste activated sludge volumes. The CEPT options were predicted to generate 65 percent greater primary sludge mass than conventional primary treatment thereby resulting in approximately 55 percent less WAS. BEPT was predicted to generate about 30 percent additional primary sludge mass resulting in approximately 15 percent less WAS.

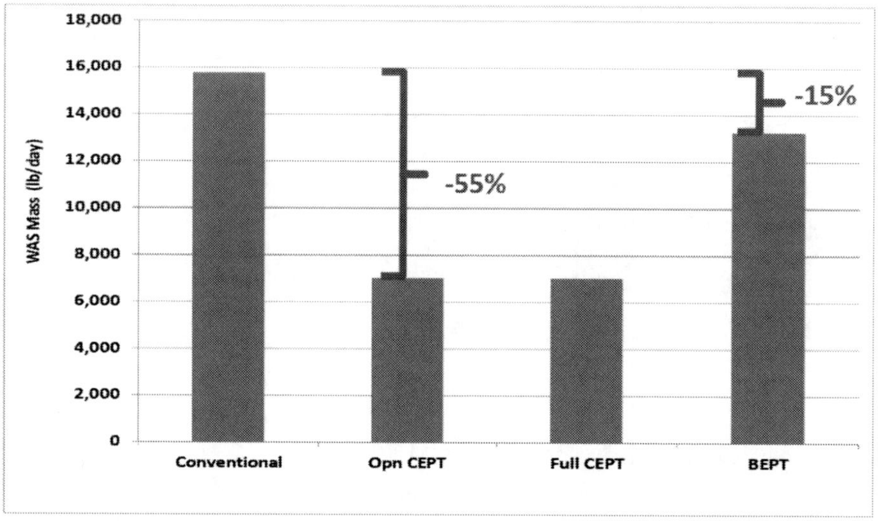

Figure 15. Predicted Waste Activated Sludge Production

Total Sludge Production

Figure 16 outlines the total sludge production of the four scenarios. The CEPT options result in an increased total sludge mass of approximately 20 percent, although the overall composition is much more biodegradable in anaerobic digestion due to the large fraction of primary sludge. BEPT results in approximately 10 percent more sludge. However, these differences do not translate into differential sizing of the co-thickening system; it is anticipated all options will require the same number of rotary drum thickeners.

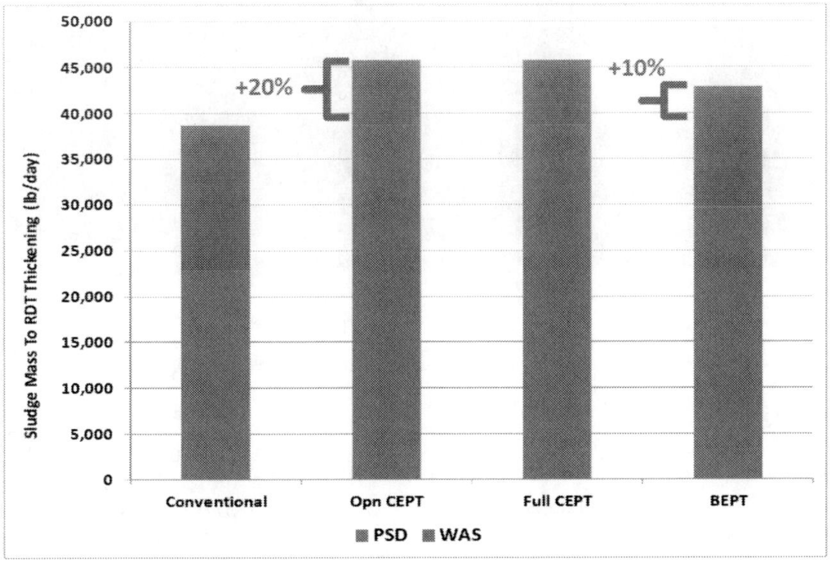

Figure 16. Predicted Combined Sludge to Thickening Units (Primary and Waste Activated Sludge)

Anaerobic Sludge Digestion

Figure 17 illustrates the anaerobic digester volumes required for each scenario. The digestion volumes were chosen based on achieving a hydraulic retention time of 22 to 23 days, except for Operational CEPT where the digester is the same size as Conventional which results in an HRT of 19 days (and therefore correspondingly slightly lower VSR and biogas production relative to Full CEPT). The larger volumes required by Full CEPT and BEPT are a result of greater sludge production.

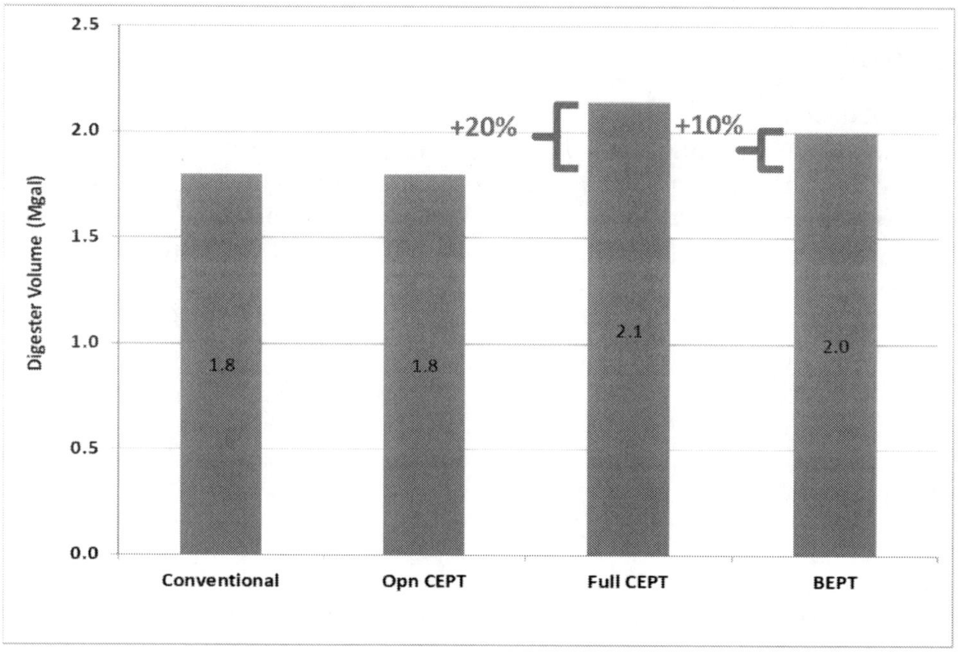

Figure 17. Required Digester Volume to Maintain Equal Digestion Retention Time

Digester Volatile Solids Reduction

Figure 18 illustrates the predicted volatile solids reduction (VSR) through anaerobic digestion. Higher VSR in the CEPT and BEPT options is a result of higher portions of primary sludge. As noted above under Figure 17, the lower HRT for Operational CEPT results in a slightly lower VSR relative to Full CEPT.

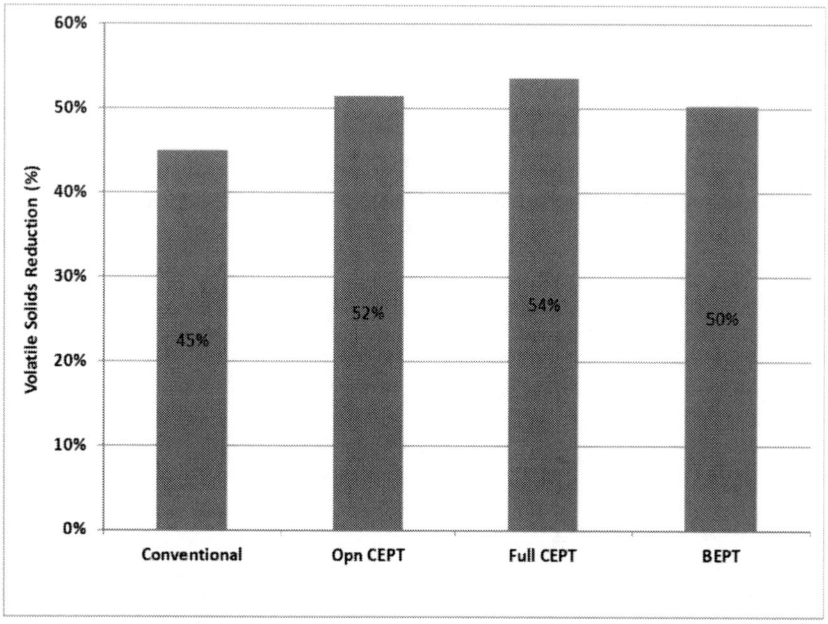

Figure 18. Predicted Volatile Solids Reduction through Digestion

Biogas Production

Figure 19 illustrates predicted biogas production. The higher biogas production of the CEPT and BEPT options are a result of higher total sludge volumes being sent to digestion and also increased biodegradability (i.e. higher VSR). Again, the lower VSR for Operational CEPT relative to Full CEPT results in a slightly lower biogas production.

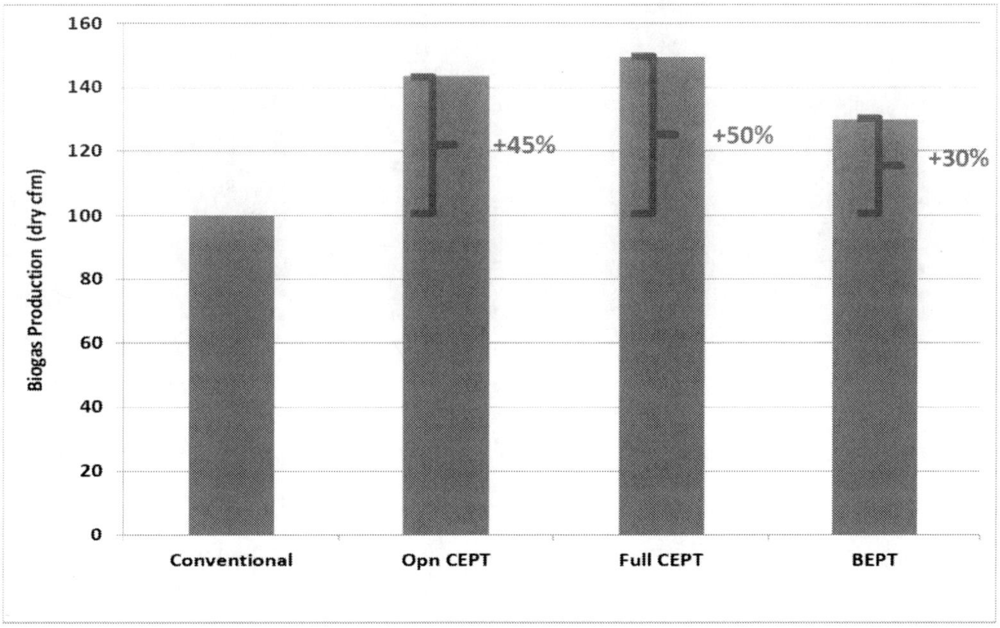

Figure 19. Predicted Digestion Biogas Production

Energy Generation

Figure 20 illustrates that Operational CEPT has a near-neutral energy balance. Full CEPT and BEPT are similar in that a significant portion of the secondary electricity requirement can be met by cogeneration, while the Conventional option has the largest difference between electricity requirement and recovery. It should be noted that the electricity consumption can be decreased through selection of energy efficient equipment and process design.

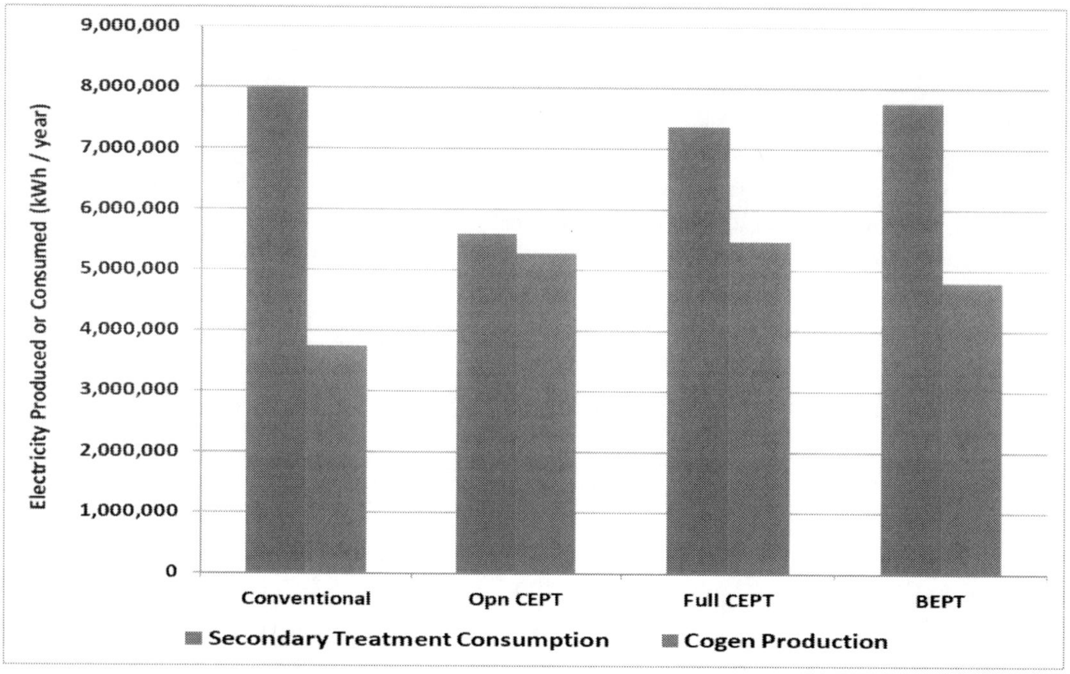

Figure 20. Predicted Secondary Treatment Energy Usage and Digestion Biogas Production

Digested Sludge Disposal

Figure 21 illustrates the predicted mass of biosolids requiring disposal. The CEPT and BEPT options, which increase the recovery of carbon by primary treatment, do result in greater mass of sludge requiring disposal. However, the impact is not proportional, due to the increased biodegradability of the sludge to digestion which mitigates the resulting mass leaving the facility.

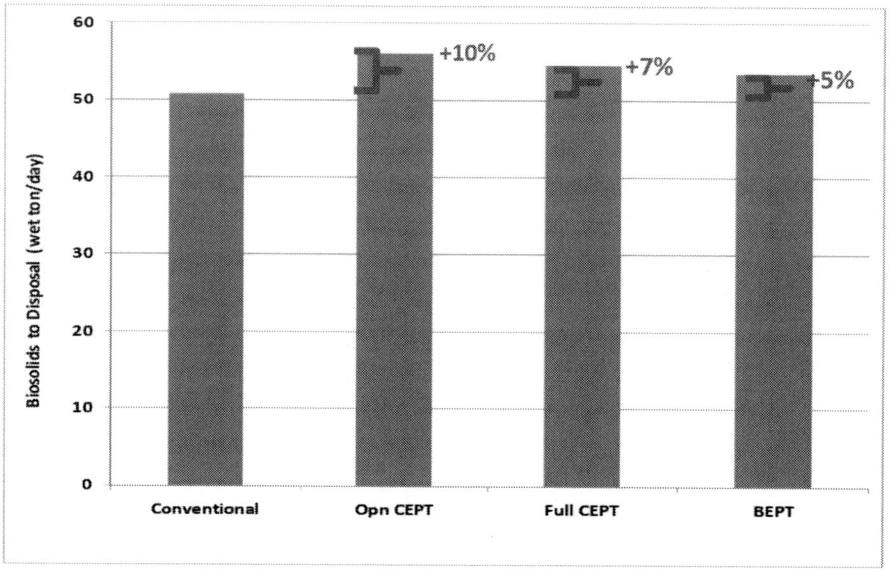

Figure 21. Predicted Mass of Biosolids Requiring Disposal Assuming a Cake Solids Concentration of 22 Percent

Polymer Usage

Figure 22 summarizes the predicted polymer usage for the four scenarios. Dewatering polymer is quite similar for all the options. RDT thickening polymer is greater for the CEPT and BEPT options as a result of PSD and WAS production. The CEPT options include a third type of polymer which is used in the CEPT process.

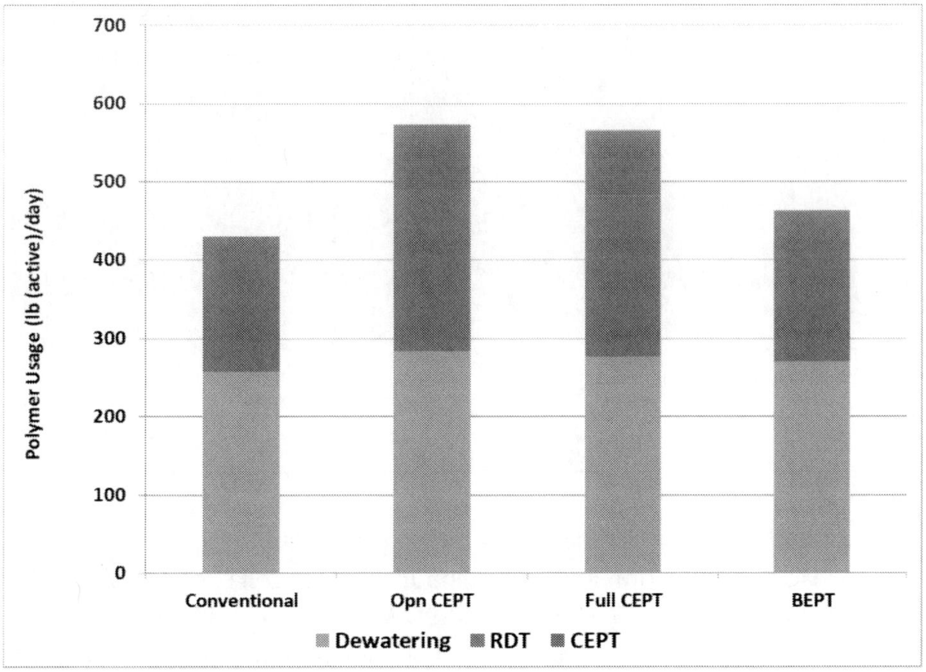

Figure 22. Predicted Polymer Requirements

COST COMPARISON

Relative Capital Cost

A cost comparison was generated comparing the capital costs associated with the four scenarios. The following cost factors were utilized to evaluate comparative capital costs:

1. Primary Clarifiers = \$1,200 / ft^2
2. Bioreactors = \$7 / gallon
3. Anaerobic Digesters = \$13 / gallon

Table 5 presents the predicted alternative capital costs (\$millions) relative to the Conventional option. It was not necessary to consider cost differences for other elements of the plant because they were essentially the same for each option. Operational CEPT has an equivalent capital cost relative to Conventional, other than some additional chemical management systems which were not included as they would be relatively minor compared to the capital costs for clarifiers, bioreactors and digesters. CEPT was found to result in a net capital saving of approximately \$25.7 million relative to Conventional principally due to savings on primary clarifiers and bioreactors, even considering increasing the digester volume to account for greater sludge quantities. For the CEPT option, due to the lack of EBPR, it was further assumed the anaerobic zone could also be deleted relative to Conventional. BEPT resulted in a net capital savings of about $2 million.

Table 5. Relative Capital Costs

Parameter	Primary Treatment Alternative			
	Conventional	Operational CEPT	CEPT	BEPT
A-Stage Bioreactor Basin	–	–	–	$1.9
Primaries	–	–	($8.0)	–
Bioreactor Basin	–	–	($15.8)	($6.3)
No Anaerobic Zone	–	–	($6.3)	–
Digesters	–	–	$4.4	$2.6
Total	**$0**	**$0**	**($25.7)**	**($1.8)**

However, for the CEPT option, the deletion of the anaerobic zone results in less fermentation of $cBOD_5$ thereby resulting in less denitrification and correspondingly higher TN of 14 mg/L. The additional costs to meet a TN of 10 mg/L were not considered in the above capital savings but would require additional denitrification volume and/or supplemental carbon addition that would reduce the apparent cost savings.

Relative Operation and Maintenance Cost

A second cost comparison was prepared to compare the O&M costs associated with the four alternatives. The following assumptions were made to develop the O&M costs:

1. Purchased Electricity = $0.08 / kWh
2. Cogeneration electricity credit= $0.08 / kWh
3. Cogeneration heat credit = $3 / MMBTU
4. Ferric Chloride = $1.34 / gal (40%)
5. Polymer (all types) = $3.85 / lb active
6. Biosolids Disposal = $42 / wet ton

Table 6 outlines the predicted O&M costs ($ per year) for each alternative relative to the Conventional option.

Table 6. Relative O&M Costs

Parameter	Primary Treatment Alternative			
	Conventional	**Operational CEPT**	**CEPT**	**BEPT**
A-Stage RAS	–	–	–	$27,000
Aeration	–	($85,000)	($51,000)	($27,000)
RAS Pumping	–	($107,000)	–	–
Ferric	–	$88,000	$88,000	$65,000
CEPT Polymer	–	$117,000	$117,000	–
RDT Polymer	–	$45,000	$45,000	$27,000
Dewatering Polymer	–	$37,000	$27,000	$19,000
Cake Disposal	–	$80,000	$57,000	$41,000
Cogeneration Electricity	–	($121,000)	($138,000)	($84,000)
Cogeneration Heat	–	($17,000)	($20,000)	($12,000)
TOTAL	**$0**	**+ $37,000**	**+$125,000**	**+ $56,000**

Present Worth Cost

The capital and O&M cost estimates were combined to compare the present worth costs of the alternatives as shown in Table 7. O&M costs were assumed to be yearly over twenty years, discounted at 5 percent.

Table 7. Relative Present Worth Costs

Parameter	Primary Treatment Alternative			
	Conventional	**Operational CEPT**	**CEPT**	**BEPT**
Capital	$0	$0	($25.7)	($1.8)
PV of O&M	$0	$0.5	$1.6	$0.7
NPV	$0	+ $0.5	($24.1)	($1.1)

FINDINGS & CONCLUSIONS

When planning new wastewater treatment facilities or upgrading existing facilities it is beneficial to consider and investigate process options through the use of whole-plant process modeling to quantify the relative energy and treatment benefits of different plant configurations aimed at satisfying objectives of energy reduction and producing low levels of effluent nutrients. Balancing the sometimes competing aspects of nutrient removal, energy conservation, and water reclamation, while considering financial and operational constraints, is crucial to assessing the total net environmental benefit of a treatment facility.

The need for advanced water reclamation to meet stringent effluent requirements led to selection of membrane technology and UV disinfection for the liquid stream. Evaluation of influent wastewater carbon recovery strategies led to selection of anaerobic digestion and a subsequent biogas combined heat recovery and power generation scheme. This combined with a goal of minimization of energy required for liquid treatment led to the evaluation of several primary treatment approaches, including conventional primary clarifiers, chemically-enhanced primary treatment, and biologically-enhanced primary treatment through the "A/B Process" which has been employed at many European facilities since the early 1980's.

Table 8 summarizes a comparison evaluation of the four scenarios. A value of 1 was assigned to the least favorable alternative; a score of 2 indicates average, while a score of 3 indicates the alternative is considered the best for that factor.

Table 8. Comparison Evaluation Summary

	Primary Treatment Alternative			
Factor	**Conventional**	**Operational CEPT**	**Full CEPT**	**BEPT**
Capital Cost	1	1	3	1
O&M Cost	3	2	2	2
Reliability	3	2	2	1
Ease of Maintenance	2	1	1	1
Ease of Operation	3	2	2	1
Technology Familiarity	3	3	2	1
Energy Balance	1	3	2	2
TOTAL	**16**	**14**	**14**	**9**

BEPT was not recommended because it requires additional treatment processes thereby increasing O&M complexity without significant energy reduction or biogas production benefits.

Full CEPT was also not recommended because the successful operation of the system is predicated on good and consistent operation of a chemical treatment system. Although there were capital savings, these were not significant enough to outweigh the operational issues.

The Conventional alternative is well understood by DWR and was recommended for initial years at reduced capacity operation. However, Operational CEPT can operate with lower MLSS concentrations with a corresponding increase in aeration efficiency and carbon capture for biogas generation. Therefore, the preferred ultimate flow sheet selected for the Crooked Creek WRF was Operational CEPT.

Key findings from the evaluation included:

- There are competing demands for the available organic matter in wastewater when considering concurrent goals of energy efficiency and nutrient removal
- Unit processes must be considered in their various combinations to understand holistic effects on overall sustainability goals
- Maximizing energy production potential through efficient capture of colloidal and soluble BOD_5 in the A/B Process can be detrimental to downstream biological nutrient removal (particularly phosphorus removal)
- Sizing MBR aeration basins for typical MLSS concentration but providing the ability to reduce the concentration and load through CEPT provides opportunities to save significant aeration energy through a combination of increased alpha factor and decreased load
- Balancing influent organic usage in anaerobic digestion and subsequent energy production with needs for driving BNR processes is key to realizing overall net energy savings
- Biogas production can be significantly enhanced through more efficient capture of pollutants in primary treatment with only a modest increase in the quantity of biosolids for disposal
- Although BEPT can improve the efficiency of primary treatment without chemical addition, because of the limiting effect on biological phosphorus removal, higher chemical usage is needed for removal of phosphorus in secondary treatment to achieve low level phosphorus requirements

The evaluation demonstrated that sustainability principles can be incorporated into selection of a treatment technology/approach that appropriately balances the sometimes competing aspects of nutrient removal, energy efficiency, and water reclamation needs.

High Efficiency Bio-Diesel Production from a Municipal FOG Facility

Michael D. Curtis, RPM
John Lanzoni, Siemens BT
Richard Parnas, RPM

RPM Sustainable Technologies
800 Village Walk, Ste 101
Guilford, Connecticut 06437

Siemens Corporation
85 John Road, Unit 1, Canton, MA 02021

ABSTRACT

Production of Biodiesel from yellow grease has long been a known commodity. Little effort has been spent to develop a viable fuel product from brown grease. The Torrington, CT Water Pollution Control Authority (WPCA) has installed and maintained a well-designed FOG receiving and pre-treatment station. The goal of the design was to concentrate FOG to a point where it is able to be used as fuel in regional sludge incinerators. RPM Sustainable Technologies (RPM), in collaboration with Torrington has performed preliminary analyses of the waste source and the site and determined that a high yield of biodiesel can be developed from FOG with the RPM 'distributed biodiesel' technology. Numerous samples taken over months from the Torrington FOG receiving station have continually shown approximately 50% mass-to-mass yield of ASTM grade Biodiesel.

The State of CT has developed far-reaching regulations requiring installation and maintenance of FOG removal equipment at nearly all Food Preparation Establishments (FPE's). Centralized collection and treatment of FOG wastes has developed is underway. The city of Torrington CT has developed a FOG facility designed to receive up to 2 M gallons per year of grease trap and Automatic Grease Recovery Units (AGRU's). This high-strength waste is approximately 90% water. Gravity separation at the plant results in 200,000 gallons per year of concentrated FOG and the potential to produce nearly 100,000 gallons a year of ASTM Bio-Diesel or Fuel Oil. This represents a significant percentage of the city's annual transportation fuel budget.

The RPM Technology detailed herein is an exciting new technology application. It involves low energy conversion of FOG to ASTM bio-diesel. It will result in millions of dollars of fuel production from a nuisance waste.

INTRODUCTION

FOG represents an immense world-wide environmental problem. It is well-documented that FOG in sewer systems results in thousands of sewer overflows annually, representing significant health threat arising from these overflows, and results in operational mayhem in sewer collection and wastewater treatment systems. Nationwide, sewer authorities and regulatory agencies have increasingly regulated the discharge and control of FOG waste. The State of Connecticut Department of Energy and Environmental Protection (DEEP – formerly DEP) has developed and implemented far-reaching regulations mandating the installation, maintenance and general permitting of AGRU's and/or passive grease traps for the vast majority of Food Preparation Establishments (FPE's) within the state. Time frames for compliance for new and existent FPE's were established with a measurable result being the generation, hauling and disposal of millions of gallons of FOG to an increasing number of Centralized FOG Receiving Facilities.

The Torrington WPCA has installed and maintained a well-designed FOG receiving and pre-treatment station. The goal, achieved in the design, was to concentrate FOG discharged to the station to a point where it is able to be used as fuel in regional sludge incinerators. Concentrated by gravity and minor heat application, a 10,000 – 12,000 BTU per gallon fuel is created that supplants the need for a portion of the fossil fuel purchase in in-state sludge incinerators.

Up to 2 Million gallons per year can be discharged to the FOG receiving system at the Torrington facility. Repeated testing, performed bi-monthly over the last year, has shown the FOG waste to be consistently nearly 90% water. With pre-treatment / separation nearly 200,000 gallons per year of concentrated FOG results. RPM can produce nearly 100,000 gallons a year of ASTM Bio-Diesel or #2 Fuel Oil from this waste. The fuel / centralized collection system provides a beneficial re-use for a true nuisance material and is being replicated throughout the northeast.

Performance Contracting at treatment facilities, resulting in Energy and Operational Savings and Upgrades is growing in sheer numbers as well as the diversity and innovation of 'at-plant' upgrades.

The proven technology of a FOG to BioDiesel conversion, the urgent need for alternate disposal mechanisms for FOG and the financial advantage of converting a waste product into a $4.00 per gallon resource makes this technology development an easy and obvious choice for inclusion in the Performance Contracting suite of technology applications.

As is the case with many capital projects in water utilities today, the need for funding and the available options to acquire monies have dwindled. As a result, many municipalities have implemented an alternative delivery and financing strategy called Performance Contracting (PC) The successes of this alternative strategy are numerous and well documented.

It is well known that wastewater facilities are the largest users of energy in a municipality. These plants typically represent approximately 30% of a municipality's utility budget. These facts combined with aging infrastructure and lack of capital creates a "perfect storm" for the PC delivery model. The partnering approach to PC delivery can and will mesh very well with a FOG to Bio-Diesel application. Producing substantial quantities of fuel at a WWTF, with no new capital input is a revenue outcome that would be welcome at many plants nationwide.

BIODIESEL PROCESSING AND PERFORMANCE CONTRACTING

As stated, Biodiesel Production from vegetable oils has been successfully performed for decades. However, comparatively little effort has been spent to develop a viable, cost-effective fuel product from FOG. Some longer-term successful efforts have been undertaken, but these usually involve glycerolysis, an energy / capital intensive process that can increase brown grease to bio-diesel operational costs to unacceptable levels. RPM Sustainable Technologies (RPM), in collaboration with the Torrington WPCA has performed extensive analyses of this concentrated waste source and the site and determined that a high yield of biodiesel can be developed from FOG with the RPM 'distributed biodiesel' technology. Numerous samples taken over months from the Torrington FOG receiving station and from FOG skimmed from primary clarifiers at four other CT waste water treatment plants have consistently shown 45-50% volumetric yield of ASTM grade Biodiesel from FOG.

Innovative pre-processing of the FOG has been shown to be critically important in efficient extraction of fuel from the waste source. RPM has developed a proprietary means of fuel separation from FOG concentrating on physical separation in addition to chemical transformation (Figure 1). Figure 1 illustrates a dramatic improvement in yield by using highly efficient separation technology to remove the methanol, water and catalyst from the pretreated FOG.

FIGURE 1

LAB SCALE SEPARATION OF BIO-DIESEL FROM FOG

Numerous 'treatments' of FOG at the lab scale and extensive operation of the full-scale Bio-Diesel plant, located at the University of Connecticut (Figure 2) have led to the design and implementation

FIGURE 2

RPM 'DISTRIBUTED SCALE' BIO-DIESEL PLANT

of the full-scale pilot plant to be located at the Torrington, CT WWTF. The RPM technology consists of a patented, continuous flow, complete-mix biodiesel processor. Resulting from this break-through technology, smaller tankage is needed as batch processes have been replaced by a small reactor system. Methanol storage is minimized and a 'distributed scale' plant can be applied at numerous smaller locales. With the ever-increasing control of FOG, the potential for numerous applications nation/world-wide.

The Torrington FOG receiving station (Figure 3) is designed such that withdrawal of FOG and 're-discharge' of non-converted materials is simplistically made with quick connect / dis-connect piping. The Pilot Plant will be trailer or container skid-mounted with tankage at or very near the skid. It is envisioned that 3-6 months of operation will ensue after on-site commissioning.

FIGURE 3
TORRINGTON WWTF FOG RECEIVING STATION

The project is being funded from public and private sources. The CT Energy Finance and Investment Authority (CEFIA) has committed $0.5 M to the project. Additional private sources of funds will bring this full-scale pilot plant on-line in the early third quarter, 2013. The municipality, an aggressive energy-conscious government, is aggressively working towards turning the proposed pilot operation into a permanent installation. Lastly, this development is being tracked by numerous corporations interested in implementing the RPM technology in a performance contract.

Work to date has been very promising and highly 'repeatable'. While not expected when testing was initialized, the results are quite predictable and as a result could be of great interest inside of a Performance Contract. Siemens Building Technology is working with RPM in project development. The ultimate level of Siemens' involvement is yet undetermined, but the project is of great interest in developing numerous municipal clients. Numerous advantages influence the positive cash model associated with implementation of this technology. These include

1) Tipping Fees Associated with FOG Delivery,
2) Reduction of 'Mass' through conversion to BioDiesel,
3) Potential Sale of BioDiesel or #2 Fuel Oil either 'behind the meter' within a municipality or on the market, and
4) Lastly, the potential for the 'waste' materials to be further converted to 'electron donors' highly useful in denitrification and ENR applications.

Rapidly degradable Hydrocarbons, some Glycerin and Volatile Fatty Acids (VFA's) are a by-product of the FOG processing system. These findings, while preliminary, are immense. Concerted efforts to collect and concentrate FOG could result in significant municipal fuel production (both biodiesel and heating oil). Waste organics might be readily utilized to enhance biological nutrient removal processes. Full-scale implementation could offer a very attractive rate of return to Torrington (Figure 4).

FIGURE 4

PLANT RATE OF RETURN

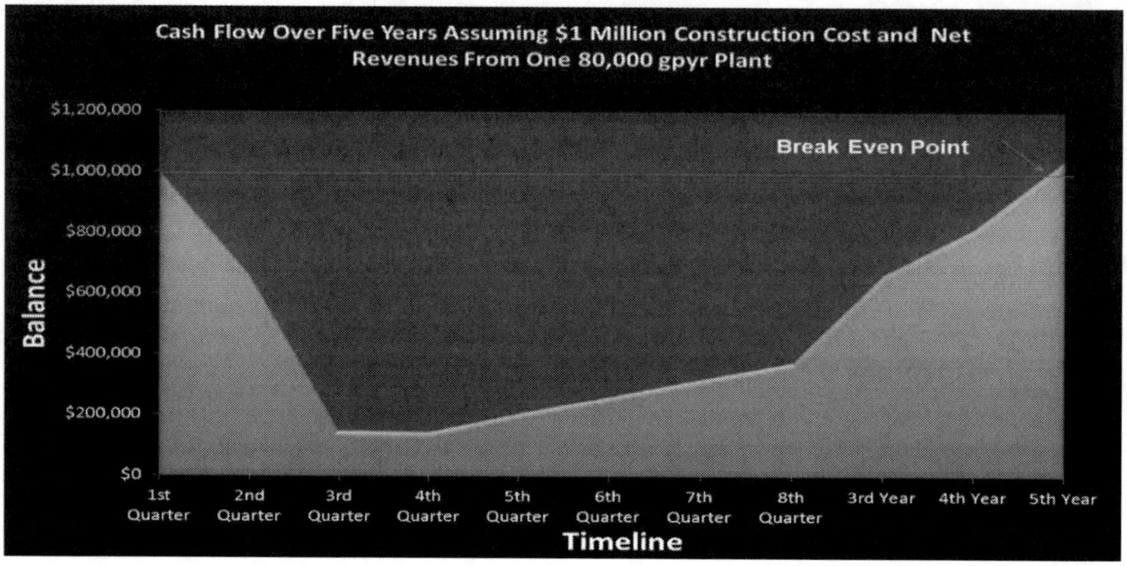

CONCLUSIONS

The RPM Patented Technology to be pilot-tested in Torrington, CT, using an over-used phrase, is a 'game-changing' technology. In a permanent application in Torrington, a 7 mgd facility, low energy conversion of FOG to ASTM bio-diesel can result in millions of dollars of fuel production from an otherwise nuisance waste. The consistency and predictive nature of the process fits well into Performance-Based contracts and will allow an accelerated 'spread' of the technology where applicable. The current structure of tipping fees and production of an expensive fuel further reinforces the cost model associated with this project.

The RPM Process, a continuous flow reactor system, has been markedly reduced in size from more conventional systems allowing distributed source biodiesel deployment and not requiring huge capital investment on behalf of a municipality. It is envisioned that numerous facilities will embrace this technology in the next decade.

Sustainable Energy Management Systems for Wastewater Treatment Facilities

James C. McCaughey, PE, BCEE
Narragansett Bay Commission
One Service Road
Providence, RI 02905
401-461-8848 x 352

ABSRACT

In the State of Rhode Island wastewater is collected and treated by 19 individual Wastewater Treatment Facilities (WWTFs) capable of receiving and treating up to 200 million gallons per day (MGD) of municipal wastewater. Based on standardized energy consumption data and assumed peak energy efficiency performance, these facilities can consume more than 78,000 MWH of electricity annually at a cost of more than $9 million. Recognizing the environmental and economic benefits of improving WWTF energy management practices the Narragansett Bay Commission, the State of Island's largest multi-community WWTF municipality, developed and launched a Sustainable Energy Management Program to help improve the energy use performance of these RI WWTFs. As a result of these efforts an estimated 4,470 MWh/year of potential energy savings have been identified and two renewable energy projects with the potential of creating more than 11,000 MWh/year of clean renewable energy are being implemented. Project efforts have been partially funded through grants from the EPA and the U.S. ARRA Program, and funds obtained through National Grid's Energy Efficiency Incentives Programs.

KEYWORDS

Energy efficiency, renewables energy, biogas, wind energy, energy conservation, biodiesel glycerin.

Introduction

Municipal wastewater contains dissolved and suspended organic matter, pathogens, nutrients (such as nitrogen and phosphorus), and chemicals from residential and commercial and industrial manufacturing processes and services and as such when released untreated can have detrimental health and environmental impacts. Municipal Wastewater Treatment Facilities (WWTF), designed to collect and treat wastewater play an important role protecting human health and the environment. WWTF operations are however large energy users and create other environmental impacts particularly in the form of air pollutant emissions (e.g., greenhouse gasses) from the consumption of fossil fuels.

To help improve WWTF energy efficiency in the State of Rhode Island the Narragansett Bay Commission applied for and received grant funding from EPA's 2008 State Innovation Grant Program to develop and implement a systematic, low cost and common-sense approach to reducing the energy demand of Rhode Island's WWTFs.

Through the resulting "Rhode Island Sustainable Energy Management Practices for Wastewater Treatment Facilities Project" (RISEMP-WWTF) the Narragansett Bay Commission (NBC), the Rhode Island Department of Environmental Management (RIDEM), the University of Rhode Island (URI), and the Rhode Island Manufacturers Extension Service (RIMES) (the Project Partners) in cooperation with EPA Region I addressed two important concerns common to all WWTF operations – energy consumption and controlling the discharge of waste Fats, Oil and Grease (FOG) from the Food Service Industry Sector (i.e. restaurants). The primary goals of this four year project were: 1) Reduce the fossil fuel energy demand of Rhode Island's 19 WWTFs and 2) Reduce the amount FOG being discharged to NBC's sewer system by promoting both the use of FOG Best Management Practices and the production of bio-diesel from waste "yellow" grease.

Purpose and Goals of the Project

As part of the RISEMP-WWTF Project the Project Partners developed and implemented two distinct but related efforts: 1) a Sustainable Energy Focused Environmental Management System (EF-EMS) Program for WWTFs, based on the well-established ISO 14001 Environmental Management System (EMS) "Plan-Do-Check-Act" approach, and 2) an Environmental Results Program (ERP) designed to improve the management and collection of FOG waste generated by local food service establishments.

I. WWTF Energy Focused Environmental Management Systems (EF-EMS)

In the State of Rhode Island municipal wastewater is managed by nineteen individual WWTFs capable of receiving and treating a collective 757 million liters per day (200 mgd) of wastewater along with some combined stormwater. These WWTFs are large energy users and, based on standardized energy consumption data, can annually consume as much as than 78,000 MWH of electricity at a cost of more than $9 million dollars per year.

Figure 1 Electrical Energy Cost Comparison: New England/U.S. 1990 through 2006
Source: U.S. Energy Information Administration http://www.eia.gov/

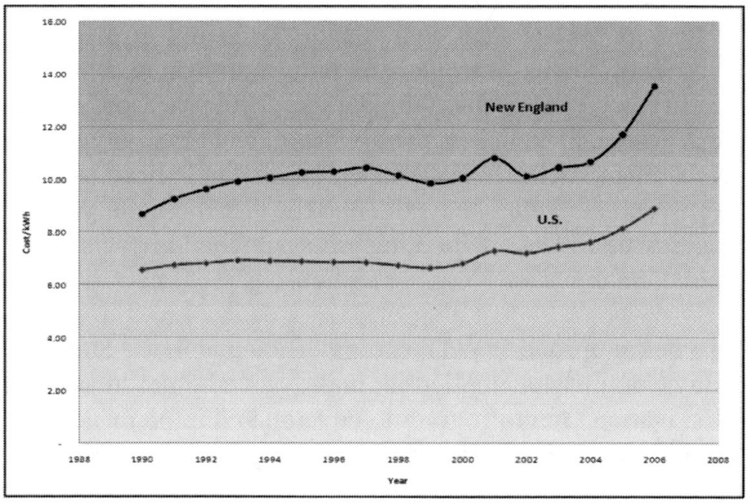

The fuels used to generate electricity vary by region and the bulk of electrical energy produced in the Northeast Unites States is obtained from the burning of fossil fuels, primarily natural gas, resulting in air pollution emissions including the greenhouse gases CO_2, CH_4, N_2O and water-vapor. Based on the average New England power industry's fuel mix every 1 MWH of electrical used results in the release of about 341 kg (750 lbs.) of $CO_{2(e)}$. Thus a 5% reduction in electrical demand by Rhode Island WWTFs will result in a reduction of more than 1,330 metric tons (1,400 tons) of $CO_{2(e)}$ annually.

The cost of generating electricity also varies by region, again mostly due to the use of available fuels. As noted above, in the Northeast expensive natural gas is the primary fuel used to generate electricity thus resulting in ever increasing and higher than average electrical energy costs.

The Project Partners set as a goal for this project the development of an Energy Focused Environmental Management System (EF-EMS) for all of Rhode Island's nineteen WWTFs, utilizing EPA's energy efficiency guidance document "Ensuring a Sustainable Future: An Energy Management Guidebook for Wastewater and Water Utilities" (the "Guidebook"): http://www.epa.gov/region6/water/energymgt/energy_mgt_guidebook_wastewater.pdf. In order to measure project outcomes, the energy performance of each participating WWTF was tracked using EPA's Energy Star Portfolio Manager http://www.energystar.gov.

The Guidebook, modeled on the ISO 14001 EMS "Plan-Do-Check-Act" (PDCA) process which includes the following:

- Plan – Establish and prioritize energy targets to meet energy goals
- Do – Implement specific practices to meet these targets
- Check – Periodically review progress
- Act – Make adjustments as needed to help meet energy goals

The Guidebook is designed to help identify energy efficiency and conservation opportunities and potential applications for the use of available renewable energy resource at WWTFs and help assure energy goals are achieved.

II. Fats, Oil and Grease (FOG) ERP for Food Service Establishments

An Environmental Results Program (ERP)[i] is an integrated approach to addressing environmental problems associated with various business sectors and other groups having large numbers of small facilities. While individual facilities within these groups may release small amounts of pollution, the aggregate impact can be significant.

The Massachusetts Department of Environmental Protection developed ERP over 10 years ago, in the midst of department-wide budget cuts, as a cost-effective strategy to improve and measure the performance of such groups. More than one-third of U.S. states have now developed or are implementing at least one ERP. Increasingly, states are also exploring new ways of applying ERP components. A typical ERP uses a combination of the following tools to address environmental problems:

- Plain-language outreach on key environmental issues;
- Voluntary or mandatory facility self-certification regarding performance on those issues;

- Agency site visits to assist facilities and verify performance; and
- Statistically based performance measurement.

Fats, Oil and Grease (FOG) that enters the sewer system from Food Service Establishments (FSEs) is a continuing problem for WWTF operations. FOG causes sewer line obstructions, problems with pumps and other WWTF equipment and can inhibit overall wastewater treatment effectiveness.

FSEs (e.g. restaurants) generally generate two forms of FOG – "Yellow Grease" consisting of used vegetable oil often contaminated with fried food residue and "Brown Grease" consisting of residual FOG removed from the wastewater stream through the use of Grease Removal Units (GRUs). The NBC limits FOG levels in commercial and industrial wastewater to125 mg/l. Yellow Grease can be potentially used as a source of renewable energy though the production of bio-diesel and collected Brown Grease can potentially be used in the anaerobic digestion of wastewater bio-solids to boost methane production.

NBC's Pretreatment Program has an aggressive regulatory FOG control program that has helped to reduce the discharge of FOG through the use Grease Removal Units, regulatory inspections and monitoring. On average however, NBC still removes as much 60 dry tons per month of FOG from the wastewater received at the Field's Point WWTF.

A noted above the goal of the FOG ERP was to reduce the amount FOG discharged to NBC's WWTFs through the voluntary use of FOG Best Management Practices while encouraging the reuse of waste FOG through local biodiesel production.

Project Approach and Results

I. EF-EMS Project - General Approach

Figure 2 EF-EMS Organizational Chart

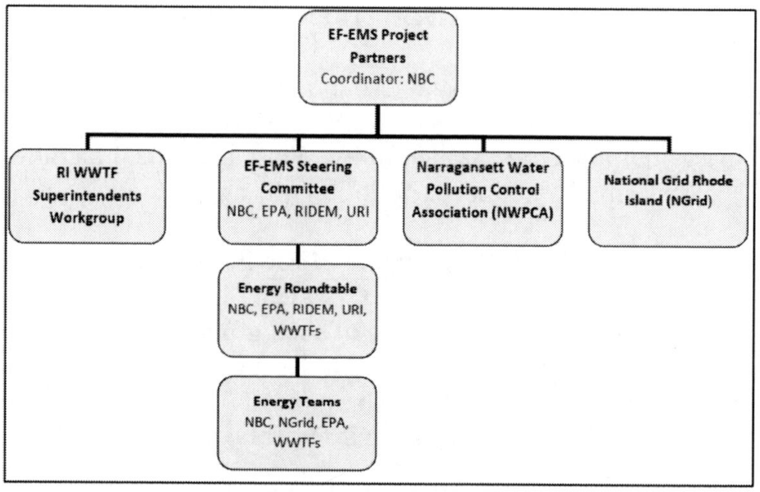

The Project Partners recognized prior to beginning this project that an important aspect of developing individual WWTF EF-EMSs will be the formation of EF-EMS Teams (in this case "Energy Teams" for each individual WWTF). In order to form effective and functional Energy Teams the Project Partners looked for members from within two existing Rhode Island based WWTF organizations: the Narragansett Water Pollution Control Association (NWPCA) - the Rhode Island "chapter" of the New England Water Environment Association, and the Rhode Island WWTF

Superintendent's Workgroup - a RIDEM headed workgroup whose membership consists of the Superintendents from all 19 Rhode Island WWTFs. Both organizations have crossover membership and meet periodically, separately and combined, to address and discuss issues pertinent to the Rhode Island WWTF community.

In forming the Energy Teams the Project Partners first established an EF-EMS Steering Committee made up of leadership members from NWPCA and the WWTF Superintendents Workgroup. As a result of a Project Kick-Off Meeting held in February 2009, attended by representatives of each Rhode Island WWTF, and several follow-up meetings and discussions, the EF-EMS Energy Steering Committee formed the Rhode Island WWTF Energy Roundtable (Energy Roundtable) consisting of key members (Superintendents and facility Managers) from each WWTF.

Through the Energy Roundtable individual WWTF Energy Teams were formed for each of the nineteen WTTFs. Each Energy Team consisted of a representative from NBC who was responsible for identifying renewable energy opportunities, a representative from National Grid's professional energy management contractor who helped identify and quantify potential Energy Efficiency Measures (EEMs), a RIMES representative responsible for EF-EMS guidance, and representatives(s) from the individual WWTF that had firsthand knowledge of the WWTF's processes and had the authority to implement identified EEMs. The Energy Team was the backbone of the EF-EMS Planning Stage process; their work established the EF-EMS Fence-Line, the baseline energy use performance level from which to measure all energy improvements, and identified all potential EEMs.

The WWTF members of WWTF Energy Team are charged with the responsibility of implementing the identified energy improvements at their individual WWTF. The Project Partners coordinate Roundtable efforts and guide the Energy Team members through the EF-EMS Plan-Do-Check-Act (PDCA) process.

Plan-Do-Check-Act

Policy Statement: Establishing an Energy Policy Statement was the first formal step taken in the development of an EF-EMS. The Policy Statement sets the level of commitment and the direction the overall organization will take on addressing the issue at hand, in this case energy management. The energy policy is looked at as a "living" document that will be updated and modified as the developing organization becomes more sophisticated about, and knowledgeable of their energy use and resulting impacts.

Guided by a RIMES instructor Energy Policy statements were put into writing collectively representing the energy management thoughts of the entire Energy Roundtable. These Policy Statements were found to contain the following set of mutually identified commitments:

- Provide quality wastewater services at a fair cost,
- Improve/maintain compliance with operating permit requirements,
- Lessen reliability on fossil fuels,

- Minimize operational costs and environmental impacts without compromising sewage collection/treatment processes or effluent quality,
- Improve energy efficiency,
- Utilize available renewable energy resources (where technically and economically feasible)
- Lessen carbon footprint

These commitments can be condensed into the following EF-EMS Energy Goals:

1. Reduce energy costs
2. Reduce use of fossil fuels
3. Improve compliance
4. Improve energy efficiency
5. Utilize renewable energy resources
6. Reduce carbon footprint

At this point no attempt is made to establish objectives for these goals. As noted in Session 4 Module 1 of the Energy Guidebook there are no standard objectives that can apply uniformly to all WWTF, thus each individual WWTF needs to develop their own specific Energy Objectives. For example, one WWTF may seek to reduce energy costs by 5% by a certain date while another WWTF may establish an objective of 10% energy use reduction by a different date. A detailed analysis of each WWTF's current energy management practices is needed to help establish achievable objectives.

This Planning exercise, in conjunction with the training given on the overall PDCA process, allowed for the identification of the following Planning Stage tasks for the Energy Roundtable:

Task 1: Establish a quantified energy use baseline for each RI WWTF,
Task 2: Conduct energy use assessments (Scoping Studies) of each RI WWTF to identify potential Energy Efficiency Measures (EEM) that address the above listed commitments,
Task 3: Quantify potential energy savings associated with each identified EEM, and
Task 4: Establish Targets for implementing identified EEMs

At this point it was the goal of the Energy Roundtable, at the completion of Task 4, to have each WWTF in the position to clearly state "what they will do" and the process to "do what they say" with respect to their ultimate energy management plans and policies.

Baseline Measurement: Energy demand within a particular WWTF will be a function of several independent variables including: volumetric flow rates, BOD and TSS loadings, ambient air temperatures and weather conditions, as well as the specific treatment processes being employed. EPA's Energy Star Portfolio Manager was found to be an excellent tool for establishing a baseline and for comparing the energy demand at a specific WWTF to other WWTFs of similar size and with similar treatment processes.

Portfolio Manager as described on the Energy Star webpage is an interactive computer based energy management tool that allows for tracking and assessing energy and water consumption across an entire "portfolio" of buildings and operations. Portfolio Manager has an application developed specifically for wastewater treatment facilities. Establishing a Portfolio Manager account is free of charge and is a relatively easy matter requiring the following basic information:

- Zip code
- Average influent flow
- Average influent biological oxygen demand (BOD_5)
- Average effluent biological oxygen demand (BOD_5)
- Plant design flow rate
- Presence of fixed film trickle filtration process
- Presence of nutrient removal process

Upon entering annual energy use data (kWh of electricity, gallons of diesel fuel, therms of natural gas, etc.) into the Portfolio Manager the program will calculate a normalized "Current Source Energy per Flow" in kBTUs/gallon/day. As noted above, this value can be compared to a nationally obtained average Energy Use Indicators (EUI) for similar facilities generating a "Rating" for your WWTF. A more detailed description of the Energy Star Program and how the Portfolio Manager rates a WWTF is beyond the scope of this paper, however this and additional information on Portfolio Manager can be found at: http://www.energystar.gov/. Figure 3 is a sample of just one of several reports that can be created using Portfolio Manager:

Figure 3 Sample Portfolio Manager WWTF Performance Report

12 Months Ending	Current Rating (1-100)	Average Flow (MGD)	Current Weather Normalized Source Energy per Flow (kBtu/gpd)	National Average Source EUI (kBtu/gpd)	Current Site Electric Use (kWh)	Annual Energy Cost (US Dollars ($))	Current Total GHG Emissions (MtCO2e)
December 2009	62	48.7	2.7299	3.1	10,972,289.1	$1,368,951.99	4,556.84
December 2008	79	47.1	2.9156	4.2	11,409,885.9	$1,315,884.69	4,668.10
Change	17	-1.6	0.1857	1.1	437,596.8	($53,067.30)	111.26

Definitions of the column headings are as follows:

Current Rating (1-100) - A benchmark rating for a facility on a scale of 1-100 relative to similar type facilities nationwide using EPA's national energy performance rating system. The benchmark rating is based on the facility's source energy use, level of business activity, and geographical location. A rating of 1 means the least energy efficient facility and a rating of the 100 most energy efficient facility in the database.

Average Influent Flow - Daily average actual flow of wastewater into the facility, measured in million gallons per day (MGD). The Average Influent Flow should be entered on a monthly basis, with the figure reflecting the average for the time period specified. This Portfolio metric is only available for Wastewater Treatment Plants.

Current Source Energy per Flow - The source energy intensity (kBtu/gallon per day) for the current 12-month period. This Portfolio metric is only available for wastewater/water treatment facilities.

National Average Source EUI (Water/Wastewater) - For a facility with a rating, this defines the source energy intensity (kBtu/gallons per day) this facility would consume if it had a national energy performance rating of 50, the rating associated with average performance. For a facility not eligible for a rating, this defines the national average source EUI calculated in kBtu/gallons per day as a weighted average across all facility of a similar type. This metric is only available for Water and Wastewater facilities.

Current Site Electric Use - Total electricity consumption (kWh) for the current 12-month period.

Annual Energy Cost - The sum of the user-entered meter costs in US Dollars ($) for the current 12 month evaluation period.

The data in this table shows that this facility reduced its combined normalized energy demand from 2.9156 kBtu/gpd in 2008 to 2.7299 kBtu/gpd in 2009. During the same time period however this facility's overall "Current Rating" dropped 79 to 62. This tells us that while this facility realized some energy efficiency improvements between 2008 and 2009 other similar type facilities did an even better job of reducing energy demand. This warrants some research into what these other facilities are doing and assessing whether or not such changes and/activities can be implemented here.

Table 1 contains Portfolio Manager Data from those Rhode Island WWTFs willing to share their data. Unavailable data was estimated using standard WWTF energy use information obtained from the Water and Environment Federation's Manual of Practice No. MFD-2 (1997), "Energy Conservation in Wastewater Treatment Facilities":

Table 1 Rhode Island WWTF Energy Use Data

Facility ID	**Data Source**	Current Rating	Current Weather Normalized Source Energy per Flow (kBtu/gpd)	National Average Source EUI (kBtu/gpd)	*Current Site Electrical Use (kWh/yr)	Current Total GHG Emissions (MtCO2e)
1	Portfolio Manager	47	4.8	4.6	1,432,299	607.1
2	Estimated	N/A	N/A	N/A	324,246	139.4

Facility ID	Data Source	Current Rating	Current Weather Normalized Source Energy per Flow (kBtu/gpd)	National Average Source EUI (kBtu/gpd)	*Current Site Electrical Use (kWh/yr)	Current Total GHG Emissions (MtCO2e)
3	Portfolio Manager	54	9.4	9.8	10,860,000	4,430.4
4	Estimated	N/A	N/A	N/A	364,745	156.8
5	Portfolio Manager	72	5.6	7.2	3,136,320	1,410.9
6	Portfolio Manager	N/A	7.7	10	267,277	131.7
7	Portfolio Manager	56	7.1	7.5	12,405,600	5,341.5
8	Portfolio Manager	62	2.7	3.1	10,855,573	4,506.7
9	Estimated	N/A	N/A	N/A	283,740	122.0
10	Estimated	N/A	N/A	N/A	40,555	17.4
11	Estimated	N/A	N/A	N/A	3,222,278	1,385.6
12	Estimated	57	11.1	11.9	734,717	278.0
13	Portfolio Manager	93	6.8	13.5	1,136,000	520.1
14	Estimated	N/A	N/A	N/A	1,213,610	521.9
15	Estimated	N/A	N/A	N/A	648,044	278.7
16	Estimated	N/A	N/A	N/A	1,657,021	712.5
17	Portfolio Manager	50	10.0	10.1	5,007,000	2,121.1
18	Estimated	N/A	N/A	N/A	890,598	383.0
19	Estimated	N/A	N/A	N/A	3,741,662	1,608.9

* Includes electrical energy used at the WWTF and excludes energy used to collect and pump wastewater through the collection system to the WWTF

Energy Assessments: An energy assessment, performed by an Energy Team, is an essential part of the EF-EMS Planning process. As noted in the EPA Energy Guide many energy utilities offer audit services as part of energy conservation/efficiency programs. In the State of Rhode Island National Grid offers these services to large industrial customers, free of charge in the form of "Scoping Studies."

A Scoping Study consists of first collecting information on baseline energy use and operational performance (accomplished here through the use of Portfolio Manager) and then conducting an on-site walkthrough of the WWTF by the Energy Team with the objective of identifying and documenting potential Energy Efficiency Measures (EEMs). A National Grid Scoping Study is overseen by a professional energy management company with experience in the specific operations being studied - in this case wastewater treatment.

In preparation for the walk through portion of the Scoping Study the Energy Team (consisting of representatives from NBC, the WWTF being audited, URI, and RIDEM) conduct a detailed review the facility's operations, collecting operational data on average daily flow, BOD, nitrogen, and TSS loadings, overall removal efficiencies, and average energy use. This collected data, as noted above, is analyzed and benchmarked using EPA's Portfolio Manager. The walkthrough portion of the Scoping Study identifies EEM opportunities ranging from light fixture replacements to blower upgrades and focuses on those particular EEMs that can be funded through National Grid's Energy Efficiency Incentives Program.

As part of each Scoping Study, NBC's Pollution Prevention Engineer performed a site specific renewable energy use assessment looking for solar and wind energy resources, biogas production opportunities, and hydro-electric energy applications.

All Scoping Study findings are summarized in a Scoping Study Report issued to the participating WWTF, National Grid and the NBC and all renewable energy assessment findings are summarized in a Renewable Energy Report issued directly to the WWTF.

Scoping Studies were completed for 18 of the 19 Rhode Island WWTFs (New Shoreham on Block Island had a combined Scoping Study/TA). Table-2 contains a summary of general EEM categories identified as part of these Scoping Studies:

Table 2 Summary of Scoping Study EEMS

EEM Category	Number of WWTFs with Identified EEMs	Number of EEMs Identified
Aeration Control	11	17
Heating/Cooling	15	39
Insulation	17	3
Lighting	17	21
Miscellaneous	1	2
Motor Replacements	8	9
Process Control	5	7
Process Monitoring	3	3
Pumping	8	13
Solids Handling	3	3
VFD	4	8
Renewable Energy	2	3
Total	84	128

Identified EEMs are classified as either "Prescriptive" or "Custom." Prescriptive EEMs can directly apply for National Grid Energy Efficiency Incentives while Custom EEMs require additional study in the form of an Energy Efficiency Technical Assessment (TA) in order to fully quantify project costs and energy savings. Technical Assessments were completed for 13 WWTFs.

EEM Implementation and Measurement: Once EEMs are identified and potential energy savings quantified the Energy Team must establish targets for implementing the EEMs. Energy Targets are measurable achievements such as installing an energy efficient piece of equipment or achieving a quantified energy savings by a targeted date. In any case an action plan needs to be developed with measurable targeted dates of completion. Here cost factors and manpower priorities come into play. Fortuitously, the Rhode Island Office of Energy Resources (RIOER) in March 2011made available $3,100,000 in American Recovery and Reinvestment Act (ARRA) funding under the Rhode Island Wastewater Utility Energy Efficiency Program to:

- Establish an Energy Efficiency Wastewater Treatment Facility Program in Rhode
- Island,
- Reduce energy costs of Wastewater Treatment Facilities through the, implementation of energy efficiency measures and deploying renewable energy and cogeneration systems,
- Assist in the stabilization of commercial and residential sewer rates,
- Realize environmental benefits to air and water quality
- Reduce the reliance on energy resources from outside of Rhode Island

Three million dollars of this funding was made available specifically for the implementation of prescriptive EEMs identified in the Project Scoping Studies. The ARRA grant will fund the EEM costs not covered by National Grid's incentive program, on a reimbursement basis, with an established installed and operational project deadline of March 31, 2012. A second round of ARRA EEM grant funds was again released in July 2011. Table 3 below contains a listing of EEMs which were implemented for one WWTF:

Table 3 Rhode Island WWTF EEMs Implementation

Project	Implementation Date	Annual Energy Savings (kWh)	Annual Renewable Energy Generation (kWh)
Electric Heater Efficiency Modifications	March 2012	12,000	
VFDs for WWTF Water Pumps	March 2012	388,000	
VFDs for WWTF Primary Sludge Pumps	March 2012	55,000	
VFD and Controls for WWTF Primary Sludge Pump	March 2012	38,000	
WWTF Lighting Upgrade	March 2012	15,000	
WWTF Administrative Building Lighting Upgrade	March 2012	90,000	

Project	Implementation Date	Annual Energy Savings (kWh)	Annual Renewable Energy Generation (kWh)
Wind Turbine Renewable Energy	October 2012		7,000,000
Total		598,000	7,000,000
% of WWTF Energy Use		2.5%	29 %

II. FOG-ERP - General Approach

The NBC in cooperation with the RIDEM, URI and EPA developed an Environmental Results Program (ERP) and a FOG Best Management Practices Workbook to assist Food Service Establishments (FSEs), serviced by the NBC, improve their FOG management practices. The workbook contains NBC's regulatory compliance inspection checklist and an explanation of the reason/necessity of each item outlined in the checklist along with a best management practices (BMP) checklist with similar explanations for each listed BMP.

The NBC FOG-ERP has been designed to improve the management of FOG by the local food service industry through a combination of efforts and actions:

1. Compliance Assistance: Pollution Prevention Engineers from the University of Rhode Island and the NBC meet with restaurant owners and managers to help implement sound and sustainable FOG Best Management Practices and better comply with NBC regulations.

2. Self-Evaluation: Restaurant owners and/or operators are given training through Compliance Assistance efforts to self-evaluate their facility in order to certify their FOG management practices utilizing the NBC Oil & Grease Compliance and Best Management Practices Workbook.

3. Regulatory Inspections: The NBC Pretreatment Program continues to inspect all restaurants on a regular basis. Participation in the FOG-ERP helps firms prepare for regulatory inspections and helps them comply with NBC regulations.

4. Certification: Restaurants that demonstrate a superior FOG management performance level are issued a Certification of Best Management Practices which may be displayed in their place of business.

A Self-Certification form is included at the end of the workbook for use by those FSEs that choose to certify their participation in the ERP. FSEs certified in this program will also be recognized as meeting the grease management requirements of the RI Hospitality Green Restaurant Certification Program[ii]. The FOG-ERP also helps each FSE comply with Rhode Island General Law (R.I.G.L 23-18.17) which requires waste cooking grease in the State of Rhode Island be recycled.

Each FSE located within the NBC service district was notified by mail about the FOG-ERP and information on the ERP program, along with electronic copies of the workbook and checklists, which are available on NBC's website, www.narrabay.com under Programs and Projects/Pretreatment Program/Grease Control Program.

As part of standard monitoring activities NBC collects and analyzes samples from the head-works of both the Field's Point WWTF and the Bucklin Point WWTF for oil and grease content. The following charts of FOG Loadings at both WWTFs show a marked decline in the amount of FOG being received at the WWTFs between the years 2006 and 2012. The impact of the FOG ERP on this decline is, at this point, inconclusive with respect to the impact from the FOG ERP efforts. On-going follow-up data collection at FSEs may give some insight as to what if any effect the FOG ERP effort had on this FOG loading reduction. It should be noted that no project grant funds where expended on this data collection and/or analysis. The increase in FOG levels in 2009 at both WWTFs is most likely due high stormwater flow for that year.

Figure 4-a NBC FP WWTF FOG Loadings 2006 - 2012

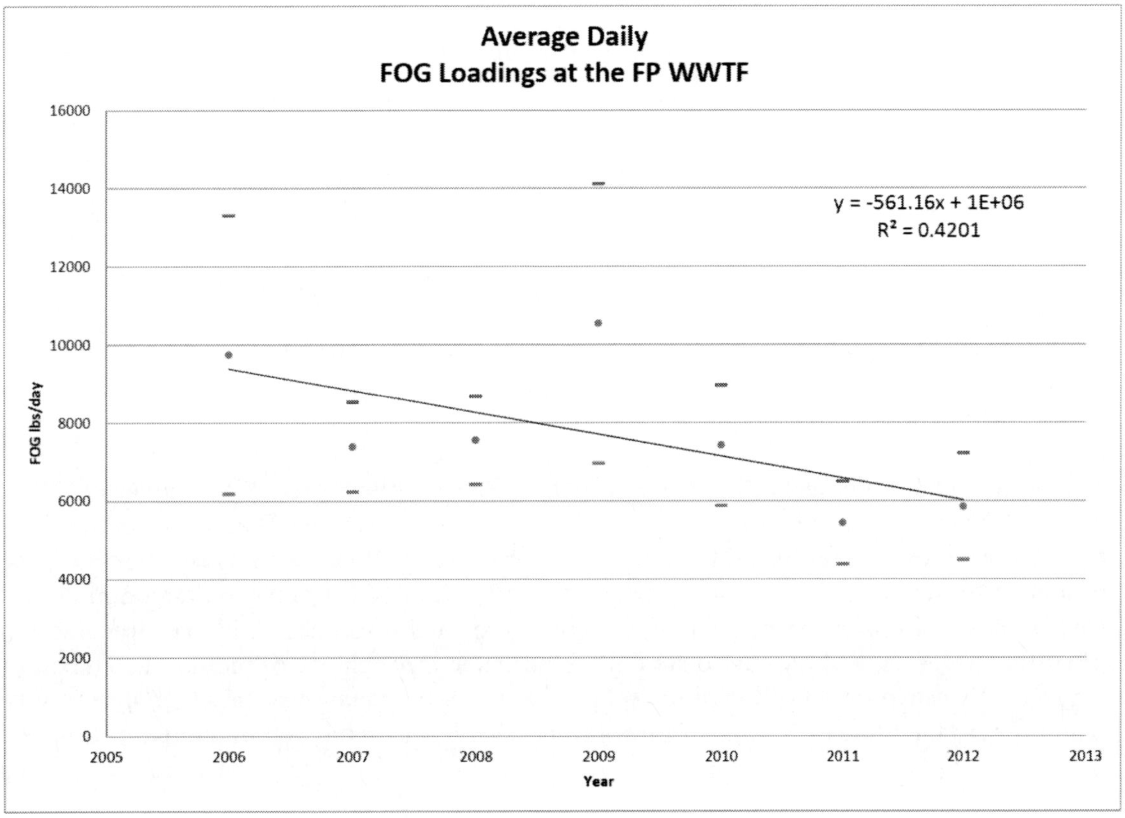

Figure 4-b NBC BP WWTF FOG Loadings 2006 - 2012

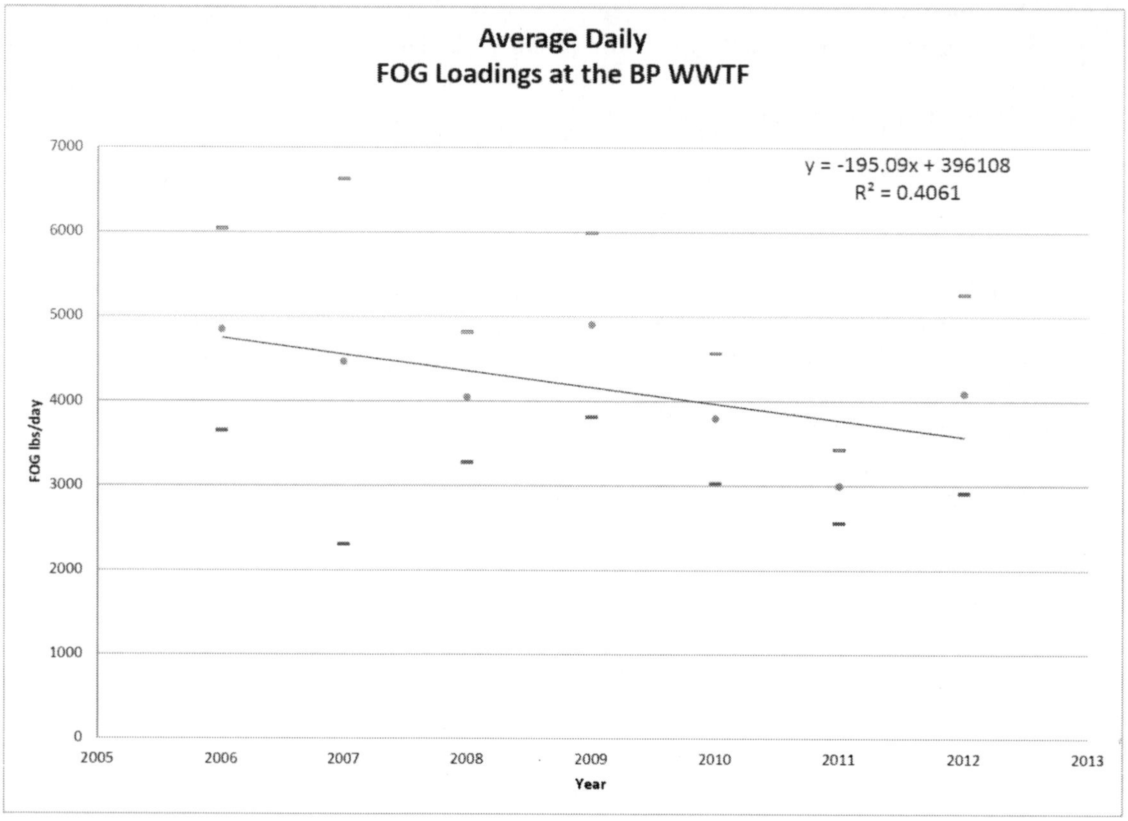

Biodiesel Manufacturing Coordination: Prior to the development of the FOG ERP the Project Partners contacted and met with the State of Rhode Island's only commercial biodiesel production facility in order to learn more about the biodiesel production process and to better understand the benefits of encouraging FSEs to send their waste "yellow grease" for biodiesel production.

The production of biodiesel from waste vegetable oils ("yellow grease") removes a potential source of landfill contamination, and reduces the release of carbon from the burning of diesel fuel produced from fossil fuel resources. It has been shown that biodiesel produced from recycled waste vegetable oils (WVO) results in an 85% reduction in the carbon intensity relative to diesel derived from fossil fuel resources. Additionally emission of sulfur oxides and sulfates are essentially eliminated through use of biodiesel. The only significant waste product of the biodiesel production process is glycerin which is currently being sent to the State of Michigan for refining. NBC is investigated in the use of this material as a potential carbon source with municipal wastewater de-nitrification operations.

At the start of the FOG ERP this biodiesel manufacturer was producing 500,000 gallons per year of biodiesel made from waste vegetable oils (yellow grease). In the summer of 2012 this facility projected a production of 1.5 million gallons. The biodiesel manufacturer employs about 20 people and collects waste vegetable oil from more than 1,000 restaurants in Southern New England.

CONCLUSIONS

The PDCA model as employed through an EF-EMS using EPA's Energy Management Guidebook for Wastewater and Water Utilities has been shown to be readily applicable to the development of a multi-jurisdictional WWTF Energy Focused Environmental Management System in the State of Rhode Island. The methodologies used to coordinate these efforts should be readily adaptable by other states and/or state political subdivisions.

As a result of this project Rhode Island's nineteen WWTFs, working together, are expected to achieve measurable reductions in energy use and subsequent reductions in the generation of greenhouse gases while accomplishing the same or improved levels of wastewater treatment. Project outcomes include improved working relationships between Rhode Island's WWTFs, improved overall environmental conditions, and sustainable energy efficient wastewater treatment operations.

The ERP approach has been proven to be a very effective way of achieving positive environmental results. At this early stage of the FOG ERP NBC has found that many of the FSEs are reluctant to participate / cooperate until they are informed that a certificate of participation can be earned. While some restaurant managers cooperated and were happy to allow NBC to assist them in completing the BMP checklist, others seemed hesitant to answer questions about their FOG practices. When they were told that their cooperation, and ultimately compliance with the identified FOG best management practices, could earn them a certificate to display within their place of business, several of these managers seemed to be far more receptive to the idea.

Having a local end-use repository for the waste yellow grease contributed greatly to the overall recycling effort of the project. It is highly unlikely that the participating FSEs would pay to have their waste yellow grease shipped to an out-of-state biodiesel production facility.

REFERENCES

EPA (2008), "Ensuring a Sustainable Future: An Energy Management Guidebook for Wastewater and Water Utilities", U.S. Environmental Protection Agency, Washington, DC, January 2008

EPA Energy Star Portfolio Manager
http://www.energystar.gov/index.cfm?c=evaluate_performance.bus_portfoliomanager

ISO New England http://www.iso-ne.com

Bickerstaff, R., (2009), "Effective Utility Management Deployment – How do We Get Started?", Proceedings from WEFTEC 2009 von Zhaern, W.M., (1996), "ISO 14000 Understanding the Environmental Standards", Government Institutes, Inc., Rockville, MD, 1996

WEF Manual of Practice No. MFD-2 (1997), "Energy Conservation in Wastewater Treatment Facilities", Water and Environment Federation, Alexandria, VA

---

[i] http://www.epa.gov/erp/
[ii] www.dem.ri.gov/programs/benviron/assist/grncert/index.htm

eVortex: A Novel Energy Harvesting Mechanism for Waste and Storm Water Conveyance Systems

Mungtra Chusilp
Arun Jaganathan, Ph.D.
Erez Allouche, Ph.D., P.E.
Trenchless Technology Center
Department of Civil Engineering
Louisiana Tech University
Ruston, LA 71270

ABSTRACT

Energy recovery from wastewater is gaining significance as utilities work toward achieving energy neutrality and sustainability in their wastewater collection and treatment systems. In this paper, an energy harvesting mechanism to harvest the excess kinetic energy available from drop structures that are installed within sewer and storm water conveyance systems is proposed. There are thousands of sewer drop structures installed across North America with heights up to 90m transporting billions of gallons of wastewater year around. An energy harvesting device called "eVortex" which was specifically designed to be retrofitted within an advanced version of drop structure called "vortex drop structure" (VDS) is described. The device consists of a low cost self-cleansing turbine integrated with an axial flux electrical generator. This paper presents the results obtained from a numerical modeling exercise undertaken to design and optimize the turbine geometry. Various performance characteristics including the mechanical power generation capacity are presented.

KEYWORD

Sewer drop structure, energy harvesting, water turbine

INTRODUCTION

Energy sustainability has been gaining significance because of the growing demand for energy and environment concerns. According to the U.S. Energy Information Administration (2012), the long term projections for U.S. energy consumption would grow at an annual rate of 0.3 % from 2010 through 2035, and the share of renewable (non-hydropower) electricity generation would increase from 4% in 2010 to 9% in 2035. Wastewater is considered as a source of renewable energy and according to the Energy Management report published by WERF (2011) the total energy available from wastewater could be as high as ten times (typical 2-4 times) the energy used for its treatment which presents a significant potential for energy recovery. Various forms of energy are available from wastewater including chemical (from bio-residuals), thermal and mechanical (kinetic and potential). Thus, energy recovery from wastewater gains significance in order to achieve energy neutrality and sustainability in wastewater industry (WEF, 2012). Several energy harvesting technologies are available to recover chemical (Ahn and Logan 2010), thermal (Baek et al., 2005) and mechanical energy (Davis 2008; PRWeb 2012; Leviathan Energy Inc.) from wastewater. In this paper, a novel kinetic energy recovery device is proposed for installation within drop structures that are available in sewer and storm water conveyance systems transporting billions of gallons of wastewater year around.

VORTEX DROP STRUCTURES

Drop structures are vertical conduits that transport wastewater from shallow collection systems to deeper tunnels. There are thousands of sewer drop structures available across North America with heights ranging from about 1.5m to over 90m conveying flow rates in excess of 127 m^3/s (2900 MGD) annually. Table 1 provides a list of drop structures installed at selected locations around the world. In a traditional drop structure called also called as "plunge drop structure", the incoming flow drops freely due to gravity (Rajaratnam et al., 1997) and continuous pounding of free falling water against the concrete floor leads to abrasive wear of concrete and other mechanical equipments. Also, the turbulence caused during the impact releases dissolved gases including hydrogen sulphide (H_2S) which eventually escapes into the atmosphere causing odor problems at the location and public health concerns. An advanced version of drop structure called "vortex drop structure" (VDS) was developed to mitigate the odor and abrasion problems associated with plunge drop structures (Vischer and Hager, 1995). A VDS consists of three main components, namely, an inlet structure, a vertical drop shaft, and a mixing reservoir. Figure 1 shows a photograph of a VDS during installation (left) and in operation (right). The incoming wastewater entering the inlet is accelerated through a spiralling path in the form of a vortex which induces negative pressure within the air core at its centre. The negative pressure drags atmospheric air into the structure which prevents odor causing gases to escape. The flow within the drop shaft rotates along a helical path and occupies only a thin layer closer to the pipe wall leaving major portion of the cross-section empty through which atmospheric air passes through freely. This helical flow is termed "wall-hugging" flow and it serves to dissipate kinetic energy through friction minimizing abrasive damage to the concrete floor. Detailed numerical analysis of VDS including its flow characteristics are reported elsewhere (Chusilp and Jaganathan, 2013).

Table 1. Summary of existing drop structures installed at selected locations

Location	Quantity	Depth (m)	Flow Rate (m^3/s)
Chicago, IL	250	-	Up to 133
Milwaukee, WI	24	Up to 91	6 - 70
Pittsburgh, PA	165	Up to 27.5	-
Singapore	18	21 - 40	Up to 19.5
Montreal, Quebec	68	Up to 30	-
Toronto, Ontario	-	Up to 27.5	-
Phoenix, AZ	7	-	-
Rochester, NY	60	-	-
Minneapolis, MN	2	Up to 16	-
Sydney, Australia	Several	39.5 - 110	-
Cleveland, OH	12	Up to 61	Up to 6.5
Richmond, VA	Several	21	-
Dearborn, MI	3	-	-
Toledo, OH	Several	-	-
Austin, TX	Several	-	-
Dorchester, MA	6	-	-

Figure 1: (Left)- A vortex drop structure during installation; (Right)- Top view of a vortex drop structure installed within a manhole.

KINETIC ENERGY HARVESTING FROM WASTEWATER

Mechanical energy from wastewater is available in terms of potential and kinetic energy. Several water turbine based mechanisms have been reported in the past to extract kinetic energy from municipal pipelines operating under partial and full flow conditions. A paddle wheel like micro turbine that can be installed within a manhole of gravity sewer have been reported (Davis, 2008). Lucid Energy has developed Water-to-Wire system using spherical vertical axis turbine for gravity sewer and potable water pipes, and a demonstration of the technology was carried out recently at San Antanio Water System (PRWeb, 2012). Benkatina hydroelectric turbine by Leviathan Energy is another commercially available device for harvesting energy from freshwater and wastewater pipelines. However, installation of such turbines within a VDS is not convenient because of its atypical wall-hugging flow that occupy only a thin layer closer to the pipe wall which requires placing turbine blades in close proximity to the pipe wall. Possibility of frequent clogging because of debris found in wastewater is another major concern in deploying water turbines within sewers. In this paper, a novel energy harvesting mechanism named "eVortex" is proposed for installation within a VDS. The turbine is specifically designed to operate effectively within a VDS without compromising its original purpose of odor control, and the blades are arranged in such a way to minimize blockage.

DESIGN AND MERITS OF eVORTEX

eVortex device consists of a low profile water turbine with an integrated electrical generator that can be easily retrofitted within the vertical shaft of a VDS. Conventional water turbines such as Francis and Kaplan turbines are generally designed for full flow conditions found in hydroelectric projects and pressure pipes. However, in a VDS flow is partial and occupies only a thin layer closer to the pipe wall leaving major portion of the pipe empty. Also, the flow spirals around the shaft. Thus, the turbine designed for VDS consists of radial arrangement of blades attached to the inner pipe wall such that major portion of the flow are intersected by the blades. The center core of the turbine was left open for air and other debris to pass through freely. Figure 2 shows the schematic of eVortex turbine in comparison with a conventional water turbine having blades located at the center of the pipe. It is expected that this open turbine with an empty space at the center would minimize the possibility of blockage caused by debris.

The section of vertical shaft containing turbine is allowed to rotate freely by the use of bearings on its either ends. In a conventional turbine the mechanical torque is usually transferred using mechanisms involving gears and shafts increasing the complexity. However, in eVortex, the rotating shaft is converted into the rotor for the electrical generator with magnets arranged on the outer wall. Figure 3 shows the schematic of eVortex showing axial flux generator integrated with the turbine. Axial flux electrical generators are suitable for low RPM applications such as this and a detailed description is provided elsewhere (Chamer et al., 1999). Thus, the advantages of eVortex design includes: a) an efficient, compact and low profile turbine architecture designed specifically for the wall-hugging flow conditions found in VDS, b) self-cleaning and low maintenance structure designed to avoid blockage, c) turbine integrated with an electrical generator which eliminates the need for additional torque transfer mechanisms such as gears, and d) a low cost unit which can be readily retrofitted within existing drop structures. In the next section, a brief description of numerical modeling carried out to design the turbine is provided.

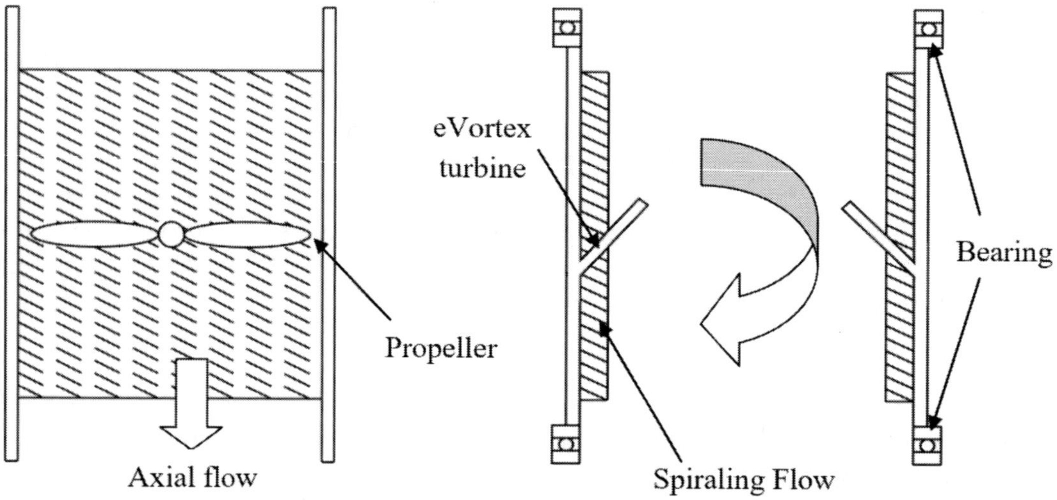

Figure 2: Schematic showing the basic concept of eVortex turbine. (Left)- Conventional water turbine with blades situated at the pipe center where the flow velocity is maximum; (Right)- eVortex turbine with blades located on the inner pipe wall where the flow is maximum (a section of vertical shaft is turned into turbine and is allowed to rotate freely employing bearings).

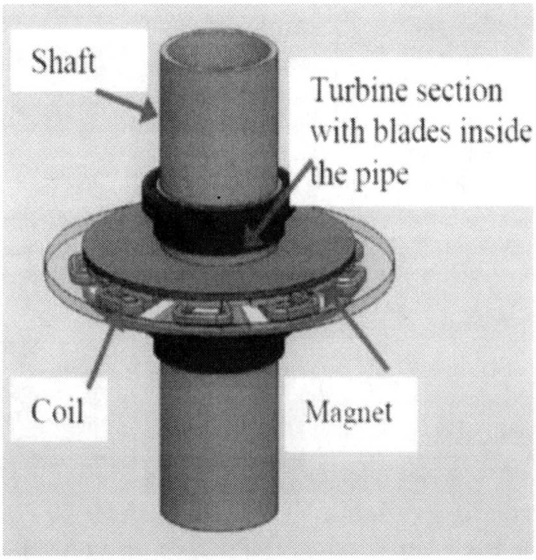

Figure 3. Schematic of eVortex showing the turbine section integrated with an axial flux electrical generator.

NUMERICAL MODELING OF EVORTEX

Numerical modeling was used to design and optimize the geometry of eVortex turbine. Numerical modeling was carried out in two phases. In the first phase, the flow characteristics of VDS were studied numerically and the model was validated by comparing the results against analytical results and experimental measurements. In the second phase, the validated model was employed to design the turbine and optimize its geometry. This section provides a brief overview of the numerical model developed during this project. Detailed numerical analysis of VDS is provided elsewhere (Chusilp and Jaganathan, 2013).

Phase 1: Study of flow characteristics through VDS

In Phase I, two separate numerical models employing single-fluid and two-fluid theories were created to study the flow characteristics of VDS. The model employing single-fluid theory was developed using Flow3D (Flow Science, 2013) and it was used to study the flow of water alone through VDS. The second model employing two-fluid theory was developed using OpenFOAM (2013) and it was used to study the flow of both air and water through VDS. Free surface flow condition in Flow3D used volume of fluid (VOF) approach (Hirt and Nichols, 1981). The domain was meshed using hexagonal cells by Fractional Area Volume Obstacle Representation (FAVOR) method. In the two-fluid model, the interface between water and air was computed using interface-capturing technique (Rusche, 2002). The two-fluid model presented an accurate representation of the problem than the single-fluid model. However, it required larger computational resources. The models were executed on the Louisiana Optic Network Initiative (LONI) Supercomputer. In Phase I, the numerical models were used to simulate flow through a commercially available VDS with a shaft 10cm in diameter and height of 1m. Figure 4 shows the photograph of VDS installed at the hydraulics laboratory of Louisiana Tech University.

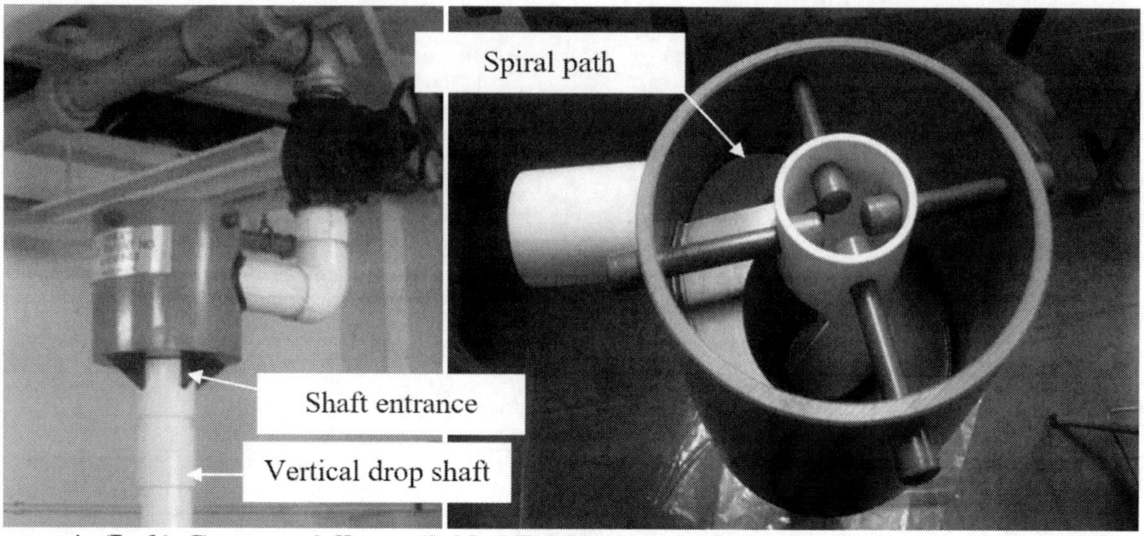

Figure 4: (Left)-Commercially available VDS having shaft 10 cm in diameter and height 1m installed at Louisiana Tech University; (Right)- Top view showing the spiral path inside the inlet.

Figure 5 shows the steady state fluid fraction within VDS for a flow rate of 0.01 m^3/s obtained from single-fluid and two-fluid models. As seen in Figure 5, the liquid portion occupies only a thin layer close to the pipe wall. The numerical models were validated by comparing the results against analytical solution (Zhao et al., 2006) and experimental measurements. Pressure transducers were mounted on the pipe wall to measure the pressure generated by the flowing water at the pipe wall. Figure 6 shows the comparison of wall pressure obtained using two numerical models, analytical and experimental measurements. As seen in Figure 6, the numerically obtained results showed a close agreement with analytical and experimental results. Figure 6 also presents the comparison of tangential and axial velocity of water obtained from numerical and analytical methods. As seen in Figure 6, velocities obtained numerically showed good agreement with analytical solutions.

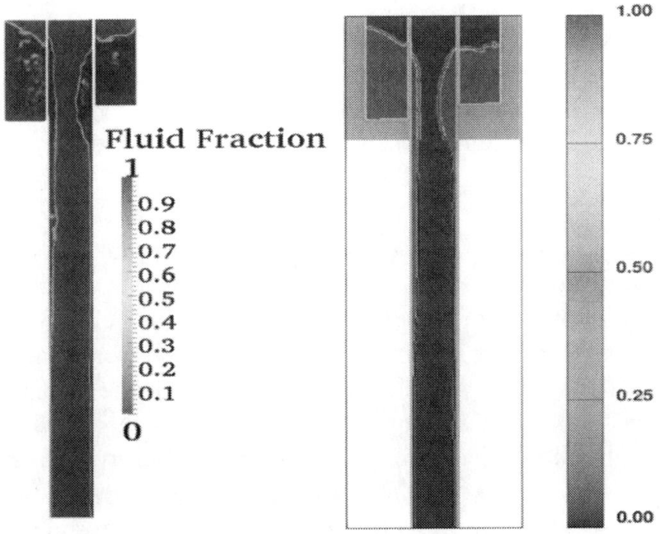

Figure 5: Numerically obtained fluid fraction for a steady state flow through VDS. (Left)- Two fluid model; (Right)- Single fluid model.

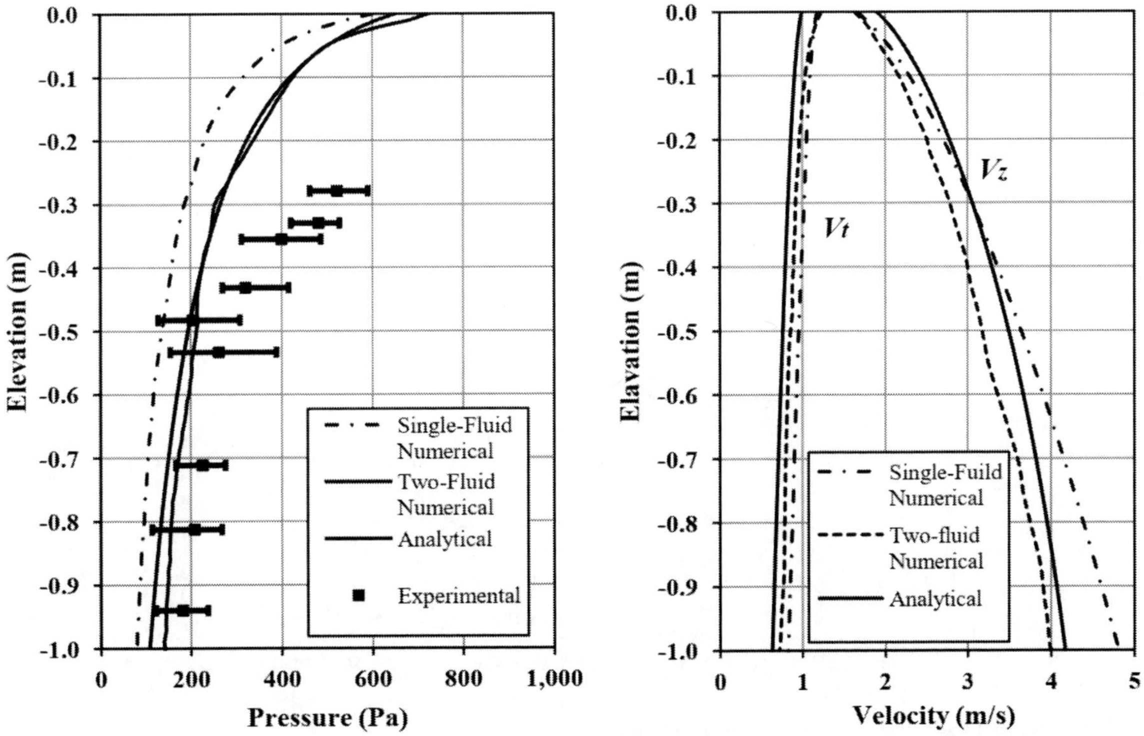

Figure 6: (Left)- Comparison of average wall pressure along the drop shaft obtained by numerical, analytical and experimental methods; (Right)-Comparison of average tangential (V_t) and axial (V_z) velocity along the drop shaft obtained by analytical and numerical modelling.

Phase 2: Modelling of eVortex turbine

Following validation, the numerical models were used to simulate a turbine located within the vertical shaft. The dimensions of VDS installed around the country vary depending on the site topography. As mentioned in Table 1, some of the largest drop structures have shaft length exceeding 90m (Milwaukee, WI) and diameter as large as 3m. However, a relatively small sized VDS with shaft ~15cm in diameter and height ~7.6m was selected for this project to keep required computational and experimental efforts within available resources. Simulations were carried out for flow rates up to 0.0252 m^3/s (400 GPM). The numerical model containing the turbine was created using single-fluid free surface model that was validated in Phase 1. The dynamics of rigid body coupled with flow was simulated using 'general moving objects' (GMO) method available in Flow3D. Figure 7 shows the geometry of turbine modeled which consists of four blades arranged on the inner pipe wall. The total height of the turbine section was 15cm and width of each blade was 4cm. The curvature of the blade was optimized such that it intersects water in perpendicular direction. The turbine section of the shaft was given free rotational degree of freedom similar to a bearing supported structure that is allowed to rotate freely. As seen in Figure 7, a major portion of the turbine section is open which allows air and other debris to pass through freely. As mentioned earlier, in VDS the flow occupies only a thin layer closer to the wall and the blades widths were adjusted accordingly.

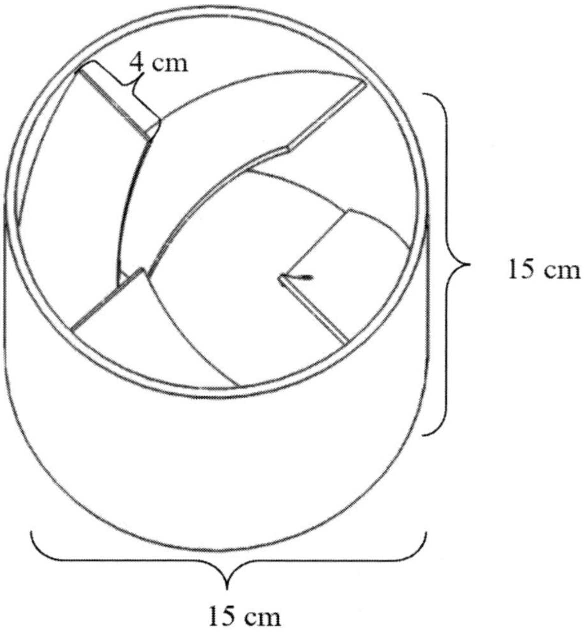

Figure 7: Dimensions of the turbine geometry.

Performance characteristics of the turbine

In this section, various performance curves of the turbine generated numerically using models developed in Phase 2 are presented. Figure 8 shows the distribution of water velocity along the entire length of VDS. As seen in Figure 8, the section where the turbine is placed is highlighted at a depth of 6m from the shaft entrance. Along the length of VDS the tangential velocity of water decreased due to drop in momentum and the vertical velocity increased aided by gravity and reached a maximum of about ~6 m/s just before hitting the blades. Power generated by the turbine was calculated numerically by incrementally applying a counter torque to the model and recording the corresponding angular velocity (Equation 1). Figure 9 gives snapshots from simulations showing the flow through the turbine at different torque loads (and angular velocities). As seen in Figure 9, during the operation of the turbine, a significant portion of its cross-section remains open for air and debris to pass through.

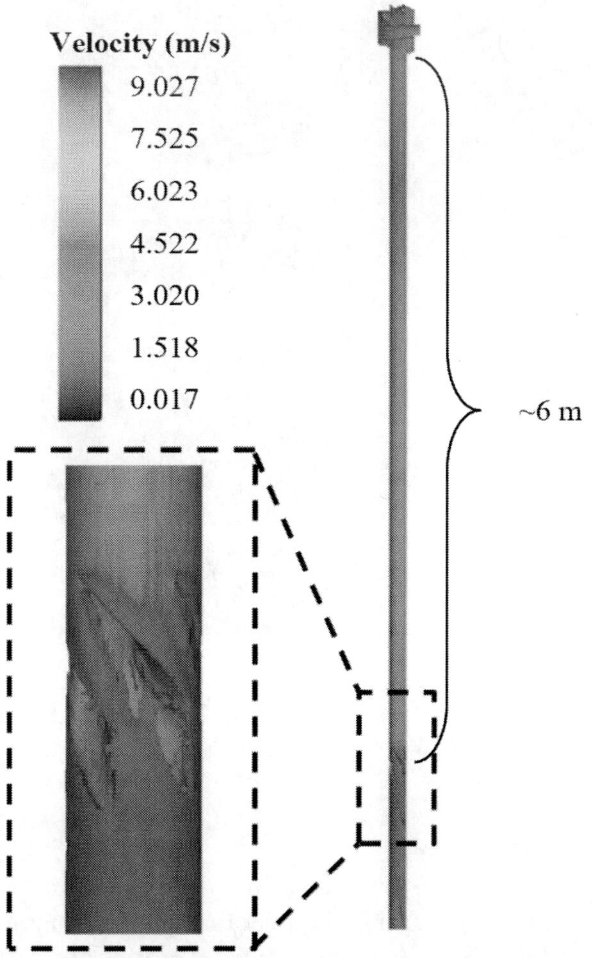

Figure 8: Velocity profile of flow through VDS with a turbine installed at a depth of 6m from the top. The picture in the inset shows a closer view of the turbine section.

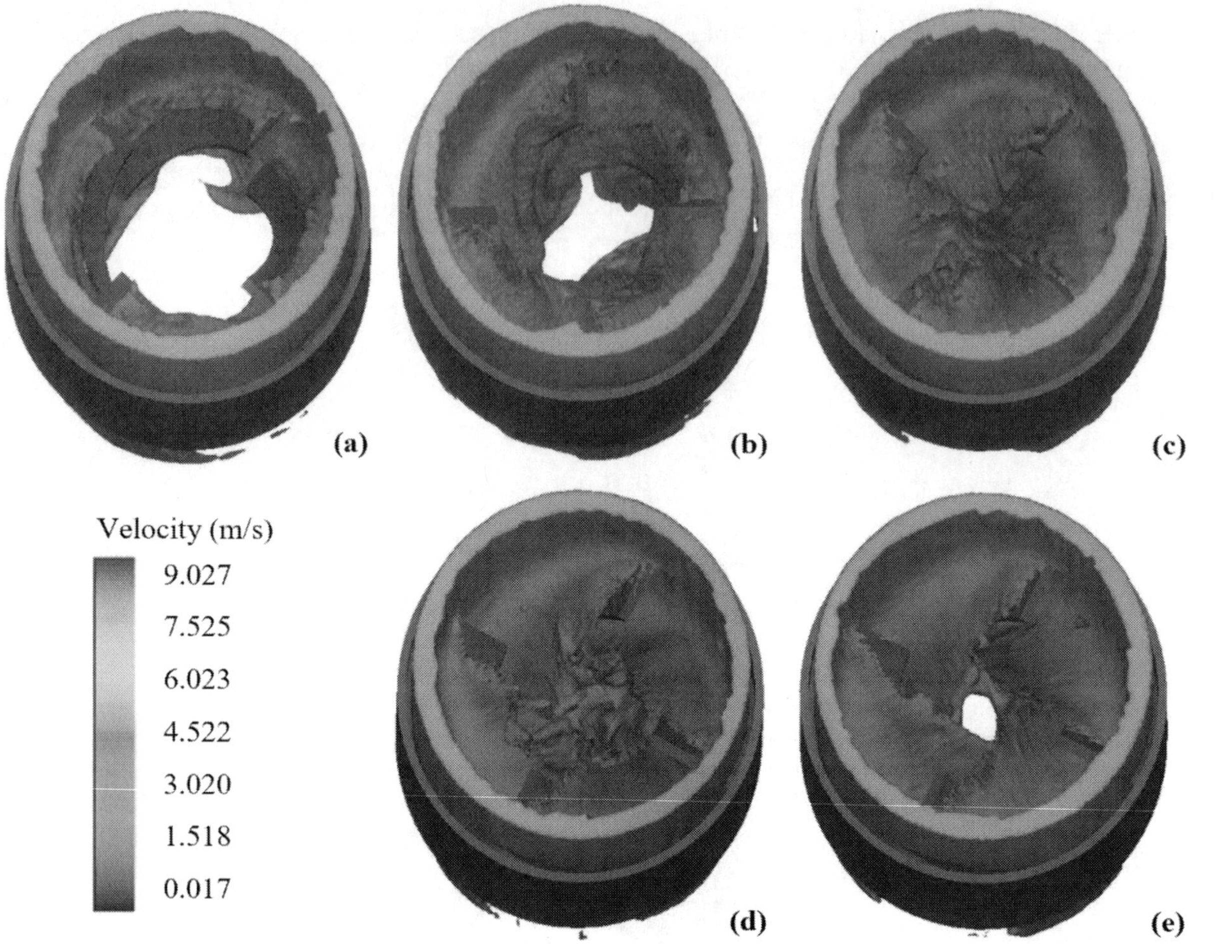

Figure 9. Shapshots from simulations showing flow through the turbine at torque loads: (a) 0 Nm, (b) 1 Nm, (c) 2 Nm, (d) 3 Nm, and (e) 4 Nm.

Estimation of mechanical power

In this section, estimation of mechanical power that can be generated from the VDS (15cm diameter, 6m drop and 0.0252 m^3/s) under consideration is presented. Power calculations were carried out for three different scenarios: 1) free falling flow in a regular plunge drop structure (where flow does not spiral around) with minimal kinetic energy dissipation, 2) spiraling flow within VDS where maximum kinetic energy is dissipated through friction, and 3) flow through a turbine installed at the bottom of a VDS. The power (*P*) generated by the turbine was calculated using the following equation:

$$P = T.\omega \quad \text{Eq. 1}$$

where '*T*' is the numerical counter torque applied to the turbine and 'ω' is the corresponding angular velocity. Power dissipated by VDS was estimated numerically. While VDS dissipates energy by creating spiraling motion of fluid, maximum power possible for similar flow rate and drop through a regular plunge drop structure was calculated using the following two equations for comparison purposes:

$$\frac{p_1}{\rho g} + \frac{V_1^2}{2g} + Z_1 = \frac{p_2}{\rho g} + \frac{V_2^2}{2g} + Z_2 + h_f \qquad \text{Eq. 2}$$

$$\mathrm{P} = \frac{1}{2}\rho Q V_2^2 \qquad \text{Eq. 3}$$

where p, V, Z, Q *and* h_f are pressure, velocity, elevation, flow rate and head loss, respectively. Table 2 provides a summary of power estimated for all three scenarios. The maximum available power for the given flow rate and elevation was approximately 1100 watts taking into account the frictional losses. VDS dissipates a significant portion of the kinetic energy through friction during spiraling flow leaving approximately 500 watts to be recovered. Maximum power that could be generated by the turbine was numerically calculated to be around 90 watts with an efficiency of around 20%. Figure 10 shows the curves for power and torque obtained from numerical simulations. While the power available from a small drop structure such as the one used in this project might not be economical, the flow rates available in some of the larger drop structures are orders of magnitude higher than what was used in this project, and it is expected that a significant amount of power could be generated by extending the eVortex technology to those large sized drop structures. The research team is currently working on an experimental program to test the eVortex and data collected from experimentation would be reported in future. Figure 11 shows the photograph of VDS (with dimensions similar to the numerical model) retrofitted with turbine constructed at Louisiana Tech University for experimental testing purposes.

Table 2: Summary of mechanical power estimations

No.	Method of estimation	Power availability (Watts)
1	Analytical estimation of available power from free falling wastewater (inside a regular plunge drop structure) at 0.0252 m^3/s for a total drop of 6m including frictional losses	~1100
2	Analytically and numerically estimated power available for recovery after energy dissipation within VDS	~500
3	Numerically estimated power recovered using eVortex turbine installed within VDS	~90

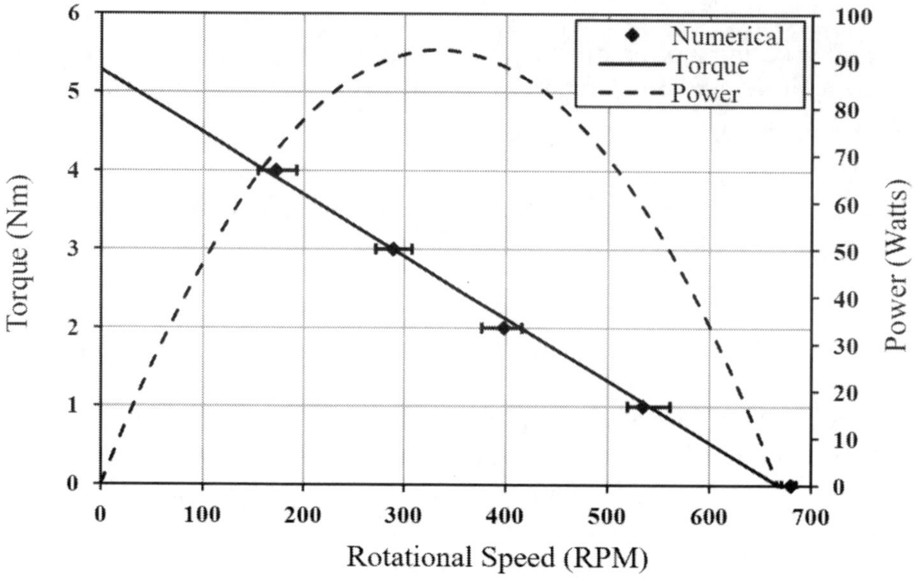

Figure 10. Numerically calculated torque and mechanical power curves for the eVortex turbine operating at a flow rate of 0.0252 m^3/s (400 GPM).

Figure 11: VDS with shaft 15cm in diameter and height 6m constructed at Louisiana Tech University; a turbine with dimensions similar to the one used in numerical model is retrofitted within the vertical shaft.

SUMMARY AND CONCLUSIONS

A novel energy harvesting mechanism developed to harvest kinetic energy from vortex drop structures was presented. A detailed numerical modeling of VDS was carried out to study it's flow characteristics using single-fluid and two-fluid theory, and the models were validated by comparing the numerical results against analytical and experimental results. After model validation, simulations were carried out to design a turbine that could effectively recover kinetic energy from VDS without compromising its original purpose of odor control. Various design and performance characteristics of the turbine along with mechanical power estimations were presented. The flow rates available in some of the larger drop structures are orders of magnitude higher than what was used in this project and it is expected that a significant amount of power could be extracted by extending the eVortex technology to those large sized drop structures. The turbine proposed has low profile and capability to self-clean because of its open nature which allows air and debris to pass freely through its core. The turbine could be easily retrofitted within the existing drop structures. Besides drop structures, the proposed concept could be used in other places including effluent discharge pipes present in industrial and wastewater treatment facilities.

ACKNOWLEDGEMENT

The funding for this research work was provided by the U.S. Department of Commerce (and EPA supplemental) through i6 program under the award number 08-79-04731. We would like to acknowledge the U.S. Department of Commerce and the EPA for their support. We would also like to thank Mr. Richard St-Aubin from IPEX, Inc. for providing additional support for this project.

REFERENCES

Ahn, Y; Logan, B. E (2010) *Effectiveness of domestic wastewater treatment using microbial fuel cells at ambient and mesophilic temperatures;* Bioresource Technology, 101(2), 469-475.

Baek, N. C; Shin, U. C; and Yoon, J. H (2005) *A study on the design and analysis of a heat pump heating system using wastewater as a heat source;* Solar Energy, 78, 427-440.

Chamer, B. J; Wu, W; Spooner, E (1999) *An axial-flux permanent-magnet generator for a gearless wind energy system;* IEEE Transactions on Energy Conversion, 14(2), 251-257.

David, R (2008) *Sewer line power generating system;* U.S. Patent 7,429,803 B2, filed August 1, 2006, and issued September 30, 2008.

Flow Science, Inc (2013) *FLOW3D v. 9.4 user's manual;* Flow Science, Inc., Santa Fe, NM.

Hirt, C. W; Nichols, B. D; (1981) *Volume of Fluid (VOF) method for dynamics of free boundaries;* Journal of Computational Physics, 39, 201-225.

Chusilp, M. and Jaganathan, A. P; (2013) Numerical *modelling of swirling flow in vortex drop structure;* (To be Published).

Leviathan Energy Inc. *Benkatina Hydroelectric Turbine;* Leviathan Energy, Clean Power from Wind Hydro Waves. Viewed February 27, 2013. http://www.leviathanenergyinc.com/benkatina.html

Munson, B. R; Young, D. F; Okiishi, T. H (2006) *Fundamentals of Fluid Mechanics, Fifth Edition;* John Wiley & Sons, Inc.

OpenCFD, Ltd (2013) *OpenFOAM, The Open-Source CFD Tool Box, User Guide v. 1.7.1;* OpenCFD, Ltd., Berkshire, UK.

PRWeb (2012) *Lucid Energy Signs Deal with San Antonio Water System to Host First LucidPipe Demonstration Site in Texas;* PRWeb, Online Visibility from Vocus, August 21, 2012. Viewed February 27, 2013. http://www.prweb.com/releases/2012/8/prweb9810189.htm

Rajaratnam, N; Mainali, A; Hsung, C. Y; (1997). *Observation on flow in vertical dropshafts in urban drainage systems;* Journal of Environmental Engineering, 123 (5), 486-491.

Rusche, H (2002). *Computational fluid dynamics of dynamics of dispersed two-phase flow at high phase fractions;* Doctoral Thesis, University of London.

U.S. Energy Information Administration (2012) *Annual energy outlook 2012*, Report no. DOE/EIA-0383.

Vischer, D. L; Hager, W. H (1995) *Vortex drops;* Energy dissipaters, IAHR hydraulic structure design manual, No. 9, Chap. 9, A. A. Balkerma, Rottendam, The Netherlands, 167-181.

WERF, Water Environment Research Foundation (2011) *Energy management exploratory team report executive summary*. Viewed February 27, 2013. http://ww2.werf.org/AM/Template.cfm?Section=Home&CONTENTID=19270&TEMPLATE=/CM/ContentDisplay.cfm

Zhao, C. H; Zhu, D. Z; Sun, S; Liu, Z (2006). *Experimental study of flow in a vortex drop shaft;* Journal of Hydraulic Engineering, 132 (1), 61-68.

Use of Alternative Water Sources for a Trigeneration System: a Case Study

P. Grace*, D. Myers*, E. Lopez Calva*, S. Zhang*, P.Mehta*

CDM, 1 Maritime Square #12-19, Harbourfront Centre, Singapore 099253

Summary

Trigeneration is the simultaneous production of electricity, cooling and heating energy. Water is used in a trigeneration system for cooling tower makeup and blow-down. The trade-offs of using alternative water sources – graywater and harvested rainwater – instead of potable water to meet these water demands are studied in the context of a hypothetical university campus. For each water supply source, the total water demand, energy requirements, utility costs and associated greenhouse gas generated for the campus is calculated. The calculations are also repeated for a conventional district supply system using electric chillers and natural gas fired boilers for comparison purposes.

Keywords

Trigeneration; rainwater harvesting; graywater reuse

Introduction

A trigeneration system utilizes waste heat generated during electricity production to provide heating and cooling energy. Recovered waste heat is used to produce hot water, which in turn can be directly used to meet heating and hot water needs or directed to an absorption chiller to generate chilled water. **Figure 1** illustrates the basic components of a simplified trigeneration system. Trigeneration is of interest as an efficient and environmentally desirable system.[1] These systems have been implemented in various settings (commercial complexes, schools, hospitals) internationally.[1]

Trigeneration systems with cooling towers for heat rejection require water for cooling tower makeup and blow-down. Alternative water sources can be used to meet cooling tower water requirements, but the use of lower quality water may lead to a faster build up of concentrated solids in the cooling towers and hence a greater water volume requirements. The diversion of alternative water sources for cooling tower requirements also means that additional alternative sources or potable water will be needed to meet other non-potable demands.

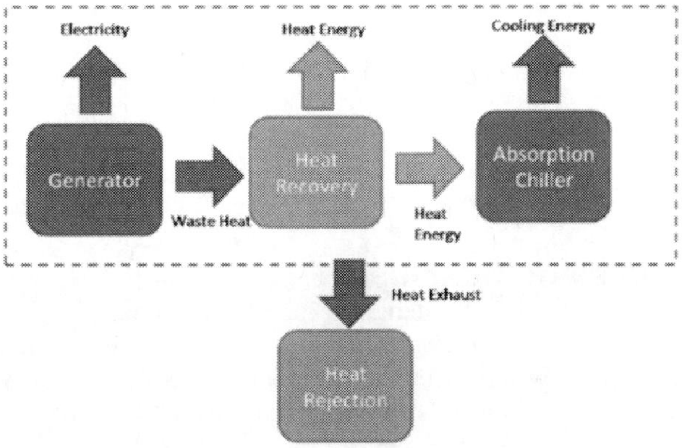

Figure 1 Simplified Overview of Trigeneration System Basic Components

This paper studies the trade-offs of using two alternative water sources – graywater and harvested rainwater – instead of potable water to meet cooling tower demands of a trigeneration system.

Approach and Methodology

This study looks at trade-offs of using potable water, graywater, or harvested rainwater to meet cooling tower demands of a trigeneration system by means of a case study featuring a hypothetical university campus with a student population of 10,000 students. Three scenarios featuring the three different water supply portfolios to the

[1] Lai & Hui. Applied Energy 87(2010) 2868-2880.

campus are set up as shown in **Figure 2.** Other non-potable demands that compete for the use of alternative sources in the campus include toilet flushing and irrigation needs in the various academic buildings and campus dormitories.

For each scenario, the total water demand, total energy consumption, utility costs and associated total greenhouse gases generated for the campus are calculated. Total energy consumption encompasses both on-campus energy use (which influences the amount of energy production, and hence the amount of heat rejection and water required in the cooling tower), as well as the energy footprint to distribute and supply water. Utility cost is calculated for both electricity and natural gas usage on campus, as well as for campus potable water use. Greenhouse gases tracked are those associated with the burning of natural gas in the trigeneration system, and use of energy in buildings, water distribution and treatment.

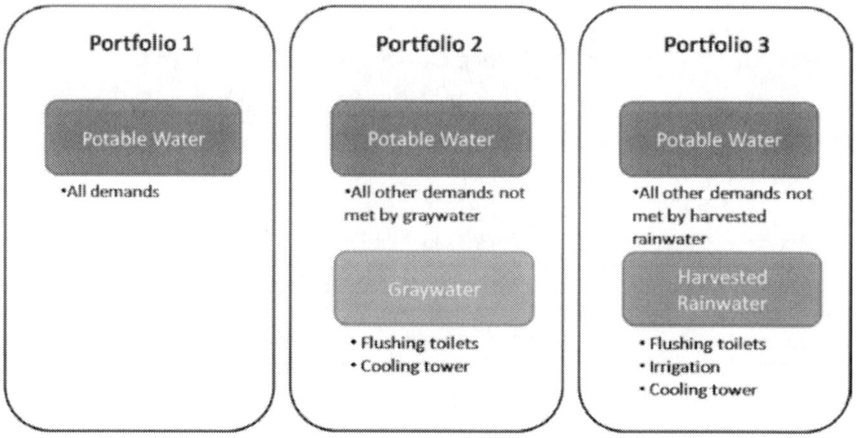

Figure 2 Modeled Campus Water Supply Portfolio Scenarios

The urban systems model developed by CDM Smith's Neysadurai Centre for Integrated Urban Solutions is used to perform the calculations. The urban systems model is a multi-sector tool that can be used to analyze impacts of implementing various green technology options in the urban environment. The model is chosen for several reasons: (1) dynamic simulation that enables accounting of diurnal and seasonal variation in rain and graywater supply with corresponding variation in water demand, (2) inter-linkages between the water and energy sectors to keep track of associated energy costs and savings, and (3) greenhouse gas layer that tracks the amount of greenhouse gas generated based on energy use.

Additional calculations are performed for the three campus water portfolio scenarios assuming that in each scenario, the trigeneration system is replaced by a conventional district supply system utilizing electric chiller and natural gas fired-boilers for comparison purposes. In comparison with absorption chillers used in a trigeneration system, a similar sized electric chiller requires about 20% less water for heat rejection.[2] Electric chillers are also more efficient at producing cooling energy for the equivalent amount of input energy.

Results and Conclusion

The annual total water demand, energy consumption, utility cost, and associated greenhouse gases generated for each of the three different campus water portfolio scenarios will be presented graphically. The results for trigeneration system will also be compared with that for the conventional district supply system to evaluate the relative environmental footprint of the trigeneration system.

Besides the trade-offs considered in this study, there are other consideration in the use of alternative water supply sources for cooling tower use such as corrosion or health concerns due to potential bacteria build-up.[3] These additional considerations will also have to be taken into account in the overall decision making process prior to use of any alternative water sources to meet trigeneration cooling tower needs.

[2] Private communication, 29 August 2012

[3] Victoria Department of Human Services (2008). Cooling towers and recycled water. <http://www.health.vic.gov.au/environment/downloads/cooling_towers.pdf>

Demonstrating the Business Case for Biogas-Fueled Stationary Fuel Cell Power Systems at Wastewater Treatment Facilities

Peter Devlin
U.S. Department of Energy
EERE Fuel Cell Technologies Office
Gregory D. Moreland
SRA International

Abstract
The U.S. Department of Energy Fuel Cell Technologies Office (FCTO) presentation will discuss its collaborative efforts with public and private stakeholders to develop and demonstrate the business case for biogas-fueled stationary fuel cell power systems at wastewater treatment facilities. Commercial deployments of biogas-fueled stationary fuel cell systems have been completed in several locations, including wastewater treatment facilities at the City of Tulare and at the Orange County Sanitation District, both in California. The integration of fuel cells with bio-methane from anaerobic digester systems at wastewater treatment facilities has been used to generate power and heat for onsite needs, power for export back to the grid, and hydrogen to fuel fuel-cell powered vehicles. These biogas fuel cell projects are being deployed in real world operations and provide a foundation for growth through additional installations at other wastewater treatment facilities. The U.S. Department of Energy (DOE) has been exploring the challenges and developmental pathways for accelerating and expanding the commercialization of biogas and stationary fuel cells at wastewater treatment facilities. The DOE's overall objective is to identify strategies and actions that could enable wastewater treatment facilities to couple renewable bio-methane with highly efficient fuel cells to produce electricity, heat, Combined Heat and Power (CHP), and Combined Heat, Hydrogen and Power (CHHP) for stationary and motive applications. FCTO's presentation will provide a status report and share findings on the technical and economic challenges and benefits of using fuel cells at wastewater treatment facilities.

Introduction
An industry-government work shop was held in June 2012 to discuss biogas and waste-to-energy technologies for fuel cell applications. The overall objective was to identify opportunities for coupling renewable bio-methane with highly efficient fuel cells to produce electricity, heat, combined heat and power (CHP); or combined heat, hydrogen and power (CHHP) for stationary and motive applications. The workshop focused on biogas sourced from wastewater treatment plants (WWTPs), landfills, and industrial facilities that generate or process large amounts of organic waste, including large biofuel production facilities, i.e., biorefineries. The following text in this manuscript reflects the presentations and discussion from this work shop.[1]

While both technical and non-technical challenges exist to the widespread use of biogas fueled fuel cells, commercial deployments of these technologies do exist and have been successful. Biogas from wastewater treatment, landfills, and commercial food processing have all been utilized and successfully demonstrated in fuel cells. Using fuel cells and other energy conversion devices, these renewable fuels have been employed to generate power and heat for onsite needs, power for export back to the grid, and hydrogen for fuel-cell powered vehicles. Figure 1 provides details on some notable examples of operating projects.

Figure 1. Biogas and fuel cell project success stories

Tulare, California, Municipal Wastewater Treatment Plant with fuel cells and ancillary equipment in foreground
Photo from Frank Wolak, FuelCell Energy

- The Tulare WWTP processes 11.5 million gallons per day (GPD) of wastewater, producing 500,000 cubic feet of digester biogas/day, which was previously flared
- Four 300 kW fuel cells convert biogas to electricity (1.2 MW capacity) and heat is used for digester heating
- Supplies close to 50% of the WWTP's electricity demand, saves approximately $3,500/day in electricity cost (estimated 4.5 year payback period, with credits), and reduces CO_2 emissions by 6,200 tons/year
- Electric efficiency of 47% and CHP efficiency of 90%

Source: Fuel Cells 2000, "Fuel Cell System Turns Waste into Electricity at the Tulare Wastewater Treatment Plant," www.fuelcellsorg/info/TulareCaseStudy.pdf

Renewable liquefied natural gas (LNG) from landfill gas plant in Altamont, CA
Photo from Mike McGowan, Linde, North America

- Produces up to 13,300 GPD of LNG from landfill gas captured from the Altamont landfill, enough to fuel 300 Waste Management, Inc. refuse hauling trucks.
- The LNG liquefaction plant operates using renewable electricity generated on site
- Displaces 2.5 million gallons of diesel fuel/year and reduces approximately 30,000 tons CO_2, 200 tons of NO_x, and 4 tons of particulates/year
- $15.5 million total capital cost (approximately $2 million in public funding from multiple agencies)

Sources: Mike McGowan, "Renewable LNG: Update on the World's Largest Landfill Gas to LNG Plant," (presentation, Biogas and Fuel Cells Workshop, Golden, CO, June 11-13, 2012); and Waste Management, Inc., "Case Study: Altamont Landfill and Resource Recovery Facility," www.wm.com/federal/case-studies/altamont.html

World's first trigeneration plant at the Orange County Sanitation District's (OCSD'S) WWTP in Fountain Valley, CA
Photo from Jack Brouwer, UC-Irvine

- Produces combined heat, hydrogen, and power (CHHP) from anaerobic digester gas collected from the OCSD wastewater treatment plant
- Fuel cell plant has a capacity of 250kW and excess hydrogen is refined to fuel up to 50 vehicles/day at a nearby hydrogen fueling station
- To date, the plant has produced more than 1.2 million kWh of electricity and close to 3,000 kg of hydrogen
- Total demonstrated efficiency of 54% (effective hydrogen production efficiency has reached 87% with electric efficiency up to 47%)

Source: Jack Brouwer, National Fuel Cell Research Center, University of California, Irvine, personal communication, September 17, 2012

Other Successes:

- ***BMW Manufacturing, Spartanburg, SC:*** Uses delivered hydrogen to fuel 86 fuel cell powered forklifts and LFG from a nearby landfill for CHP in a combustion engine, saving 1.8 million kWh and 1.2 million tons of CO_2 per year
- ***Sierra Nevada Brewery, Chico, CA:*** 1.2 MW fuel cell CHP using digester biogas and natural gas
- ***Gills Onions, Oxnard, CA:*** 600 kW fuel cell CHP using digester biogas produced from onion waste meets 95% of the food processing plant's electricity needs
- ***East Bay Municipal Utility District (EBMUD), Oakland, CA:*** Digesters at the municipal WWTP process post-consumer food waste, food processing waste, restaurant grease, and other waste streams and the biogas-fired CHP plant provides 90% of the WWTP's energy needs

Biogas-Powered Fuel Cells – Considering the Opportunity

The U.S. consumed 98 quadrillion (quads) British Thermal Units (BTUs) of total energy in 2011, with over 40 quads used to generate electric power. Of this, almost 27.5 quads (or 70%) was from fossil fuel sources such as coal and natural gas.[2] At the same time, heat for hot water and space heating of buildings must be supplied through combustion of additional fossil fuel. Generating electricity for commercial, industrial, and residential applications using CHP technologies would greatly improve the overall efficiency of both electricity generation/delivery and heating by moving the electricity generation close enough to the point of use that waste heat energy can be used for heating and hot water.

Each year, about 0.74 quads of biogas is produced in the U.S. – about 0.6 quads at landfills, 0.11 quads from animal waste management operations, and 0.03 quads at industrial and domestic wastewater treatment plants.[3] Of this amount, less than half is captured and used. Much of the gas captured at landfills is flared rather than being used for power generation or other uses. Of the 7.6 million metric tons of biogas captured at landfills in 2010, more than half was flared.[4] As a whole, only 19 percent of the nation's wastewater treatment plants use biogas produced in their anaerobic digesters for power generation.[5]

Biogas is naturally produced from the anaerobic digestion of organic waste materials. Although its raw composition varies depending on the type of source material (Table 1), it is chiefly composed of methane (CH_4) and carbon dioxide (CO_2), with small amounts of oxygen, nitrogen, hydrogen sulfide, and other trace components not shown here. This mixture can be used as a fuel to produce energy, most commonly in an internal combustion engine to produce heat, electricity, or a combination of both (CHP).

Table 1. Biogas composition from various sources

Composition	Natural Gas	Biogases			
		Waste Water	Food Waste	Animal Waste	Landfill
Methane (Vol%)	80-100	~50-60	~50-70	45-60	40-55
Carbon Dioxide (Vol%)	<3	30-40	25-45	35-50	35-50
Nitrogen (Vol%)	<3	<4	<4	<4	<20
Oxygen (Vol%)	<0.2	<1	<1	<1	<2
H_2S, ppm	<0.1	<400	<10000	<300	<200
Non-H_2S Sulfur, ppm	<10	<1	<1000	<30	<30
Halogens, ppm	<0.1	<0.2	<0.2	<0.2	<100
Moisture, %	<0.02	~3	~3	~3	~3

Source: Wolak, Frank. "Fuel Cell Power Plants: Biofuel Case Study – Tulare, CA." Biogas and Fuel Cells Workshop. Golden, CO. June 11–13, 2012. http://www1.eere.energy.gov/hydrogenandfuelcells/pdfs/june2012_biogas_workshop_wolak.pdf.

The biogas can be used as-is, but is often cleaned up to the quality and heating value of natural gas (termed "bio-methane"). As with natural gas, bio-methane can be compressed and stored in gaseous tanks or liquefied and stored in liquid tanks, and can be transported via truck or pipeline.

Bio-methane, with appropriate clean-up to remove trace contaminants, can also be used to power a fuel cell. A fuel cell converts chemical energy contained in a fuel directly into electric power without combustion. Like engine systems, fuel cell systems can be configured to produce heat as well as power. Certain types of fuel cell systems can also be configured to produce pure hydrogen. The production of CHHP is just beginning to be explored, and is sometimes referred to as "tri-generation." Figure 2 shows the major biogas energy pathways.

Figure 2. Biomass to electricity, heat, and hydrogen pathways

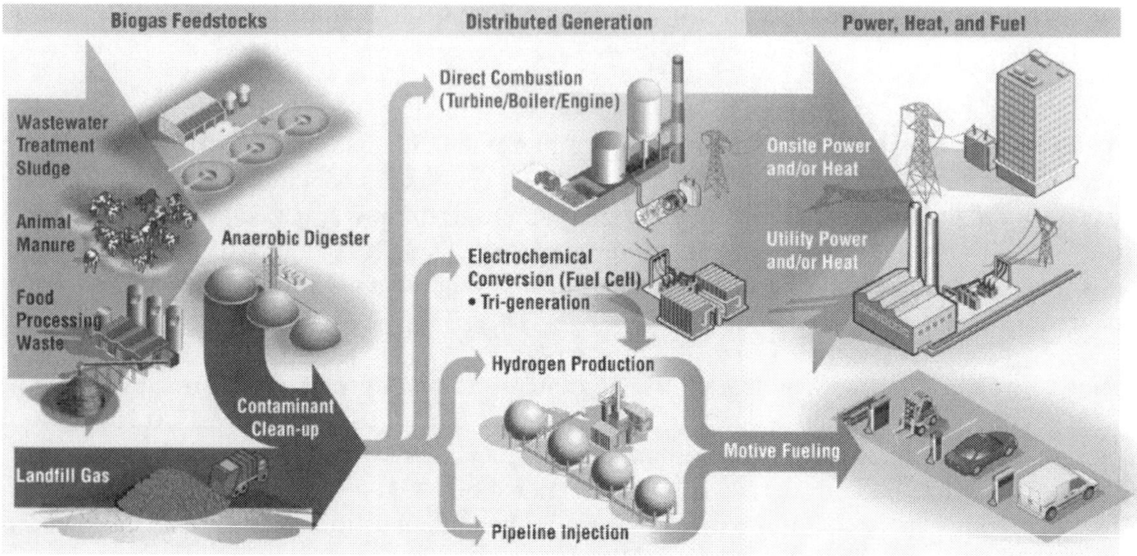

The use of bio-methane in fuel cells has the potential to provide heat and power more efficiently than conventional combustion-based technologies (Figure 3). Fuel cells running on bio-methane provide between 30 percent to 65 percent electrical efficiency in CHP applications versus approximately 15 percent to 40 percent for engine and turbine technologies.[6]

Figure 3. Comparison of electrical efficiency of power generation equipment

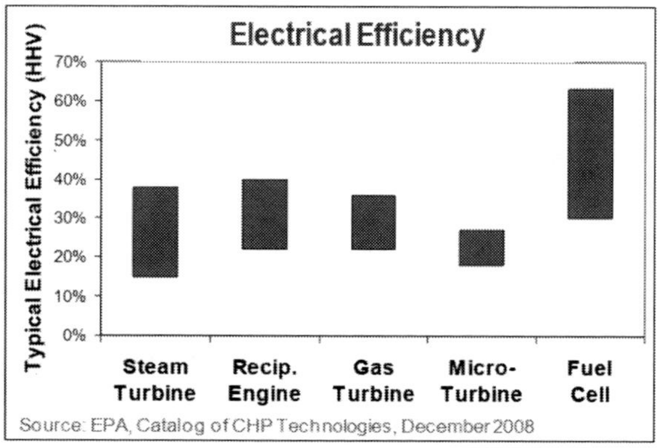

Fuel cells also generate almost no criteria air pollutant emissions (Table 2). These are certain pollutants that are known to be hazardous to human health and are regulated by EPA under the Clean Air Act. Criteria pollutants include sulfur dioxide, particulate matter, carbon monoxide, nitrogen oxides, ozone, and lead.

Capturing biogas for energy has other environmental benefits as well. Methane is a potent greenhouse gas (GHG), with twenty-one times the global warming potential of CO_2. Methane emissions from landfills and wastewater treatment plants account for about 30 percent of all methane emissions in the United States,[7] and about 10 percent of total national GHG emissions in CO_2 equivalents.[8] Generating power with biogas also displaces GHG emissions from conventional power generation. The benefit of displacing other sources of power generation depends both on the amount of biogas available in a particular region and on the fuel mix used in different locations. For example, California has a very "green" power generation fuel mix, so the benefit of displacing this electricity is low. However, California has a large amount of biogas available in comparison to other states. The combination of these two factors makes California a good opportunity for GHG reduction using biogas.

Table 2. Comparison of criteria air pollutant emissions from power generation technologies

Technology	Air Pollutant (lb/M Wh)	
	NO_X	CO
Reciprocating Engine	1.51	2.52
Combustion Turbine	0.83	3.45
Microturbine	0.17	0.29
Fuel Cell	0.0001	0.0015

Source: Hamilton, Steve. "Biogas from Municipal WWTPs: Fuel Cells Viewed as a Value Proposition." Biogas and Fuel Cells Workshop. Golden, CO. June 11–13, 2012. http://www1.eere.energy.gov/hydrogenandfuelcells/pdfs/june2012_biogas_workshop_hamilton.pdf.

The economics of biogas fuel cell projects depends on a number of factors. Biogas competes with conventional sources of natural gas as a fuel for the fuel cell. High prices for the competing fuel i.e. conventional natural gas are favorable for biogas CHP/CHHP. Likewise, power produced from biogas in a CHP application must compete with grid electricity. High prices for grid electricity also are favorable for biogas CHP and CHHP economics.

Anaerobic digestion with capture and use of biogas for the management of waste products like manure or food processing waste reduces odor, waste stream pathogens, and waste volume, which may have economic or public relations value. Waste heat from power generation can also be used to improve digester efficiency, especially in cold-winter climates.

Available Biogas Resources

The major biogas resource segments for anaerobic digestion are landfills, wastewater treatment plants, farm operations, and business and industry sites that are involved in food and beverage production. Anaerobic digestion is a long-established and proven technology, with extensive markets in East Asia and Europe. Germany is considered the leading European country, with more than 7,000 facilities and 18 terawatt hours (TWh) of electricity generated from biogas.[9]

In the United States, there are more than 2,200 biogas-producing sites that are currently operational, including about 1,500 digesters at WWTPs, 186 digesters on farms, and 576 landfill-based energy projects.[10] Substantial opportunities exist for increasing the number of biogas projects in all resource segments, with a large GHG reduction potential from the capture of biogas. The American Biogas Council estimates that there are over 120,000 sites available for development, not including private or industrial wastewater treatment and landfill sites. Table 3 summarizes current biogas production and potential production in the United States.

Table 3. Summary of current and potential biogas capture and use for farms, wastewater treatment facilities and landfills in the United States

	Number of Locations	Biogas Resource	Notes
			Operational Digesters
Wastewater Treatment	1,500		• Only 250 use the biogas they produce.
Landfills	576		
			Potential Resources
Farms	8,200	1,700 MW	• Only counting dairy and swine farms. About 70-100 cubic feet per day of digester gas is produced per milking cow; a dairy farm of 500 cows can generate 100 kW of electricity.*
Wastewater Treatment	3,250	750 MW	• 2,000 WWTPs that process more than 1 million gallons per day (GPD) of wastewater do not have a digester. 1,250 WWTPs are producing but not using biogas. • A typical WWTP processes 100 GPD for every person served (1 cubic foot of digester gas can be produced per 100 gallons of wastewater). • 100 kW can be generated from 4.5 million GPD of wastewater.
Landfills	510		• Every 1 million tons of municipal solid waste equals 432,000 cubic feet per day of LFG for 20 years, or 1 MW of electricity.*

** assuming 30% conversion efficiency*

Sources: Serfass, "Biogas Markets" and Ahmed, "Biogas Impurities."

In 2012, total methane emissions from these facilities was 16.6 million metric tons (MMT), 13.3 MMT from landfills, 2.5 from manure management operations, and 0.8 from WWTPs.[11] If this methane was captured and used for CHHP in fuel cells, up to 1.9 MMT of hydrogen and 81,000 gigawatt hours (GWh) of electricity could be produced.[12] In addition, by generating power from the biogas, up to 300 MMT of CO_2 equivalent emissions would be avoided, equal to about 13% of national CO_2 equivalent emissions from electricity generation.

Biogas resources vary widely across the United States. California, New York, and Texas are examples of states with the richest resources for wastewater. California, Pennsylvania, and Texas have the richest resources for landfill gas. California and many states in the Midwest and Southeast have the richest biogas resources from agriculture. Figure 4 shows the combined biogas resources from WWTPs, landfills, and livestock operations in the U.S.

Figure 4. Biogas combined resources from landfill gas, wastewater treatment, and manure management

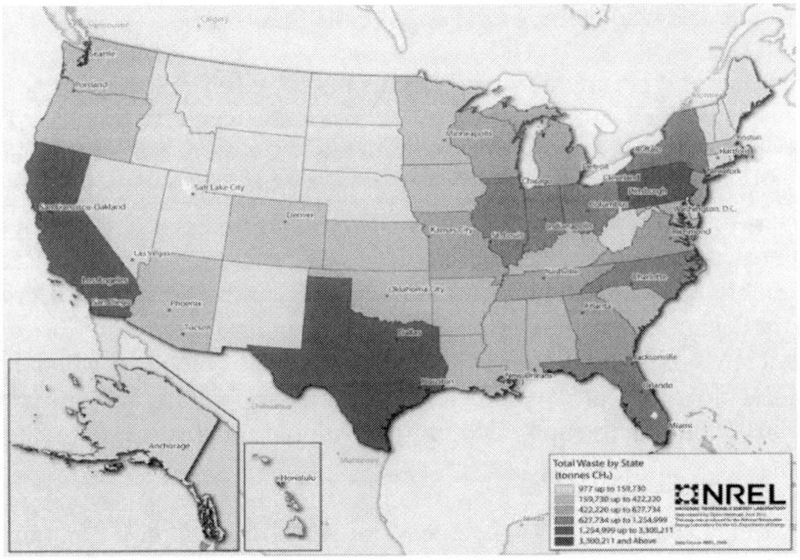

Source: Steward, Darlene. "Biomass Resources Overview and Perspectives on Best Fits for Fuel Cells." Biogas and Fuel Cells Workshop. Golden, CO. June 11–13, 2012. http://www1.eere.energy.gov/hydrogenandfuelcells/pdfs/june2012_biogas_workshop_steward.pdf.

As mentioned before, the cost of grid electricity can significantly impact the value potential for a biogas-to-electricity project, and these costs vary a great deal across the country. California, a state with significant biogas resources and relatively high electricity prices, is an example of a state with a relatively attractive potential for biogas project opportunities.

In terms of biogas production potential, Figure 5 shows that fats and greases and food wastes from bakeries, restaurants, and groceries will produce the highest amount of biogas per ton digested. However, because collecting these types of wastes presents logistical difficulties, the net energy benefit may be substantially reduced.

Figure 5. Cubic meters of biogas production per ton of substrate

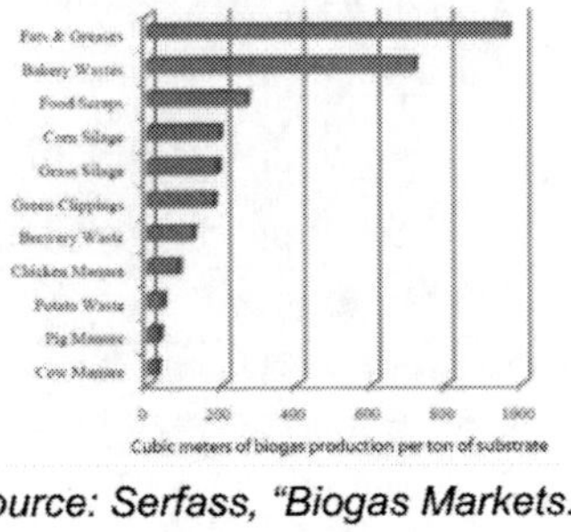

Source: Serfass, "Biogas Markets."

Biogas and Fuel Cell Technologies – Current Status

A range of technologies is important for the effective use of biogas to power fuel cells, including creating the biogas from waste, cleaning or upgrading the biogas to meet the fuel purity specifications required for fuel cells, and the fuel cells themselves.

Anaerobic digestion (AD) is generally a mature technology for treatment of sewage, manure, some food wastes, and other low solids wastes.[13] For wastewater treatment, AD reduces the volume of solids and volatile content, reducing sludge disposal costs, while creating biogas. Adding various food wastes to wastewater sludge, or "co-digestion," can increase the output of biogas in some cases.[14] AD for higher solids waste streams is less well developed, especially in the U.S., though research on the development of new processes is ongoing. R&D needs for AD include improving the understanding of the biological organisms used in different AD processes, improving reactor design and optimization, and understanding co-production and optimization of useful by-product streams.[15] An issue with AD as a source of biogas for fuel cells is the stability in mass flow of the biogas stream, which can vary with temperature and with changes in the waste feed stream. Blending the upgraded biogas with natural gas is one method to overcome this pitfall.

Biogas from landfills is essentially a special case of AD, where carbon-containing solid waste materials break down over time, and, in the generally anaerobic environment of the sealed landfill, generate methane. Other gaseous constituents are also generated, however, and given the variability of landfill waste, LFG often has more varied contaminants than from wastewater or industrial digestion. The typical, although not universal, components of biogas from various sources were shown previously, in Table 1, and include CH_4, CO_2, nitrogen, oxygen, water vapor, and trace amounts of hydrogen sulfide (H_2S), other sulfur species, and halogens. Siloxane species are also prevalent in both landfill and wastewater gases. Landfills can be expected to produce LFG over lifespans that typically last about 20 years, with peak production rates between years 5 and 7.[16] In addition to use of LFG for power generation, liquefaction of methane from LFG has been demonstrated at industrial scales.[17]

Many cleanup or upgrading technologies are being used in bio-methane production tmeet the requirements of various Combined Heat and Power systems including fuel cells. Table 4 shows a variety of processes and their relative cost. Most of these technologies are well known. However, the variability of the types and concentrations of contaminants for each application and, indeed, each site, is high. Thus, while cleanup technologies may be well known, each installation generally requires a custom designed suite of cleanup equipment, which drives up cost. The complexity and cost of cleanup systems can also increase due to variability in the biogas

Table 4. Common cleanup processes and relative cost

Upgrade Method	Contaminant Removal	Cost to Upgrade ($/1000 cu ft)
Biological	H_2S	1.86
Iron Oxide (sulfa treat)	H_2S	0.79
Iron Oxide (Sulphur Rite)	H_2S	1.49
Membrane	CO_2, H_2O	2.13
Water Scrubber	H_2S, CO_2	0.38
PSA	CO_2	2.53
Activated Carbon	H_2S	0.45
Amine	H_2S, CO_2	4.58

Source: Handley, Ian. "Biogas Technologies and Integration with Fuel Cells." Biogas and Fuel Cells Workshop. Golden, CO. June 11–13, 2012. http://www1.eere.energy.gov/hydrogenandfuelcells/pdfs/june2012_biogas_workshop_handley.pdf production rate. Cleanup can generally be expected to add 2 cents per kWh to electricity production costs.[18]

High temperature fuel cell systems, including phosphoric acid, molten carbonate, and solid oxide fuel cells would typically be considered for use with bio-methane fuels because of the potential to generate large quantities of useful heat for CHP. These fuel cells are currently demonstrated and commercially available, in modules from 100 – 400 kW in output. However, initial costs are high, and fuel purity requirements demand complex cleanup systems. Fuel purity requirements for these high temperature fuel cells are given in Table 5. Fuel cell systems capable of hydrogen co-production are currently in demonstration.

Table 5. Fuel purity requirements for high temperature fuel cells (values in ppm)

Type of Fuel Cell	PAFC	MCFC	SOFC
H_2S	2.0	0.1 – 5.0	1.0
COS, CS_2, mercaptan		1.0	
Organic sulfur		6.0	
H_2S, COS, CS_2		0.5 – 10.0	
HCl, ppm		0.1	"few"
Halogens	4.0	0.1 – 1.0	1.0 – 5.0
Halogenated organics		0.1	
NH_3	1.0	10,000	5,000
Siloxanes		1.0	0.01
Tars		2,000	

Source: Ahmed, "Biogas Impurities."

The majority of current biogas-to-energy projects are using the biogas in a conventional combustion-based system to produce heat and/or power for use on site or, in some cases, for sale back to the electric grid. However, as a result of generational improvements in commercial stationary fuel cell technologies, the costs of these technologies are declining, and their reliability and performance levels are improving in real-world environments. Accordingly, fuel cell technologies are becoming viable alternatives to the incumbent combustion-based technologies.

The Potential Operating and Financial Benefits from a Stationary Fuel Cell System

Fuel cells operating on biogas offer a pathway to renewable electricity generation that will deliver critical power, while providing electric and thermal energy to facilities on a continuous basis. This reliability results in daily operating cost savings. Furthermore, interruptions in electricity from grid power outages can disrupt operations, which can cost hundreds of thousands of dollars, and also create safety and health risks that cannot be monetized.

Fuel cell systems are highly reliable. Recent energy service contracts have been based on a 95% availability, which includes scheduled downtime for maintenance. Unscheduled downtime is typically much less. Local grid reliability varies considerably, but fuel cell systems can meet a facility's energy requirements with reliability that can meet or exceed the performance of the grid. Compared to other clean energy sources that generate intermittently (i.e., wind or solar) fuel cell CHP provides reliable, steady power levels.

The devastation Superstorm Sandy caused on the eastern seaboard in November 2012 has brought more focus on the need for better systems, such as micro-grids, to keep power up-and-running during natural disasters. Stationary fuel cells can be a major contributor to micro-grids because they can provide clean power close to where it's needed, and they have the flexibility to do load following, which is important when the grid goes down and loads fluctuate.

The largest stationary fuel cell systems are base load systems, including the phosphoric acid fuel cell systems from Clear Edge Power, the solid oxide fuel cell systems from Bloom Energy, and the molten carbonate systems from Fuel Cell Energy. However, because they're able to island, they are also serving the role of backup power during outages. As an example, during Super Storm Sandy, 22 or the 23 fuel cells operated by Clear Edge Power and located in Sandy's path stayed in operation. The one failure was caused by a hardware failure that was quickly remedied, and the fuel cell system returned to service. For another example, during an outage in San Diego during August 2011, an Albertson's supermarket with a Clear Edge Power fuel cell system stayed in operation, providing food and supplies to the public and without losing any of its perishables.

Fuel cell CHP systems can help overcome power quality issues. Voltage variations, frequency or waveform distortions are examples of power quality issues faced by many facilities relying on grid power. While most equipment tolerates these power quality deficiencies, critical equipment (including computing and communications equipment) may be disrupted or damaged by poor quality electricity.

Fuel cells generate DC power that is delivered to the power conditioning unit of the CHP system. The power conditioning unit raises the voltage and converts it to a high-quality AC power supply. Because voltage and waveform are managed by the power conditioning unit, the CHP system can deliver higher quality power than the grid.

Electricity consumption by facilities can vary considerably with fluctuating cooling, lighting, and office equipment power demands. High utility demand charges can be reduced through effective energy management and fuel cell CHP technologies that can reduce demand for grid power. Because CHP systems operate most efficiently in base load operations, they reduce power demand during all periods. However, the reduction in peak power demand achieved by the fuel cell CHP system should be included in the analysis of system benefits.

Applying federal and state financial incentives is a way to maximizing the financial benefits of fuel cell CHP projects. Federal incentives in the form of the Energy Investment Tax Credit can provide up to $3,500/kW or 30% of the project costs (whichever is lower). Most states have energy incentive programs that help offset capital equipment costs and promote innovative and renewable energy technologies, including fuel cells. Financial incentives for CHP systems are available in the form of direct financial grants, tax incentives, low-interest loans, or utility and environmental policies that increase the financial prospects for a project.

While Department of Treasury regulations prevent tax-exempt entities, e.g., not-for-profit organizations, from directly taking advantage of tax benefits for property that they own, the Internal Revenue Code (IRC) and Treasury regulations allow these entities to enter into contract-for-services agreements with private-sector project developers such as fuel cell vendors and energy service providers. Structured to comply with IRC rules, these contract-for-services agreements can be negotiated so that both the tax-exempt customer and the project developer can benefit from the tax credit. Understanding the ability of federal/private parties to take advantage of these incentives is a key consideration in deciding whether to fund the project through direct appropriations or alternative financing.

The Potential for CHHP from a Stationary Fuel Cell System

One of the immediate challenges in the development of hydrogen as a transportation fuel is finding the optimal means to roll out a hydrogen-fueling infrastructure concurrent with the deployment of hydrogen fueled vehicles. The low-volume hydrogen requirements expected in the early years of fuel cell vehicle deployment make the economic viability of stand-alone, distributed hydrogen generators challenging. A possible solution to this "underutilized asset" problem is to evaluate the potential for generating multiple products, thus improving the capital utilization of the system.

DOE sponsored an evaluation and validation of a project using a high-temperature fuel cell to co-produce electricity as the primary energy stream, with hydrogen as a higher-value co-product.[19] As shown on Figure 6, at the high-temperature fuel cell, methane is internally reformed at the fuel cell anode to hydrogen and carbon dioxide. The fuel cell operates near 600°C and uses molten carbonate electrolyte as the charge carrier. Heated air is combined with the waste gas from the hydrogen purification system and oxidized. These resultant waste gases are fed to the cathode. The fuel cell cathode converts waste gas carbon dioxide to the carbonate charge carrier to complete the fuel cell circuit. The fuel cell stack generates a direct current voltage, which is then converted to alternating current by an inverter in the electrical balance of plant. The system produces 480 VAC, 60 HZ, and a nominal 300 kW without hydrogen co-production. Excess carbon dioxide and water leave the cathode as exhaust, and heat can be recovered from these exhaust gases.

Figure 6: CHHP Power Plant Schematic

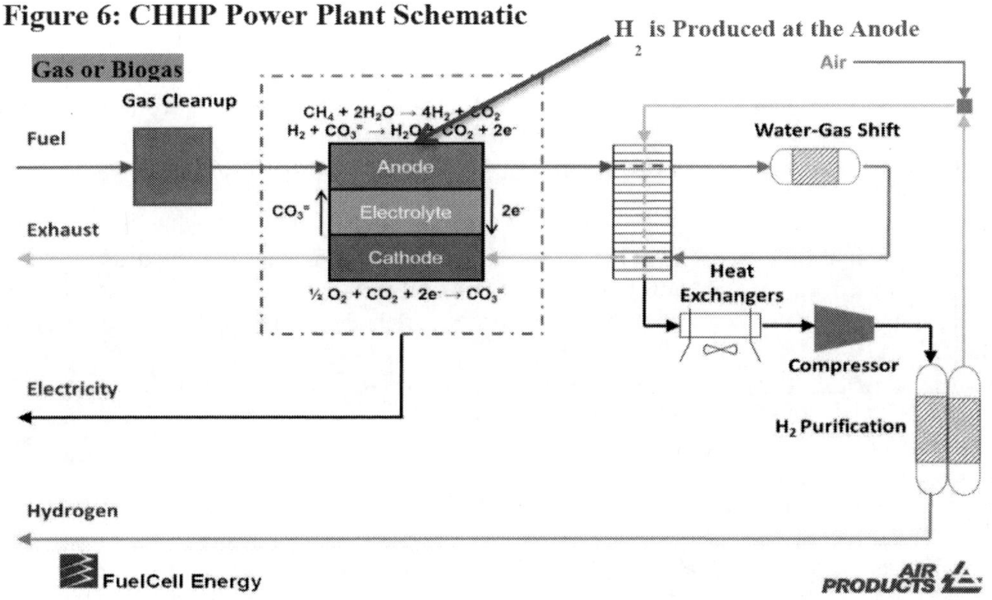

About 70 to 80% of the hydrogen is converted to power, and some hydrogen remains available for recovery. The anode exhaust gas is cooled and sent to a water-gas shift catalytic reactor to convert most of the carbon monoxide present in the stream to hydrogen and carbon dioxide. After an additional cooling step, this gas stream is then compressed and sent to the pressure swing absorption (PSA) system. The PSA uses adsorbents to remove carbon monoxide, carbon dioxide, and water to produce a high-purity hydrogen stream. The waste gas from the PSA is catalytically oxidized and returned to the cathode. The PSA system can also be placed in stand-by mode to stop hydrogen production and allow for maximum power production by the DFC® system, thereby improving the system efficiency and economics.

During operation on AD, a detailed heat and material balance was performed to determine the overall efficiency of the Hydrogen Energy Station. The calculated efficiency of 53.3% exceeded the target of 50%. More than 5 million standard cubic feet of AD gas was processed, and over 1 million kWh of power was exported to the power grid in Orange County, CA.

Overall, the CHHP power plant at this site:

- generates approximately 250 kW of electricity daily
- has 100 kg/day hydrogen capacity (350 and 700 bar), enough hydrogen fuel to power 25 to 50 vehicles.

Based on the knowledge gained from the design, installation, and initial operation at this site, an economic analysis on the cost of hydrogen fuel was performed. Hydrogen production rates up to 6,000 kilograms per day were considered. At this level of production, the analysis estimates that hydrogen can be produced at (1) near-term pricing of \$6 to \$8 per kilogram utilizing PSA for hydrogen purification and (2) longer-term pricing of \$3 to \$5 per kilogram using next generation, electrochemical separation processes. Because fuel cell electric vehicles (FCEVs) are much more energy efficient than conventional vehicles with gasoline internal combustion engines, consuming more than 50% less fuel for the same power output, these prices will be competitive with gasoline prices.

---

[1] "Biogas and Fuel Cells Workshop Summary Report: Proceedings from the Biogas and Fuel Cells Workshop, Golden Colorado, June 11-13, 2012" http://www.nrel.gov.docs/fy12osti/56523.pdf

[2] U.S. Energy Information Administration, *Annual Energy Review* (Washington, D.C.: U.S. Energy Information Administration, October 2011).

[3] EIA, Annual Energy Review.

[4] U.S. Environmental Protection Agency, *Inventory of U.S. Greenhouse Gas Emissions and Sinks: 1990–2010*, 430-R-12-001, (Washington, DC: U.S. Environmental Protection Agency, April 15, 2012).

[5] Shabbir Ahmed, "Biogas Impurities and Cleanup for Fuel Cells" (presentation, Biogas and Fuel Cells Workshop, Golden, CO, June 11–13, 2012), http://www1.eere.energy.gov/hydrogenandfuelcells/pdfs/june2012_biogas_workshop_ahmed.pdf.

[6] Satyapal, "Expanding the Use of Biogas with Fuel Cell Technologies."

[7] Ibid.

[8] EPA, Inventory of U.S. Greenhouse Gas.

[9] Ian Handley, "Biogas Technologies and Integration with Fuel Cells" (presentation, Biogas and Fuel Cells Workshop, Golden, CO, June 11–13, 2012), http://www1.eere.energy.gov/hydrogenandfuelcells/pdfs/june2012_biogas_workshop_handley.pdf.

[10] Patrick Serfass, "Biogas Markets and Federal Policy" (presentation, Biogas and Fuel Cells Workshop, Golden, CO, June 11–13, 2012), http://www1.eere.energy.gov/hydrogenandfuelcells/pdfs/june2012_biogas_workshop_serfass.pdf.

[11] EPA, Inventory of U.S. Greenhouse Gas Emissions.

[12] Ibid.

[13] Ruihong Zhang, "Biogas Production Technologies" (presentation, Biogas and Fuel Cells Workshop, Golden, CO, June 11–13, 2012), http://www1.eere.energy.gov/hydrogenandfuelcells/pdfs/june2012_biogas_workshop_zhang.pdf.

[14] Steve Hamilton, "Biogas from Municipal WWTPs: Fuel Cells Viewed as a Value Proposition" (presentation, Biogas and Fuel Cells Workshop, Golden, CO, June 11–13, 2012), http://www1.eere.energy.gov/hydrogenandfuelcells/pdfs/june2012_biogas_workshop_hamilton.pdf.

[15] Duff, "Biomass Program Perspectives."

[16] Ahmed, "Biogas Impurities."

[17] McGowan, "Renewable LNG."

[18] Ahmed, "Biogas Impurities."

[19] Final Report on DOE Cooperative Agreement DE-FC36-01GO11087 "Validation of An Integrated Hydrogen Energy Station." http://www.osti.gov/bridge/servlets/purl/1053779/

Carbon Free Water – A Successful Case Study in Sonoma County, California

Cordel Stillman, Sonoma County Water Agency

Amy Bolten, Sonoma County Water Agency

ABSTRACT

Once identified as the largest energy user in Sonoma County, the Sonoma County Water Agency (Agency) has implemented a series of measures to reduce our impact on the environment with the goal of net zero emissions from our water and wastewater electricity supply. These measures, implemented under the umbrella program Carbon Free Water by 2015, include water efficiency, energy efficiency, development of renewable energy projects, and procurement of renewable energy contracts. Combined, these efforts have resulted in a 97 percent reduction in the Agency's emissions related to our purchased electricity since 2006.

In addition to greenhouse gas reduction efforts related directly to the Agency's water and wastewater electricity supply, the Agency has assumed a leadership position by implementing innovative efficiency and renewable energy projects for the region as a whole. Lessons learned from the Agency's work towards achieving Carbon Free Water by 2015 served to develop the expertise necessary to implement these community-wide projects. Overall, Sonoma County Water Agency's experience in working towards Carbon Free Water by 2015 serves as a valuable case study for other utilities looking to reduce their greenhouse gas footprint.

KEYWORDS

Sonoma County, carbon footprint, water efficiency, energy efficiency, renewable energy, greenhouse gas emissions, photovoltaics, biogas, Farms to Fuel, Sonoma Clean Power, community choice aggregation, financing, electric vehicles, Applied Solutions

INTRODUCTION

The Agency has a special interest in energy matters that arise from a unique set of circumstances, the primary being that the Agency is among the region's largest users of electricity. Guiding our efforts is the Sonoma County Water Agency Energy Policy (Policy) adopted by our Board of Directors on March 22, 2011.[1] The Policy focuses on the Carbon Free Water by 2015 program, which implements programs that reduce the Agency's greenhouse gas emissions, with the goal of achieving a net zero carbon energy supply by 2015. Carbon Free Water by 2015 is being pursued through a combination of water efficiency, energy efficiency, and renewable energy purchase and development as outlined below.

ENERGY EFFICIENCY THROUGH WATER USE EFFICIENCY

Recognizing that the first step towards achieving Carbon Free Water by 2015 was to reduce the overall energy demand within the system, and that the largest use of system power was to pump water, the Agency implemented a series of aggressive water conservation programs. The Agency pumps and distributes water to serve more than 600,000 residents in two counties. Through implementing programs to reduce end-use water demand, the agency can pump less water and therefore use less energy. Water conservation initiatives include public awareness campaigns, programs targeting conversion through low water use landscaping, and rebates and direct install programs for low water use fixtures.

Some examples of public awareness efforts conducted by the Agency include annual summer water conservation public outreach and education campaigns through multimedia advertising and public meetings. Leveraging resources to achieve greater impact the Agency teamed up with NBC Bay Area and several other water agencies to produce the public education television series, *Water Matters*, which aired in 2009 and 2010. This series featured the Sonoma County Water Agency and its Russian River water supply system. In addition, the Agency offers tours of the Russian River water supply system and some of its sanitation districts.

Another approach to increasing water conservation on the demand side is ingraining water conservation into the local culture. Each year, the Sonoma County Water Agency's Water Education Program reaches thousands of elementary school students with lessons and materials on water conservation and stewardship of our local watersheds. This comprehensive regional program incorporates teacher workshops, curriculum materials, contests, assemblies, student incentives, classroom presentations, and field trips – all of which are free to schools within our service area in the California counties of Sonoma and Northern Marin.

The Agency also provides information on rebates and conservation education for homes, businesses, and agriculture. In 2012, the Agency and the Sonoma-Marin Saving Water Partnership successfully held its second annual Eco Friendly Garden Tour. A dozen gardens were featured, educating more than 700 visitors in water efficient garden practices such as drought tolerant plants and low water use irrigation systems. The Agency also developed and manages the Qualified Water Efficient Landscape program, a bilingual program to train landscapers in water efficient landscaping practices. Through these efforts, ongoing water conservation initiatives have helped reduce water use throughout the region approximately 28 percent since 2002.

ENERGY EFFICIENCY THROUGH OPERATIONS EFFICIENCY

The second pathway to reducing the embedded energy use in our water system is the implementation of operational energy conservation measures wherever possible. Energy efficiency measures include replacing old electric motors and pumps, optimizing aeration systems at treatment plants, and other system operation fine-tuning. Through these changes, the Agency has improved the efficiency of the water transmission system itself by 18 percent (figure 1). The energy intensity of water in our system went from utilizing 2,618 kWh/MG (kilowatt-hours per million gallons) of electricity for water delivered in 2005 to using 2,142 kWh/MG of electricity for water delivered in 2011.

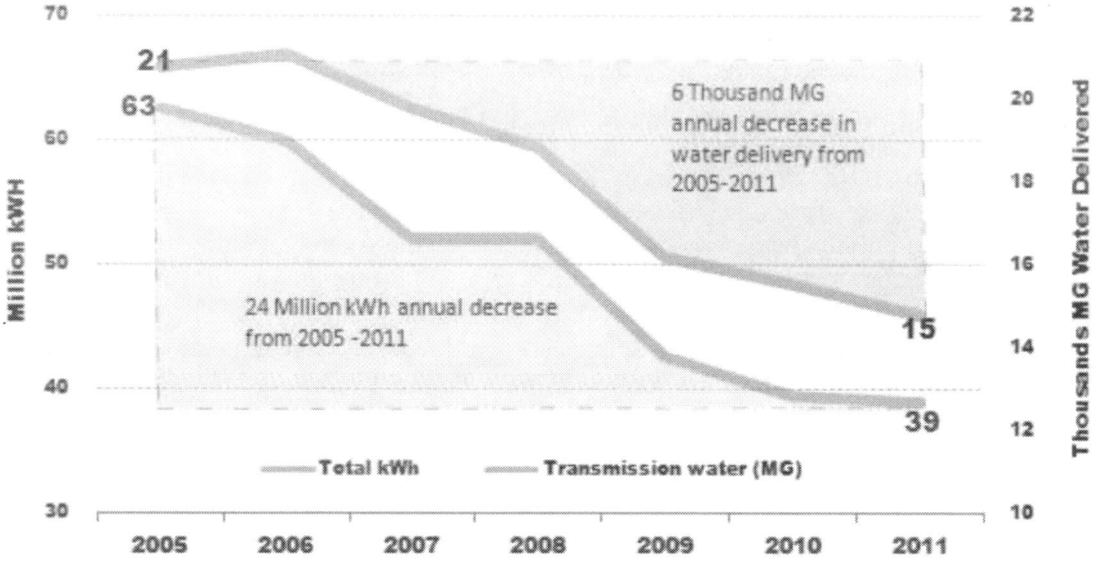

Figure 1. Sonoma County Water Agency annual energy use and water delivery

Additionally, we have piloted two types of geothermal heating and cooling systems. At our Operations and Maintenance building, a ground source heat pump system has been installed to deliver highly efficient heating, ventilation, and air conditioning. This system is expected to reduce our heating and cooling energy use at that location by 50 percent by utilizing the thermal mass and constant temperature of the earth. A variation of this system is being installed at our newly constructed Service Center where the heating and cooling system will be a loop located in a 100 million gallon wastewater treatment pond. This system will capitalize on the constant temperature and thermal mass of the large body of water.

RENEWABLE ENERGY

The development and purchase of renewable energy sources is the final piece of the Carbon Free Water by 2015 Program. As of 2011, approximately 94 percent of the Agency's energy consumption demands were met through renewable energy sources including solar electric, landfill gas and hydroelectric power (figure 2).

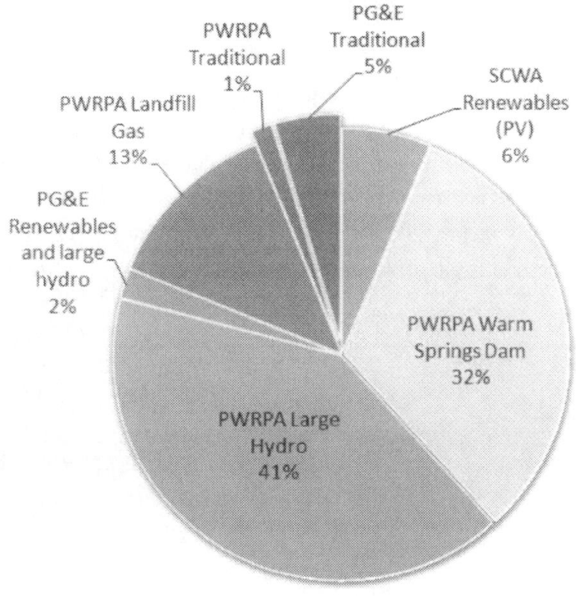

Figure 2

There are two mechanisms utilized by the Sonoma County Water Agency to develop our renewable energy supply. The first is building our own power generation assets to power our operations. These include three photovoltaic systems to power our administration building and two wastewater treatment plants that total an approximate capacity of 2 megawatts (MW).

In addition to developing our own power supply facilities, the Sonoma County Water Agency is a member of the Power and Water Resources Pooling Authority, a joint powers authority created to take advantage of the power resources and needs of various water agencies in California. This allows the Agency the freedom to wheel power from specific sources for our own use. Through this mechanism, the Agency has been able to purchase renewable energy from local power production facilities in Sonoma County. For example, the Agency currently purchases 100 percent of the electricity from Sonoma County's Landfill Gas to Energy project. We are also able to wheel power generated at our own hydroelectric facility. Moving forward we are developing a diverse portfolio of utility-scale renewable energy projects. The electricity from these efforts may be sold as a revenue stream or used to supply Sonoma Clean Power, a community choice aggregation program in formation in Sonoma County. At this time, these projects are structured as public-private partnerships. However in the future, the Agency may be building revenue generating assets on its own. For example, the Agency is developing a 20 MW solar photovoltaic system on unused land at the regional airport as well as evaluating other open spaces for large solar installations. Other projects include biomass, geothermal and Farms to Fuel, a project that will produce renewable energy through anaerobic digestion of local poultry waste.

GREENHOUSE GAS REDUCTION

In 2012, the Agency achieved The Climate Registry's (TCR) Platinum Status, the highest achievement level recognized by TCR.[2] The Agency is one of three entities nationwide to receive this recognition. Platinum status is based on meeting three requirements: the adoption of a carbon reduction plan, achieving greater than 20 percent reduction in GHG emissions and demonstrating implementation of five or more best practices for climate protection. The Agency has registered its GHG emissions since 2006. Since that time, we have reported a 97 percent reduction in emissions related to our purchased electricity supply (figure 3).

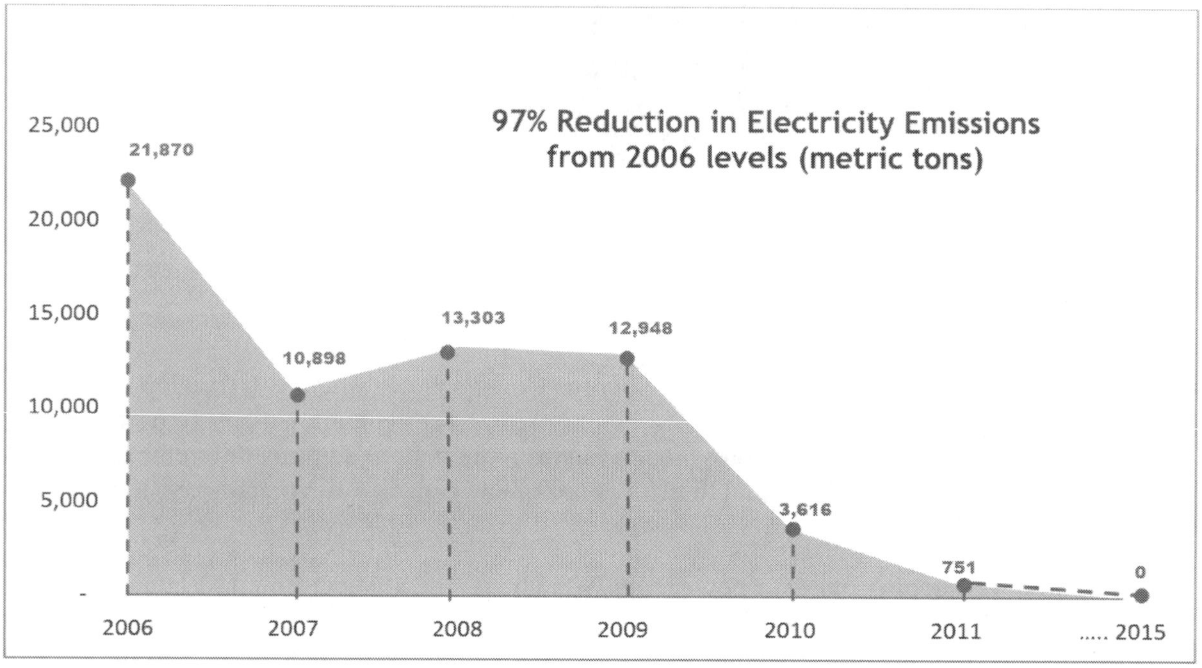

Figure 3

DEMONSTRATING LEADERSHIP

In addition to the Agency's dedication to producing Carbon Free Water by 2015, the Agency's Energy Policy includes the ability to develop projects of regional benefit. Over the past five years, the Agency has taken a leadership position in launching a number of programmatic efforts for the benefit of the entire community, within and beyond Sonoma County's borders. Some examples are described below.

Sonoma County Energy Independence Program

The County of Sonoma collaborated with the Agency in 2009 to launch the Sonoma County Energy Independence Program (SCEIP), an innovative water efficiency, energy efficiency, and renewable energy installation financing mechanism for home and building owners. As of January 2013, almost $60 million has been disbursed to 1,807 projects to fund efficiency and renewables on private buildings. The program has been a huge success and has helped promote the growth of local green jobs. Other counties in California and across the nation are making efforts to replicate this program.

Sonoma County Efficiency Financing Program

The Agency has recently launched another program to finance energy efficiency and water conservation retrofits through aggregating public and non-profit projects for municipal bond financing.[3] This program is a scaled down version of Delaware's Sustainable Energy Utility (SEU) model, which aggregates large numbers of energy and water retrofits to be collectively funded through a joint municipal bond. Contractually guaranteed cost savings through a pre-qualified set of energy service companies back the bond. We are working to secure a multi-million dollar bond to finance these retrofits, while creating a positive impact on the local economy.

Sonoma Clean Power

The ability to develop projects of regional benefit has allowed the Agency to investigate Community Choice Aggregation (CCA). CCA allows one or more cities and/or counties to form a service area that provides for the purchase of power generation of customers within that service area. Sonoma Clean Power, the Sonoma County CCA, would include the goals of reducing greenhouse gas emissions, cost competitive power rates, developing local jobs in renewable energy and developing long-term rate stability and energy reliability for Sonoma County residents through local control.[4] On December 4, 2012, the Sonoma County Water Agency's Board of Directors approved the formation of the joint powers authority for Sonoma Clean Power, and the project continues to move forward.

Electric Vehicles

The Agency is a leader in electric vehicle infrastructure development, currently collaborating with local governments to develop a regional electric-vehicle charging infrastructure. Not only is the Agency participating in a regional program to fund and install electric vehicle chargers throughout the county but the Water Agency also organizes reoccurring meetings for local governments to discuss electric vehicle deployment and infrastructure.

Applied Solutions

On a broader level, the Agency birthed Applied Solutions, a national organization of counties and cities with the goal of identifying and sharing the most innovative local government programs.[5] This non-profit network now boasts 150 county and city members across the country as well as universities and members from the Middle East, Africa, and Canada.

CONCLUSION

Recognizing the need to reduce carbon emissions, the Agency has successfully undertaken multiple efforts with the goal of achieving Carbon Free Water by 2015. As a result, the Agency produced only 751 metric tons of CO_{2e} in 2011 for its purchased electricity supply, a 97 percent reduction since 2006. This monumental reduction of 21,119 metric tons of CO_{2e} is equal to removing 4,141 passenger vehicles from the road, or electricity supplied to 2,633 homes annually. As we near the milestone of becoming a net zero carbon electricity operation, the Agency remains dedicated to effectively and sustainably managing water resources for people and environment.

REFERENCES

1) Sonoma County Water Agency. Energy Policy. http://www.scwa.ca.gov/files/Energy%20Policy.pdf

2) Sonoma County Water Agency. Sonoma County Water Agency Achieves Climate Registry Platinum Status. http://www.scwa.ca.gov/lower.php?url=press-releases&article=sonoma-county-water-agency-achieves-climate-registry-platinum-status-2012-05-08

3) Sonoma County Water Agency. Sonoma County Efficiency Financing (SCEF) Program. http://www.scwa.ca.gov/scef/

4) Sonoma County Water Agency. Community Choice Aggregation (Sonoma Clean Power). http://www.scwa.ca.gov/cca/

5) Applied Solutions. http://www.appliedsolutions.org

Estimating Carbon Emissions in Water Supply Portfolio Evaluations Including Recycled Water and Greywater

E. Lopez Calva*, S. Zhang*, D. Myers*, P. Grace*, P. Mehta*

CDM Smith Neysadurai Centre for Integrated Urban Solutions, 1 Maritime Square #12-19, Harbourfront Centre, Singapore 099253

Summary

Carbon emissions assessments and carbon footprint analysis are becoming more common in water supply planning studies. In most cases analysis are made by defining carbon emissions at the supply option level for large-scale supplies (i.e., conventional water filtration, desalination, etc). After assembling a water supply strategy with a defined portfolio, the carbon emissions of the supply options involved are combined to define the carbon emissions of the portfolio. This approach is less applicable when the supply portfolio includes smaller scale supply options that can be implemented in a decentralized fashion and in many times on-site (building level). Water supply analysis can better capture the real benefits of options when water supply is analyzed as a portfolio in a systemic way. This paper describes the application of a systems model to evaluate the actual use of sources in a portfolio. The model provides the opportunity to estimate effectiveness in the reduction of demand from municipal systems, annual costs savings, and energy use and indirect GHG emissions associated with the sources. Two portfolios are evaluated in two separate cities with large variations in the energy footprint of the municipal system. Comparison will be made on the benefits of the additional sources, and GHG emissions (driven by the current municipal water footprint). The use of a systems tool allows us to better account for the real GHG impacts on water supply portfolio decisions.

Keywords

Water supply; carbon footprint; carbon emissions, supply portfolio

Introduction

Carbon emissions assessments and carbon footprint analysis are becoming more common in water supply planning as part of the metrics that decision-makers use in deciding on investments for capital projects associated with water supply. These assessments are more common and widespread in academic studies where the case is made for accounting for the energy footprint of water supply (NRDC, 2004).

These studies in most cases are based on large-scale supply infrastructure (i.e., conventional water filtration, desalination, centralized wastewater recycling, etc.) by taking into account the energy used to produce and distribute a unit volume of water (m^3) and also taking into account the emissions associated with that energy use. Carbon emissions of the supply options involved are combined to define the carbon emissions of the supply portfolio. This approach is less applicable when the supply portfolio includes smaller scale supply options that can be implemented in a decentralized fashion (decentralized recycling) and in many times at the building level (i.e., greywater) due to greater level of analytical complexity where the small scale options can present many different configurations and can have a utilization that is not as uniform as a large-scale source. Water supply analysis can better capture the real benefits of options when water supply is analyzed as a portfolio in a systemic and dynamic way, where the use of a small scale source triggers a reduction in a municipal source. With this approach, the use of outdoor sources, for example, can be linked to precipitation data for better estimates of the sources' yield.

This paper describes the application of a dynamic systems model to evaluate the actual use of sources in a portfolio. The model provides the opportunity to estimate effectiveness the reduction of demand from municipal systems, annual costs savings, and energy use and indirect GHG emissions associated with the sources.

Approach and Methodology

The use of two on-site supplies is evaluated for two different theoretical cities A and B, with relatively different annual precipitation and precipitation patterns and where the energy and carbon footprint of the main municipal supply maybe significantly different. Data from two actual cities is used for cities A and B corresponding to Melbourne, Victoria, Australia (as of today, prior to the operation of the Wonthaggi Desalination Plant, Melbourne has a low energy footprint for conventional surface water supply) and San Diego, California, U.S.A. (high energy footprint for the main water supply).

Two decentralized supply options, greywater and small-scale recycling are theoretically introduced in the water supply portfolio of the cities and assumed to reduce the use of the main municipal source. A systems dynamics model developed by CDM Smith's Neysadurai Centre for Integrated Urban Solutions is used to simulate the performance of the on-site supplies, the reduction in use of the main municipal source, and the corresponding new greenhouse gas emissions for the new supply portfolio. The model simulates the hydrology and the impacts of that hydrology on outdoor water demand (relevant for recycled water use) on a daily basis. It also simulates the energy use associated with recycled water and greywater and estimates the indirect carbon emissions associated with the energy use, based on actual data for power generation emission factors for the two reference cities (Melbourne and San Diego). The total annual use of greywater, recycled water and corresponding reductions to use of municipal supply are estimated, along with the net carbon emissions change triggered by the implementation of the two new decentralized options.

Results and Conclusion

The annual total water use for each of the three sources, energy consumption, and associated greenhouse gases will be presented. Estimates of capital costs associated with the two decentralized options will be presented and related to the net change in greenhouse gas emissions. The net change in greenhouse gas emissions for the cities will be compared and the factors driving the difference discussed. Discussions will include considerations on hydrology, energy footprint of the main supply, and electricity generation emissions for the local energy use. The net benefits in reductions of greenhouse gas result from the framework presented in Figure 1.

Besides the trade-offs considered in this study, there are other consideration in the use of alternative water supply sources. These additional considerations will also be discussed given that the dynamics model used for the study generates outputs on a number of additional relevant metrics.

References

National Resources Defense Council (NRDC). 2004. Energy Down the Drain. Pacific Institute, Oakland, California.

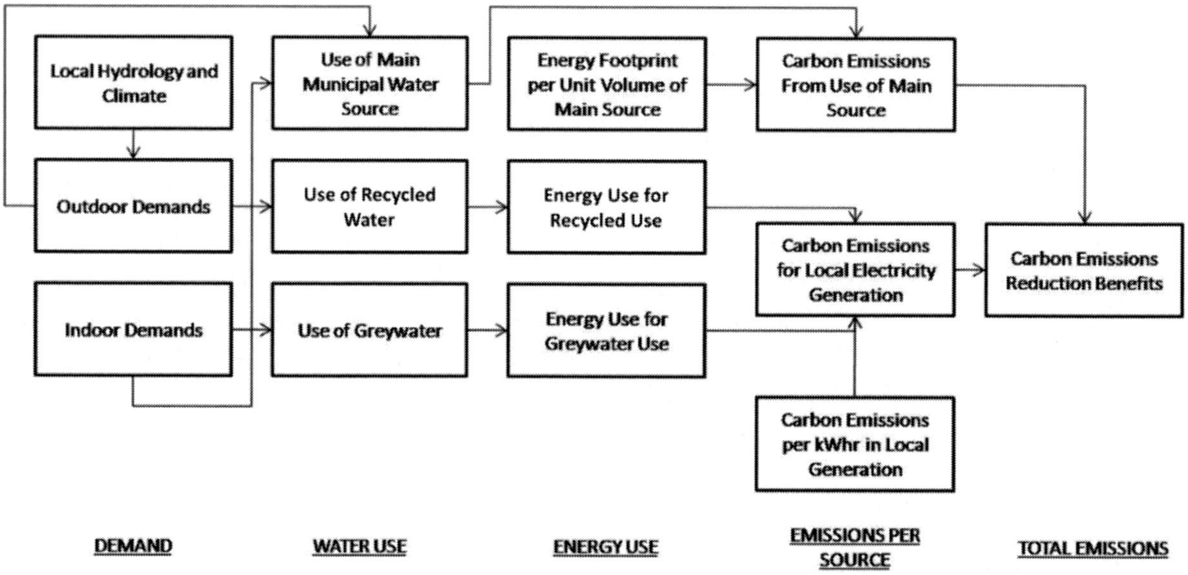

Figure 1. Systems Structure for Greenhouse Gas Reduction.

Climate Ready Water Utilities: Helping the Water Sector Prepare for and Adapt to a Climate Change

Curt Baranowski, US Environmental Protection Agency

ABSTRACT

Drinking water and wastewater utilities face many challenges due to variable climate conditions and resulting impacts on water resources. These challenges include infrastructure damage due to flooding and sea level rise, water quality degradation, drought, combined sewer overflows due to intense precipitation events, and other factors. Each challenge can compromise operations and threatens the ability of utilities to fulfill their mission of providing clean water to communities. Responding to this challenge the U.S. Environmental Protection Agency developed the Climate Ready Water Utilities initiative to help the water utility sector adapt to climate change by providing a clear understanding of climate science and adaptation options. The CRWU initiative has several tools and resources which advance climate adaptation planning and actively engage the water sector, including a risk assessment tool and community workshops. These resources allow utilities and stakeholders to begin the climate change adaptation process.

KEYWORDS: Climate ready, climate change, water utility, adaptation, risk assessment, workshop, estuary.

INTRODUCTION

In 2011, the National Drinking Water Advisory Council delivered the Climate Ready Water Utilities (CRWU) report to the U.S. Environmental Protection Agency's (EPA) Administrator Jackson. The report contains 11 finding and 12 recommendations, including the creation of a CRWU initiative (NDWAC, 2010). Based on the report, the CRWU initiative was established to provide resources to help the water sector adapt to climate change impacts by promoting a clear understanding of climate science and adaptation options. Integrated water resources management is a key theme within CRWU, as the tools and projects use a holistic approach to collaborate with partners and promote effective watershed planning with the utility sector.

To address a number of the recommendations in the report, CRWU has several ongoing activities to advance climate adaptation planning, raise awareness and encourage proactive action from utilities. These activities include easy-to-understand informational materials, such as the Adaptation Strategies Guide, collaborative workshop guides such as the Extreme Events Workshop Planning Tool, and a comprehensive risk assessment tool, the Climate Resilience Evaluation and Awareness Tool (CREAT). Whether utilities choose to combine these activities or implement only one, CRWU tools enable them to begin the process of considering how climate change will affect their resources and ability to fulfill their missions. CRWU can guide them through the process of creating an adaptation plan that will not only make water systems more resilient, but communities as well.

TOOLS

The newly updated Adaptation Strategies Guide for Water Utilities is a PDF document that provides region-specific climate science information and adaptation options to help utilities begin climate adaptation planning. It is an interactive, "clickable" guide that walks users through easy-to-understand, regionally relevant climate information, as well as specific challenges and adaptation options based on their location. The updated sections provide new information on how drinking water and wastewater utilities can approach sustainable practices, specifically through green infrastructure and energy management activities. As the guide walks users through the process of identifying climate challenges, it incorporates sustainable practices into a detailed, categorized list of adaptation options to help utilities develop planning, operations and capital/infrastructure strategies. The guide also includes worksheets users can use to track their progress and decisions throughout the document. The Adaptation Strategies Guide serves as a useful introduction to the climate change and adaptation planning fields, and shows users how to begin the initial stages of implementation.

The Extreme Events Workshop Planning Tool takes the principles of the Adaptation Strategies Guide one step further and helps water sector utilities hold extreme events workshops. The tool allows users to customize their own workshop based on five different scenarios: extreme flooding, extreme drought, sea level rise, intense wildfire and reduced snowpack. During the workshop, participants openly discuss the potential impacts of more frequent and more intense extreme events to not just the utility, but the community as a whole. The tool is built around the concepts of integrated and adaptive water resources management, by encouraging collaboration with stakeholders outside of the utility. It incorporates local decision makers and stakeholders into climate change adaptation planning, which gives the utility and the community a more robust approach to managing impacts, and can help build community based support networks. The workshop tool provides everything a utility would need to plan a workshop, including an introduction to climate science presentation, a facilitator's guide and matching PowerPoint presentation, discussion points for participants, and breakout group sessions to identify both impacts and adaptation options.

The second version of the Climate Resilience Evaluation and Awareness Tool, CREAT 2.0, was recently released and is now available online. CREAT represents the first effort to provide practical, easy-to-use software that translates the most recent climate change science into actionable information for drinking water, wastewater and storm water systems. The tool assists users in conducting climate change risk assessments and promotes a localized understanding of climate change impacts for utility owners and operators.

CREAT 2.0 includes new information for extreme events analysis, location-specific climate change scenarios, energy efficiency considerations, and the ability to compare multiple analysis files for greater flexibility in evaluation. Updated training is also available within the tool, with new training modules that include automated presentations to walk users through the step-by-step processes within CREAT. Specifically, the modules assist the water sector with inputting data to identify and assess threats, identifying adaptation options and developing an implementation plan. Utilities in Las Vegas, Nevada, Oakland, California and Wilmington, Delaware have already piloted CREAT 2.0, and CRWU initiative continues to work with utilities around the country to implement risk assessments. Large-scale training workshops are scheduled in collaboration with New York and New Jersey local and state governments, with multiple dates set for the spring of 2013.

The CRWU initiative continues to build on its relationship with the Climate Ready Estuaries (CRE) program. EPA's CRE program and the CRWU initiative have partnered to encourage watershed-scale adaptation planning. Initial applications of this partnership have paired National Estuary Programs (NEP) with utilities in their watershed to perform CREAT assessments, specifically in Hoboken, New Jersey, Moro Bay, California, and the Albemarle-Pamlico NEP in North Carolina. In New York and New Jersey, the utility and the National Estuary Program identified current and future climate threats and adaptation options, completed the CREAT risk assessment, and released a report documenting the process, which is now available on EPA's website. In North Carolina, two utilities from neighboring communities worked with the NEP to help manage difficult coastal conditions and ensure they can continue providing clean water while protecting the Sound. Their report will be available online in the near future.

CONCLUSION

For more information on EPA and CRWU activities, please visit the website at www.epa.gov/climatereadyutilities. The resources and tools can be accessed via the website. For information regarding training on CRWU products, please visit the "training" tab and register for the upcoming webinars on specific tools and pilot activities. For any questions, please contact CRWU at CRWUhelp@epa.gov.

References

National Drinking Water Advisory Council; 2011; Climate Ready Water Utilities: Final Report of the National Drinking Water Advisory Council.

QUANTIFYING FUGITIVE CH_4 EMISSIONS FROM FACULTATIVE SLUDGE LAGOONS

Anup Shah[1*], Lauren Fillmore[2], Angela Detweiler[3], John Willis[1], Brad Bebout[3], Adrienne Frisbee[3], Cheryl Kelley[4], Jeff Chanton[5], Leslie Bebout[3]
[1] Brown and Caldwell, [2] Water Environment Research Foundation, [3]NASA-AMES Research Center, [4]University of Missouri-Columbia , [5]Florida State University
* Corresponding Author: Anup Shah (Email: ashah@brwncald.com)

ABSTRACT

As part of WERF's U2R08 project, NASA Ames Research Center (ARC) and Brown and Caldwell researchers performed liquid and gaseous phase sample collection at two different sites to focus on the methane (CH4) emissions from facultative sludge lagoons (FSLs) and Photo-synthetic oxidation ponds (PSOPs). This paper presents a case study on the CH_4 emissions from wastewater treatment in facultative sludge lagoons. The Owner of the site preferred to remain anonymous for this paper.

The effort under U2R08 was limited to one sampling event. These estimates are based on this sampling event and largely represent an order of magnitude estimate for FSLs receiving and storing anaerobically digested sludge during fall.

Overall CH_4 fluxes were highly variable for the four FSLs (mean 3.6 ± 3.1 g/m^2/d) with the highest fluxes coming from the two lagoons being fed digested biosolids on a daily basis. High CH_4 bubble concentrations indicated that methanogens are quite active in these lagoons in the fall. Ebullitive fluxes are significant; contributing at least 75% of the total flux into the atmosphere. However, Since CH_4 escaping in bubbles substantially avoids oxidation, ponds with higher oxygen concentrations will be characterized by higher oxidation rates, lower diffusive fluxes, and relatively higher proportions of flux attributable to ebullition. Considering the large surface areas, fugitive CH_4 emissions from FSLs can be significant, especially for the lagoons where undigested solids are stored.

This investigation was a step in the direction of providing a better understanding of the wastewater treatment industry's true impacts from a GHG standpoint by taking into consideration the emissions from FSLs and PSOPs. Further comprehensive research and year-round measurements will provide much better understanding of the full impact and may lead to develop better solutions to remediate their impact as warranted by their significance.

KEYWORDS

Methane, CH_4, Greenhouse Gases, WERF, Facultative Sludge Lagoons (FSLs), Fugitive greenhouse gas emission, Photosynthetic Oxidation Ponds (PSOPs).

INTRODUCTION AND BACKGROUND

FSLs represent one of the lowest energy processes available and as such are attractive options from energy use and indirect greenhouse gas (GHG) emissions perspectives. These area-intensive processes are cryophilic-to-mesophilic treatment and storage units that are uncovered, except for a layer of aerobic liquid. The literature suggests that CH_4 evolution is mitigated by methanotrophic bacteria in the water column that can aerobically oxidize CH_4, as occurs naturally in lakes and other environments (Bastviken et al., 2008; Kelley, 2003; Bastviken et al. 2002). Because each system has potentially significant anaerobic volumes, quantifying the levels of CH_4 emissions provides insight into the significance of direct CH_4 emissions and whether improvements are warranted to address fugitive emissions in these otherwise very sustainable technologies. The primary objective of this research was to provide an order of magnitude estimate of CH_4 emission from FSLs.

For this research, NASA-ARC performed liquid- and gaseous-phase sample collection at four FSLs at a wastewater treatment facility. The utility operating the FSLs opted to remain anonymous in this case-study. The FSLs at the site received anaerobically digested sludge (15 days SRT @ 35°C) and provided up to five years of storage. According to the facility data, the FSLs provided up to 40% additional volatile solids reduction during the storage before being "harvested" by dredges for ultimate land disposal.

The four FSLs on the site were at varying stages of operation during the site visit and sample collection for this research on November 3, 2010. This paper summarizes the results of that sampling and estimates annual CH_4 emissions from these process units on a per unit loading basis.

Summary of CH_4 Evolution and Processing

CH_4 evolves and is processed in a system using a variety of mechanisms. This discussion summarizes those mechanisms so as to provide context from the results reported later in this paper.

Biogenic CH_4 evolves when organic matter is mineralized by anaerobic microbes, known as methanogens, which belong to the domain Archaea. These methanogens are quite diverse, widely distributed, and able to use different types of substrates as a source of energy, the most common being hydrogen, where $CO_2 + 4\ H_2 \rightarrow CH_4 + 2\ H_2O$ (Ehrlich, 1996). Other substrates include acetate and methylated compounds, where CH_4 is always produced as a byproduct. In aquatic systems, CH_4 production will occur in the anoxic region just below the sediment surface. When pressure builds up, bubbles form and are released at the sediment-water interface. As the bubbles travel through the water column, some will be oxidized by methanotrophic bacteria, whereas others will escape directly into the atmosphere. In the oxidation process, CH_4 is oxidized to CO_2, the latter having a global warming potential (GWP) 25 times lower than CH_4 over a 100-year time frame (IPCC, 2007). The methanotrophy in these organic-rich lagoons and ponds is the 'high capacity-low affinity oxidation' type where CH_4 oxidation is aerobic; therefore, the availability of oxygen is the limiting factor. Figure 1 provides a schematic overview of these processes.

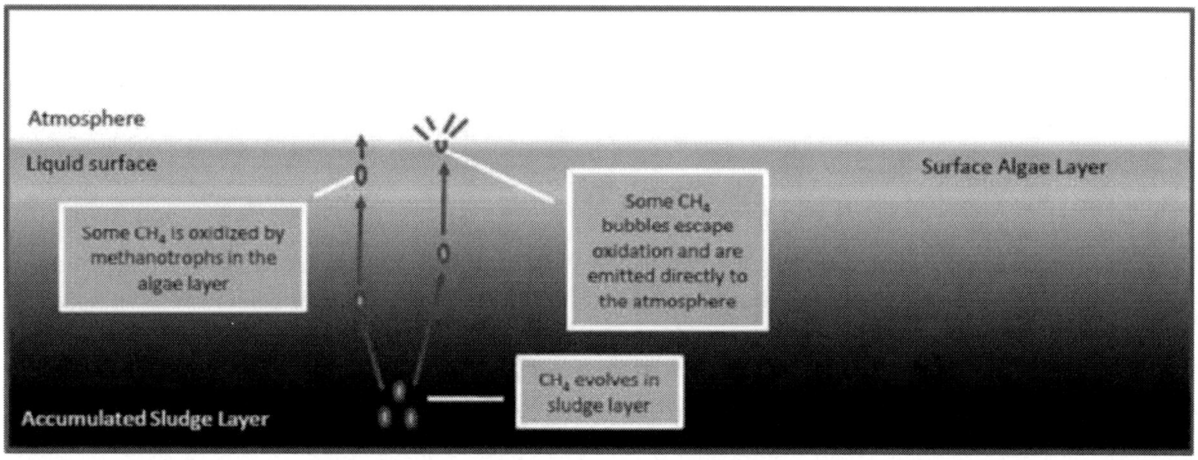

Figure 1. Schematic of CH_4 Evolution from Facultative Sludge Treatment Lagoons.

SAMPLING METHODOLOGY

CH_4 Gas Sampling. The concentration of CH_4 in the lagoons was determined by measuring the dissolved CH_4 in the water column and the CH_4 in gas bubbles trapped in the sediments. Dissolved CH_4 was sampled, in triplicate at each FSL by filling up 60 ml syringes with approximately 30 ml of lagoon water and 30 ml of nitrogen gas (Airgas NI UHP35). The exact volumes of the gas phase and aqueous phase were recorded for each syringe to calculate CH_4 concentrations based on Henry's Law for air-water partitioning. Syringes were shaken vigorously for about two minutes to equilibrate the CH_4 previously dissolved in the water (~99%) with the added nitrogen. Gas samples were stored in glass containers (exetainers) containing a hypersaline solution (200 g/L NaCl), and brought back to the lab for analysis. Storing the samples with the hypersaline solution inside the containers decreases leakage rates; the hypersaline solution (in which CH_4 is exceedingly insoluble) forms a long diffusive path length between the gas sample and the rubber septum. Dissolved CH_4 samples were also injected into 38 ml evacuated serum bottles for $\delta^{13}C$ stable isotope analysis. Potassium Hydroxide (KOH) pellets (~0.13 g) were added to serum bottles to inhibit microbial CH_4 oxidation from occurring between the time the gas was sampled and analyzed.

Gas bubbles containing CH_4 were collected from the FSLs by disturbing the bottom of the lagoons with a rod, and trapping the rising bubbles with an inverted funnel connected to a plastic tube stretched out on a pole. The bubbles displaced the water in the initially-filled funnel and were drawn up through the tube and into a 60 ml syringe. Gas samples were then injected into a 10 ml evacuated serum bottle containing a KOH pellet for storage and transport. Samples were collected in triplicate for each site.

These samples were taken for two reasons: 1) to assess the CH_4 concentrations of the gas present in the sediments, and 2) to determine the carbon isotopic signature of the CH_4 in that gas.

CH_4 Flux Measurements. CH_4 fluxes from the FSLs were determined using two different methods. The diffusive CH_4 flux from the lagoons was estimated based on the CH_4 concentration differences between the atmosphere and the lagoon water, the gas transfer coefficient, and the wind speed over the surface of the pond, as described in Wanninkhof (1992) and Barber et al. (1988). The second method was to quantify the fluxes directly by using floating flux chambers, which account for the diffusive (dissolved) and ebullitive (bubble) flux, the latter known to be significant in shallow CH_4 saturated aquatic environments (Martens and Klump, 1980; Chanton and Martens, 1988; Chanton *et al.*, 1989).

Three acrylic flux chambers designed according to EPA standards (Kienbusch, 1986) were deployed at each FSL over a period of one hour. A foam collar around the chamber kept the chambers afloat on the surface of the lagoon (Figure 2). A two-way valve on top of the chamber was opened so that the pressure inside the chamber (headspace volume ~12L) would equilibrate with the outside pressure. An initial time point sample of the headspace gas was collected from each chamber for later analysis. Two more time points were taken during the following hour of the incubation. Samples of the last time point were injected into 10 ml evacuated serum bottles, containing a KOH pellet, for CH_4 isotope analysis. The flux was calculated using a linear regression of CH_4 concentration within the headspace of a chamber over time, after correcting for the volume of the headspace and the area covered by the chamber.

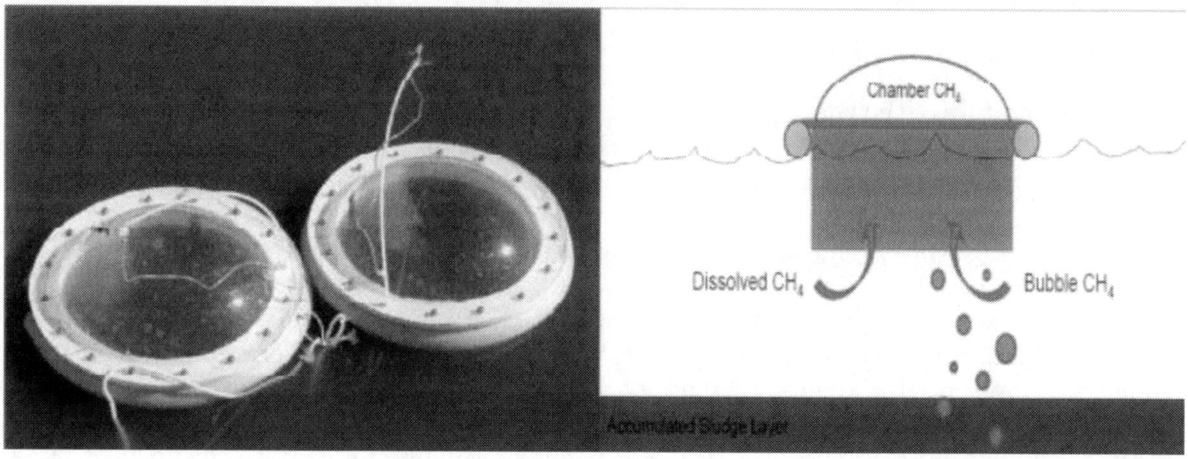

Figure 2. Floating Flux Chambers Capture Diffusive and Ebullitive CH_4 at the Water-Air Interface.

These two methods acted as checks on each other, since floating chambers may not adequately maintain the turbulent regime of the boundary layer (Broecker and Peng, 1984) and the determination of a realistic gas transfer coefficient can be difficult (Raymond and Cole, 2001).

Determination of CH_4 Concentrations. All gas samples collected in the field were brought back to the lab at NASA ARC and analyzed for CH_4 on a Shimadzu 14A gas chromatograph, using a flame ionization detector, and a two-meter Porapak N column. The oven temperature was held at 40°C, the injector at 120°C and the detector at 125°C. The estimated flow rate for nitrogen was ~30 cc/min, 500 cc/min for air, and 50 cc/min for hydrogen.

Stable Isotopes. Gas samples collected in serum bottles were analyzed for $\delta^{13}C$ values at Florida State University and the University of Missouri-Columbia. Samples for stable isotope analysis were injected into a gas chromatograph coupled via a combustion interface to a Finnigan Delta Plus XL Isotope Ratio Mass Spectrometer (IRMS). Samples with low CH_4 concentrations (<1,000 ppm) were injected into a cryogenic focusing unit to concentrate the samples before running them through the IRMS.

Stable isotope values are expressed as the ratio of the heavy isotope (^{13}C) to the light isotope (^{12}C), relative to a standard. The departure from a standard is expressed as δ (delta) in parts per thousand (‰), and can be calculated by (equation 1):

$$\delta X = [(R_{sample} / R_{standard}) - 1] \times 10^3,$$

where R_{sample} is the C^{13}/C^{12} ratio of a sample, and $R_{standard}$ is the ratio of C^{13}/C^{12} of PeeDee Belemnite.

The isotopic analysis of $\delta^{13}C\text{-}CH_4$ is a useful tool to quantify the contribution of diffusive and ebullitive flux into the chamber, based on the two-source mixing model (equation 2):

$$f_{dissolved}=(\delta_{chamber}-\delta_{bubbles})/(\delta_{dissolved}-\delta_{bubbles})$$

CH_4 Oxidation. The oxidation efficiency of a system, or the amount of CH_4 that is oxidized in the water column by methanotrophic bacteria before reaching the surface, was also calculated using stable carbon isotopes (Chanton et al., 2008; Chanton et al., 1999). The fraction of CH_4 oxidized is based on the discernible isotopic fractionation due to the preferential uptake of ^{12}C over ^{13}C by methanotrophs, making the oxidized (dissolved) CH_4 more enriched than the source (bubble) CH_4 (Coleman et al., 1981; Barker and Fritz, 1981). Assuming that the lagoons more closely resemble a closed system, the amount of CH_4 oxidized can be calculated according to the equation below (Chanton et al., 2008) (equation 3):

$$f_{oxidized} = 1 - \{[(\delta_{oxidized}+1000)/(\delta_{source}+1000)]^\wedge(\alpha/1-\alpha)\};$$

Other Parameters. Water temperature was measured at each lagoon with a FisherBrand M113849 thermometer. Average wind speeds were obtained from the NOAA National Weather Service website (www.weather.gov). Due to wind gusts during the field sampling date, the atmospheric CH_4 concentration was not sampled on site. Instead the global CH_4 average of 1.77 ppm (IPCC, 2007) was used in the Wanninkhof and Barber flux models. A sensitivity test using higher atmospheric CH_4 levels (8 ppm) was performed and found that the results only deviated by less than <0.01 mg/m^2/d from the Wanninkhof and Barber flux models

RESULTS AND DISCUSSION

For the purpose of this paper, the identification (Figure 3) and general descriptions of the operational status of the four FSLs are as follows:

FSL-A is thought to be representative of well operated units and had been dredged earlier in 2010. The unit had been fed 15–30 minutes prior to sampling but it is unlikely that the feed translated into increased gas production as the feed and sampling locations were a considerable distance apart.

FSL-B represents the longest "sludge age" with visible mounds of sludge and a thick surface layer of algae. This FSL had not been dredged in a long time (last dredged in 2007); however, it was still being 'fed'.

FSL-C represents an extremely overloaded condition under 'abnormal operation' and possibly 'worst case' emissions.

The surface of the FSL was green tinged. Secondary effluent flushing was turned off to FSL-C in 05/19/2009. This FSL was last dredged during the 2008 harvest season. FSL-C had been fed continuously over a period approaching 48 hours due to a valve failure.

However, samples were collected to provide a good contrast to the other three lagoons that were well managed.

Samples from FSL-C should serve to demonstrate that FSL-C was still emitting CH_4; likely at a higher rate despite of being in an 'abnormal operation'. From the green house gas management standpoint, the data point from an 'abnormal case' also demonstrates the importance of managing the lagoons in a 'normal' operational state.

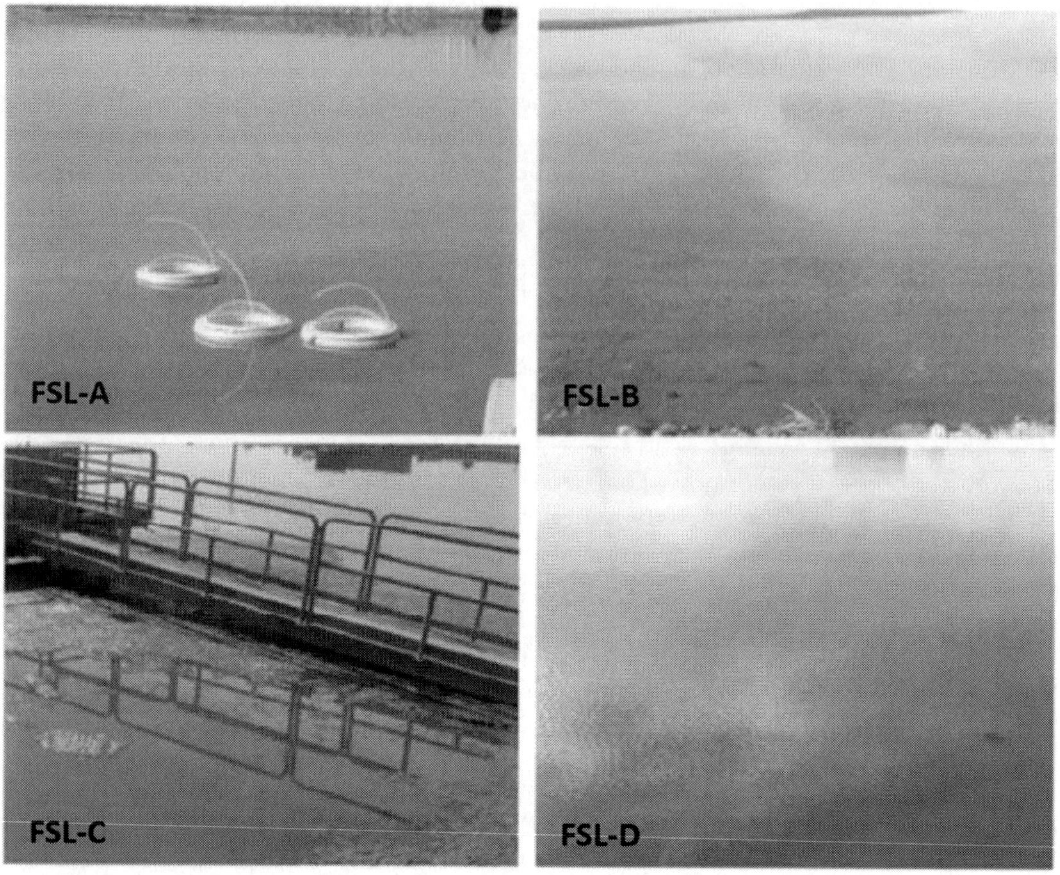

Figure 3. Facultative Sludge Lagoons Have 10 to 13 Feet of Biosolids Storage and Two to Five Feet of Water, Depending on Length of Time Since Dredging.

FSL-D represents the 'normal' operation. The surface of this FSL was covered in green algae. The surface of this particular unit was normally flushed with secondary effluent to elutriate.

Bubble CH_4 concentrations were high for all four FSLs, ranging from 76–83% (v/v). These concentrations did not differ significantly from one another (ANOVA: $F=0.54$; $p=0.67$) (Fig. 4A), reflecting an active methanogenic community late into the fall for these lagoons. Dissolved CH_4, on the other hand, was significantly different for the four FSLs (ANOVA: $F=509.69$; $p<0.0001$), except between FSL-B and FSL-D (T-TEST: $p=0.08$), which presented extremely low CH_4 concentrations of 1.7±1.7 μM and 5.7±2.5 μM, respectively. The higher dissolved CH_4 concentrations for FSL-A (167.3±26 μM) and C (400.6±12 μM) may be explained by the fact that these two lagoons are actively being fed digested solids on a daily basis. The latter was accidentally fed for 48 hours straight around the time that we sampled it. Overall, the mean dissolved CH_4 concentration for the lagoons sampled was 143.8±170.3 μM.

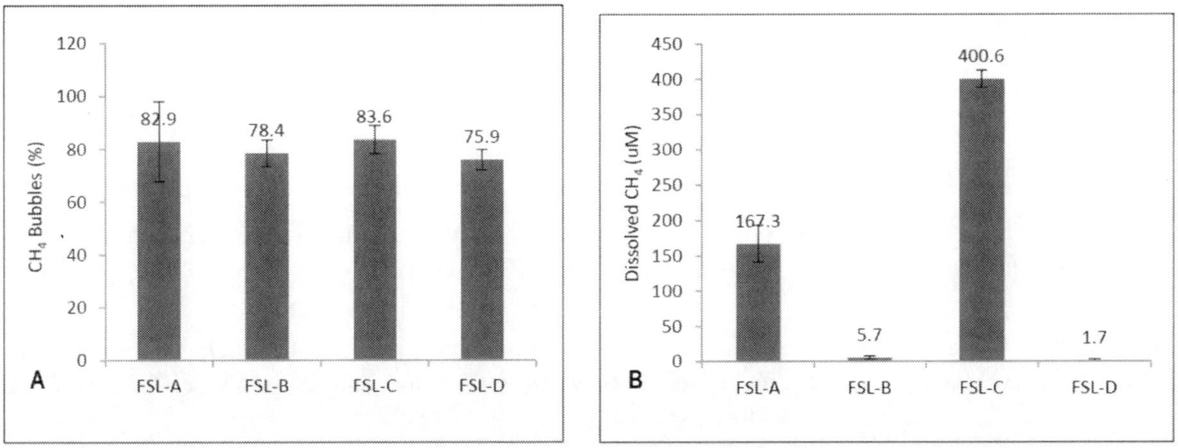

Figure 4. CH_4 Bubble Concentrations (%) (A), and Dissolved CH_4 Concentrations (µM) (B) for Four FSLs, Sampled on November 3, 2010.

Measured CH_4 fluxes at the water-surface interface were highly variable (ANOVA: $F=15.32$; $p=0.001$), presenting a tenfold difference between the lowest flux of 0.8±0.5 g/m^2/d for FSL-D and the highest flux of 7.7±2.3 g/m^2/d for FSL-A, with a mean daily CH_4 flux of 3.6±3.1 mg/m^2/d (or 3.6x10^{-7} kg/ha/d) (Figure 5). Such high deviation from the mean is due to the fact that these lagoons are at different operational states (see site descriptions above), with the highest fluxes coming from the lagoons actively being fed biosolids, as was observed for the dissolved CH_4 data. Preliminary data from oxidation ponds sampled year-round at another site show that CH_4 fluxes are highly correlated to seasonal changes, with higher fluxes during the summer and fall, and lower fluxes in the winter and spring. Therefore, the use of one time point in the fall to extrapolate the average annual CH_4 flux of this system will most likely lead to an overestimation.

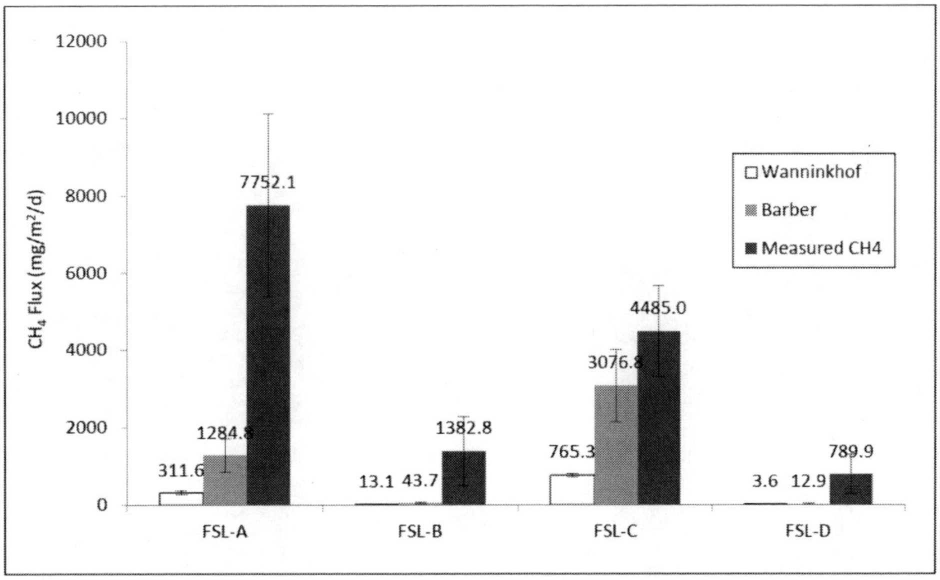

Figure 5. Measured CH_4 Flux Compared to the Calculated CH_4 Flux Based on Flux Models Described in Wanninkhof (1992) and Barber et al. (1988).

The calculated CH_4 fluxes based on Wanninkhof (1992) and Barber et al. (1988) were much lower than the measured CH_4 flux determined by the floating chambers (Figure 5). This was no surprise considering that these flux models do not take into account ebullition, which in this case contributed 75–100% of the total CH_4 measured in the chambers (Table 1). These models may become more useful in reflecting actual CH_4 fluxes during the winter and spring months, when ebullition flux is reduced due to lower methanogenic activity. A sensitivity test was performed using a higher atmospheric CH_4 concentration of 8 ppm (observed at another oxidation pond site) in the flux models, but this only showed a decrease in the diffusive fluxes by less than 0.01 $mg/m^2/d$ (data not shown). Due to this minute change, the data reported was based on a background global atmospheric CH_4 concentration of 1.77 ppm.

Table 1. Contribution (%) of Diffusive and Ebullitive Fluxes to Chambers Based on a Two-Source Mixing Model (eq. 2), and Oxidation Efficiency (eq. 3) of the FSLs Based on a Fractionation Factor of 1.02.

Site	Avg chamber $\delta^{13}C-CH_4$ (%)	Avg dis $\delta^{13}C-CH_4$ (%)	Avg bub $\delta^{13}C-CH_4$ (%)	$f_{diffusive}$ (%)	$f_{ebullitive}$ (%)	Oxidation efficiency (%)
FSL-A	-51.2±0.4	-49.6± 0.2	-51.1±0.3	25	75	8
FSL-B	-49.3±0.2	-47.5±0.5	-49.5±0.2	-2	102	10
FSL-C	-49.2±0.3	-47.7±0.1	-47.9±0.8	N/A*	N/A*	1
FSL-D	-49.2±0.2	-41.1**	-47.9±0.6	-3	103	30

* *The two-source mixing model did not produce any results for FSL-C, since the $\delta^{13}C$ values for dissolved and bubble CH_4 were too similar, and the $\delta^{13}C$ value for the chamber was more depleted than both (Figure 6).*

***Two replicates of dissolved CH_4 samples for FSL-D presented extremely enriched $\delta^{13}C$ values, indicating that CH_4 might have oxidized in these bottles between the time they were sampled and analyzed, therefore, only one replicate was used.*

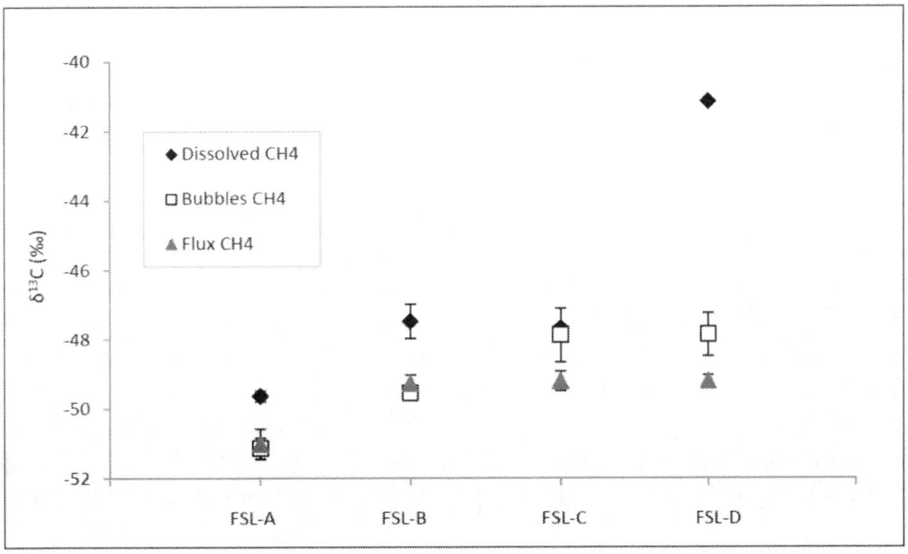

Figure 6. $\delta^{13}C$-CH_4 Values for Gas Samples Collected in all Four FSLs.

To the researchers' knowledge, the fractionation of carbon by CH_4 oxidizing bacteria has not been empirically determined for sludge lagoons. Therefore, in order to calculate the oxidation efficiency in the lagoons an assumed fractionation factor (α) of 1.02 was used based on a review of α values calculated for oxidation activity in stream cultures by Barker and Fritz (1981), lakes by Bastviken et al. (2002), and landfills by Chanton et al. (2008). The oxidation efficiency for the four FSLs ranged from 1–30% (Table 1). Overall, the oxidation efficiency observed during this research was lower than that observed in another site in which open oxidation ponds were in use. Possible explanations are that oxygen concentrations in the lagoons are extremely low limiting aerobic CH_4 oxidation activity and the CH_4 loading is higher in FSLs compared to theoxidation ponds. Oxygen production in the ponds is likely dependent on the health of the algae. The highest oxidation efficiency occurred in FSL-D, which appeared to have the most "green" and "healthy" algae, whereas the lowest oxidation occurred in FSL-C, which was subjected to overfeeding of biosolids and presented a murky color the day that we sampled it. Dissolved oxygen (DO) data provided by the facility also supports this hypothesis, depicting DO concentrations at FSL-D to be on average three times higher compared to the other three lagoons sampled (Table 2). Interestingly enough, the FSL-D liquid layer was flushed with secondary effluent, unlike the other lagoons, as a mechanism to reduce struvite buildup, odor control, and to improve algal blooms.

Table 2. Average Dissolved Oxygen (DO) Concentrations (mg/l) in the FSLs for Oct. 2009 – Oct. 2010, Provided by the Wastewater Treatment Facility.

Site	Avg DO (mg/l)	Min DO (mg/l)	Max DO (mg/l)
FSL-A	3.2	0	19.7
FSL-B	3.1	0.5	13.6

FSL-C	2.6	0.3	9.2
FSL-D	10.1	0.8	29.1

CONCLUSION AND FURTHER RESEARCH

Overall CH_4 fluxes were highly variable for the four FSLs (mean 3.6±3.1 g/m^2/d) with the highest fluxes coming from the two lagoons being fed digested biosolids on a daily basis. High CH_4 bubble concentrations indicated that methanogens are quite active in these lagoons in the fall. Ebullitive fluxes are significant; contributing at least 75% of the total flux into the atmosphere. The oxidation of CH_4 dissolved in the water could be significant. Since CH_4 escaping in bubbles substantially avoids oxidation, ponds with higher oxygen concentrations will be characterized by higher oxidation rates, lower diffusive fluxes, and relatively higher proportions of flux attributable to ebullition. Considering the large surface areas, fugitive CH_4 emissions from FSLs can be significant, especially for the lagoons where undigested solids are stored.

Facultative Sludge Lagoons are area-intensive processes and remain uncovered. Except for a layer of aerobic liquid on the surface, majority of the FSL volume represent cryophilic-to-mesophilic treatment and storage of sludge. Pond/lagoon treatment technologies represent some of the lowest energy processes available and as such are attractive options from an indirect GHG perspective. However, quantifying the fugitive GHG emissions from these process is required to provide a more complete and accurate estimate of their overall GHG footprint.

While it is intuitive that sludge treatment lagoons have significant anaerobic zones, oxidation ponds have also been shown to develop anaerobic layers, particularly in the deeper portions. Literature also suggests that CH_4 evolution is mitigated by methanotrophic bacteria in the water column that can aerobically oxidize CH_4, as occurs naturally in lakes and other environments. Because each system has potentially significant anaerobic volumes, quantifying the levels of CH_4 emissions would provide insight into the significance of direct CH_4 emissions and whether improvements are warranted to address fugitive emissions in these otherwise very sustainable technologies.

The effort under U2R08 was limited to one sampling event. These estimates are based on this sampling event and largely represent an order of magnitude estimate for FSLs receiving and storing anaerobically digested sludge during fall.

Due to shallow depths and large surface areas, meteorological conditions play a major role in determining the characteristics and behavior of facultative lagoons. Distinct seasonal variations occur throughout the year. The diurnal changes in the lagoon are minimal during the winter months compared to those observed during the summer. Considering this, additional sampling under various seasonal conditions throughout the year is recommended to further refine these estimates.

ACKNOWLEDGEMENTS

This research was possible through the funding support provided by the Water Environment Research Foundation (WERF), and the helpful guidance of Lauren Fillmore, WERF Senior Program Director. The project team gratefully acknowledges the active participation and "in-kind" contributions from the NASA AMES Research Center, City of Sunnyvale, California, and the other utility member participating in this research by permitting access to their facultative sludge lagoons. Dedication and contributions from Dr. Brad Bebout and Angela Detweiler of NASA AMES were critical for this research project. Access to field data from NASA's previous research efforts at Sunnyvale was essential to the success of this project.

REFERENCES

Barber, T.R.; Burke, R.A.; and Sackett, W.M. (1988). Diffusive Flux of Methane from Warm Wetlands. *Global Biogeochemical Cycles*, 2: 411-425.

Barker, J.F. and Fritz, P. (1981). Carbon Isotope Fractionation during Microbial Methane Oxidation. *Nature*, 293: 289-291.

Bastviken, D.; Cole, J.J.; Pace, M.L.; and Van de Bogert, M.C. (2008). Fates of Methane from Different Lake Habitats: Connecting Whole-Lake Budgets and CH_4 Emissions. J. Geophy. Res.,113: G02024.

Bastviken, D.; Ejlertsson, J.; Tranvik, L. (2002). Measurement of Methane Oxidation in Lakes: A Comparison of Methods. *Environmental Science and Technology*, 36: 3354-3361.

Broecker, W.S. and Peng, T-H (1984). Gas Exchange Measurements in Natural Systems. In: Brutsaert W, Jirka GH (eds) Gas Transfer at Water Surfaces. Reidel Publishing Co., Dordrecht, pp 479-495.

Chanton, J.P.; Powelson, D.K.; Abichou, T.; and Hater, G. (2008). Improved Field Methods to Quantify Methane Oxidation in Landfill Cover Materials Using Stable Carbon Isotopes. *Environmental Science and Technology*, 42: 665-670.

Chanton, J.P.; Rutkowski, C.M.; and Mosher, B.M. (1999). Quantifying Methane Oxidation from Landfills Using Stable Isotope Analysis of Downwind Plumes. *Environmental Science and Technology*, 33: 3755-3760.

Chanton, J.P.; Martens, C.S.; and Kelley, C.A. (1989). Gas Transport from Methane-Saturated, Tidal Freshwater and Wetland Sediments. *Limnol. Oceanogr.*, 34: 807-819.

Chanton, J.P. and Martens, C.S. (1988). Seasonal Variations in Ebullitive Flux and Carbon Isotopic Composition of Methane in a Tidal Freshwater Estuary. *Global Biogeochem. Cycles*, 2: 289-298.

Ehrlich, H.L. (ed) (1996). Geomicrobiology – 3rd edition. Marcel Dekker, Inc., New York, pp 628-642.

IPCC (2007): Climate Change 2007: The Physical Science Basis. Contribution of Working Group I to the Fourth Assessment Report of the Intergovernmental Panel on Climate Change [Solomon, S.; Qin, D.; Manning, M.; Chen, Z.; Marquis, M.; Averyt, K.B.; Tignor, M.; and Miller, H.L. (eds.)]. Cambridge University Press, Cambridge, United Kingdom and New York, NY, USA, 996 pp.

Kelley, C.A. (2003). Methane Oxidation Potential in the Water Column of Two Diverse Coastal Marine Sites. *Biogeochemistry* 65: 105-120.

Kienbusch, M. (1986). Measurement of Gaseous Emissions Rates from Land Surfaces Using an Emission Isolation Flux Chamber. User's Guide, EPA Users Guide, (EPA 600/8-86/008).

Martens, C.S. and Klump, J.V. (1980). Biogeochemical Cycling in an Organic-Rich Coastal Marine Basin - I. Methane Sediment-Water Exchange Processes. *Geochim. Cosmochim. Acta*, 44: 471-490.

Raymond, P.A. and Cole, J.J. (2001). Gas Exchange in Rivers and Estuaries: Choosing a Gas Transfer Velocity. *Estuaries* 24:312-317.

Wanninkhof, R. (1992). Relationship Between Wind Speed and Gas Exchange Over the Ocean. *Journal of Geophysical Research* 97:7373-7382.

Peak Amplification of Energy Demand, Energy Cost, and Carbon Emission in Water Reclamation and Purification Processes during Diurnal Cycles

Reza Sobhani [1,*], Lory E. Larson[2], Diego Rosso[1]

[1] Civil & Environmental Engineering, University of California, Irvine, CA 92697
[2] Southern California Edison, CSUB, Irwindale, CA
*Corresponding author (rsobhani@uci.edu; T: 949-943-6906; F: 949-824-3672)

ABSTRACT

Due to the fluctuations of different environmental parameters, the energy footprint of water reclamation and purification processes has a dynamic behavior during diurnal periods. Among the influencing factors, hydraulic load and constituent load have the highest diurnal impact on the energy footprint of the biological activated sludge process during water reclamation, and energy intensive membrane processes (e.g. reverse osmosis) during water purification. Besides the diurnal energy footprint variations, the energy-associated carbon emission also is facing a variation among the portfolio of energy sources employed by power supply agencies in different hours of a diurnal period. During the night, the carbon intensity of the unit power generation is lower due to a higher contribution of clean-source energies; while during the daytime, fossil fuel-source power has a bigger contribution in fulfilling the power augmented demand.

KEY WORDS: Energy footprint; carbon emission; environmental parameters; peak amplification; diurnal cycle.

INTRODUCTION

Water reclamation and purification (WRP) implies processing wastewater to meet potable standards (Metcalf & Eddy, 2007). Energy-conservative, cost-effective, and reliable treatment technologies are vital to the successful implementation of WRP projects (Sobhani, 2011). The employed treatment technologies are typically energy intensive and are associated with direct and indirect release of greenhouse gases (Côte et al. 2004; Monteith et al., 2005; Foley et al., 2010). Improving energy efficiency during the operations provides an opportunity for minimization of energy footprint (eFP) on a large scale, and concurrently enhances the mitigation of greenhouse gas emissions to the already evident public and environmental health benefits of WRP processes (Gori et al., 2011; Rosso and Stenstrom, 2008).

Many studies discussed the dynamics of water reclamation processes (WRP) due to the abrupt or gradual variations generated by many environmental parameters (*inter alia,* Ekama and Marais, 1979). Among those parameters, we can mention hydraulic load (Q), constituent load concentrations (C) including biochemical oxygen demand (C_{BOD}), ammonia (C_{NH4}), and total dissolved solids (C_{TDS}), water temperature (T_w), and ambient temperature (T_a). Furthermore, the carbon intensity of the unit power generation (g_{CO2}/kWh), named κ, varies in the diurnal period independently from the water characteristics listed above. The aim of this study is to understand the dynamics of eFP in WRP processes caused by variations of environmental parameters, and scrutinize the dynamic behavior of its associated cost and carbon emission caused by variations among the energy demand and different diurnal contributions of power sources.

Two treatment plants (Plant A and Plant B) were selected for this study. The plants are located in the United States in the same warm area ($T_{avg} = 20^oC$). Plant A is a large municipal water reclamation plant with Q ~ 360,000 m^3 d^{-1}, and inclusive of headworks, primary treatment and secondary activated sludge processes. A larger portion of Plant A secondary effluent is transferred to Plant B, which is a water purification plant with Q ~ 260,000 m^3 d^{-1}, and inclusive of MF, RO, and AOP processes for deep aquifer recharge. The discharged water is extracted from the downstream groundwater wells for potable purpose. The smaller portion of Plant A effluent is discharged into a water body.

Overall, for both plants, the studied processes are inclusive of biological activated sludge (AS) process, microfiltration (MF), reverse osmosis (RO), and advanced oxidation process (AOP). The AS process is employed for dissolved organic matter, suspended solids, and nutrient removal. The MF process is employed for residual suspended and colloidal solid removal. The RO process is utilized for residual dissolved constituent removal. Finally, AOP includes an ultraviolet (UV) radiation coupled with H_2O_2 injection used for residual trace constituent removal, in this case N-Nitrosodimethylamine (NDMA). Among the highest energy intensive processes of WRP treatment, AS and RO are the two processes selected here for discussion.

ENVIRONMENTAL PARAMETERS

Different environmental parameters fluctuate through the treatment plants' influent during diurnal periods, including Q, C (i.e. C_{BOD} and $C_{NH4\text{-}N}$), and T_w. Henze et al. (2002), Metcalf and Eddy (2003), and Hammer and Hammer (2008) showed similar diurnal patterns of Q fluctuation observed at treatment plants. In general, minimum flow occurs during the early morning hours when water consumption is lowest and when the base flow consists of infiltration and small quantities of sanitary wastewater. The first peak flow generally occurs in the late morning when wastewater from the peak morning water use reaches the treatment plant. The second peak flow generally occurs in the early evening between 7 and 9 pm. The time of occurrence and the amplitude of the peak flow rate vary with the size of the community, the length of the collection system, and even the days of the week (Leitão et al., 2006).

C_{BOD} and C_{NH4} also are significantly changed during the course of a day (Metcalf and Eddy, 2003). In general, C_{BOD} fluctuation follows the Q variation. When lower Q contains lower C_{BOD}, higher Q generally contains higher C_{BOD}. The first C_{BOD} peak often occurs before noon, when the second one occurs during the late evening. Meanwhile, $C_{NH4\text{-}N}$ has different diurnal behaviour. Henze et al. (2002) and Hammer and Hammer (2008) showed that the peak of influent NH_4^+-N often occurs in the morning and stays low during the afternoon and drops significantly during the night. Figure 1 presents the average daily water characteristics collected from the studied plant (i.e. Plant A) and distributed by diurnal variation patterns provided in the aforementioned literature.

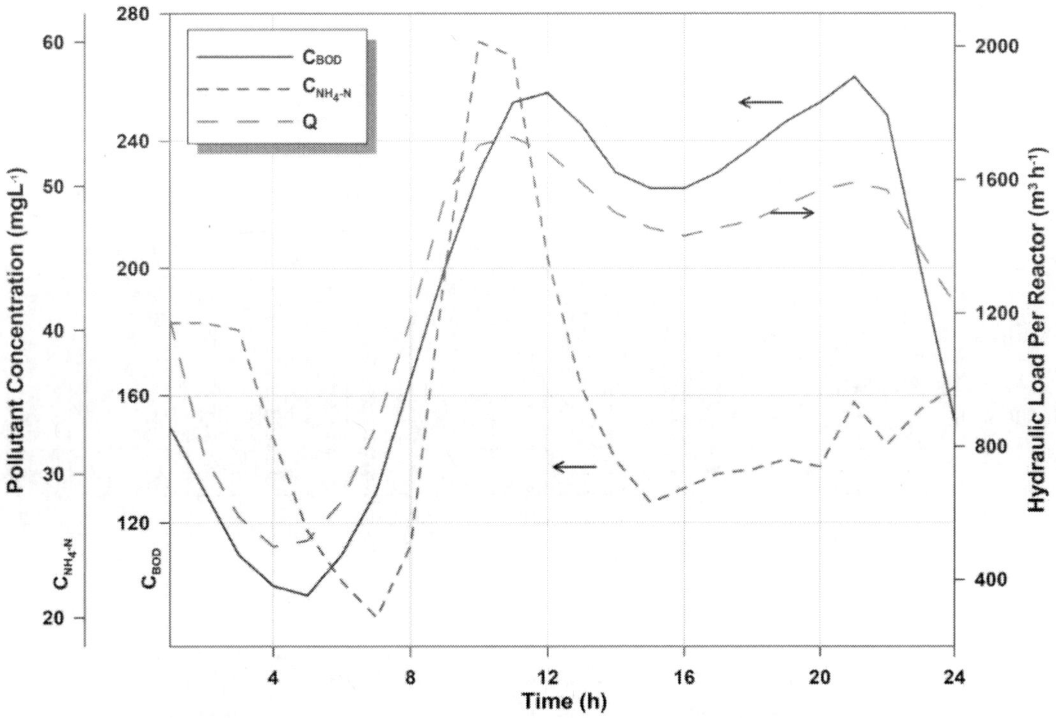

Figure 1. Variation of water characteristics (C_{BOD}, $C_{NH4\text{-}N}$ and Q) plotted using the daily average from the studied plant (Plant A) and the daily variation patterns provided in the available literature (Henze et al., 2002; Metcalf and Eddy, 2003; Hammer and Hammer, 2008).

Besides the water-based environmental parameters, the energy cost (USD/kWh) and the greenhouse gas emission of unit energy generated, κ (g_{CO2}/kWh) are the other parameters that independently change in a diurnal period. Both aforementioned parameters show more amplified variation in the summer season compared to the winter time. These parameters vary due to the changes across the diurnal power demand that forces the power supply companies to utilize different contributions of energy sources (e.g., hydroelectric, nuclear energy, gas or coal) in their diurnal power generation portfolios.

Power utilities employ a higher contribution of fossil power sources (e.g., natural gas or coal) to meet the augmented power demand during the diurnal peak hours, especially during the hot summer time, resulting in higher overall κ values (664 and 690 g_{CO2}/kWh, respectively), while power generated from nuclear and hydroelectric sources with smaller κ values (63 and 15 g_{CO2}/kWh, respectively) dominates the nightly power supply.

The overall κ value is calculated by adding the diurnal contributions of each power source, multiplied by their specific carbon emissions available in literature (Fridleifsson et al., 2008; Lenzen et al., 2008). Figure 2 presents the results of our study in Southern California and reveals that during the summer, κ varies between 385 and 495 gCO_2/kWh, while during the winter it varies between 355 and 465 gCO_2/kWh. Meanwhile, the energy cost varies between 0.04

USD/kWh and 0.13 USD/kWh during the summer season, and 0.04 USD/kWh and 0.06 USD/kWh during the winter time (SCE, 2011).

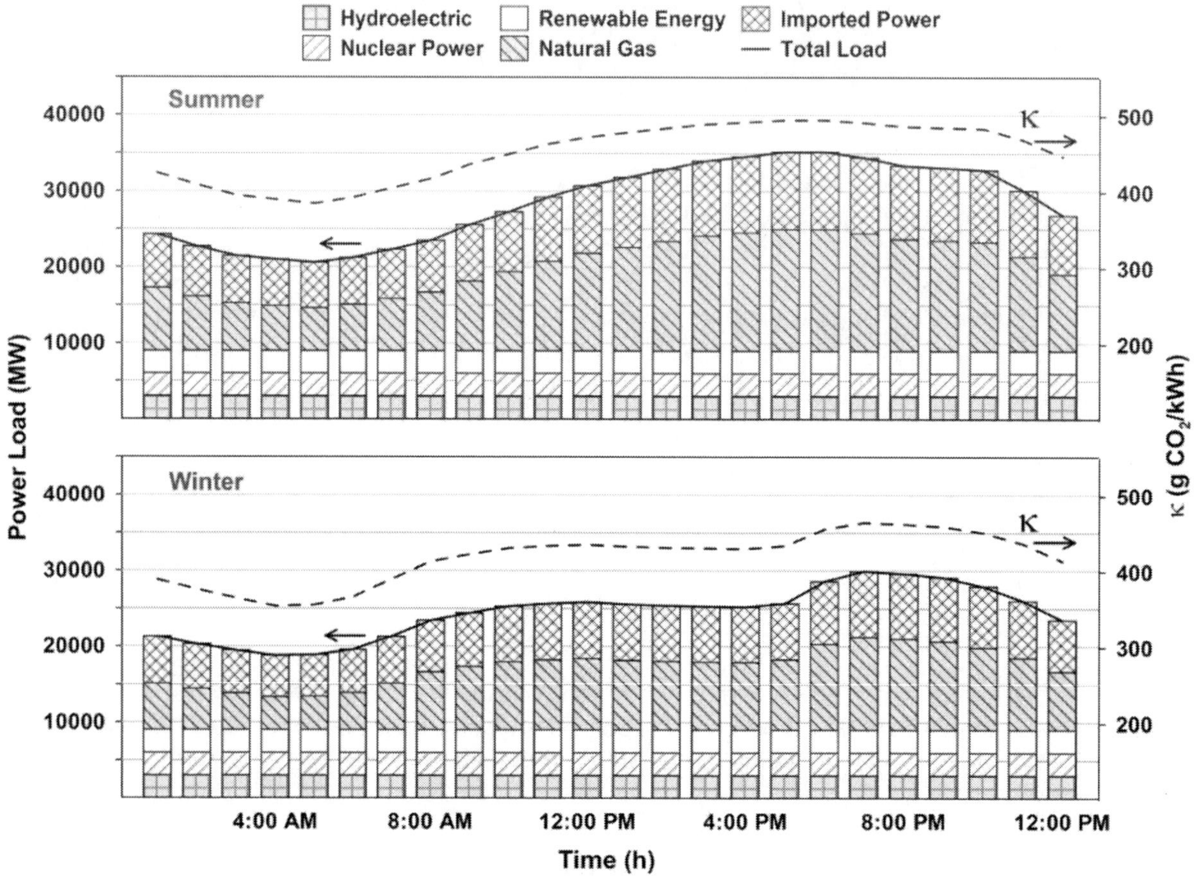

Figure 2. Diurnal power load (FERC, 2012) and power generation portfolio (CAISO, 2012) for the considered electrical grid, and calculated greenhouse gas emission of unit power generated (κ) during summer and winter in California, USA.

RESULTS AND DISCUSSION

We compiled the environmental parameters in a short-term dynamic model that outputs the diurnal variations of eFP and energy-associated carbon emission (eCE) through the WRP processes. The data set provided by Plant A was based on daily measurements that do not carry the time-sensitivity required for this study. As usual for the majority of treatment plants, at Plant A the daily average flow is the only flow information collected. As a result, the environmental parameters collected from this plant were distributed diurnally based on the variation patterns taken from the available literature (i.e., discussed in the previous section).

The results reveal that in the AS process, the intertwined C and Q variations concurrently affect both oxygen requirements (R_o) and oxygen transfer efficiency (OTE). During the peak daytime hours of AS operation, providing more air to fulfil the augmented oxygen demand increases the size of fine bubbles, due to the higher discharge pressure in diffusers. Accordingly, OTE declines and eFP variation is amplified in the AS operation tanks. We assumed that the AS process operates a fine-bubble diffuser system with the same dynamics and response to organic load as in Leu et al. (2009). On the other side, the higher energy demand in AS process is concurrent with the higher κ value during the peak day-time hours especially during the hot summer days. Figure 3 depicts how the concurrency of different environmental parameters' variations significantly amplifies eFP and eCE changes in the AS process during a typical diurnal period.

Oxygen requirements is divided in two compartments: 1) oxygen requirement for carbonaceous biodegradation; 2) oxygen requirement for nitrogen removal. The results of our model show that these compartments retain the diurnal variation patterns of C_{BOD} and C_{NH4}. The carbonaceous R_o increases significantly during C_{BOD} peaks (i.e. the late morning and the late evening), while the R_o required for nitrogen removal increases considerably during the C_{NH4} early morning peak. Furthermore, both compartments decrease during the night and increase during the day due to diurnal flow variations. The results of our model reveal that the intertwined Q and C variation amplifies the diurnal R_o variation over 97%.

The variation in oxygen transfer rate (OTR) is formed by the diurnal fluctuations observed in R_o and OTE, which are formed by C and Q variations. The results of our model show that OTR abates during the night, then starts to rise during the morning and reaches its peak at noon. It has a mild decease after noon and has another peak during the evening. Meanwhile, Q and consequently R_o present similar variation behaviors. The results (Fig. 3) show that the air flow rate (AFR) variation is 45% higher than R_o variation at the same diurnal period. This difference is because of the reciprocal impacts of OTE and AFR.

High C_{BOD} during the daytime peak hours decreases the α value (i.e., $\alpha = \frac{k_L a_{WW}}{k_L a_{TW}}$). At the same time, higher Q and C necessitate higher AFR. It causes higher inlet pressure in the aeration system and increases the size of the fine bubbles aerated from the submerged membrane disc diffusers. Higher bubble size results in lower standard OTE (SOTE). The concurrency of the decrease in oxygen transfer parameters (i.e., α and SOTE) produces a compounding steep OTE decrease during the peak daytime hours. On the other hand, to meet the treatment requirements when the lower transfer efficiency occurs, the AS process requires higher AFR that exacerbates the OTE depression. Conversely, the nightly concurrence of lower Q and C values requires lower R_o, improves OTE value, and consequently alleviates AFR.

Based on selected patterns for the environmental parameters' variations, the results (Fig. 3) reveal that eFP decreases after the midnight and rises during the daytime with two peaks: the first peak occurs before noon and the second occurs during the early night (~9 pm). The mutual impacts of R_o and αSOTE variations increase the diurnal eFP variation by 48% (i.e., compared to the R_o variation alone). The diurnal eCE variation is exacerbated by the regional κ variations. The peaks of eFP are accompanied by κ value at peak hours during the daytime and early evening. As a result, compared to eFP, the eCE diurnal variation is 22% higher during the summer and 20% higher during the winter time.

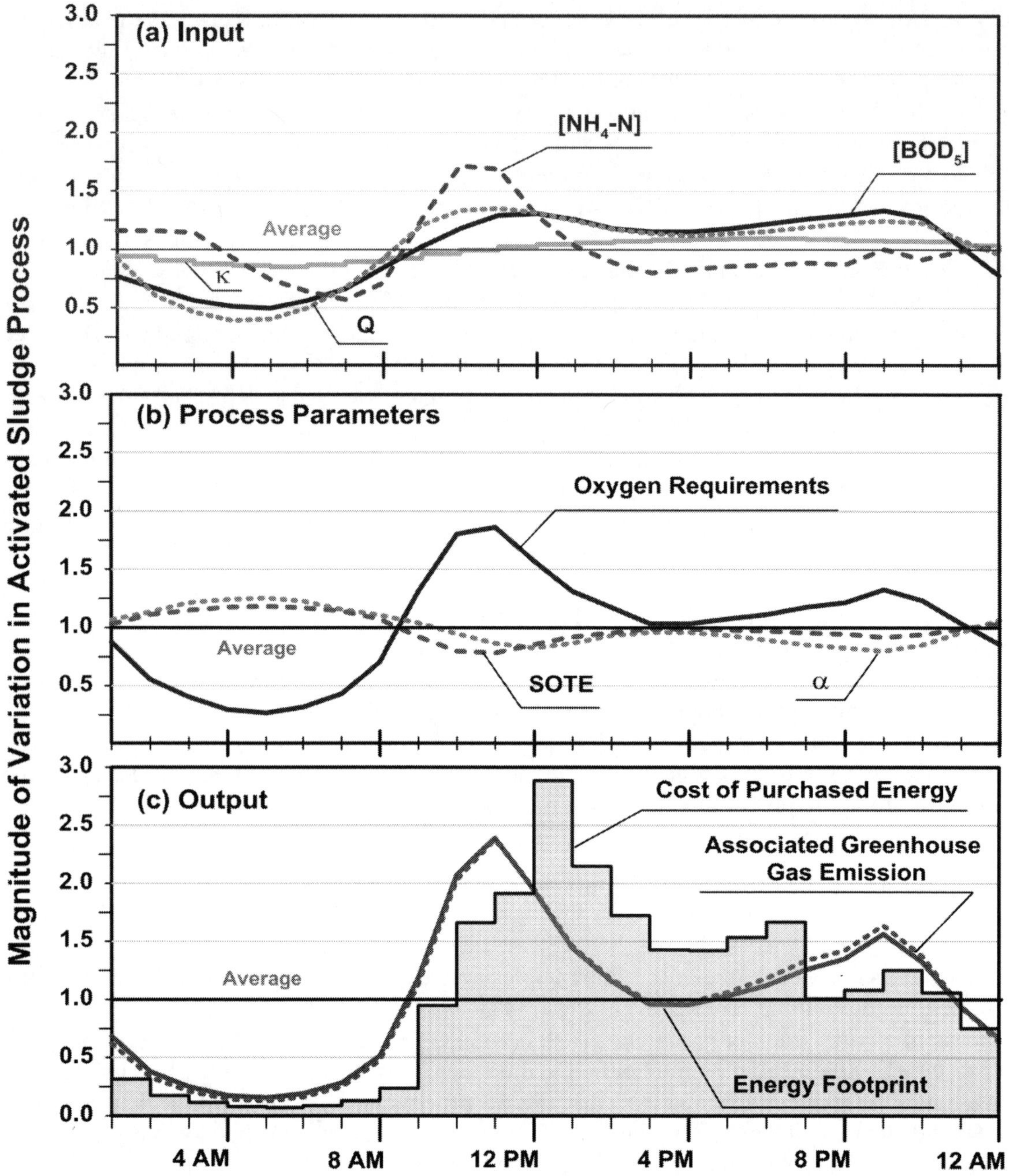

Figure 3. Magnitude of variation in activated sludge process across (a) input parameters, (b) process parameters, and (c) output parameters. Note that while the input parameters (hydraulic and constituent load) have peaking factors of 1.3-1.7, the energy parameters (energy footprint, energy carbon emission, and instantaneous energy cost) have peaking factors of 2.4-2.9.

In Plant B, the employed data was provided by the plant SCADA system. In this plant, RO is the most energy intensive process. Fluctuations among the dissolved solid concentration of RO feed water cause the variations among the osmotic pressure. When the RO influent faces higher constituent load (TDS concentration), to keep the RO flux at a stable level, higher applied pressure is required. An increase among the applied pressure requires higher power consumption by the RO feed pumps. During the off-peak hours, when the constituent load is lower,. lower applied pressure and consequently lower power consumption would be needed. In addition to the constituent load, the power consumption of the RO feed pumps is varied because of the fluctuations in the hydraulic load. Other environmental parameters (e.g., water temperature) could affect the RO power consumption as well. However, in the studied plants they were considered either with inconsiderable variations (i.e., water temperature) or negligible effect during a diurnal period (e.g., long-term membrane fouling).

The variations of the power consumption can be simplified and calculated based on the diurnal hydraulic load variation and solid load (i.e. represented by the conductivity) variation. In Plant B, the diurnal power variations are mostly the results of the load fluctuations in the RO feed water. The results (Fig. 4) indicate that when the hydraulic and constituent loads (i.e., represented by conductivity) fluctuate 56% and 13%, respectively, the required power for the RO feed pumps varies 74% during a typical diurnal period. Studying the required power reveals that between midnight and noon, the power variation is dominated by the flow fluctuation, while between afternoon and midnight the power variation is dominated by the conductivity variation.

The results show that the required power for RO pumping decreases from midnight until 7am, due to the simultaneous decrease among the RO feed flow and conductivity. After 8am, because of the increases in flow and conductivity, the required power starts to rise. During the afternoon, when the feed flow looks stable, because of the conductivity increase, we observe a steady increase in required power until it reaches the peak around 5 pm. After that, the required power decreases until midnight, when the feed water conductivity declines.

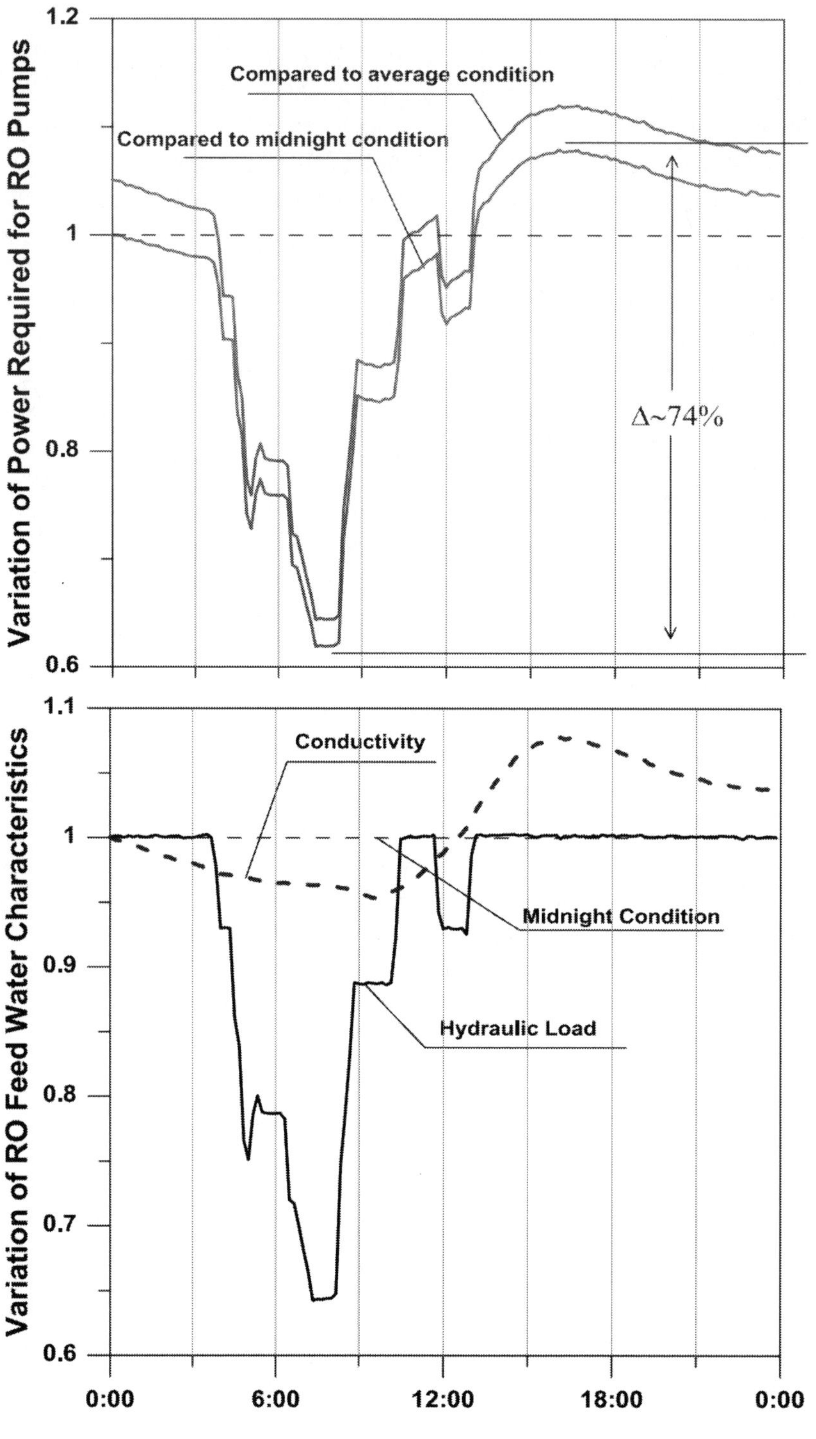

Figure 4. Diurnal variations of RO required power caused by fluctuations of RO hydraulic and constituent (i.e., conductivity) loading. Before noon power requirement variations are dominated by flow fluctuation, while in the afternoon the variations are dominated by the conductivity fluctuations.

SUMMARY AND CONCLUSSIONS

In the water reclamation and purification processes, energy footprint and its associated carbon emission have dynamic behavior in the diurnal period. The variable environmental parameters have a non-linear effect on energy footprint and its associated carbon emission.

Among the influential factors in the activated sludge process, constituent concentration and flow have the highest impact on the diurnal energy footprint, while aeration oxygen transfer efficiency has the highest reciprocal impact.

The water purification processes in Plant B inherit the water characteristic variations that are encountered by plant A during the diurnal periods. In this plant, RO process is the most energy intensive process. Between midnight and noon, the variation of required power for RO feed pumping is dominated by the flow fluctuation, whereas in the afternoon until midnight the power variation is dominated by the conductivity fluctuation. During the latter period of time, the feed flow is stable and the fluctuations observed are largely in feed water conductivity.

Elevated values of carbon intensity for unit power generation (κ) during the daytime, especially at peak hours, turn the time-of-use of power drawn to a critical element for the energy-associated carbon emission. In some energy-intensive processes (e.g. AS process), the shift of portions of the operations from peak hours to off-peak hours could significantly decrease the energy-associated carbon emission.

The results of this analysis can be used for retrofitting existing treatment plants or improving the energy and carbon emission efficiency in current reclamation processes. This paper demonstrates how dynamic energy modeling can be used as a quantitative tool to assess such time-of-use energy savings, which is the necessary element for a cost-benefit analysis when implementing alternatives for flow management (e.g., equalization basins). Furthermore, the curbing or elimination of peak power demand charges may enhance the value of modeling energy usage dynamically and make a stronger case for flow management alternatives.

REFERENCES

CAISO (2012) Renewables Watch, California ISO, http://www.caiso.com/green/renewableswatch.html (last accessed on August 2012).

Côte, P., Masini, M., Mourato, D. (2004) Comparison o f membrane options for water reuse and reclamation. Desalination, 167, 1-11.

Ekama, G.A.; Marais, G.v.R. Dynamic behavior of the activated sludge process. (1979) Water Pollution Control Federation, , 51(3), Part I, 534-556.

FERC(2012) California ISO (CAISO) Daily Report, Federal Energy Regulatory Commission: Sacramento, 2012, http://www.ferc.gov/market-oversight/mkt-electric/caiso/caiso-iso-archives.asp (last accessed on August 2012).

Foley, J., de Haas, D., Hartley, K., Lant, P. (2010) Comprehensive life cycle inventories of alternative wastewater treatment systems. Wat. Res. 44 (5), 1654-1666.

Fridleifsson, I.B.; et al. The possible role and contribution of geothermal energy to the mitigation of climate change. In: IPCC Scoping Meeting on Renewable Energy Sources. Hohmeyer, O., Trittin, T., Eds.; Proc. Luebeck, Germany, January 2008, 59-80.

Gori, R.; Jiang, L-M.; Sobhani, R.; Rosso, R. (2011) Effects of soluble and particulate substrate on the carbon and energy footprint of wastewater treatment processes. Water Res., 45 (18), 5858-5872.

Hammer and Hammer (2008) M.J. Water and Wastewater Technology. 6th ed., Pearson Education, Inc.: New Jersey.

Henze, M.; Harremoes, P.; Cour Jansen, J. la; Arvin, E. (2002) Wastewater Treatment. 3rd ed., Springer: Germany.

Leitão, R.C.; van Haandel, A.C.; Zeeman, G.; Lettinga, G. (2006) The effects of operational and environmental variations on anaerobic wastewater treatment systems: A review. Bioresource Technology, 2006, 97 1105–1118.

Lenzen, M. Life cycle energy and greenhouse gas emissions of nuclear energy: A review. Energy Conversion and Management, 2008, 49, 2178-2199.

Leu, S-Y.; Rosso, D.; Larson, L.E.; Stenstrom, M.K. (2009) Real-time aeration efficiency monitoring in the activated sludge process and methods to reduce energy consumption and operating costs. Water Environment Research, 81(12), 2471-2481

Metcalf & Eddy, Inc – AECOM (2007) Water reuse: issues, technologies, and applications, McGraw-Hill, NY.

Metcalf, & Eddy, Inc.(2003) Wastewater engineering: treatment and reuse, McGraw-Hill, NY.

Monteith, H.D.; Sahely, H.R.; MacLean, H.L.; Bagley, D.M. (2005) A rational procedure for estimation of greenhouse-gas emissions from municipal wastewater treatment plants. Water Environ. Res., , 77(4), 390-403.

Rosso, D.; Libra, J.A.; Wiehe, W.; Stenstrom, M.K. (2008) Membrane properties change in fine-pore aeration diffusers: Full-scale variations of transfer efficiency and headloss. Water Research, 42(10–11), 2640–2648

SCE (2011) Southern California Edison, General Service - Industrial Rate Schedules, Schedule TOU-8

Sobhani, 2011. Analysis of Diurnal Variations in Energy Footprint and Its Associated Carbon Emission for Water Supply and Reuse in Arid and Semi-Arid Areas, PhD Dissertation.

Energy Consumption Trends in BNR Mixing Technologies

Lamya Badr, Hazen and Sawyer, P.C.
Joe Rohrbacher, PE, Hazen and Sawyer, P.C.
Patricia Drummey Stiegel, PE, Hazen and Sawyer, P.C.
Don Howard, City of Greensboro
Lori Cooper, City of Greensboro

ABSTRACT

Secondary treatment is typically the largest energy consumer at municipal wastewater treatment plants (WWTPs), with aeration representing most of this demand. The energy used in the activated sludge process generally increases with enhanced levels of treatment. As WWTPs face more stringent nitrogen and phosphorus effluent limits, energy requirements for pumping and mixing will continue to increase relative to aeration. Although much of the energy reduction efforts in activated sludge treatment have been focused on reducing aeration energy and demand, comparatively, limited attention has been focused on reducing the energy associated with recycle pumping and mixing.

Biological nutrient removal (BNR) processes generally include multiple unaerated anoxic and/or anaerobic zones to remove nitrogen and phosphorus, respectively. Greater unaerated tankage requirements are associated with increased levels of treatment; for example five-stage BNR processes require anaerobic, pre-anoxic and post-anoxic zones. These unaerated zones need to be mixed in order to maintain solids in suspension and provide efficient treatment. Unaerated tankage may account for more than 50% of the total activated sludge volume in a five-stage BNR process, and mixing requirements for these zones can exceed aeration requirements, particularly in underloaded systems. For example, **Figure 1** illustrates a case in which the energy consumed by the mixing process was significant in comparison to the aeration process. The Figure illustrates the breakdown of energy consumption in the biological treatment system at a 20 million gallon per day (mgd) 5-stage BNR plant located in North Carolina. At this facility, the jet aeration mixing system consumed 32 % of the total energy consumption.

Submersible mixers, vertical turbine mixers and jet mixing systems have provided decades of reliable mixing of unaerated zones at municipal WWTPs. These technologies have typically been applied at a range of mixing energy densities. Typical energy inputs have ranged from 0.25 horsepower per 1,000 cubic feet (cf) to over 1 hp per 1,000 cf. Promising new technologies including hyperboloid mixers and large bubble mixing have resulted a reduction in mixing energy requirements to energy densities of 0.1 horsepower per 1,000 cf and below. Advances in the understanding and design of traditional mixing systems have also allowed for a reduction in energy input using conventional mixing. The results of this study compare energy inputs for different mixing technologies based on data from various pilot studies and papers. Furthermore, the results highlight the variability of energy consumption within one mixing technology and between the different types of mixing technologies.

This paper will provide a historical overview of mixing technologies used in activated sludge treatment with an emphasis on the development and application of energy requirements. Energy efficient mixing design will be discussed. The evaluation of mixing technologies for the City of Greensboro's (North Carolina) T.Z. Osborne Water Reclamation Facility (TZO WRF) Expansion and BNR Upgrade Project will be presented as a case study to illustrate the technical, energy and cost considerations for selecting mixing technologies for anaerobic and anoxic zones. Energy consumption information and present worth costs that were developed for four mixing technologies (jet mix, vertical mixers, hyperboloid mixers and large bubble mixing) at the TZO WRF are presented.

KEYWORDS

Mixing, energy consumption, cost analysis, EnviroMix, Invent, large bubble, Pulsair, vertical turbine.

INTRODUCTION

Adequate mixing is required to keep mixed liquor in suspension and, subsequently, promote denitrification processes in anoxic zones, and phosphate release in anaerobic zones. It is important to provide adequate mixing while minimizing oxygen transfer in both the anoxic and anaerobic environments. Thus, low-oxygen transfer mixing equipment is a critical element of the biological nutrient removal (BNR) process. As more mixing technologies enter the market, the decision of what mixing technology to implement is increasingly difficult, as it should consider cost, maintenance, reliability, and energy consumption.

Biological nutrient removal (BNR) processes generally include multiple unaerated anoxic and/or anaerobic zones to remove nitrogen and phosphorus, respectively. Greater unaerated tankage requirements are associated with increased levels of treatment; for example five-stage BNR processes require anaerobic, pre-anoxic and post-anoxic zones. These unaerated zones need to be mixed in order to maintain solids in suspension and provide efficient treatment. Unaerated tankage may account for more than 50% of the total activated sludge volume in a five-stage BNR process, and mixing requirements for these zones can exceed aeration requirements, particularly in underloaded systems. For example, **Figure 1** illustrates a case in which the energy consumed by the mixing process was significant in comparison to the aeration process. The Figure illustrates the breakdown of energy consumption in the biological treatment system at a 20 mgd 5-stage BNR treatment plant located in North Carolina. At this facility, the jet aeration mixing system consumed 32 % of the total energy consumption.

The purpose of this paper is to provide an overview of mixing technologies used in activated sludge treatment with an emphasis on energy requirements. Furthermore, the mixing evaluation that was developed for the City of Greensboro will be presented as a case study to illustrate the technical, energy and cost considerations for selecting mixing technologies for anaerobic and anoxic zones.

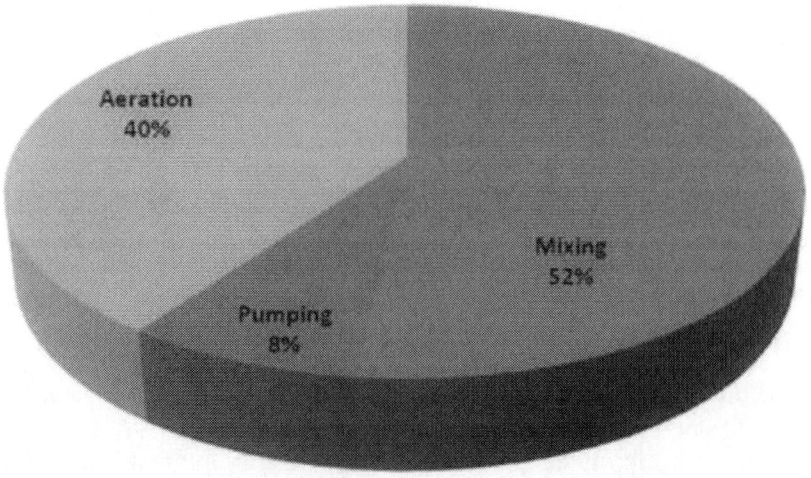

Figure 1 – Example of Mixing Energy Consumption in Biological Treatment Process

MIXING BACKGROUND

Mechanical Mixing

Mechanical mixers may either be platform-mounted (vertical turbine) mixers or submersible mixers. Vertical turbine mixers consist of a drive unit, vertical shaft, and hydrofoil impeller. This type of equipment has a long, proven history of being able to meet process mixing requirements when specified correctly.

The following characteristics are typical of vertical turbine mixers:

- Hydrofoil impellers provide an axial flow pattern (from top to bottom of tank floor).
- Hydrofoil impeller consists of 3 or 4 blades.
- Radial impellers are used for high shear applications (oxygen dispersion).
- Installation is such that the impeller is located approximately one impeller diameter above the tank floor.
- Wetted parts are typically constructed of 316 stainless steel for wastewater applications.

Mixers can be provided with parallel shaft or right-angle drive units. Right-angle drives tend to be more expensive and slightly less efficient than parallel shaft drives; however, right-angle drives are more robust and have proven to be more reliable than parallel shaft drives. **Figure 2** illustrates a schematic of a vertical turbine mixer with a hydrofoil impeller (Hazen and Sawyer 2009).

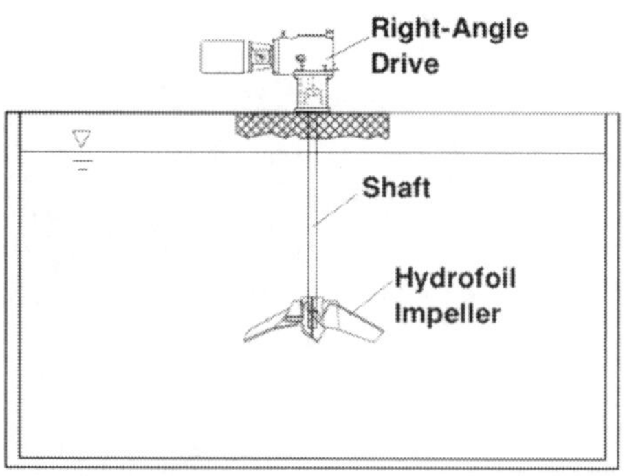

Figure 2 – Illustrative Section of a Vertical Turbine Mixer (Hazen and Sawyer 2009)

While vertical turbine mixers have traditionally been designed to provide a mixing intensity of approximately 0.3 hp per 1,000 cubic feet of mixed liquor volume (Oton et al 2009), recent studies indicate sufficient mixing can be provided at a mixing intensity of approximately 0.10 to 0.15 hp per 1,000 cubic feet of mixed volume using the hydrofoil (axial) impeller. A mixing study developed by Philadelphia Mixing Solutions at the Derry Township WWTP (Pennsylvania) in 2009 illustrated a 90 percent energy savings in using hydrofoil impellers in comparison to specifying an energy consumption of 0.3 hp per 1,000 cubic feet. Additionally, the hydrofoil impellers were found to consume less than 50 percent of the energy consumed by radial impellers, which are typically used for air/gas dispersion. Furthermore, vertical turbine mixing impellers may be designed to minimize rag build-up which leads to decreased operational efficiency and damage to the mixer shaft and blades. Impellers with smooth surfaces that are angled instead of straight-edged help prevent rag collection in the mixing basin (Philadelphia Mixing Solutions).

Submersible mixing is a form of mechanical mixing in which the mixer motor is rail-mounted, or mounted on the basin floor, and submersed within the treatment tank. Submersible mixers have been implemented in wastewater treatment for several decades and are considered to provide effective mixing in biological treatment. While submersible mixers have traditionally been the small-blade/high-speed type with relatively high energy requirements (up to greater than 1 hp per 1,000 cubic feet), the large-blade propeller standard is a lower energy option. The large-blade design was popularized in Europe in the mid-1990s and operates at low speeds (up to 100 rpm), requiring a fraction of the horsepower of the small-blade design. **Figure 3** illustrates a low-speed submersible mixer. Submersible mixer manufacturers have claimed that large-blade mixers are as energy efficient as emerging technologies such as hyperbolic and large bubble mixing (Cliburn 2008).

Figure 3 – Low-Speed Submersible Mixer (Flygt 2012)

Hyperboloid Mixing

Hyperboloid mixers, such as Invent's Hyperclassic mixer, are specifically designed with a hyperboloid shaped mixer body that results in a manufacturer-claimed lower required energy input compared to vertical turbine mixers. Invent is the only well-established manufacturer of hyperboloid mixers, which have been implemented in Europe for more than a decade and are now gaining recognition in the United States with over 50 municipal installations. According to Invent, hyperboloid mixers require an energy input of around 0.1 hp per 1,000 cubic feet of mixed liquor volume and have a higher allowable length to width (L/W) ratio (3:1 at side water depths greater than 10 feet) than conventional mixers (Hazen and Sawyer 2009).

The following characteristics are typical of the hyperbolic Invent Hyperclassic Mixers:

- Installation is such that mixer body is approximately 1/10 of the mixer diameter above the tank floor.
- Unlike vertical turbine mixers, air diffusers cannot be installed directly underneath the hyperboloid mixers body due to the hyperboloid mixer design.
- Non-clogging hyperboloid mixer body.
- Radial flow pattern is generated by the mixer, parallel to the floor and towards the basin walls.
- Mixer body is constructed of fiberglass-reinforced plastic (FRP); shaft may be constructed of FRP or stainless steel.
- Shaft diameter is 4.5" for 1.5 hp mixers and 6.6" for 2, 3, and 5 hp mixers.
- All mixer drives are parallel shaft helical geared.

Hyperboloid mixers can be implemented as standard mixers, as described above, or as mixer aerators. The mixer aerators are installed with ring spargers placed below the mixer body on the basin floor. The air sparger diffuses air under the mixer which distributes across the tank as the mixer rotates. Mixers equipped with ring spargers require motors that are about 1.2 to 3.8 times larger than typical hyperboloid mixers since additional energy is required to effectively aerate. These mixers can be installed with variable frequency drives (VFDs) to reduce power consumption when operating in a mixing-only mode. **Figure 4** presents a schematic of a hyperboloid mixer with an air sparger on the basin floor.

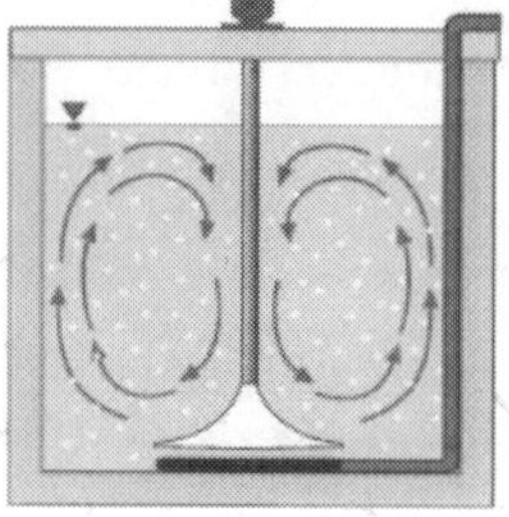

Figure 4 - Section of a Hyperboloid Mixer with Air Sparger (Invent, *Hyperclassic Mixing and Aeration System*)

There have been several full-scale pilot studies that have compared hyperboloid mixing with mechanical mixers, two of which took place at the Bowery Bay Water Pollution Control Plant (WPCP) in New York and the DC Water Blue Plains Facility. In 2004, two Invent hyperboloid mixers were installed as part of a pilot study in two anoxic zones at the 150 mgd-Bowery Bay WPCP. The hyperboloid mixers were compared with submersible mixers that were installed in 1996 as part of a nitrogen removal upgrade project. The results of the pilot study indicated that both types of mixers achieve similar uniform distribution with minimal oxygen transfer. The Invent mixers demonstrated a specific energy consumption of 0.1 hp per 1,000 cubic feet. By 2008, all of the submersible mixers were replaced with hyperboloid mixers for a total of 92 units at the Plant (Oton et al 2009).

A more recent full-scale pilot study took place at the 370 mgd DC Water Blue Plains facility. In 2008, six hyperboloid mixers were installed in anoxic/swing zones and compared with conventional vertical turbine/hydrofoil mixers. Field testing confirmed that both hydrofoil and hyperboloid installations achieved homogenous mixing conditions with the hyperboloid mixers achieving a slightly better mixed liquor distribution. The conventional hydrofoil mixers operated with an actual energy consumption of 0.29 hp per 1,000 cubic feet while the hyperboloid mixers operated at 0.15 hp per 1,000 cubic feet. DC Water has continued to add Invent mixers in the BNR basins at the Blue Plains Facility (Oton et al 2009).

Jet Mixing

A jet mixing system consists of a pipe header (manifold) with a series of jet nozzles that are used to mix the tank liquid utilizing a recirculated flow at a high velocity. Self-priming, submersible or dry pit centrifugal pumps are required for liquid recirculation. Furthermore, an air pipe can be located along the header in order to add oxygen to an individual cell if required. Jet mixing headers can be provided with nozzles projecting in one direction, opposing directions, or alternating directions. **Figure 5** illustrates a jet mix system installation.

One significant advantage to jet mixing is that in-tank components are minimized and moving or mechanical parts can be placed such that accessing the BNR tank for maintenance is not required. Overall power requirements for jet mixing are typically greater than other mixing technologies due to friction losses in the jet aeration piping and pump losses.

Figure 5 - Jet Mix and Aeration (K_La Systems)

Large Bubble Mixing

The large bubble mixing system consists of stainless steel piping, nozzles or accumulator plates, air compressors, dryers, air injection valves, PLC based controllers, and associated equipment. Compressors supply air through small diameter piping to nozzles that are mounted on the floor of the BNR tank. Compressed air is released through the nozzles in a short burst, creating a large bubble that mixes the contents of the tank as the bubble rises. The large bubble mixing system achieves mixing throughout a specified volume by allowing nozzles to fire short bursts of air for a specified duration and frequency which can result in varied mixing patterns. Manufacturers of large bubble mixing systems claim energy use to be 0.09 to 0.15 hp per 1,000 cubic feet of mixed liquor volume. **Figure 6** illustrates a large bubble mixing system.

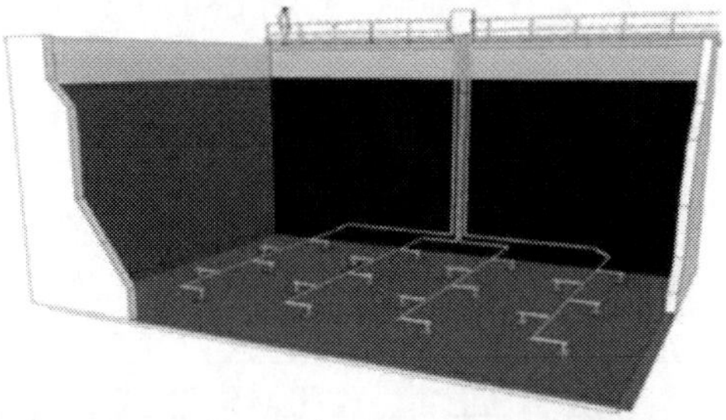

Figure 6 - Schematic of a Large Bubble Mixing System (EnviroMix)

In 2009, the BioMix (by EnviroMix) system was installed in the F. Wayne Hill Water Resources Center (WRC) in Buford, Georgia and compared to submersible mechanical mixers as part of an independent study. The mixing system was installed in two anaerobic zones of one of the plant's BNR treatment trains. The results of this study established that BioMix is as effective in mixing as submersible mechanical mixers, is compatible with anaerobic and anoxic environments, and presents a 60% energy savings based on horsepower per volume of liquid mixed (Randall 2010).

Similarly to the F. Wayne Hill WRC, in 2011, the Renewable Water Resources (ReWa) Mauldin Road WWTP in Greenville, SC had the BioMix system installed as part of a pilot testing project. The purpose of this pilot test was to implement large bubble mixing in the anoxic zones of the plant's biological treatment basins and to compare nutrient removal, mixing performance, and energy consumption with that of submersible mixers. BioMix equipment was installed in one anoxic basin alongside mechanical mixers. The results of the pilot test indicate that the BioMix mixing system can provide adequate mixing with negligible oxygen transfer, resulting in effluent phosphorus concentrations averaging 0.34 mg/l. The results also illustrated 75% energy savings in comparison to the propeller submersible mixer (McManus).

Table 1 summarizes large bubble mixing pilots and projects in the process of obtaining large bubble mixing technologies. It should be noted that **Table 1** only includes BNR applications, as there are several large bubble mixing installations for other applications such as sludge and chemical tank mixing.

Table 1 – Current BNR Mixing Applications of Large Bubble Mixing

Plant	Mixing Manufacturer / Status	Activated Sludge Process	BNR Mixing Zone	Total Mixing Volume (mg)	Overall Plant Capacity (mgd)
Mauldin Road WWTP, Greenville, South Carolina	BioMix (EnviroMix)	A^2/O	One (1) anoxic zone.	0.13	70
F. Wayne Hill WRC, Gwinnett County, Georgia *	BioMix (EnviroMix)	5 – Stage Bardenpho	Two (2) anaerobic zones in series.	0.81	60
South Granville WASA WWTP, North Carolina	BioMix (EnviroMix) Expected to be operational by February 2014.	5 – Stage Bardenpho	One (1) anaerobic zone, three (3) pre-anoxic zones (series), two (2) swing zones (parallel), two (2) post anoxic zones (parallel).	2.70	5.5
Mount Pleasant Center Street WWTP, South Carolina	BioMix (EnviroMix)	MLE	Two (2) anoxic zones.	0.38	3.7
Village Creek WWTP Plant 001, Jefferson County, Alabama	Design is complete.	Activated sludge with upstream anoxic selector.	One (1) anoxic zone per basin, six (6) basins total	2.0	40 peak
Village Creek WWTP Plant 002, Jefferson County, Alabama	Design is complete.	Activated sludge with upstream anoxic selector.	One (1) anoxic zone per basin, four (4) basins total.	2.6	60 peak

Plant	Mixing Manufacturer / Status	Activated Sludge Process	BNR Mixing Zone	Total Mixing Volume (mg)	Overall Plant Capacity (mgd)
Valley Creek WWTP, Jefferson County, Alabama	Preliminary design is complete.	Activated sludge with upstream anoxic selector.	One (1) anoxic zone per basin, ten (10) basins total.	0.7	44 average
Marysville WRF, Ohio	Pulsed Hydraulics, Inc.	Modified A^2/O	Anaerobic and anoxic zones of four (4) BNR basins.	1.2	8
Village of Cridersville WWTP, Ohio	Pulsed Hydraulics, Inc.	Activated sludge.	Anoxic/aerobic swing zones.	Unknown	0.8
Redhook WWTP, New York	Pulsed Hydraulics, Inc.	Unknown	Unknown	Unknown	Unknown

*Working on extension across all 10 trains in both anaerobic and anoxic zones.

Energy Consumption

Specific energy consumption from various pilot studies and papers were consolidated for different mixing technologies. **Figure 7** illustrates the collected specific power consumption (in hp per 1,000 cubic feet) from the various studies. The values illustrated in **Figure 7** were obtained from the following studies and are referenced, by number, in the figure:

1) "Mixing Things Up." Treatment Plant Operator. (Wray 2012).
2) "Invent Mixer Evaluation Technical Memorandum." Boyle. (Cliburn 2008).
3) "The Fine Line Between Thorough Mixing and Energy Consumption." (Oton et al 2009).
4) "Evaluation of Energy Conservation Measures." EPA. (USEPA 2010).
5) "A Comparative Analysis of a BioMix System and a Submersible Propeller Mixer". Clifford W. Randall and William O. Randall. (Randall 2010).
6) TZO WRF Mixing Alternatives Study. Hazen and Sawyer. July 2012.
7) Metcalf & Eddy, "Wastewater Engineering Treatment and Reuse." (Metcalf 2003).
8) Philadelphia Mixing Solutions BNR Installations at PMS's recommended HP (average of 11) (Philadelphia, b).

It is important to note that manufacturer-claimed energy consumption can differ from what is applied in the field. Furthermore, energy consumption may vary from plant to plant based on site-specific conditions. The error bars illustrated in **Figure 7** represent a reported range of values instead of a single value.

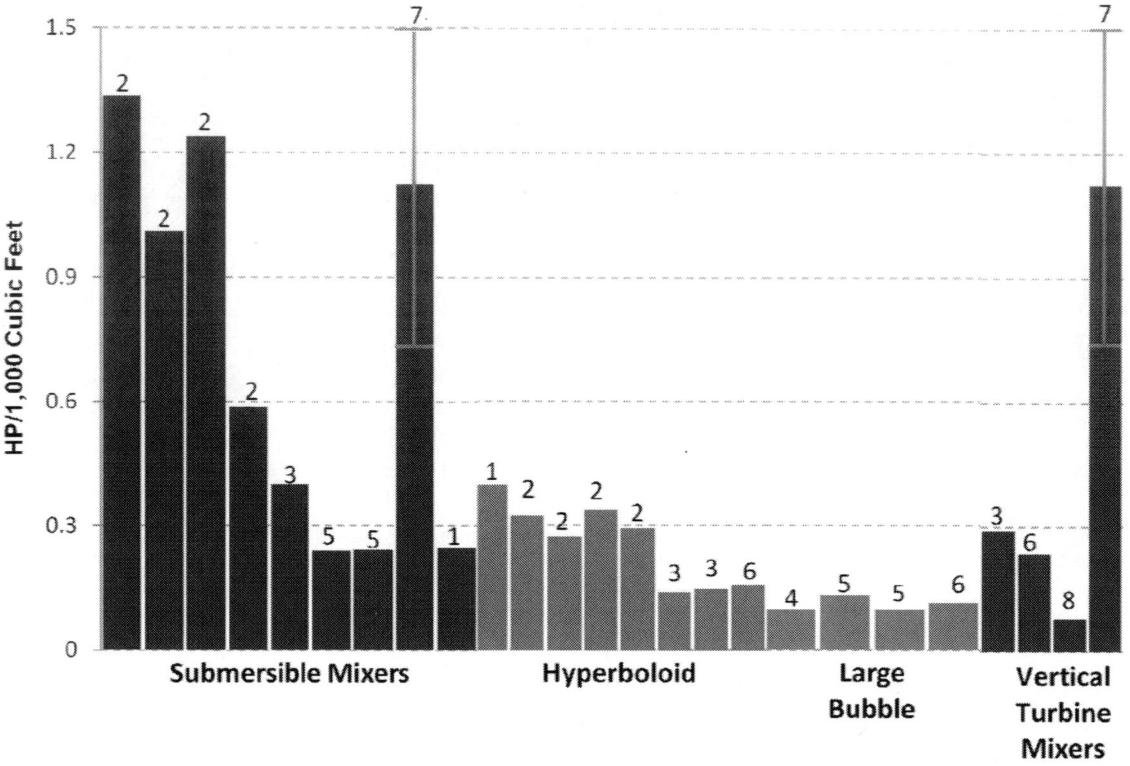

Figure 7 - Specific Energy Consumption of Various Mixing Technologies (Numbered References Listed Above)

CITY OF GREENSBORO CASE STUDY

As part of the Water Reclamation Facility (WRF) Expansion and Biological Nutrient Removal (BNR) Upgrade Project for the City of Greensboro, the existing two-stage activated sludge process will be converted to a five-stage biological nutrient removal (BNR) process. The existing aeration tanks at the T.Z. Osborne WRF will require mixing equipment in the post-anoxic/swing zones. Four (4) new aeration tanks will also require mixing in the pre- and post-anoxic zones. Various low-oxygen transfer mixing technologies were compared based on capital cost and operational efficiency to determine the best mixing technology to be incorporated into the BNR basins at the TZO WRF.

Currently, the TZO WRF contains twelve (12) two-stage BNR basins which are configured with 7 to 10 treatment cells depending on the BNR basin. The first three cells of each of the BNR basin can be operated in aerobic or anaerobic modes. These cells are equipped with pumped jet mixing and aeration systems which can provide mixing with and without air. When the first three cells are not aerated, they provide anaerobic volume for biological phosphorus release prior to the aerobic cells that provide for the oxidation of the organic matter and ammonia.

As part of the expansion and BNR upgrade, one cell of each basin will be subdivided to create swing cells that will typically be operated as anoxic zones. Four additional BNR basins will also be constructed. The swing zones will be capable of operating aerobically if necessary. The last cell of each BNR basin will be equipped as a reaeration zone by utilizing fine bubble diffusers only and will not require additional mixing. Nitrified recycle (NRCY) pumping will also be installed to return nitrate from the end of the aerobic zone to the first or second unaerated cell. The first three cells of the BNR basins will typically operate as pre-anoxic zones for denitrification. The post-anoxic cells in the existing BNR basins, and both the pre- and post-anoxic cells in the proposed BNR basins will be equipped with mixing that keeps mixed liquor suspended solids (MLSS) in suspension and maintains a homogeneous mixed liquor.

The following low-oxygen transfer mixing technologies were evaluated as part of the mixing alternatives evaluation: jet mixing, vertical turbine mixing, hyperboloid mixing, and large bubble mixing. Costs for each of these mixing technologies were developed and compared to one another with respect to: equipment costs, operation & maintenance requirements, and structural requirements (i.e. concrete for support and access).

RESULTS & DISCUSSION

Costs

A number of assumptions were made to calculate the capital, operations and maintenance (O&M), and 25-year total present worth cost for each of the mixing alternatives evaluated for the TZO WRF. The general assumptions applied to all of the mixing alternatives are as follows:

1) All capital costs include 15% for contractor overhead and profit.
2) All capital costs include 25% of equipment costs for labor to install the equipment.
3) Use 3% for rate of inflation.
4) Use 5% for the interest rate in calculating the total present worth.
5) Use $0.06/kW hour for power costs.

Figure 8 illustrates the equipment and capital costs of each mixing technology evaluated for the TZO WRF. While the equipment costs for each alternative are similar, the additional costs associated with jet mix increases the overall capital cost of the jet mix option. The non-equipment capital cost includes the cost of concrete for pump support and access, as it is assumed that concrete slabs will cover the entire cells with the jet mix option, similar to the existing jet mix installation at the WRF.

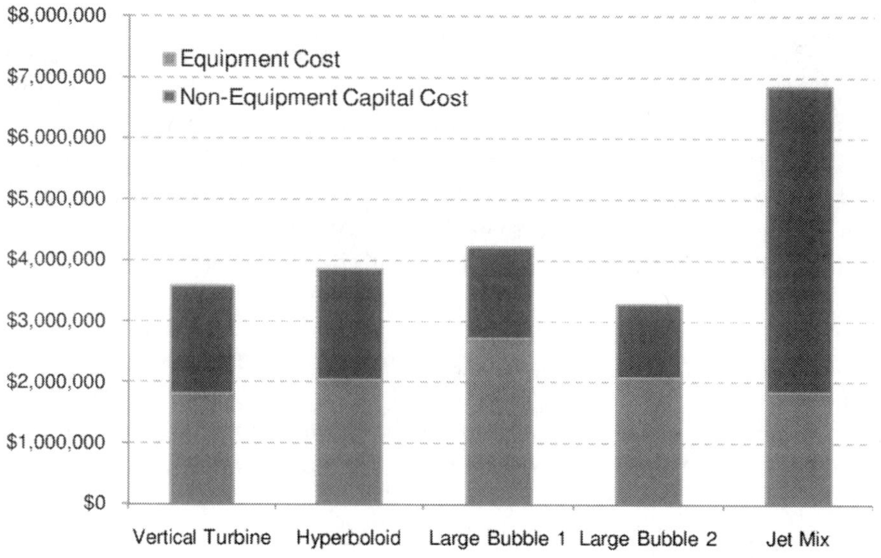

Figure 8: Equipment & Capital Cost of Mixing Alternatives

Figure 9 illustrates the net present worth of each mixing alternative. Maintenance costs were determined based on a percentage of equipment costs for each alternative based on previous experience. The percentages used for annual maintenance fees are 1, 1, 3, and 4 percent for vertical turbine, hyperboloid mixers, jet mix, and large bubble mixing. The percentages are higher for jet mix and large bubble mixing as they incorporate the maintenance associated with mixing pumps (for jet mix) and compressors, dryers, filters, and other equipment (for large bubble mixing). In general, the O&M cost associated with maintenance is minor compared to capital and power costs. **Figure 10** illustrates the net present worth without maintenance costs considered.

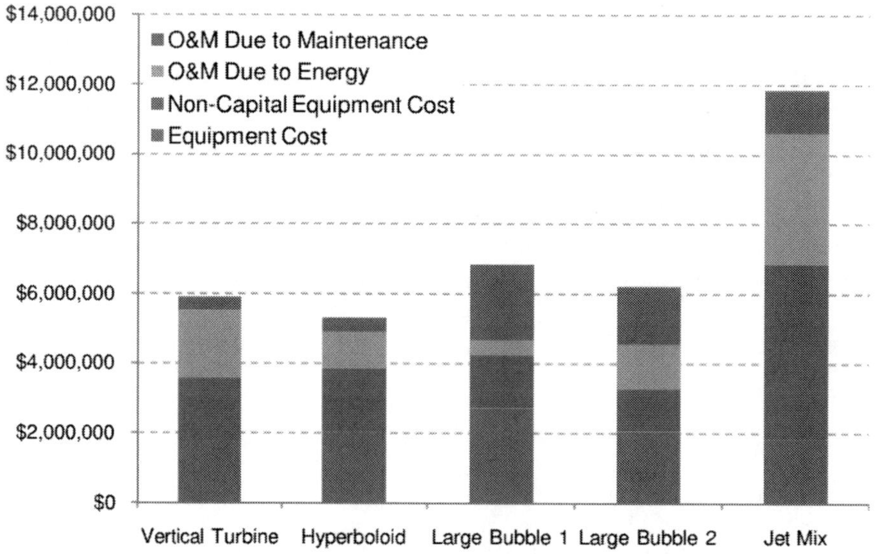

Figure 9: Net Present Worth of Mixing Alternatives

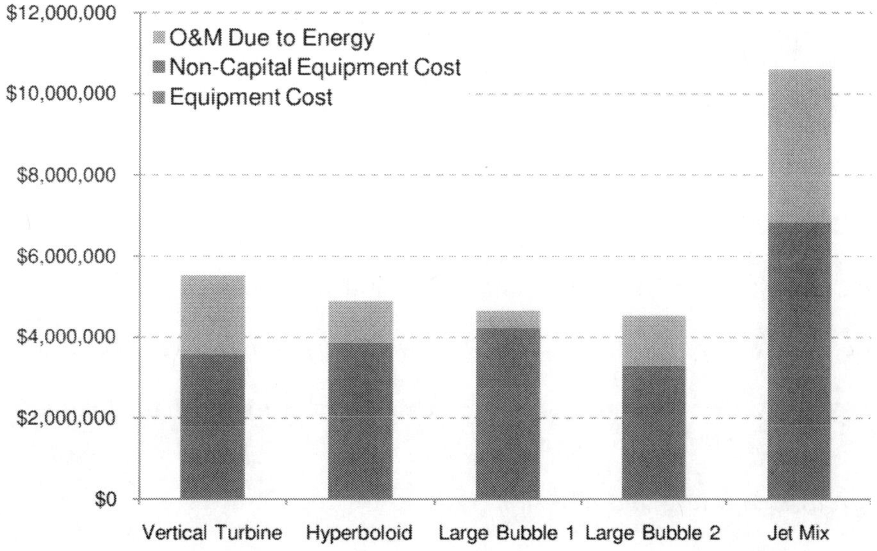

Figure 10: Net Present Worth of Mixing Alternatives Without Maintenance Costs

CONCLUSION

The results of the mixing evaluation performed for the TZO WRF BNR Upgrade and Expansion project indicate that conventional vertical turbine, hyperboloid, and large bubble mixing net present worth costs are competitive with one another. Energy costs for jet mix and conventional vertical turbine technologies are expected to be higher than for hyperboloid and large bubble mixing applications.

Several pilot tests and papers have been developed to compare energy use in hyperboloid and large bubble mixing with conventional mixing (submersible and vertical turbine mixing). The results of these studies indicate that energy consumption may vary from site to site and that manufacturer-claimed energy consumption may differ from actual energy consumption. Furthermore, conventional vertical turbine mixers have evolved to be more energy efficient and can be within the same energy consumption range as hyperboloid mixers. Although large bubble mixing is claimed to consume the least amount of energy when compared to various mixing technologies, there are a limited number of large (>1 MG of mixing volume) full-scale BNR applications with large bubble mixing currently in operation. However, there are several full-scale projects with large bubble mixing currently in the design or construction phase.

ACKNOWLEDGEMENTS

The authors would like to thank the following for their assistance in developing this paper:

Mark Messere, Hazen and Sawyer

Greg Everhart, Combs & Associates, Inc.

Gary Broome, Heyward Incorporated

Roger Wright, Atlantic Process Sales, LLC

Stuart Humphries, EnviroMix

Skip Brunhaver, Pulsair

Tony Combs, Combs & Associates, Inc.

Jason Beck, CDM Smith

REFERENCES

Cliburn, Mike; 2008; *Invent Mixer Evaluation*; Technical Memorandum; Boyle.

Flygt low-speed mixers. 2012. Flygt. <http://www.flygtus.com/3424270.pdf>.

Hazen and Sawyer; 2009; *Technical Memorandum No.8F-FINAL Mixer Evaluation for BNR Basins 1-6*. City of Tallahassee.

Invent; *Hyperclassic Mixing and Aeration System.*

Randall, C.W. and W.O. Randall; 2010; *A Comparative Analysis of Bio^{M}x System and a Submersible Propeller Mixer: Mixing in Aerobic Zones at the F. Wayne Hill Water Resources Center, Buford, Georgia.*

McManus, Glen and Randy Boyette; *A Comparative Biological Treatment Analysis of a BioMix System and a Submersible Propeller Mixer Operating in an Anoxic Zone*; Renewable Water Resources.

Metcalf and Eddy; 2003; Wastewater Engineering Treatment and Reuse; Fourth Edition; McGraw Hill.

Oton, Sonia M.; Gidugu, Srinivas D.; Carroll, Janice R.; Saurer, Paul D.; Fillos, John; Ramalingam, Krish; Kharkar, Salil M.; Minassian, Rouben Der; Osit, Matthew S.; Solimando, Daniel A.; Deur, Allen; 2009; *The Fine Line Between Thorough Mixing and Energy Consumption;* Hazen and Sawyer, P.C. DC Water and Sewer Authority; NYC Department of Environmental Protection.

Philadelphia Mixing Solutions (a); *Anoxic Mixing White Paper.*

Philadelphia Mixing Solutions, Ltd (b); Partial Installation List.

Pulsair System, INC; 2011; *Municipal Wastewater Applications.*

USEPA; 2010; *Evaluation of Energy Conservation Measures for Wastewater Treatment Facilities*; 832-R-10-005.

Wray, Jared; 2012; *Mixing Things Up: Large Blade, Slow Speed Wastewater Mixing Technologies Gain Recognition for Providing Solid Performance with Low-Energy Usage*; Treatment Plant Operator.

Making Energy Efficiency Happen – Experience and Lessons Learned in the EPA Region 4–TN Department of Environment & Conservation Energy Management Initiative

Robert J. Freeman, P.E., US Environmental Protection Agency, Region 4
Jennifer Dodd, Tennessee Department of Environment and Conservation

ABSTRACT

In early 2011 the EPA Region 4 Office in Atlanta (EPA R4) and the Tennessee Department of Environment & Conservation (TDEC) established a partnership to encourage and assist water/wastewater utilities in adopting energy efficiency and conservation measures. Water/wastewater energy use is typically the largest energy cost of most municipal governments, and energy is the largest controllable cost of operation for water/wastewater systems. In spite of these facts, most public water and wastewater systems have not seriously considered energy efficiency and conservation. Region 4 and TDEC developed a strategy for an energy management initiative (EMI) that would focus on a relatively small group of water/wastewater utilities and work with them to encourage the implementation of energy efficiency/conservation measures. That effort lasted approximately 18 months and consisted of assembling a team of volunteer partners with specific skill-sets and experience to conduct the EMI, identifying the specific utilities to be invited to participate, development of a structured process of energy assessment site visits and a series of day-long workshops supporting activities to work with the utilities. The results of the EMI was energy savings at the seven participating utilities of over 5,000,000 kWh per year, cost savings exceeding $400,000 per year, and overall percentage energy savings of 16.9%. The participating utilities ranging in size from 31 L/s (0.7 mgd) actual flow to 745 L/s (17 mgd) mgd actual flow and all the utilities identified significant energy efficiency and conservation opportunities. The overall energy savings achieved was 16.9%, from both operational modifications and equipment improvements. Lessons Learned are presented that will help improve the conduct of an EMI in the future and help to identify key issues that should be addressed.

KEY WORDS

Energy efficiency, energy conservation measures, energy management, public water/wastewater systems, integrating energy management into water/wastewater management.

INTRODUCTION

Where are we and how did we get here?

Energy efficiency is a concept whose history dates back as far as written history itself. Some of the earliest examples of energy efficiency were the improvements in wood burning devices such as wood stoves and fire boxes with draft controls to gain more benefit from a supply of firewood than possible with an uncontrolled fire. Designs of sailing ships were improved to capture more of the energy of wind and increase the size and speed of those vessels. The process of getting more work or benefit from the existing supply of resources led to steam powered machinery, internal combustion engines, and the explosion of mechanical and electrical inventions during the industrialization period of the 1800's and 1900's. The seemingly inexhaustible supply of cheap energy fueled much of that growth. The modern concepts of energy efficiency and energy conservation can be traced back to two dramatic developments of the late 1960's and the early 1970's. Awareness was growing about the serious environmental consequences associated with both the extraction of resources to produce energy and the lack of concern about the byproducts of the use of energy – the dramatic increase in pollution. Liquid wastes that caused rivers to burn and made Lake Erie into the "dead sea", and smokestack emissions so great that a Federal Court shut down the steel mills in Birmingham, AL in January, 1971. The shock of the 1973 Arab Oil Embargo demonstrated the dangerous dependence of our energy appetite on the availability of limited resources used to produce energy. The 1970's saw the passage of the landmark environmental legislation, the Clean Air Act, the Clean Water Act, and others that began the process of assessing the environmental consequences of our lifestyles.

As the cost of energy and pollution increased, a new approach began to emerge that focused on conservation of resources, more efficient use of energy, and recognizing the need to protect our environment. Today we see incandescent light bulbs as archaic, hybrid automobiles are top sellers, high efficiency heating/air conditioning systems are common, water saving plumbing fixtures are required, appliances wear the Energy Star label, and buildings and schools designed to be energy efficient earn the Leadership in Energy and Environmental Design (LEED) certification. Energy audits of buildings, factories, and residential homes are encouraged and offered by numerous private companies who have entered the efficiency/conservation business as well as the electric utilities, often at no charge.

In the midst of this growing recognition of the significant energy and cost savings opportunities that result from efficiency and conservation, the public water/wastewater sector has only begun to see widespread interest. The opportunities for energy/cost savings at public water/wastewater systems are significant – the largest energy users in most cities are the water/wastewater systems, typically 30-40 percent of their total energy budget, and energy is the largest controllable cost of operations in most systems. The obvious question is why haven't more water/wastewater systems taken advantage of this reality?

The Disconnect

As more ideas and information came available in recent years regarding the proven technology for energy efficiency and conservation in water/wastewater systems, the continued lack of widespread embrace of those approaches in the public water/wastewater sector was confounding. A major contributor to this conundrum has been identified as the "disconnect" between the consumers of the energy for water/wastewater systems – the operations management/staff – and the people who pay the bill for the energy used at those systems – the city financial people who are located "downtown". Several studies have shown that the number of water/wastewater plant superintendants and operators who see their energy bills is close to zero. Consequently, it is hardly surprising that water/wastewater facilities do not see energy management as an issue, much less a priority. Likewise, most city governance officials spend little time, if any, focused on water/wastewater treatment (other than permit compliance), so they are usually unaware of the opportunities for energy efficiency/conservation. The two key success measures in the typical water/wastewater system management approach are consistent permit compliance and minimal or no customer complaints related to operations or service. In order for energy management to become an integral part of the normal management approach, both groups, operations and governance, need a better understanding of the relationships between energy use, cost and management of energy, and the operating requirements of their systems. The desired result will be a common commitment focusing on both optimizing the present performance of their water/wastewater infrastructure and assuring its sustainability in the future.

EPA Region 4 – Tennessee Department of Environment and Conservation First Steps

In January, 2008, EPA published "Ensuring a Sustainable Future: An Energy Management Guidebook for Wastewater and Water Utilities" (EPA Guidebook), a manual focused on the management process of energy efficiency and conservation instead of the technology. Numerous technology focused manuals, design guides, textbooks, and other references had been published on energy management but only limited implementation of those concepts were evident in the public water/wastewater sector. Addressing "The Disconnect" was increasingly recognized as a significant barrier to more widespread embrace of efficiency/conservation approaches. The EPA Guidebook and its promotion by the EPA Office of Wastewater Management (EPA OWM) around the country sparked a renewed energy management focus on the public water/wastewater sector and encouraged those efforts already underway.

Workshops featuring the EPA Guidebook were held in all ten EPA Regions around the country, with a workshop in Nashville, TN on October 14, 2008, jointly sponsored by EPA OWM, the EPA Region 4 Office (EPA R4), and the Tennessee Department of Environment and Conservation (TDEC). That one day workshop was attended by approximately 20 water/wastewater utilities from Tennessee and laid the foundation for the EPA R4 – TDEC energy management partnership.

TENNESSEE WATER & WASTEWATER ENERGY EFFICIENCY PARTNERSHIP

Strategy and Plan

Shifting budget priorities within EPA and national economic problems delayed EPA R4's ability to build on the Nashville workshop until early 2011. In February, 2011, EPA R4 and TDEC established a partnership whose goal was to assist a select group of Tennessee water/wastewater utilities recognize the opportunities for energy and cost savings at their facilities and implement the necessary changes/improvements to achieve those savings. During the spring of 2011 EPA R4 and TDEC engaged other stakeholders to participate in the partnership, beginning with the Tennessee Valley Authority (TVA), the power supplier for most of the state who has been an advocate for and supporter of energy efficiency and conservation. Other partners included the Tennessee Department of Economic and Community Development (TDECD), the Municipal Technical Advisory Service of the University of Tennessee (MTAS), the University of Memphis – Civil Engineering Department, the Environmental Finance Center – University of North Carolina (EFC), and Schneider Electric USA. The partner makeup was key to achieving the goals and hoped for success of the effort; expertise of the partners in the areas of process operation, equipment efficiency, and cost estimating provided the ability to identify and present energy management opportunities in a credible way. Additionally, partners had to be willing and able to make a significant time commitment to the effort to see the process through. The partners adopted the name Tennessee Water & Wastewater Energy Efficiency Partnership (Partnership).

The typical model for energy efficiency/conservation efforts for public water/wastewater facilities is to conduct an energy audit, present the utility the results of that audit with recommendations for improvements, and wish them well. That model works fairly well with industrial/commercial utilities that recognize energy use as a cost and energy savings as equal to a profit – those utilities have a significant financial motivation for action. Public utilities rarely have that motivation, in part due to "The Disconnect" described earlier. EPA-R4 and TDEC agreed in the beginning that this effort would not follow the typical model but would be a structured program expected to take at least 12 months, characterized by frequent interaction with the utilities individually and as a group to develop relationships with both the governance and the operations parts of the utilities. Participation by utilities would be by invitation, contingent on their commitment to be involved for the duration of the effort.

By the end of June, 2011, the Partnership had developed a program outline based on a successful public water/wastewater utility energy management effort conducted by the EPA Region 7 – Kansas City Office. The Energy Management Initiative (EMI), as the effort was called, was to be a two phase approach. The first phase would be an energy assessment at each utility identifying potential energy efficiency/conservation measures; the second phase would be a series of four one-day Energy Management Workshops to encourage integration of energy into the management approach of the utilities and develop and encourage implementation of specific energy improvements.

In order to clarify our energy savings expectations, the Partnership set a goal of 10-15% overall energy savings for the EMI.

Phase 1 – Preliminary Energy Assessment

The Preliminary Energy Assessment plan was as follows:

- Initial site visits by the partnership team to each utility to meet the governance and operations authorities, present the plan and goals of the EMI, walk through the facility to conduct a preliminary energy assessment to identify possible operational and equipment improvements, collect operational and energy use data, and set up an Energy Star Portfolio Manager account for each wastewater facility to benchmark and track energy use.

- Preparation of a Preliminary Energy Assessment Report (Report) for each utility visited. The Report will focus on both modifications to process operations and equipment related energy efficiency/conservation opportunities.

- Return visit to each utility to present their Report and discuss its results. If the Report identified potentially significant energy saving opportunities, the utility would be encouraged to participate in the second phase of the EMI.

Phase 2 – Energy Management Workshops

The second phase of the EMI was a series of four one-day Energy Management Workshops held at the TDEC Training Center in Murfreesboro, TN, five to eight weeks apart. Those Workshops were structured to include presentations by team members and invited outside experts and facilitated sessions where the utilities worked together as groups. The goal was to produce energy management plans for each utility that included a management strategy to integrate energy into their existing management approach, identification of specific energy improvements, and an implementation schedule for those improvements. Workshops were planned as follows:

- Workshop #1 topics:
  - ✓ The importance of a management approach that incorporated an energy focus as presented in the 2008 EPA Guidebook
  - ✓ The need to see and understand the energy costs of the utility – understanding the information in an electric bill – usage charges, demand costs, peak & off peak charges
  - ✓ Examples of different opportunities for energy efficiency/conservation – both operational and equipment
  - ✓ Homework assignment – develop inventory of major energy using equipment and priority for improvement (EPA Guidebook worksheets)

- Workshop #2 topics:
  - ✓ Review Workshop #1 & present results of homework
  - ✓ Importance of energy goals and benefits of creating an energy champion or team
  - ✓ Opportunities for harvesting the energy in waste streams – biosolids, Fats-Oil-Grease
  - ✓ Overview of funding/financing alternatives for energy projects
  - ✓ Facilitated group session to discuss potential energy projects and barriers to implementation
  - ✓ Homework – update Portfolio Manager data entry, finish energy project ranking

- Workshop #3 topics:
  - ✓ Review Workshop #2 and present results of homework
  - ✓ Presentations on different funding/financing alternatives – grants, State Revolving Fund loans, TVA energy efficiency incentives, performance contracting
  - ✓ Facilitated group session to work on energy project list, identify problems or assistance needed, identify accomplishments
  - ✓ Homework – focus on energy management plan preparation

- Workshop #4 topics:
  - ✓ Utilities present energy management plans
  - ✓ Utility experiences in getting support for energy management projects
  - ✓ Breakout sessions for utilities to get specific help from partners to identify energy management projects and strategy for funding and implementation

Plan Implementation – Successes & Disappointments

The Partnership identified candidate utilities to be invited to participate, focusing on wastewater facilities, and chose a range of sizes based on design wastewater flow located across the state. The ten utilities selected ranged from 39.4 L/s (0.9 mgd) to 1095.8 L/s (25.0 mgd) design wastewater plant flow. The location of the initial utility candidates were three in east and west Tennessee and four in central Tennessee including Nashville. In July, 2011, EPA R4 sent letters of invitation to the ten selected utilities outlining the potential energy and cost saving opportunities at most water/wastewater facilities, the overall EMI concept, and the plan to go forward. Our anticipation of an enthusiastic acceptance of our invitation to benefit from a Partnership of highly experienced professionals at no cost to the utilities came to a halt when we only received one positive response from the utilities – the others did not respond at all. Our first "Lesson Learned" was harsh – when EPA's name is on the invitation the likelihood of a positive response is questionable. Fortunately our other team members with relationships with the utilities intervened personally – TDEC and the MTAS representative – resulting in six more positive responses for a total of seven participating utilities. The Partnership also received a request to be included from a utility that was not on our original invitation list. They were added bringing our total to eight utilities participating. The Preliminary Energy Assessments were conducted at the utility's facilities in October and November, 2011. Each site visit required between four and six hours to conduct the briefing, walk through their facilities, and set up a Portfolio Manager account for each utility. The Reports were prepared and the return visits to the eight utilities were in January and February, 2012. Based on the results of those site visits all but one utility was encouraged to continue to the phase 2 Workshop portion of the EMI. One of the eight utilities visited had already completed construction projects at their wastewater system that incorporated significant energy efficiency improvements. Our Report concluded that the benefit to proceed to the phase 2 Workshop part of the EMI would not be worth the time and effort required. Another utility that we recommended to participate in the phase 2 Workshops was in a reorganization situation and could not commit to participating in the Workshops. The result was seven utilities committed to the Workshop process. Those utilities and their actual wastewater plant flows are shown in Figure 1.

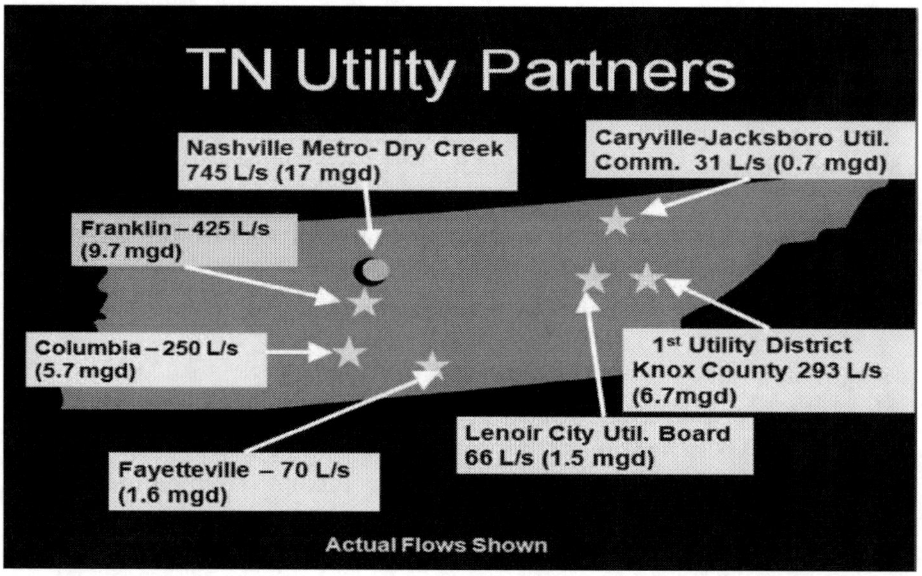

Figure 1. Location and Actual Flows of Tennessee Utilities Participating in the EMI

The importance of having wastewater treatment process expertise in our Partnership was important for all of our Preliminary Energy Assessment visits, but it was dramatically demonstrated at three of our utilities wastewater plants, Columbia, Franklin, and Fayetteville where our team members were able to determine that a significant reduction in aeration energy could be achieved with no detriment to the performance of the treatment system. Those savings were achieved in very little time with no costs to the utility and are shown as annual rates as follows:

- Columbia – 1,300,000 kWh saved – Over $100,000 cost of energy saved per year
- Fayetteville – 450,000 kWh saved – Over $35,000 cost of energy saved per year
- Franklin – 1,500,000 kWh saved – Over $130,000 cost of energy saved per year (additional savings identified in Workshop Phase of EMI)
- Total Reduction of Over 2800 metric tons CO_2 equivalent

ENERGY MANAGEMENT INITIATIVE RESULTS

The Workshop phase of the EMI began on March 15, 2012, and the last workshop was held August 20, 2012. All seven of the utilities had at least two people at each workshop and participated in all of the workshop activities and homework tasks.
Through the process of the Preliminary Energy Assessment Phase and the Workshop Phase of the EMI all of the utilities identified significant energy efficiency/conservation improvements for their facilities. The following list summarizes the energy efficiency and conservation measures identified by the utilities, including those already implemented and those planned for implementation in the near future, along with the energy and cost savings from those measures. Several of the utilities are showing cost savings from the EnerNOC program, a TVA incentive program that pays utilities who can reduce their power demands during peak demand periods. EnerNOC provides cost savings but no energy savings.

Caryville-Jacksboro Utilities Commission –actual wastewater flow 31 L/s (0.7 mgd)- results shown in Figure 2:

Energy Conservation Measures—Wastewater Plant			
Action	**Project Cost**	**Projected Annual Energy Savings**	**Projected Annual Cost Savings**
Infiltration/Inflow Reduction	TBD	79,000 kWh	$7,000
Aeration systems improvements	TBD	109,000 kWh	$8,750
	TOTAL:	**188,000 kWh** **19.4% reduction**	**$15,750**

Figure 2. Caryville-Jacksboro Utility Commission EMI Results

Lenoir City Utility Board – one of the utilities with both water and wastewater system energy savings identified -actual wastewater flow 66 L/s (1.5 mgd) – EMI results shown in Figure 3:

Energy Conservation Measures - Water & Wastewater Plant				
Location	**Action**	**Project Cost**	**Projected Annual Energy Savings**	**Projected Annual Cost Savings**
Wastewater Plant	Optimize operation of aeration and sludge return system	$0	348,000 kWh (**21.8%** wastewater energy savings)	$28,000
Water Plant	Install VFDs - Motor/Pump Systems-High Service and Source Water	$80,000	175,000 kWh (**11.7%** water energy savings)	$14,000
Water & Wastewater TOTAL:			**523,000 kWh**	**$42,000**

Figure 3. Lenoir City Utility Board EMI Results

Fayetteville Public Utilities – actual wastewater flow 70 L/s (1.6 mgd) – EMI results shown in Figure 4:

Energy Conservation Measures - Wastewater Plant				
Action	**Project Cost**	**Projected Annual Energy Savings**	**Projected Annual Cost Savings**	**Overall Energy Reduction**
Optimization of aerobic digester operation	$0	517,000 kWh	$34,000	**27%**

Figure 4. Fayetteville Public Utilities EMI Results

Columbia – actual wastewater flow 250 L/s (5.7 mgd) – EMI results shown in Figure 5:

Energy Conservation Measures-Wastewater Plant				
Action	**Project Cost**	**Projected Annual Energy Savings**	**Projected Annual Cost Savings**	**Overall Energy Reduction**
Reduced aeration and modified UV control	$0	1,300,000 kWh	$100,000	**18%**

Figure 5: Columbia EMI Results

First Utility District of Knox County – actual wastewater flow 293 L/s (6.7 mgd) – EMI results shown in Figure 6:

Energy Conservation Measures -Wastewater Plant			
Action	Project Cost	Projected Annual Energy Savings	Projected Annual Cost Savings
Optimize return sludge pumping and aeration system	$0	650,000 kWh	$50,000
Solar array installation	TBD	60,000 kWh	TBD
EnerNOC program	$0	Demand response only	$18,000
	TOTAL:	**710,000 kWh** **15% Reduction**	**$68,000**

Figure 6. First Utility District of Knox County EMI Results

Franklin – actual wastewater flow 425 L/s (9.7 mgd) – EMI Results shown in Figure 7:

Energy Conservation Measures - Wastewater Plant			
Action	**Project Cost**	**Projected Annual Energy Savings**	**Projected Annual Cost Savings**
Aeration optimization	$0	1,490,440 kWh	$120,000
Lighting replacement	TBD	11,000 kWh	$3000
Solar array installation	$0	168,000 kWh	$7500
Electric Meter Change	$0	0 kWh	$48,000
EnerNOC Program	$0	Demand response only	$16,000
	TOTAL:	**1,669,440 kWh** **13% reduction**	**$194,000**

Figure 7. Franklin EMI Results

Nashville Metro – Dry Creek Wastewater Plant – actual wastewater flow 745 L/s (17 mgd) – EMI results include the energy efficiency project that was implemented at the plant in 2009 and went online in 2010, before the EMI began. These energy savings are not counted in the total energy savings shown later in this paper. The Dry Creek EMI results are shown in Figure 8.

<table>
<tr><th colspan="4">Energy Conservation Measures-Wastewater Plant</th></tr>
<tr><th>Action</th><th>Year Implemented</th><th>Estimated Annual Energy Savings</th><th>Estimated Annual Cost Savings</th></tr>
<tr><td>Installation of fine bubble diffusers</td><td>2009</td><td>2,400,000 kWh</td><td>$145,000</td></tr>
<tr><td>EnerNOC Program</td><td>2009</td><td>Demand response only</td><td>$65,000</td></tr>
<tr><td colspan="2" rowspan="2">TOTAL:</td><td>2,400,000 kWh</td><td rowspan="2">$210,000</td></tr>
<tr><td>16.7% reduction</td></tr>
</table>

Figure 8. Nashville Metro – Dry Creek Plant EMI Results

All of the utilities met or exceeded our original goal of 10-15% energy savings, some by a significant margin; four of the utilities identified saving of almost or exceeding 20% at their wastewater plant. The total energy savings identified by all the utilities, not including the Nashville Metro-Dry Creek Plant energy savings, is shown as follows:

- Total Energy Savings – 5,000,000 kWh per year
- Overall Percentage Savings – 16.9%
- Cost of Energy Savings – over $400,000 per year
- Carbon Emission Reduction – Over 4300 metric tons CO_2 equivalent

All of the utilities are continuing to pursue additional energy efficiency and conservation measures, and EPA R4 will continue our partnership with TDEC to encourage and assist these utilities to implement more energy and cost saving measures. TDEC has also requested EPA R4 to participate in a 'Round 2' of the EMI with a new set of utilities.

The entire EMI process took 18 months, longer than expected, and substantially longer than the typical energy initiatives with specific utilities that have been conducted in the southeast in the past. The EMI results have been substantially better than most typical energy audits or energy initiatives achieve as well. <u>If energy savings half the level achieved so far with the EPA R4 – TDEC EMI could be achieved at all of the wastewater plants over 44 L/s (1 mgd) in Region 4, over 1 billion kWh could be saved per year, with cost savings approaching $100,000,000 per year.</u>

On October 11, 2012, a recognition ceremony was held at the Bicentennial Capitol Mall in Nashville to celebrate the accomplishments of all of the participating utilities. The Regional Administrator, EPA R4, and the Commissioner of TDEC were in attendance, along with several Mayors of the participating cities. The local television and news media covered the event and several news outlets throughout Tennessee covered the story. A photograph of the utility participants and dignitaries is shown in Figure 9 below:

Figure 9. EPA R4 – TDEC Energy Management Initiative Recognition Ceremony in Nashville

CONCLUSIONS AND LESSONS LEARNED

The Lessons Learned from the EPA R4 – TDEC EMI are still being recognized and collected. The 18 month long EMI effort had no predecessors or examples in this part of the country to follow or learn from. The only example of something like this was the EPA Region 7 energy initiative in Missouri. As a result, many unanticipated learning opportunities and challenges surfaced and are still being evaluated. The following list includes some of the more noticeable Lessons Learned that will help improve both the process of a public utility energy initiative and the effectiveness in terms of integrating an energy approach into traditional water/wastewater system management:

- The EPA R4 – TDEC EMI demonstrated the substantial opportunities that exist for energy savings at public water/wastewater utilities of all sizes.
- 'The Disconnect' is the major barrier to more wide spread adoption of an energy management approach at public water/wastewater utilities, and it significantly hinders reaping the benefits available from energy efficiency and conservation.
- The most important element required for an EMI success is to collaborate with state partners who have a positive relationship with the utility personnel and the governance authorities. This excludes most permitting and enforcement staff from participating.
- The EPA R4 – TDEC EMI started well in its inclusion of the governance authorities with the utility operations personnel, but we did not or could not get consistent involvement of those governance authorities through the EMI process. Utility operations personnel are critical to the implementation and operation of energy saving improvements, but many of

those improvements may require upper level support and approval. If those authorities do not understand the value and benefits of energy efficiency and conservation measures, it may be difficult or impossible to move them to implementation. It is also much easier to get utility operations personnel seriously interested in energy efficiency if a directive comes down from top management to carefully consider energy in normal decision making. This lesson about the inclusion of governance authorities is actually several lessons, and they are among the most important. This is part of 'The Disconnect' problem.

- An EMI team must make an effort to include at least one wastewater treatment process specialist. Many energy audits and energy initiatives are primarily equipment focused. Equipment efficiency can provide significant energy savings, but no equipment improvement can beat a no-cost operational improvement like shutting off an unneeded aeration blower that operates to maintain an excessively high dissolved oxygen concentration in the wastewater aeration basins. Operational improvements can be the "low hanging fruit" that provide quick energy and cost benefits, and may encourage interest in other energy efficiency opportunities. Our EMI team had two knowledgeable wastewater process people who helped identify substantial energy savings related to operational improvements.

- An EMI team should also include someone with real world experience in the cost of energy upgrades and improvements. Credible cost estimates and energy savings estimates are the key elements needed to develop payback period numbers – the common tool used to rank and prioritize projects.

- A more comprehensive focus on the importance of developing an energy management plan should be part of the first workshop that is conducted, and attendance of the governance authorities should be strongly encouraged at that workshop, or part of the workshop that will help them understand the opportunities and challenges of energy efficiency at water/wastewater systems.

- An inadequately staffed EMI team and an inadequate amount of time committed to an EMI is worse than not having an EMI at all. Failure to produce results or achieving only marginal success due to inadequate planning or time and expertise commitment poisons the water for future EMI efforts.

- Do not begin the EMI process if you do not have the time and/or ability to track and document actual energy savings. Projections and estimates must be verified with actual numbers.

- Recognize that local government utilities are usually much slower than private water/wastewater systems in their speed of decision making, particularly when significant sums of money are involved. Also, smaller governments make decisions faster than larger governments, but they have fewer resources.

REFERENCES

"Ensuring a Sustainable Future: An Energy Management Guidebook for Wastewater and Water Utilities", January, 2008, US Environmental Protection Agency, Washington, DC.

Herndon, Kerry. EPA Region 7, Kansas City, KS,Personal Communication, February 2011-June 2012

Energy Efficient, Sustainable Improvements and Technologies Implemented at Wastewater Treatment Plants

Brian M. Sibiga, PE[1] and Frank A. Nerone, PE[1,2]

[1]Wendel, 140 John James Audubon Parkway, Amherst, NY 14228

[2]Former Administrative Director, Niagara County Sewer District No. 1

The goal to operate economically sustainable wastewater treatment facilities has pushed entities throughout New York State to identify and implement energy efficient facility and technology improvements. This paper will paper will provide a case study of results from projects and programs at nine wastewater treatment systems. The paper will provide energy savings, construction costs and return on investment information on technologies including aeration diffuser upgrades, blower upgrades, dissolved oxygen and pumping control systems, digester mixing systems, anaerobic digester gas collection, belt press upgrades and replacement, lighting upgrades, HVAC upgrades and energy management systems.

INTRODUCTION

Wastewater treatment plants in New York State have continued to experience challenges related to balancing ever increasing energy costs while meeting stricter discharge limits. This paper will provide real life examples of eleven wastewater entities in New York State that have developed economically sustainable retrofit programs that reduced their impact on the environment while also saving money.

The goal to reduce energy costs and improve existing treatment processes resulted in the undertaking of energy efficiency improvements at the following wastewater treatment plants in New York State:

1. Town of Newfane, Newfane, NY
2. Oneida County Sewer District, Utica, NY
3. Town of Grand Island, Grand Island, NY
4. Niagara County Sewer District No.1, Niagara Falls, NY
5. City of Kingston, Kingston, NY
6. Town of Webster, Webster, NY
7. Town of Chenango, Chenango, NY
8. City of Beacon, Beacon, NY
9. Erie County DEP Big Sister Creek WWTP, Lakeshore, NY
10. Washington County Sewer District No. 2, Fort Edward, NY
11. Village of Silver Creek, Silver Creek, NY

The eleven treatment systems presented above provide sewage treatment for approximately 310,000 residents throughout New York State. The treatment plants have a combined design average flow rate of 96 MGD and detailed parameters are presented in Table 1 below. A figure is provided that presents the geographical location of the various wastewater treatment plants that were evaluated in this case study.

Table 1 – Wastewater Treatment Plant Parameters

Wastewater Plant	Capacity	Treatment Processes	Population Served
Newfane	1.6 MGD	Primary, Secondary and Tertiary	8,996
Oneida County	48.0 MGD	Primary and Secondary (Nutrient Removal)	110,000
Grand Island	3.5 MGD	Primary and Secondary (Unox)	15,000
Niagara County	14.0 MGD	Primary, Secondary, and Tertiary	40,000
Chenango	0.8 MGD	Primary and Secondary	8,000
Beacon	6.0 MGD	Primary and Secondary	13,243
Erie County Big Sister	7.7 MGD	Primary and Secondary	35,420
Kingston	6.8 MGD	Primary and Secondary	24,000
Webster	7.5 MGD	Primary and Secondary	33,000
Washington County	2.5 MGD	Primary and Secondary	15,000
Silver Creek	0.75 MGD	Primary, Secondary and Tertiary	1,900

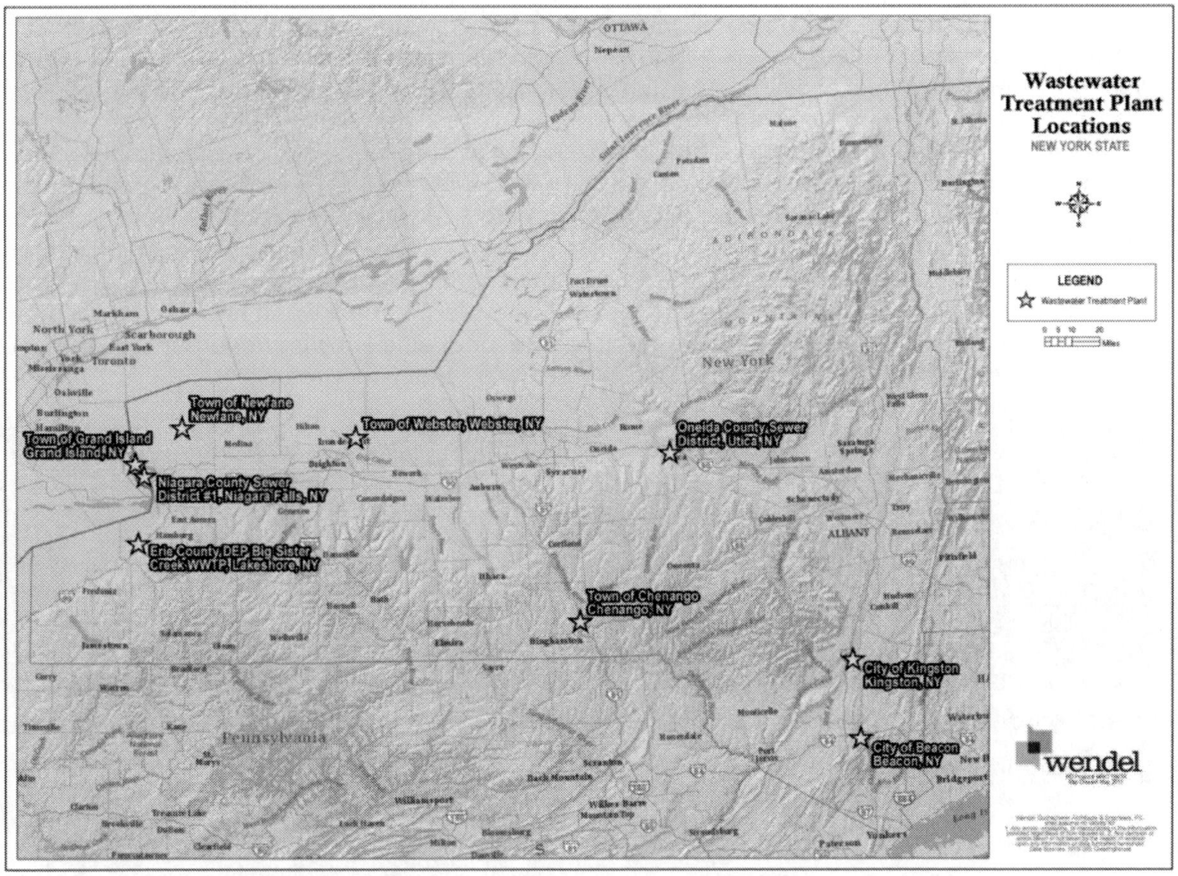

Figure 1 – Geographical Location of Wastewater Treatment Plants

METHODOLOGY – DETAILED STUDY

Each project included a detailed study/planning phase that included the completion of an investment grade energy audit. As part of the energy audit a baseline was calculated utilizing methods approved by the US Department of Energy's International Performance Measurement and Verification Protocol. Energy and efficiency conservation measures were identified through a walkthrough of the existing facilities and through interviewing the operations staff. Baseline data for the energy and efficiency conservation measures identified during the walkthrough was determined using remote and existing logging equipment to determine equipment run hours and/or equipment kW. Plant specific loading characteristics, historical utility data, and operator interviews were utilized to normalize the calculations for facility usage and significant facility seasonal variations. The data was utilized to form the baseline from which energy and operational savings was calculated.

PROJECT IMPROVEMENTS

The detailed study and planned phase provided recommended energy and operational efficiency improvements for the various wastewater systems studied in this paper. The recommended improvements met the individual entities payback and return on investment criteria to provide an economically sustainable implementation model. The recommended improvements for the various wastewater systems are presented in the following paragraphs.

Town of Newfane

The recommended energy and operational efficiency improvements for the Newfane project focused on the largest energy users which are the aeration system and the tertiary filtration system. Operational efficiency improvements were focused on the solids dewatering process. The recommended aeration system improvements included conversion of the existing coarse bubble diffused aeration system to a fine bubble ceramic aeration system, replacement of the existing 200 hp aeration blower which operated at full output with a 75 hp positive displacement blower with a VFD, utilization of a dissolved oxygen control system to match blower output to process air and mixing requirements. The recommended tertiary filtration system improvements included the replacement of the existing mixed media sand filtration system with a coarse media sand filtration system and automation of filter control system to backwash filter based on head loss instead of time of day. The recommended operational efficiency improvements to the solids dewatering process included rehabilitation of the existing belt filter press.

Figure 2 – Proposed Aeration Diffusers

Figure 3 – Proposed Filter Feed VFDs

Oneida County
The recommended energy efficiency improvements for the Oneida County project included upgrades to the aeration control system, lighting facilities, wastewater treatment plant influent pump station and the Sauquoit Creek pumping station. The aeration control system improvements included automation of the control system to modulate blower output from existing single stage blowers with by adjusting inlet vanes based on dissolved oxygen levels and to adjust air supply feed valves to the various aeration basins. Pump station improvement included the replacement of standard efficiency motors with premium efficiency motors, replacement of rheostat variable frequency drives with digital VFDs and the incorporation of an automated control system at both the influent pump station of the wastewater treatment plant and Sauquoit Creek pump station. Lighting system upgrades included the replacement of existing T-12 bulbs with T-8 bulbs and the installation of occupancy sensors.

Figure 4 – Proposed base flow pump

Figure 5 – Existing PSA system towers

Town of Grand Island
The recommended energy efficiency improvements for the Grand Island project included upgrades to the surface aerators for the existing high purity oxygen system, air supply blower improvements for the existing PSA oxygen generation system, influent pump station and lighting improvements. The surface aerator improvements included replacement of the existing surface aerators which were deteriorating due to age with new high efficiency surface aerators. The air supply blower improvements included modifications to the existing PSA oxygen generation system to allow for the oxygen generation rates to be reduced to match demand requirements. The air supply blower was reduced from 200 hp to 75 hp and equipped with a vfd to allow for the air supply rate to match the process requirements and eliminate much of the venting to atmosphere of excess oxygen. The influent pump station was upgraded to incorporate a base flow pump that can be utilized at low flow conditions to reduce energy usage.

Niagara County Sewer District No. 1
The recommended energy efficiency improvements for the Niagara County Sewer District project included upgrades to the existing aeration system, incorporation of an automated dissolved oxygen control system, upgrade of the existing mixed media filtration system, and the existing filter feed system. The existing aeration system was a mechanical aeration system at the end of its useful life that utilized manual adjustment of speeds, manual monitoring of dissolved oxygen control levels, and provided limited mixing capabilities.

The recommended aeration system improvements included installation of fine bubble ceramic diffusers, new centrifugal blowers, air supply piping, a new blower building. The recommended automatic dissolved oxygen control system included eight total dissolved oxygen probes (one per each contact and stabilization basin), air feed valves per each contact and stabilization basin, air supply header pressure transmitter, electric actuated blower inlet valves, and an algorithm driven automated control system. The existing tertiary filtration system included three dual media sand filters with a surface wash and water backwash system and constant speed filter feed pumps. The recommended system included the installation of a dual air-water backwash system, coarse mono-media sand, a new filter underdrain system, variable speed drives on the filter feed pumps, an automated PLC control system with dry/wet weather operating scenarios.

Figure 6 – Proposed tertiary filters

Figure 7 – Proposed aeration system blowers

City of Kingston

The recommended energy efficiency improvements for the City of Kingston project included replacement of the UV disinfection system, upgrades to the digester mixing system, replacement of the sludge thickening system, lighting improvements, upgrades to the process water system, and upgrades of the existing settled sewage pumps. In addition, several operational upgrades were recommended that resulted in additional savings including modifications to the existing belt filter press and upgrading of the polymer system for the belt filter press. The UV disinfection system improvements included the replacement of the existing constant light intensity system with a low pressure high output UV system with flow pacing and a self cleaning feature. The upgrades to the digester mixing system included cleaning of the existing digesters, installation of a nozzle based mixing system and rehabilitation of the existing gas collection piping. The existing dissolved air flotation system was replaced with a gravity belt thickener which was more energy efficient and easier to operate and maintain. Various interior lighting systems throughout the wastewater plant were upgraded with high efficiency light bulbs and ballasts and operating sensors were installed to eliminate unneeded lighting. The existing process water system was equipped with pumps at the end of their useful life and discharge throttling valves to regulate pressure. The process water system was upgraded with new pumps with VFDs, pressure sensors and an automated control system to maintain pressure at the required set point. Upgrades to the existing settled sewage pumps recommended rehabilitation of existing pump interiors or replacement of the existing pumps with new pumps to enhance the wire to water efficiency.

Town of Webster

The recommended energy efficiency improvements for the Town of Webster project included incorporation of a methane recovery system, pump, motor and control upgrade for the process water system and the incorporation of an automated dissolved oxygen control system. The methane recovery system work included the replacement of the existing steel covers which were leaking and not ballasted correctly with fiberglass covers and rehabilitation work on the existing digester gas boiler. The pump, motor and controls upgrades for the process water system included the replacement of the existing oversized pump with a new pump with a VFD and premium efficiency motor sized for their current demands and equipped with an automated control system. The automatic dissolved oxygen control system work included the incorporation of a PLC based control system that automatically adjusts blower air output to the aeration by opening and closing the inlet valve based on dissolved oxygen levels.

Figure 8 – Proposed digester cover

Figure 9 – Installation of digester cover

Town of Chenango

The recommended energy efficiency improvements for the Town of Chenango project included replacement of the SBR diffusers, replacement of the air supply blowers, and incorporation of an automated dissolved oxygen control system. An additional operational efficiency improvement to the existing solids dewatering system was also recommended. The SBR diffuser replacement work included the replacement of existing coarse bubble diffusers with ultra fine bubble membrane diffusers. The air supply blower replacement included the replacement of two existing 50 hp blowers with two new 25 hp positive displacement blowers with VFDs. The automated dissolved oxygen control system allows for blower output to be adjusted based on oxygen levels in the three existing SBRs versus the current system which monitors DO levels and the blower output is manually adjusted. The existing belt filter press and polymer system was replaced with a new three belt press and the polymer system was also upgraded.

City of Beacon

The recommended energy efficiency improvements for the City of Beacon project included replacement of the existing centrifugal air supply blowers, replacement of the influent and return activated sludge pumps, lighting and boiler upgrades, and improvements to the dissolved oxygen control system. Additional operational efficiency improvements to the existing solids dewatering system were also recommended. The air supply blower work included the replacement of two 200 hp centrifugal blowers with two new turbo blowers and to the existing dissolved oxygen control system. The influent and return activated sludge pump work included the replacement pumps that were at the end of their useful life and had lost both pumping capacity and efficiency due to wear. The lighting and boiler upgrades included the replacement of low efficiency lighting with new ballasts and fixtures while the boiler work included the replacement of an existing boiler that was at the end of its useful life with a more efficient model.

Figure 10 – Existing RAS pumps

Figure 11 – Existing air supply blowers

Erie County DEP Big Sister Creek

The recommended energy efficiency improvements for the Erie County DEP Big Sister Creek project included an automated dissolved oxygen control system, return activated sludge pump VFD and controls upgrades, aerobic digester VFD and controls upgrades, lighting upgrades and an automated energy management system. The automated dissolved oxygen control system work included the incorporation of one new dissolved oxygen probe in each aeration basin, the installation of new electric actuators on the existing blower inlet valves and the installation of new PLC based control panel to modulate blower output based on dissolved oxygen levels. The return activated sludge pump VFD and controls upgrades included the installation on the existing RAS pumps and the installation of flow meters and total suspended solids meters on the pump discharge to pace the return sludge flows on influent flows to the plant and to allow for modified MCRT control to take place. The aerobic digester VFD and control system upgrades included the installation of new VFDs on two existing blowers, the installation of new DO probes and automation of the blower valves. The lighting upgrades included the replacement of low efficiency ballasts and T12 bulbs with high efficiency fluorescent T8 and LED bulbs. The automated energy management system included the automation of the following wastewater treatment systems:

- Raw sewage pumps
- Grit system blowers
- Re-aeration system blowers
- UV disinfection system
- Effluent process water system
- WAS pumps
- Plate and frame filter press
- Dissolved air flotation (DAF) units
- Chemical sludge pumps
- Ferric chloride pumps
- Thickened sludge pumps
- Filter press feed pumps
- Solids contact clarifiers

IMPLEMENTATION METHODS

Various methods of installation were utilized to implement the various energy efficiency improvements identified at the nine wastewater systems. Two of the projects (Town of Newfane and Niagara County Sewer District No. 1) included energy efficiency improvements that were implemented by employee installation directly and additional improvements that were implemented through the traditional design, bid, build model. The remaining seven projects were implemented through energy performance contracting where the savings generated by the improvements are utilized to cover the equipment procurement and installation costs. Financing of the energy performance contracting was either provided through municipal leasing or through conventional bonding.

MEASUREMENT AND VERIFICATION

The measurement and verification phase is an important step to an energy efficiency improvement project because it provides validation of the original energy savings projections and provides an opportunity for further commissioning of the improvements to enhance savings. On several of the projects covered in this paper, the measurement and verification phase was utilized in conjunction with training of the operations staff to enhance the effectiveness of the equipment that was installed further reducing the time for the savings to pay for the improvements.

CASE STUDY RESULTS

Energy savings for six of the wastewater systems were compiled after the construction was completed and the results are compared to the savings projected in the preliminary study phase and return on investment data is presented for the implemented improvements. The results for the remaining three wastewater systems are based on the projected savings from the preliminary study and the project costs based on bidding of the work. The results of the implemented projects are as follows:

Town of Newfane

The efficiency improvements implemented provided energy savings totaling approximately $117,850 per year. An additional $153,600 of energy incentives further reduced the total return on investment of the improvements to 5.6 years. The measured energy and operational savings were $22,400 above what was projected in the preliminary study phase. The performance of the individual measures is presented below. It should be noted that the filtration system improvements were implemented to comply with a consent order and any energy savings were utilized to try to offset the total cost of the required improvements.

Efficiency Improvement Measures	Projected Energy/Oper. Savings	Measured Energy/Oper. Savings	Project Cost	Energy Incentives	Return on Investment
Aeration System	$65,000	$84,000	$250,000	$86,000	1.95 years
Filtration System *	$2,000	$3,600	$250,000	$7,600	67.0 years

* Project was completed to comply with consent order requirements.

Efficiency Improvement Measures	Projected Energy/Oper. Savings	Measured Energy Savings	Project Cost	Energy Incentives	Return on Investment
DO Control System	$50,000	$57,000	$150,000	$110,000	0.70 years
Influent Pump Station	$27,500	$35,000	$425,000	$60,000	10.4 years
Sauquoit Creek PS	$18,450	$20,000	$325,000	$39,000	14.3 years
Lighting Upgrades	$15,000	$13,750	$200,000	$88,660	8.1 years

Oneida County WPCF

The efficiency improvements implemented provided energy savings totaling approximately $125,750 per year. An additional $297,660 of energy incentives further reduced the total return on investment of the improvements to 6.4 years. The measured energy and operational savings were $14,800 above what was projected in the preliminary study phase. The performance of the individual measures is presented below.

Efficiency Improvement Measures	Projected Energy/Oper. Savings	Measured Energy Savings	Project Cost	Energy Incentives	Return on Investment
Surface Aerator Upgrade	$11,275	$17,275	$314,000	$13,200	17.4 years
Influent Pump Station	$5,500	$4,500	$79,900	$6,650	16.3 years
PSA and Blower System	$20,000	$24,000	$62,500	$29,000	1.4 years

Town of Grand Island

The efficiency improvements implemented provided energy savings totaling approximately $45,775 per year. An additional $48,850 of energy incentives further reduced the total return on investment of the improvements to 8.9 years. The measured energy and operational savings were $9,000 above what was projected in the preliminary study phase. The performance of the individual measures is presented below.

DO Control System	$23,450	$25,250	$260,000	$60,000	7.9 years
Belt Press Modifications	$5,000	N/A	$50,900	NA	10.0 years

Niagara County Sewer District No. 1

The NCSD No.1 commenced a capital project to replace aging equipment that was failing due to age. A program was undertaken to attempt to minimize the effect of the capital project on the rate payers by maximizing the energy efficiency of required improvements. The efficiency improvements implemented provided energy savings totaling approximately $48,000 per year. An additional $100,000 of energy incentives further reduced the total return on investment of the improvements to 55.8 years. The measured energy and operational savings were $13,000 above what was projected in the preliminary study phase. The performance of the individual measures is presented below.

Efficiency Improvement Measures	Projected Energy/Oper. Savings	Measured Energy/Oper. Savings	Project Cost	Energy Incentives	Return on Investment
Filtration System Upgrade	$20,000	$23,000	$680,000	$70,000	26.5 years
Aeration System Upgrade	$15,000	$25,000	$2,100,000	$30,000	82.8 years

City of Kingston
The efficiency improvements implemented provided energy savings totaling approximately $135,595 per year. An additional $108,590 of energy incentives further reduced the total return on investment of the improvements to 8.7 years. The measured energy and operational savings were $22,400 above what was projected in the preliminary study phase. The performance of the individual measures is presented below.

Efficiency Improvement Measure	Projected Energy/Oper. Savings	Measured Energy Savings	Project Cost	Energy Incentives	Return on Investment
UV System Upgrade	$35,750	$37,200	$430,000	$20,640	11.0 years
Digester Mixing	$45,460	$43,200	$356,000	$48,900	7.1 years
Settles Sewage Pumps*	$11,570	NA	$135,000	$11,500	10.7 years
Sludge Thickener	$42,350	$38,500	$325,000	$25,000	7.8 years
Lighting Upgrades	$5,125	$5,125	$48,500	$2,550	9.0 years

*The settled sewage pump upgrade work was implemented outside of the energy project and measured energy savings are not available.

Town of Webster
The efficiency improvements implemented provided energy savings totaling approximately $56,711 per year. An additional $17,100 of energy incentives further reduced the total return on investment of the improvements to 9.0 years. The measured energy and operational savings were $1,974 above what was projected in the preliminary study phase. The performance of the individual measures is presented below.

Efficiency Improvement Measure	Projected Energy/Oper. Savings	Measured Energy Savings	Project Cost	Energy Incentives	Return on Investment
Pump Motor Upgrades	$5,197	$9,296	$54,100	$8,200	4.9 years
DO Control Upgrades	$18,400	NA*	$311,750	$15,000	16.1 years
Digestion Methane Recovery	$55,900	NA*	$330,000	NA	5.9 years

Town of Chenango

The efficiency improvements implemented provided energy savings totaling approximately $79,170 per year. An additional $18,800 of energy incentives further reduced the total return on investment of the improvements to 8.8 years. The performance of the individual measures is presented below.

Efficiency Improvement Measure	Projected Energy/Oper. Savings	Measured Energy Savings	Project Cost	Energy Incentives	Return on Investment
DO Control System	$4,870	NA*	$73,500	$3,800	14.3 years
SBR and Blower Upgrade	$18,400	NA*	$311,750	$15,000	16.1 years
Solids Dewatering Upgrade	$55,900	NA*	$330,000	NA	5.9 years

* Measured energy and operational savings are currently not available because the improvements are being constructed.

City of Beacon

The efficiency improvements implemented provided energy savings totaling approximately $121,103 per year. An additional $59,900 of energy incentives further reduced the total return on investment of the improvements to 10.5 years. The performance of the individual measures is presented below.

Efficiency Improvement Measure	Projected Energy/Oper. Savings	Measured Energy Savings	Project Cost	Energy Incentives	Return on Investment
Pump and Motor Upgrades	$27,160	NA*	$695,000	$11,875	25.2 years
Boiler, HVAC and Lighting	$7,000	NA*	$107,000	$4,000	14.7 years
Air Supply Blowers	$38,315	NA*	$478,000	$44,025	11.3 years
Belt Press and Polymer	$48,628	NA*	$51,846	NA	1.1 years

* Measured energy and operational savings are currently not available because the improvements are being constructed.

Erie County Big Sister Creek

The efficiency improvements implemented provided energy savings totaling approximately $130,685 per year. An additional $142,325 of energy incentives further reduced the total return on investment of the improvements to 7.3 years. The performance of the individual measures is presented below.

Efficiency Improvement Measure	Projected Energy/Oper. Savings	Measured Energy Savings	Project Cost	Energy Incentives	Return on Investment
Lighting Upgrades	$6,370	NA	$59,750	$3,465	8.8 years
DO Control System	$24,670	NA	$150,500	$57,450	3.8 years
Energy Mgmt. System	$58,200	NA	$654,500	$30,000	10.7 years
RAS Pump Upgrades	$26,970	NA	$54,550	$15,210	1.5 years
Digester Aeration Blower	$14,475	NA	$182,550	$36,200	10.1 years

* Measured energy and operational savings are currently not available because the improvements are being constructed.

Washington County Sewer District No. 2

The efficiency improvements implemented provided energy savings totaling approximately $56,711 per year. An additional $17,100 of energy incentives further reduced the total return on investment of the improvements to 9.0 years. The measured energy and operational savings were $1,974 above what was projected in the preliminary study phase. The performance of the individual measures is presented below.

Efficiency Improvement Measure	Projected Energy/Oper. Savings	Measured Energy Savings	Project Cost	Energy Incentives	Return on Investment
Lighting Upgrades	$2,600	NA	$41,000	$5,000	13.8 years
Blower Upgrades	$60,000	NA	$150,500	$41,000	3.8 years
Diffuser Upgrades	Above in Blower Savings	NA	$654,500	$17,000	10.7 years
DO Control System	Above in Blower Savings	NA	$175,000	$1,000	1.5 years
Solids Dewatering	$143,000	NA	$1,450,000	$12,000	10.1 years
Digester Mixing and Septic	$145,000	NA	$2,300,000	NA	16.0 years
Microturbines	$64,000	NA	$472,000	NA	7.8 years

* Measured energy and operational savings are currently not available because the improvements are being constructed.

Village of Silver Creek

The efficiency improvements implemented provided energy savings totaling approximately $74,970 per year. An additional $68,500 of energy incentives further reduced the total return on investment of the improvements to 8.86 years. The performance of the individual measures is presented below.

Efficiency Improvement Measure	Projected Energy/Oper. Savings	Measured Energy Savings	Project Cost	Energy Incentives	Return on Investment
Lighting Upgrades	$3,500	NA	$45,000	$2,500	8.8 years
Aeration Diffusers	$22,000	NA	$250,000	$25,000	3.8 years
Blower Upgrades	$22,500	NA	$288,000	$26,000	10.7 years
Influent Pumps	$26,970	NA	$150,000	$15,000	1.5 years

SUMMARY AND CONCLUSIONS

In conclusion, the projects described in this case study allowed the wastewater treatment plants to undertake necessary capital improvements while improving energy efficiency and taking advantage of available incentives. The case study highlighted that various implementation methods could be successfully utilized to incorporate energy and operational efficiency improvements. Seven of the nine wastewater systems were able to implement an energy and operational efficiency improvement project with a return on investment of less than 10 years while one other system had a return on investment of less than 11 years.

Minimizing Energy Consumption in Decentralized Wastewater Collection and Treatment Systems

Kartiki S. Naik[1*] & Michael K. Stenstrom[2]
University of California, Los Angeles[1&2]
knaik@ucla.edu[1], stenstro@seas.ucla.edu[2]

ABSTRACT

Centralized wastewater treatment, widely practiced in developed areas, involves transporting wastewater from large urban or industrial areas to a large capacity plant using a single network of sewers. Alternatively, the concept of wastewater collection, treatment and reuse at or near its point of generation is called decentralized wastewater treatment (Tchobanoglous and Crites, 1998). Smaller decentralized plants with energy-efficient reclaimed water pumping, modularization of expansion and minimum capital investment can meet the increasing need for reclamation and wastewater management accessibility in rapidly developing regions. Hence, a feasibility analysis and comparison methodology subject to user-defined constraints for a decentralized configuration of several treatment plants and collection networks is necessary. We devised a methodology to preliminarily assess the feasibility of installing decentralized systems, based on local needs. It evaluates the following parameters and their combinations to develop various scenarios: Pump-back energy consumption, Sewer construction cost, Capacity expansion cost. We applied this methodology to the case study of Hollywood and a typical central treatment plant. The reclaimed water distribution network, sewer expansion and pump station expenditure, and capital expansion for completely centralizing wastewater collection and reclamation exceeded the cost of the same for a decentralized system. Also, the energy consumption for pumping back reclaimed water is about twice the aeration energy consumption and significant fraction of the treatment energy consumption. In this study, decentralization was evidently the more favorable solution.

KEYWORDS: Decentralization, reclamation, wastewater, energy, pumping

INTRODUCTION

Centralized wastewater treatment consists of a large treatment plant connected to an elaborate wastewater collection system. Wastewater collection, treatment and reuse at or near its point of generation are called decentralized wastewater treatment (Tchobanoglous and Crites, 1998). Decentralized configurations generally use shorter collection networks of relatively smaller size. Such systems are sometimes used for relieve sewers leading to larger central treatment plants. For example, in Los Angeles, the D.C. Tillman and Glendale wastewater treatment plants, located upstream of the Hyperion treatment plant avoided overloading trunk sewers and reclaimed water near potential users (City of Los Angeles, 2011).

With reclamation on the rise, it is very necessary to have a simple system not only to reclaim the wastewater but also to convey it to the respective region for use. Since reclaimed pipelines flow in the direction opposite of that of sewers, they usually need to gain some elevation in their course. Reducing the length of the pipeline is the principle solution to minimize pumping energy consumption. Centralized plants generally cannot be located close to all the potential reclaimed water users. It is especially difficult to establish such a large treatment facility in any residential neighborhood. Thus, for centralized plants, returning recycled water can incur large energy consumption due to pumping. Adapting the design of large centralized treatment plants to local reclamation needs to reduce pumping cost can be complicated. Decentralized treatment systems have a smaller scale, thus simplifying modification according to local needs. With smaller collection-reclamation loops, decentralized wastewater treatment systems can reduce pumping back energy consumption.

Decentralization is also an alternative to the expensive construction of larger trunk sewers to convey the added flow of newly developed areas. With smaller sewers and treatment plants, these systems can also provide wastewater treatment and collection access to remote communities. Many developing regions do not have any access to wastewater collection and treatment rendering them susceptible to disease outbreaks. Satellite or decentralized treatment systems can be an excellent solution to reduce public health risk.

Both treatment concepts have pros and cons; for instance, large treatment plants have more economy of scale. Satellite treatment plants are usually assumed to be limited to primitive treatment systems such as septic tanks, and in turn difficult to optimize for energy and cost. (Massoud et al., 2008) But to take advantage of improved public health, energy-efficient reclamation and conveyance, and modular expansion, it is necessary to devise an optimization algorithm to obtain a customizable, cost-effective, energy-efficient wastewater collection and treatment configuration for such regions.

This study involves optimizing sewer networks and decentralized treatment plant locations in an urban area to minimize pump-back energy consumption for recycled water. We computed pump-back energy consumption as a fraction of treatment and aeration energy for a hypothetical but realistic example (Hollywood). We optimized the decentralized wastewater collection and treatment configuration for Hollywood for minimum energy consumption and cost.

METHODS

Local and treatment process characteristics: We applied this analysis to Hollywood, CA. To assess the preliminary feasibility of a decentralized system, we quantified its benefits and compare them with those of a centralized system. The trunk pipes of the sewer and reclaimed water lines that we designed for this study bifurcated based on topography. Fig. 1 shows the reclamation pipeline direction and the bifurcation. It also shows the pipe lengths and the elevations at different points.

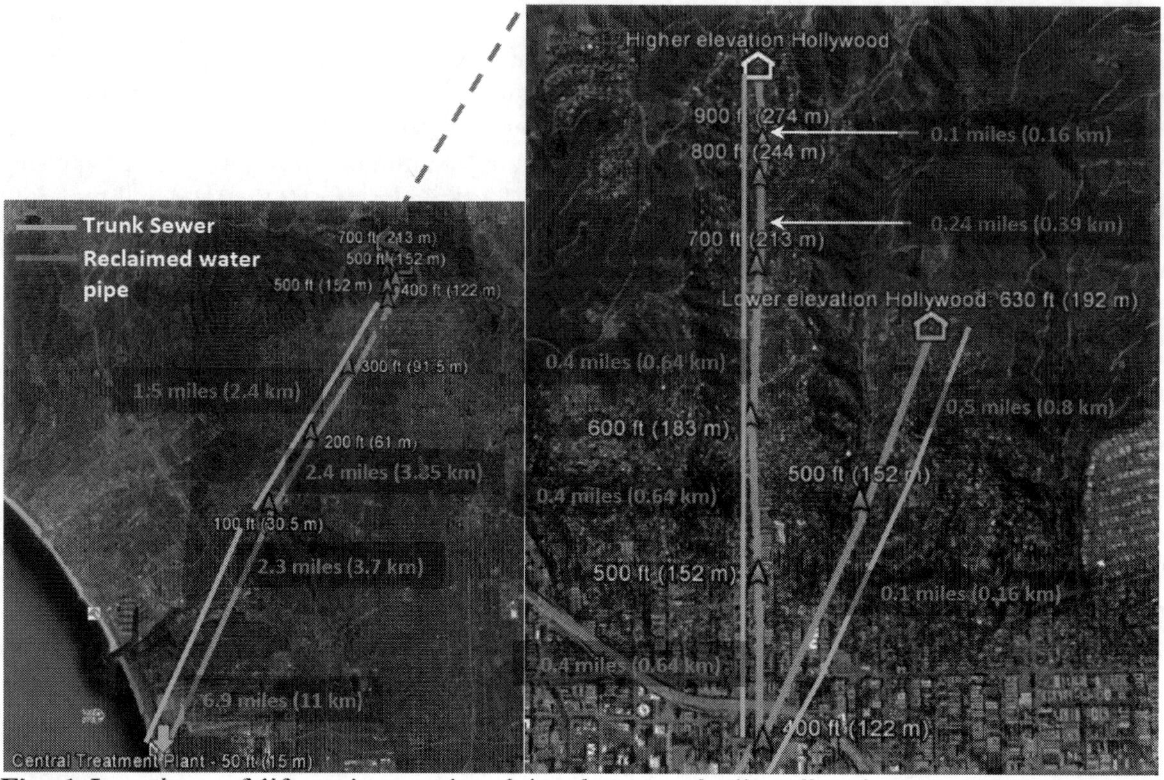

Fig. 1 Locations of lift stations and reclaimed water pipeline direction (Google Earth© 7.0.2 2013)

Table 1 describes the relevant demographics and flow characteristics. The distance from the central treatment plant is equivalent to the maximum distance of Hollywood's furthermost extremities from the plant. We assumed a uniform population distribution. \

Table 1 Regional characteristics and assumptions for case study (references: United States Census, 2011 Google Earth, Metcalf and Eddy, 2003)

Regional characteristics	**Value**
Population of Hollywood	123,000
Flow rate per capita	66 Liters/day (0.066 m^3/day)
Distance from central treatment plant	15.8 miles (25260 km)
Percent wastewater recycled	90%
Population density	7540 per sq. mile (2950 per sq. km)
Average Elevation for Hollywood	1219 ft (372 m)
Average Elevation for Central Treatment Plant	50 ft (15 m)
Flow treated	45 MGD (1.97 m^3/s)

Fig. 2 shows the typical treatment train for which the treatment energy was calculated. We selected this treatment plant configuration to avoid skewing of the relation between the pump-back and treatment energy consumption because of expensive treatment. We designed each process according to required standards and calculated the cumulative energy consumption.

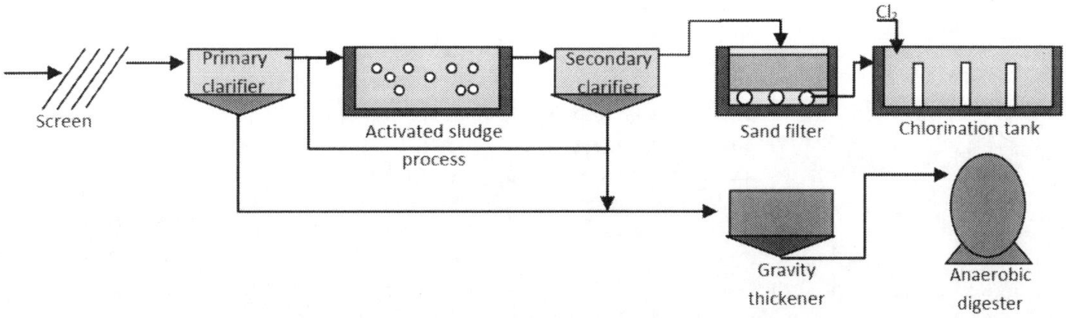

Fig. 2 Hypothetical wastewater treatment configuration for analysis

We also compared it with the expected capital cost of decentralized treatment plant. Aeration energy is usually 45-75% of the entire treatment energy consumption (Rosso et al., 2008). Hence we analyzed the pumping back energy consumption as a fraction of the aeration energy consumption.

Sewer Design: To design the trunk sewer, we computed the flow generated at every mile along the sewer direction. We estimated the wastewater generation per mile by calculating the flow on a per capita per unit length (miles) basis and designed each unit length for that flow. Sewers are commonly designed using empirical curves relating the depth of flow ratio (partial to full flow) to other ratios such as discharge, velocity, wetted area, wetted perimeter and hydraulic radius. But such curves are tedious to use in a sewer design model without a large amount of interpolation and assumption of linearity. Hence, we designed the sewer pipe for this distance, using two methods: iteration and theoretical design (Ott and Jones, 1988). For the iteration method, we used suitable depth of flow and velocity of flow ratios (for partial to full flow) to iterate between different solutions. This takes several manual trials and an approximate estimate of the sewer size for the initial guess. The theoretical design involves calculating the sewer size and depth of flow from theoretical equations relating the sewer diameter and depth of flow.

For the iteration method, we use sewer design curves to determine the velocity of flow ratio from the maximum diameter ratio. We iterated the sewer size to meet the constraints for velocity of flow and depth of flow ratios. After obtaining the desired pipe size and depth of flow ratio, we design the pipe for that section. We applied this iteration technique to sections of entire sewer trunk at every mile.

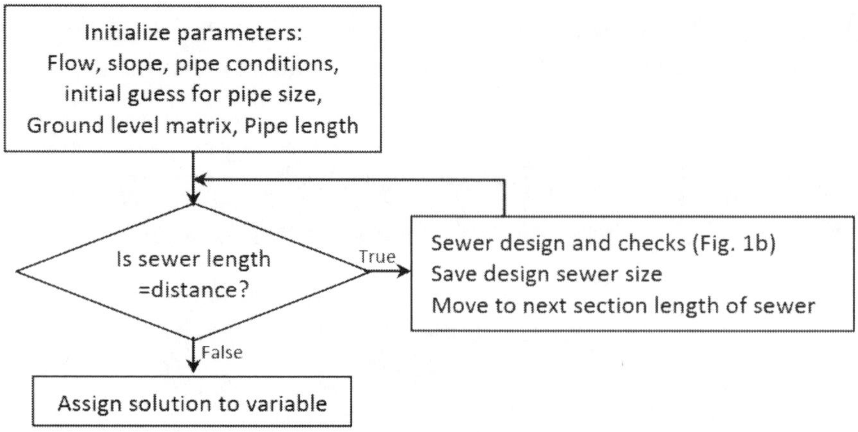

Fig. 3a Basic algorithm for sewer design

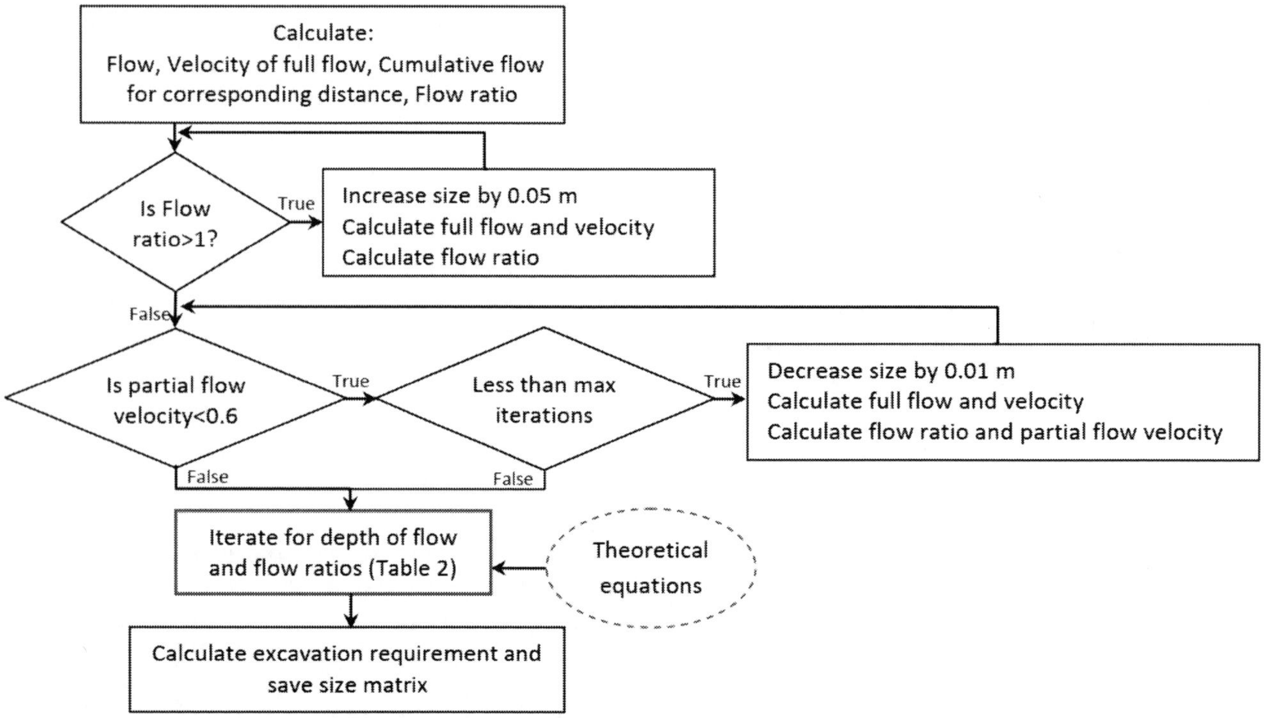

Fig. 3b Detailed algorithm for sewer design

For the theoretical design, we used explicit equations derived by Ott and Jones (1988). We created a routine in Matlab® to assume a sewer size and run a loop to iterate the size as the flow depth ratio and velocity ratio requirement changes. Fig. 3 describes the algorithm for designing an entire trunk sewer. Fig 3a denotes how we divided the trunk sewer, and designed each section sequentially. Fig. 3b shows the detailed design procedure.

To design each link we used the theoretical equations derived by Ott and Jones. This is the only published literature which derives explicit theoretical equations for partial flow velocity, the angle that the depth of flow subtends and the relation between flow and depth of flow ratios. The theoretical model considers two cases of flow: (1) Depth of flow ≤ Radius of pipe (2) Depth of flow > Radius of pipe. Figs. 4a and 4b consider both these cases and show corresponding schematics of the sewer pipe.

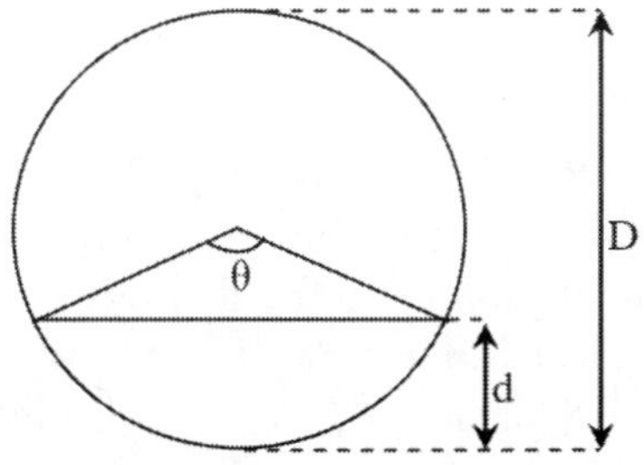

Fig. 4a Sewer with partial flow (d/D ≤ 0.5)

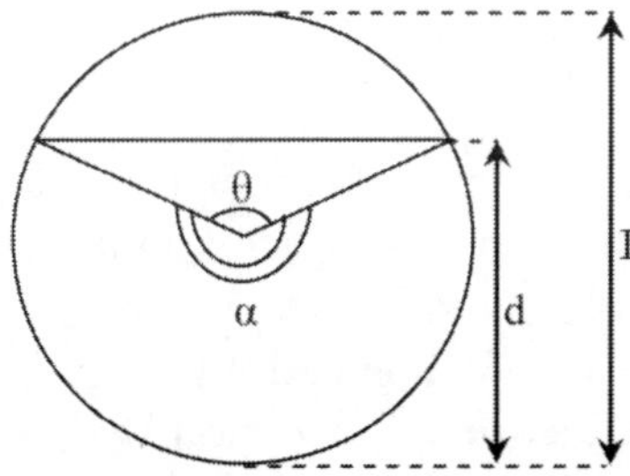

Fig. 4b Sewer with partial flow (d/D ≤ 0.5)

Table 2 describes the theoretical equations and the corresponding variables for the two cases of flow. The equations for the higher depth of flow have been simplified by changing the use of the subtended angle. These equations were used in the iteration part of the model after choosing a potentially suitable pipe size using the hydraulic parameter constraints.

Table 2 Theoretical equations for Sewer Design Model (Ott and Jones, 1988)

Equation (d/D) ≤ 0.5	**Equation (d/D) > 0.5**	**Variables**
$\frac{d}{D}=\frac{\left[1+\cos\left(\frac{\theta}{2}\right)\right]}{2}$	$\frac{d}{D}=\frac{\left[1-\cos\left(\frac{\theta}{2}\right)\right]}{2}$	d: Depth of flow D: Pipe size θ: Angle subtended at center (refer to Fig. 2) α: (2π- θ) R_{PF}: Hydraulic radius of wetted area A_{PF}: Wetted area of pipe Q_{PF}: Partial flow rate Q_F: Flow for completely filled pipe V_{PF}: Velocity at partial flow
$\theta=2\arctan\left\{\frac{\sqrt{1-\left[2\left(\frac{d}{D}\right)-1\right]^2}}{\left[2\left(\frac{d}{D}\right)-1\right]}\right\}$	$\theta=2\arctan\left\{\frac{\sqrt{1-\left[1-2\left(\frac{d}{D}\right)-1\right]^2}}{\left[1-2\left(\frac{d}{D}\right)-1\right]}\right\}$	
$R_{PF}=\frac{D}{4}\left[\frac{2\pi-(\theta-\sin\theta)}{(2\pi-\theta)}\right]$	$R_{PF}=\frac{D}{4}\left[\frac{(\theta-\sin\theta)}{\theta}\right]$	
$A_{PF}=\frac{D^2}{8}[2\pi-(\theta-\sin\theta)]$	$A_{PF}=\frac{D^2}{8}(\theta-\sin\theta)$	
$\frac{Q_{PF}}{Q_F}=\left[\frac{(\theta-\sin\theta)^{5/3}}{2\pi(\theta)^{2/3}}\right]$	$\frac{Q_{PF}}{Q_F}=\left[\frac{(\alpha-\sin\alpha)}{2\pi(\alpha)^{2/3}}\right]$	
$V_{PF}=\left[\frac{8Q_{PF}}{D^2(\theta-\sin\theta)}\right]$	$V_{PF}=\left[\frac{8Q_{PF}}{D^2(\alpha-\sin\alpha)}\right]$	

Pump-back energy consumption: We calculated the pump-back energy based on the topography of the region. As given in Table 1, the distance from Hollywood city to the central treatment plant is 16 miles (25.6 km). Hollywood city is at an elevation about 360 meters higher than the location of the central plant. We had to design each section of the reclaimed water pipe for full flow. We divided the reclaimed water pipeline into sections based on the placement of the pumping stations. We computed the pump-back energy consumption for each section, calculating the pipe size for each section depending on the flow. Fig. 3 shows the placement of pumping stations and trunk pipeline for reclaimed water supply.

We used a maximum design head of 100 ft (30.5 m) for lifting stations for the reclaimed water. From elevation data of the points in the direction of the reclaimed water pipeline from the central plant to Hollywood, we determined the positions of the pumping stations and the distances of the sections. We calculated the pumping head using the power equation. To account from deviations from the assumed direction, we assumed that the actual length of each section of the pipe was 20% more than the straight line measurement.

$$P = \frac{\rho g Q h}{\eta}$$

P: Power (kW), ρ: Density of water (N/m^3); Q: Discharge in pipe (m^3/s);
h: Head (m); η: Pump efficiency

Capital investment: We have considered the capital cost of the decentralized treatment plant assuming the same treatment configuration. This cost was calculated based on capacity of the plant, using literature cost relationships that include economies of scale. This assumption simplified the determination of the capital cost of two plants of different capacities and identical treatment configuration. We conducted a review of literature establishing and describing relations of cost with the capacity of the plant corresponding to various years. Table 3 summarizes these relations from different sources.

The entire set of equations was assessed for consistency by projecting the values from one year to the rest of the years. We used Producer Price Indices calculated by the Bureau of Labor Statistics (U.S.A) to project the values (U.S. Dept. of Labor, 2012). The projected costs were not consistent with the reported values over the time period of the references. Hence it was necessary to compute the construction costs from the correlations from all the years and project them to 2012. Hence, these costs are reported as a range of values.

The sewer cost is computed based on the (LADWP, 1983) and the Producer Price Index developed by the Bureau of Labor Statistics, U.S.A. The reported construction cost per unit length of the sewer followed an exponential relation with the diameter (Fig. 2). The Producer Price Index is a parameter which is used to project the production cost of a commodity to any desired year for which this index is available. (U.S. Dept. of Labor, 2012).

Table 3 Literature review of relations between construction cost and flowrate of wastewater treatment plants (EPA,1978; EPA, 1980; Qasim, 1992; UN, 2003; USEPA, 2007)

Reference	Region	Source of data	Type of cost	Unit process	Regression	Variables
EPA, 1978	U.S.A	Winning bids for Construction Grants Program for EPA	Construction cost, piping, excavation	Site work excluding excavation	$C=(1.12 \times 10^5)(3780Q)^{0.97}$	C = cost (dollars) Q=flow rate (m^3/day)
				Preliminary treatment	$C=(5.79 \times 10^4)(3780Q)^{1.17}$	
				Influent pumping	$C=(1.47 \times 10^5)(3780Q)^{1.03}$	
EPA, 1980	U.S.A	Winning bids for Construction Grants Program for EPA	Construction cost, piping, excavation	Equalization	$C=(6.76 \times 10^4)(3780Q)^{0.60}$	C = cost (dollars) Q=flow rate (m^3/day)
				Oxidation ditch	$C=(4.68 \times 10^5)(3780Q)^{0.57}$	
				Rotating biological contactor	$C=(6.09 \times 10^5)(3780Q)^{0.77}$	
Tsagarakis et al., 2002	Greece	National survey of MWTPs	Construction	Conventional ASP (whole plant)	$C=0.116X^{0.954}$	C = cost in 10^6 dollars per 1000 population equivalent X = plant size in 1000 population equivalent
United Nations, 2003; Qasim et al., 1992	U.S.A.	Manufacturers, conceptual design and published data from USEPA report	Construction (1996 dollar values)	Screening and grit removal with bar screens	$CC = 674Q^{0.611}$	CC: Capital cost in dollars Q: average design flow in m^3/day
				Conventional activated sludge with diffused air	$CC = 72Q + 368{,}043$	

We calculated the cost for small treatment plants based on the EPA Report on Biological Nutrient Removal Processes and Costs, 2007 and the EPA cost-flowrate correlations from the years 1978, 1980 and 1992 (1996 dollar values) (EPA,1978; EPA, 1980; Qasim, 1992; UN, 2003; USEPA, 2007). We used this relation to compute a construction cost and projected it to 2012. As the decentralized plant belonged to the category of small treatment plants similar to those reported, one value of construction cost was computed using this relation.

RESULTS AND DISCUSSION

We compared the centralized and decentralized configurations for three types of scenarios:

- Pump-back energy consumption
- Sewer expansion
- Capacity expansion and Sewer expansion

Reclaimed water pumping and construction costs

We computed the pumping energy consumption at every 100 feet of elevation. Fig. 5 shows the pump-back energy consumption at every point. It also depicts the pump-back energy consumption as a percent of the total and aeration energy consumption. Based on average values of energy consumption for wastewater treatment, we calculated the treatment energy consumption as $101,840. We assumed that the aeration energy consumption using the influent BOD ($45,828). Both these values correspond to wastewater generated in Hollywood.

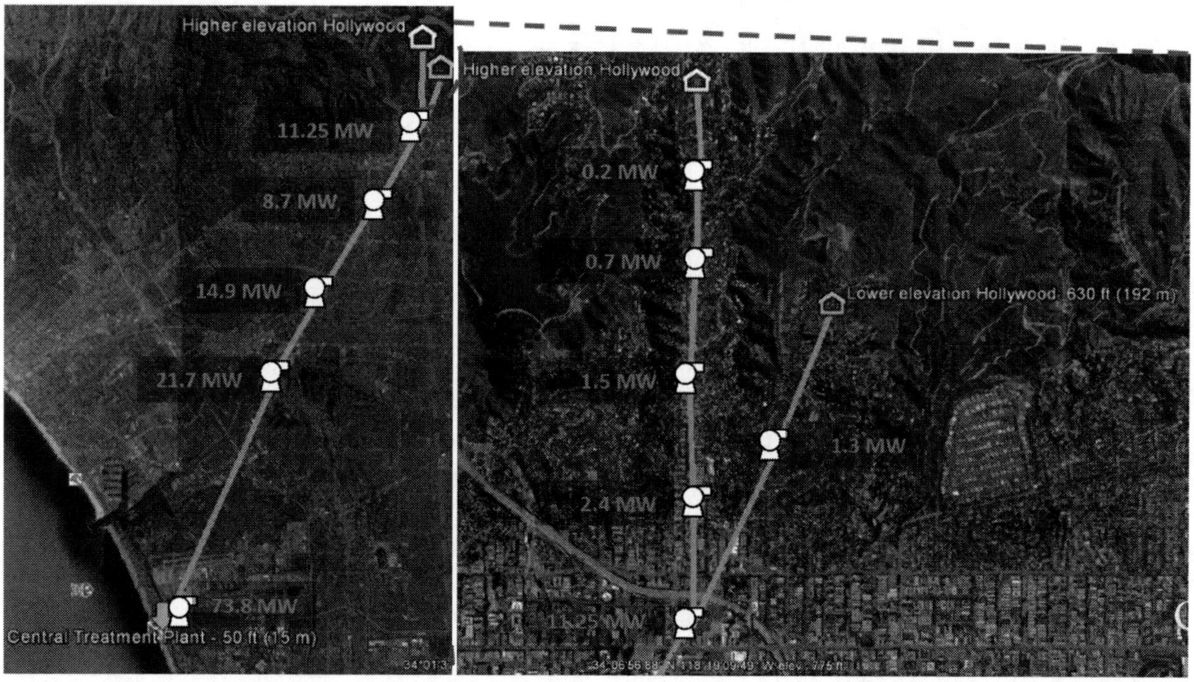

Fig. 5 Pump-back Energy consumption for Hollywood, CA (Google Earth© 7.0.2 2013)

We also calculated the cost of installing the reclaimed water pipeline. We referred to several reclaimed water installation reports (LA City Sanitation District, 2010) and arrived at pipe sizes and an average cost of $243 per linear foot of installation. This cost includes both material and installation. This cost has been converted to 2012 USD. Thus, the total cost for the reclaimed water pipeline is **$67,692,000.**

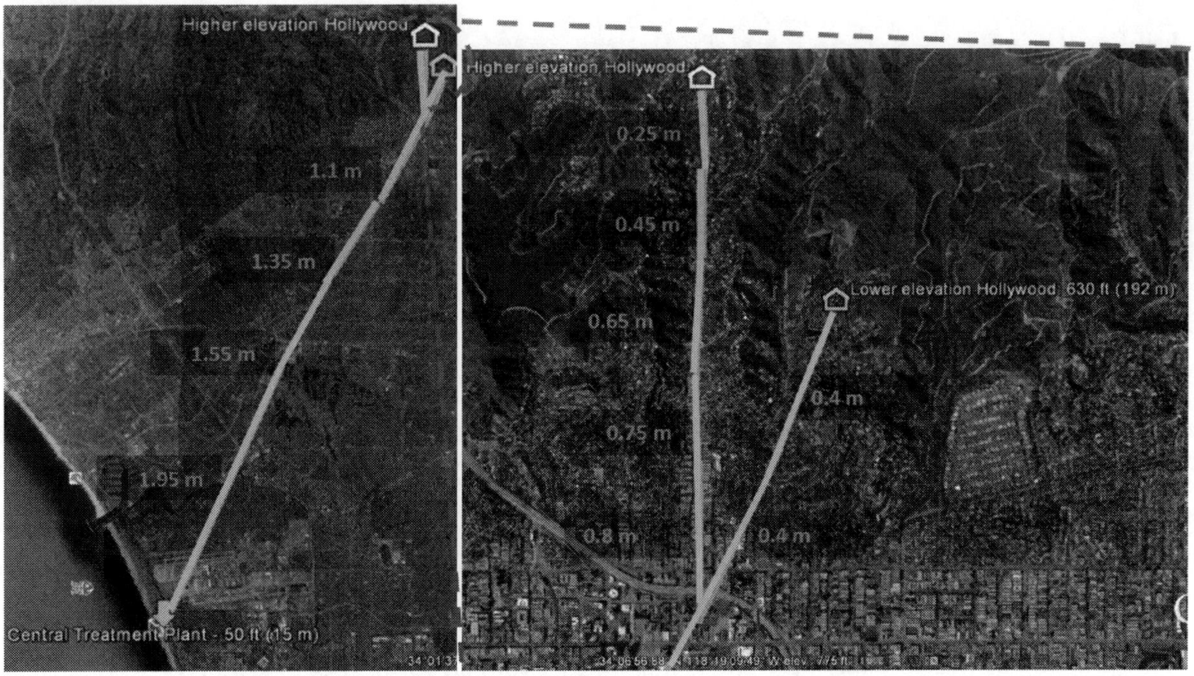

Fig. 7 Sewer size for collection network (Google Earth© 7.0.2 2013)

We assumed that the cost of lifting stations is 10% of that of a treatment plant. We considered the cost of multiple lift stations is 30% higher than that of a single station. Thus we computed a total construction cost of all the lifting stations as **$7,259,633-$13,819,501.**

Sewer and reclaimed water pipe construction without capacity expansion

To connect Hollywood to the central treatment plant, the sewer diameter has to be enlarged to accommodate the flow. Fig. 6 shows the increase in sewer diameter with flow for the same case. Near the central plant, the design diameter approaches **2 meters**. In case of decentralized treatment plant for Hollywood, the design trunk sewer diameter is **0.45 meters.**

Expansion of old sewers is more expensive than building smaller new sewers. The cost for the entire sewer is computed to be **$865,972,105** (LADWP, 1983) and the Producer Price Index developed by the Bureau of Labor Statistics, U.S.A.) (U.S. Dept. of Labor, 2012). Fig. 7 shows the relation of construction cost of a sewer with its diameter for extrapolated till 2012. The construction cost for a main sewer trunk connecting the region to a decentralized plant at a radius of 5 miles is calculated to be **$6,004,499.**

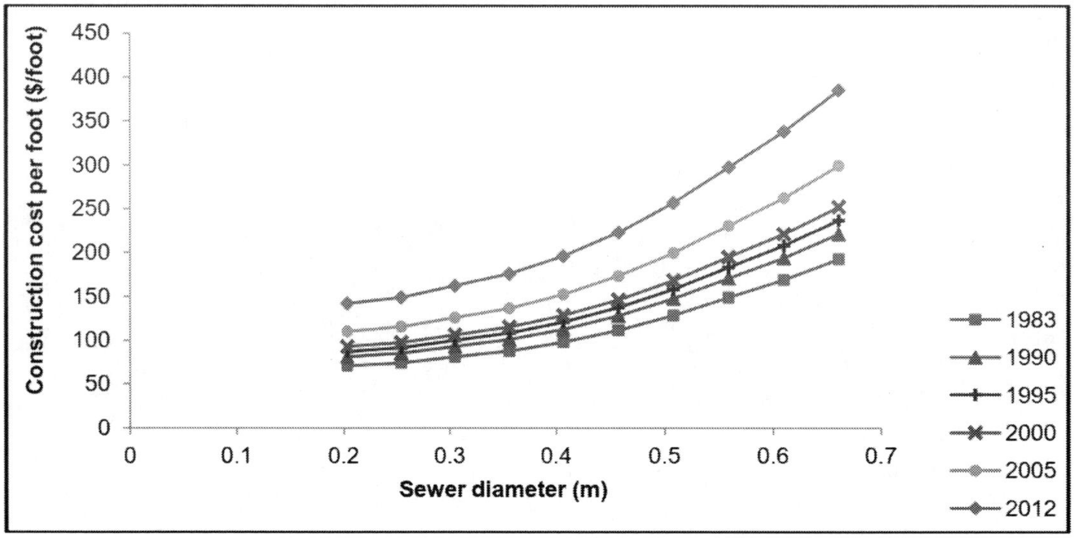

Fig. 7 Variation of construction cost of sewer per length with sewer diameter and time (LADWP, 1983)

The construction cost of the decentralized treatment plant of the required capacity of 2.5 MGD was calculated to be in the range of **$5,942,092.68 - $22,498,583** (EPA, 1978; EPA, 1980; LADWP, 1983; Qasim, 1992; UN, 2003; USEPA, 2007). The upper limit of this range is a conservative estimate, since it estimates the construction cost of a BNR-based system rather than a conventional ASP-based system. Thus, construction cost for the sewer connecting to the central plant exceeds the combined cost of the decentralized treatment plant and the sewer connecting to it.

Comparison using capacity expansion cost for central treatment plant

The capital cost for central plant was computed to be in the range **$55,843,334.45 - $106,303,852.96.** The minimum expansion cost can be assumed to be 50% of the minimum construction cost, i.e. **$27,921,667.23**. As compared to the combined cost of the decentralized plant **($5,942,092.68 - $22,498,583)** and the connected sewer **($6,004,499)**, the combined cost of expanding the sewer connecting to the central plant **($67,839,165)** and expansion of the central plant facilities **($27,921,667.23)** is much higher (EPA, 1978; EPA, 1980; LADWP, 1983; Qasim, 1992; UN, 2003; USEPA, 2007).

In this case study, the decentralized plant was assumed to have biosolids treatment. A decentralized treatment system can be optimized and controlled more efficiently by utilizing the central treatment facilities for biosolids treatment, similar to the Los Angeles City D.C. Tillman and Glendale wastewater treatment plants. At these plants and other upstream plants operated by the Los Angeles County Sanitation Districts, primary sludge, waste activated sludge, screenings and grit are all returned to the trunk sewers. This simplifies the distributed plant design and construction, and does not overload the trunk sewers, since these flows are typically less than 1% of the treated flow.

CONCLUSION

Pump-back power consumption of the recycled water, expansion cost of sewer lines and central treatment plant facilities determine the feasibility of installing a decentralized system for a new/remote region. Pump-back power consumption is a large fraction of the total treatment power consumption and can surpass aeration energy consumption. With the demand for recycled water on the rise, pump-back power consumption has to be controlled to make water reclamation more practical and economical. To support long reclaimed water distribution networks, pump stations require to be installed which can be a significant part of the capital investment. The enlargement of sewers also had a very high cost as compared to installing a decentralized plant and a local sewer. If the expansion of central facilities to accommodate the added flowrate is required, it tips the scale further in favor of decentralized wastewater treatment. Thus, decentralized treatment can be an extremely practical and economic alternative for expanding communities and urban areas. Furthermore, it ensures access to wastewater treatment for remote regions. This study establishes the need to model decentralized treatment systems, in terms capacity, number of units and location.

REFERENCES

City of Los Angeles, Dept. of Public Works, http://www.lacitysan.org, Accessed 11/27/2011

Environment Protection Agency, (1978), Technical report - Construction Costs for Municipal Wastewater Treatment Plants: 1973 – 1977, EPA/430/9-80-003

Environment Protection Agency, (1980), Technical report - Construction Costs for Municipal Wastewater Treatment Plants: 1973 – 1977, EPA/430/9-80-003

Google Earth© 7.0.2, 2013, Accessed (7/2012-2/2013)

LA City Sanitation District, (2010), Wastewater Capital Improvement Program http://www.lacitysan.org/fmd/WCIP/WCIPbook10-11.pdf

LADWP, (1983), Design Manual, Cost and Quantity Estimating Manual, http://dpw.lacounty.gov/des/Design_Manuals/CostQuantityEstimatingManual.pdf (Accessed, 01/30/2012)

Massoud, M. A., Tarhini, A., and Nasr, J. A., (2008), Decentralized approaches to wastewater treatment and management: Applicability in developing countries, Journal of Environmental Management 90(1): 652–659

Metcalf and Eddy Inc., (2003) *Wastewater Engineering: Treatment, Disposal, Reuse*, 4th Ed. McGraw-Hill, New York, N.Y.

Ott, C. and Jones, P. (1988), Partial Flow Properties in Circular Sewers, J. Trans. Eng., 114(1): 113-119

Qasim, S. R., Siang, W. (D.) L., Motley, E. M., Heung, K. G., (1992), Estimating Costs for Treatment Plant Construction, Journal AWWA, 84(8):56-62

Rosso D., Stenstrom M. K., Larson L. E., (2008), Aeration of large-scale municipal wastewater treatment plants: state of the art, Water Science and Technology, 57(7):973-8.

Tchobanoglous, G. and Crites, R. W. (1998) *Small and Decentralized Wastewater Management Systems,* 0-07-289087-8, McGraw-Hill, USA

United Nations Economic and Social Commission for Western Asia, (2003), Waste-water treatment technologies: a general review, United Nations

United States Census Bureau website, http://quickfacts.census.gov/qfd/states/06000.html Accessed, 01/15/2012)

United States Department of Labor, Bureau of Labor Statistics, http://data.bls.gov/pdq/SurveyOutputServlet (Accessed, 01/24/2012)

USEPA, (2007), Biological Nutrient Removal Processes and Costs, EPA: 823R07002

Ten Years and Counting in the Wastewater Process Energy Efficiency Industry: Northern California's CalPOP Program

Edward Myers*, MSChE, Lescure Engineers, Inc.,
John Bidwell, Quantum Energy Services and Technology, Inc.
4635 Old Redwood Highway
Santa Rosa, CA 95403
(707) 575-3427

ABSTRACT

An energy efficiency industry operating programs within the water and wastewater sector can speed the reduction of energy consumption. We examine the experience to date of a utility sponsored Northern California wastewater energy efficiency program started in 2001. This program, called California Wastewater Process Optimization Program, or CalPOP, has been quite successful, and generated a wealth of experience and lessons learned. The primary objective of CalPOP is finding and implementing Energy Efficiency (EE) capital improvements that significantly reduce electricity or natural gas demands. CalPOP has found many unique EE measures in wastewater processing. Some, such as optimizing pumps and blowers, yield significant, reliable energy savings when properly evaluated and implemented. Other methods of saving energy, particularly those with less of a track record, are riskier and do not always perform as predicted. Non-technical factors have been critical to the success of the program, in the form of addressing organizational barriers and drivers that impact the probability of implementation. To date, the great majority of EE measures proposed by CalPOP have been, or are expected to be, implemented. CalPOP has been deemed a success by the objective standards used to measure California EE programs. The lessons learned are presented within a strategy of risk management.

KEYWORDS

Energy Efficiency, Wastewater, Energy Efficiency Program, Energy Efficiency Industry

INTRODUCTION

To promote an energy efficient economy, it is vital to have a viable and fully engaged energy efficiency industry. Specialized practitioners within this industry can develop a body of knowledge, tools, and best practices to assist all other sectors of the economy in improving energy efficiency (EE) in their operations. However, for an EE industry to mature and prosper, it must itself have a successful business model, and be able to justify its own existence economically. Is the water – energy nexus a niche that can support a segment of the emerging EE industry? In this presentation, we will look at the experience of a Northern California program which promotes EE in one sector of the energy-water nexus – publicly owned and operated wastewater treatment facilities.

Northern California is primarily supplied with electricity and natural gas by Pacific Gas and Electric (PG&E). PG&E operates a portfolio of programs to promote EE improvements, including the California Wastewater Process Optimization Program (CalPOP). CalPOP started in 2001 in response to the emerging California energy crisis of that period. CalPOP was developed and is managed by Quantum Energy Services and Technologies (QuEST) of Berkeley, California, a private engineering firm specializing in implementing EE programs and services. CalPOP provides no-cost high-quality energy audits, and cash incentives for completed EE measures identified within the audits. QuEST's Implementation of CalPOP as a utility (PG&E) program lends both credibility and objectivity to the program as viewed by the program's customers, namely municipal and special district wastewater facilities that are customers of PG&E. Lescure Engineers is an engineering subcontractor to QuEST, and has performed ten of the CalPOP facility audits, primarily within the 2010-2012 three-year program cycle.

With the success of the statewide EE industry, and other EE policy efforts, California has been able to hold overall electric energy consumption to 1990 levels, while adding 7 million new residents since 1990. When the California Public Utilities Commission (CPUC) sets electricity and natural gas rates, they include a public goods charge on all customer energy bills that funds EE programs statewide. California's investor owned utilities (IOU), such as PG&E, receive EE funding from the CPUC to administer a portfolio of EE programs, some of which are contracted to third party implementers such as QuEST's implementation of CalPOP.

The power savings these EE programs yield are treated as another resource to balance energy supply with demand, whereby installed EE projects are a reliable means of supplanting the need for constructing and fueling new power plants. Specifically, EE programs are evaluated in terms of the "avoided costs" of power generation. Accordingly, EE programs are closely scrutinized by both the IOUs and the CPUC to insure they are cost effective. The EE programs are typically authorized for 2 to 3 year cycles, and must apply for renewal. If a program underperforms against its EE goals and program budget, and is determined not to be cost effective, the program is at risk of not being renewed. Cost effectiveness translates to verified energy savings, in sufficient quantity, and at reasonable costs within a given time horizon. As a result, programs such as CalPOP develop operational strategies that maximize the yield of energy savings, contain costs and minimize the risk of program termination.

It has been widely recognized that wastewater processing has abundant opportunities for energy savings, although there are formidable market barriers to overcome in the pursuit of installed EE projects. This has been chronicled in the California Market in an market characterization study for the water and wastewater sector recently prepared for PG&E and Southern California Edison(SCE) and the California Measurement Advisory Council (CalMAC)[1]. The Environmental Protection Agency has published a guide to wastewater EE measures[2], and established a benchmarking methodology with an Energy Star rating system for wastewater facilities[3]. The New York State Energy Research & Development Authority (NYSERDA) has developed a Best Practices Handbook for Water and Wastewater Energy Management[4], and the Water Environment Federation (WEF) published a very helpful EE practitioner's guide, Manual of Practice 32 (MOP 32)[5]. Typically, for local governments, wastewater facilities represent the largest utility bill account among all municipal facilities and operations. Subsequently, the pursuit of energy efficiency in the wastewater sector requires engagement with local governments and utility districts as is depicted in these studies.

METHODOLOGY

The objective of the CalPOP program is finding and implementing wastewater process energy savings through a mix of capital improvements and process control modifications. The CalPOP audits are targeted to specific processes and equipment whereEE improvements offer a compelling business case, and are most likely to be implemented. The capital improvements recommended typically require $10,000 to $1,000,000 in capital expenditures, and require the support of Staff, Councils and Boards within the municipality or district.

A CalPOP audit is not a comprehensive audit in terms of producing an exhaustive list of EE measures for the wastewater facility. Instead, the processes are reviewed, and discussions held with staff to identify targeted EE measures – discrete improvements that can be implemented together, or separately as desired, without extensive re-design of the facility. Proposed EE measures must fit within the customers' risk tolerances in terms of minimizing the probability of upsetting plant processes and ultimately, discharge permit violations.

Power logging is typically performed for two weeks on the targeted equipment, to capture facility baseline energy use and the effect of daily variations in equipment use. Typically, pertinent historical data from Supervisory Control, Alarm, and Data Acquisition (SCADA) historical files are combined with logged power data. Then, a spreadsheet and/or process computer model is developed to predict the energy savings from proposed improvements. The calculations are checked by a third party technical reviewer contracted by PG&E.

The primary deliverable of a CalPOP investigation is a high-quality engineering audit report that accurately identifies EE measures and their expected outcomes in terms of project costs, energy and energy bill savings along with an estimated CalPOP incentive. Engineering audits have been provided at no cost to wastewater facilities that have been previously screened to maximize the probability of program participation and project installation. The standardized audit report format presents highly accurate energy saving predictions of these computer models on the targeted processes. A preliminary design description and capital budget estimate is supplied, and the benefit/cost ratio is expressed for each measure as a simple payback period (Capital costs divided by first year energy savings). This report also provides the technical basis to quantify expected rebate incentives available upon project completion, and to qualify the project for zero and low interest loans from PG&E or the California Energy Commission (CEC). CalPOP program participants value the engineering audit reports which typically would not have otherwise been commissioned by the participant, and yield EE opportunities that the customer was not previously aware of, or had no basis for making a decision to pursue them.

The audit report also projects the dollar amount of the CalPOP program rebate incentive, which is roughly equivalent to the first year cost savings from the installed EE measures. Power logging is again performed when the installation is complete to verify the actual energy savings. Again, calculations are checked by a third party technical reviewer contracted by PG&E.

The incentive is calculated based on installed and measured natural gas savings (measured in Therms, equivalent to 100,000 BTUs), and total electric energy, measured in both Megawatt-hours (MWh) savings and peak demand Megawatt (MW) reductions. In many instances of completed CalPOP projects, the project features significant improvement in the control and operation of treatment processes. While participant administrators appreciate the program incentive and operation al cost savings, plant operators very much appreciated improvements in the control of treatment processes.

RESULTS

Case Study Summaries

As to the potential for cost effective energy savings in wastewater treatment, we have found EE opportunities in many aspects of the treatment processes. The movement of fluids – pumping of water and supplying the air for aeration – is the biggest single area where opportunities abound, and savings can be realized with low risk. Four successful fluid movement projects with the CalPOP program are as follows:

1. The City of Santa Rosa's Reclamation Pumps Retrofit. This project was the focus of a presentation at the 2011 Energy/Water conference[6]. Two large reclaimed water pump stations owned and operated by the City of Santa Rosa, California, were thoroughly optimized for improved energy efficiency. Capital costs were $720,000, and the project installation was publicly bid. Electricity consumption was nearly cut in half. The project yielded an 895 MWh annual reduction in electrical energy, saving an estimated $107,000 per year. The City pursued the project vigorously, and it took just over three years from the delivery of the CalPOP audit to project completion. The savings were greater than predicted in the audit, as a result of further refining the retrofit during final design.

New 200 HP Variable Speed Pump (Center)

New 75 HP Variable Speed Pump

Figure 1 – City of Santa Rosa Reclamation Pump Retrofits

2. Vallejo Sanitation and Flood Control District (SCFD) added a small variable speed air compressor to two existing larger fixed speed compressors. The project was implemented in-house for $17,000, yielding a measured cost savings of $13,800 annually. Annual energy consumption of the air compressor system dropped 70% from 233 MWh to 71 MWh, more than predicted in the audit. The implementation was complete one year after the delivery of the audit report.

Figure 2 – Vallejo SCFD Air Compressor Retrofit

3. The Town of Yountville optimized speed on Roots style positive displacement blowers at an aeration basin, where a constant airflow is maintained. The retrofit consisted of changing the pulleys on the belt drive, reducing annual energy consumption from 91 MWh to 51 MWh, and yielded $5,000 annual power bill savings on a $2,400 investment. The savings were nominally less than predicted in the audit.

Figure 3 – Yountville Aeration Blower Retrofit

4. Two high-speed turbo blowers replaced two existing multi-stage centrifugal blowers project at the Napa Sanitation District's Soscol Water Recycling Facility. Prior to installation, the project was modified beyond original project recommendations set forth in the Facility Audit Report. The modified project was designed to obtain additional energy savings by separating the channel air system from the secondary aeration basin system, and adding two smaller blowers to service the channel air system demand thereby allowing for optimal sizing and operation of the replacement blowers installed for the secondary aeration basin system. The installed aeration system improved the Plant staff's ability to control and stabilize the aeration system, which periodically suffered wide swings in blower pressure and airflows. The installed project the delivered more than 1.0 MWh in annual energy savings to the District, and measured savings slightly exceeded estimates provided in the QuEST's Audit.

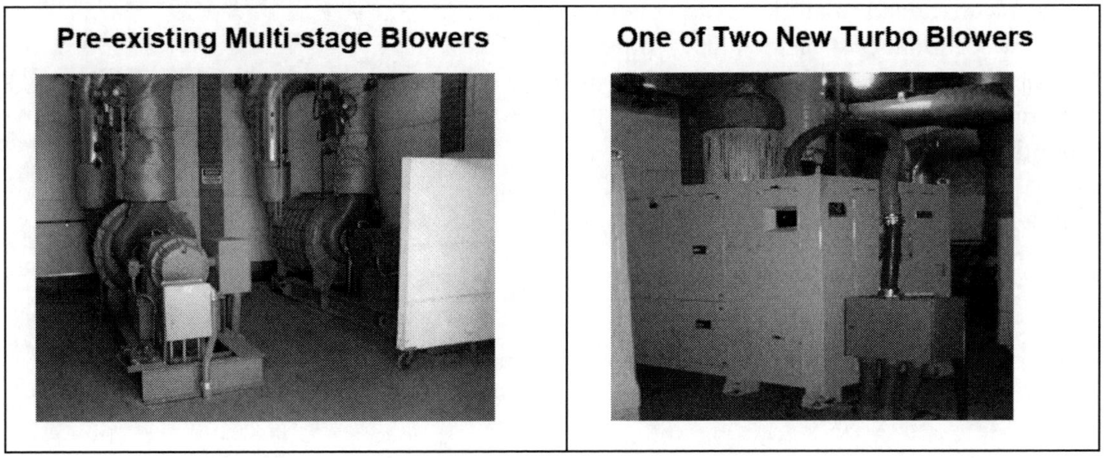

Figure 4 – Napa Sanitary Turbo Blower Aeration Retrofit

These four examples represent relatively low risk projects. Pumps, blowers, and compressors are well understood, and if accurate and detailed process information is obtained, and the existing and proposed equipment is carefully modeled, energy savings can be closely predicted.

However, there are many other areas for improving the EE in wastewater. Since there are many types of wastewater processes, there are many EE opportunities that are unique, and as a result, less proven. Of the following seven examples, two have been completed, with the others undergoing design, or still in the planning stage, or declined by the owner:

1. At the Town of Yountville, the anaerobic digester gas collection and flare system had fallen into disrepair. Originally designed to burn digester gas and supplement with natural gas to heat the digester, corrosion and plugging in the digester gas collection piping resulted in burning natural gas exclusively. Further, the waste digester gas flare had corroded to the point of failure, and digester gas was vented to atmosphere. The CalPOP audit showed the natural gas cost savings could justify the refurbishment of the system. An improved flare was installed, along with new digester gas flow meters. The project was made possible by a zero interest on-bill financing (OBF) loan provided by PG&E, and took two years from audit report completion to finish. Capital costs were $128,500; natural gas consumption dropped two-thirds, translating to natural gas savings of $10,200/year. Predicted savings from the audit were $13,700/year. The efficiency of the sludge heater in utilizing digester gas was over-estimated in the audit phase, since it could not be observed in this mode of operation. In the future, more accurate estimates will be possible when this type of opportunity is encountered.

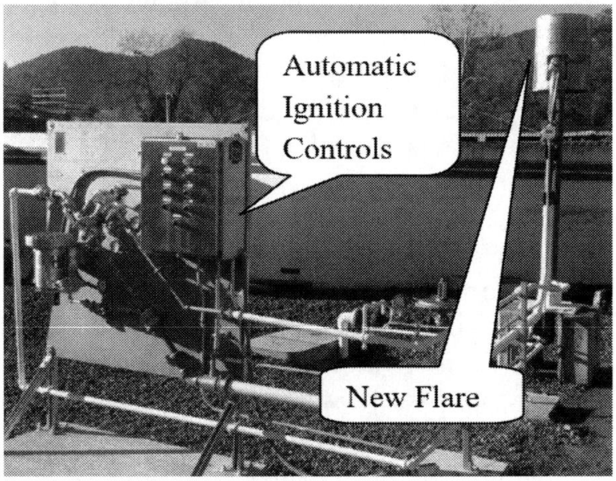

Figure 5 – Yountville Digester Gas Collection Flare Retrofit

2. At Vallejo SFCD, an aeration basin had fixed speed positive displacement blowers and Dissolved Oxygen (DO) sensors, but lacked automatic DO control of aeration. Variable speed drives were added to all three blowers by District staff, at a capital cost of $19,000. Annual electrical energy was reduced 51 MWh, a $4,300 annual cost savings. The project implementation took 15 months. The estimated energy savings in the audit were twice as high, based on an empirical savings model which was subsequently abandoned. While the project remained a financial plus for Vallejo with a 4.4 year simple payback, the CalPOP program did not receive the full compensation that was anticipated based on its performance fee structure.

Figure 6 – Vallejo SCFD DO Control Aeration Retrofit

3. The City of American Canyon installed one of the first Membrane Bio Reactor (MBR) activated sludge treatment systems on the West Coast. It uses considerable submerged air, for both supplying dissolved oxygen, and agitating the membranes to minimize solids buildup and the associated pressure drop across the membranes. Since the system was installed, the MBR vendor has recommended methods to significantly reduce agitation aeration. Furthermore, the dissolved oxygen can only be manually controlled. Reducing agitation aeration, and adding DO control, will require considerable improvement in field instrumentation and controls, totaling $217,000. But with estimated savings of $59,000 per year, the simple payback is attractive. However, as with Vallejo DO control example above, there is a greater risk of under-performance, due to the complexity of biology, operational needs, and limitations of the available models for predicting DO control energy savings. Final design has begun, and completion is expected by the end of 2013.

4. Also at the City of American Canyon, the UV disinfection system has three banks of lamps which currently run at full power all the time. A prior UV dose study determined that only two banks were required most of the time. The audit estimated annual savings at $14,000, and the controls upgrades needed to automate the dose control would cost at least $100,000. American Canyon is moving forward with final design, with completion planned for late 2103. However, prior CalPOP experience in improving UV energy efficiency at other sites has shown mixed results as there are examples of UV measure successes and one failure, albeit unrelated to performance of the EE equipment installed. The relationship between UV dose and disinfection efficacy is complex. The cost of disinfection violations could easily negate electricity savings, and may cause the operators to fall back to full power operation.

5. At the City of Yuba City, a purified oxygen activated sludge process uses Pressure Swing Adsorption (PSA) to generate oxygen enriched air. Yuba's PSA reciprocating compressors have become increasingly expensive to maintain, and site staff are motivated to replace them with newer compressors. Our audit showed the screw compressors that have been used as replacements at other sites are only marginally more energy efficient, and replacement could not be cost justified on compressor energy efficiency alone. The audit did find an apparent opportunity to improve the PSA controls and increase the yield of oxygen, in turn reducing air demand on the compressors. However, this improvement has never been tested elsewhere, and therefore the savings are uncertain. Even with the potential boost in overall PSA energy efficiency from both new compressors and controls, the simple payback would be over 10 years. The project is being weighed by the owner. Unfortunately, following the audit, another northern California wastewater facility with a PSA unit that retrofitted with screw compressors has found them less reliable, and more expensive to maintain than anticipated. In retrospect, this PSA project may not have been appropriate to the CalPOP program, given its novelty, and a considerable risk of under-performing.

6. The Livermore Amador Valley Water Management Agency (LAVWMA) operates a large (ten pumps totaling 4,800 HP) pump station to lift treated wastewater over a 400' grade for disposal in San Francisco Bay. When any of the pumps are off, the check valves must hold back 180 psi or more pressure. Eight of the ten check valves had significant leaks, and replacing them is estimated to save 146 MWh per year, or $15,100, at a total capital cost of $64,300. LAVWMA has replaced the two valves with the highest leak rates. The remaining six check valves are slated for replacement by the end of 2013.

7. The Sonoma County Water Agency, Airport Wastewater Treatment Facility uses membranes to filter secondary wastewater. Upstream of the filters, screens are installed to prevent larger particles from potentially clogging the filters. When screening, the screens have a few psig pressure drop. When backwashing, these screens required a pressure drops of 20 psig to dislodge solids. The upstream pressure is controlled by variable speed pumps. The upstream pressure is maintained as required for backwash, and while screening the excess pressure is wasted downstream through pressure relief valves. We proposed a more sophisticated control strategy that would allow the upstream pressure to stay at its current setting during backwash, then reduce pressure during screening. The pump heads would also be optimized for this strategy. Although estimated savings of 187 MWh and $15,200 per year were predicted, with a three year simple payback, the Agency eventually declined to implement the measure, saying that their operational mode had changed. In retrospect, this was an unproven measure, and perhaps too complicated.

Energy Savings Summary in Wastewater Processes

As a subcontractor within the CalPOP program, Lescure Engineers performed audits on ten wastewater facilities, and proposed eighteen separate EE measures. The EE cost savings, including verified and estimated, are concentrated in two areas: 1) secondary treatment, and 2) effluent pumping. Figure 7 below illustrates this.

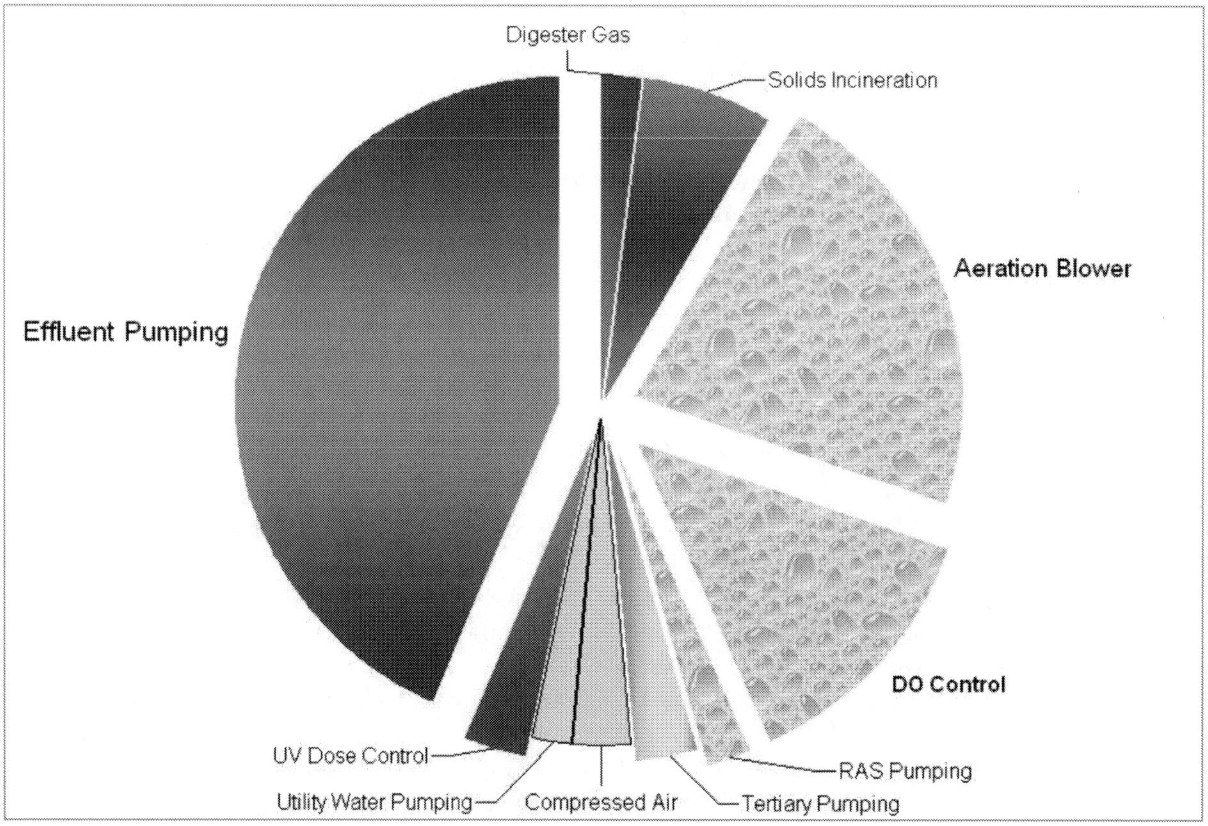

Figure 7 – Breakdown of Cost Savings by Process (Lescure Engineers' Projects)

Secondary treatment EE savings result from optimizing the delivery and control of aeration air in the activated sludge process. High speed turbo blowers move air more efficiently than fixed speed multistage blowers, and the turbo blowers have built-in turndown capabilities that respond to varying air demand without energy wasting control valves. DO control monitors the biological air requirements and delivers only what is needed, which typically reduces the overall quantity of air delivered under manual control, and can significantly reduce aeration costs.

Effluent pumping costs are considerable at California wastewater plants that cannot discharge effluent year round, due to limited surface water during the dry summer season. The effluent pumps may deliver water to irrigation sites, or transport the treated water many miles (over hilly terrain) to ocean disposal sites. These pumps can often be optimized with variable speed controls, or other means, to move water with less energy demand.

The remaining wastewater process areas – including site utilities – have been found to have less potential for energy savings, but may be worth pursuing. If measures with smaller savings potential are blended with larger savings measures, it can create an attractive package.

Speed of implementation impacts the CalPOP program funding. Part of the funding is contingent on successful implementation. Figure 8 below shows the correlation between the capital cost and time between the completion of the audit and startup of the improvements for the Lescure CalPOP projects to date. Smaller projects that can be done in-house with staff, such as two on the left, can complete in roughly one year; at the other extreme, larger publicly bid projects, shown on the right, can have much longer completion times.

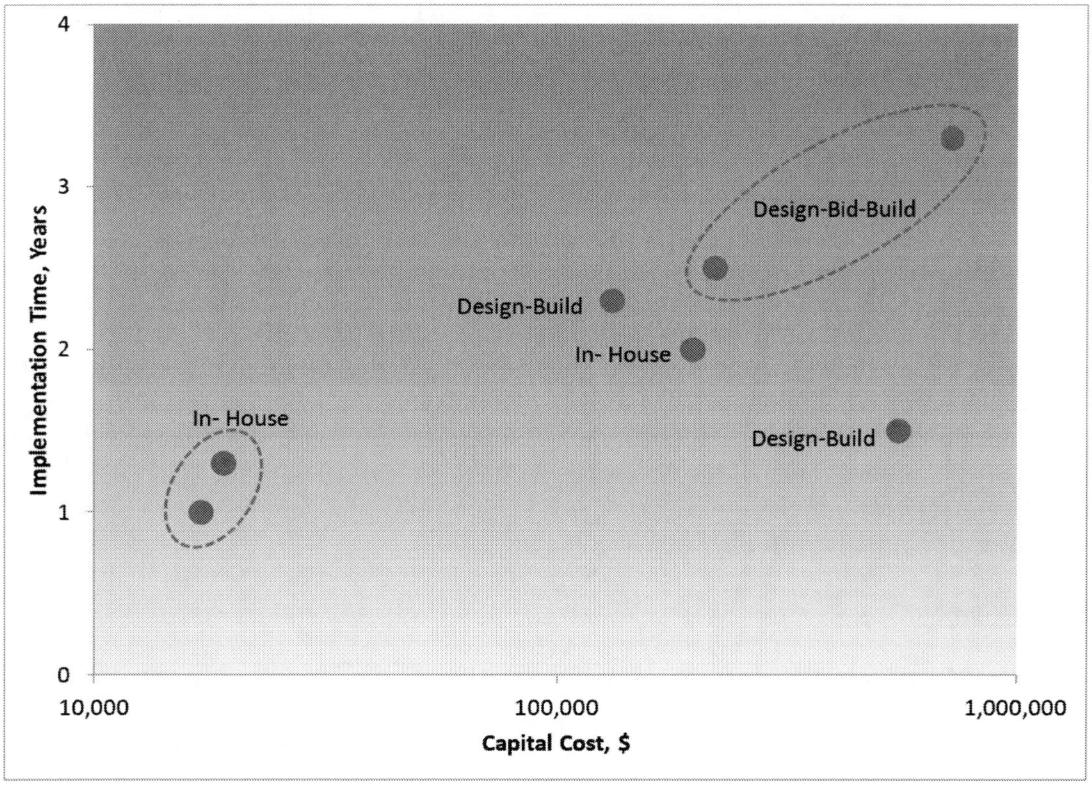

Figure 8 – Implementation Time vs. Capital Costs (Lescure Engineers' Projects)

Performance Overview of the CalPOP Program to Date

Over twelve years, CalPOP has both audited and completed projects in more than 65 municipal wastewater facilities throughout California, though most have these have been concentrated in the PG&E service territory in the northern half of the State. These projects have total verified savings of 51,800 MWh/year, and have been undertaken a total cost of $ 15.3 Million, with the assistance of $3.3 million in program incentives to program participants.

Of all the 48 customer sites audited by the CalPOP program in the last five years, 46% have completed installations and an additional 25 % of are expected to complete installations before the end of 2014. About 29% of audited sites have either declined or are unlikely to complete projects complete projects within before the end of 2014. The CalPOP program, as administered by QuEST as a Third Party Implementer under contract with PG&E, has a renewed contract to implement the program for 2013 and 2014; in this time several additional projects will be added with the intent to complete installations for these customers within this two-year program cycle.

As a subcontractor within the CalPOP program over the last 5 years, Lescure Engineers has experienced a 30% project completion rate for audited projects to date, and 45% of projects are expected to complete before the end of 2014. On balance, 10% of projects are unlikely to complete, and 15% of the project customers have declined to implement. Overall, Lescure Engineer's 75% success rate leaves room for improvement, but is comparable to the CalPOP program overall, and is about the best that is expected in the real world.

DISCUSSION AND CONCLUSIONS

Lessons Learned

There are, of course, many facets to developing and maintaining an EE program in the wastewater sector. The pursuit of energy efficiency is not typically among the highest priorities of wastewater facility operators and municipal goverments, as the primary focus of plant operators is to maintain treatment processes in a manner that minimizes the risk of very costly permit violations. However, in recent years, the pursuit of operational cost savings in municipal operations coupled with the emerging local government interest in climate action plans has elevated EE as a priority in their wastewater facilities. Consequently, an EE niche has been slowly emerging within the wastewater engineering services sector, although it is still considered to be a nascent practice within this sector. Overall, given the predominant interest of facility operators in minimizing the operational risks, the risk management aspects of a projects becomes a useful lens to encapsulate many of the non-technical lessons learned with the CalPOP program. There are three main outcome risks to be managed: 1) Non-implementation; 2) Slow implementation; 3) Under-performing improvements.

Non-Implementation

To keep the CalPOP program cost effective from the utility's perspective, and in turn for the EE industry, a high percentage of the audits must proceed to implementation in a timely fashion. However, barriers to implementation often emerge that fall outside of the technical and financial feasibility of projects as presented in the audit reports. We have found the following barriers to implementation:

1. Capital funds are not available, or take a long time to be internally allocated;
2. The measures proposed appear risky, particularly if they have not been proven elsewhere;
3. All decision makers were not brought on board prior to starting the audit, and do not support the plan of action;
4. Other facility upgrades or priorities demand the attention of staff;
5. The owner does not have the capacity to assume additional debt.

The first barrier, available capital funds, is being addressed at the California state policy level. The California Energy Commission has had a low interest loan program for public agencies since the early 2000's. More recently, PG&E has initiated a zero interest "on-bill financing" (OBF) Program, up to $250,000 per project, which has been well received. If the staff at the site seem unwilling to risk unproven measures, then propose only proven measures. The other barriers have to be addressed at the outset: identifying and involving all decision makers; understanding other projects or demands on staff; and assuring debt capacity exists.

On the positive side, we found there are drivers within the staff and community – aside from strictly financial motivation - that can increase the odds of implementation. These are:

1. Operational benefits aside from energy cost reduction, such as improved controls, fewer upsets and breakdowns, better ability to control the processes and achieve compliance;
2. Interest in replacing worn and obsolete capital equipment as part of an energy efficiency improvement;
3. The concepts for the efficiency measures are proposed by staff;
4. Community and local government support for greenhouse gas reductions associated with reduced energy demand.

The barriers and drivers should be assessed at the outset, before resources are committed for an investment grade audit. If on balance the prospects look unfavorable, it may be wise to decline to participate with the site, or delay until circumstances improve. However, a degree of non-implementation must be anticipated in planning the EE program finances; just as no one in baseball bats a thousand, some projects simply won't implement.

Slow Implementation

Speed of implementation has been an ongoing concern. The CalPOP program has typically run in 2-3 year cycles, with the goal of completing all projects within each cycle. Major capital projects in the wastewater industry take two or more years from inception to startup. As shown previously, the greater the capital cost, the longer implementation takes. However, the larger projects also yield larger energy savings, and can improve the cost effectiveness of the EE program. Programs such as CalPOP must plan for this length of time for implementation, and be patient for the up front investment in outreach and audits to pay off. In addition to the time it takes to implement once the audit is complete, further up-front time is required for marketing and performing the audit. It takes a long term commitment by utilities and policymakers for these programs to bear fruit in the wastewater sector.

Under-Performance

Under-performing improvements present a risk to an EE program that can be far reaching. It undermines the credibility of the EE team's ability to predict savings and deliver a high quality engineering audit. When a project under-performs significantly, not only will the owner involved be less receptive to additional measures, but bad word of mouth may discourage other wastewater facilities from getting involved with the program. The extreme of this – when a project is deemed a failure – a non-performing improvement that saves no energy – can be a far reaching disaster for all involved. The under-performance risk is mitigated up front by using measures proven elsewhere. One other key – for larger, more complex projects - is a final design team with prior experience in the particular type of measure. It further helps if the audit team can retain influence, if not control, over the final design process.

Formal Evaluation of CalPOP Performance

As the CalPOP program is ultimately funded by ratepayers, it must meet performance goals set for all ratepayer funded efficiency programs. In effect, it must be able to compete economically with new supply (power plants, gas purchases, and infrastructure) as a method of balancing electricity (and natural gas) supply and demand. Accordingly, there are cost effectiveness metrics that programs must meet in order to be renewed in successive three year program cycles. These apply to the program cost to the utility vs. savings delivered, and also the combined cost of the program plus the capital cost for the ratepayer that implements efficiency measures. The intent of the latter is to insure EE programs are cost effective to the economy as a whole, not just for the utility.

The cost-effectiveness of EE programs that are funded through a public goods charge are measured and regulated by the California public Utilities Commission (CPUC), which seeks to provide objectively derived estimates of avoided costs to the utility and society at large. The CPUC has developed and deployed a transparent and repeatable costing methodology that does not rely on either proprietary data or models. The inputs to this costing model are extensive, and are beyond the scope of this paper to discuss, although they span everything from attributes of climate zones and utility planning areas, sector load shapes, transmission and distribution systems, system reliability, load shapes, energy losses, environmental externalities, market price forecasts, and attributes of the programs and the ordained EE measures that are deploy within the program.

The key outputs of the cost-effectiveness measurement process are expressed in terms of a "Total Resource Cost" or TRC, and a Program Administrator Cost (PAC) test, and in both cases the net present value (NPV) and Benefit-Cost (BC) ratios are the ultimate expression of program performance. The TRC seeks to measure programs cost-effectiveness in terms of net societal costs and benefits, whereas the PAC test is the cost-effectiveness of program delivery.

A Program's TRC costs include total program budgets and net incremental (participant) costs of all the measures that get installed under the program, less the total program rebates. TRC net benefits, for both gas and electricity, are the sum of the present value of benefits from all of the programs installed measures within the given prescribed duration of the program; an adjustment is made to gross benefits by applying a CPUC specified-Net-to-Gross ratio. The PAC costs only include total program budgets and customer incentives, yet account for benefits in the same manner as the TRC.

Summary Conclusions

Through the TRC and PAC tests, the CPUC and PG&E have concluded the CalPOP program delivery has been cost effective and has provided a clear net benefit to the California economy. In the final analysis, the CalPOP program, and much of the California EE industry as a whole, relies on the policy support from the CPUC, and program administration from the IOUs. Implementation of CalPOP as an utility (PG&E) program, combined with the program's track record of successful projects, establishes the credibility, objectivity and viability of CalPOP in the eyes of the program's potential customers. However, the customers' predominant drivers for engaging the services of the CalPOP are to obtain reliable, no-cost technical expertise and to garner the financial assistance offered in the form of program incentives and qualifying projects for zero and low-interest loans offered by the utility and the State.

While there are free-standing, privately financed Energy Service Companies (ESCOs) that are able to implement energy efficiency improvements in buildings, they have yet to penetrate water and wastewater facilities to any degree, in our experience. There have been instances where CalPOP has received ESCO requests for wastewater EE services to supplement the development of thier broader portfolio of EE measures in other types of facilities belonging to municipalities that the ESCO is contracted to serve . Given elevated levels of fiscal stress in recent years, many municipalities will be driven to reduce operational costs, seek financial incentives and attractive lending options to fund EE projects, and it is likely that both ESCOs and CalPOP will be in a position to serve this type of municipality. However, equipped with the financial, technical and institutional support services needed to overcome the market barriers to EE that are unique to the municipal wastewater sector, CalPOP will continue to be well positioned to serve this segment of the municipal market.

In the final analysis, the CalPOP program, and much of the California EE industry as a whole, relies on the policy support from the CPUC, and program administration from the IOUs. Given the direct program experienceof CalPOP, we believe it would be difficult to establish an ongoing energy efficiency industry in the energy-water nexus without the policy support afforded by the ratepayer funded efficiency programs, and an ongoing effort by policymakers to address market barriers.

REFERENCES

[1]Kema, Industrial Sectors Market Characterization: Water and Wastewater Industry, January 2012,http://www.calmac.org/publications/Final_Industrial_Sector_Market_Characterization_Water_Wastewater.pdf

[2] U.S. Environmental Protection Agency, September 2010, Evaluation of Energy Conservation Measures for Wastewater Treatment Facilities, EPA 832-R-10-005

[3] U.S Environmental Protection Agency, October 2007, ENERGY STAR® Performance Ratings Technical Methodology for Wastewater Treatment Plant

4 The New York State Energy Research & Development Authority (NYSERDA), September, 2010, Water and Wastewater Energy Management: Best Practices Handbook http://www.nyserda.ny.gov/Commercial-and-Industrial/Sectors/Municipal-Water-and-Wastewater-Facilities/MWWT-Tools-and-Materials.aspx

[5] Water Environment Federation, 2010, Energy Conservation in Water and Wastewater Facilities, MOP 32, McGraw-Hill

[6] Edward Myers, John Rosenblum, John Bidwell, August 2011, Case Study: Retrofits Yield Dramatic Energy Savings at California Reclaimed Water Pump Stations, WEF Energy/Water 2011 Conference Proceedings

THE WATER IMPLICATIONS OF GENERATING ELECTRICITY: WATER CONSUMPTION ACROSS THE UNITED STATES BASED ON DIFFERENT ELECTRICITY PATHWAYS THROUGH 2050

Jordan Macknick, National Renewable Energy Laboratory
Sandra Sattler, Union of Concerned Scientists
Kristen Averyt, Cooperative Institute for Research in Environmental Sciences
Steve Clemmer, Union of Concerned Scientists
John Rogers, Union of Concerned Scientsts
National Renewable Energy Laboratory
15013 Denver West Parkway
Golden, CO 80401
(303) 275-3828

ABSTRACT

The power sector withdraws more freshwater annually than any other sector in the U.S. The current portfolio of electricity generating technologies in the U.S. has highly regionalized and technology-specific requirements for water. Water availability differs widely throughout the nation. As a result, assessments of water impacts from the power sector must have a high geographic resolution and consider regional, basin-level differences. The U.S. electricity portfolio is expected to evolve in coming years, shaped by various policy and economic drivers on the international, national and regional level; that evolution will impact power sector water demands. Analysis of future electricity scenarios that incorporate technology options and constraints can provide useful insights about water impacts related to changes to the technology mix. Utilizing outputs from the Regional Energy Deployment System (ReEDS) model, a national electricity-sector capacity expansion model with high geographical resolution, we explore potential changes in water use by the U.S. electric sector over the next four decades under various low carbon energy scenarios, nationally and regionally.

KEYWORDS

Energy-water nexus, electricity, cooling system, renewable, low-carbon

INTRODUCTION

The electricity sector in the United States has a significant impact on national and regional water resources and is highly dependent on water. The United States Geological Survey (USGS) has estimated that 41% of all freshwater withdrawals in the United States in 2005 were for the electricity sector, primarily for thermoelectric cooling needs (Kenny et al., 2009). Although water consumption for electricity generation accounts for a much small portion of total water consumption (3%) (Solley et al., 1998), it can have impacts in places with low water availability or high water temperatures (Averyt et al. 2013). The future development of the power sector will have important impacts on regional water resources, while the availability of water resources can impact the types of power plants and cooling systems that are built. Consequently, the power sector may also be vulnerable to variability in water quantities (e.g. drought), especially those that may result from potential climatic changes (Dai, 2010; van Vliet, 2012; Averyt et al., 2013). The 2007 drought in the southeast exposed many thermal generators, including Browns Ferry nuclear plant, to water-related shut downs and curtailments due to high discharge temperatures and shallow or exposed cooling water inlet locations (NETL, 2009b).

Power plants have two primary quantity-related impacts on water resources: withdrawal and consumption. According to the USGS, "withdrawal" is defined as the amount of water removed from the ground or diverted from a water source for use, while "consumption" refers to the amount of water that is evaporated, transpired, incorporated into products or crops, or otherwise removed from the immediate water environment (Kenny et al., 2009). Both water withdrawal and consumption values are important indicators for water managers and other stakeholders determining power plant impacts and vulnerabilities associated with water resources. Once-through cooled facilities withdraw large amounts of water, increase the temperature of the water when it is returned, and can impact aquatic ecosystems, depending on local water availability (Reynolds, 1980; EPA, 2011). Recirculating cooling systems (or closed-loop) withdraw less, but consume more water per unit of generation than once-through systems on similar power plants, which could have important consequences in times of low water availability (Macknick et al., 2011).

Low carbon energy technology choices can have different impacts on water resources, depending on energy source and cooling system decisions. Renewable energy sources, including non-thermal renewables (e.g., solar photovoltaics and wind) and thermal renewables (e.g., geothermal and concentrating solar power), along with non-renewables (e.g., nuclear, fossil technologies with carbon capture and storage, or CCS) have a wide range of water impacts (Macknick et al., 2011). Some state agencies, as well as the Environmental Protection Agency, have recognized the connections between energy and water and have proposed policy actions to address the impacts of power plants' water use and the environmental impacts of their cooling systems (CSLC, 2006; Kyl, 2010; NYSDEC, 2010; EPA, 2011).

Prior efforts have examined water-related impacts of future electricity production on national and North American Electric Reliability Corporation (NERC) regions. This study goes beyond these efforts through consideration of highly detailed regional analysis of water impacts based on a regionalized electricity model. EPRI (2011) and Roy et al. (2012) consider regional impacts of water withdrawal in the electricity sector, but the regional electricity demand is developed at the NERC level and disaggregated according to existing generation and transmission, not taking into consideration the impacts of new generation and transmission or water consumption (EPRI, 2011; Roy et al., 2012). Other studies that consider electricity generation at the NERC region or sub-region level (NETL, 2009a; Elcock, 2010; Chandel et al., 2011; Cooley et al., 2011) do not have sufficient geographic resolution to adequately evaluate highly localized water impacts. Van Vliet et al. (2012) assessed the relative vulnerability of existing power plants (select plants in the United States and Europe) to projected changes in cooling water source temperatures and flows, but not on scales that are policy relevant for state water managers (van Vliet et al., 2012).

This paper evaluates the water implications of a range of future electricity generation mixes in the United States expected to have appreciably different water profiles and impacts at a regional level that is relevant for policymakers and water managers (Clemmer et al., 2013). Because of the large role of the power sector in contributing to and potentially mitigating climate change, several of our scenarios incorporate deep cuts in carbon emissions in the electricity sector. Generating diverse electricity mixes allows for an examination of water use at broad scales and of how mixes emphasizing different technologies might fare, particularly during droughts and considering competing uses. The results of this work can be used in conjunction with other estimates of future water availability (EPRI, 2011; Harto and Yan, 2011; Roy et al., 2012) to identify areas that may see competition for water resources between the energy sector and other sectors.

METHODS

As described in Clemmer et al. 2013), we consider four electricity scenarios: one reference scenario and three carbon-constrained scenarios emphasizing different low carbon technologies. The reference scenario (Scenario 1) is patterned off of EIA's Annual Energy Outlook 2011 (AEO 2011) reference case. Scenario 2 assumes that the United States meets the electricity sector's share of a cumulative economy-wide carbon budget of 170 gigatons of CO2eq from 2012-2050 through economic competition of low-carbon technologies. Scenarios 3 and 4 include the electricity sector's share of the U.S. CO2eq emissions budget plus additional targets for specific low--carbon technologies. For scenario 3, we assume nuclear generation would grow from approximately 20% of the U.S. electricity mix today to 29% in 2035 and 36% in 2050, while coal with CCS would grow to 15% of the generation mix by 2035 and 30% by 2050. For scenario 4, we assume aggressive deployment of energy efficient technologies and buildings would reduce U.S. electricity demand 20% by 2035 and 35% by 2050 versus the reference case, while generation from renewable energy technologies (wind, solar, geothermal, biomass, and hydropower) increases from 10% in 2010 to 50% in 2035 and 80% by 2050

The national electricity capacity expansion model used to analyze these scenarios, the Regional Electricity Deployment System (ReEDS), forecasts the deployment of supply-side generation and transmission capacity for the power sector in the contiguous United States in two year increments to the year 2050 (Short et al., 2009). ReEDS is a long-term capacity expansion and dispatch model that represents all major generation technologies, including coal, natural gas combined cycle, natural gas combustion turbines, fossil fuels with carbon capture and storage (CCS), nuclear, hydropower, wind, solar, geothermal, biopower, and storage. It is unique among capacity expansion models for its highly discretized regional structure and statistical treatment of the impact of variability of wind and solar resources on capacity planning and dispatch. Further details of the electricity scenarios and modeling assumptions can be found in Clemmer et al., 2013).

Water withdrawal and consumption coefficients are applied in a consistent functional unit of cubic meters per MWh of electricity generated and are adapted from Macknick et al., (2012a), as utilized in Averyt et al., (2011). Only operational freshwater uses are analyzed by region, but use of saline water is also tracked. Water use rates are assumed to be constant over the duration of the analysis, though any changes in technologies' thermal efficiencies could alter those rates. We aggregate fuel and cooling type technologies from existing plants to match ReEDS technologies along with fuel and cooling system categories discussed in Macknick et al., (2012a). Median withdrawal and consumption factors are utilized for all technologies. All existing coal and nuclear plants are matched with the Generic category in Macknick et al., (2012a), which includes estimates of water usage for a variety of different types and vintages of coal plants. For concentrating solar power (CSP) technologies, which include both parabolic trough and power tower systems in ReEDS, we use dry-cooled parabolic trough systems water use rates for this analysis, with exception of the cooling system sensitivity analysis, where we considered wet-cooled parabolic trough systems. Geothermal technologies have large ranges of water use values, depending on the specific technology and whether they use externally sourced freshwater or the on-site geothermal fluids for cooling. For purposes of this study, we assume all new geothermal technologies are dry-cooled binary systems that required additional freshwater for makeup due to operational geofluid loss, per assumptions in Clark et al., 2011. For the cooling system sensitivity analysis we consider wet/dry hybrid-cooled binary systems, which use more freshwater as part of their cooling systems. Due to data constraints, we use dedicated biopower plant values for plants that co-fire biomass with coal. ReEDS does not consider combined heat and power (CHP) systems. Water use from CHP technologies are thus omitted, though CHP could play an important role in meeting energy efficiency reductions such as those envisioned under Scenario 4. Whereas substantial evaporation can occur from reservoirs that produce hydroelectricity, we have elected to not consider this water usage (withdrawal or consumption) due to the complexities in attributing water use to particular demands on these reservoirs, such as drinking water supply, recreation, and flood control (Gleick, 1992; Torcellini et al., 2003).

Water use coefficients were applied to technology-specific generation output from each ReEDS power control area (PCA) region, based on 2008 generation data by technologies with specific cooling systems, as compiled and described in Averyt et al., 2013. We consider only volumes of inland freshwater resources and existing wastewater utilization for regional analysis. Saline water volumes utilized in electricity production are tracked but excluded from the regional analysis. In cases where EIA does not report cooling systems for certain fuel-types within a particular PCA region, the cooling system makeup for the fuel-type with most generation is applied. If there is no existing generation in a PCA, new thermoelectric generation (with the exception of geothermal and CSP technologies) is assumed to be cooled with recirculating cooling systems. In addition, reflecting Environmental Protection Agency guidelines (EPA, 2011), all new thermal generation is assumed to be equipped with recirculating cooling towers or dry-cooled systems, depending on existing structures in each PCA. For retirements of thermal generation, once-through cooled thermal plants within a PCA are assumed to retire prior to plants with recirculating cooling technologies. In coastal areas, retiring once-through cooled facilities utilizing saline water are assumed to be replaced by power plants utilizing freshwater in recirculating cooling systems.

Water withdrawal and consumption values are displayed according to USGS Hydrologic Unit Code 2 levels (HUC-2) (Seaber et al., 1987). Water values are calculated based on electricity generation by PCA region as calculated by ReEDS and then distributed to HUC-2 regions on a land area basis. Figure 1 shows the mapping of ReEDS PCA regions to USGS HUC-2 regions. For a PCA entirely within a HUC-2 region, all generation and water use in the PCA is attributed to the HUC-2. For a PCA that overlaps with multiple HUC-2 regions, generation and water use from the PCA are apportioned to overlapping HUC-2 regions based on the percent of the HUC-2 region inside each PCA region. No attempts are made to project water demands from other sectors or any metrics of water availability.

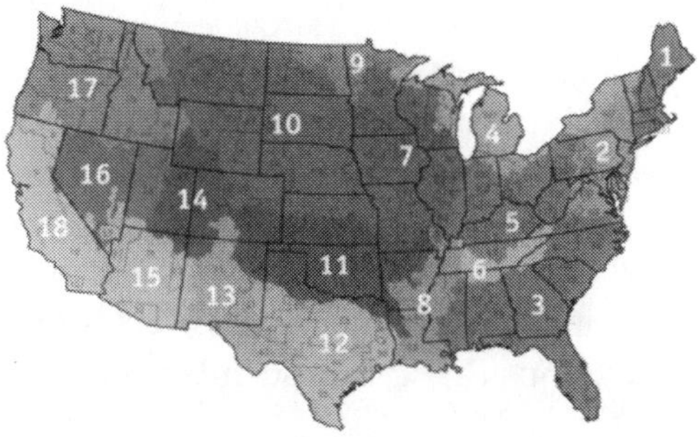

Figure 1. Mapping of ReEDS PCA regions (grey outlines) to USGS HUC-2 regions (shaded regions).

RESULTS

Results from the ReEDS modeling show a range of electricity generation technologies deployed between 2010 and 2050 under the four scenarios (figure 2). Under the reference case (Scenario 1), natural gas becomes the dominant fuel for generating electricity, while renewable energy experiences more modest growth, to meet the projected increase in electricity demand and replace retiring coal and nuclear plants. Under the carbon budget scenarios (Scenarios 2-4), conventional coal generation is largely phased-out by 2030. Under Scenario 2, conventional natural gas generation increases in the early years to replace coal and reduce power plant carbon emissions, while renewable generation (particularly wind and solar) and natural gas with CCS make significant contributions in the last half of the forecast. Under Scenario 3, coal with CCS and nuclear steadily increase after 2020, providing approximately two-thirds of total U.S. generation by 2050, while renewable energy technologies provide most of the remaining generation. Under Scenario 4, energy efficiency more than eliminates the projected growth in electricity demand, while wind and solar increase to meet a large share of the renewable energy target of 80% by 2050.

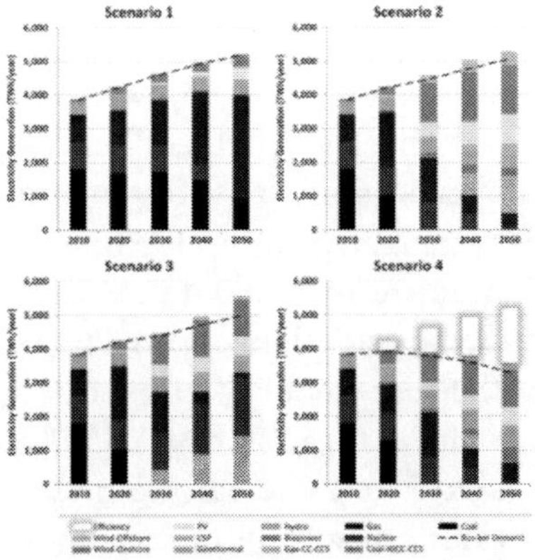

Figure 2. National electricity generation by scenario. Results from the ReEDS model indicate a variety in total electricity generation values and deployed electricity generation technologies in 2030 and 2050. Scenario 1, reference case; Scenario 2, carbon budget, no technology targets; Scenario 3, carbon budget with coal with CCS and nuclear targets; Scenario 4, carbon budget with efficiency and renewable energy targets.

National level water consumption trajectories vary widely depending on energy scenario (figure 3). Compared with 2010, Scenario 1 shows an increase in national water consumption of 320 million cubic meters (0.6%) by 2030, but a decrease of 1.7 billion cubic meters (34.2%) by 2050. Scenario 3 leads to a 1.8 billion cubic meter (35.0%) reduction by 2030, though subsequent increases in consumptive uses lead to a net increase of 700 million cubic meters (21.7%) from 2010 values. Scenarios 2 and 4 follow a similar decreasing trajectory until 2030, reducing consumptive uses by 3.1 (60.0%) billion cubic meters by 2030, yet diverge from 2030 to 2050. In 2050, total reductions in consumption for Scenario 2 are 2.8 billion cubic meters (55.4%), whereas Scenario 4 leads to reductions of 4.2 billion cubic meters (85.2%) from 2010 values.

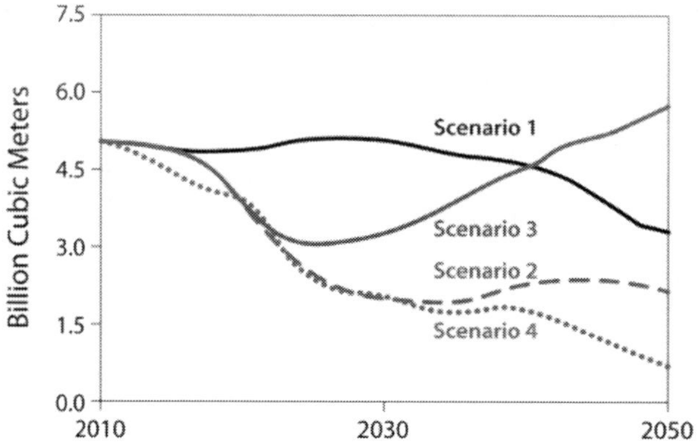

Figure 3. National level water consumption results for four electricity scenarios. Scenario 1, reference case; Scenario 2, carbon budget, no technology targets; Scenario 3, carbon budget with coal with CCS and nuclear targets; Scenario 4, carbon budget with efficiency and renewable energy targets.

National consumption trends are slightly different than withdrawal trends due to different relative withdrawal and consumption factors for the energy technologies and cooling systems deployed. Under Scenario 1, consumption in 2030 increases from 2010 as a result of increased electricity demand being met by primarily natural gas combined cycle plants, with no substantial reduction in coal and nuclear generation. By 2050, coal and nuclear generation is substantially reduced and replaced with natural gas combined cycle generation, which has a lower consumption rate than coal and nuclear generation. Scenario 2 consumption decreases greatly from 2010 to 2030 due to coal plant retirements, and then increases from 2030 to 2050 as a result of building new natural gas combined cycle plants with CCS. Scenario 3 consumption declines sharply from 2010 to 2030 due to the retirement of conventional coal facilities and additional natural gas combined cycle generation. From 2030 to 2050, consumption increases due to increased deployment of coal with CCS and nuclear facilities utilizing recirculating cooling technologies, which have higher water consumption rates than other technologies. Scenario 4 national water consumption declines steadily due to a reduction in total energy demand and increased penetration of renewable technologies.

Similar to withdrawal impacts, regional changes in consumptive uses vary greatly by energy scenario and may differ from national trends (figure 4).

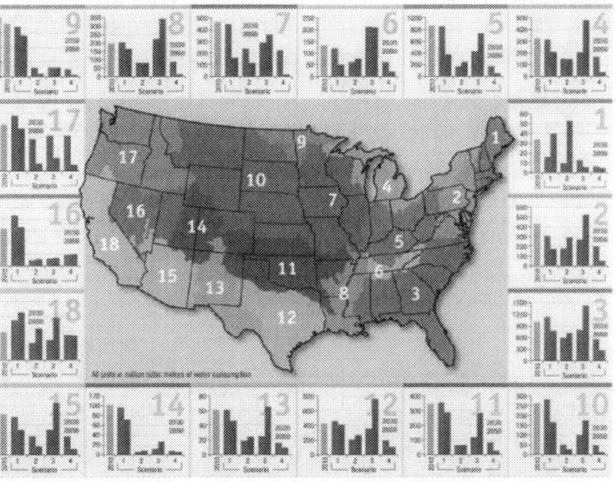

Figure 4. Water consumption results (in million cubic meters per year) for 2030 (blue bars) and 2050 (red bars) by HUC-2 region for electricity Scenarios 1-4; Scenario 1, reference case; Scenario 2, carbon budget, no technology targets; Scenario 3, carbon budget with coal with CCS and nuclear targets; Scenario 4, carbon budget with efficiency and renewable energy targets; y-axes have different scales and are for intra-region comparison purposes.

In particular, Scenarios 1 and 3 lead to changes in water consumption in many regions (figure 4) that may differ from national trends (figure 3). Compared with 2010 values, Scenario 1 leads to a slight increase in consumptive uses in 2030; this national-level increase is a result of increases in consumption for certain HUC-2 regions representing the southeast (3, 8), southwest (16, 18), northwest (17), and central (10, 11, 12) parts of the nation. The majority of regions in the U.S. show reductions in expected consumptive use. By 2050, although national levels of consumptive uses decrease, HUC-2 regions in the northeast (1) and the southwest (16, 18) show increases in consumptive uses. Increases in consumption in the southeast and central regions are largely due to the retirement of once-through coal and nuclear facilities, which are replaced by new power plants operating with recirculating cooling systems. The switch from technologies utilizing once-through systems to technologies utilizing recirculating systems contributes to the decrease in withdrawals in the southeast and central regions, yet increased consumption levels. Increases in consumption in the southwest and the northwest regions in Scenario 1 are largely due to the construction of new recirculating cooling systems in areas that currently have no generation or have predominantly hydropower generation. Under Scenario 3, national levels of consumption decrease in 2030 (figure 3), yet HUC-2 regions 6 and 8 in the southeast show increases in consumptive uses (figure 4). By 2050, after substantial increases in nuclear and coal with CCS generation, consumption increases from 2010 values in a variety of regions in the Mid-Atlantic (2), Great Lakes (4), southeastern (3, 6, 8), central (12), and southwestern (13, 15, 18) parts of the country. All other regions (1, 5, 7, 9, 10, 11, 14, 16, 17) show decreases in consumption in 2050 from 2010 values. Scenario 2 shows reductions in consumption for all regions except regions 1 and 18 in 2050. Scenario 4 leads to reductions in consumption for all regions in both 2030 and 2050.

DISCUSSION

Reductions in national water consumptive uses (figure 3) appear in each electricity scenario, yet higher resolution results provide more insight for water management. Water resources are managed on relatively small spatial scales that may cross national and international political boundaries. Providing projections of future water use from the electric sector on a watershed-level (HUC-2 region) may be more useful for water managers than providing estimates at a national-, state-, or FERC region-level; examining results by region may (as in our cases) provide results that differ importantly from national ones. A watershed-level approach also facilitates analysis with estimates of current and projected water availability metrics. Switching from once-through cooled systems to recirculating systems may lead to an increase in consumption for some regions. Under Scenario 1, total national level consumptive uses increase by the year 2030 (figure 3), driven by increases in eight regions (3, 8, 10, 11, 12, 16, 17,18) in the southeast, central and southwest parts of the nation (figure 4). Many of these regions have experienced droughts in recent years; increased droughts or other reductions in water availability could affect availability of water for the power sector. National levels of consumption increase from 2030 to 2050 in 9 regions (2, 3, 4, 6, 8, 12, 13, 15,18) in the east, southeast, midwest, and southwest parts of the nation (figure 4), driving increases in national consumption totals (figure 3). In cases where national level water consumption decreases, many regions may still show increases in consumption. For example, from 2030 to 2050, Scenario 1 shows decreases in national levels of water consumption (figure 3), yet during this same period consumptive uses in some northeast and southwest regions (1, 16, 18) rise above 2010 values (figure 4). In addition, from 2010 to 2030, Scenario 3 consumptive uses decline on a national level (figure 3), yet increase in two regions (6, 8) in the southeast (figure 4).

High renewable energy penetration scenarios lead to the most substantial reductions in water consumption. Scenarios 2 and 4 both show greater than 50% renewable penetration by 2050 and only two regions (1 and 18, representing the northeast and California) in Scenario 2 show increases in consumption in 2050 (figure 4). In contrast, for Scenarios 1 and 3, which have non-renewable technologies providing more than 50% of generation in 2030 and 2050, 14 of the 18 regions show an increase in consumption in 2030 or 2050. In regions 16 and 18 in the west, freshwater consumption in 2030 is greater under Scenario 4 (with high renewable penetration) than Scenarios 2 and 3 (figure 4). This is largely driven by the deployment of geothermal and solar technologies into these areas in Scenario 4. Scenario 4 levels of consumption in 2030 for regions 16 and 18 are, however, lower than Scenario 1 and 2010 baseline values (figure 4). By 2050, consumption in region 16 remains higher than that in Scenarios 2 and 3 and below that of Scenario 1 and the 2010 baseline. For region 18, consumption in 2050 for Scenario 4 is the lowest of all scenarios.

A high degree of spatial resolution in this study assists in targeting specific areas for future analysis that may be overlooked when analyzing data on a national or state level. National level reductions in withdrawals or consumptive uses may be a result of large changes in some regions, while increases may be occurring in other regions, such as in the southwest and southeast. Future, more granular studies could be conducted in the southeastern and southwestern parts of the nation, which historically have seen increasing energy-water competition and which consistently show potential water usage increases under the electricity scenarios considered in this study. Electricity generation and deployment in these areas could be limited by water availability, increasing demands from other sectors, or legal availability related to the water transfers between water rights holders (e.g., agriculture and municipal). In some areas, large increases in water withdrawals or consumptive uses by the power sector or any other sector by 2030, combined with potential climatic changes such as droughts, could prompt policy changes that affect future energy and water management decisions. However, we did not evaluate water availability as a constraint, nor did we evaluate climatic changes and how those could affect decision-making; future analyses could evaluate these issues.

Freshwater usage associated with many renewable and non-renewable technologies could be reduced by utilizing alternative water resources (e.g., shallow brackish water) or by utilizing dry cooling technologies, yet these technologies and options incur additional costs and performance penalties and may affect power plant siting decisions.

CONCLUSIONS

This work provides an initial assessment of the water impacts of various electricity scenarios at a USGS HUC-2 region level. National level electricity sector consumptive uses decline in 2030 for all low carbon scenarios, while the reference case, Scenario 1, shows a slight increase. By 2050, all scenarios have lower consumption levels than 2010 except for Scenario 3, which includes high penetration of coal with CCS and nuclear technologies.

Regional trends may differ substantially from national level results. Scenario 1 leads to increases in consumption in 2030 for areas in the southeast, southwest, and central parts of the nation, and Scenario 3 leads to increases in Mid-Atlantic, Great Lakes, southeastern, Texas, and southwestern states. In general, regional increases in consumption are associated with Scenarios 1 and 3, whereas the Scenarios 2 and 4 show consistent reductions.

This high level analysis provides insight to decision-makers on water-relevant spatial scales related to the impacts of future electricity scenarios on water resources. Future research of this nature is recommended, including a more detailed study on the impacts for specific geographic regions that may see increases consumption, analysis of additional electricity generation scenarios, evaluation of different cooling system policies and technologies, and an assessment of how total projected water usage for power plants and other sectors compares with projected water availability in each region. For further information, see Macknick (2012b).

REFERENCES

Averyt K, Fisher J, Huber-Lee A, Lewis A, Macknick J, Madden N, Rogers J and Tellinghuisen S (2011) *Freshwater use by U.S. power plants: Electricity's thirst for a precious resource* (Cambridge, MA: Union of Concerned Scientists)

Averyt K, Fisher J, Huber-Lee A, Lewis A, Macknick J, Madden N, Rogers J and Tellinghuisen S (2013) *Developing a comprehensive database of power plant cooling water use: Lessons learned, challenges, and opportunities.* Environmental Research Letters 8:1. 015001

Chandel M K, Pratson L F and Jackson R B (2011) *The potential impacts of climate-change policy on freshwater use in thermoelectric power generation.* Energy Policy 39 6234-6242

Clark C, Harto C, Sullivan J and Wang M (2011) *Water Use in the Development and Operation of Geothermal Power Plants* (Argonne, IL: Argonne National Laboratory)

Clemmer S, Macknick J, Mai T, Rogers J and Sattler S (2013) *Modeling low-carbon U.S. electricity futures to explore impacts on national and regional water use.* Environmental Research Letters 8:1. 015004.

Cooley H, Fulton J and Gleick P (2011) *Water for Energy: Future Water Needs for Electricity in the Intermountain West* (Oakland, CA: Pacific Institute)

Dai A (2010) *Drought under global warming: a review.* Wiley Interdisciplinary Reviews: Climate Change 2 45–65

California State Lands Commission (CSLC) (2006) *Resolution by the California State Lands Commission regarding once-through cooling in California power plants.* Proposed April 13, 2006

California Energy Commission (CEC) (2009) *Best Management Practices & Guidance Manual: Desert Renewable Energy Projects* (Sacramento, CA: California Energy Commission)

Department of Energy (DOE) (2012) *SunShot Vision Study Energy* (Washington, DC: U.S. Department of Energy)

Elcock D (2010) *Future U.S. water consumption: the role of energy production.* Journal of the American Water Resources Association, 46(3)

Electric Power Research Institute (EPRI) (2011) *Water Use for Electricity Generation and Other Sectors: Recent Changes (1985-2005) and Future Projections (2005-2030)* (Palo Alto, CA: EPRI)

Energy Information Administration (EIA) (2010) *Form 906/920/923: Utility, Non-utility, and Combined Heat And Power Plant Database. 2007 Monthly Time Series.* (Washington, DC: U.S. Energy Information Administration)

Energy Information Administration (EIA) (2011) *Form 860: Annual Electric Generator Report* (Washington, DC: U.S. EIA)

Environmental Protection Agency (EPA) (2011) *Cooling Water Intake Structures--CWA 316(b), Basic Information* (Washington, DC: U.S. EPA)

Gleick P (1992) *Environmental consequences of hydroelectric development: the role of facility size and type.* Energy 17 735–747

Harto C B, Yan Y E, Demissie Y K, Elcock D, Tidwell V C, Hallett K, Macknick J, Wigmosta M S and Tesfa T K (2011) *Analysis of Drought Impacts on Electricity Production in the Western and Texas Interconnections of the United States* (Chicago, IL: Argonne National Laboratory)

Kenny J F, Barber N L, Hutson S S, Linsey K S, Lovelace J K and Maupin M A (2009) *Estimated Use of Water in the United States in 2005* (Reston, VA: U.S. Geological Survey Circular 1344)

Kyl J (2010) *Deploying Solar Power in the State of Arizona: A Brief Overview of the Solar-Water Nexus* (Washington, DC: Office of Senator Jon Kyl)

Macknick J, Newmark R, Heath G and Hallett K C (2011) *A Review of Operational Water Consumption and Withdrawal Factors for Electricity Generating Technologies* (Golden, CO: National Renewable Energy Laboratory)

Macknick J, Newmark R, Heath G and Hallett K C (2012a) *Operational Water Consumption and Withdrawal Factors for Electricity Generating Technologies: A Review of Existing Literature.* Environmental Research Letters 7:4. 045802

Macknick J, Sattler S, Averyt K, Clemmer S and Rogers J (2012b) *The water implications of generating electricity: water use across the United States based on different electricity pathways through 2050.* Environmental Research Letters 7:4. 045803

National Energy Technology Laboratory (NETL) (2009a) *Estimating Freshwater Needs to Meet Future Thermoelectric Generation Requirements* (Pittsburgh, PA: National Energy Technology Laboratory)

National Energy Technology Laboratory (NETL) (2009b) *Impact of Drought on U.S. Steam Electric Power Plant Cooling Water Intakes and Related Water Resource Management Issues* (Pittsburgh, PA: National Energy Technology Laboratory)

National Renewable Energy Laboratory (NREL) (2012) *Renewable Electricity Futures Study.* ed Hand M M, Baldwin S, DeMeo E, Reilly J M, Mai T, Arent D, Porro G, Meshek M and Sandor D (Washington, DC: U.S. Department of Energy)

New York State Department of Environmental Conservation (2010) *Best Technology Available (BTA) for Cooling Water Intake Structures*. Proposed March 4, 2010.

Reynolds J Z (1980) *Power plant cooling systems: policy alternatives.* Science 207 367–372

Roy S B, Chen L, Girvetz E H, Maurer E P, Mills W B, Grieb T M (2012) *Projecting Water Withdrawal and Supply for Future Decades in the U.S. under Climate Change Scenarios.* Environmental Science and Technology 46 (5) 2545-2556

Short W, Blair N, Sullivan P and Mai T (2009) *ReEDS Model Documentation: Base Case Data and Model Description* (Golden, CO: National Renewable Energy Laboratory)

Torcellini P, Long N and Judkoff R (2003) *Consumptive Water Use for U.S. Power Production* (Golden, CO: National Renewable Energy Laboratory)

Van Vliet M T H, Yearsley J R, Ludwig F, Vögele S, Lettenmaier D P and Kabat P (2012) *Vulnerability of US and European electricity supply to climate change*. Nature Climate Change 2 676-681

Solley W B, Pierce R R and Perlman H A (1998) *Estimated Use of Water in the United States in 1995* (Reston, VA: U.S. Geological Survey Circular 1200)

Seaber P R, Kapinos F P and Knapp G L (1987) *Hydrologic Unit Maps* (Denver, CO: United States Geological Survey, accessed most recent data from ftp://ftp.ftw.nrcs.usda.gov/wbd/)

WATER AND ENERGY UTILITITES: IMPROVING COLLABORATION

Cindy Dyballa, independent consultant
Takoma Park MD
301-220-9041
c_dyballa@yahoo.com

ABSTRACT

Recent years have seen a burst of interest in the relationship of energy and water use and savings. This nexus has been defined and even quantified in many ways. The technical aspects of this nexus are becoming well understood, and technologies are emerging to address both energy savings in water systems and water savings in the energy sector. The number and success of programs that save both end-use water and energy is growing.

But the water and energy sectors operate in different worlds. It seems we don't speak the same language, even when we are trying to. Without a way to find common economic and political ground and create opportunities to work together for mutual benefit, significant progress in water and energy savings will remain elusive.

This paper focuses on water and energy utility collaboration for end-user water and energy efficiencies. It, explores some of the key differences in the water and energy utility sectors, identifies barriers to collaborative program success, and suggests opportunities for greater collaboration. Besides differences in the dominant management systems (public or private), utilities differ in pricing and rate structures, utility size and numbers, and state regulatory structures, organizations and constraints. This affects utility objectives, finances, and how demand management and efficiency are treated within each sector.

Discussion is based on interviews with leading practitioners in the water and energy fields, institutional obstacles from program experience, and barriers and solutions identified by a workshop of water and energy conservation practitioners, as well as available literature summarizing each of the two sectors. The goal of this work is to articulate and understand these differences, in order to improve communication and contribute to more and better joint program efforts.

KEYWORDS

Water efficiency, water savings, partnerships, urban water use, energy efficiency, energy-water nexus, programs, energy savings

INTRODUCTION

One key aspect of the energy-water nexus--the role of water and energy efficiency in water and energy production, distribution, use and treatment--has received recent attention. Technologies are emerging to improve efficiencies. The impacts of greater efficiencies on water and energy providers, their customers and broader environmental goals are becoming clearer. Energy (electric and gas) and water (water and wastewater) utilities are exploring various methods to achieve both direct and embedded energy and water savings. Wastewater systems in particular are paying closer attention to improving operational energy efficiencies that reduce their energy costs.

Both water and energy utilities and their regulators have begun to estimate and target end-use, and in some cases embedded, water and energy savings in their programs. But despite recent program successes, few recent efforts have focused specifically on one specific approach that can contribute to water and energy efficiencies: collaborative end-user programs between water and energy utilities to achieve joint efficiencies.

Joint utility programs that reach end users have potential to create benefits for both parties and contribute to energy and water savings. But the water and energy sectors have historically not worked well or often together, and instead have generally created separate but somewhat parallel efficiency efforts (AWE and ACEEE, 2011). Industry differences between the water and energy utility sectors get in the way of finding common economic and political ground. Challenges have emerged during recent experience in joint efficiency programs.

Some have called for stronger collaboration on energy and water efficiency programs. A national workshop of water and energy experts, looking broadly at how to build joint water-energy efficiency efforts, called for several steps to strengthen cross-sector collaboration in their action "blueprint" (AWE and ACEEE, 2011). Analyses in several states, including Colorado, California and Texas, have urged states to address joint water and energy saving efforts. And a recent nationwide search for exemplary programs yielding energy and water savings highlighted several joint utility efforts (Young and Mackres, 2013).

A Colorado analysis of savings potential from water-energy efficiency programs urged state energy and water agencies to encourage joint efforts, noting residential and commercial program opportunities (Tellinghuisen, 2009). A Texas analysis concluded that implementing end-user efficiencies is one of the keys to sustainable use of both energy and water (Hardberger et al, 2009). Some California water and energy experts believe that despite extensive state experimentation with joint programs, the two sectors must still fully explore what's possible within and beyond existing institutional constraints (Parker, 2012). Another California paper urges further state support for energy-water partnerships of electric and water utilities, mechanisms to better estimate embedded efficiencies, and comprehensive guidelines for managing water and energy together (White, 2013). Broader calls for a greater focus on the water-energy nexus have also recognized the role of joint end-user programs, including the US Department of Energy (2006) and the US General Accounting Office (GAO) (2011).

This paper addresses the issues presented by utility collaboration, with a focus on end-user efficiencies, not energy and water savings in utility operations. The goal of this work is to articulate and understand these challenges and differences, along with possible solutions, in order to improve collaboration and contribute to more effective joint program efforts that save both water and energy.

METHODOLOGY

This discussion is based on several sources, published and unpublished, from both the water (water and wastewater) and energy (electric and gas) fields. Available literature summarizing each of the two sectors was consulted for comparative industry profiles. Barriers to utility collaboration and possible solutions have been considered by experts in at least one forum (AWE and ACEEE, 2011). Programmatic challenges were extracted from documented program experience of exemplary water-energy efficiency programs (Young and Mackres, 2013), unpublished background material for that effort, relevant program efforts uncovered through an online Water-Energy Program Directory (ACEEE, 2013), California's PUC pilot programs (EcoNorthwest, 2011) and elsewhere.

Structured telephone interviews with water and energy efficiency experts and practitioners supplemented this information. The 17 interviewees were selected with attention to filling gaps, not presenting a comprehensive view of each sector. The nine program practitioners consulted have worked in several states in the eastern, western, northwestern and midwestern US. Nonprofit, research and government interviewees brought a state, regional or nationwide perspective and were familiar with the range of program experience in their field. Individual interviewees are not listed by name or quoted, giving them freedom to speak frankly about challenges.

In this paper, water refers to both water and wastewater; energy refers to electric and gas. The terms providers and utilities are used interchangeably. Efficiency here includes both efficiency (same service, reduced use) and conservation (reduced use and service) programs aimed at the end uses of water and energy.

RESULTS

This section summarizes challenges to implementing joint water-energy utility programs for combined water and energy end-use and embedded savings. These include differences in the water and energy industries, regulatory structures, economic factors, communications and information, and approaches to end use efficiency programs.

Challenges Identified by Others

Lack of existing cross-sector working relationships was high on the list of barriers to success for a collection of energy and water experts gathered by the Alliance for Water Efficiency (AWE) and American Council for an Energy Efficient Economy (ACEEE) to develop a joint water-energy action plan "blueprint". The group observed that the two sectors operate under different regulatory and business models and with structures that don't recognize efficiency benefits the same way. They put forward ways to improve collaboration on joint programs, research and policies and codes: "just add water" to integrate water and energy utilities' planning and implementing programs; learn from and replicate best practice programs; and pursue partnership in educational opportunities (AWE and ACEEE, 2011).

They also called for state and federal level policy support in the forms of: a platform to enable communication of energy and water regulators; increased federal and state coordination in grant funding, research, regulation, and technical assistance; regulatory structures and incentives that reward water and energy efficiency; specific energy-water elements added to existing federal legislation; state and federal tax incentives; and federal collection of water and energy end-use data across sectors (AWE and ACEEE, 2011). Participants in a 2010 western workshop, focused on increasing electric and water use efficiencies and on energy recovery from water operations, made similar suggestions for promoting end-use efficiencies, including a research collaborative and attention to ways to transfer the concept of decoupling, separating revenue from sales volume, from the electric to the water industry (Liner, 2012).

Perhaps the most robust discussion on end-user programs has taken place in California. A recent white paper on the water-energy nexus in California considers state level policy and programmatic issues that arise when contemplating programs (Parker, 2012). A paper prepared for the California PUC, in addressing the complexities and economic incentives of embedded energy and water, identifies the key issue as how to manage the water-energy portfolio together (White, 2013).

In considering federal approaches to reduce both energy use for water production and water use for energy development, GAO (2012) concluded that regulatory systems and economic feasibility inhibit widespread adoption of useful technologies and approaches. GAO (2011) noted water industry barriers to achieving end-user water efficiencies that reduce water-related energy uses, among them: water pricing that doesn't reflect all costs; how water utilities operate; competing operational priorities; and lack of public awareness about the energy demands of urban water.

Water and energy industry differences

Number, size and ownership in the water and energy industries differ substantially, despite a wealth of individual variation. Water and wastewater providers are likely to be smaller, more disparate, and publicly managed. These three structural factors have implications for the success of any efficiency program. While consistent precise numbers are difficult to obtain, approximate comparisons are still useful.

Start with the sheer number of providers: some 52,000 separate community water systems for drinking water,[1] and over 15,000 wastewater treatment facilities (EPA, 2009 and 2008), compared with just over 3,200 electric providers. Several of those interviewed commented on this disparity, on fragmentation in the water sector generally, and on the ability of smaller utilities to cope.

The larger electric providers have huge geographic service territories of entire metropolitan areas and even regions, which dwarf the scale of all but the very few largest water wholesalers. And only 145 electric providers represent nearly ¾ of total US electric sales (Nadel, 2013). Some 400 very large drinking water systems (each serving over 100,000 population), representing the top 1%, provide water for nearly half the served US population, and about 4,000 water systems (each with over 10,000 population) together serve more than 245 million (EPA, 2010). A Colorado analysis of savings potential from water-energy efficiency programs flagged mismatched service territories as one hurdle (Tellinghuisen, 2009). One practitioner interviewed noted the difficulties in bridging the scale at which the typical energy and water utilities operate. This vastly different scale is reflected in sales and revenues. Electricity dwarfs the others, with some $368.2 billion in revenue (APPA, 2012). US water utilities collected $42.1 billion in revenue in 2010; wastewater treatment revenue in 2010 was $46.9 billion, mostly from municipal sewage or wastewater treatment (Statista, 2013).

Ownership of the largest energy utilities is virtually all private, and of the larger water providers public. Nearly 7 in 10 of US electric customers (residential and commercial) are served by investor-owned utilities (IOUs), and over ¾ of these IOUs are large. The vast majority of larger gas utilities are also investor owned. Electric and gas IOUs dominate both energy sales and revenue (APPA, 2012). By contrast most Americans get their water and wastewater services from large, publicly managed utilities (EPA, 2008). Virtually all wastewater systems, regardless of size, are publicly managed. Only 7% of the US population with sewer service is provided by private entities (Pinsent

Masons, 2012). These ownership tendencies hold even though the total number of community water systems is evenly split public-private (EPA, 2011) and only a fraction of total electric utilities are investor-owned (APPA, 2012). This picture is complicated by the presence of public electric cooperatives, water wholesalers, federal water and power producers, and joint municipal utilities providing drinking water, wastewater and/or electricity services.

Regulatory Structure
These ownership tendencies affect regulatory structure. Public water providers and private energy IOUs dominate their respective sectors. These largely operate under different state regulatory schemes, managed by separate agencies, imposing different constraints. IOUs are regulated by state utility commissions, and water utilities are often governed by state agencies with other concerns. For public water and wastewater providers the overriding regulatory concern is health and water quality; public safety is a regulatory concern for energy, and more importantly costs and rates (shareholders) (also for water IOUs). Almost all those interviewed stressed the challenges this creates for joint programs. The regulatory scheme affects utility objectives, finances, focus on the customer or ratepayer, and most importantly how demand management and efficiency are treated.

Table 1.**Water and Energy Sectors Compared Nationwide (approximations)**

Characteristic	**Water**	**Wastewater***	**Electricity**	**Gas**
Number	52,000	15,000	Over 3,000	Nearly 2,000
Size and service	400 serve nearly half of customers	1770 produce 92% of volume	145 make ¾ of sales	140 make 90% of sales
Dominant ownership	Public (majority)	Public (over 90%)	Investor owned	Investor owned
State regulation	Drinking water quality	Water quality	Rates and profits (for IOUs)	Rates and profits (for IOUs)
Rate setting	Local or regional	Local or regional	State PUC approved	State PUC approved
Revenues	$42 billion	$47 billion	$368 billion	$115 billion
Approach to efficiency	Best management practices, conservation planning	Operational efficiencies	Energy efficiency Resource standards	Energy efficiency Resource standards

* Wastewater statistics are available in terms of facilities, not managing entities. Other numbers in this table refer to entities.

Economic Factors
Investor-owned energy utility rates are approved by state regulators; IOUs must justify rate increases to recover costs of operations and earn a fair rate of return on investments. For public water and wastewater providers, rate setting is primarily local and not reviewed by the state (Wisconsin is one exception). This can affect accounting for the costs, financing and benefits of efficiency programs, a major concern for most of those interviewed.

Most of those interviewed noted the ability of IOUs in nearly half the states to separate, or decouple, costs of efficiency programs from rate setting (ACEEE, 2012). Many raised the unclear authority of state-regulated energy utilities to invest in water efficiency programs that save energy in many states. One state expert interviewed urged a permissive, rather than legislative or rule making, state approach to this issue. Several noted that there's no other good way to economically estimate and apportion regional environmental and other benefits. Recognizing and allocating embedded savings should help efficiency programs to be cost-effective for individual utilities, but making such estimates is difficult and often requires incompatible or unavailable data (White, 2013).

Simple financial comparisons are difficult because public utilities and IOUs take different budgeting and rate-setting approaches. Most municipal utilities work on an annual basis, taking a cash-needs approach for each year. Energy IOUs, and many public energy utilities, employ a multiyear approach (Aldinger, 2010).

Recent declines in sales and revenue concern both industries. In water this is often attributed to reduced demand from successful efficiency programs. In reality, reduced demand for many reasons, combined with increasing operating costs, has led to increased water rate pressures, and a "new normal" for both the water and electricity industries (Beecher and Chesnutt, 2012; Faruqui, 2012). And while financial needs are an overarching concern of the water sector, sustainability challenges are intertwined; energy efficiency was identified as the most important sustainability issue facing the water sector by nearly half the respondents to a survey of water and wastewater managers (Black & Veatch, 2012).

The effects on collaboration for efficiency programs can be powerful. One practitioner commented that area water utilities may want to participate, but can't, since they have no separate budget for efficiency programs and no way to recoup investment in the short run. On the other hand, another practitioner saw reducing their program costs as a motivation to attract partners. One expert interviewee noted that due to accounting differences, some energy utilities tend to assume that the bulk of savings will show up in the customer's bill. A state expert noted that while the benefits of embedded energy are widely accepted, the estimation of these is fraught with problems. Of note, few of those interviewed mentioned significantly expanding outside sources of funding as a way to address this.

Communications, Stove-piping, and Information Sharing

Perceived sector differences can influence communications and collaboration. Even the terms differ in each industry--water providers speak in terms of water deliveries and households; electric utilities measure sales; wastewater managers talk about facilities, not customers. Those interviewed pointed to weak understanding of how the other sector uses information, makes decisions, sets goals and formulates budgets. Several described the silos in which water, wastewater, electric and gas providers each tend to operate, even within one combined utility, industry norms of talking across these silos only when required, and a corresponding default attitude of mistrust. Several practitioners hit stumbling blocks during program design with time consumed by satisfying several internal bureaucracies, competing concerns over public credit, legal cautions, or the need to elevate to senior management just to get started. Mistrust takes many forms. One practitioner noted that some public water utilities don't want to widely share flaws in their operations (for example a high water loss percentage) as they are directly responsible to the public. Another offered that when the focus is on required program costs, not the customer, it's easy to assume the other utility would do the program without you.

The challenges inherent in data and information sharing for program design and evaluation were highlighted by several interviewees. Several practitioners noted that IOUs (primarily energy) have more restrictions on sharing customer information. This complicates both the delivery of customer program services and program evaluation. At least two successful programs found they needed extra time to spell out formal processes for multiple utility partners to work with customers, exchange and report program information, and get approvals.

Effective data collection is also a collaboration issue, and can require a formal agreement to spell out roles and resolve privacy issues. Collecting data from water and wastewater utilities to evaluate pilot programs was, in itself, a barrier to success in pilot joint programs requested by the California PUC (EoNorthwest, 2011). Consistent water and energy data reporting standards are being sought in California, to help identify gaps, establish best practices, and develop better programs (White, 2013). Texas researchers have made a similar call for better data to solidify estimates (Stillwell et al, 2011).

Approaches to End Use Efficiency Programs

Water and energy utilities take their approach to end use efficiency to some degree from their state regulators. And state policy approaches for water and energy efficiency tend to differ.

Many state water policy makers have focused on promoting or requiring water efficiency planning and/or best management practices (BMPs) for water efficiency. More than half of US states require some form of water efficiency planning, and 19 require some level of plan implementation. Most of the state planning frameworks stress measures or practices rather than percentage water use reduction goals. Conservation activities are required in 15 states as part of a water permitting or rights process[2] (AWE and ELI, 2012).

State energy regulators are moving toward adoption of statewide resource standards--energy savings performance targets that utilities must work towards. The 24 states with energy performance standards provide a strong impetus for local energy efficiency programs that go beyond the easy savings. Several of these states have legislated requirements for utilities to acquire "all cost-effective" energy efficiency resources; one state (California) requires electric utilities to look to energy efficiency as the first provider of new energy. Six more states use other policies to promote significant energy efficiency program savings, as reflected in reduced electricity sales of at least 0.5%/year (Foster et al, 2012). Some additional states (not mapped) include energy efficiency programs as part of an energy integrated resource planning process.

One result is seen in the scale of investment in end use efficiency and the ability to track results. Annual savings from energy efficiency programs increased significantly, topping 18 million mwh in 2010, roughly the amount of electricity used in Wyoming. Utility budgets, 85% for electric utility programs, rose to almost $7 billion in 2011 (Foster et al, 2012). No similar nationwide numbers are readily available for the water sector, although the California Urban Water Conservation Council estimates water savings from implementation of a few drinking water utility BMPs (CUWCC, 2013).

Both water and energy efficiency end-use efforts may get a boost from the emergence of local and state policies addressing other issues. Now 37 states have some form of planning for climate change, and 19 states have set greenhouse gas emissions targets (CCES, 2012). California is not unique in the convergence of several separate efforts to motivate water and energy utility collaboration to achieve greater efficiencies:

- A state assessment that water and wastewater production, distribution, use and treatment are a significant energy use statewide, leading to state policies emphasizing all sources of energy efficiency;
- Statewide water reduction goals of 20% by 2020 tied to urban water management plans that help meet water use reduction targets (AWE, 2012); and
- Climate change legislation (CA AB 32) mandating attention to actions that produce greenhouse gases.

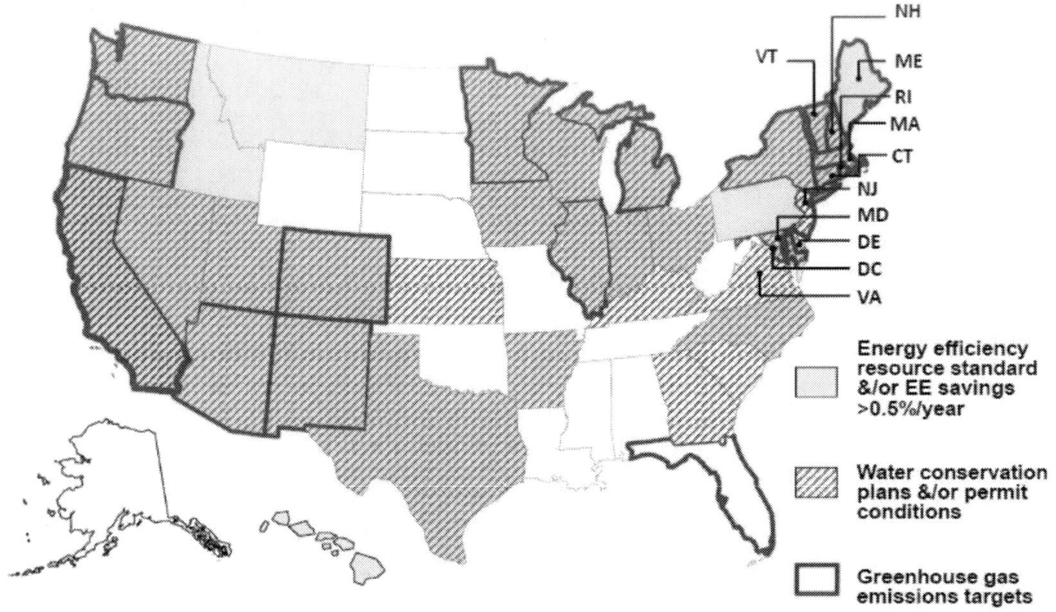

Figure 1. **Selected Water and Energy Efficiency and Climate Policies by State**

Taken together, the convergence of water conservation, energy efficiency and greenhouse gas targets in a state can both enable a regional perspective on benefits and encourage water and energy utilities to pursue joint efficiency programs. Consider a California proposal to link these systems, by meshing water efficiency best practices with energy efficiency program requirements, and designing a more comprehensive approach that builds on both water BMPs and energy resource standards to provide guidelines for managing the water-energy efficiency relationship as one integrated portfolio[3] (White, 2013).

Program Experience

Program experience across the US affirms both these challenges to joint programs and presents creative options to address them.

Evaluation results for several California PUC-requested joint utility water-energy saving pilot programs between three large energy IOUs and their water providers yielded significant energy savings. This was despite somewhat limited scope calculations of water and energy savings (individual utility, not regional perspective) and limited data availability, particularly for wastewater savings (EcoNorthwest, 2011). The PUC is moving ahead to integrate water into utilities' energy efficiency portfolios, and is considering changes to its cost-effectiveness procedures for energy-water efficiency measures to include energy embedded in water and wastewater. But they appear so far to be stopped short of a broader inclusion of energy embedded in all water and wastewater activities from source to use and treatment, regardless of service territory (Parker, 2012).

Several of the most successful energy-water savings programs selected by a group of water and energy experts resulted from joint utility collaboration. A closer look at these reveals both the challenges in joint programs and ways to overcome them. These programs went to great lengths to solidify joint utility commitment, ensure collaboration, and clarify benefits to partners up front. Several efforts started out narrowly before expanding to broader goals. Getting multiple organizations to understand their roles, trust each other, and communicate effectively was in itself a major accomplishment. Some programs had to establish structured decision making processes, rules of operation, or draft legal agreements. Once established, program partners found that collaboration helped mitigate issues of funding, staff time, and organizational commitments (Young and Mackres, 2013). The collaboration aspects of several of these are described below.

One exemplary multifamily efficiency program, an effective three way collaboration of public and private water and energy utilities in Austin Texas, began with an interest in deeper building energy savings and a desire to reach a difficult customer segment with one-stop services. The utilities designed and integrated complementary programs and policies. One utility manages the program, others contribute to rebates; utilities staff are involved, but outsourcing the actual work can free up staff time and resources.

Some of the exemplary programs have similar outside partner involvement. The Denver area Watts to Water program delivers one stop services for hotels and large office buildings. Utility partners Denver Water (public) and Xcel (investor-owned) provide strong support and funds, with a collaborative marketing and outreach strategy, but this program is managed by a separate group with excellent contacts in the target customer group. Getting the many other partners all on board did take time upfront.

The successful Windsor Pays program, partnering the Town and two separate energy and water Sonoma county agencies agency to implement the program, brought a broader regional climate protection authority into the program's design. Similarly, a broader regional air quality management district helped with seed funding for the commercial ozone laundry program of a major electric IOU and the City of Santa Rosa Utilities Dept. (water and wastewater), was initiated as one of the PUC energy-water pilots. And for Living Wise programs that have signed on multiple sponsors, a turnkey contractor manages the program and seeks contributions (funds, staff time or other resources) from water and energy utilities beyond the initial utility sponsor. The Living Wise resource action program in southern California was highlighted as exemplary due to its commitment of both energy and water utilities (Young and Mackres, 2013).

Broader city climate change goals (beyond energy savings) allowed a bigger upfront investment for Boulder's public buildings program, designed as an energy performance contract. Since the City does not pay itself as supplier for water, water savings measures are bundled with other measures in packages that together show good payback. The Sonoma County Water Agency committed in 2011 to pursue a carbon free water supply; this provides another rationale for conservation programs to achieve long term water use reductions that also conserve power.

Joint end-user programs are not always appropriate or useful to an individual utility. MWRA, an award-winning program, has collaborated extensively with area energy utilities for many years, but does not currently pursue aggressive end user programs, with or without partners, at this time. Instead it takes advantage of state renewable energy standards and a regional greenhouse gas initiative to reduce its own energy use and produce power.

ACEEE has also recently posted an online water-energy program directory with summaries of over 450 energy saving programs nationwide that also produce water savings. Most do not involve collaboration between water and energy programs. Of those that do, several utilize a nonprofit group, existing or formed solely for program delivery, that is not restricted by state regulations for water or energy providers. These groups can deliver both water and energy savings and can pursue partnership commitments without actively involving all utilities in day to day decisions (ACEEE, 2013).

Outside this documented program experience are other working water and energy utility partnerships, some in place for years, among them San Antonio, Fort Collins, the San Francisco East Bay area (East Bay, 2013), and West Basin in suburban Los Angeles (West Basin, 2011). Interviews show that many of these partnerships were originally motivated by mutual concern for customer convenience and service, long before the current policy discussion about the energy-water nexus. In at least one case, the partners agreed they did not want to be embarrassed by customers asking why they could not coordinate similar programs and services.
For example, the Seattle area residential WashWise and other area utility end-user programs, targeting both water and energy, partnered water, wastewater, energy utilities and other groups in a long-standing set of programs. Begun as a pilot and focused on measures most cost effective for all, the program design directly addressed equity and collaboration among the partners and built in evaluation of savings and costs. Keys to partnership success were leveraging partner costs, identifying mutual benefits, using regional climate change and environmental goals as justification, and including private business and community group partners (Dietemann, 2008).

DISCUSSION AND CONCLUSIONS

What does all this suggest about ways to strengthen collaboration? Despite the many hurdles involved, water and energy utilities across the US, both public and private, have already joined together across their differences because they see benefits to customer service as well as immediate efficiencies for their system. All but one of those interviewed was very positive about the usefulness of and potential for collaborative end-user efficiency programs as one useful tool for reducing both water and energy use. There were several comments that this area needs more attention to fulfill its savings potential.

Key differences between the water and energy industries have significant implications for collaboration, and can inform what works and what doesn't. The larger energy utilities tend to be investor-owned, the larger water utilities public. The water industry is more fragmented.

Regulatory structure and rate-setting influence how utilities account for efficiency activities. While potential savings are real, how they are accounted for and who gets to count them inhibit collaboration. Programs that clearly define this upfront and seek contributions proportionate to immediate benefits stand a better chance of success. In recognition of the difficulties in estimating and apportioning costs and benefits, especially embedded savings, some successful programs bring in an outside party for seed funding to get started and to acknowledge regional benefits.

The regional economic and environmental benefits of joint programs can be significant, as water becomes more scarce more often in more places throughout the US, and as pressures of climate change force both energy and water utilities to adapt. The three trends prompting state attention in California—climate change, water and energy—are not unique, and can motivate action elsewhere.

Most practitioners described workarounds to address built-in hurdles of communications, differing perspectives, attributing costs and benefits, sharing customer information, and satisfying bureaucratic requirements. These methods can help obtain better results:

- Get the right folks talking at the start.
- Clarify common goals up front.
- Take the time to work through mechanics of collaboration.
- Start small, even with joint messaging and education or a pilot project.
- Pool resources proportionately.
- One utility can take the lead programmatically, and others contribute what they do best.
- Acknowledge difficulties in estimating costs and benefits and apportion contributions.
- Keep it simple for smaller scale utilities.
- Bring in another party (nonprofit) to provide the glue and attract outside funding.
- Agree that service and convenience to customers is an important mutual goal.
- It never hurts to ask (well, almost never).

Opportunities for utility collaboration on end-use efficiencies may be most immediate in situations with:

- Municipally owned joint utilities;
- Energy cooperatives with similar scale as mid-size municipal water providers;
- State energy efficiency resource standards and/or other policies to promote robust utility energy efficiency programs;
- Utilities in any sector that need to go beyond low hanging fruit;
- States and regions with persistent drought and water shortages; and
- Water and wastewater utilities facing significant energy costs or capital costs for expansion.

State policies can strongly support stronger program collaboration, although these policies are not easily adopted. Interviews and formal studies have suggested an array of potential methods to encourage partnership efforts:

- Individual agency actions that address cost-benefit issues;
- Regular communication among energy and water regulators;
- Water-saving targets for water utilities, just as many states have energy-saving targets;
- Integration of state goals for water and energy efficiency and climate change;
- In-state coordination of various agency mechanisms that support water or energy efficiency such as best management practices, planning, and energy efficiency resource standards;
- Water and energy end-use data collection efforts that cut across sectors;

- Use of existing funding mechanisms to jumpstart increased collaboration; and
- Formal mechanisms to recognize the broader benefits of programmatic water and energy savings.

Any and all of these actions can provide water and energy utilities with another avenue to increase efficiency of their service delivery, reduce their own costs, and achieve regional environmental benefits.

REFERENCES AND ENDNOTES

Aldinger, C, Oct. 2010. An Analysis of the Differences in Water Rates of Investor-Owned Water Utilities and Government-Owned Water Utilities. Peasley, Aldinger & O'Bymachow for California Water Association. At http://www.calwaterassn.com/wp-content/uploads/2010/11/CWA-Water-Rates-Differences-White-Paper.pdf.

Alliance for Water Efficiency (AWE) and American Council for an Energy Efficient Economy (ACEEE), May 2011. Addressing the Energy-Water Nexus: A Blueprint for Action and Policy Agenda. At http://www.allianceforwaterefficiency.org/blueprint.aspx.

Alliance for Water Efficiency (AWE) and Environmental Law Institute (ELI), Sept. 2012. The Water Efficiency and Conservation State Scorecard: An Assessment of Laws and Policies. At http://www.allianceforwaterefficiency.org/final-scorecard.aspx

American Council for an Energy Efficient Economy (ACEEE), 2013. Water-Energy Program Directory. At http://www.aceee.org/w-e-programs, accessed February 2013.

American Public Power Association (APPA), 2012. 2012-2013 Annual Directory & Statistical Report. At 111.publicpower.org.

Beecher, J and T Chesnutt, Dec. 2012. Declining Water Sales and Utility Revenues: A Framework for Understanding and Adapting. For Alliance for Water Efficiency and Johnson Foundation.

Black & Veatch, 2012. Strategic Directions in the US Water Utility Industry. At http://brstage.bv.com/docs/management-consulting-brochures/2012-water-utility-report.pdf.

California Energy Commission, Sept. 2009. 2009 Integrated Energy Policy Report-Draft Committee Report. CEC-100-2009-003-CTD.

California Urban Water Conservation Council (CUWCC) 2013. Urban water supplier best management practices are listed online at http://www.cuwcc.org/bmps.aspx?ekmensel=b86195de_24_0_7794_2.

Chesnutt, T, et al, March 2012. A Balanced Approach to Water Conservation in Utility Planning. For Water Research Foundation.

Center for Climate and Energy Solutions (CCES), 2012. Climate Action by State. Online at http://www.c2es.org/us-states-regions.

Department of Energy, US, 2006. Energy Demands on Water Resources: Report to Congress on the Interdependence of Energy and Water.

Dietemann, A, March 2008. Water and Energy Efficiency Partnerships. Seattle Public Utilities unpublished presentation to EPA Water and Energy Meeting.

East Bay Municipal Utility District, 2013. High-Efficiency Residential Clothes Washer Rebates. Online at http://www.ebmud.com/water-and-wastewater/water-conservation/high-efficiency-residential-clothes-washer-rebates, accessed Feb. 2013.

ECONorthwest March 2011. Embedded Energy in Water Pilot Programs Impact Evaluation. Prepared for California Public Utilities Commission Energy Division.

Environmental Business International at Statista 2013. Revenue of the U.S. water utility industry between 2000 and 2010. Online at http://www.statista.com/statistics/192890/revenue-of-us-water-utilities-since-2000/.

Environmental Business International at Statista 2013. Revenue from U.S. Wastewater Treatment-since Since 2000. Online at http://www.statista.com/statistics/192838/revenue-from-us-wastewater-treatment-since-2000/.

Environmental Protection Agency, US, 2006. Community Water Systems Survey.

Environmental Protection Agency, US, 2008. Clean Watersheds Needs Survey 2008.

Environmental Protection Agency, US, 2009. FACTOIDS: Drinking Water and Ground Water Statistics for 2009. Office of Water. EPA 816-K-09-004.

Environmental Protection Agency, US, June 2011 (updated February 2012). Fiscal Year 2010 Drinking Water and Ground Water Statistics. EPA 817K11001.

Faruqui, A and E Shultz, December 2012. "Demand Growth and the New Normal." Public Utilities Fortnightly. At www.fortnightly.com.

Foster, B et al 2012. The 2012 Energy Efficiency State Scorecard. ACEEE Report number E12C. At http://aceee.org/research-report/e12c.

General Accounting Office, US, March 2011. Energy-Water Nexus: Amount of Energy Needed to Supply, Use, and Treat Water Is Location-Specific and Can Be Reduced by Certain Technologies and Approaches. GAO-11-225.

General Accounting Office, US, September 2012. Energy Water Nexus: Coordinated Federal Approach Needed to better Manage Energy and Water Tradeoffs. GAO- 12-880.

Hardberger, A and A Stillwell et al, April 2009. Energy-Water Nexus in Texas. University of Texas at Austin and Environmental Defense Fund. At http://ssrn.com/abstract=1873566.

ICF, 2008. Water and Energy: Leveraging Voluntary Programs to Save Both Water and Energy. For EPA Office of Water.

Klein, G, 2005. California's Water-Energy Relationship. California Energy Commission CEC-700-2005-011-SF. At http://www.energy.ca.gov/2005publications/CEC-700-2005-011/CEC-700-2005-011-SF.PDF.

Liner, B et al, Nov. 2012. "Appendix C, Western Electric Power Sector/Water Sector Collaboration Workshop (2010)" in: Water Energy Future Workshop, Nov. 27 2012. Water Environment Federation, Danfloss and Alliance to Save Energy.

Nadel, S, February 2013. Unpublished ACEEE analyses of data from US Energy Information Agency 2004 Natural Gas Annual and 2013 Annual Energy Outlook.

Park, L, September 2012. California's Water-Energy Nexus: Pathways to Implementation. GEI Consultants.

Pinsent Masons, 2012. Pinsent Masons Water Yearbook 2012-2013 (14th edition). Pinsent Masons LLP, London.

Stillwell, A et al 2011. "The Energy-Water Nexus in Texa*s."* Ecology and Society*, 16*(1), 2. At http://www.ecologyandsociety.org/vol16/iss1/art2/.

Tellinghuisen, S June 2009. Water Conservation = Energy Conservation, A Report for the Colorado Water Conservation Board. Western Resource Advocates.

West Basin Municipal Water District, 2011. Water Efficiency Master Plan Highlights 2011-2015. At http://www.westbasin.org/files/water-efficiency-master-plan/we-masterplanhighlights-final.pdf.

White, R and M Zafar, Jan. 2013. Rethinking the Water Energy Nexus: Moving toward Portfolio Management of the Nexus Policy and Planning Division. CA Public Utilities Commission.

Young, R and E Mackres, January 2013. Tackling the Nexus: Exemplary Programs That Save Both Energy and Water. ACEEE & AWE. ACEEE Report number E131. At http://www.aceee.org/research-report/e131.

---

[1] EPA defines a community water system as one that provides water for human consumption through pipes or other constructed conveyances to at least 15 service connections, or regularly serves at least 25 people year-round.

[2] In Great Lakes Basin states, an interstate compact with water conservation requirements covers parts, not all, of these states (AWE and ELI, 2012).

[3] White (2013) concludes that in some cases, balancing the water-energy portfolio together for greater water and energy efficiencies may lead to more water or energy use in a particular situation.

The Importance of Water for Power Generation

Mike Preston, PE, Black & Veatch
Black & Veatch
11401 Lamar Ave
Overland Park, KS 66211
(913) 458-2626

ABSTRACT

The importance of water for modern day power generation is well known and becoming the focus of more intense study of late. Latest survey data available from the USGS "Estimated Use of Water in the United States in 2005" estimates that thermoelectric power generation accounted for 41 percent of all freshwater withdrawals, roughly 201 billion gallons per day. This is striking compared to the 44.2 billion gallons per day of water withdrawn for public supply documented in the same survey. How is all of this water used in the power generation process? This paper will provide an overview of various power generation technologies, their unique water requirements, water and wastewater treatment processes used in power generation, affects of recent regulatory changes on power generation water use, and power generators evolving attitudes towards water and wastewater. Hopefully this will provide insight and education for those unfamiliar with this industry and possibly foster opportunities for greater cooperation between water and energy sectors.

KEYWORDS

Power generation, water use, energy-water nexus, wastewater treatment.

INTRODUCTION

Due to its abundance and some unique physical properties, water is the fluid of choice for power generation processes across the spectrum of fuels. Whether it be fossil, nuclear, or renewable (such as hydroelectric, biomass or solar thermal), the majority of power generation relies on water as the primary medium for transferring energy from raw natural resources to useable electric energy that can be reliably distributed to our businesses and homes.

As noted above, to those that are unfamiliar with the various power generation processes, the water demands can seem extreme. As water resources become increasingly more precious, it is helpful to understand how water is used in the generation process. As this is better understood, it may present more opportunities to better manage available water resources and to facilitate meaningful discussions across water and energy utilities to make the best use of the water resources available. This paper provides an overview of water use in the power industry.

POWER GENERATION AND WATER USE

Figure 1 is a graphic of the electric power consumption in the United States for 2011. Total power generation in 2011 was estimated at nearly 4,000 Terawatt hours (10^{12} watts) in 2011. Fossil and nuclear power generation comprise approximately 85 percent of the power generated in the United States. On average, a typical home in the United States will use 11 – 12,000 kilowatts hours per year.

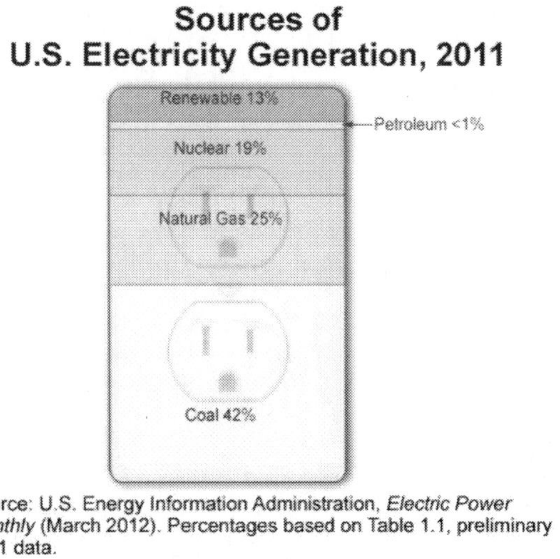

Figure 1 – US Electric Generation (Credit US Energy Information Administration)

Figures 2 and 3 indicate the amount of water withdrawn in the United States and its various uses according the USGS Estimated Use of Water in the United States in 2005 publication. Obviously much has changed since 2005, but the overall water use for power generation would be expected to be similar or slightly lower than shown in this figure for 2012. Any reduction in water withdrawal would be based on an overall reduction in total power generation and an increase in less water intensive power generation processes. However, power generation is still the largest category for both total water withdrawal and fresh water withdrawal.

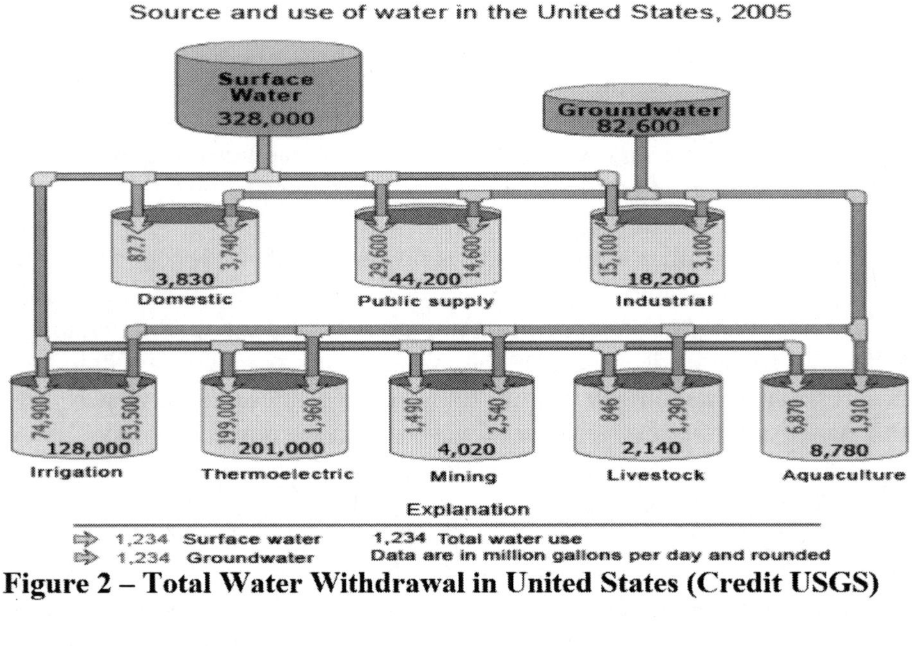

Figure 2 – Total Water Withdrawal in United States (Credit USGS)

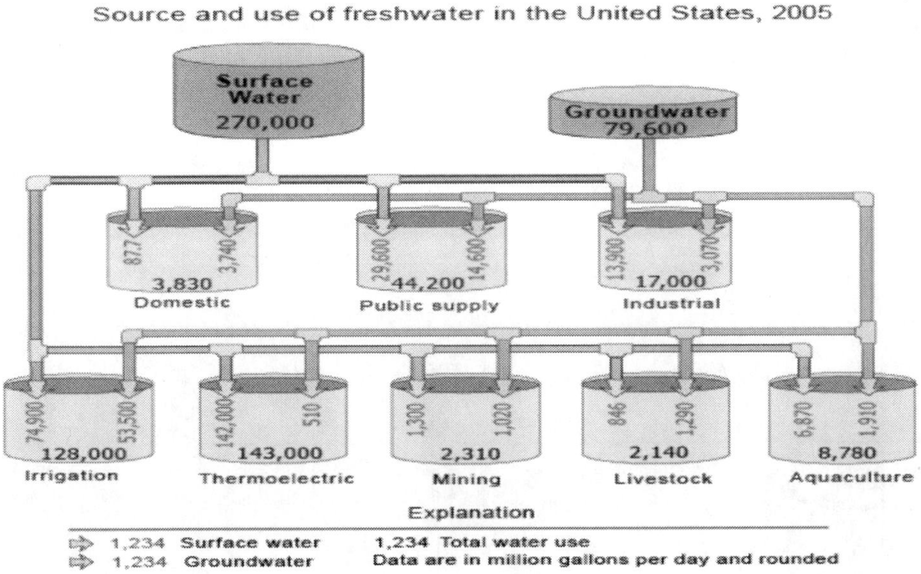

Figure 3 Fresh Water Withdrawal in United States (Credit USGS)

It is important to distinguish between water withdrawal and water consumption. Actual water consumption (primarily in the form of evaporation) estimates for thermal power generation are only about 2 - 3 percent of the fresh water withdrawn. So, while the power generation water withdrawal figures are quite daunting, actual water consumption is very modest. By comparison, irrigation accounts for 85 percent of the fresh water consumption in the United States.

It is also interesting to note that power generation water use peaked in the 1980's and has since declined and held steady in the last 30 years despite an increase in population and power consumption. This is the result of increasing efficiencies in power generation and the use of less water intensive power generation technologies (primarily natural gas combined cycle). Figure 4 exhibits this trend.

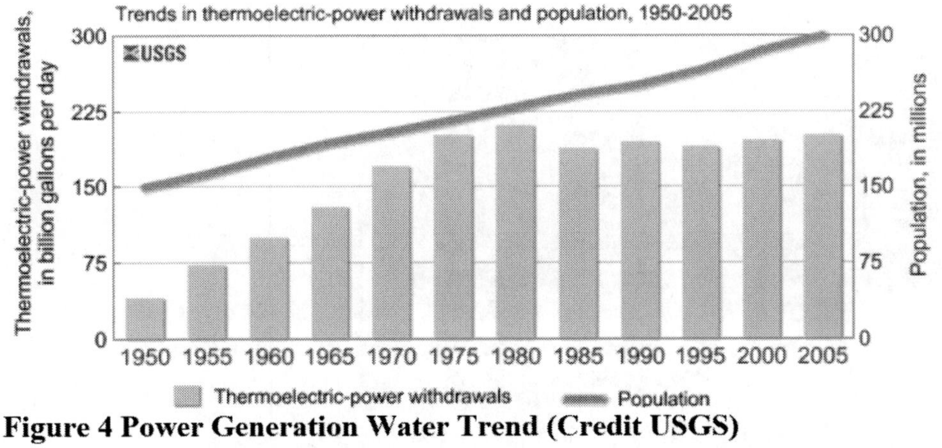

Figure 4 Power Generation Water Trend (Credit USGS)

Figure 5 provides an approximate breakdown of water use by power generation technology and process use. It is worth noting here that while many renewable forms of power generation may have beneficial impacts to air quality, several use quantities of water similar to thermal technologies. In studying the diagram, it is readily apparent that cooling water is the largest user of water in the power generation process. Other significant users are ash handling, boiler makeup, and scrubbing (air quality control). The rest of this paper will discuss these processes and how water is used.

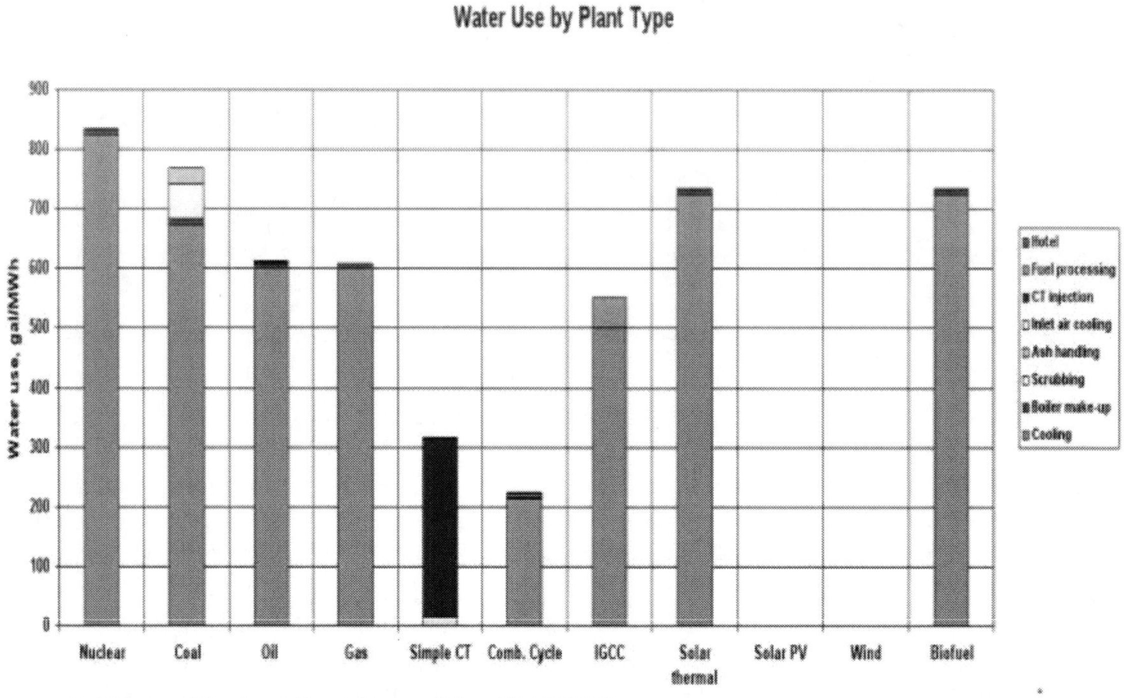

Figure 5 Water Use by Plant Type (Credit EPRI)

Cooling Water

Water for cooling is the major water use in thermal (nuclear and fossil) power generation processes. Figure 6 is a simple schematic of a coal fired power station.

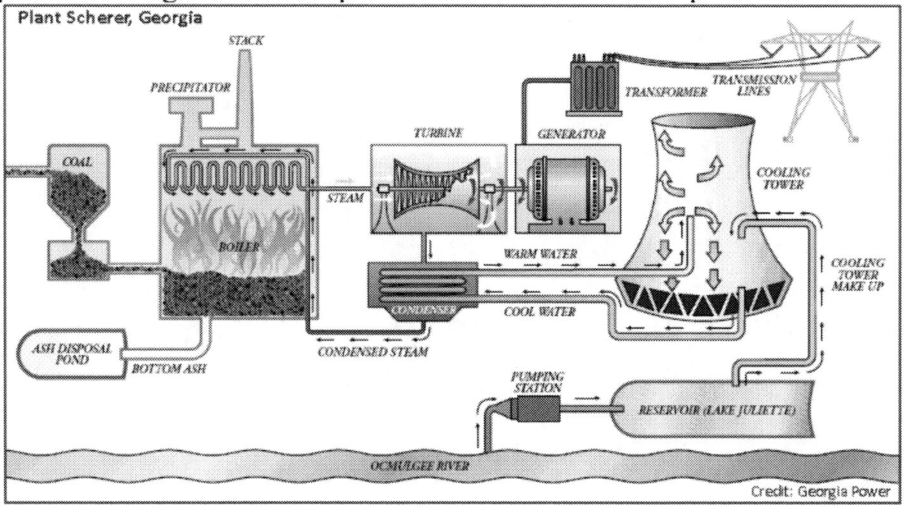

Figure 6 Power Plant Schematic

Figure 5 above illustrates an approximation of cooling water consumption for various power generation technologies and cooling methods. Many conditions can impact the actual value for a given location.

In light of the current "dash to gas" in the power industry, it is also notable that combined cycle power stations (which are primarily gas fired) use significantly less water per Megawatt-hour than coal and nuclear stations. This is due to the fact that combined cycle power stations utilize the Brayton Cycle, meaning that typically over 50 percent of the power generated is by a combustion turbine and does not require cooling water.

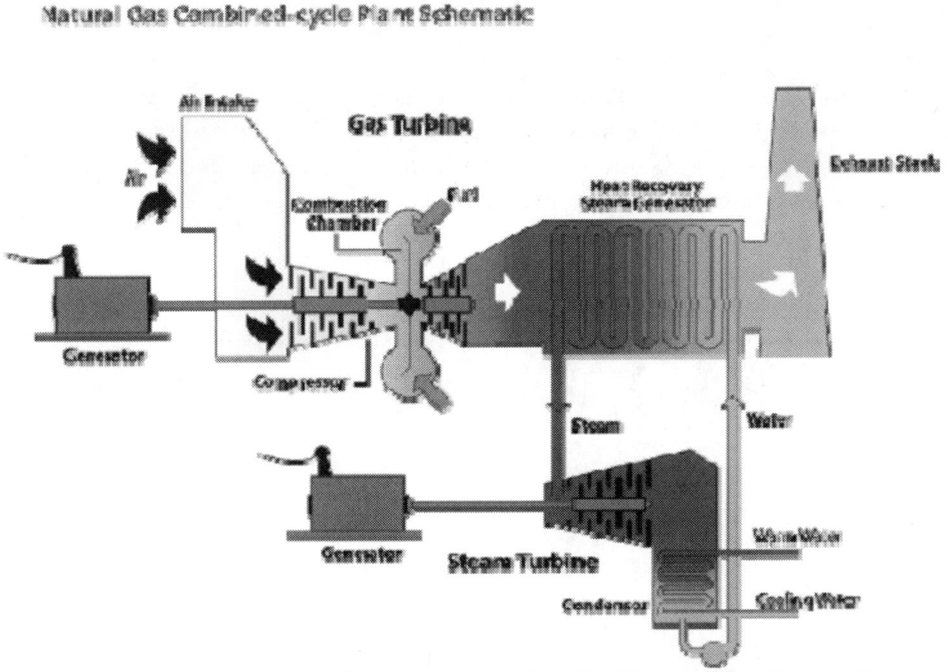

Figure 7 Combined Cycle Power Plant (Credit Consumers Energy)

It is also worth noting that plant efficiency is another contributing factor to lower water use per MWh in combined cycle facilities. Generally, coal plants and nuclear plants have an efficiency (defined as Btu in a kWh/plant heat rate in Btu's) in the range of 30 – 40 percent. Combined cycle plants are generally in the range of 40 – 55 percent.

There are two primary types of cooling water systems employed in the majority of power stations. Figure 8 is a simple schematic of a once through cooling system (open loop). The once through cooling system utilizes the raw water directly in the steam condenser. These types of systems are typically used in locations where an abundant water supply exists. Due to recent regulations, once-through cooling systems for new facilities in the United States are rare. The benefit of this approach is generally a slightly more efficient plant resulting from a cooler water supply and a lower auxiliary electric load to accomplish the cooling. These types of systems generally require only chemical conditioning of the water to minimize the potential for scale formation or corrosion of heat transfer surfaces and microbiologically induced corrosion (MIC).

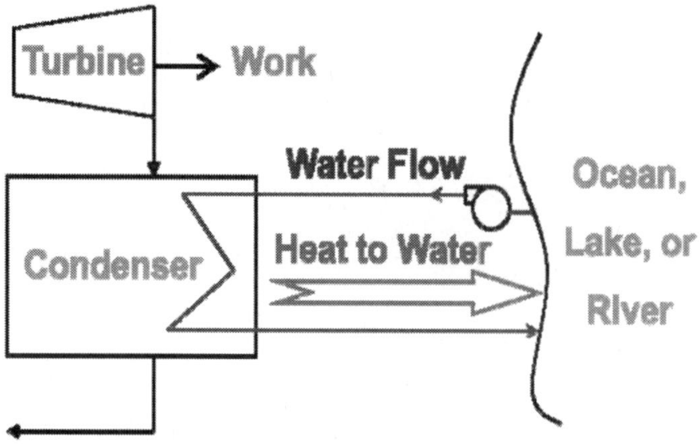

Figure 8 Once Through Cooling

The other primary cooling system is a recirculating cooling water system (closed loop) as shown in Figure 9. The recirculating cooling system utilizes evaporation as the primary mode of cooling. The evaporative cooling is optimized by use of a cooling tower. The cooling tower promotes evaporation by increasing the air-water interface.

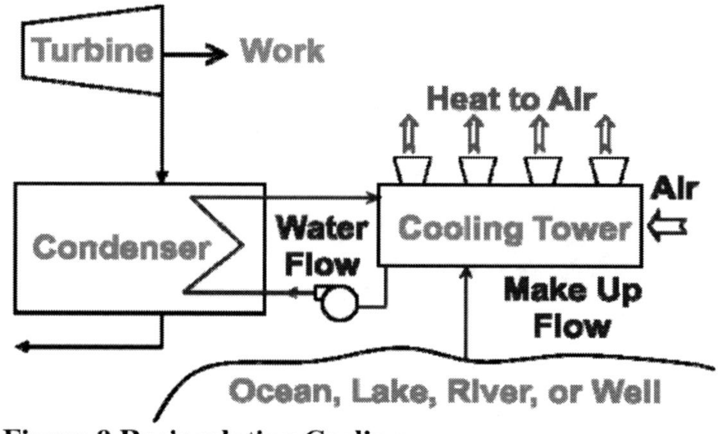

Figure 9 Recirculating Cooling

The evaporative cooling process results in a concentration of soluble salts in the recirculating water. This concentrating affect must be controlled using a blowdown stream to minimize scale formation and subsequent material degradation and efficiency loss of heat transfer surface. The blowdown stream flow is calculated to maintain a "Cycles of Concentration" (COC) in the cooling water. Essentially, this is a concentrating factor. Salts present in the makeup water at "X" concentration will be present in the cooling water at X x COC in the cooling water (provided the saturation point is not surpassed).

The COC is determined by the quality of the makeup water available and is set to avoid concentrating certain salts beyond their solubility point. Of particular scaling concern are hardness, alkalinity, silica, and phosphate. General criteria for the concentration of these salts in circulating water systems are shown below and assume the use of chemical scale/corrosion inhibitor and acid feed for alkalinity control.

Calcium, mg/l as $CaCO_3$	900
Silica, mg/l as SiO_2	150
Phosphate, mg/l as PO_4	30
Alkalinity, mg/l as $CaCO_3$	150 – 200

Other dissolved salts can affect material selections and must also be considered in the design of cooling water systems. Sulfate levels can impact the selection of concrete and chloride levels can impact the condenser tube metallurgy. Ammonia, phosphate, iron and manganese levels are also limited to minimize fouling and nutrient concentrations. Additionally, the level of total dissolved salts (TDS) can be affected by the plant's air permit as it regulates drift from the tower.

In order to promote more efficient evaporation, the cooling tower contains a fill material. There are several types of fill available from splash (open) fill to film (structured) fills. Film fills result in the highest evaporation efficiency; however they are the most prone to clogging with precipitated salts or suspended solids. Therefore, the impact of COC to suspended solids must also be considered. Suspended solids not only enter via the makeup water but are also scrubbed from the air.

Makeup to power plant cooling towers may also be supplied by treated effluent (grey water) from Publically Owned Treatment Works (POTW). The level of treatment can vary but secondary treatment is generally the minimum considered acceptable and tertiary treatment is preferred. Many states encourage this type of industrial water reuse through publication of guidelines and regulations. Use of this type of water for cooling tower makeup can present some unique challenges and various industry groups are now working to try to identify best industry practices and bring some consistency to this application to address these challenges.

In some cases, pretreatment of the cooling tower makeup water may be justified or required. The various types of pretreatment common in power stations are lime/soda softening, clarification, and filtration. Alternately, treatment of the cooling tower blowdown may be required to meet discharge requirements or to allow a portion of this water to be reused. Blowdown treatment systems can tend to be more complex and costly including high recovery reverse osmosis or thermal treatment (ie brine concentrators, crystallizers, etc).

Dry cooling and hybrid cooling technologies are gaining more acceptance in the industry as water supply issues become more of a concern. However, these types of cooling result in notable efficiency reductions and are still relegated to use only when deemed necessary due to water restrictions.

High Purity Water

High purity (demineralized) water is used in power stations primarily as makeup water for the steam generation cycle. The steam generation process utilizes demineralized water to make steam that is low in impurities in order to satisfy steam purity requirements imposed by the turbine manufacturers. The use of demineralized water also maintains efficient boiler operation by eliminating scaling of heat transfer surfaces in the boiler and the systems feeding the boiler. Demineralized water requirements for a typical utility type boiler (pressure > 1500 psi) are as follows:

Conductivity	0.1 mircomho/cm
Sodium	3 ppb as Na
Hardness	0
Chlorides	3 ppb as Cl
Sulfates	3 ppb as SO_4
Silica	10 ppb as SiO_2
Total Organic Carbon	100 ppb

Demineralized water is also used in combined cycle plants to control air emissions and boost power production from the combustion turbine. Demineralized water can be injected into a combustion turbine to control combustion temperature and consequently the formation of nitrogen oxides which are regulated. This is typically termed NOx injection water. Demineralized water (or steam) can also be injected into the turbine to increase the mass flow through the turbine and consequently the power generation of the turbine. This is termed power augmentation and is typically utilized for short periods of time.

Most equipment in the steam generation process is constructed predominately of carbon steel. Demineralized water presents a challenge in this application to minimize corrosion of mild steel materials. Corrosion is minimized by controlling the chemistry of the water in the steam cycle to minimize ongoing corrosion.

Demineralized water is produced using ion exchange or a combination of membrane and ion exchange processes. Ion exchange was the treatment of choice until the development and acceptance of reverse osmosis (RO) in the power industry. A basic ion exchange train is shown in Figure 10.

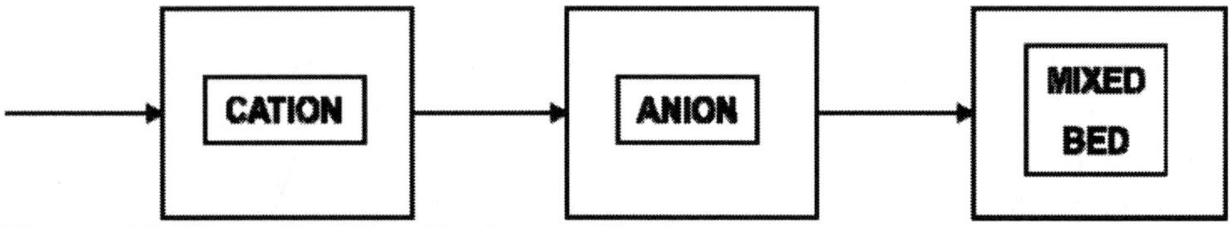

Figure 10 Three Bed Ion Exchange Train

The ion exchange process utilizes ion exchange resin to remove ionic impurities from the raw water. The resin bead provides a structure for attaching exchangeable ions. In the case of power generation demineralizers, the exchangeable ions are H+ (cation resin) and OH- (anion resin). As the water passes through the bed of resin, ionic impurities such as calcium, magnesium, sodium, chloride, sulfate, silica, iron, etc are exchanged in the cation and anion exchangers for H+ and OH- ions. The mixed bed exchanger includes both cation and anion resin and is used to polish the water to a final purity consistent with demineralized water. The demineralizer is a batch process. When the cation and anion exchangers are exhausted, they are regenerated in place using sulfuric or hydrochloric acid (cation) and sodium hydroxide (anion). Well designed demineralizer trains are about 90 percent water efficient and the cation and anion regeneration wastes will be self-neutralizing (or require minimal pH adjustment to fall within the pH range of 6 -9).

As membrane processes evolved and gained acceptance, the ion exchange train has evolved into a RO – mixed bed process. Most recently this has developed into a two pass RO and resin polishing process. The RO itself is not capable of producing demineralized quality water so a mixed be is added to polish the RO effluent. This process is now the dominate demineralizer process used in the power industry.

The RO process relies on a semi-permeable membrane to reject dissolved solids. Figure 11 is a simple schematic of an RO process. The RO product water is passed through a mixed bed exchanger to remove the final traces of dissolved solids to make demineralized water. The RO-mixed bed process is roughly 60 – 75 percent water efficient. Generally the RO reject can be reused within the power station.

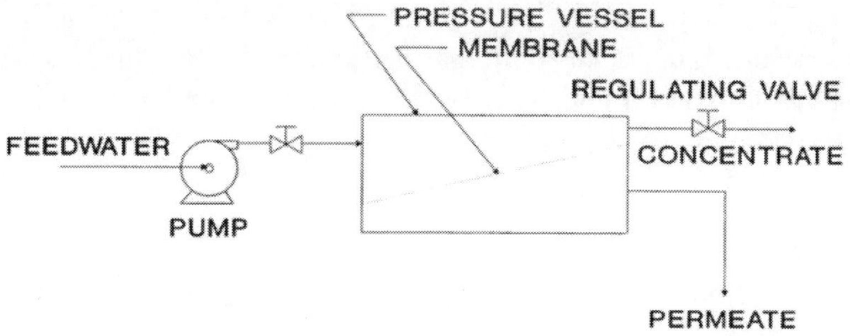

Figure 11 RO Process

Material Handling

Coal fired power stations generally use water in the handling of solid materials used or generated at the power station. Large quantities of coal are handled daily to support operations. Water is used in the coal unloading and storage process to control fugitive dust.

Water is also integral to the ash handling processes. Two types of ash are produced in the combustion process. Bottom ash is a heavy ash that falls to the bottom of the boiler furnace. It is very course and has the consistency and appearance of volcanic rocks. The ash is removed in one of two ways. Historically, ash has been sluiced from the bottom of the boiler. Ash is collected in large water filled hoppers at the bottom of the furnace and periodically sluiced out. Sluiced ash is often directed to a wet impoundment (pond or lagoon) where the ash settles and is stored and the water is recovered and reused for sluicing. Alternately, it is sluiced to large tanks where the ash settles and is loaded into trucks for landfill disposal or sale and the water is recovered and reused. This is a very water intensive process however it generally does not consume a large amount of water. This process has also come under much scrutiny recently with high profile issues related to wet ash impoundments.

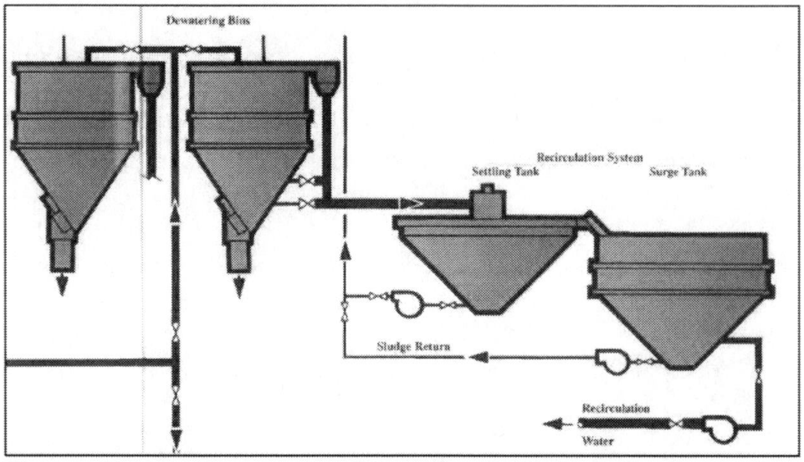

Figure 12 Ash Dewatering Bins (Credit United Conveyor Corp)

More recently, boilers have been designed to convey bottom ash out of the boiler. In this case the bottom ash falls into a water filled conveyor and is conveyed to a bunker adjacent to the boiler. The ash can be loaded from there in trucks for disposal. This process uses much less water and is generally the process of choice for new power stations. In either case the bottom ash can be sold for use in construction materials.

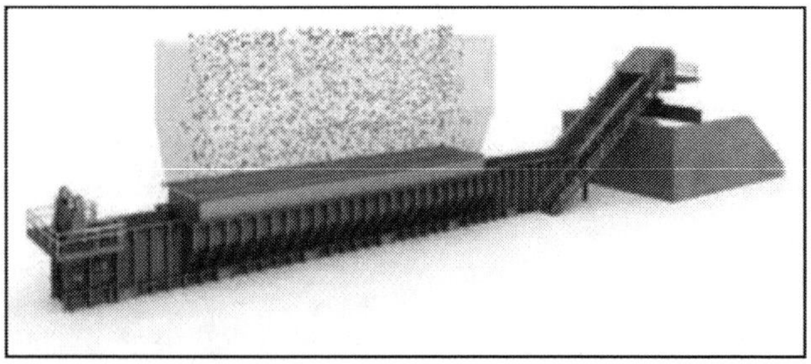

Figure 13 Bottom Ash Conveyor (Credit United Conveyor Corp)

Fly ash is the light ash that leaves the boiler with the combustion gas. It is a very fine ash and has the consistency of talcum powder. It is removed in fabric filters or electrostatic precipitators installed in the combustion gas stream as it heads to the stack. Historically, the fly ash that was collected was sluiced to wet ash impoundments for disposal. More recently, the value of fly ash as an additive to construction materials has been realized and has resulted in switching from wet ash handling to dry ash handling. In this case the fly ash is pneumatically conveyed to hoppers where it can be loaded into trucks and sold.

Air Quality Control

Air quality control is the most high profile issue facing power generation facilities today. There are many processes involved in the cleanup of the combustion gas primarily employed at coal fired power stations. Fly ash removal systems were mentioned in the previous section. In addition to reducing particulate, acid rain and smog producing constituents must also be reduced. The most water intensive process in the flue gas cleanup stream is the flue gas desulfurization (FGD) process. This process is used to remove sulfur compounds from the flue gas that would otherwise produce acid rain if released to the atmosphere.

Historically, the FGD process is designed around a gas liquid absorbtion process. The absorber utilizes an alkaline reagent in liquid form (typically a lime or limestone based slurry) to react with the sulfur in the flue gas according to the following reactions.

$$CaCO_3 \text{ (solid)} + SO_2 \text{ (gas)} \rightarrow CaSO_3 \text{ (solid)} + CO_2 \text{ (gas)}$$
$$CaSO_3 \text{ (solid)} + H_2O \text{ (liquid)} + \tfrac{1}{2}O_2 \text{ (gas)} \rightarrow CaSO_4 \text{ (solid)} + H_2O$$

The absorber is designed to promote gas liquid contact with the most prevalent design being the spray tower design. The slurry is sprayed into the gas in the absorber vessel through several stages of packing. Residual solution is collected in the bottom of the absorber and recycled. A portion is bled off and the gypsum solids are dewatered. The gypsum formed can be sold or landfilled. Figure 14 shows a high level overview of the process.

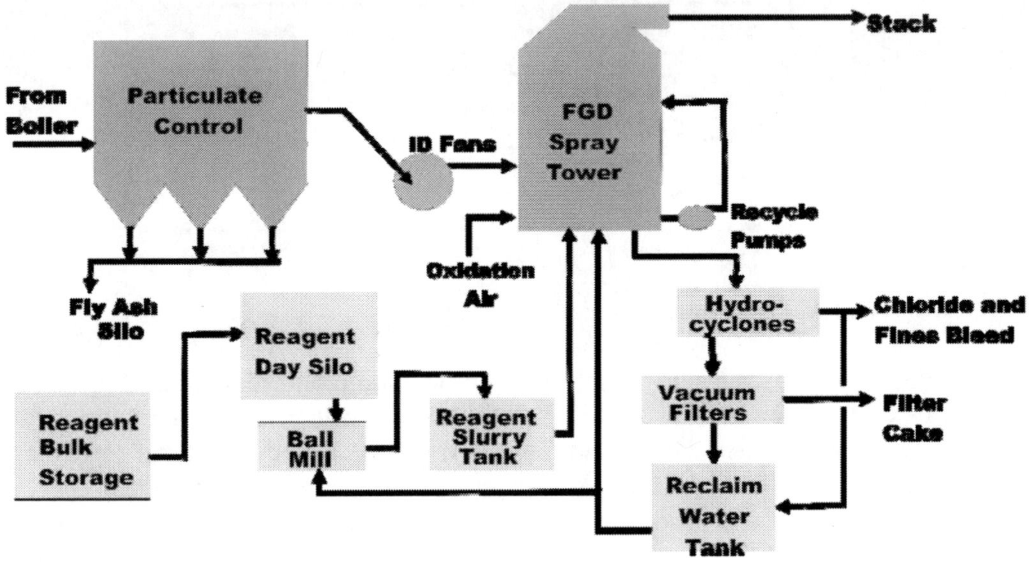

Figure 14 Wet FGD Process

Makeup water is supplied to the absorber to account for water losses in the gas leaving the absorber, in the bleed stream used to control chemistry in the absorber, and in the filter cake. Makeup water is added in the form of reagent preparation water, mist eliminator wash water, and general makeup. Makeup supplies for reagent preparation water are generally fresh water sources. The quality of this water is regulated to assure that the reagent capability, and consequently the absorber performance, is not compromised. Mist eliminator wash and general makeup water quality is not as closely regulated and can be sourced from internal plant recycle streams or from treated municipal effluent.

The filter cake in this process can be high quality gypsum solid. This solid is often sold for use in wall board production. The chemistry of this gypsum is closely regulated and therefore the inputs to the absorber process are closely regulated, including water supply.

The primary wastewater stream in this process is the chloride and fines bleed stream. This stream is used to control the buildup of chlorides and fines in the absorber. This stream is characterized by high TDS (3 – 6 percent), high TSS (2 – 4 percent), high COD (100 – 200 mg/l), and high heavy metals concentrations. Currently, the primary treatment approach is to reduce TSS and heavy metals. This is accomplished in a physical chemical process as depicted in Figure 15 below. It is a two stage precipitation process where the bulk of the TSS is precipitated in the first stage and the heavy metals are precipitated in the second stage. Heavy metals are precipitated using iron co-precipitation or metal sulfide precipitation.

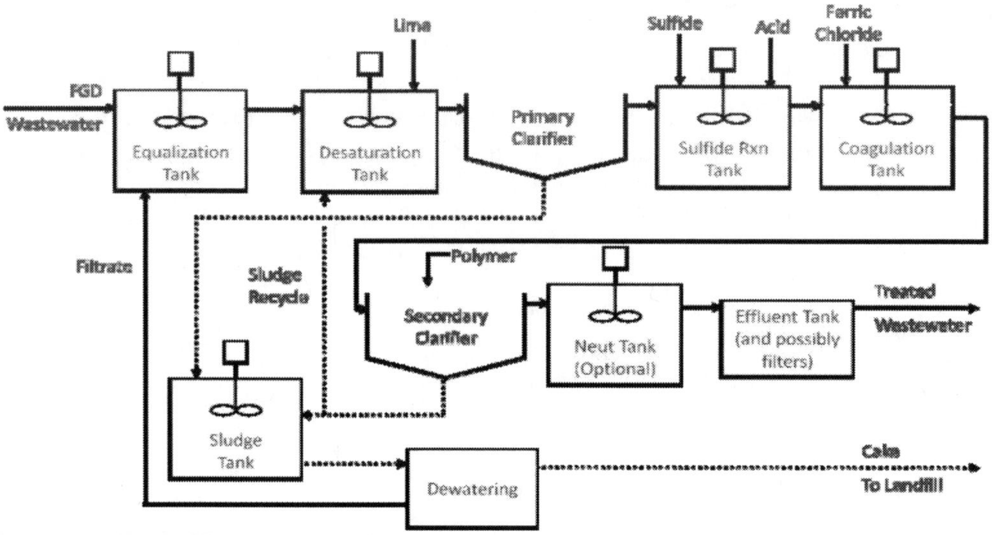

Figure 15 FGD Wastewater Treatment

Another common FGD design uses a spray drier to accomplish pollutant reduction. Figure 16 is a simplified flow diagram of a spray drier process. Similar to the wet FGD, the spray drier uses an alkaline slurry stream to introduce the reagent into the flue gas. However, in this case, the entire reagent stream is evaporated and solids are created by reaction with the flue gas. The solids are essentially dry and are removed from the spray drier. Virtually no wastewater is created.

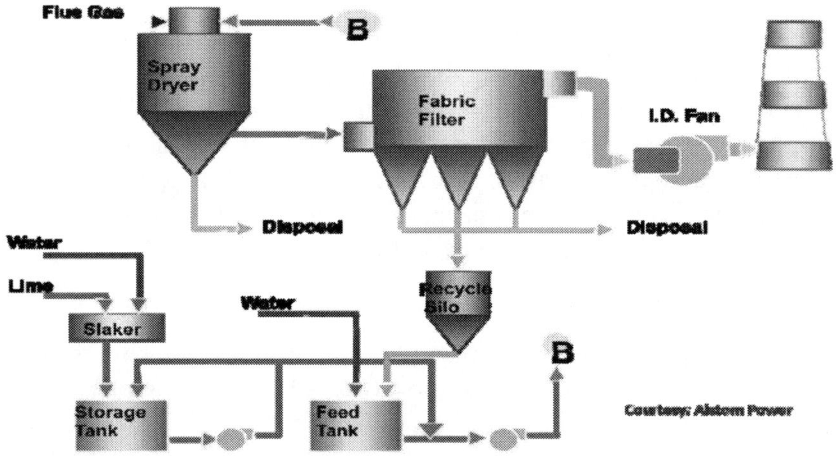

Figure 16 Semi Dry Absorber (Credit Alstom Power)

Wastewater Treatment

Wastewater treatment at a power station can take a variety of forms. It can be as simple as oil removal and neutralization to very complex and costly thermal systems that eliminate all wastewater discharge (zero liquid discharge, ZLD). Primary sources of wastewater include the following

Cooling Tower Blowdown
Steam Cycle Blowdown
Water Treatment System Wastewater (ie RO reject, Demineralizer Regeneration Wastewater)
Ash Handling Water (coal plants)
FGD Wastewater (coal plants)
Landfill Leachate (coal plants)
Metal Cleaning Wastewater (coal plants)
Coal Pile Runoff (coal plants)

Wastewater treatment processes commonly found in power stations are TSS reduction and heavy metals reduction processes. Both of these processes use physical chemical clarification treatment equipment to precipitate and settled solids. Settled solids are dewatered and land filled as non-hazardous materials. In many cases this treated wastewater is discharged to a receiving stream or body of water. In some cases it is discharged to a local POTW facility.

As mentioned above, some power stations completely eliminate their process liquid discharge creating a ZLD facility. This is generally accomplished by a well thought out recycle of certain plant wastewaters and treatment and reuse of the remaining wastewater. The two technologies that have garnered a majority of these applications are the brine concentrator/crystallizer and high efficiency, or high rejection, RO (ie HERO™, OPUS, etc).

The brine concentrator/crystallizer is a thermal technology that produces a distillate stream (TDS < 50 mg/l) and a dewatered solid waste. Figure 17 is simplified schematic of the process.

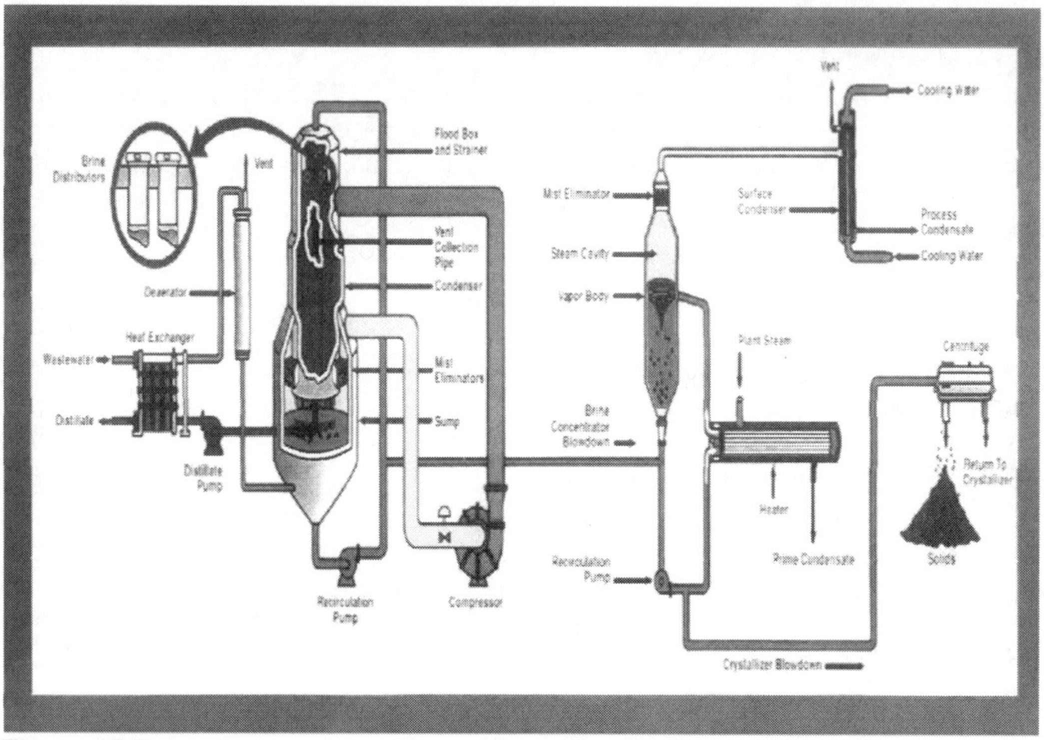

Figure 17 Brine Concentrator/Crystallizer (Credit General Electric)

Although the process has been optimized, it is requires large amounts of thermal and electric energy to drive the process, on the order of 80 – 100 kwh/1000 gal of treated water in the brine concentrator and 1.1 lbs/hr of steam per pound of distillate in the crystallizer. Special materials of construction and complex equipment design result in higher capital costs.

The high rejection RO process is based on pretreating the RO feed water so that the RO recovery can approach 90 percent. The objective is to remove scaling constituents in the RO feed, with exception of silica. Then adjust the RO feed to pH 10.5 - 11 and treat through the RO. The higher pH prevents the silica from fouling the RO membranes and allows for increased recovery. To be true ZLD, the RO reject must be treated in a crystallizer or sent to an evaporation pond for disposal. Figure 18 is a simplified schematic.

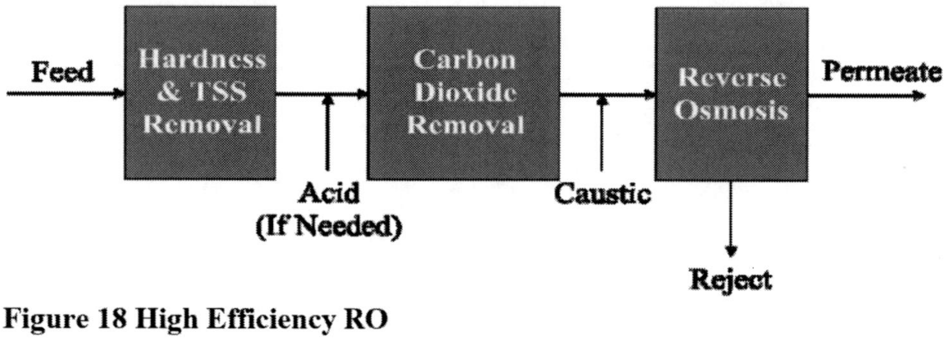

Figure 18 High Efficiency RO

This process is designed to offer an alternative to the brine concentrator that reduces the operating costs for the owner.

Power Plant Water Balance

Figure 19 is a cartoon version of a power plant water balance. The balance includes the water intensive processes at the site and provides a means for accounting for water use at the plant. Generally a balance calculation is performed using a spreadsheet software and will track not only water flows but also the project water quality through each process. These balances are necessary as a basis for NPDES permits and are helpful in providing an overview of the plant water management plan. They can also be used to consider various water reuse scenarios, anticipate wastewater treatment concerns, and prepare water management philosophies.

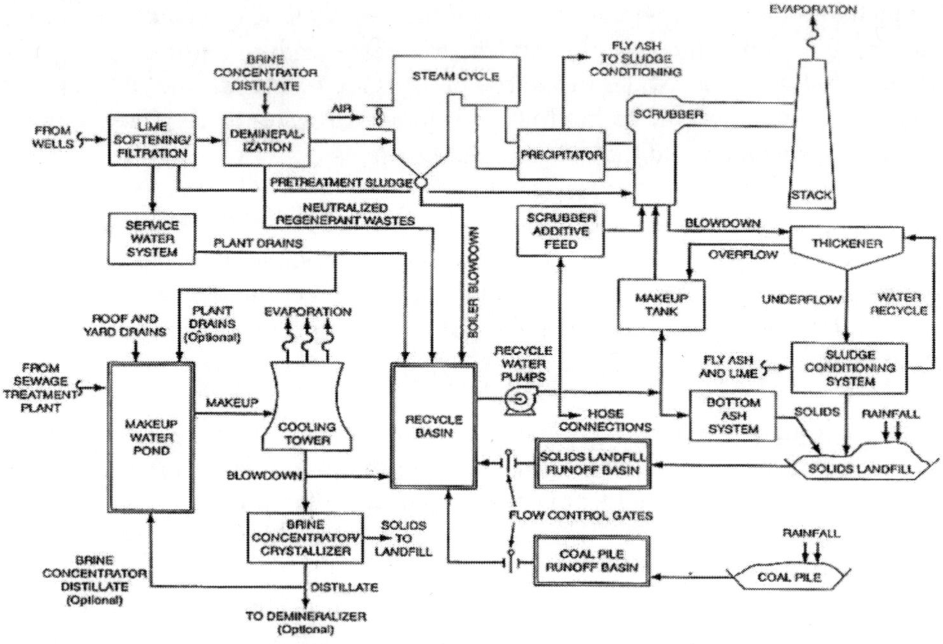

Figure 19 Power Plant Water Balance

Changing Regulations

Power utilities today face an onslaught of evolving regulations from air to water to solid waste. This creates a tremendous amount of uncertainty for power utilities. All of these regulations have one thing in common, water. Whether it is water supply, water use within the generation facility, or wastewater, many utilities are being forced to focus more on water related issues. Air regulations have been the most visible issues recently. Many of the air regulations have been resolved and the impacts are making their way through the system, however there are still some significant issues in the future (ie carbon dioxide regulations). Other, lower profile regulatory changes are also occurring that add to the uncertainty. These are not as high profile, but may have significant consequences nonetheless. These are summarized below.

<u>Clean Water Act Section 316(b)</u> Section 316(b) of the federal Clean Water Act (CWA) requires the location, design, construction and capacity of cooling water intake structures to reflect the best technology available (BTA) for minimizing adverse environmental impacts to aquatic

species. EPA released a proposed Phase II rule in March 2011 to establish BTA for all existing facilities withdrawing more than 2 million gallons per day (MGD) for cooling purposes based on impingement mortality standards and case-by-case entrainment reviews. One approach to reducing mortality is limiting the intake screen velocity to 0.5 fps through screen velocity.

The impact of this new rule may require many power stations utilizing once through cooling to modify their intake structures to meet new requirements, or convert their systems to cooling towers to reduce surface water intake.

Effluent Limitation Guidelines (40CFR 423) EPA entered into a consent agreement with various environmental organizations in which it agreed to propose revisions to power plant wastewater effluent limit guidelines by July 23, 2012, and finalize this rulemaking by January 31, 2014. These dates have changed, but the final revisions are scheduled for May 2014. Once the rule is finalized, plants would need to come into compliance with the new standard within 3 to 5 years of the final rule or as their NPDES wastewater discharge permits are renewed. The revised limits will likely include minimum best available technology (BAT) and/or water quality based limits for metals (arsenic, mercury, selenium, etc.) in power plant effluents.

The impact of this new rule will vary depending upon the final effluent standards and the water quality of the waterbody receiving the wastewater discharges. It is anticipated that additional treatment systems, likely physical/chemical and/or biological, will be required to meet new limits. Alternatively, water recycling/reuse or costly zero discharge systems may be required in some power generation facilities.

Coal Combustion Residual Management EPA published a proposed rulemaking in June 2010 that would regulate coal combustion residues (CCRs) - i.e. fly ash, bottom ash, boiler slag and flue gas desulphurization materials generated from coal-fired power plants – as either "special wastes" subject to regulation under Subtitle C of the Resource Conservation and Recovery Act (RCRA), or solid waste under Subtitle D of RCRA. The practical result of either action will force coal-fired power plants to eventually convert their collection, handling, transport, and storage of CCR from wet to dry systems, and close or retrofit existing wet surface impoundments (ash ponds). EPA is expected to finalize this rulemaking in 2012, in which case coal-fired power plants will be forced to discontinue wet ash handling operations within 5 to 7 years depending upon how quickly the requirements are adopted and take effect in each state.

The impact of this new rule is uncertain at this time but could require nearly all coal-fired power generation facilities that currently use wet sluicing and/or ash ponds to convert to a form of dry handling and disposal system. Current impoundments may have to be closed or retrofitted and existing water and wastewater management systems reconfigured.

Summary

The predominant forms of power generation (coal, gas, nuclear) are water intensive processes. While power generation facilities withdraw large quantities of water, much of it is returned to the environment so that water consumption is only about 3 percent of the total water consumed in the United States for agriculture, public, and industrial supply. Up to 70 – 90 percent of the process water used in power generation is for cooling. Cooling water is supplied from three predominant sources, surface water (lakes, rivers), ground water, and treated wastewater. Additional process water is used for high purity makeup water requirements, material handling, and air pollution control. Wastewater generated in power stations is treated in a variety of ways, including complete elimination of process water discharge (ZLD). Power utilities currently face a very turbulent regulatory environment creating tremendous uncertainty and challenging their traditional business model. Many facilities will reevaluate their water management plans and may look for alternative water supplies and wastewater discharge options. There may be opportunities for power utilities to work with water utilities to resolve future water supply issues.

Watergy in the U.S.: Energy Management, Policy Mechanisms and Technologies to Optimize Energy Use in Pumping and Aeration Systems

Robert Bruce Lung, Alliance to Save Energy

Lisa Tryson, Danfoss North America

John Masters, Danfoss North America

Ryan Graf, Policy Navigation Group

A recent report published by the U.S. Department of State and prepared by the National Intelligence Council cites that, without deliberate changes in conservation and water use, the demand for water globally will exceed current sustainable water supplies by 40% -- reaching 1,800 trillion gallons annually by 2030[1]. Today, approximately 13% of total U.S. electricity consumption is attributed to activities necessary to extract, treat, convey and heat water. In addition, approximately 35% of municipal energy use is specifically attributed to water and wastewater treatment[2]. As a result, improving water and energy efficiency will be increasingly indispensable for water and wastewater utility managers and end users.

This paper will provide valuable insights into the Alliance to Save Energy's Watergy approach to assessing water and energy efficiency opportunities in water and wastewater agencies. By using this approach, the Bucks County Water and Sewer Agency (BCWSA) was able to identify significant energy savings opportunities, establish an energy management system based on ISO 50001 to achieve continual improvements in energy efficiency and uncover financing mechanisms to facilitate implementation of energy savings improvements. This approach can easily be replicated in many other U.S. water and wastewater agencies. In addition, establishing energy management systems can not only enable persistent water and energy efficiency gains, but can also provides the context for implementing more capital-intensive energy efficiency optimization projects. Special attention is paid to VFDs as an underutilized technology in the United States.

Water and Wastewater Energy Issues

Treating water for human consumption, conveying treated water to consumers, and treating wastewater are all extremely energy intensive processes. Every gallon of water that passes through a potable water or wastewater treatment system represents a significant energy cost. Globally, energy is among the top three cost items to water and wastewater utilities, accounting for 40–60% of the operating costs for supplying water in many developing countries[3].

Water supply and wastewater treatment are also infrastructure-intensive, requiring expensive pumping, treatment, distribution, sewers, fan, motor, and disinfection systems. Water access and wastewater treatment capacity can be expanded much more quickly and inexpensively through water and energy efficiency improvements than through new infrastructure, deferring the need for additional infrastructure investments. Also, water and wastewater investment decisions that neglect energy efficiency have a domino effect that increase investments in other sectors, such as power plants, investments to extract and transport the additional fuel, and investments necessary to offset the environmental costs associated with air emissions and declining water and hydrocarbon reserves.

Part of what makes water supply such an energy intensive process is that a large proportion of distribution systems around the world are leaky. Even cities in fully industrialized countries can have high leakage rates due to aging infrastructure. In addition, cracks in sewer lines can lead to infiltration of wastewater treatment systems, leading to higher wastewater loads. Whenever water is lost or collected due to leaks, excess energy must be consumed for both pumping and treating water. When potable water is lost before it reaches the end user, the associated water production costs cannot be recovered.

In developing countries, often one-third to one-half of the volume of water produced is lost to leaks and system inefficiencies. In many cities in Mexico, for example, one-third of the water that enters distribution systems is lost before it reaches the consumer. In Brazil, it is not uncommon for such losses to reach 40%. Indian cities lose about half of their water to leakage on average, with losses ranging from 40% in large metropolitan cities to 50–60% in smaller cities. In South Africa, most big cities also lose an average of approximately 40%. The United States is also not immune to large water losses. In 2003, 35 U.S. municipal water agencies recorded a water distribution loss rate of more than 15% at a cost of approximately $103 million[4].

The Watergy Approach

Realizing that a combined approach to water and energy efficiency can yield greater savings than focusing on either water or energy efficiency alone, the Alliance to Save Energy developed a program called Watergy in 1997 to help water and wastewater treatment authorities improve their efficient delivery and treatment of both resources. The Watergy program contains three main elements: 1) an assessment of energy and water using applications, 2) training/capacity-building for staff and/or customers, and 3) identification of financing mechanisms to facilitate implementation of water/energy efficiency measures. Since its inception, the Watergy program has achieved significant water and energy savings in more than 100 municipalities in 16 developing countries. Through technical and managerial improvements, the program realizes immediate improvements in water service, increased water delivery, reduced water and energy consumption, and more revenue for system upgrades and new customer connections[5].

The Watergy approach internationally involved assessing water and energy efficiency throughout all parts of the municipal water delivery systems and addressing opportunities found throughout these systems. These opportunities frequently included:

- Pumping system improvements (repairs and retrofits)
- Leak detection, repair and pressure reduction
- Automation of system operations
- Regular monitoring & metering of energy and water end uses

Many of the projects that were identified had rapid payback periods; often within one year. Rates of return depended on local conditions, but approximate paybacks for various energy efficiency measures are as follows:

- Immediate to several months: using capacitors to optimize electric installation power factors, improving operations and maintenance (O&M) procedures; managing pressure within the system.

- 1 – 2 Years: installing metering equipment; retrofitting worn pumps, configuring system controls to operate equipment during off-peak demand periods.

- 2 - 3 Years: installing new automation systems, variable speed drives, or premium efficiency motors.

Based on this success, the Alliance introduced the Watergy approach in the United States in 2010. Working with a large water and sewer agency on the east coast, the Bucks County Water and Sewer Agency (BCWSA), the Alliance in partnership with Policy Navigation Group and Process Energy Services implemented a Watergy project to assess water/energy efficiency opportunities in BCWSA's facilities. The project also involved training on the water/energy nexus, research into financing mechanisms and outreach materials to showcase BCWSA's achievements. The Watergy project began with an energy assessment of seven facilities (three pumping stations and four wastewater treatment plants) that represented approximately 75% of BCWSA's energy footprint. The assessment identified 16 cost-effective energy efficiency measures in three separate categories that could yield annual energy savings of 4.1 million kWh per year, or roughly 20 percent of the agency's annual energy use. The categories of projects included energy management practices, operational measures and capital equipment measures.

In addition, the Watergy project included training for BCWSA staff on water/energy efficiency, research into financial mechanisms that could reduce the payback periods from implementation of the assessment recommendations and some outreach resources to inform outside stakeholders about the steps BCWSA is taking to operate sustainably.

The training session followed the assessment and introduced BCWSA's future energy management team to key concepts related to the water/energy nexus. This included the energy intensity of water and wastewater treatment, resource uncertainties for both water and energy, the condition of water/wastewater treatment infrastructure and opportunities for energy efficiency, and the benefits of addressing such opportunities. The training also outlined the steps for establishing an energy management system including developing a policy, a management plan, selecting the energy manager and team, establishing an energy use baseline, data collection and reporting, and communication. These concepts were used to underscore the importance of addressing water and energy efficiency continuously. Lastly, the training included examples of successful water/energy efficiency projects to showcase what is possible when a forward-looking water/wastewater agency makes water and energy efficiency a strategic priority.

The research into financial mechanisms was aligned with the assessment recommendations and integrated into BCWSA's energy management system. In addition to identifying incentive programs, the research methodology and results were shared with the energy management team and included five key aspects:

1. Identifying the major incentive programs for which BCWSA is eligible,
2. Understanding the requirements to qualify for various incentives,
3. Integrating incentives and their qualification requirements in the project planning process,
4. Integrating financial incentives within the financial analysis of energy efficiency projects, and
5. Building the capacity to capture payments for eligible projects once they are completed.

The Watergy team helped BCWSA identify the major incentive programs for which they were eligible, including Federal, state, and electric utility. While this was initially straightforward, it helped BCWSA understand the variety of incentive programs, how they are structured and the different sources for incentives.

Next, the research helped BCWSA understand the requirements for qualification and the priorities behind the different sources of incentive programs. These priorities determine the requirements for each program. Because these requirements determine project eligibility for incentives, understanding them could enable BCWSA to target different incentives based on what the projects can accomplish.

Then, once the incentives and their qualification requirements were identified, they were integrated into the project planning process of the assessment recommendations. This involved matching the baseline and ongoing energy use of end use equipment to various incentives in order to be able to plan and qualify for them more effectively. Also, the amounts of the identified financial incentives were integrated into the financial analyses for each project. Because the funds from incentive programs have the potential to positively affect the return on investment for energy efficiency projects, they can facilitate implementation, particularly for projects that require extensive capital investments.

Finally, the financial mechanism research helped BCWSA build the capacity to capture payments for their eligible projects. Capturing such incentive payments requires two fundamental items: the technical expertise to explain how a project fits the criteria for the incentive program and the relationship with the funding organizations. Because energy efficiency projects in water/wastewater treatment plants can vary in scope and complexity, they don't always correspond to the typical projects that incentive program administrators encounter. Therefore, it was necessary to provide technical explanations to ensure that the projects would meet the incentive program requirements. This also involved developing relationships with incentive program administrators to demonstrate the quality and trustworthiness of the projects. Once program administrators are confident that energy efficiency projects will meet or exceed expectations, they are more likely to view an organization that submits incentive applications more favorably. By developing goodwill with incentive program administrators, water and wastewater agencies can improve their chances of getting funding for new capital projects.

The financial research identified the potential sources of funding for projects from Federal, state, and private sources. The sources of funding included utility rebates, Federal tax credit and discretionary programs, and state funding programs. The financial analysis identified the available sources from the potential sources and applied those to the identified projects resulting in the real rate of return of these projects. With full implementation BCWSA can see a 229% return on investment over 10 years.

The training enabled BCWSA to establish an energy management team and get personnel on the ground to evaluate their motor-driven systems for energy efficiency potential. This training and subsequent energy management team activity is yielding a large number of ideas beyond the scope of the original energy assessment to improve energy performance. The energy management team is implementing a plan that will support a corporate energy policy developed after the assessment. The team has implemented most of the operational measures including lowering thermostats, shutting off lighting when personnel are not present, cycling pumps and aerators to avoid redundant operation and installing pump timers to avoid continuous operation that was unnecessary. These measures are yielding annual energy savings of approximately 300,000 kWh. In 2012, BCWSA began implementing some of the capital-intensive measures in the assessment. One such project involved installing new Variable Frequency Drives on two wastewater pumps, which is saving approximately 150,000 kWh per year. In addition to other measures identified in the assessment, BCWSA's energy team is evaluating low impact hydroelectric generation as part of an infrastructure expansion and is acquiring energy management software to better monitor existing SCADA-controlled equipment.

The outreach included producing a video of BCWSA's approach and some of the measures they undertook to improve their energy performance. In addition, senior management from BCWSA participated in conferences to share their lessons learned from the Watergy project and emphasize their perspective to policymakers and energy efficiency program developers.

An integrated approach to water and energy efficiency can yield greater energy and water savings than focusing separately on water or energy efficiency -- and can result in improved system reliability. Based on the successful Watergy project at a U.S. agency that improved water and energy efficiency at several pumping stations and wastewater treatment plants, other water and wastewater utilities and municipalities can learn how they can implement similar projects and achieve significant water and energy efficiency gains, thereby addressing both water scarcity and energy costs simultaneously. This approach can be replicated in other parts of the world including industrialized states, newly industrialized countries and developing countries.

Technologies for Water & Energy Efficiency

Addressing the water/energy nexus is paramount in the development of infrastructure and water/wastewater treatment demand due to urbanization. In fact, the U.S. Department of Energy estimates that pumping systems in water, wastewater treatment and irrigation applications jointly serve as the United States' third largest consumer of electricity. According to the Consortium for Energy Efficiency, the estimated annual potential savings in municipal water & wastewater agencies is 31 billion kWh.

However, it also has been estimated that the potential exists to save as much as 11-18% of this energy consumption through innovative, easily implemented and readily available technologies. One example of such technology is variable frequency drives (or VFDs), which can improve the energy efficiency of water and wastewater plants, while also minimizing water losses.

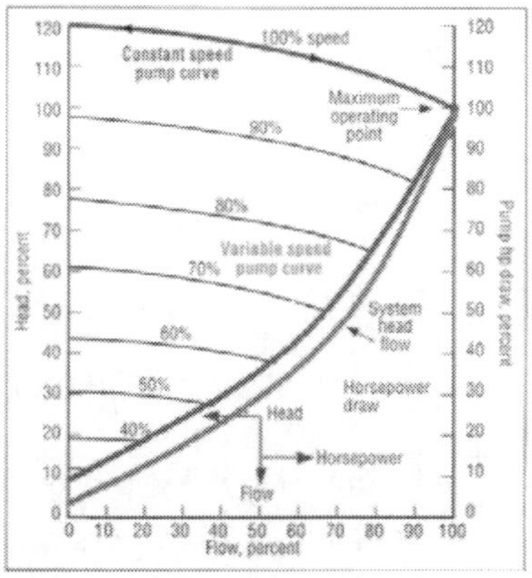

Variable frequency drives (VFDs) are an electronic type of adjustable speed drive (ASD) that control the speed of an induction motor. VFDs control motor speed by varying the frequency and voltage of alternating current (AC) from utility-supplied electricity while maintaining torque in response to power demand from the application served by the motor. Controlling motor speed is beneficial for industrial motor-driven applications because it can lead to reduced energy use and costs, improved system reliability, lower maintenance needs, and better process control. By adjusting motor speed to match the demands of motor-driven systems and applications, VFDs can yield significant energy savings as well as other benefits.

VFDs act as liaisons between the electric grid and the motor to control flow and speed. The primary benefit of VFDs is improved operating efficiency, which often results in significant energy and energy cost savings for many motor-driven systems. Energy savings can range from 20% to 50% of application-specific energy consumption. Through Affinity Laws, it is possible to predict how changing the speed of a motor will affect flow, pressure and power. The drive will control the speed of electrical motors by varying the frequency and voltage while maintaining torque. As a result, even a 10% reduction in motor speed can yield a 27% decrease in energy use.

In the United States, VFD deployment occurs at a slower rate than in other industrialized states. In the United States, VFD saturation in industrial applications is estimated at approximately 9%[6], compared to Europe and Japan where approximately 19% and 38%[7] – respectively - of existing industrial applications are served by VFDs. One study estimated that VFD penetration in the U.S. could increase to 14%, which could yield aggregate annual energy savings of 78,000 GWh[8]. Incentives to install VFDs may help increase VFD penetration. Currently, several states and utilities offer incentives for plants that install VFDs on motor-driven applications.

Technologies for water and wastewater systems are subject to complex regulation and approval processes on both the federal and state levels. It is here that education is needed. VFDs are a proven, readily adaptable technology with established and proven design standards that can save energy without compromising water quality or delivery.

Saving Energy in Plant Retrofits

VFDs can be retrofitted onto existing motor systems or built with original equipment and can control both standard and high-efficiency motors ranging in size from fractional horsepower (hp) motors to units rated at several thousand hp

There are several easily implemented design options for retrofitting motor-driven applications in water and wastewater plants including process adjustments, blower and pump sizing improvements and motor efficiency upgrades. Yet, comparatively, VFDs offer the easiest implementation and serve as the largest opportunity for saving energy in these applications because of the precise load control afforded by the drive's variable speed control.

Another example of a wastewater treatment retrofit with VFDs was done by Wendell Engineers of a wastewater treatment plant in upstate New York comprised the following:

- Rebuilding two 450 HP finished water vertical turbine pumps
- Rebuilding one 350 HP finished water vertical turbine pump
- Installing new motors with variable speed drives at the finished water and raw water pump stations (7 motors from 125 to 450 HP)
- Installing a programmable logic control (PLC)-based SCADA system with remote telemetry system to modulate pump speed, improve pump energy data collection, and maximize energy efficiency
- Upgrading filter valve actuators and the coagulant chemical feed system
- Lighting system replacement

Prior to the retrofit, the plant was using 5.8 million kilowatt hours annually, at an estimated cost of more than $694,000. The retrofit significantly improved energy use – reducing electricity consumption by 1.47 million kilowatt hours per year (approximately 25%) and providing energy savings of approximately $96,000 annually.

VFDs and Water Conservation

Distribution system water losses increase as system pressure also increases. Studies show that facilities can experience net revenue water losses up to 30%, which, because of the interdependency of water and energy, also result in wasted energy during pumping and treating. Since net revenue water losses are partly due to pipe leaks employing VFDs to maintain constant pressure in the system has a positive effect on reducing net revenue water losses.

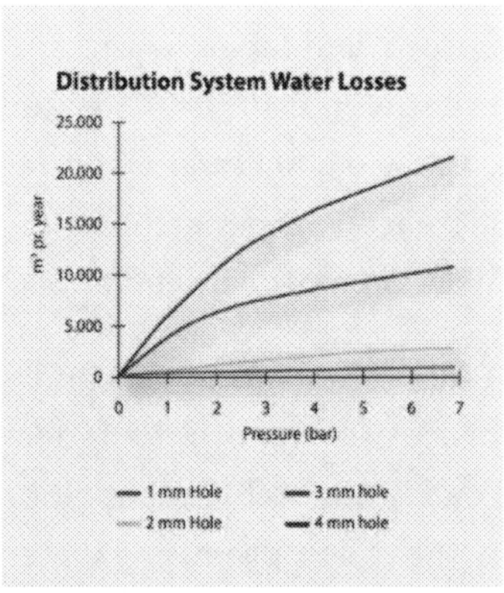

This is because, whereas constant speed pumps operate at maximum performance and efficiency for only a percentage of the day, a VFD with a built-in cascade controller receives the pressure feedback signal, initiates additional pump drives as needed and provides a master speed reference signal to all the pumps, ensuring that they provide matched pressure and flow.

This intelligent pump control functionality of VFDs can, thereby, minimize water loss. For example, if a 1/8-inch leak occurs in a water main, simply reducing the pressure from 75 psi to 60 psi would result in the conservation of approximately 530,000 gallons of water per year otherwise lost in a single leak.

Wastewater System Control

A typical wastewater treatment plant can experience load fluctuations by as much as a factor of five over a 24-hour period. As a result, a systems approach is required to achieve accurate and reliable input signals. Manually controlling the aeration process can cause excess energy consumption. However, when a VFD-based control system is utilized in an aeration blower application, energy costs can be cut by 60% or more.

In West Virginia, a wastewater treatment plant saved between $50,000 and $60,000 annually when it replaced three 75 HP centrifugal blowers with one 100 HP and one 50 HP rotary lobe blower – each controlled by a variable frequency drive. One significant parallel benefit was that the dissolved oxygen levels also improved. In the original installation, dissolved oxygen levels varied from 0.5 ppm to 6.0 ppm over a 24-hour period. In the updated installation with VFDs, the dissolved oxygen levels were maintained steadily at 2.2 ppm throughout the day.

Together, the BCWSA, New York water treatment plant and West Virginia wastewater treatment plant projects help validate the effectiveness of variable frequency drives as but one example of a technology that could improve energy use and leakage in water, wastewater and irrigation activities throughout the country, especially when implemented in as part of a comprehensive Watergy strategy. However, because of a culture of focusing on primary mission, lack of energy literacy and financing mechanisms for the water/wastewater treatment sector, both education and policy measures are required to help further encourage the use of these technologies to save energy, costs and improve the water/energy infrastructure.

Policy Mechanisms

Several barriers currently exist that prevent widespread adoption of energy efficiency practices within the water and wastewater treatment sector. This includes many hurdles obstructing the rapid deployment of variable frequency drives and other proven technologies that harbor tremendous opportunities for improving energy and water use.

To begin with, there is little guidance between the federal, state and municipal levels. A framework that is useful for local officials, operators, and other stakeholders is needed. This framework should be interwoven with the permit process and actions like an energy efficiency analysis could be a state driven requirement. In addition, regulations are not streamlined and often conflict with other priorities related to water quality and delivery. Some regulations such as emissions permits and FERC regulations present barriers to energy efficiency measures such as CHP. At the state level, energy generated from biosolids and biogas is not included in renewable energy or energy efficiency portfolio standards. One of the biggest impediments is narrowly defined utility incentive programs. When water and wastewater treatment agencies seek incentives for energy efficiency measures, they often need to go through a customized incentive process that doesn't often yield the desired funding. Lastly, state definitions of energy service companies (ESCOs) vary widely and many don't identify water and wastewater treatment plants as eligible for service provider opportunities.

Additionally, the following barriers inhibit the use of readily available energy-saving technologies: first cost investment challenges of technology upgrades in both retrofit and new facility projects, constraints of municipality budgets and regulatory acceptance.

Proposed solutions to overcome these obstacles include: standardizing legislations and codes regarding ESCOs and developing standards and education programs to encourage the design of new facilities with modern equipment; conducting policy outreach with a common voice among industry stakeholders; fast-tracking the permitting process for CHP and energy generated from biogas; developing rebates, credits or other financing programs for retrofit projects; and defining and articulating both the economic value of water/energy efficiency. In addition, maintaining the current 20% green project reserve that applies funds distributed via the EPA's Clean Water State Revolving Funds would be a huge boon. Applying the successful green project reserve to other federal wastewater treatment grants could yield results that are even more significant.

The opportunities for energy saving in this sector are vast. The path forward requires a holistic approach: the right financing mechanisms and policy inducements to deploy proven technologies.

---

[1] National Intelligence Center, Global Trends 2030.

[2] Consortium for Energy Efficiency, "Water/Wastewater Systems," 2007.

[3] Alliance to Save Energy, "Achieving Energy Efficiency in Water Supply & Wastewater Treatment Utilities through Watergy Strategies," 2012.

[4] Ibid.

[5] Ibid.

[6] Motor System Market Opportunities Assessment, 2002.
[7] NEEA, report 00-048, 2000
[8] MSMOA, 2002

Energy Savings From Advanced Process Control: Two Case Studies

Steve Kestel[1]*, Matthew Gray[1], Weixin Shi[2], Junquan Yan[3]

1: BioChem Technology, King of Prussia, Pennsylvania
2: BioChem Technology, Shanghai, China
3: Jieyuan Draining Co., Ltd, Yangzhou, Jiangsu Province, China
*Email: skestel@biochemtech.com

ABSTRACT

As energy costs continue to increase and effluent permits become more stringent, process control has become one of the most important components of the operations of activated sludge processes. Utilizing a process controller that adjusts the dissolved oxygen (DO) set points based on real time influent loading can result in significant energy savings at a biological nutrient removal wastewater treatment plant (WWTP). This paper presents two case studies of plants that implemented such a control system. At the Metropolitan Wastewater Reclamation District's Robert W. Hite Treatment Facility in Denver, Colorado, an average of 11% less air was used when using the process controller compared to fixed DO set-point control. At the Liuwei WWTP, located in Yangzhou, Jiangsu Province, China, the plant achieved an energy savings of 21.0% using a process controller compared to fixed aeration.

KEYWORDS: Process Control, Process Optimization, Energy Savings

INTRODUCTION

With continuing increases in energy costs and the requirements of biological nutrient removal (BNR), process control has become one of the most important components of the operations of activated sludge processes.

The energy use of the aeration system for the activated sludge process accounts for approximately sixty percent of the total energy use of the plant (EPA, 1989). A well designed and implemented process control system can save significant amounts of energy.

The purpose for process control at wastewater treatment plants (WWTPs) is to respond to disturbances. WWTPs are continuously subject to disturbances caused by the inflow side of the process. The wastewater influent varies in flow, concentration, and composition throughout the day and year. The WWTP must still maintain consistent effluent quality despite these diurnal fluctuations. A process control system with proper instrumentation will minimize the effect of diurnal fluctuations on the effluent quality while achieving treatment goals.

Dynamic dissolved oxygen (DO) set-point control has started to become popular with the increased cost of energy and the emergence of better online instrumentation. The basic concept of dynamic DO set-point control is to vary the DO set-point with the influent loading to meet an effluent goal. Successful implementation can provide aeration energy savings of 10 to 20% while maintaining the effluent quality. There are two control principles, feedback and feed forward, that are used with dynamic DO set-point control.

There have been several case studies (Liu, 2005; Walz, 2009) of facilities using ammonium analyzers to enhance the aeration control system. This paper discusses the implementation and results of a feed forward process controller used in two energy savings projects.

METHODOLOGY

Control Theory

Wastewater treatment plants (WWTP) do not typically operate at the maximum design conditions during most of the day, most of the year, and lifetime of the plant, which provides flexibility to optimize the operations. The aeration system of an activated sludge process is the highest energy consuming system within the treatment train. With the use of DO and ammonia control the energy consumption of the aeration system can be lowered tremendously by taking advantage of the additional treatment capacity during diurnal lows.

The purpose of the DO set-point and ammonia controller is to meet effluent ammonia targets while minimizing the amount of oxygen supplied to the process. Minimizing the supplied oxygen reduces the loading on the air blowers, resulting in significant energy and monetary savings.

Residual DO in each aerobic stage of the process can be used to control the specific nitrification rate of the biomass. Increasing the DO concentration increases the nitrification rate by increasing the availability of oxygen to the autotrophic biomass. The increase in the nitrification rate results in a decrease in the effluent ammonia concentration. The converse is also true; lowering the DO concentration will increase effluent ammonia.

Feed-Forward Process Controller Description

The DO set-point controller uses a feed-forward control algorithm to match the DO set points in the aerobic stages to the variations in diurnal loading. The control algorithm utilizes a biological mathematical model that is customized to the plant and the specific process to calculate an ammonia concentration profile throughout the bioreactor. The model is used with an optimization algorithm to determine the effect different DO set points will have on the effluent ammonia concentration. The optimization algorithm ensures the DO set points calculated are the lowest possible to maintain effluent ammonia targets. Figure 1 is a diagram that visually explains the feed-forward DO set-point controller.

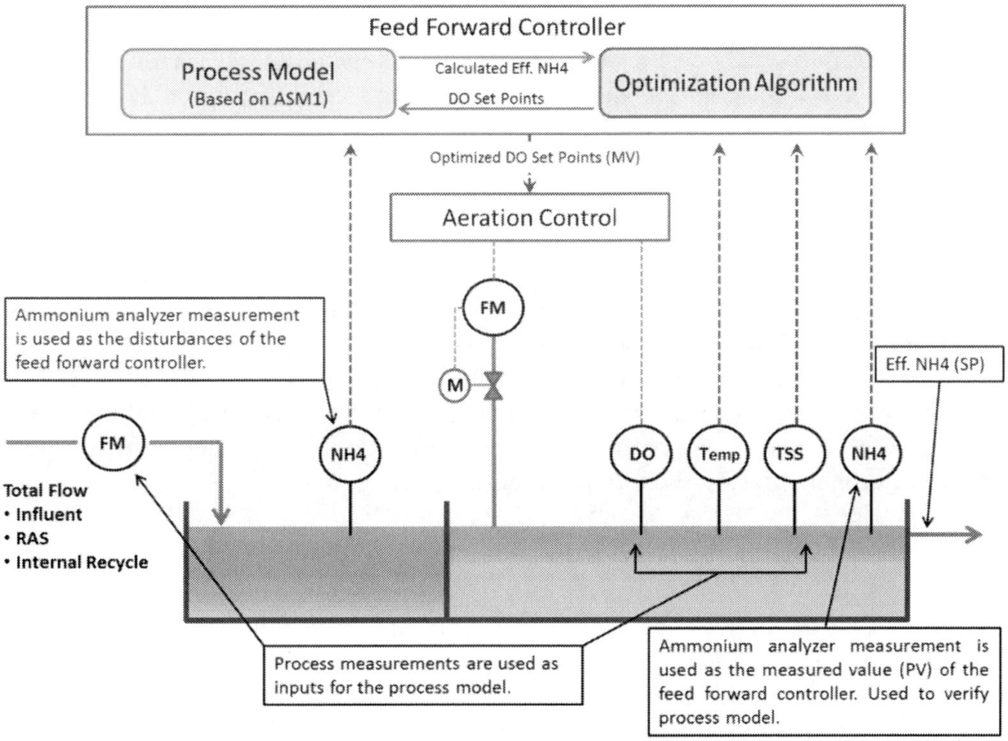

Figure 1. General Dissolved Oxygen and Ammonia Control Diagram

Summary of Measured Values

The feed-forward control algorithm requires the following inputs:

- Effluent Ammonia (mg/l) (End of Aerobic zone)
- Influent Ammonia (mg/l) (End of Pre-Anoxic zone)
- Dissolved Oxygen (mg/l) (one in each aeration zone)
- MLSS (mg/l) within the bioreactors
- Temperature (C°) within the bioreactor
- Influent flow rate (MGD)
- Return Activated Sludge flow rate (MGD)
- Mixed Liquor Return flow rate (MGD)

At a minimum only one set of ammonium analyzers is required; the controller is capable of calculating the influent ammonia concentrations of multiple trains if the flows are known.

Controlled Values:

The residual ammonia concentration at the end of aeration is used as a control set-point for all trains. This effluent ammonia concentration target set-point is selected by operators. The ammonia set-point can be changed on the user interface at any time.

Manipulated Values:

The DO set points for the individual aerobic stages are used to control the bioreactor effluent ammonia concentration. The controller DO set-point calculations are limited to a DO set point range and maximum step change selected by the operator.

Activated Sludge Model Description

The controller utilizes a mathematical model based upon the International Water Association (IWA) Activated Sludge Model #1 (ASM1) that is customized to the plant specific process. The purpose of the model is to calculate the ammonia concentration within the bioreactor and to work with the process optimization algorithm to calculate the lowest DO set points to achieve the ammonia effluent goal. The model incorporates both the biological kinetics of the activated sludge process and physical characteristics of the specific plant and the treatment process. While it is not a complete hydraulic model, the model calibration process indirectly accounts for the impact of hydraulic heterogeneities on the treatment performance.

Biological Kinetics

The purpose of the model is to simulate the concentrations of ammonia within the aerobic reactors, so the model only incorporates process rates from ASM1 that remove or add ammonia to the process. The IWA model process rates include several parameters that are difficult to measure, making it difficult to tune. Below is the method that was used to modify the original ASM1 process rates.

$$-\left(f_{NH}+\frac{1}{Y_A}\right)\mu_A\cdot\frac{S_{NH}}{K_{NH}+S_{NH}}\cdot\frac{S_{DO}}{K_{DO_A}+S_{DO}}\cdot X_A+k_a\cdot S_{ND}\cdot X_H-f_{NH}\cdot\mu_H\cdot\frac{S_S}{K_S+S_S}\cdot\frac{S_{DO}}{K_{DO_H}+S_{DO}}\cdot X_H$$

Autotrophic Uptake | Nitrification | Ammonification | Heterotrophic Uptake

$$-\left(\cancel{f_{NH}}+\frac{1}{Y_A}\right)\mu_A\cdot\frac{S_{NH}}{K_{NH}+S_{NH}}\cdot\frac{S_{DO}}{K_{DO_A}+S_{DO}}\cdot X_A+\cancel{k_a\cdot S_{ND}\cdot X_H-f_{NH}\cdot\mu_H\cdot\frac{S_S}{K_S+S_S}\cdot\frac{S_{DO}}{K_{DO_H}+S_{DO}}\cdot X_H}$$

Insignificant | Modified to use measurable variables | Incorporated into measured NH_4 adjustment

$$\rightarrow\quad SARR\cdot\frac{S_{NH}}{K_{NH}+S_{NH}}\cdot\frac{S_{DO}}{K_{DO_A}+S_{DO}}\cdot X_{MLSS}$$

Specific Ammonia Removal Rate (mg NH4 removed per g of TSS) | Mixed Liquor Suspended Solids (g of TSS)

The specific ammonia removal rate is essentially the nitrification process rate with modifications to be able to use measured values instead of the IWA model variables such as growth rate, biomass, and yield for autotrophs. Like the IWA model, the actual ammonia removal rate is dependent upon limiting variables ammonia, and dissolved oxygen.

Feed Forward Controller Program

The feed forward program consists of two parts, the user interface and the process control engine.

The user interface is programmed and compiled utilizing the development software Visual Basic by Microsoft. The operational modes and set points can be accessed, implemented, and modified directly through the user interface. Additionally, soft-coded tuning parameters can easily be accessed through the Human Machine Interface (HMI), which allows for on-the-fly control tuning, giving plant operators and engineers the ability to adjust the system as necessary. All data necessary for controller operation is easily visible on the user interface with the train overview screens and historical graphing

The process control engine is where the process model and optimization algorithm calculations are implemented. The mathematical software MatLab is used to implement the process control engine.

CASE STUDY 1

The Metropolitan Wastewater Reclamation District, Denver, Colorado

The Metropolitan Wastewater Reclamation District (Metro District) in Denver, Colorado, treats wastewater from Denver and parts of northern Colorado at its Robert W. Hite Treatment Facility (RWHTF), with an average daily flow of about 530,000 m^3/day (140 MGD). The RWHTF currently consists of two distinct secondary treatment plants. The South Secondary (SSEC) is a pure oxygen plant, and the North Secondary (NSEC) is an activated sludge plant. The NSEC is set up as four "quads" with three trains each. The trains are configured as a Modified-Ludzak-Ettinger (MLE) process, with three anoxic zones and one aeration zone in the first pass, and one aeration zone in each of the second and third passes. The NSEC is also equipped with centrate and return activated sludge re-aeration basins (CaRRB), but this process was not the primary focus of the study. The NSEC treats approximately 303,000 m^3/day (80 MGD).

In 2009, the Metro District spent over $4.8 million in electrical charges, of which approximately one third was for operating the blowers in the North Secondary. In 2010, the Metro District created the NSEC Energy Optimization Initiative. The initiative was formed to allow operating staff to identify and implement operational changes and technologies to reduce energy consumption.

As part of NSEC Energy Optimization Initiative, the Metro District's Model User Group (MUG) worked with consultants to conduct model simulations to calculate potential aeration savings at different DO set points. BioChem Technology (BioChem) also modeled the process to quantify the potential benefits of using feed-forward dynamic DO set point control. The simulations showed that feed-forward control was able to save 12 to 23% in aeration compared to current operating parameters without affecting the effluent quality. Based upon the simulation results, the Metro District contracted with BioChem to demonstrate the energy savings potential of a feed forward controller.

NSEC Pilot Installation

The controller installed for the demonstration project was the Bioprocess Intelligent Operating System (BIOS) as manufactured by BioChem. The BIOS uses a feed-forward control and optimization algorithm to determine the optimal set points for DO to achieve treatment goals while minimizing energy consumption. The demonstration started on September 1, 2011 and ended in March of 2012.

For the NSEC pilot installation, the controller was programmed to control one quad (Trains 7, 9 and 11). Train 9 is instrumented with ammonium analyzers in the first anoxic zone and at the end of the last aeration basin. The instrumented train is considered the "master train", and all three trains were controlled based on ammonia measurements in the master train. The controller calculated DO set points for each of the three aeration zones (Pass A, Pass B, and Pass C) in each of the three trains. A parallel quad (Trains 8, 10 and 12) was instrumented identically to the controller quad and acted as a baseline for the pilot. The DO set points for the baseline trains were set by the plant operators. Figure 2 is an aerial view of the NSEC that depicts the feed-forward controller and baseline quads, and Figure 3 is a schematic of the demonstration.

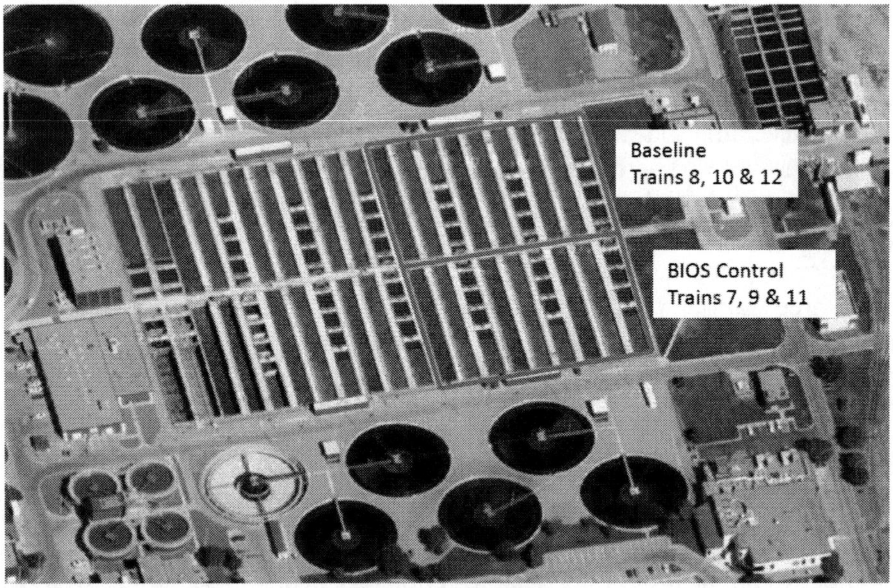

Figure 2. Aerial View of the NSEC at the RWHTF

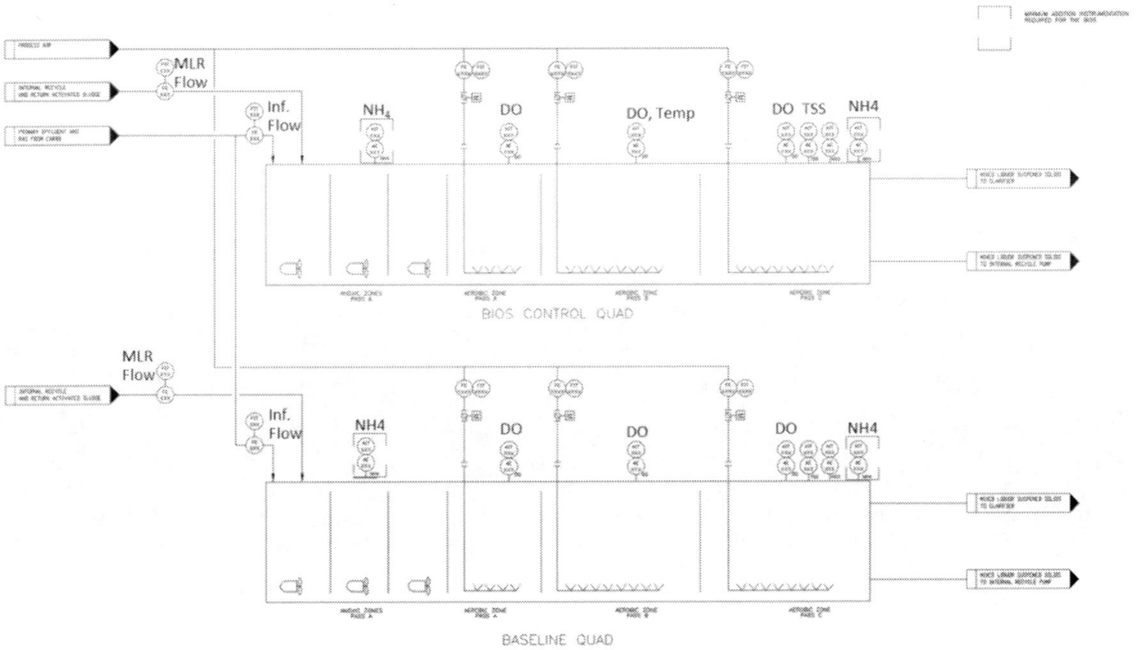

Figure 3. Process Flow Diagram of the Feed-Forward Controller Demonstration

The model only considers the aerobic section of each NSEC train. The aerobic section is modeled as eleven (11) continuously stirred tank reactors (CSTR) in series. The reactor influent (primary effluent), RAS, and mixed liquor recycle (MLR) flows are totalized and used as the flow input to the first aerobic CSTR. The output conditions of the CSTR become the input conditions of the subsequent CSTR. The model's configuration is summarized in Figure 4 below.

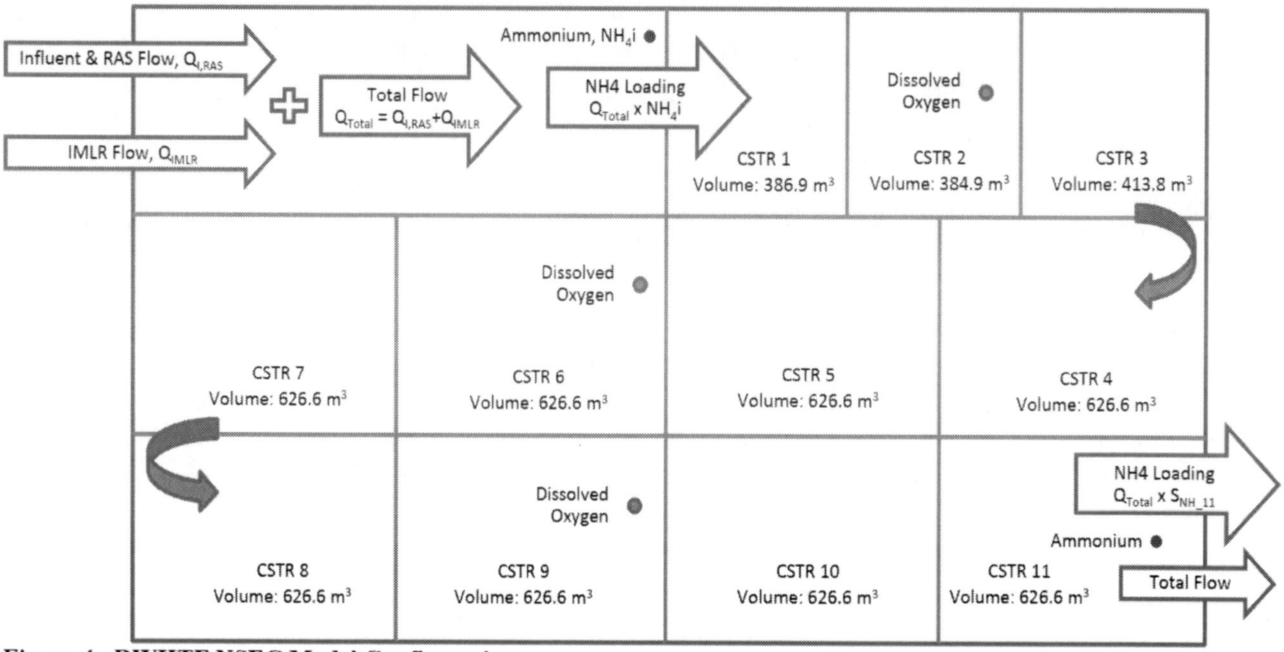

Figure 4. RWHTF NSEC Model Configuration

The controller runs on a virtual Windows 7 machine embedded in the existing Metro District server. It communicates with the plant Distributed Control System (DCS) through a Structured Query Language (SQL) Server Express database. The controller retrieves plant operating data from the database, and transmits DO set points for each aeration zone to the database for implementation by the plant DCS. For the pilot installation, the controller user interface allows operator input of set-point limits, operating data and manual overrides. The controller was installed in the second half of August, and it began operation on September 1, 2011.

NSEC Pilot Results

The controller uses a feed-forward control algorithm to provide the required DO set points to meet the variation in diurnal loadings. The aeration energy savings are generated by lowering the DO within the aeration tanks as needed by the process. As DO levels drop, the oxygen transfer efficiency increases, requiring less airflow to meet the same oxygen demand.

A typical week of DO set points for each pass of Trains 9 (Feed-forward) and 10 (baseline) is shown in Figure 5. In general the DO set points for Train 9, the feed-forward controlled train, were below the DO set points for Train 10, the baseline train. The lower DO set points resulted in decreased airflow usage, as shown on Figure 6. Despite the decreased airflow usage, nitrification performance of the two trains was similar, as shown in Figure 7. The two trains had similar influent ammonia concentrations, and similar effluent ammonia (NH_3-N) and nitrate (NO_3-N) concentrations. This indicates that the lower DO set points did not have a negative effect on nitrification performance of the secondary treatment.

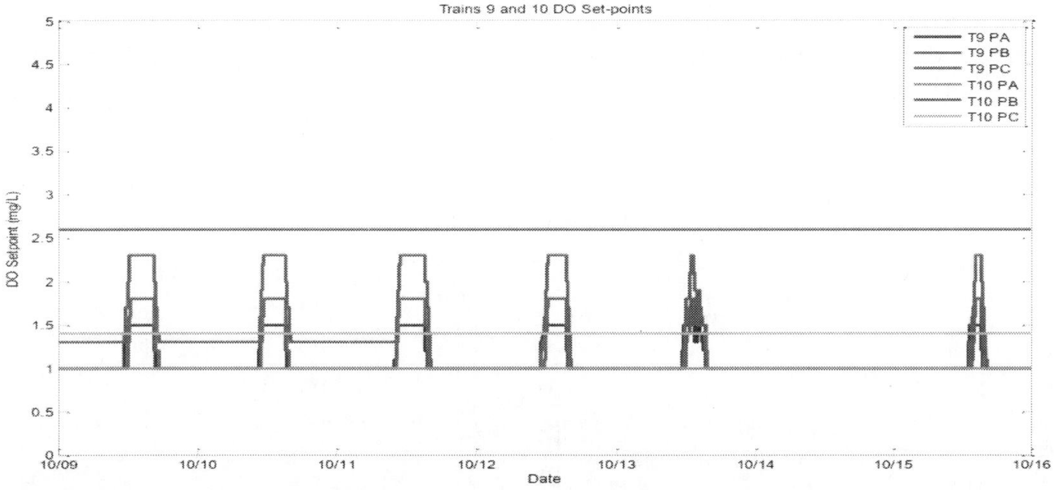

Figure 5. Train 9 and 10 DO Set points

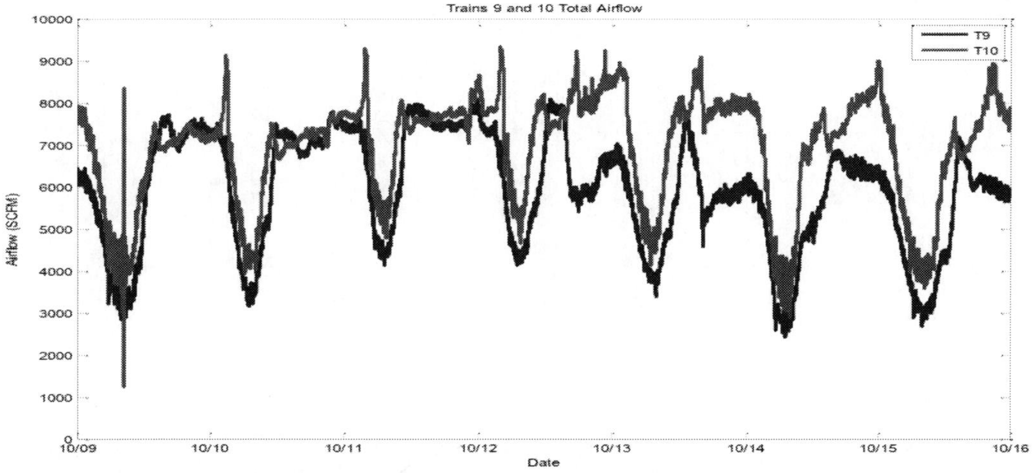

Figure 6. Train 9 and 10 Total Airflows

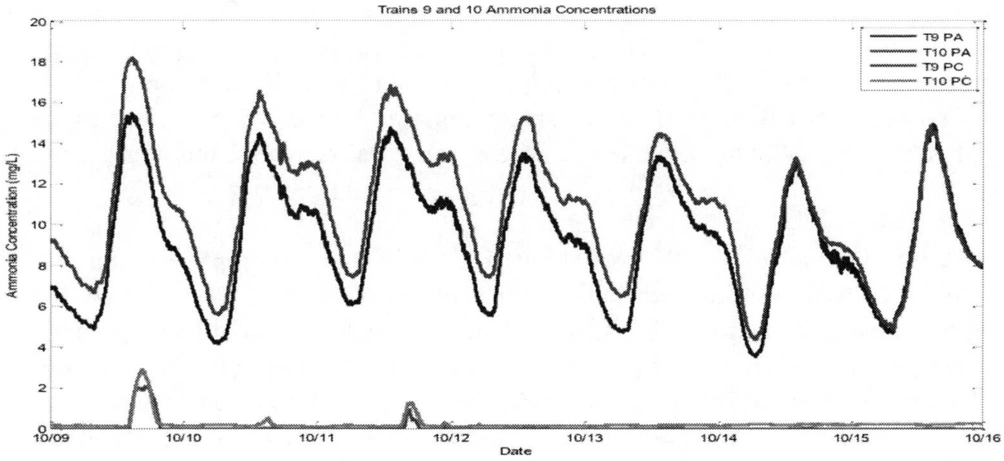

Figure 7. Train 9 and 10 Ammonia

The total airflow to each train was used to compare the feed-forward controlled quad to the baseline quad. The results of the comparisons are summarized on Figure 8. The airflow reduction between the baseline and feed-forward quads was up to 20%. During the six months of the demonstration project the BIOS controlled quad used 11% less air than the baseline quad.

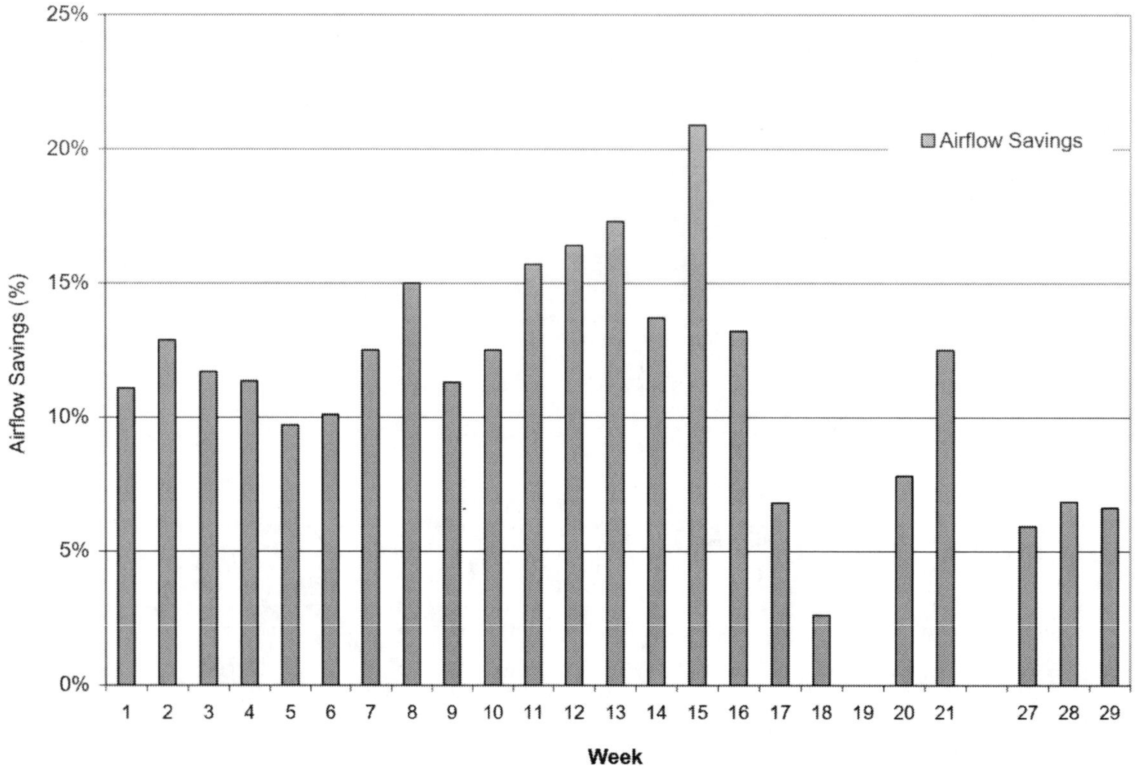

Figure 8. BIOS Aeration Savings

Prior to week 17 the average savings was 13.5%, but during week 17 the operator lowered the DO set-point for the Baseline quad and the other two quads not controlled by the controller. Because the controller DO set points were consistently lower than the baseline while meeting the effluent ammonia limit, operations felt comfortable dropping the DO set points for the rest of the plant to take advantage of the potential savings. As expected, the relative savings generated by the controller dropped, but it was still able to achieve savings of 6%. With further optimization of the selected DO range and effluent goal, the future savings can be increased. Note that the pilot was stopped during weeks 22 to 26 while the DCS system was down for maintenance and upgrades.

The feed-forward controller was part of the Metro District's NSEC Energy Optimization Initiative. Prior to the pilot installation and as part of the Initiative, the Metro District limited the number of blowers in operation to two. This limitation of blowers generated a DO sag during the peak of the day, but reduced the airflow usage by an estimated 10%. During the pilot the feed-forward controller was not affected by the limits on the number of blowers, and only required two blowers independent of the blower restriction. Comparing the DO set points calculated by the feed-forward controller to the DO set points utilized by the plant before the NSEC Energy Optimization Initiative was started shows a theoretical savings of 22-30% using the feed-forward controller.

To place a monetary value to the observed savings, the observed airflow differentials were entered into the following equation, which includes the Metro District specific power cost in 2010 ($0.075/kWh) and conversion factor for airflow to power (0.000423 kilowatt hour per standard cubic foot [kWh/scf]):

$$\text{Daily Aeration Cost}\left(\frac{\$}{d}\right) = Airflow\left(\frac{scf}{\text{min}}\right)\cdot 0.00423\left(\frac{kWh}{scf}\right)\cdot 1440\left(\frac{\text{min}}{day}\right)\cdot 0.075\left(\frac{\$}{kWh}\right)$$

The estimated aeration savings during the demonstration are approximately $18,635. With a full installation of the BIOS, the estimated yearly savings would be $134,026 at the 11% observed average savings.

To fully realize the benefits of the feed-forward controller, it is essential to have an aeration system that is capable of controlling the airflow to meet the dynamic DO set points calculated by the controller. In its default setting the controller limits the DO set-point change to 0.3 mg/l per calculation cycle of 15 minutes. The NSEC aeration control system shows an acceptable level of performance for control. Aeration control over a five day period in Train 9 Pass B is shown in Figure 9.

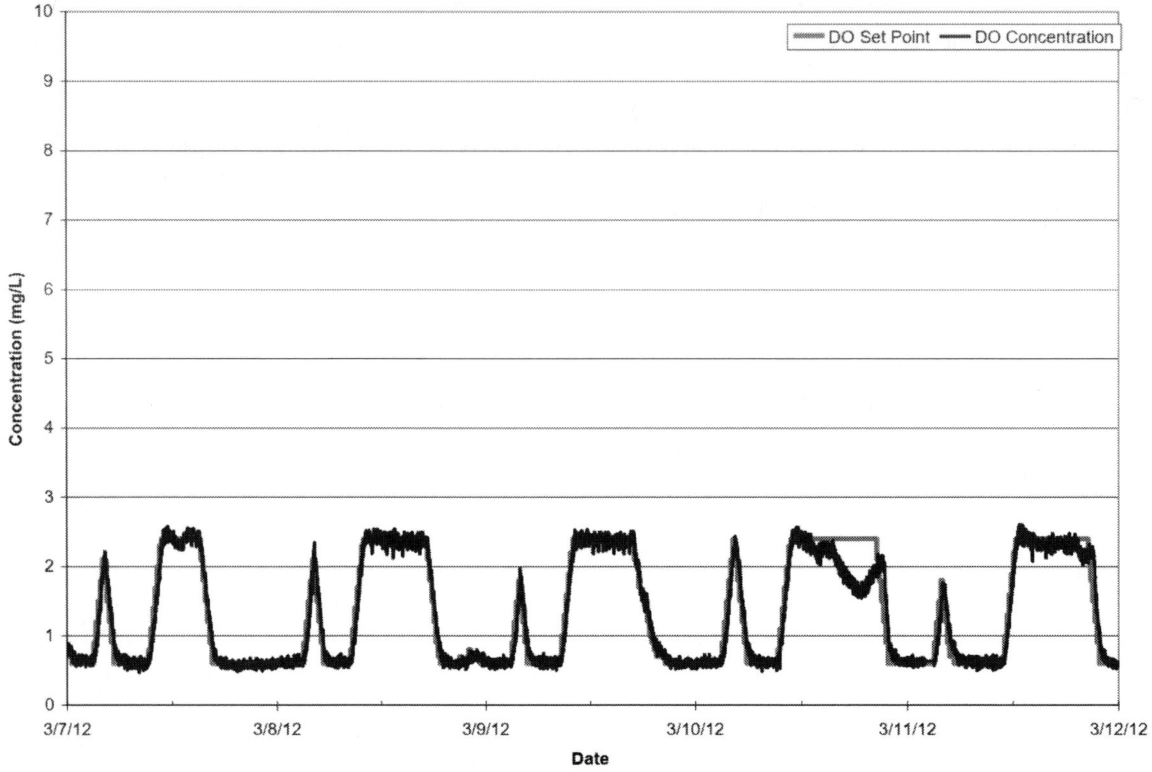

Figure 9. DO Set-Point Tracking for Train 9 Pass B

With any control system the control is only as good as the instrumentation that is used. For the feed-forward control system there are several instruments required: flow, DO, temperature, MLSS, and ammonium analyzers. Most of the instrumentation is well established within the wastewater community and requires limited maintenance. In recent years ammonium analyzers have made great strides in regards to robustness, but they still require frequent maintenance and calibration. The Metro District checked the analyzers once a week with grab samples and adjusts the analyzer calibrations curves as needed.

The Metro District provided and installed four (4) Endress & Hauser (E&H) ammonium analyzers based upon BioChem's recommendations. Overall, the ammonium analyzers worked well. Train 9 and 10 analyzer readings closely tracked each other, with only few periods of questionable readings. Figure 10 shows ammonium readings during a typical day.

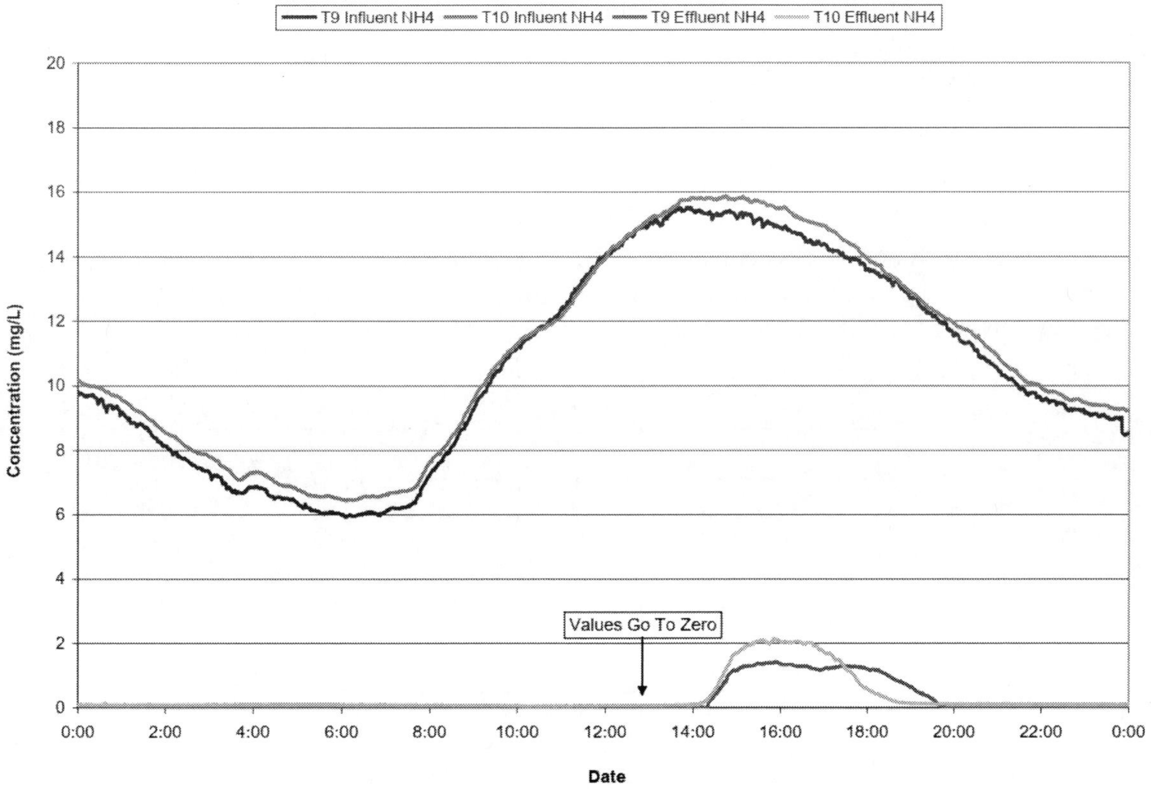

Figure 10. Train 9 and 10 Ammonium Analyzer Measurements

CASE STUDY 2

Liuwei Wastewater Treatment Plant, Yangzhou, Jiangsu Province, China

Liuwei WWTP (Liuwei) is located in Yangzhou, Jiangsu Province, China. Phase I of Liuwei has been operating since April 2005 utilizing an oxidation ditch process with a capacity of 50,000 m^3/day (13.2 MGD). By April 2007, design work for the Phase II expansion project had begun. In July 2008, the construction of the Phase II expansion began. The Phase II expansion added an A^2O process to the plant with a capacity of 100,000 m^3/day (26.4 MGD). Phase II start-up began in October 2010. In total, the Liuwei WWTP is responsible for serving an area of 19.16 square miles and a population of 390,000 persons. In order to ensure the Phase II expansion produces high quality effluent, a BIOS control system was included in the design and has been operating since February 2011.

The Phase II expansion is a completely separated process from Phase I. The liquid treatment train consists of bar screens, primary clarifiers, hydrolysis treatment, secondary treatment, secondary clarifiers, a high density sedimentation tank and microfiltration prior to final effluent discharge. The designed influent conditions and effluent permit are listed in Table 2.

Table 2: Influent Conditions and Effluent Permit for Liuwei WWTP

	COD	BOD_5	SS	TN	NH3-N	TP
Influent Conditions	450	200	230	45	35	4.0
Effluent Permit	50	10	10	15	5	0.5

The secondary treatment process consists of two trains in an anaerobic-anoxic-oxic (AAO) configuration. A diagram of the process can be found in Figure 11 showing how the flow through the major processing units.

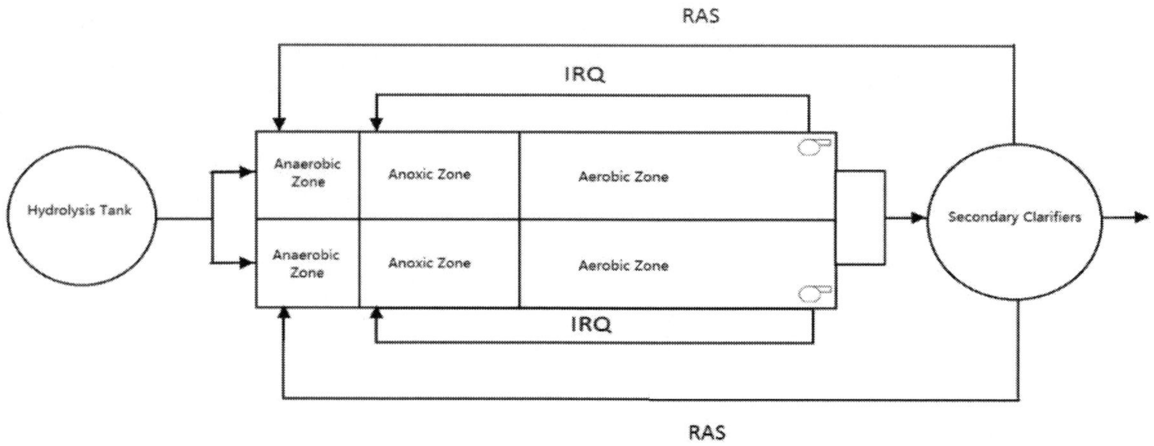

Figure 11. Major Process Units at Phase II of Liuwei WWTP

In order to implement the BIOS and maximize its benefits to the plant, installation locations of analyzers and required equipment including air valves and airflow meters were carefully selected. A diagram of one train is shown in Figure 12 indicating how these instruments were installed at Liuwei.

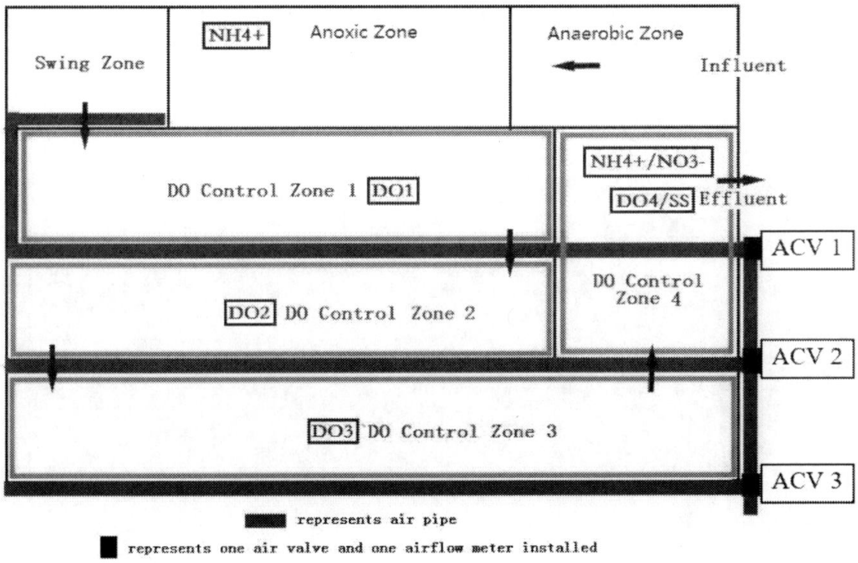

Figure 12. Layout of One Process Train and Installation Positions for Equipment

One ammonia analyzer was installed at the end of the anoxic zones and one ammonia/nitrate analyzer was installed at the end of the aerobic zones to measure the influent and effluent ammonia concentrations. For the first three DO control zones, three sets of air valves/airflow meters were installed in order to control the air entering into these zones. For DO control zone 4, manual valves were installed at each drop-leg. These manual valves are adjusted periodically to maintain a low DO in the effluent of the aeration tank in order to make sure the internal nitrate recycle doesn't return too much DO back to the anoxic zone.

Liuwei WWTP Results

The advanced control system optimizes the DO set points based on the influent flow rate and the influent ammonia concentration, as shown in Figure 13.

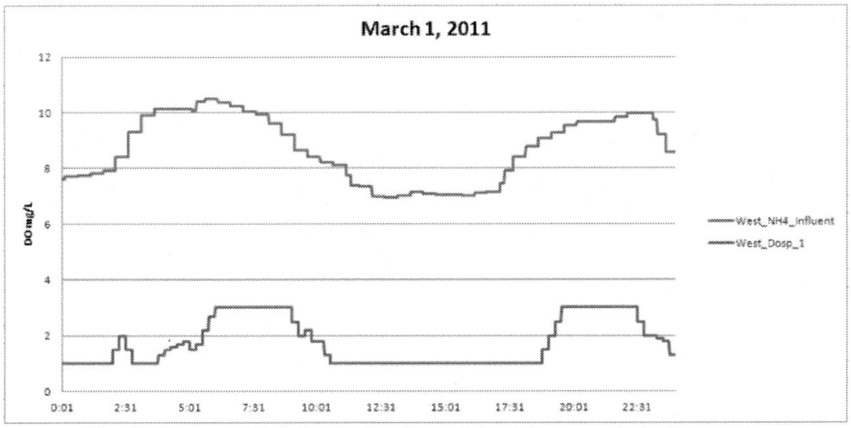

Figure 13: Calculated DO Set-Point Based On Changes in Influent Ammonia

One of the secondary objectives of implementing the advanced control system was to maintain tight DO set-point tracking under changing influent flow rates and ammonia. Figure 14 demonstrates how the influent loading changed in the period of April 8 to 18, 2012. The incoming flow rate varied from 72,225 m^3/day (19.08 MGD) to 92,288 m^3/day (24.38 MGD) and the ammonia concentration varied from 4.7 g/m^3 to 37.2 g/m^3.

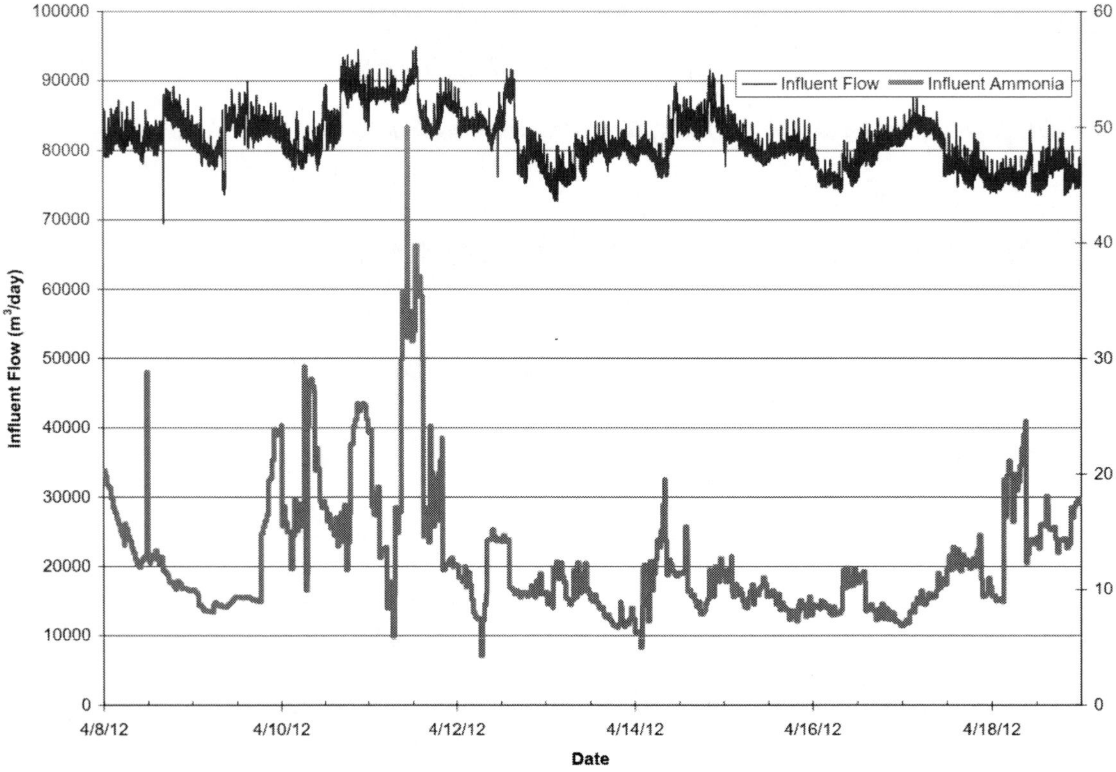

Figure 14: Changing influent flow rate and concentration over an 11-day period

DO control performance of one pass over the same period is shown in Figure 15. During this period, the DO was within ±1.0 mg/L of the set-point 99.1% of the time, and within ±0.5 mg/L of the set-point 93.9% of the time. The DO set-point was changing to match the changes in the influent loading.

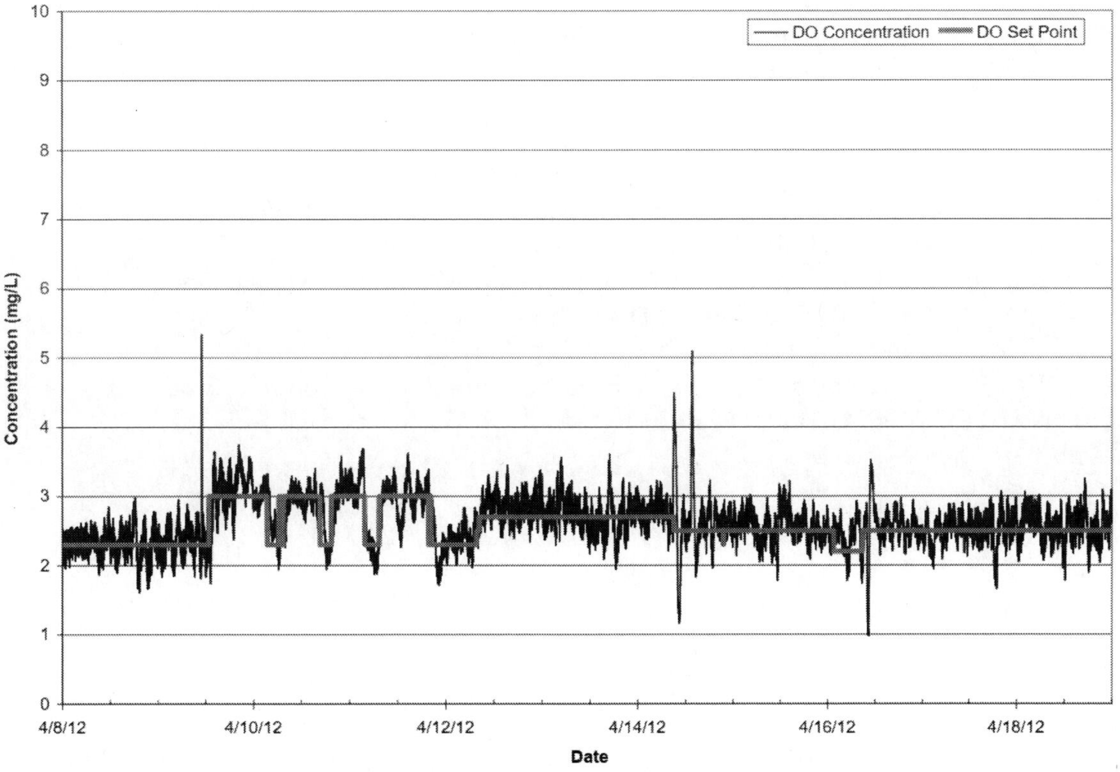

Figure 15: 11-day DO Control Curve Under Changing DO Set points

Under manual aeration control the blowers were set to supply a constant 16,000 m^3/hr of airflow which consumed an average of 195 kVA. While utilizing the advanced control system, the plant is now consuming an average of 154 kVA, an energy savings of 21.0%.

CONCLUSIONS

Overall the installations have been a success. The objective of the Metro Distric pilot project was to verify that dynamic dissolved oxygen control of the aeration system based on feed-forward control could reduce the aeration demand while meeting treatment requirements. Over the 6 months of the pilot, the feed-forward controlled quad used 12% less air than the baseline quad. With further optimization of the selected DO range and effluent goal, the savings can be increased. The estimated aeration savings during the demonstration, as a direct result of reduced airflow requirements, is approximately $18,635. With a full installation of the control system, the estimated yearly savings is $134,026. Based on these results, the Metro District proceeded with a full installation of the control system.

Likewise, the process control system at Liuwei WWTP has been a success. By implementing this control system, the plant has reduced its energy requirement by 21.0% while consistently meeting its effluent requirements.

REFERENCES

Endress Hauser (2011) ISEmax CAS40D, Technical Information.

Fry, C., and Langley, S. Ion-Selective electrodes for Biological Systems. Amsterdam: Harwood Academic , 2001.

Liu, W., G.J.F. Lee, P.E. Schloth, M.E. Serra (2005) Side by Side Comparison Demonstrated a 36% Increase of Nitrogen Removal and 19% Reduction of Aeration Requirements Using a Feed Forward Online Optimization System. Proceedings of WEFTEC 2005, Washington, DC, CD-ROM.

U.S. Environmental Protection Agency (1989) Fine Pore Aeration System, Design Manual; EPA/625/1-89/023; Washington, D.C.

Walz, T., J.R. Coughenour, T. Stahl (2009) Energy Savings at Phoenix 23rd Avenue Wastewater Treatment Plant Using Feed-Forward Process Control. Proceedings of WEFTEC 2009, Orlando, FL, CD-ROM.

DYNAMIC ENERGY FOOTPRINT MONITORING OF WASTEWATER AERATION SYSTEMS VIA FULL-SCALE DEPLOYMENT OF REAL-TIME OFF-GAS ANALYZERS

Matthew K. Jeung[1], Lory E. Larson[2], Michael K. Stenstrom[3], Diego Rosso[1,*]

[1] Civil and Env. Engineering Dept., University of California, Irvine, CA 92697-2175
[2] Southern California Edison, CSBU, Irwindale, CA 91706
[3] Civil and Env. Engineering Dept., University of California, Los Angeles, CA 90095-1593
* Corresponding author (e-mail: bidui@uci.edu)

ABSTRACT

Wastewater aeration is a highly complex and dynamic process for which site-specific analysis and control are needed. Off-gas testing is a method that can determine the aeration process requirements and conditions in real-time, so that appropriate operational changes and control strategies can be implemented. For this research project, off-gas analyzers were installed at five wastewater treatment plants in Southern California to continuously measure standard oxygen transfer efficiency (SOTE) and off-gas flux and perform a subsequent energy analysis. The results were as expected and show that there is an inverse relationship between SOTE and off-gas flux that follows a distinct diurnal pattern. The energy analysis showed that the energy usage during peak demand is amplified due to compounding factors of increased COD and volumetric wastewater load. Utilizing off-gas monitoring to determine process requirements and performance, many energy saving strategies (e.g. blower type, variable frequency drives, control strategies) that can lower peak and overall energy demand are feasible and can be implemented in a energy rebate program.

KEY WORDS – Aeration; Oxygen Transfer Efficiency; Off-gas; Optimization; Blower Energy; Process Control

INTRODUCTION

Aeration is an essential and energy-intensive process for most wastewater treatment plants, posing as prime energy footprint contributor with 45 to 75% of process energy drawn (Reardon, 1995). Fine-pore diffusers, the most common municipal aeration technology today, are marred by fouling and scaling problems associated with rapid declines in performance and increases in process energy demand and costs (Rosso and Stenstrom, 2006). However, by measuring oxygen transfer efficiency (OTE, %) with continuous off-gas monitoring (Figure 1), significant saving margins in energy and maintenance costs can be quantified and achieved (Leu et al, 2009). Dynamic OTE data can be obtained by the off-gas testing method (Redmon et al, 1983), which was used previously to compare aeration system efficiencies (Libra et al, 2002) and process energy footprints (Rosso et al, 2011), and has been shown to be a preferred method to study oxygen transfer in membrane bioreactors (Krause et al, 2003) as well as biofilters (Gillot et al, 2005). Continuous off-gas monitoring has also been applied to optimize aeration by introducing feedback dissolved oxygen (DO) control with feed-forward off-gas control strategies (Trillo et al, 2004).

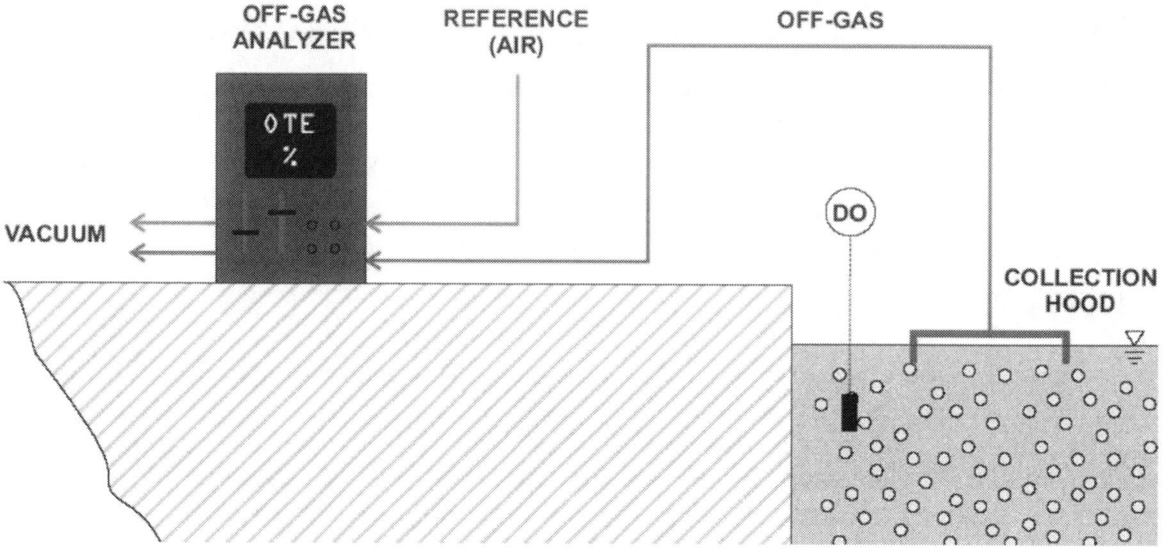

Figure 1. Schematic of continuous off-gas testing set-up.

Control strategies for aeration systems aim to maintain relatively stable DO by utilizing dissolved oxygen (DO) probes in the aeration basin that obtain DO signals which are input into a feedback loop that controls the blowers. DO has a great effect on the performance and efficiency of biological treatment processes and is a strong indicator of process conditions. When DO is below optimal conditions, microbial respiration and metabolism may be inhibited and the sludge composition may change, which may lead to reduced treatment efficiency and process failures. Above optimal DO may result in denitrification issues in anoxic reactors or an indication of excess energy through over-aeration. (Ferrer, 1998; Serralta et al., 2002). Even though several control strategies have been developed to improve the DO control system (Ferrer, 1998; Stare et al, 2007; Holenda et al, 2008).

The DO control system is typically tuning the air flow discharge to the aeration tanks employing a simple feedback loop feeding a PID controller. The application of air flow control to individual basins or individual diffuser grids and the type of blower control is selected based on process requirements and economic considerations (U.S. EPA, 1989). To optimize the energy consumption of aeration systems, the goal is to provide the minimal amount of air needed to meet substrate removal requirements. Through the application of low-cost on-line off-gas measurements, oxygen transfer rate and air flow rate can be measured, providing a new tool for process analysis and accurate energy calculations. In addition, continuous long-term off-gas measurements can be used to monitor the decline in efficiency due to diffuser fouling.

In this research project, existing off-gas monitoring prototypes were redesigned to become permanent installations for continuous off-gas monitoring. This paper presents the latest off-gas technology as well as the energy analysis from continuous off-gas monitoring.

METHODS

Fully-automated off-gas monitoring analyzers with their collection hoods were developed and permanently installed in 5 wastewater treatment plants in Southern California to monitor process efficiency for at least nine months (Figure 2). These plants include smaller (~40000 m^3/d) to larger (~400000 m^3/d) facilities. The analyzers and collection hoods are developed to be affordable and able to be assembled from readily available components. The collection hood capture areas ranged from 1.70 to 2.97 m^2. From these installations, OTE and off-gas flux ($m\ h^{-1}$) data was collected hourly in order to determine dynamic variations in process performance. Hourly measurements are appropriate because large-scale aeration systems have time-constants of the same order, i.e. an action with a consequent reaction cannot be taken more frequently than hourly (seldom, every 30 minutes).

Figure 2. Installed off-gas analyzer (right) and hood (right).

The off-gas analyzers were visited approximately once every 10 days for periodic data collection. The raw data collected from the instrument are 4-20 mA signals which are typically used in plant data systems. OTE is determined by comparing the oxygen mole fraction in the off-gas with the oxygen mole fraction of the reference gas used in the aeration process (e.g. 20.95% for air). By using a reference gas for each hourly measurement of OTE, the off-gas analyzer is self-calibrating. OTE was corrected to standard conditions (i.e., αSOTE, %) using site-specific temperature and pressure data, and process DO values (the salinity correction in these plants never exceeds 1% departure from ideal conditions). The off-gas velocity readings from the off-gas analyzer are used to calculate the off-gas flow rate through the instrument. Based on the off-gas flux (i.e. the analyzer off-gas flow rate per capture area of the collection hood), the off-gas flow rate for the entire aeration basin can be estimated. Since off-gas flux is directly proportional to the air flow rate (AFR) through the aeration system, the power drawn by the blowers and their energy footprints can be quantified compared between different processes.

To incorporate off-gas flux data into an aeration energy model, the equation for blower power requirements involving adiabatic compression is modified for the wastewater aeration process (Metcalf and Eddy, 2003):

$$BHP = \frac{W_{air}RT}{c\gamma\eta}\left[\left(\frac{P_d}{P_i}\right)^{\gamma} - 1\right]$$

$$= \frac{\rho_{air}\cdot AFR\cdot RT}{c\gamma\eta}\left[\left(\frac{Z + h_L + DWP + P_i}{P_i}\right)^{\gamma} - 1\right]$$

$$= \frac{\rho_{air}\cdot A_{TANK}\cdot J_{air}\cdot RT}{c\gamma\eta}\left[\left(\frac{Z + h_L + DWP + P_i}{P_i}\right)^{\gamma} - 1\right]$$

where BHP is the power drawn by the blower(s), W_{air} is the mass air flow rate, R is the universal gas constant, T is the inlet (ambient) temperature, P_i is the inlet pressure, P_d is discharge pressure, c is the molecular weight of air, γ is the ratio of specific heats (0.283 for air) η is the combined motor and blower efficiencies (typically ranging from 0.7 to 0.9), ρ_{air} is the air density, AFR is the volumetric air flow rate, Z is the hydrostatic pressure from the diffuser submergence, h_L is the head loss in the air distribution line, DWP is the dynamic wet pressure, A_{TANK} is the tank bottom area, and J_{air} is the air flux. The energy footprint of the aeration process is calculated by normalizing the blower power drawn with the site specific wastewater plant volumetric load.

RESULTS AND DISCUSSION

The results verify that off-gas testing can be applied to continuously monitor the efficiency of wastewater aeration processes for extended periods (Figure 3). The depth of the aeration basin was approximately 6 m, therefore the αSOTE values are similar to typical process conditions (~3 to 5 %/m). As predicted, hourly changes in off-gas flux result in inverse αSOTE patterns; at higher off-gas flux, αSOTE is lower. This is largely due to phenomena such as decreased specific area of the bubbles (i.e. bubble surface area per bubble volume) and decreased bubble residence time within the aeration basin. For the aeration process, higher off-gas flux is demanded when increased carbonaceous oxygen demand (COD) levels and/or increased volumetric wastewater loads enter the reactor. For most wastewater treatment plants, peak COD and loading events occur simultaneously and in a diurnal pattern. This results in a daily increase in oxygen uptake rate (OUR), which lowers the DO and triggers the DO feedback control system.

In Figure 3, elevated off-gas fluxes are observed each day from approximately 9am to 6pm. The timing, duration and amplitude of this diurnal peak are highly dependent on the quantity and quality of the wastewater and the characteristics of the wastewater collection system. Large wastewater treatment plants typically receive water from large wastewater collection systems that carry wastewater for relatively long residence times before arriving at the plant. For these larger systems the diurnal off-gas flux peaks will be seen later in the day and are generally less dramatic than for smaller wastewater treatment plants.

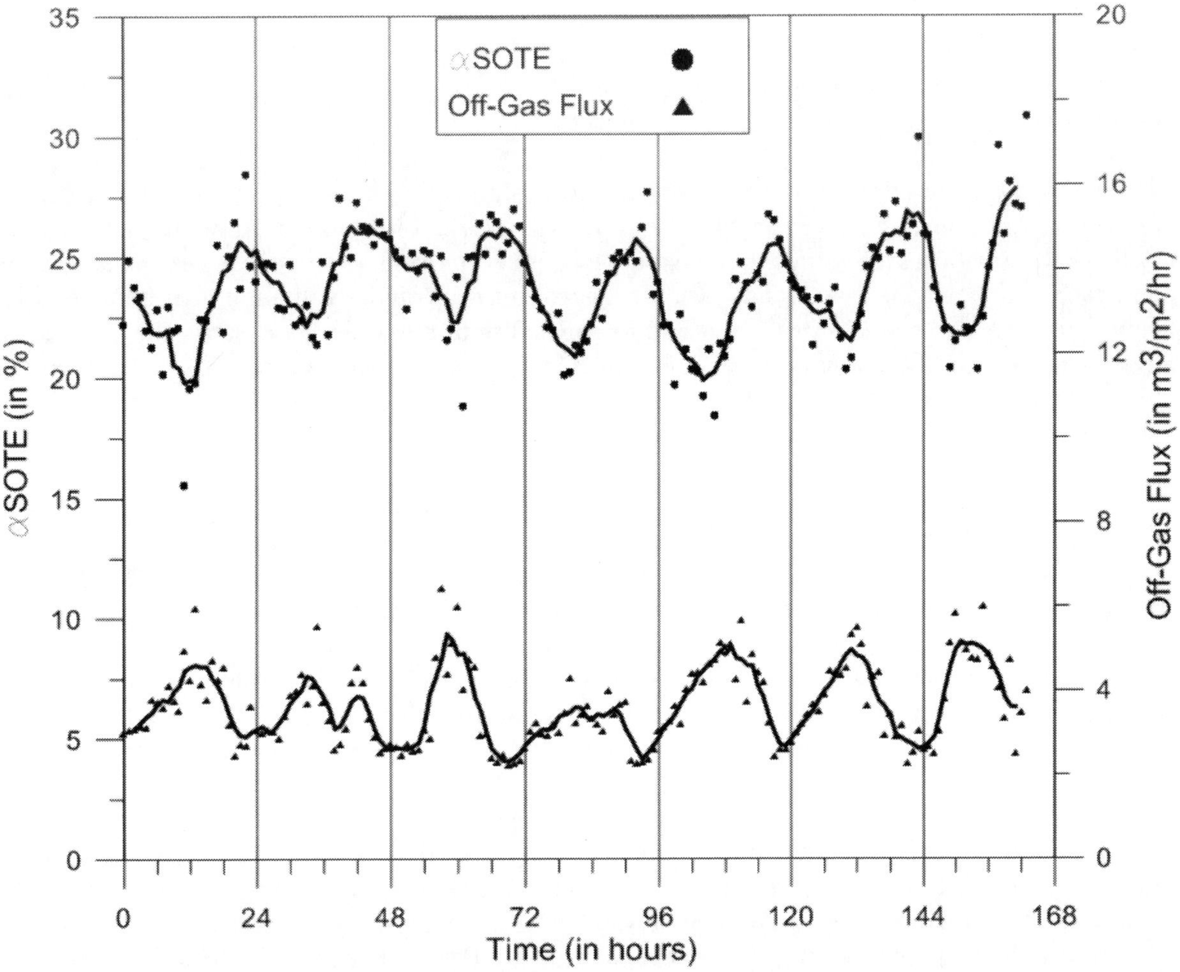

Figure 3. Standard oxygen transfer efficiency in process water (αSOTE) and off-gas flux vs. time for a one-week period; The curves represent a running average of the hourly measurements of αSOTE and off-gas flux.

In addition, differences may be seen between separate and combined sewers. Wet-weather flows would introduce greater total daily load into a wastewater plant, however COD concentration levels could be lower and more distributed, which would reduce the relative amplitude of the off-gas flux peak. These wet weather patterns can play an important role in wastewater aeration energy usage and operational cost and should be considered in the process design and operation.

Historical time-series of measured OTE and calculated blower power drawn and energy usage can be used to calculate the excess energy wasted due to peak demand and diffuser fouling. By comparing the cost of cleaning to the energy wastage due to diffuser fouling, an optimal cleaning frequency can be recommended for each plant. Also, the effectiveness of cleaning for each plant can be quantified in order to compare traditional and emerging cleaning methods (e.g., in-line chemical cleaning, flexing/reverse flexing of membranes, etc.). Another strategy to limit energy usage during peak times is the implementation of equalization basins, where possible. By distributing the wastewater load from the peak to off-peak times, the overall energy and operational cost to treat the water can be much lower and off-set the cost of the equalization basin. Moreover, burdensome peak power demand charges would be limited if not entirely avoided, and since these charges are typically billed in longer cycles (e.g., quarterly) the overall energy cost savings would be further pronounced.

Through off-gas monitoring, appropriate process design and operation decisions can be made to reduce energy usage and operational costs. For example, in order to determine the most suitable options for aeration system design or expansion, other analysis on a cost-benefit basis, such as comparing the use of fixed and variable speed blowers and evaluating the benefits of an equalization basin, can be considered and quantified with dynamic OTE data. The benefits and disadvantages of the various types of centrifugal blowers and positive displacement (PD) blowers can be quantified. Also, the benefits of variable frequency drive (VFD) retrofits, which enable the change of the rotational speed of the blower, can be quantified.

The dynamics and conditions in typical wastewater aeration processes exacerbate the energy footprint to perform wastewater treatment. As seen in Figure 4, when the oxygen uptake rate is highest and the most oxygen is needed, the oxygen transfer within the aeration basin is least efficient and the power and energy curves amplify. When DO is constant, off-gas data can be used to calculate the OUR as the ratio of the oxygen transfer rate (OTR, $mass_{O2}$/time) to the water volume below the off-gas hood. The cost of energy may also follow the same diurnal pattern, because of varying energy sources at different times of the day. Therefore there may be large diurnal differences in operational costs between different wastewater aeration processes.

To limit the diurnal fluctuations in the aeration process performance and to reduce overall energy consumption, a feed-forward control strategy using signals from off-gas monitoring may be implemented (Trillo et al, 2004). If process requirements and oxygen transfer performance are determined in real-time by off-gas analysis, this information can be utilized to predict and control the air required to meet process requirements. By implementing a feed-forward control system based on off-gas signals, an adaptive and energy efficient control of the aeration system can be achieved.

Based on the energy and cost saving capabilities mentioned, an energy rebate program can be implemented to provide cash incentives to improve the energy efficiency of wastewater aeration systems. With off-gas monitoring of wastewater aeration, energy savings for various design and operation improvements can be quantified. Based on the energy savings, a cash refund would be provided to the wastewater utility. Therefore, in addition to reducing the energy burden of wastewater aeration, current energy saving techniques would result in further investment toward sustainable wastewater treatment and infrastructure and possibly the re-investment for further energy and cost savings.

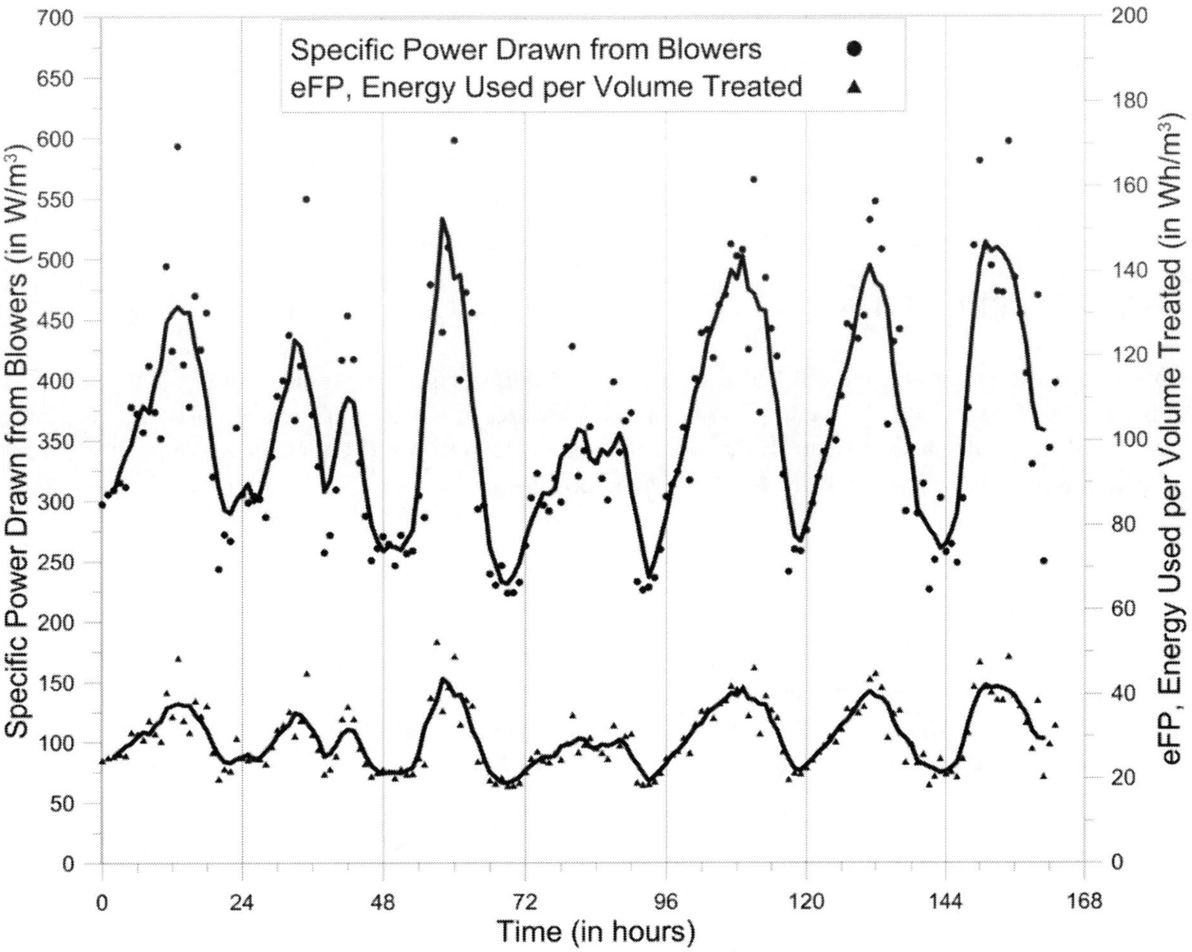

Figure 4. Specific power drawn (W m^{-3}) and specific cumulative usage, also referred to as energy footprint (Wh m^{-3}) for a treatment plant performing biological nutrient removal in Southern California; The curve represents a running average of the hourly measurements.

SUMMARY AND CONCLUSIONS

The objectives of this research project were to develop and apply an off-gas analyzer for continuous OTE monitoring and energy analysis of the wastewater aeration process. Below are the main conclusions from this research project:

- An automated off-gas analyzer was developed for continuous OTE monitoring. The instrument was affordable, self-calibrating, and built for permanent installation.
- Continuous OTE monitoring if applied for several months can be used to optimize cleaning schedules by comparing energy wastage to cleaning cost.

- Dynamic OTE data and the subsequent energy analysis were used to provide insight into the costs and benefits of various aeration processes and upgrades and expansions to aeration processes.
- If connected to plant data systems, real-time analysis of OTE signals can have dynamic operational implications, such as feed-forward off-gas control and energy minimization.
- Lastly, real-time OTE monitoring can be used for benchmarking the energy efficiency of aeration technologies, to aid the establishment of rebate programs for energy savings from the wastewater aeration process.

ACKNOWLEDGMENTS

We gratefully acknowledge the California Energy Commission, for their financial support of this project. Also, we would like to thank the wastewater utilities and particularly the operators involved for their participation in this research project. Without their generous support and help the success of this project would not have been possible.

REFERENCES

Ferrer, J., Rodrigo, M.A., Seco, A., Penya-roja, J.M. (1998) Energy saving in the aeration process by fuzzy logic control. Water Science and Technology 38 (3): 255-263.

Gillot, S., Capela-Marsal, S., Roustan, M., Héduit, A. (2005) Application of the off-gas method to the measurement of oxygen transfer in biofilters. Chemical Engineering Science. 60: 6336-6345.

Holenda, B., Domokos, E., Redey, A., Fazakas, J. (2008) Dissolved oxygen control of the activated sludge wastewater treatment process using model predictive control. Computers and Chemical Engineering. 32: 1270-1278.

International Water Association (2008) Biological Wastewater Treatment – Principles, Modelling and Design. IWA Publishing, London, UK.

Krause, S., Cornel, P., Wagner, M. (2003) Comparison of different oxygen transfer testing procedures in full-scale membrane bioreactors. Water Science Technology. 47 (12): 169-176.

Leu, S.Y., Rosso, D. Larson, L.E., Stenstrom, M.K. (2009) Real-time aeration efficiency monitoring in the activated sludge process and methods to reduce energy consumption and operating costs. Water Environment Research. 81: 2471-2481.

Libra, J.A., Schuchardt, A., Sahlmann, C., Handschag, J., Wiesmann, U., Gnirss, R. (2002) Comparison of the efficiency of large-scale ceramic and membrane aeration systems with the dynamic off-gas method. Water Science and Technology. 46 (4-5): 317-324.

Metcalf and Eddy, Inc. (2003) Wastewater engineering: treatment and reuse. 4th ed., McGraw-Hill, New York.

Reardon, D.J. (1995) Turning down the power. Civil Engineering, ASCE. 65 (8): 54-56.

Redmon, D., Boyle, W.C., Ewing, L. (1983) Oxygen Transfer Efficiency Measurements in Mixed Liquor Using Off-gas Techniques. Journal of the Water Pollution Control Federation. 55 (11): 1338-1347.

Rosso, D. and Stenstrom, M.K. (2006) Economic implications of fine-pore diffuser aging. Water Environment Research. 78 (8): 810-815.

Rosso, D., Lothman, S.E., Jeung, M.K., Pitt, P., Gellner, W.J., Stone, A.L., Howard, D. (2011) Oxygen transfer and uptake, nutrient removal, and energy footprint of parallel full-scale IFAS and activated sludge processes. Water Research. 45: 5987-5996.

Seralta, J., Ribes, J., Seco, A., Ferrer, J. (2002) A supervisory control system for optimising nitrogen removal and aeration energy consumption in wastewater treatment plants. Water Science and Technology. 45 (4-5): 309-316.

Stare, A., Vrecko, D., Hvala, N., Strmčnik, S. (2007) Comparison of control strategies for nitrogen removal in an activated sluedge process in terms of operating costs: A simulation study. Water Research. 41: 2004-2014.

Trillo, I., Jenkins, T.E., Redmon, D., Hilgart, T., Trillo, J. (2004) Implementation of feedforward aeration control using on-line off-gas analysis – the grafton wwtp experience. Proceedings of the Water Environment Federation, WEFTEC 2004 91-95 27-45.

U.S. Environmental Protection Agency (1989) Design Manual: Fine Pore Aeration Systems, EPA/625/1-89/023.

ADVANCED CONTROLS MAXIMIZES ENERGY EFFICIENCY OF A LARGE EFFLUENT PUMP STATION

Curt Wendt, PE, CAP

CDM Smith, 2301 Maitland Center Parkway, Suite 300, Maitland, FL 32751

ABSTRACT
Orange County Utilities Department had an aging 40-mgd effluent pump station that needed pumping system replacements at the South Water Reclamation Facility. This included rehabilitating five existing 6,500-gpm vertical turbine pumps, installing a new 1,000 HP motor, replacing the existing variable frequency drives, and replacing the obsolete pump station control system. When this station was commissioned, it was realized that the specified replacement control scheme was too simplistic and advanced controls were required.

A better strategy was developed which incorporated the use of a second PID controller to form a cascade control scheme. The equipment changes needed to implement this strategy were minimal because the control system was software based. Although the need to reduce energy consumption was not originally recognized, new advanced controls provided the flexibility to reduce the electrical costs of the station.

KEYWORDS
Cascade Control, Large Pump Stations, PID Controller, Pump Control, Variable Frequency Drive

INTRODUCTION
Pumping is a major user of electrical power in a typical wastewater plant. In order to reduce operating costs, many plants have under taken energy efficiency upgrades. These typically start with replacing the existing motor with a high efficiency motor. After high efficiency pump motors are installed, the next step to improving energy efficiency is installing variable frequency drives (VFDs). VFDs are much more energy efficient than constant speed motor to better match pumps to changing flow conditions.

However, installing VFDs on pumps to save energy is just part of the solution. To use VFDs to their greatest benefit, they must be controlled by an energy efficient control scheme. Many energy update projects focus on using high efficiency motors and VFDs without reviewing how the pumps are controlled. The smooth control of large pump stations is required to provide efficient operation of the pumps and minimize power consumption at the station. Modifying the existing control scheme to produce smooth pump operation can yield a significant savings in energy and operation costs. The good news is that often little to no capital costs are required.

TEXTBOOK PUMP STATION CONTROL

The textbook example of a pump station is simplistic by definition. It consists of an inflow, a wetwell with a level sensor, a VFD, a pump, and a controller. The wetwell is an optimal size, the inflow doesn't change too drastically, and the starting and stopping of the pumps is ignored. Also the pump is allowed to operate at any speed.

For analog value control, the proportional-integral-derivative (PID) controller is a standard controller that has stood the test of time. It has a setpoint (SP) input, a process value (PV) input, and a control value (CV) output. The setpoint is the target value that you want to achieve. In this case, the target value is a wetwell level. This can be set manually by an operator. The process value (PV) is the field measurement that corresponds with the setpoint. In this case, the PV gets its value from the wetwell level instrument. The control value (CV) is the controller output. This output changes something in the system that impacts the process value. In this case, the CV determines the speed at which the pump runs.

The PID controller has three major configuration constants: proportional (P), integral (I), and derivative (D). Controllers can be customized by changing their configuration. This is usually configured or tuned during equipment field startup. P and PI controllers are the most often used configurations.

The proportional controller is a simple multiplier. The setpoint is subtracted from the process value, in this case level, and an error is calculated. That error is then multiplied by the proportional value and the output to the VFD is generated. The proportional band shows how far from the setpoint the PV must be to produce a 100% output on the controller at a given gain. The higher the controller gain (or P value) the smaller the band and process error will be. However, increasing the gain to very high values causes instability (i.e. oscillation) causing the output to rapidly go from 0% to 100% and then back again. Whenever this occurs, too much controller gain is the likely culprit.

The PI controller builds up a memory of how the system responds. Using only a proportional controller, there will always be a process error left over. The Integral term is used to increase the effect of the Proportional value over time to eliminate the error. This is not as intuitive as the P term alone because it depends on what has happened in the past. For instance, if the controller has had a hard time bringing the system back to the setpoint over a long time, when it finally does reach the setpoint, the pump won't just go to zero speed. The controller will keep the pump running well below the setpoint, as if it is still anticipating a future surge in demand. To the casual observer this doesn't seem right. It is important to remember that only a pure proportional controller with no I or D terms will produce a 0% output when it is below the setpoint.

PUMP STATION CONTROL CURRENT PRACTICE

A typical large pump station consists of a wetwell, level sensor, multiple pumps and VFDs, and a discharge flow meter. In the real world, it seems like the pumps and wetwells are either too big or too small. Unlike the text book example, flow coming into the wetwell is not constant and sudden changes can occur in the inflow. Running the pump dry is not an option. Neither is overflowing the wetwell. Also, pumps cannot run at all speeds equally well, only at the speeds that keep them on the pump curves.

A popular control scheme is constant level. The theory is that the wetwell level will stay constant if what comes in is pumped out. Constant level control really means controlling the level *around* a setpoint. The maximum use of wetwell level from just above low level to just under high level is the best choice because using all the available freeboard produces smoother pump station operation. A very large wetwell would allow the pump to move water at the pump's best efficiency for very long periods of time regardless of the short term changes in inflow. The constant level scheme can be misunderstood by control engineers. If implemented strictly, level will be maintained at the expense of turning pumps on and off rapidly. Although this will keep a strictly constant level, this is not really how we want to achieve our goals. Pumping water efficiently as much of the time as possible is our real goal.

In simple schemes, all pumps are controlled using a single PID controller based on wetwell level. Running all pumps at the same speed whenever multiple pumps were operating makes sense especially if all the pumps are the same size and produce roughly the same amount of flow and pressure at a given speed.

Small wetwells, common in existing PS upgrades, compound the problem of control. Small wetwells change level quickly and it may be difficult to make the pump controls respond fast enough to prevent either overflows or low levels caused by over pumping. Large wetwells change level very slowly even if a large inflow change occurs. By the time the level changes enough to increase the pump speed, there is so much additional water that it take a very long time for the pump to catch up. In a large system, the rate of level change becomes as important as the actual level change.

Large pump stations have multiple pumps to accommodate the wide range of flows. Starting a second or third pump causes a sudden sharp change in available pumping capacity. Using a fast acting controller will allow the system to quickly compensate to the change on-line pump capacity.

Using a single PID Controller therefore requires a tuning compromise to balance those needs. A slow control action is needed to react to gradual changes in wetwell level. A fast control response is needed to react to quick change in dynamics of adding or subtracting a pump. A compromise setting must be found that reacts quickly enough to correct the sudden change of adding a pump but slowly enough to make use of the available wetwell storage space.

When to start and stop pumps must be determined. The most common technique is to start an additional pump when the running pumps reach 100% speed. A time delay is often used to avoid turning on additional pumps if they are not needed. In a similar manner, then the running pumps slow to a low speed (say 70%), a pump is stopped, and the remaining pumps speed up to produce the same flow with one less pump. The low speed cut off is selected to prevent the pumps from running off their pump curve.

This is how our example project started. Orange County Utilities Department had an aging 40-mgd effluent pump station that needed pumping system replacements at the South Water Reclamation Facility. This included rehabilitating five existing 6,500-gpm vertical turbine pumps, installing a new 1,000 HP motor, replacing the existing variable frequency drives (VFDs), and replacing the obsolete pump station control system. The five 1000 HP motors and the associated liquid rheostats and energy recovery electrical controls were replaced with new motors and variable frequency drives. A sixth pump, motor, and VFD combination was also added.

The project included replacement of the existing medium voltage motor control center and wet well level control system. Modifications were made to the existing supervisory control and data acquisition (SCADA) system including programming to incorporate a new wet well level control scheme. In this case a simple pump control scheme was specified by the engineer so that all the variable speed pumps were controlled using a single PID controller based on level.

When this station was commissioned, it was realized that the specified replacement control scheme was too simplistic and that advanced controls were required.

Figure 1 - One of the six 6,500 gpm vertical turbine pumps

PROBLEMS

During the startup of the pump station, a number of anomalies were noticed including running more pumps than required, rapid pump cycling, unresponsiveness to changes in discharge conditions, and flow surging.

When using a single controller, there can be a tendency to run more pumps than required. If the controller is configured to be too fast, it will try to keep the wetwell level in a narrow band and start and stop pumps too quickly. Additional pump starts can trigger electrical demand charges, greatly increasing the electrical power cost of operation.

Rapid pump cycling can occur if the controller responds too slowly. When an additional pump starts, it initially runs at fully speed and causes a large amount of water to be discharged. It may lower the wetwell level so quickly, that it runs only a short time before being requested to stop. This occurs because the slowly responding control loop doesn't have enough time to reduce the speed of the pumps.

Discharge pressure changes also affect pump operation. The pump curves are different for one, two, or three pumps in simultaneous operation. The shutoff value shifts for the different pump combinations, so a simple low speed setpoint of say 70% doesn't cover all operating scenarios. If it is set for the maximum number of running pumps, it will not be ideal for use with a single running pump.

Discharge flow surges or sags when a pump is started or stopped. When a second pump starts, suddenly the active pumping capacity almost doubles for the same speed command. This sudden change causes the control system to overshoot, sending a pulse of water down stream and possibly causing a sudden drop in the wetwell. Over time it will recover. In the same way, when a pump is stopped, the pumping capacity is suddenly reduced causing the control system to initially undershoot, which may cause a rise in the wetwell level.

Tuning the single PID controller for the Orange county effluent pump station was difficult during startup and commissioning due to several factors. The level in the wet well normally rises slowly, but the dynamic change of adding an additional pump required quick response. A good compromise between these two requirements was not possible.

Figure 2 - The effluent pump station control system

Simple single loop controllers are limited in their ability to achieve multiple objectives. Oftentimes a good compromise between quick and slow response cannot be found. Wet wells over flow, or more pumps are started and stopped than needed. PIDs take time to respond to sudden large changes. The slower the PID, the longer it takes to compensate for a large dynamic system change.

Pump flow is not linear to its speed but the controller output is linear. As the wetwell level increases, it sends a proportional request to speed up the pump to get more flow. However, centrifugal pumps are not linear so more or less flow than desired will be pumped. If not enough flow is pumped, the controller will sense this through a rising wetwell level which will cause the controller to further increase the request. This makes the pump station system harder to control.

Other factors also added to the difficulty of controlling the Orange county effluent pump station. During startup, certain VFD run speeds were locked out, because the pumps had natural resonant frequencies that needed to be avoided to reduce mechanical vibration. If a speed was needed that was blocked, the VFD would alternate between the frequency just below the lockout speed and the frequency just above the lock out speed. The VFD will continue to alternate between these two frequencies until the pump speed increase above the lock out range.

Additionally, it was discovered that excess flow held in an equalization basin during the day was released all at once sometime after midnight. This caused a sudden increase in wet well level that was difficult to deal with. This situation was addressed by changing operating procedures to release this flow more slowly.

These anomalies demonstrated that a simple control scheme with multiple pumps, in this case, was not ideal.

SOLUTION

A better strategy was developed which incorporates the use of a second PID controller to form a cascade control scheme. With this type of control, the first (or outer) controller reacts slowly to allow the wet well to buffer the flow as much as possible by allowing the level to rise and fall around a set point. The outer loop is configured separately to gradually adjust pumping. It is tuned to use most of the wetwell capacity in order to smooth the pumping operation. This permits fewer pumps to run over time.

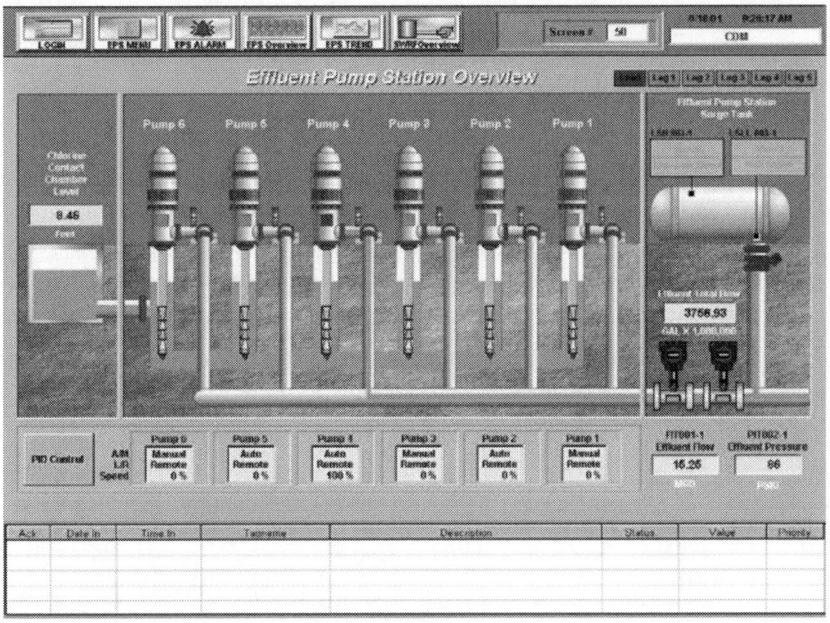

Figure 3 - The pump station control display

The second (or inner) controller changed pump speed quickly in reaction to changes in discharge flow and pressure based on the number of pumps running. The inner controller converts multiple pumps into one ideal virtual pump. The desired flow is reached by speeding up the VFDs until it is achieved. If the controller reaches 100% speed and needs more flow, the next pump is started. As soon as the additional flow is sensed, the controller quickly adjusts the speed of all the pumps to produce the requested flow. This makes a virtual pump that responds linearly with the flow request.

Traditionally pumps are turned on and off either by a level setpoint or by the speed of the VFDs. This is often set at the point where check valves close which is usually off the pump curve. If speed is used, a different setpoint is needed for different combinations of pumps. At Orange County, the control system records the flow produced by different combinations of running pumps. This information can be used to decide when to turn off a running pump to reduce capacity.

An even better scheme uses best efficiency instead of flow to start and stop pumps optimizes overall operation. This creates a virtual pump with a broad range of best efficiency versus the flow profile. This saves power and mechanical wear. Centrifugal pumps are most energy efficient at higher speeds as shown on the BEP on the pump curves. It is not desirable to run three pumps when the same flow could be produced by two at a better efficiency and lower power use.

The equipment changes needed to implement this strategy were minimal because the control system was software based and was able to be reprogrammed to add the second controller and setup a cascade control system.

SUMMARY AND CONCLUSIONS

Cascade control using two PID controllers improves control and saves money. Installing this modification saved thousands of dollars in electrical demand charges each month, which were being assessed each time pumps cycled incorrectly under the single loop control scheme. Although the need to reduce energy consumption was not originally recognized, new advanced controls provided the flexibility to reduce the electrical costs of the station.

Other benefits include:

- Saves power by running pumps at their best efficiency possible
- Saves Mechanical wear by running at their best efficiency possible
- Smooth operation prevents downstream process upsets
- Adaptable for force main use with varying discharge pressures
- Low cost – the second controller is probably already available in software and a discharge flowmeter is also probably already present

Advanced Water Reclamation Minimizes the Energy Footprint:
The Riverfront Water Reclamation Facility

Brooks Newbry, P.E., PhD, Water Technology Group
Ken Knickerbocker, P.E., Wood, Patel and Associates
Dan Lueder, City of Cottonwood
Water Technology Group
1028 N 24th Street
Mesa, Arizona 85213

ABSTRACT

The vision of the City of Cottonwood, Arizona for its Riverfront Water Reclamation Facility (RWRF) is to make it a benchmark for environmental leadership. Central themes are energy efficiency, product quality, research opportunities and public outreach. Designed to be fully powered by solar energy, it will remove refractory compounds of increasing environmental concern in order to produce a finished water of essentially drinking water quality, suitable for a wide variety of uses. Combining state-of-the-art treatment with research facilities and a public learning center, the City of Cottonwood intends to offer this facility as a vehicle for substantially advancing the art and science of water reclamation.

KEYWORDS

Advanced wastewater treatment, pharmaceuticals treatment, energy conservation, endocrine disruptor treatment, advanced oxidation, DO control, ionized air

INTRODUCTION

The City of Cottonwood, Arizona prides itself on environmental awareness. Several years ago the City Council recognized various water resource challenges, including beneficial reuse of reclaimed water to water starved parks, decreasing supplies of high-quality raw water for potable use, impending stringent limits on treated wastewater discharges and increasing public and professional concern over potential threats from "emerging" contaminants. One anticipated requirement is a limit on total nitrogen (TN) of 1 milligram per liter (mg/L) in treated effluent discharged to the local receiving water (the Verde River); this is believed to be among the most stringent TN limits in the nation. Other contaminants of particular concern were endocrine-disrupting compounds (EDCs) and pharmaceuticals and personal care products (PPCPs). The City Council also recognized the need to minimize the City's carbon footprint associated with water use and wastewater treatment.

Based on the need to address these issues, the City Council directed staff to develop concepts and a plan for a satellite water reclamation facility that would meet these challenges as well as providing reclaimed water to a City park that includes old stand cottonwood trees; the namesake of the City. The Council specifically directed staff to meet the following four overarching goals:

- Produce a very high quality finished water to maximize reuse/discharge opportunities
- Minimize nonrenewable energy consumption
- Create a facility that is recognized by both the public at large and the technical community as a major asset
- Minimize project costs, both construction and life-cycle

A project team was selected and embarked on the planning process. What emerged was a preliminary plan for a unique project, the Riverfront Water Reclamation Facility (RWRF). The RWRF will be a 0.3 million gallon per day (mgd) (1.1 megaliter per day, mLd) facility, unique in its design, operation and performance. It is expected to be a major asset to the local community and to the water management profession at large.

The reaction to the preliminary plan from many in the professional community was "these goals cannot be met affordably." The project is not yet complete, but the approach taken by the project team and conclusions reached to date indicate that the goals set out by the City Council are fully achievable.

DEVELOPING PROJECT CONCEPTS

The goals established for the project led to four basic concepts upon which initial planning was based:

- Advanced treatment would be required
- Energy optimization and renewable energy use would both be critical
- The facility would need to be inviting to the public
- Cost-effectiveness would need to guide all project decisions

Advanced Treatment to Achieve Finished Water Objectives

The first concept addressed the finished water objectives. Removing PPCPs and EDCs has been the subject of research, but no full-scale applications (to the authors' knowledge) to-date, although several projects are in development. Various studies have examined the occurrence of these materials in municipal wastewater and their transformation through various types of wastewater treatment processes. It has been demonstrated that they are always present in municipal wastewater. Conventional biological treatment processes degrade them in varying degrees, with some completely destroyed, many partially degraded and many more passing through biological treatment essentially unchanged (Joss et al., 2006).

Advanced oxidation processes (AOP) have proven to be effective at degrading some PPCPs and EDCs (AWWA Research Foundation, 2007). The transformation rates for specific compounds have been shown to depend on the AOP type and operating conditions. Little is known about the products resulting from these transformations or their possible environmental implications.

Based on this review of the state of the art, the project team concluded that the treatment train would need to consist of conventional "secondary" biological treatment followed by advanced treatment processes. They determined that the secondary treatment process would use extended aeration activated sludge and the advanced treatment processes would need to include filtration and AOP. They also concluded that an additional biological treatment step might be necessary following the AOP. However, little information is available on byproducts of PPCP and EDC destruction by AOP. Therefore, the decision was made to defer implementing biological treatment downstream of the AOP until performance data have been collected and evaluated.

The "heart" of the treatment train will be the secondary treatment process and the AOP. Options for both were evaluated. A brief summary of those evaluations and the basis for selecting the options to be implemented are presented in the following sections.

Secondary Treatment Process Selection

After a detailed evaluation of available technologies, the project team determined that extended aeration activated sludge would best meet the nitrogen removal goal as well as the goal for EDC and PPCP removal. Numerous options are available for implementing extended aeration activated sludge. Those considered for possible application for the RWRF, along with qualitative evaluation criteria, are presented in Table 1.

Each of the options considered has positive features and drawbacks relative to the project goals (detailed discussion of the evaluation process is beyond the scope of this paper). Based on this evaluation, the "SEQUOX" process was selected for implementation. Among the reasons for selecting this process were:

- It has been demonstrated to consistently remove TN to very low concentrations (consistent with the quality goal).
- It minimizes power consumption (consistent with the energy goal); power minimization is discussed below in more detail.
- It combines several unit operations into a single common–wall structure, minimizes the need for mechanical equipment (other than blowers), does not require large internal recycles and does not require chemical addition for treatment (all of which are consistent with the cost minimization goal).
- It requires a minimum of operator attention and maintenance.

Advanced Oxidation Process Selection

"Advanced oxidation" is based on using the "hydroxyl radical" as the primary oxidant. The hydroxyl radical is one of the strongest oxidants known to exist, second only to fluorine in its oxidizing potential. AOPs have been recognized as potential tools for water treatment and wastewater remediation for decades (Parsons, 2004). However, only a relatively few have been applied full-scale.

Table 2 lists some of the AOPs considered by the project team, along with the criterion used to evaluate each alternative. This is an area of active research and development, and other technologies are likely to become commercially viable in the future.

Fenton's Reagent has been used for bench-scale laboratory work for decades and has been applied full-scale in a few instances. It is by far the least energy-intensive of the options considered. However, because of the relative lack of full-scale experience it was eliminated from consideration.

Ultrasound processes (alone and in combination with O_3, UV and other reactants) have been studied and applied for many years. They are attractive because of the relative simplicity of the equipment required. However, they have not been applied full-scale and they require considerable amounts of energy. For these reasons they were eliminated from consideration.

The project team concluded that an O_3-based technology would best meet the project's goals. Although proven at full-scale and effective, the UV/O_3 option requires large amounts of power. Therefore, it was eliminated. The process selected was O_3/H_2O_2. This process offers the additional advantage of being able to operate without the use of H_2O_2, if operational experience shows that this operating mode is suitable (either continuously or during certain periods).

Table 1 – Extended Aeration Treatment Option Screening Summary					
Criterion	Treatment Option				
	Oxidation Ditch	Sequencing Batch Reactor	Membrane Bioreactor	Integrated Fixed Film/ Activated Sludge	SEQUOX
Quality Goal					
Effluent Quality	◓	◓	○	◓	○
Performance Reliability	◓	●	○	◓	○
Operational Ease	◓	●	●	◓	○
Operational Flexibility	●	●	●	●	◓
Energy Goal					
Power Consumption	●	◓	●	◓	○
Power Turndown	●	●	●	●	○
Public Value Goal					
Land Area Required	●	○	○	◓	◓
Noise Potential	●	◓	◓	◓	○
Odor Potential	◓	◓	◓	◓	◓
Cost Minimization Goal					
Equipment Requirements	◓	●	●	◓	○
Ease of Maintenance	◓	◓	●	◓	○
Chemical Requirements	◓	●	◓	◓	○

Key:

○ = Option has positive attributes relative to criterion

◓ = Option is neutral relative to criterion

● = Option has negative attributes relative to criterion

Table 2 – Advanced Oxidation Process Option Screening Summary

Criterion	Treatment Option				
	Ozone (O_3)	Ozone/ Hydrogen Peroxide (O_3/H_2O_2)	UV Radiation/ Ozone (UV/O_3)	Fenton's Reagent (Fe^{2+}/H_2O_2)	Ultrasound
Quality Goal					
Full-Scale Demonstration	○	○	○	●	●
Performance Reliability	◓	○	○	○	●
Operational Ease	○	◓	◓	○	○
Operational Flexibility	◓	○	●	◓	◓
Energy Goal					
Power Consumption	◓	◓	●	○	●
Power Turndown	◓	○	●	◓	●
Public Value Goal					
Land Area Required	◓	◓	◓	◓	◓
Noise Potential	◓	◓	◓	○	○
Odor Potential	◓	◓	◓	◓	◓
Cost Minimization Goal					
Equipment Requirements	◓	◓	●	◓	○
Ease of Maintenance	◓	◓	●	◓	○
Chemical Requirements	○	◓	○	●	○

Key:

○ = Option has positive attributes relative to criterion

◓ = Option is neutral relative to criterion

● = Option has negative attributes relative to criterion

Energy Optimization and Renewable Energy Use

Energy optimization focused on three concepts:

- Minimizing energy requirements within the RWRF
- Minimizing energy requirements for RWRF support functions
- Using renewable energy to the extent possible

Minimizing In-Plant Energy Use

Some of the largest energy users within the plant will be:

- Blowers, for secondary treatment operation
- Ozone generators, for the AOP
- Wastewater pumping
- Process building ventilation and odor control systems.

A number of features were included in the RWRF design to minimize energy requirements in these areas. Some of the more innovative features are briefly discussed below.

The secondary treatment process selected is uniquely suited to minimizing power requirements for blower operation. It uses a dissolved oxygen (DO) control system to modulate the blowers. Most DO control systems attempt to hold the aeration basin DO at a specific setpoint, by ramping blowers up and down in response to an instantaneous DO measurement and generally using a proportional-integral-derivative (PID) control strategy. In contrast, SEQUOX uses an operating range, consisting of a minimum and maximum DO value (D_{MIN} and D_{MAX}, respectively). The process control strategy uses DO measurements taken at one-minute intervals. It changes blower speed by one hertz (Hz) in response to the DO measurements, depending on which of three conditions. A simplified version of the control strategy is as follows:

- DO below DO_{MIN} – the plant is "oxygen transfer limited" and the blowers gradually ramp up;
- DO between DO_{MIN} and DO_{MAX} – the plant has sufficient oxygen and the blowers gradually ramp down to a minimum aeration rate required for mixing;
- DO above DO_{MAX} – the plant has excess oxygen and the blowers are turned off until the oxygen demand increases.

This blower control strategy minimizes blower operation without compromising process performance. Limited comparative analysis conducted to date indicates that this strategy reduces net aeration power requirements by thirty percent and more compared with conventional blower control strategies.

Producing ozone requires an oxygen source. The project team concluded that the most cost-effective means for providing oxygen would be on-site generation. Some systems use pressure swing adsorption (PSA), in which compressed air is supplied to a separator. The RWRF facility will use vacuum swing adsorption (VSA), which requires only compressing the oxygen from the separator. This requires only about twenty percent of the power that PSA requires.

The process flow train developed by the project team requires several unit processes. The initial process layout required as many as three intermediate pump stations to create sufficient hydraulic grade to flow through the WRF. By taking advantage of existing topography, and revising initial process selections, it appears that this can be reduced to a single intermediate pump station (to pressurize the AOP system).

Much of the RWRF will need to be enclosed to meet State regulatory requirements. Initial concepts called for buildings to house the treatment processes, together with relatively large ventilation systems to provide the required airflows through occupied spaces. This ventilated air would require treatment for odor control, with associated substantial energy (and possibly chemical) requirements. Revised concepts replaced much of the building space with sectional, easily removable covers over the process tanks. Additionally, an ionization system will be used to treat the supply air for all ventilated spaces. This system will supply air that destroys odor-causing (and corrosive) compounds within the ventilated space, improving the work environment and extending the life of equipment within the space.

Minimizing Energy Requirements for Support Functions

The treated water from the RWRF is intended to be suitable for a variety of uses. Two of the anticipated major uses are landscape irrigation during the summer, and water supply augmentation (through aquifer recharge), during the winter. The location chosen for the RWRF will minimize the energy requirements for implementing these uses. It will be adjacent to a public park with ball fields and other areas that currently use considerable amounts of potable and Verde River water for irrigation. The selected location is also in close proximity to a suitable site for aquifer recharge. Finally, the location is immediately adjacent to a main sewer interceptor and wastewater pump station, which will be the source of the wastewater for treatment. Currently, the City pumps all its wastewater to a wastewater treatment plant located at the highest elevation in the City. The RWRF is located at the lowest elevation. The location will minimize energy requirements for pumping both the influent and the treated water.

Using Renewable Energy

The City Council saw a major opportunity for using solar energy to minimize the net consumption of nonrenewable energy at the RWRF. With about 300 days of sunshine per year in Cottonwood, solar energy is a viable renewable energy source. A comprehensive study examined both the technology (including solar panel type, panel location, tracking system, panel tilt, panel orientation and energy storage) and economics (including construction costs, operating costs production based incentives and renewable energy credit). Based on the analysis and the City's preferences, it was decided to use a stationary mounting system, polycrystalline panels and direct purchase of the solar power generation system.

The yearly power consumption for the facility is conservatively estimated to be approximately 1 MW (see Table 1). The RWRF site can accommodate enough solar panels to generate approximately 340,000 kWh, or 34% of the total energy required. In an effort to provide as much renewable energy as possible, the City decided to expand the solar panel infrastructure to adjacent soccer fields, giving soccer fans a respite from the intense Arizona sun. Figure 1 presents the proposed layout of the solar infrastructure.

In all, there are 1,800 polycrystalline panels, each with the capacity of producing 295 W, for a total of 776,454 kW per year. The solar panel coverage area is estimated to be 37,650 ft^2. After balancing the proposed design of the solar panel system with the total power requirements, it was determined that the RWRF would need to be connected to the electrical grid to ensure full power

Table 1 – Estimated Power Requirements for RWRF

EQUIPMENT	Total Connected (HP)	USAGE (%)	ANNUAL POWER CONSUMPTION (kWh/yr)
Headworks	19	90%	111,748
Secondary Treatment	120	57%	447,645
Tertiary Filters	4	50%	13,070
AOX System	20	50%	65,350
Reclaimed Water Pump Station	120	10%	78,420
Plant Water Pump Station	20	50%	65,350
Odor Control System	30	85%	166,641
Plant Lighting/Visitor Center/Ops Bldg	-	-	50,000
Misc 10%	-	-	99,800
Total			**1,098,023**

is available at all times. However, it is anticipated that optimizing energy use will allow the facility to operate using solar power exclusively during at least part of the year.

The solar energy system design goes well beyond generating energy to power the plant. The panels, including mounting systems and location on the buildings, were designed to include the orientation of the buildings to take advantage of the passive solar capacity of the integrated system. The solar shading will reduce the overall building energy requirements. Elevated solar arrays also provide shaded areas for viewing soccer on the adjacent soccer fields. The location of the plant, its architectural features and its landscaping features are all designed to maximize the efficiency of the solar power system as well as to provide indirect public benefits.

Creating a Public Asset

Wastewater treatment provides obvious public benefits, but these benefits often go unrecognized. A major goal was to have the RWRF recognized as a major asset to the community. The project team identified two major strategies to achieve this goal:

- Emphasize aesthetics, to make the RWRF appealing to the local community
- Provide research and education opportunities, to make the RWRF appealing to both the local community and the broader professional community.

Emphasize Aesthetics

The City Council intends for the RWRF to be viewed as a public asset. In addition to producing very high quality finished water for use where potable water would otherwise be required, they expect the facility to be a source of civic pride.

Odor control is a very high priority. Treatment processes were selected to minimize odor generation and release. Basic controls, common to modern wastewater treatment plants, will be implemented, such as containing the headworks and bagging screenings. Biosolids, a necessary

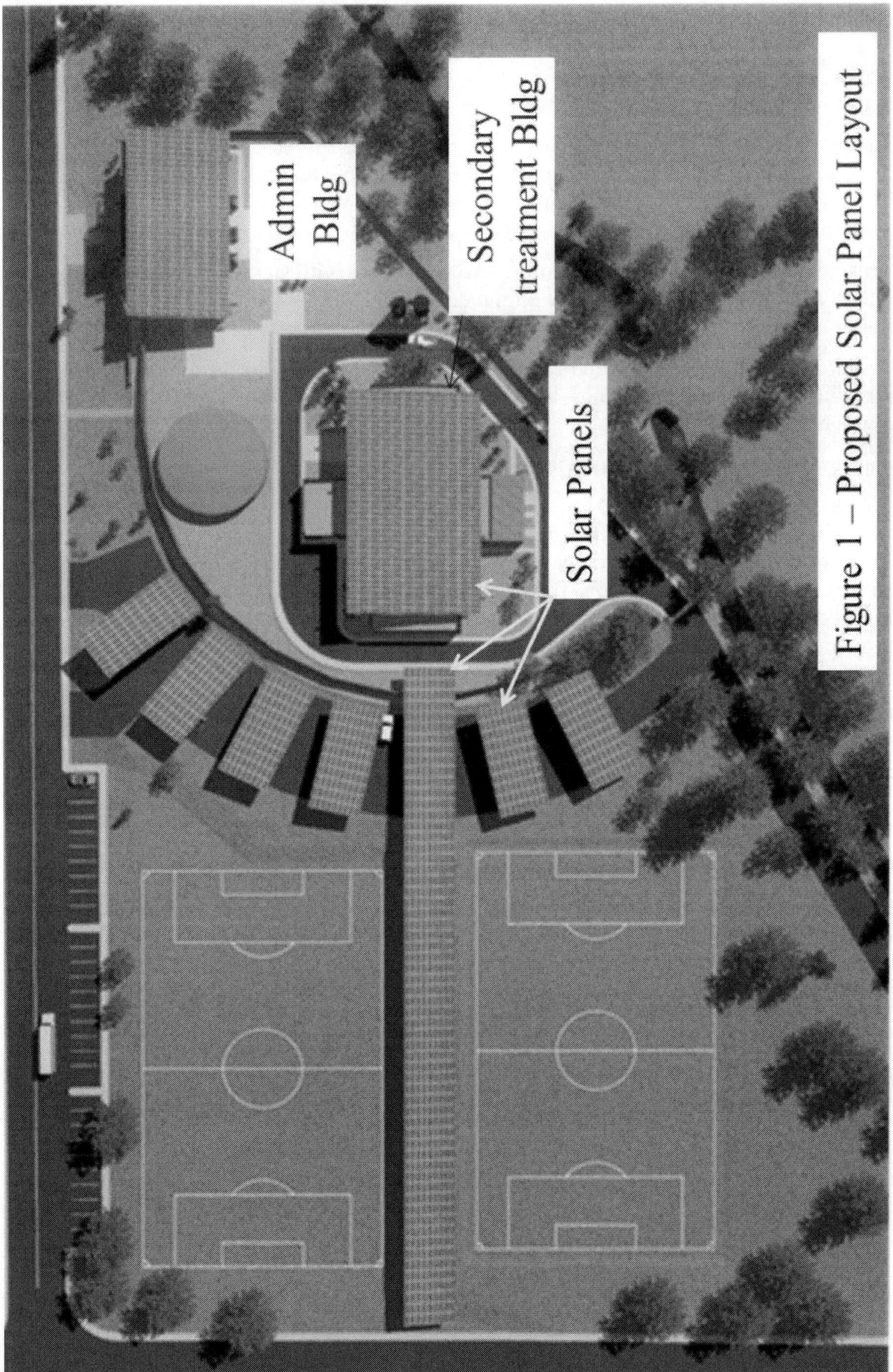

Figure 1 – Proposed Solar Panel Layout

byproduct of secondary treatment but which can be a source of odor complaints, will be returned to the sewer for conveyance to the main Cottonwood WRF for treatment. An innovative odor control process (ionized air treatment of supply air, discussed earlier) will both control odors and minimize both capital and operating costs.

Noise control is also a high priority. Sound attenuation enclosures on noise-producing equipment will keep noise levels at the plant boundary below an acceptable level (nominally 60 decibels, the sound level of normal conversation).

Truck traffic is a common complaint near treatment facilities. Processes have been selected to minimize the required truck traffic onto the site. Processes with significant requirements for chemicals and other consumables were eliminated during initial planning to minimize the delivery requirements for these materials.

The architectural concept is to blend with the surrounding community with minimal obstructions to sight lines. Low profile buildings, with architectural features consistent with the surrounding area, will be used.

Create Research, Education and Public Outreach Opportunities

The vision of the City Council is that the RWRF will be a research and environmental awareness facility, offering major public outreach and educational programs. It is anticipated that programs will be developed that attract participation from the local community and well beyond. Indeed, the state regulatory agency, Arizona Department of Environmental Quality (ADEQ), has requested the facility be made available for possible internships for new wastewater treatment plant operators.

The facility will include a Visitor/Operations Center that will include spaces and provisions for educational programs and research. The project lends itself to research by including state of the art technologies constructed to full scale, allowing the collection of empirical data to further the science of many facets of water reclamation, from advanced wastewater treatment to energy use optimization. Additionally, the facility will afford the public access to an otherwise restricted utility by integrating public friendly design into several treatment processes.

PROCESS CONFIGURATION

The RWRF will consist of (or, in some cases, have provisions for future addition of) the following unit operations and processes:

- Influent Pumping
- Raw Sewage Fine Screening
- Grit Removal
- Extended Aeration Activated Sludge with Nitrification/Dentrification
- Phosphorus Removal
- Filtration
- Advanced Oxidation
- Biological Filtration (possible future addition)
- UV Disinfection (possible future addition)
- Finished Water Pumping

A process flow diagram is shown in Figure 2.

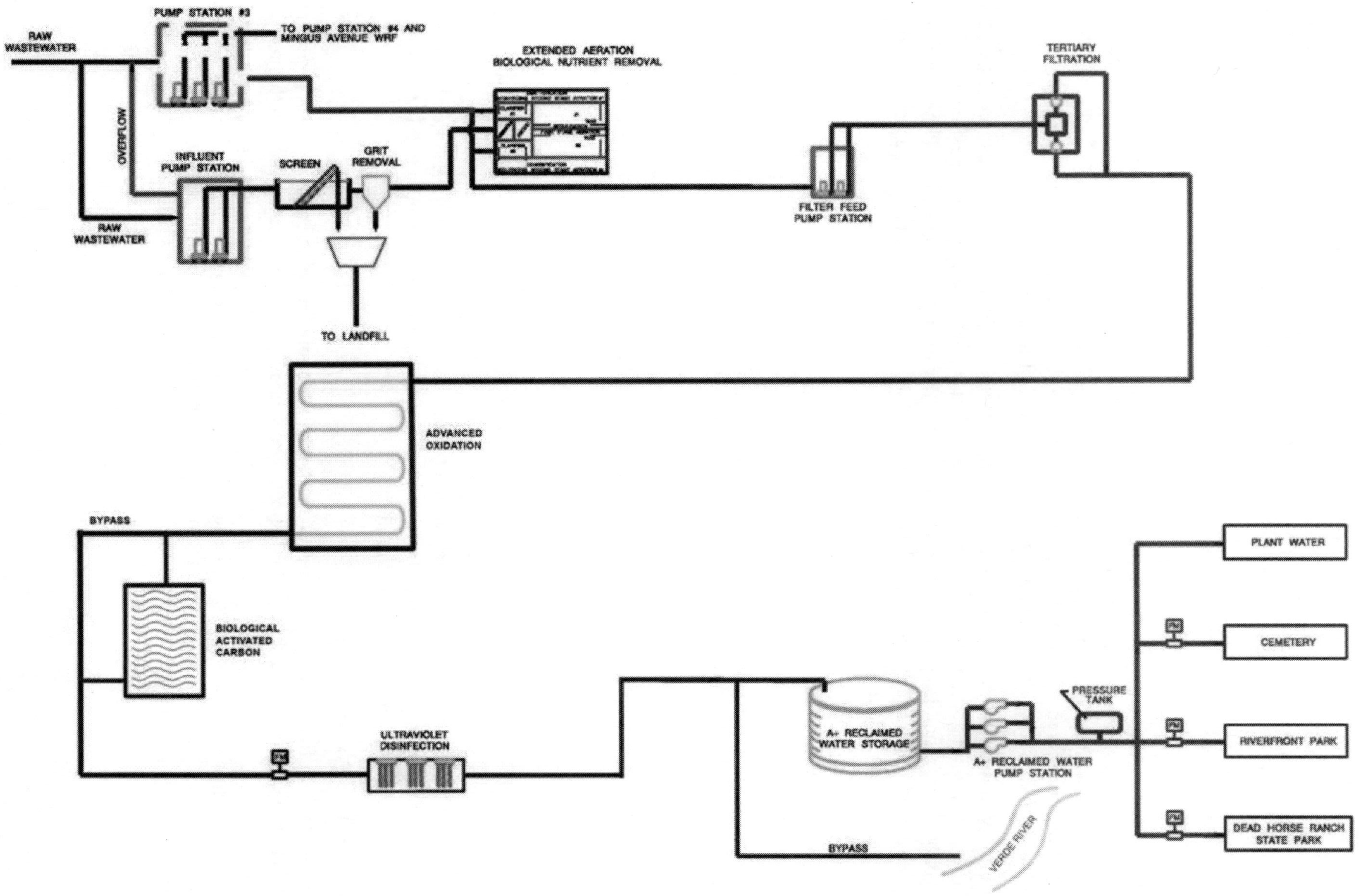

Figure 2 – Process Flow Diagram

Influent pumps will transfer raw wastewater from an existing interceptor sewer, adjacent to the plant site, to the headworks. A mechanical bar screen (6 mm nominal solids removal) will remove large debris from the wastewater. A screenings washing compactor will remove the majority of the putrescible organics from the screenings and return this material to the wastewater flow for treatment. Washed screenings will be collected in a bagger and held for periodic collection and disposal. A vortex type grit removal system will follow preliminary screening for grit capture.

The secondary treatment system will consist of aeration tanks and clarifiers within a single, common-walled structure. Return and waste activated sludge (RAS/WAS) transfer will use airlift pumps, which require no moving parts. The clarifiers also require no moving parts. This arrangement results in a system that is easy to operate and maintain and is energy efficient. A schematic of the secondary treatment system is shown in Figure 3.

Filtration will use either continuous backwash sand filters or disk filters. Evaluation of these two options is continuing. Sand filters are considered more reliable for meeting an effluent phosphorus limit (not discussed earlier, but one of the treatment requirements). However, the pumping head requirements are considerably higher than for cloth disk filters, necessitating an intermediate pump station.

Advanced oxidation will use a serpentine tube reactor for contacting O_3/H_2O_2 with filtered secondary effluent. This treatment process is illustrated schematically in Figure 4. It is expected to attract considerable research interest.

Provisions will be made for future addition of a biological filtration system. The products of advanced oxidation of EDCs and PPCPs have not been characterized to any significant extent. It is anticipated that much of the degradation will result in organic molecule fragments that are more amenable to biological degradation than the original compounds. The fate and effects of such compounds in the environment have not been determined; responsible environmental protection may call for further treatment of them prior to releasing the treated effluent.

Provisions will also be made for future addition of effluent disinfection. AOP will destroy all microbial contaminants of concern and will meet disinfection requirements. However, if a biological filtration system is added in the future, subsequent effluent disinfection may be required.

The finished water will be conveyed to different locations for reuse or discharge, depending on the season, reuse demands and other factors. A pumping facility will be included that will transfer this water to areas where it is required.

PROJECT STATUS

The RWRF is in final design. Thirty percent design is complete, value engineering is in progress and final project details are being established. The schedule is to begin construction the last quarter of 2013 with the facility to go on line approximately eighteen months later. To minimize cost and compress the construction schedule, a Construction Manager At Risk (CMAR) delivery system will be used. This will make the owner, design team and construction contractor partners in achieving the project objectives and realizing the vision of the City Council.

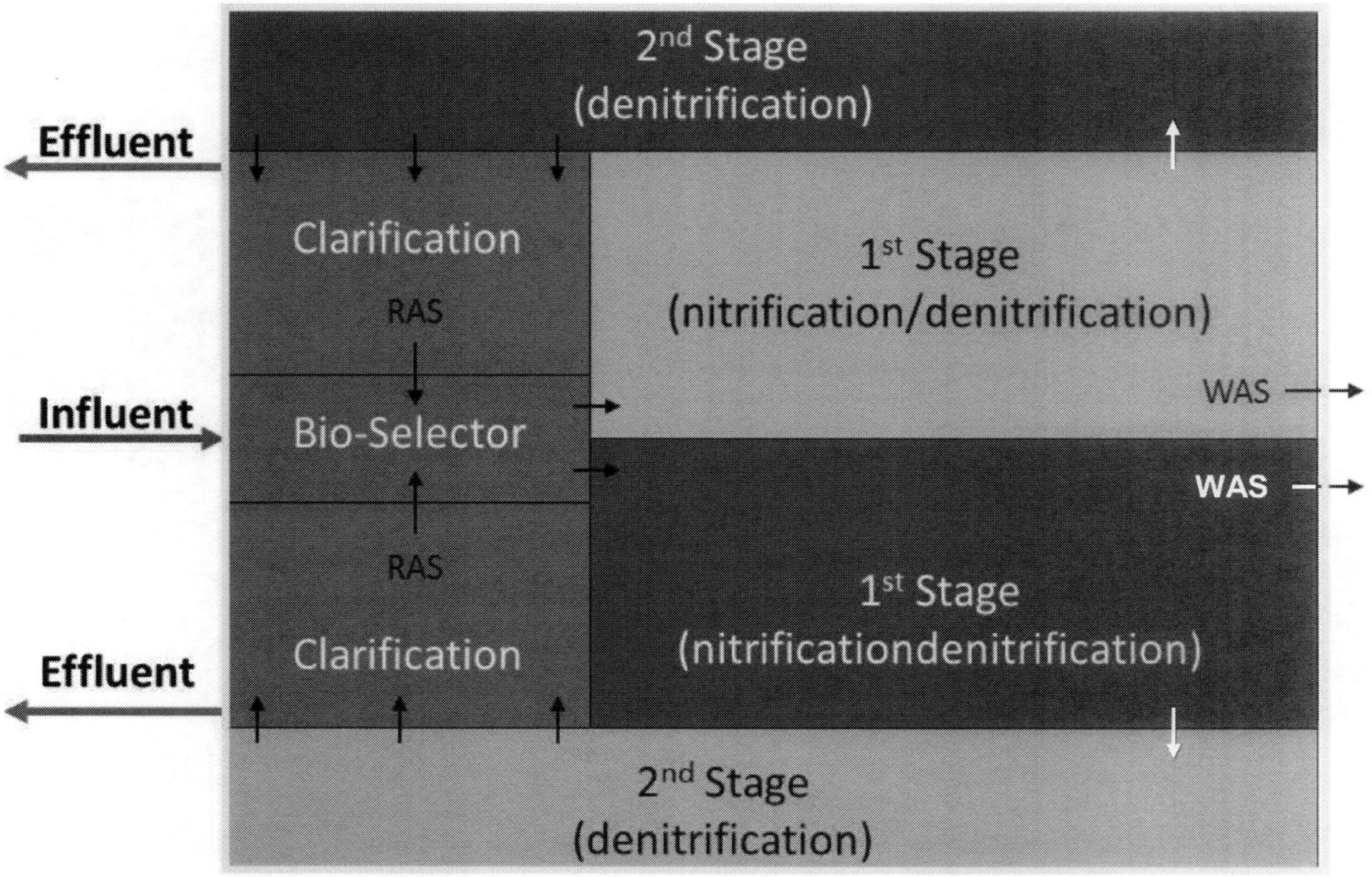

Figure 3 – Secondary Treatment Process Schematic

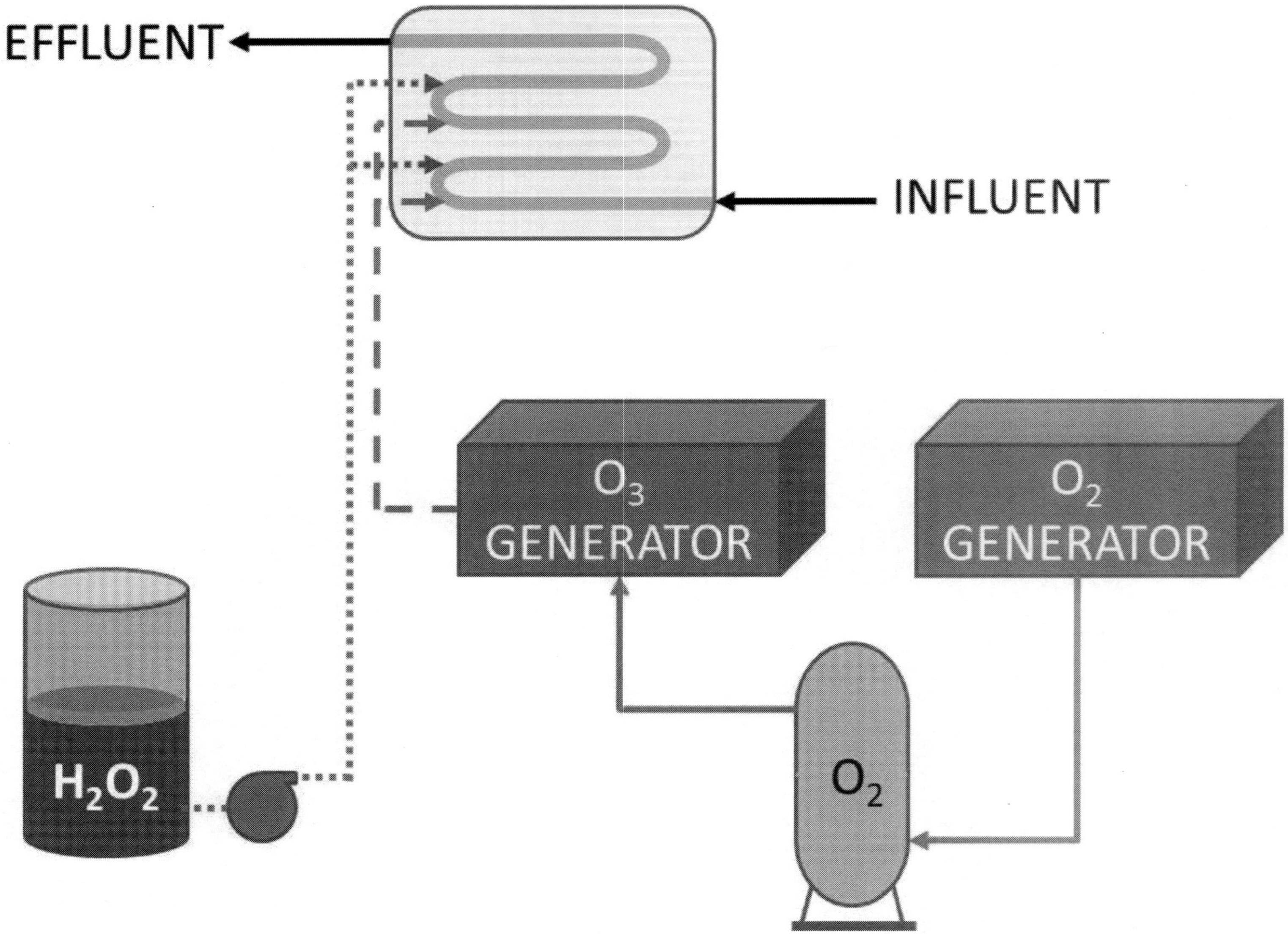

Figure 4 – Advanced Oxidation Process Schematic

REFERENCES

AWWA Research Foundation, "Removal of EDCs and Pharmaceuticals in Drinking and Reuse Treatment Processes," 2007

Joss, Adriano et al., "Biological degradation of pharmaceuticals in municipal wastewater treatment: proposing a classification scheme," Water Research, 40 (8), 2006

Parson, S. (Ed.), Advanced Oxidation Processes for Water and Wastewater Treatment, IWA Publishing (2004)

Reclaimed Water Quality Requirements for Electric Utilities

Don Vandertulip, PE, BCEE,
CDM Smith, San Antonio, TX

ABSTRACT

Water use for thermal cooling represents a significant percentage of total water use ranging from 81 percent to 15 percent depending on the geographic region in the United States. Reclaimed water offers some significant advantages to help meet cooling water demands in the electric power industry, allowing the existing source waters to be used for other purposes, including potable water use. The quality of reclaimed water is dependent on the background water quality used to transport sanitary waste to a wastewater treatment plant (WWTP) and on the processes used at the WWTP to remove contaminants from the wastewater. Perceptions of the water quality required for cooling purposes varies from the east coast to the west coast, with demands for low TDS and low hardness water typical on the east coast and facilities on the west coast in water constrained areas treating available water to meet their process requirements.

Water quality required for various cooling process applications are examined and the ability of reclaimed water to provide reliable alternate waters for power plant cooling are addressed. Advantages in using reclaimed water to address a portion of the energy-water nexus are identified and examples provided for several power plant cooling successes. Cost in meeting cooling water treatment and transmission to the power plant are identified to assist future consideration in using new industry best practices to solve the energy-water nexus.

KEYWORDS

Reclaimed water, Power plant cooling, Water quality

INTRODUCTION

One of the significant advantages in using reclaimed water is that it is usually continuously available in predictable quantities, a significant advantage in water supply planning. There is however, a difference in the quality of reclaimed water compared to potable water in each geographic area. Nationally, there are also differences in perception of the water quality required for use in cooling operations. This paper examines the historical use of reclaimed water for power plant cooling operations, differences in perceived need and desired reclaimed water quality, and future opportunities to expand its use in the power industry.

Advantages in using reclaimed water examined in this paper include:

- Continuous availability;
- Consistent quality;
- Sustainable source;
- Replacement for potable demands; and
- Reasonable, predictable cost.

CURRENT USE OF RECLAIMED WATER FOR COOLING

An Argonne National Laboratory report (DOE, 2007) identified that 136 billion gallons per day (bgd) of reclaimed water were being used for power plant cooling operations with 200 bgd projected by 2010. According to this report, 57 power plants in 16 states used reclaimed water in 2007, with most facilities in Florida, California and Texas. One of the largest individual users is the Palo Verde Nuclear Generating Facility in Arizona. Forty-six of these facilities use reclaimed water in cooling towers with the balance using reclaimed water in cooling lakes and air pollution scrubbers (Vandertulip, 2012). Several facilities using reclaimed water for plant cooling are shown in Table 1.

Table 1: Example Facilities Using Reclaimed Water for Cooling

Cooling Tower/Pond Facility	Reclaimed Water Use		Reclaimed Water Source
	L/sec	BGY	
Palo Verde Nuclear Generating Station, AZ	2,655	22.15	Phoenix, AZ 91st Avenue WRP
CPS Energy, San Antonio, TX	1,953	16.29	Dos Rios WRC, San Antonio, TX
West Basin Water District, Fountain Valley, CA	1,558	13.00	Hyperion WWTP, Los Angeles, CA
Stanton Energy Center, Orange County, FL	563	4.70	Eastern Water Reclamation Facility, Orange, County, FL
DCO Energy Marina Thermal Facility, Atlantic City, NJ	14	0.12	Atlantic County Utilities, Atlantic City, NJ

Palo Verde and Red Hawk Power Plants 31 miles west of Phoenix, Arizona use an annual average of 2,655 liters/sec (l/sec) [22.15 billion gallons per year (bgy)] from the Phoenix 91st Avenue reclamation plant. Peak summer demands reach 3,718 l/sec [85 million gallons per day (mgd)]. Palo Verde is the largest nuclear generating station using reclaimed water for thermal cooling. CPS Energy in San Antonio, Texas is using 1,953 l/sec (16.29 bgy) for power plant cooling from two lakes filled with local stormwater runoff supplemented with reclaimed water supplied under contract with the San Antonio Water System. West Basin Water District in Orange County, California provides supplemental treatment at the Edward C. Little Water Recycling Facility to secondary effluent from the Los Angeles Hyperion WWTP. A total of 1,558 l/sec (13 bgy) of reclaimed water is produced in five different qualities of water for multiple uses including cooling in industrial cooling towers, single and double pass reverse osmosis water for boilers and high pressure boilers and injection water for injection to reject sea water intrusion into the potable supply aquifer. Orange County, Florida Stanton Energy Center uses 563 l/sec (4.7 bgy) from the Eastern Water Reclamation Facility, Orange, County, Florida for cooling tower consumptive use. Another east coast facility, operated by Atlantic County Utilities/DCO Energy Marina Thermal Facility in Atlantic City, New Jersey is using 14 l/sec, or 120 million gallons per year (mgy) for cooling. The Marina Thermal Facility serves as a central utility plant providing heating, chilled water for cooling, and backup power to the Renaissance Pointe area of Atlantic City, including the Borgata Hotel and Casino (Booker, 2008).

DIFFERENCES IN DESIRED RECLAIMED WATER QUALITY

As a general rule, water used for power plant cooling operations on the United States (U.S.) east coast is soft with low total dissolved solids (TDS) while water further west tends to be harder and have higher TDS levels. Initial designs for cooling operations typically take the existing water quality into account and choices of materials of construction for facilities, piping and equipment are based on the anticipated beginning water quality and the water quality calculated when blowdown is necessary. Reclaimed water can add between 250 and 500 milligrams per liter (mg/l) of TDS to the TDS level existing in the potable water system. If an east coast utility has designed for a feed water quality based on a maximum TDS of 100 mg/l, an alternate reclaimed water supply with 500 mg/l may impact the number of cycles the water can be used or potentially affect equipment sensitive to higher TDS concentrations. In contrast, reclaimed water supplied to the Palo Verde Nuclear Generating Station in Arizona has 1,200 mg/l of TDS and Arizona Power provides onsite treatment to soften the water and obtains 25 cycles of use, where most facilities operate at five cycles (Day, 2008). The impact of increasing a cooling water source by 400 mg/l is more pronounced when starting from a datum of 100 mg/l compared to starting at 1,200 mg/l. Water quality at several cooling facilities will be discussed as examples of differing water quality perceptions, beginning on the east coast and proceeding west.

The Atlantic County Utilities/DCO Energy Marina Thermal Facility in Atlantic City requires cooling water supply meeting the parameter concentrations shown in Table 2.

Table 2: Water Quality for Cooling, DCO Energy Marina Thermal Facility

Parameter	Units	Value
Alkalinity	mg/l as $CaCO_3$	<25
Chloride	mg/l	<10
Total Hardness	mg/l as $CaCO_3$	<30
Calcium Hardness	mg/l as $CaCO_3$	<20
Magnesium Hardness	mg/l as $CaCO_3$	<10
Total Iron	mg/l	<0.2
Manganese	mg/l	<0.02
pH	pH units	6.5 - 8.5
Sulfate	mg/l	<10
TDS	mg/l	<100
Barium	mg/l	<0.5
Sodium	mg/l	<10
Nitrate/Nitrite	mg/l	<1
Ortho Phosphate	mg/l	<0.5
Total Silica	mg/l	<12
LSI/Corrosivity	S.U.	-1 +/- 0.5

Booker, Municipal – Industrial Partnership for Reuse: A Case Study in the Northeast U. S., 2008 WateReuse Symposium

Note that the total dissolved solids (TDS) limit is set at 100 milligrams per liter (mg/l) and the total hardness is limited to less than 30 mg/l expressed as calcium carbonate. While this is the water quality typical for this area of New Jersey, most western states have significant TDS and hardness concentrations to contend with, even to meet secondary drinking water standards of 500 mg/l TDS.

JEA, an electric, water and sewer agency) in Jacksonville, Florida provides a hypothetical cooling water quality example in their Best Practices Manual (JEA, 2005) shown in Table 3.

Table 3: JEA Best Management Practices Cooling Water Example

	JEA Makeup	2 Cycles	3 Cycles	4 Cycles
Calcium	150 ppm	300 ppm	450 ppm	600 ppm
Sulfate	130 ppm	260 ppm	390 ppm	520 ppm
M Alkalinity	110 ppm	220 ppm	330 ppm	440 ppm
Silica	30 ppm	60 ppm	90 ppm	120 ppm
TDS	500 ppm	1000 ppm	1500 ppm	2000 ppm
pH	7.8	8.2	8.6	8.8
LSI	0.4	1.4	2.1	3.1
pHs	7.4	6.85	6.5	5.7

Best Management Practices and Guidance Manual for Cooling Towers, JEA, August 2005

This table illustrates the concentration of parameters in the reclaimed cooling water as the number of cycles increase from makeup water to four cycles. Assuming the reclaimed water starts at 500 mg/l TDS, within four cycles of operation, the TDS will have increased to 2,000 mg/l and the other parameters will also concentrate. The pH is shown to increase with the number of cycles consistent with the increase in alkalinity. If acid were used to depress the pH, mineral saturation would occur at the pHs shown for each number of cycles. Depressing the pH to increase the mineral solubility must be balanced with the drop in Langlier Index that will occur as the pH drops and alkalinity is consumed.

West County Energy Center, a Florida Power and Light (FP & L) power plant in West Palm Beach County, Florida has established realistic reclaimed water quality criteria (Bashaw, 2009) for its cooling plants with three of the parameter values required by Florida Department of Environmental Protection (FDEP). Maximum total suspended solids concentration, pH range and minimum total chlorine concentration are required in FDEP 62-610. Table 4 identifies the FP & L reclaimed water quality criteria and key water quality parameters for cooling water quality. The range of concentrations shown for the prevailing reclaimed water quality typically meet the criteria but there are obvious occasional excursions. An FP & L criterion for TDS is 750 mg/l and the prevailing TDS range for reclaimed water is 516 to 676 mg/l, below the desired limit. Note also that the criteria for total hardness is 210 mg/l compared to the 30 mg/l total hardness at the DCO Energy Marina Thermal Facility. The prevailing range for total hardness at the West County Energy Center is 167 to197 mg/l. Compared to the hardness criteria at the DCO Energy Marina Thermal Facility the West County Energy Center water is hard, but much softer than water sources in western states.

Table 4: Key Water Quality Parameters for Cooling Water Use

Parameter	Units	Requirements	Prevailing Water Quality
Alkalinity	mg/l as $CaCO_3$	<120	74 to 118
Ammonia	mg/l as N	<1.0	0.0 to 0.5
Calcium (as Ca)	mg/l	<50	44 to 53
Chloride	mg/l	<350	196 to 251
Total Hardness	mg/l as $CaCO_3$	<210	162 to 197
Iron - Total	mg/l	<0.3	0.1 to 0.7
Nitrate	mg/l as N	<10	0.5 to 13.3
Nitrite	mg/l as N	<1	0.0 to 3.8
Ortho-Phosphate	mg/l as P	<1.67	0.1 to 1.0
Total Phosphate	mg/l as P	<1.67	0.2 to 4.4
pH[1]	SU	6 to 8.5	6.7 to 7.9
Sulfate	mg/l	<250	48 to 63
Total Chlorine[1]	mg/l	>1	--
Total Dissolved Solids	mg/l	<750	516 to 676
Total Suspended Solids[1]	mg/l	<5	2.0 to 45
Turbidity	NTU	<3.0	2.0 to 9.0
[1]These requirements are also contained in Part III of Chapter 62-610.			

Bashaw, 2009 WateReuse Symposium

San Antonio Water System has supplied reclaimed water to fill and maintain the water level in two cooling lakes operated by CPS Energy, San Antonio, Texas since 1962. Water is pumped by CPS Energy from the San Antonio River immediately downstream from the convergence of the San Antonio and Medina Rivers, a point that received the total discharge from San Antonio Water System water reclamation plants. The combined capacity of the two lakes is 110,600,100 meter^3 (89,700 acre-feet). CPS Energy has contracted for 61.65 million meter^3 (MCM) (50,000 acre-feet) per year and at this rate of use, would exchange the total volume of the cooling lakes at about a 22 month frequency. A cooling tower pilot study was completed in March 2000 using reclaimed water from the Leon Creek Water Recycling Center (LCWRC) (SAWS, 2000). Table 5 provides a comparison of the Edwards Aquifer based potable water system water quality compared to the quality of reclaimed water at the LCWRC. The Edwards Aquifer water is from a karst limestone aquifer resulting in high TDS (297 mg/l) and high hardness (263 mg/l as $CaCO_3$). TDS in the reclaimed water ranges between 710 and 720 mg/l with hardness in the range of 272 to 280 mg/l (SAWS, 2006). Results of concentrating the identified parameters with five cycles of water exchange for both potable water and reclaimed water are shown in Table 6.

Based on the Pilot Study, a workshop was held for commercial customers with cooling towers to identify results of the pilot study and how the use of reclaimed water could support local water needs. Based on the study, five cycles of operation could be maintained with reclaimed water without an increase in onsite water treatment cost for the cooling towers.

Table 5: Typical Make-up Water and Recycled Water Quality

Component	Units	Typical City Make-up (Edwards Aquifer)	Typical Recycled Water (Leon Creek)
pH	pH	7.2	7.4
Calcium as $CaCO_3$	mg/l	205	225
Magnesium as $CaCO_3$	mg/l	58	75
M-Alkalinity as $CaCO_3$	mg/l	228	215
Conductivity	mmhos	460	1000
Silica as SiO_2	mg/l	12	14
Chlorides as Cl	mg/l	25	117
Sulfates as SO_4	mg/l	26	53
Iron as Fe	mg/l	<0.2	0.05
Nitrate as NO_3	mg/l	<1.0	31-71
Nitrite as NO_2	mg/l	0.0	0.0-1.2
Ammonia as NH_3	mg/l	0.0	1.1-4.2
Total Phosphate as PO_4	mg/l	0.2	4 -11
Organics as BOD	mg/l	0.0	2
Total Suspended Solids	mg/l	<5	0.5 –1.0
Total Chlorine as Cl_2	mg/l	<1.0	1.0 – 1.5

(SAWS, 2000)

Table 6: Actual Cooling Water Chemistry at Five Cycles

Component	Units	Cooling Water with City Make-up (Edwards Aquifer)	Cooling Water with Recycled Water (Leon Creek)
pH	pH	8.0	7.4
Calcium as $CaCO_3$	mg/l	1025	1125
Magnesium as $CaCO_3$	mg/l	290	375
M-Alkalinity as $CaCO_3$	mg/l	100*	40*
Conductivity	Mmhos	2300	5000
Silica as SiO_2	mg/l	60	70
Chlorides as Cl	mg/l	125*	585*
Sulfates as SO_4	mg/l	1100*	1360*
Iron as Fe	mg/l	0.2	0.2
Nitrate as NO_3	mg/l	<1.0	155-355
Nitrite as NO_2	mg/l	0.0	0
Ammonia as NH_3	mg/l	0.0	<0.5
Total Phosphate as PO_4	mg/l	<1.0	20-50
Organics as BOD	mg/l	0.0	0.0
Total Suspended Solids	mg/l	<10	<10
Total Chlorine as Cl_2	mg/l	0.0-2.0	1.0 -4.0

*This assumes sulfuric acid is used for pH control; chlorides will be higher if Chlorine is used.
(SAWS, 2000)

The study also found:

- Acceptable control of mild steel corrosion due to nitrate and orthophosphate;
- Galvanized steel and mild steel corrosion rates should drop;

- Acceptable copper corrosion rates (<0.3 millimeters per year) achieved with increase of toly-triazole (TTA) feed to 2 mg/l;
- Good control of scale with sulfonated styrene/maleic anhydride (SS/MA) scale control polymer at normal feed rates (10 ppm);
- Cost of microbiological control somewhat higher due to increased halogen demand; and,
- Water treatment costs with recycled water similar to current treatment costs.

In Denver, Xcel Energy operates the Cherokee Station, a 717 megawatt (MW) coal-fired steam electric station that uses between 6.165 and 11.097 MCM (5,000 to 9,000 acre-feet) of water for cooling, silo washdown and fire control. Reclaimed water (Table 7) is provided by Denver Water Recycling Plant about a half-mile from Cherokee Station. Xcel has an existing water right to surface water which is lower in cost than reclaimed water and will continue in use to maintain the water right. Reclaimed water and surface water are blended in an onsite reservoir prior to cooling tower use. As part of Xcel's conservation plan, they began using reclaimed water in 2004 and its use provides reliability to plant operations as they are no longer at risks to drought induced shortages (EPA, 2012).

Table 7: Denver Water Reclaimed Water Quality

Denver Water Reclaimed Water Quality Criteria		
Parameter	Units	Typical Range
Alkalinity, Total as CaCO3	mg/l	50-150
Ammonia as N	mg/l	0-0.4
Boron	mg/l	0.2-0.4
Calcium	mg/l	40-70
Chloride	mg/l	65-170
Chlorine, Total	mg/l	1.5-4.0
Iron	mg/l	0.05-0.6
Magnesium	mg/l	5-20
Manganese	mg/l	0.003-0.08
Nitrate + Nitrite as N	mg/l	5-30
Nitrate as N	mg/l	5-20
Nitrite as N	mg/l	0.01-0.05
Ortho Phosphorous, Dissolved as P	mg/l	0.04-0.3
pH	SU	6-8
Phosphorous, Total as P	mg/l	0.04-0.4
Potassium	mg/l	10-20
Sodium	mg/l	90-200
Specific Conductance	mg/l	360-1250
Sulfate	mg/l	80-250
Temperature	^{0}C	10-30
Total Coliform	MPN/100 ml	<1.0
Total Kjeldahl Nitrogen	mg/l	0.2-2
Total Organic Carbon	mg/l	4-8

(EPA, 2012)

Palo Verde Nuclear Generating Station initiated a lease agreement in 1973 with the Sub-Regional Operating Group (SROG) formed by the cities of Phoenix, Mesa, Tempe, Glendale and Peoria to receive 110.97 MCM (90,000 acre-feet) per year of reclaimed water for power plant cooling. The initial lease price remained unchanged at $30 per acre-foot (AF) until renegotiated in 2010 with stepped increases to $159/AF in 2020, rising to $547/AF in 2050 (Carpenter, 2012). Due to the initial high TDS (1,200 mg/l) and hardness (300 mg/l) from SROG, Arizona Public Service operates a lime softening water treatment plant onsite to achieve 25 cycles of concentration through the cooling towers. Influent and effluent water quality from the onsite softening plant is shown in Table 8. Blowdown concentrate approaches 33,000 mg/l and is stored in 650 acres of onsite evaporation ponds. Concentrate from the 1060 MW Redhawk combined cycle gas plant is evaporated and concentrated in a brine concentrator, zero liquid discharge facility (Day, 2008).

Table 8: Influent/Effluent Data at Palo Verde Water Reclamation Plant

Scale Forming Constituents (mg/l)	Influent Quality	Effluent Quality
Alkalinity (as $CaCO_3$)	189	27
Calcium (as $CaCO_3$)	183	73
Magnesium (as $CaCO_3$)	123	15
Silica	19	3.5
Phosphate	10	<0.1

(Day, 2008)

DISTRIBUTION IN THERMAL COOLING DEMANDS

The 2012 Environmental Protection Agency (EPA) *Guidelines for Water Reuse* identifies water and reclaimed water usage across the U.S. and presents population trends over the last decade (EPA, 2012). Since water use for thermal cooling represent a significant percentage of total water use (81% in the Mid-Atlantic states to only 15% in the Mountain and Plains States) (EPA, 2012) conversion to using reclaimed water or selecting reclaimed water in new applications offers significant advantages for both the utility and the area customers they serve. Several figures developed for use in the EPA *Guidelines for Water Reuse* are presented below to illustrate regional differences in demand and use of water for thermal cooling. Figure 1 (Figure 5-1 *EPA Guidelines*) identifies the distribution of freshwater use by category in the U.S. with an average of forty-nine percent of the freshwater used in the thermoelectric energy sector. In the northeast region, Figure 2 (Figure 5-5 *EPA Guidelines*) identifies seventy-two percent of water use dedicated to the thermoelectric energy sector. This offers a significant opportunity to replace potable water or potable source water with reclaimed water where there is a consumptive use for power production.

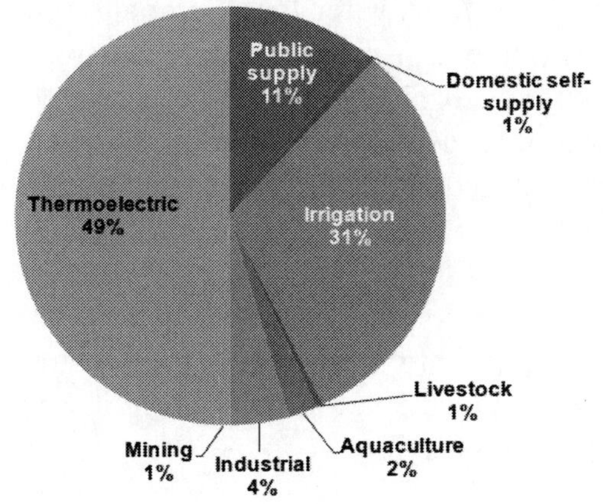

Figure 5-1
Freshwater use by category in the U.S.

Figure 1: Freshwater Use by Category

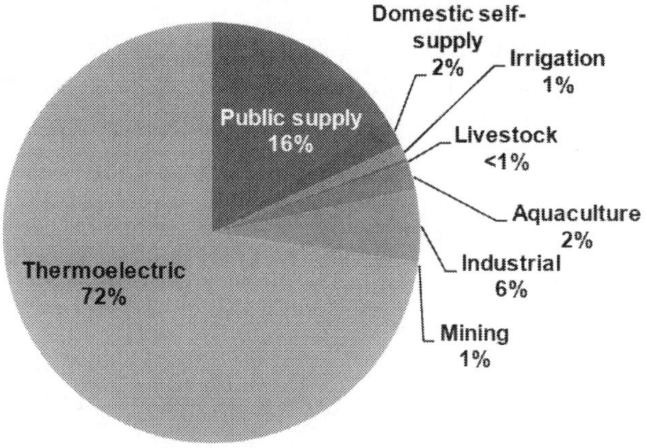

Figure 5-5
Freshwater use by sector for the Northeast region

Figure 2: Northeast Freshwater Use

Energy projects in the northeast that are using reclaimed water include: University of Connecticut campus use at the central power plant including use in a boiler; Dominion Power's Brayton Point Power Plant in Somerset, Massachusetts; a power plant at the Central landfill in Johnston, Rhode Island is the largest reclaimed water project in the northeast.

The Mid-Atlantic region includes nearly 30 million population and significant densities in the Washington, D.C area. The region has significant opportunities to increase use of reclaimed water in the energy sector with limited projects to date. Loudoun County Virginia's Broad Run Water Reclamation Facility provides reclaimed water for evaporative cooling to several data centers. Also in Virginia, Fairfax County uses reclaimed water from the Norman Cole, Jr. Pollution control Plant for cooling at a co-generation plant and Bedford City WWTP provides reclaimed water for cooling at a food packaging facility. Two wastewater treatment facilities in Chesterfield and Fauquier counties provide reclaimed water for cooling and stack scrubbing to coal-burning power facilities.

The southeast region includes Florida with the most developed water reuse program in the U.S. including numerous power plants, some discussed in this paper. Florida reclaims and reuses forty-nine percent of wastewater treatment plant effluent, the highest in the U.S. Although sixty-seven percent of the freshwater in the southeast region is used in thermoelectric generation, there are not many applications using reclaimed water outside of Florida. This indicates a significant opportunity to convert to reclaimed water as the region's water portfolio is integrated.

The Midwest and Great Lakes region uses sixty-three percent of freshwater for thermoelectric cooling, much of it for once through cooling with sixty-one percent of the freshwater use in Minnesota used for once through cooling. Historically abundant surface waters have restricted the development of more intense water use and reclamation. Increasing populations and population density are increasing the potential for use of reclaimed water for power generation and industry.

The South Central region has the second highest population growth rate, warmer climate and lower rainfall rates. Freshwater use for thermoelectric power generation represents only twenty-five percent of water use in the region. Reclaimed water is used for power plant cooling in San Antonio, El Paso, several panhandle and Rio Grande Valley communities.

The Mountain and Plains region uses only fifteen percent of the regions freshwater for thermoelectric power generation with seventy percent use in irrigation. Population growth over the last decade has been higher on a percentage basis than any other region in the U.S. Power plant cooling in Denver has been discussed above. Montana had an early entry into power plant cooling when a power company purchased a mining operation in Colstrip, Montana and used reclaimed water for power plant cooling. There are also increased uses for reclaimed water in the oil and gas industry for hydraulic fracking and reclamation of produced water.

The Pacific Southwest has seen rapid growth, especially Arizona with over twenty-four percent growth and Nevada at thirty-five percent growth. Although California's growth rate has been near the national average, it still represents an increase of 3.4 million people in the last decade with resulting increases in energy demand. As a region, only twenty-seven percent of the freshwater has been used for thermoelectric power generation and nearly half (49 percent) used for irrigation as shown in Figure 3 (*EPA Figure 5-35*).

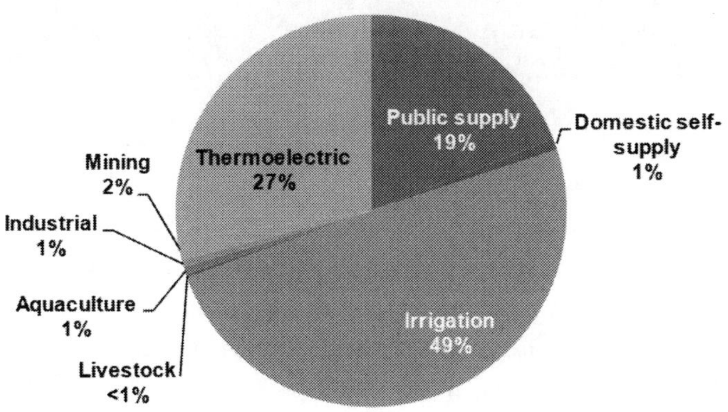

Figure 5-35
Freshwater use by sector for the Pacific Southwest region

Figure 3: Pacific Southwest Freshwater Use

Reclaimed water use at Palo Verde in Arizona has been discussed. Freshwater use in power plant cooling is not a significant use in California but two percent of reclaimed water use is used in geothermal energy production.

In the Pacific Northwest, thirty percent of freshwater is used for thermoelectric power generation but relatively small quantities of reclaimed water, again presenting significant opportunity to substitute reclaimed water for freshwater in this consumptive use.

As the power industry expands facilities to meet future power needs, one criterion in that planning process is to secure adequate water for their operations from the lowest cost source that is still available. As U.S. populations shift to the southwest and west with stressed water supplies, the next available water supply may be at considerable distance from the planned power generation facility. This potentially translates to significant cost to move the water to its point of use. Much of the current water supply in California is imported water, either from out of state on the Colorado River system of water in northern California that is referred to as state water being transferred to the arid southern region of the state. This cost is reflected in pipes, pumping, regulatory and legal constraints. Two examples of western projects will be discussed to highlight the economic advantage seen in choosing reclaimed water to supplement existing water supplies.

The Inland Empire area east of Los Angeles, California has multiple water sources. When those water supplies are compared on energy for production basis, reclaimed water requires 400 kilowatt hours per acre-foot (KWH/AF) of water; groundwater pumping requires 950 KWH/AF, Colorado River aqueduct water 2,000 KWH/AF, and ocean desalination 4,400 KWH/AF Raucher, 2009). Strictly form an energy perspective, maximizing the use of reclaimed water will reduce water supply cost and energy demand for the region. In San Diego, a similar analysis has been applied in a study of power demand by water process which includes the energy cost of importing state water to San Diego (Ataide, 2010). Surface water, ground water and non-potable recycled water each require about 1000 KWH/AF. Recycled potable water requires double the energy at 2,000 KWH/AF while imported state water requires up to 3,200 KWH/AF. As in the Inland Empire, ocean desalination would require between 4,100 and 5,000 KWH/AF. While there is much discussion about the energy-water nexus, using a lower energy requirement water for power production will save energy and reduce the cost of water.

With the reclaimed water quality and cost background, constraints to expanded use of reclaimed water were identified and potential steps forward identified in the recently released National Research Council (NRC) report *Water Reuse: Potential for Expanding the Nation's Water Supply Through Reuse of Municipal Wastewater* (NRC, 2012). This report recognizes that the water quality from many reclamation plants in the U.S. is of higher quality than approved potable water supplies. In May 2012, the American Society of Mechanical Engineers (ASME) and the Water Environment Federation (WEF) convened a workshop in Washington, D.C. to focus on expanding the use of reclaimed water within the electric power industry, producing a best practices workshop report. The ASME/WEF Workshop - Municipal Wastewater Reuse by Electric Utilities: Best Practices and Future Directions (ASME, 2012) and 2012 EPA *Guidelines for Water Reuse* (EPA, 2012) compiled and presented information on water and reclaimed water use practices in the U. S. Specific applications for reclaimed water in the power industry presented in these two reports should serve as an entry point for more power industry interest in reclaimed water and water utility interest in meeting the critical water need of power generators.

RECLAIMED WATER ADVANTAGES FOR THE ELECTRIC POWER INDUSTRY

Examples of current reclaimed water use have been presented. Even in regions of the U.S. that have been historically considered water rich, increasing populations and new demands for other domestic and industrial purposes are beginning to stress local water supplies. Changing weather patterns are or may add to concerns about the long term capability of a local water supply to meet all demands. Advantage offered by use of reclaimed water is that it will be continuously available in proportion to the water used in the local area as a fairly stable percentage of that local water use on an annual basis will be discharged to the local sanitary sewer system for treatment. The portion contributed to the sanitary sewer remains relatively stable month to month with seasonal increases for outdoor irrigation not being reflected in changing sewer flows. There are already areas within the U.S. that have already committed all of their reclaimed water for reuse, removing reclaimed water as an option for current alternate supplies.

The quantity and quality of wastewater discharges are controlled by state issued discharge permits and the quality required to meet stream quality standards has continued to increase in recent years. Therefore, the quality of water available for reuse has continued to improve and in some states, certain discharge permits require near potable quality water. For most cooling and industrial process water purposes, having an adequate quantity of water at a consistent quality allows the industrial user (power plant) to provide any incremental treatment to provide the quality of water they deem appropriate for their intended use. Reusing this highly treated water again for an additional use allows other source water to be used first for a domestic use. This integrated approach is more sustainable than a single use of a water source and development of infrastructure to add to the water footprint of a community.

By using reclaimed water instead of a community potable water or the same source water (untreated) used for the community potable water supply, a power plant allows the community to defer cost of expanding the potable water system to meet increased demands based on higher domestic use and increased demands for water by the power industry. Reuse of reclaimed water may also allow the community to delay addition of wastewater plant improvements if they were to continue discharge to a stream at the same or higher wastewater flows.

The two offsets to potable water and wastewater expansion cost that could be anticipated by a community are part of the reason that reclaimed water is typically priced lower that the local potable water supply. This lower price and the ability to contract for the supply over extended service period, allows the power industry to define their future cost for cooling water as a component of their cost of service.

Advantages in using reclaimed water discussed above include:

- Continuous availability;
- Consistent quality;
- Sustainable source;
- Replacement for potable demands; and
- Reasonable, predictable cost.

SUMMARY

Nationally, forty-nine percent of freshwater is used in the thermoelectric industry. Many electric power generating facilities are already using reclaimed water to meet all or a portion of their water demand. As competition for locally available water increases, reclaimed water offers numerous advantages to both the power industry and the community water and wastewater utility. ASME and WEF have partnered to provide information to both industries on benefits to both industries from local reuse of water provided by the wastewater industry. Examples have been provided of electric power generating stations across the U.S. that are effectively using reclaimed water and providing cost benefits to both the electric and wastewater customers. Power industry and wastewater utility providers should meet locally to discuss future use of a valuable water resource that can offer advantages to both utilities and their customers.

REFERENCES

Ataide, Randy M. (2010) *San Diego's Water Sources: Assessing the Options*, Equinox Center and Fermanian Business & Economic Institute, San Diego, CA, July 2010.

ASME/WEF Workshop (2012) *Municipal Wastewater Reuse by Electric Utilities: Best Practices and Future Directions*, ASME, Washington, DC, September 2012.

Booker, Randall (2008) *Municipal – Industrial Partnership for Reuse: A Case Study in the Northeast U.S.*, WateReuse Symposium, Dallas, Texas, September 2008.

Carpenter, Guy, David Cavazos, and Jeremy Cook (2012) *Sale of the Century: Determining the Price of Reclaimed Water to be Leased to the Largest Nuclear Power Plant in the U.S.*, WateReuse Symposium, September 2012.

Day, Henry and Robert Lotts (2008) *Water Reuse at Palo Verde Nuclear Generating Station (and Redhawk), Including Discussion of TDS, the Energy-Water Nexus and Sustainability*, WateReuse Symposium, September 2008.

Department of Energy (2007) *Use of Reclaimed Water for Power Plant Cooling,* ANL/EVS/R-07/3, Argonne National Laboratory, Argonne, IL, August 2007.

JEA (2005) *Best Management Practices and Guidance Manual for Cooling Towers*, JEA Jacksonville, FL, August 2005.

EPA (2012) *Guidelines for Water Reuse*. EPA/600/R-12/618, Environmental Protection Agency. Washington, DC, September 2012.

Raucher, Bob (2009) *Groundwater Desalination as an Integral Part of Comprehensive Regional Water Resource Management*, WateReuse Symposium, September, 2009.

San Antonio Water System (2000) *Pilot Cooling Tower Study Report,* San Antonio Water System Water Recycling Program, March 15, 2000.

San Antonio Water System (2006) *Recycled Water Users Handbook,* San Antonio Water System Water Recycling Program, May, 2006.

NRC (2012) *Water Reuse: Potential for Expanding the Nation's Water Supply Through Reuse of Municipal Wastewater*, The National Academies Press: Washington D.C, July, 2012.

Vandertulip, Don (2012) *Reclaimed Water Quality Requirements for Electric Utilities*, ASME/WEF Workshop-Municipal Wastewater Reuse by Electric Utilities: Best Practices and Future Directions, Washington, DC, May, 2012.

Maintaining NPDES Permit Compliance at Power Plants While Using Municipal Treatment Plant Effluent with Variable Ammonia and Nitrate Levels.

Christopher Varnon, PE, CDM Smith
Randi Kim, PE, CDM Smith
Joseph Viciere, PE, CDM Smith
Nathan Schmaus, PE, CDM Smith
Tina Petersen, PE, CDM Smith
Jaron Bergin, CHMM, Calpine Corporation

ABSTRACT

Calpine Corporation's Hidalgo Energy Center (HEC) is a state-of-the-art, combined-cycle power plant located in Edinburg, Texas. Fueled by clean, natural gas, the Hidalgo Energy Center has been generating up to 479-megawatts of reliable, affordable electricity for the Texas power market since 2000. To conserve natural resources, HEC uses reclaimed wastewater effluent from nearby municipalities for plant cooling, thereby minimizing the impact to local water resources.

Over the past twelve years, municipal reclaimed water has served as a reliable source of water for HEC. However, on occasion HEC has exceeded its discharge limits for ammonia-N. Since the effluent is primarily comprised of cooling tower blowdown, the suspected source of the ammonia-N is the municipal reclaimed water. HEC engaged CDM Smith to conduct a study to evaluate the occurrence of high ammonia-N and nitrate-N concentrations in its intake water and throughout the cooling water process. The study includes an evaluation of the existing processes, making an engineering judgment as to whether corrective action needs to be taken to address the occurrence of high ammonia-N and nitrate-N in the cooling water blowdown, and proposing recommendations as necessary to allow HEC to maintain compliance with their permit limits. Study components included:

- A review of the condition, design, and performance history of the treatment plants providing reclaimed wastewater to assess the reliability of these facilities. This review included calculating average and maximum values for ammonia-N and nitrate-N and developing trend analyses for each facility to estimate expected water quality.

- Using the expected water quality estimates and cooling tower operational information to develop a mass balance around the cooling tower system in order to estimate expected effluent concentrations of ammonia-N and nitrate-N across a range of cycles of concentration. Results were used to determine if HEC would exceed its permit limits based on the expected water quality in the future.

Upon review of historical data from the reclaimed water source and consideration of improvements made to the WWTP to improve process reliability, it was determined that HEC should be able to continue to operate its cooling tower while meeting the permit requirements for ammonia and nitrate-N, given the operation within recommended cycles of concentration. Additional steps to increase the operational flexibility of the facility and provide opportunities for HEC to overcome potential upsets from the McAllen WWTP were recommended for consideration with one or a combination of the following actions:

- Increasing onsite storage capacity
- Working with McAllen to improve nitrate-N removal, or
- Use of other sources of water with low concentrations of ammonia-N and nitrate-N.

KEYWORDS

Wastewater Effluent, Ammonia, Nitrogen

INTRODUCTION

Calpine Operating Services Company, Inc. (Calpine) owns and operates the Calpine Hidalgo Energy Center (HEC), an electric power generating plant, at 4005 North Seminary Road, at the northwest corner of the intersection of Monte Cristo Road (Farm-to-Market Road 1925) and Seminary Road in Edinburg, Hidalgo County, Texas. Calpine purchased the facility during its construction and it has been in operation since June 1, 2000.

HEC utilizes a recirculating cooling tower system to dissipate heat generated by its power generating plant. HEC receives treated municipal wastewater effluent from the City of McAllen's North Wastewater Treatment Plant (McAllen WWTP) for use as makeup water in the cooling tower system to offset evaporation, drift, and blowdown. Calpine also has an agreement to receive treated municipal wastewater from the City of Edinburgh, but to date has rarely utilized this as a source of water. HEC has been using the McAllen WWTP water since beginning operation in 2000.

HEC currently discharges its cooling tower system blowdown to the North Main Drain III, and ultimately to the Laguna Madre in Segment No. 2491 of the Bays and Estuaries. This discharge is permitted under the Texas Water Code Chapter 26 and TPDES Permit No. WQ004138000. Effluent limitations, monitoring requirements, and other conditions for this discharge are set forth in the TPDES Permit, as well as in the rules of the Texas Commission on Environmental Quality (TCEQ), the laws of the State of Texas, and other orders of the TCEQ. The HEC blowdown is monitored for BOD, TDS, sulfate, chloride, nitrate-N, ammonia-N, and residual chlorine.

In 2011 and 2012, Calpine detected ammonia-N concentrations in the HEC effluent above the permitted limits as shown in **Table 1-1**. Prior to 2011, Calpine had intermittent exceedances of its discharge permit, but never in consecutive months.

The purpose of this study was to evaluate operating conditions and associated data for both the McAllen WWTP and HEC to determine:

- the root cause of the higher than normal ammonia nitrogen conditions;
- expected effluent quality from McAllen WWTP in future;
- anticipated HEC effluent quality under current cooling tower operating conditions; and

- whether or not modifications to the cooling tower operating conditions at HEC would help maintain permit compliance.

Table 1: HEC Ammonia-N Permit Limit Exceedances

Outfall No. 001			
Month/Year	Ammonia Nitrogen DAV Concentration	Ammonia Nitrogen DMAX Concentration	Ammonia Nitrogen DMAX Loading
	Limit = 1.0 mg/L	Limit = 3.0 mg/L	Limit = 23 lb/d
April 2011	1.200	c	c
December 2011	1.740	4.97	23.1
January 2012	1.464	4.95	23.8
DAV = Daily Average; DMAX = Daily Maximum mg/L = milligrams per liter; lb/d = pounds per day; c = compliant			

The study methodology included:

- an evaluation of McAllen WWTP operating conditions at the time that the higher than ammonia nitrogen conditions were observed;
- trend analysis of McAllen WWTP effluent data to estimate expected effluent quality (ammonia-N and nitrate-N);
- a mass balance around the HEC cooling tower under current operating conditions to determine the expected effluent quality (ammonia-N and nitrate-N) for the HEC blowdown and the discharge;
- a mass balance around the HEC cooling tower under modified operating conditions to determine the expected effluent quality (ammonia-N and nitrate-N) for the HEC blowdown and the discharge; and
- an evaluation of optimal operating conditions to maintain compliance with permit requirements.

In addition to analyzing the current water source, CDM Smith evaluated possible alternative water sources that could be utilized if it is determined that reliable compliance with permit limitations cannot be achieved with the treated wastewater from McAllen. The analysis includes a general summary of alternative water sources that were identified in the Region M Water Plan along with a discussion of regulatory framework, availability, and quality.

Based on an analysis of the current and possible alternate water sources for use in the cooling tower system at HEC, CDM Smith provided recommendations for Calpine to meet and maintain compliance with permit limits as well as to increase the reliability of ongoing operations.

EVALUATION OF EXISTING WATER SOURCE TO HEC

The City of McAllen North Wastewater Treatment Plant (McAllen WWTP) is an extended aeration activated sludge process. The facility is currently operated under TPDES Permit No. WQ0010633004. The McAllen WWTP recently underwent a comprehensive improvements and expansion project to provide a total design capacity of 15 million gallons per day (mgd) and upgrade their treatment process with state-of-the-art equipment and controls, including a comprehensive SCADA system that allows the plant operators to more closely monitor process feedback and consistently produce a high equality effluent. The McAllen WWTP completed the transition to its expanded operation in January 2012. In discussion with McAllen WWTP operations staff, during the months of December 2011 and January 2012, the facility went through the process of start-up and commissioning of the new treatment units and had some process upsets while transitioning back and forth between the old plant and the new plant.

The reclaimed water from the City of Edinburg historically has been of lower quality than the water from the McAllen WWTP, and since the Edinburg facility is older, without the state-of-the-art equipment and controls that the upgraded McAllen WWTP now has, CDM Smith did not perform a detailed analysis on data from the City of Edinburg's WWTP.

The McAllen WWTP pumps a portion of its reclaimed effluent to HEC. The reclaimed water from McAllen is stored in a water storage pond (approximately 400,000 gallons) located in the southwest corner of the HEC site. From the storage pond, the makeup water for the HEC cooling tower system is pumped into two 500,000-gallon storage tanks, which are used to supply the cooling tower and provide fire water storage. These tanks are kept at near full capacity and any excess water from the storage pond is recirculated back to the storage pond.

The McAllen WWTP final effluent limitations for ammonia-N, which is the HEC storage pond influent water, are presented in **Table 2**. In comparison, the Calpine HEC Outfall 001 discharge limitations for ammonia-N and nitrate-N are presented in **Table 3**, which are actually more restrictive than those imposed on the McAllen WWTP.

Table 2: McAllen WWTP Ammonia-N Effluent Limitations

Effluent Characteristics	Discharge Limitations				
	Daily Average		7-day Average	Daily Maximum	Single Grab
	lbs/day	mg/L	mg/L	mg/L	mg/L
Ammonia-N	375	3	6	10	15

Table 3: HEC Ammonia-N Effluent Limitations

Effluent Characteristics	Discharge Limitations				
	Daily Average		Daily Maximum		Single Grab
	lbs/day	mg/L	lbs/day	mg/L	mg/L
Ammonia-N	7.7	1.0	23	3.0	3.0
Nitrate-N	1,052	137	2,219	289	289

The daily averages (average of all data for the entire month) and daily maximums (maximum individual sample in each month) of the ammonia-N concentrations detected in the weekly grab samples taken from the McAllen WWTP effluent during 2011 and 2012 are presented in **Table 4**.

Table 4: McAllen WWTP Effluent Ammonia-N Concentrations

Month	Ammonia-N	
	Daily Average	Daily Maximum
	mg/L	mg/L
January 2011	0.10	0.12
February 2011	1.20	1.79
March 2011	0.48	1.82
April 2011	1.10	2.10
May 2011	0.66	1.33
June 2011	0.81	2.50
July 2011	0.19	0.31
August 2011	0.60	1.84
September 2011	0.42	0.66
October 2011	0.38	0.58
November 2011	0.31	0.67
December 2011	2.89	11.0
January 2012	1.69	4.19
February 2012	0.24	0.46
March 2012	0.19	0.41
April 2012	0.09	0.21
May 2012	0.13	0.30
June 2012	0.11	0.16
July 2012	0.14	0.25
August 2012	0.13	0.18
September 2012	0.17	0.24

The McAllen WWTP currently monitors the ammonia-N concentration in its effluent per its TPDES Permit, but nitrate-N is not monitored, as the McAllen WWTP does not have a permit limit for nitrate-N.

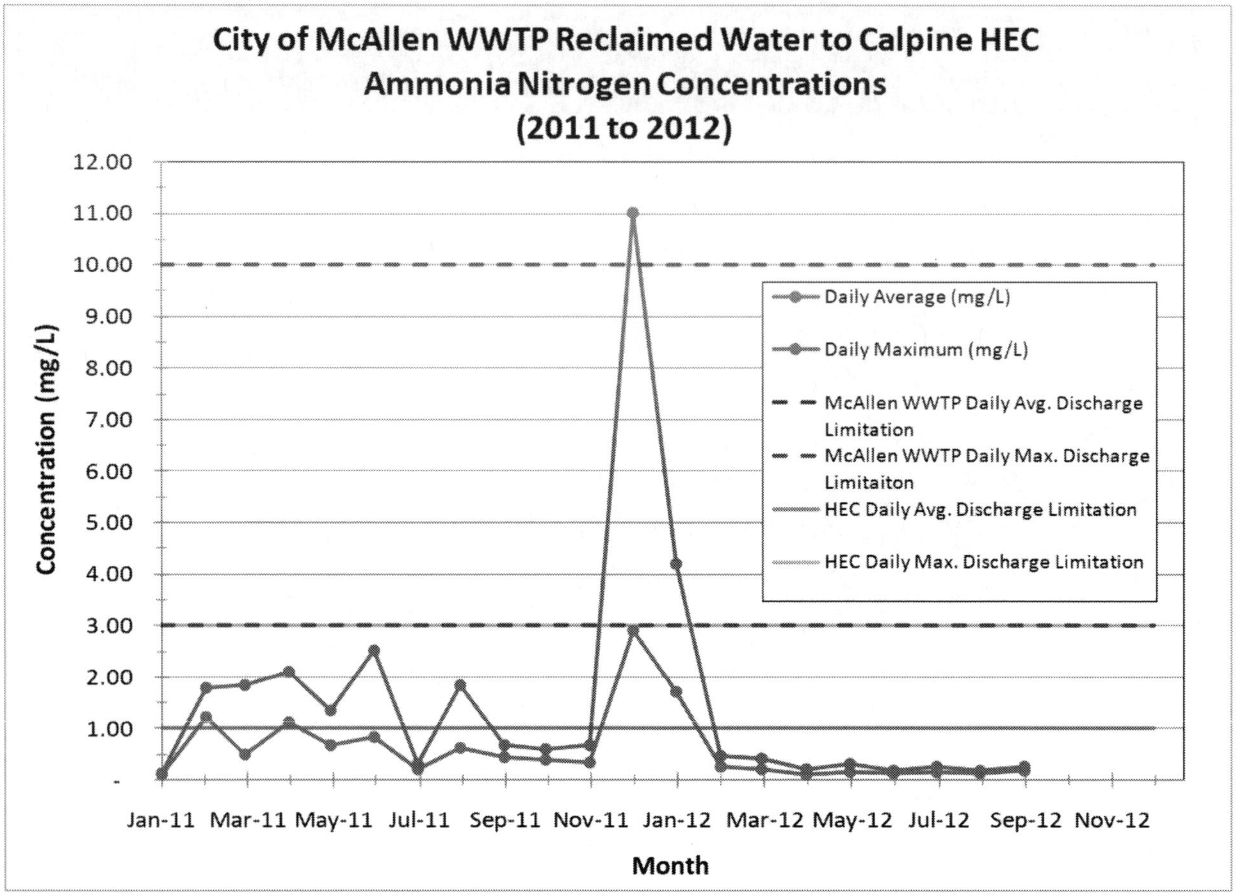

Figure 1: McAllen WWTP Effluent Ammonia-N Concentrations (2011 - 2012)

From January 2011 to November 2011, the daily average ammonia-N concentration fluctuated between 0.10 mg/L and 1.20 mg/L and the daily maximum concentration of ammonia-N fluctuated between 0.12 mg/L and 2.50 mg/L (See **Figure 1**).

During December 2011, the daily average and daily maximum ammonia-N concentrations reached a maximum of 2.89 mg/L and 11.0 mg/L, respectively. During January 2012, the daily average and daily maximum ammonia-N concentration were at 1.69 mg/L and 4.19 mg/L, respectively. The high concentration of ammonia-N from December 2011 to January 2012 is attributed to the McAllen WWTP transitioning from 8 mgd to 15 mgd operation after completion of its facility upgrades.

Since the McAllen WWTP completed its transition in January 2012, the ammonia-N concentration in the reclaimed water effluent has been relatively stable and is well below what is required for the cooling tower system makeup water at the HEC for the latter to meet its discharge limitations (See **Figure 2**).

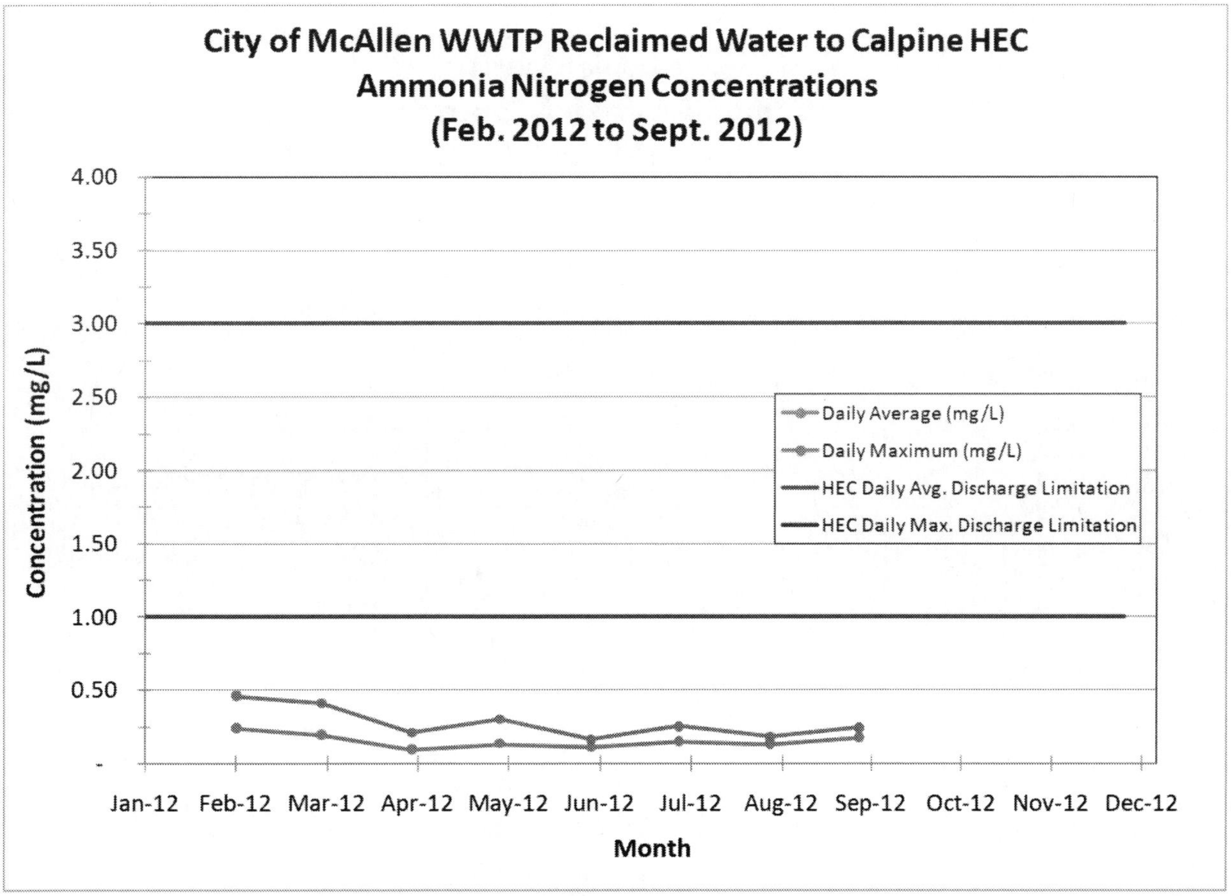

Figure 2: McAllen WWTP Effluent Ammonia-N Concentrations (Feb. 2012 – Sept. 2012)

From February 2012 through September 2012, the mean daily average concentration of ammonia-N was 0.15 mg/L with a standard deviation of 0.05 mg/L and the maximum daily average concentration was 0.24 mg/L. During this same period, the mean daily maximum concentration was 0.28 mg/L with a standard deviation of 0.11 mg/L and the maximum daily average concentration was 0.46 mg/L. Since the sample size is relatively small, the mean concentrations are reported with their respective standard deviations. The standard deviation is the degree to which the concentrations (daily average and daily maximum) for each month differ from their respective sample mean. This data is shown in **Table 5** and the cumulative probabilities of ammonia-N and nitrate-N concentration for this time period is shown in **Figure 3**.

Table 5: McAllen WWTP Ammonia-N Concentrations (Feb. 2012 – Sept. 2012)

	Average	Standard Deviation	Typical Range
Daily Average Concentration	0.15 mg/L	0.05 mg/L	0.10 mg/L to 0.20 mg/L
Daily Maximum Concentration	0.28 mg/L	0.11 mg/L	0.17 mg/L to 0.39 mg/L

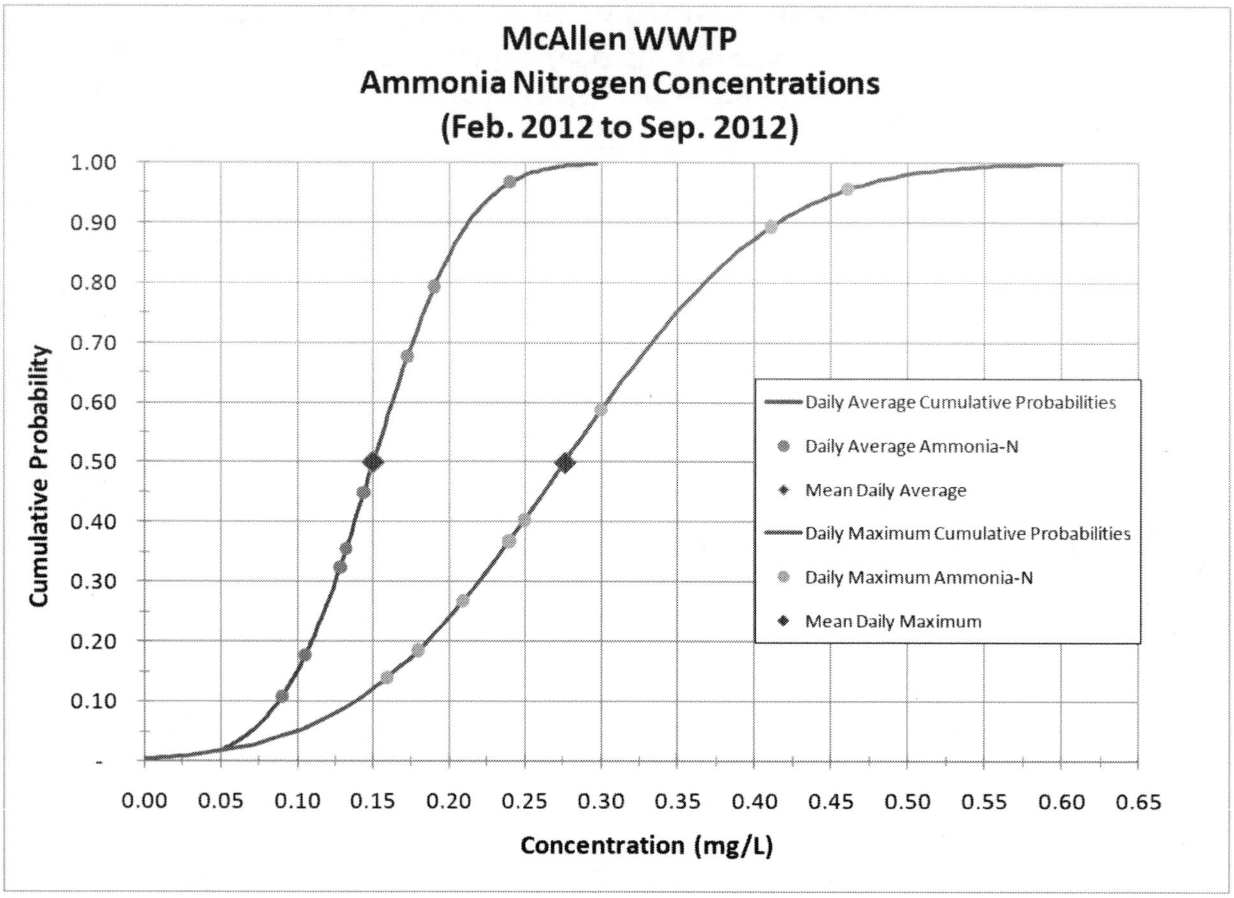

Figure 3: Cumulative Probabilities of McAllen WWTP Effluent Ammonia-N Concentrations (February 2012 to September 2012)

Expected Ammonia-N Concentration from McAllen WWTP

Since the current operation of the McAllen WWTP includes upgraded equipment and process control, the future ammonia-N concentrations in the McAllen WWTP effluent are expected to remain stable and continue to be consistent to those detected since February 2012 after the McAllen WWTP began operating their newly expanded facility.

The mean ammonia-N concentration plus one standard deviation corresponds to the 84th percentile of the data set, so based on the ammonia-N concentrations detected in the McAllen WWTP effluent since February 2012, this suggest that 84% of future daily average ammonia-N concentrations will be below 0.20 mg/L. The maximum ammonia-N concentration in the data set corresponds to the 97th percentile, which suggests that 97% of future daily average concentrations will be below the maximum recorded daily average concentration of 0.24 mg/L. Using the same statistical analysis, it is expected that 84% of future daily maximum ammonia-N concentrations will be below 0.39 mg/L and that 96% of future daily maximum concentrations will be below the maximum recorded concentration of 0.46 mg/L (See **Table 6**).

Table 6: Expected Ammonia-N Concentrations

	Mean (mg/L)	Mean Plus 1 STDEV (mg/L)	Maximum (mg/L)
Expected Daily Average Concentration	0.15	0.20	0.24
Expected Daily Maximum Concentration	0.28	0.39	0.46

Evaluation of McAllen WWTP Effluent Nitrate-N Concentrations

McAllen does not have a nitrate-N limit in its permit and therefore does not monitor the nitrate-N concentration in its effluent. The nitrate-N concentrations analyzed in this section were taken at the HEC makeup water storage pond. HEC began monitoring the nitrate-N concentration in the makeup water storage pond, taking weekly grab samples, in June 2012.

The HEC Outfall 001 discharge limitations for Nitrate-N are presented in **Table 7**.

Table 7: HEC Nitrate-N Effluent Limitations

Effluent Characteristics	Discharge Limitations				
	Daily Average		Daily Maximum		Single Grab
	lbs/day	mg/L	lbs/day	mg/L	mg/L
Nitrate-N	1,052	137	2,219	289	289

The daily average (average of all data in each month) and daily maximum (maximum individual sample in each month) of the nitrate-N concentrations that were detected in the weekly grab samples taken from HEC's makeup water storage pond are presented in **Table 8** and in **Figure 4**.

Table 8: Calpine HEC Makeup Water Storage Pond Nitrate-N Concentrations

Month	Daily Average	Daily Maximum
	mg/L	mg/L
June 2012	30.75	33.60
July 2012	35.84	41.30
August 2012	29.92	32.00
September 2012	33.85	36.50
October 2012	34.92	36.10

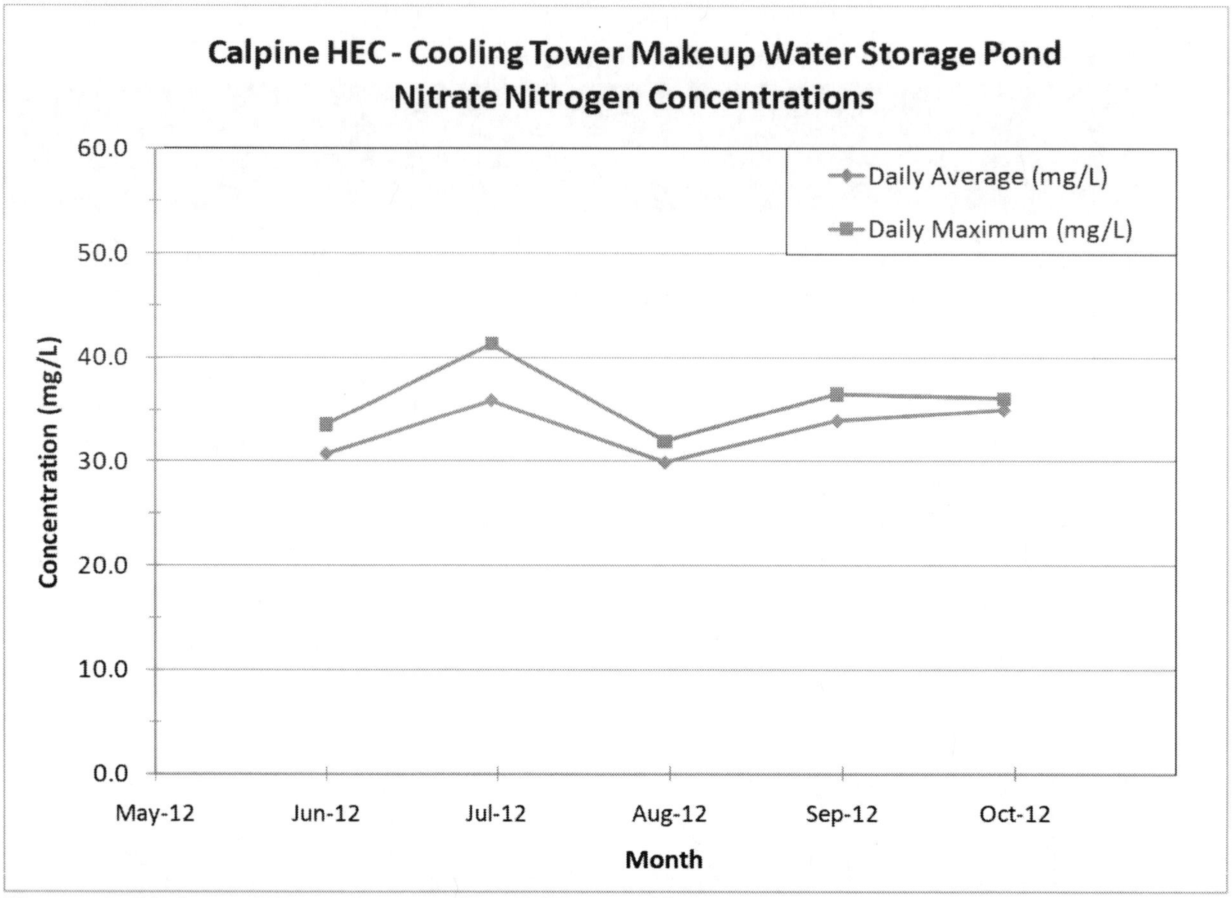

Figure 4: Makeup Water Storage Pond Nitrate-N Concentrations

Since the sample size is relatively small, the mean concentrations are reported with their respective standard deviations (see **Table 9** and **Figure 5**). From June 2012 through October 2012, the mean daily average concentration of nitrate-N was 33.06 mg/L with a standard deviation of 2.60 mg/L and the maximum daily average concentration was 35.84 mg/L. During this same period, the mean daily maximum concentration was 35.90 mg/L with a standard deviation of 3.54 mg/L and the maximum daily maximum concentration was 41.30 mg/L.

Table 9: Nitrate-N Concentrations

	Average	Standard Deviation	Typical Range
Daily Average Concentration	33.06 mg/L	2.60 mg/L	30.46 mg/L to 35.66 mg/L
Daily Maximum Concentration	35.90 mg/L	3.54 mg/L	32.36 mg/L to 39.44 mg/L

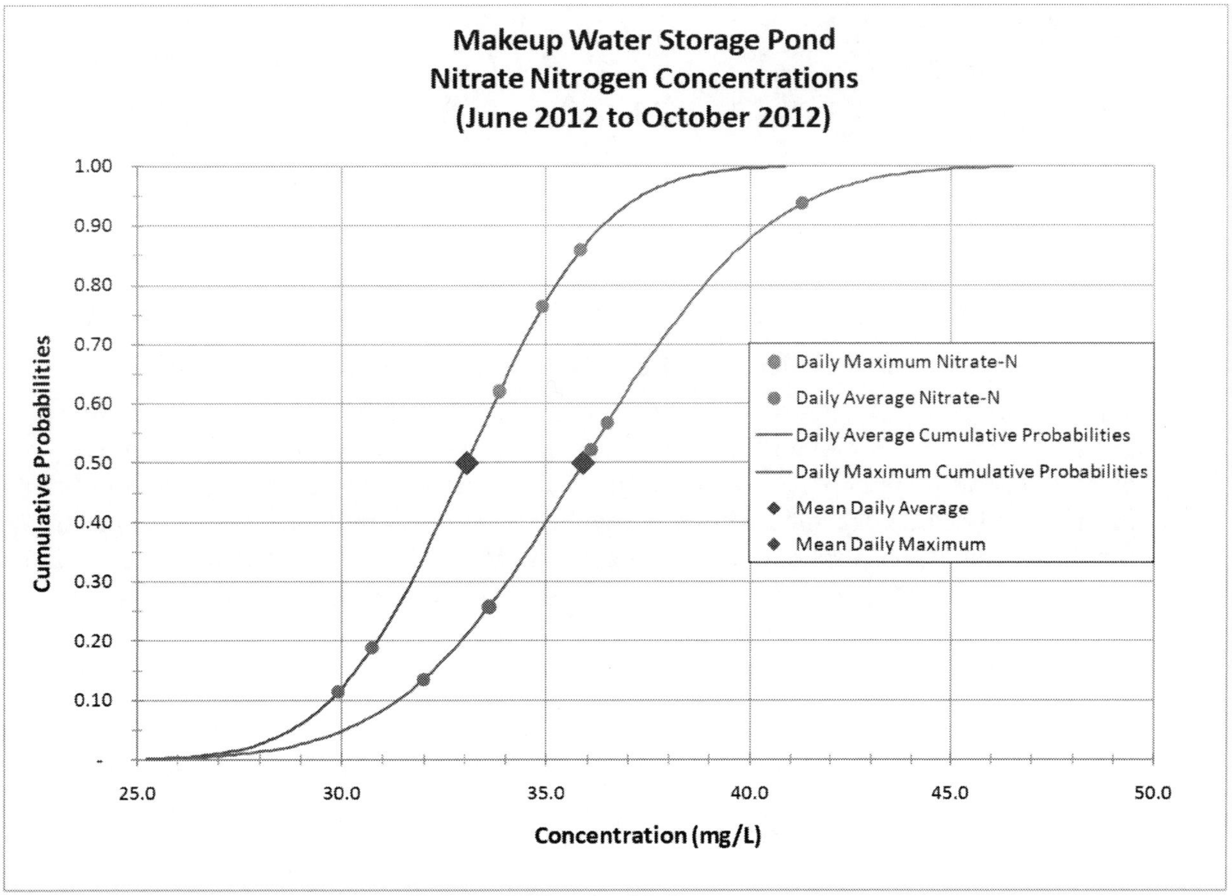

Figure 5: Cumulative Probabilities of Nitrate-N Concentrations

Expected Nitrate-N Concentration at Cooling Tower Influent

It is expected that the reclaimed water from McAllen WWTP that is stored in the makeup water pond will continue to have similar nitrate-N concentrations as detected since June 2012 as long as the McAllen WWTP continues its standard operations of their new facility.

The mean nitrate-N concentration plus one standard deviation corresponds to the 84th percentile of the data set, so based on the nitrate-N concentrations detected in the makeup water pond since June 2012, this suggest that 84% of future daily average nitrate-N concentrations will be below 35.66 mg/L. The maximum nitrate-N concentration in the data set corresponds to the 86th percentile, which suggests that 86% of future daily average concentrations will be below the maximum recorded daily average concentration of 35.84 mg/L. Using the same statistical analysis, it is also expected that 84% of future daily maximum nitrate-N concentrations will be below 39.44 mg/L and that 94% of future daily maximum concentrations will be below the maximum concentration of 41.30 mg/L (See **Table 10**).

Table 10: Expected Nitrate-N Concentrations

	Mean (mg/L)	Mean Plus 1 STDEV (mg/L)	Maximum (mg/L)
Expected Daily Average Concentration	33.06	35.66	35.84
Expected Daily Maximum Concentration	35.90	39.44	41.30

Site Observation at McAllen WWTP in Relation to Nitrate-N Concentration

The McAllen WWTP is currently operated as an extended aeration process. During CDM Smith's visit to the McAllen WWTP, personnel expressed their interest in operating the extended aeration plant in a more conventional way, due to the success of their pretreatment program in limiting organic loading below the values used in the design of the basins and the flexibility that the control and monitoring capability of their new system affords. The diffused aeration system at this facility is served by four 350-Hp high speed, direct drive blowers that utilize air bearing technology. In their current mode of operation, two blowers in service provide too much air to the aeration system, while one blower does not provide enough air.

During the site visit, CDM Smith discussed with McAllen operation personnel the possibility of operating the extended aeration plant more similar to a Modified Ludzack-Ettinger (MLE) process, and have only one blower in service to save energy and to limit the detrimental effect that too much dissolved oxygen in the mixed liquor may present to the plant's clarification process. In that mode, an anoxic zone would be created upstream of the aeration zone in which only sufficient air would be provided for mixing. An internal recycle of the nitrified mixed liquor after the aeration zone would be returned to the anoxic zone, converting the nitrates brought from the mixed liquor into nitrogen gas. The utilization of nitrates as an oxygen source would result in long-term energy savings for the McAllen WWTP, and the potential benefit for Calpine with this mode of operation is that nitrate-N in the McAllen effluent would decrease substantially. Even if an internal recycle stream was not incorporated and the front of the aeration basins were operated as anoxic zones, some denitrification would occur, although the energy savings and reduction in nitrate-N would be substantially lower.

The downside of operating the aeration basins in this manner is that the anoxic zone would not contribute to the volume required for nitrification and as the influent flow and total ammonia-N loading approaches the level for which the basins were originally designed, the ability to achieve the current low concentrations of ammonia-N in the McAllen WWTP effluent might be compromised. This may not be an issue if operating data indicates that the basins and aeration system were sized conservatively enough to sacrifice this volume for use as an anoxic zone. Based on anecdotal evidence of the reduction in plant loading due to the pre-treatment program, this may be the case.

A way that the plant could maintain its designed nitrification capacity, continuing to achieve low effluent ammonia-N concentrations, and at the same time decrease its nitrate-N effluent concentration would be to install denitrification filters to reduce the nitrate-N in the clarified water. This would require new equipment and a supplemental carbon source, such as ethanol, to be provided. Since McAllen does not have any limitations for effluent nitrate-N in their permit, denitrification filters would not need to be installed to maintain compliance. These filters could be installed to treat the 3-mgd effluent flow that is conveyed to HEC, which is only 20 percent of the total design capacity.

The new McAllen 15-mgd plant is a state of the art plant with all necessary equipment and controls to reliably perform as designed. The data collected and analyzed by CDM Smith support that effluent ammonia-N will continue to be low as currently operated. Effluent nitrate, however, is a product of the nitrification process and will vary based on the influent Total Kjeldahl Nitrogen (TKN) to the wastewater plant. Sampling performed at our request on November 21 through 26 indicate that typical TKN influent concentrations to this plant range from 47 mg/L to 74 mg/L with an average of 61 mg/L. Consequently, the limiting number of cooling tower cycles of concentration at HEC to meet the requirements of its permit is based on the nitrate-N concentration coming from McAllen, not the ammonia-N concentration.

Cooling Tower System Model

To determine the recommended cooling tower operating parameters a model was developed using the mass balance equation (**Equation 1**) and the cycle of concentration equation (**Equation 2**). The model was then executed inputting the average cooling tower water circulation flow rate, the average rate of evaporation, rate of drift loss, and the influent nitrate-N concentration (See **Table 11**). The cooling tower influent nitrate-N concentration used in the model was 35.66 mg/L. Because of the relatively small sample size of the influent nitrate-N concentrations, the mean plus one standard deviation, which corresponds to the 84th percentile, of the daily average nitrate-N concentrations is used in the model instead of the mean to provide a more conservative estimate of the future water quality. The model was executed in Microsoft Excel using the solver analysis tool at varying blowdown flow rates to determine the corresponding cycles of concentration.

Cooling Tower Operation

CDM Smith evaluated the operation of the recirculating cooling tower system with respect to the cycles of concentration to determine if current operation will enable HEC to meet permit limits or if operational adjustments are needed. The average cycles of concentration presented in **Table 12** are based on the ammonia-N concentrations and the nitrate-N concentrations that were detected in the samples taken from the McAllen WWTP Reclaimed Water, the HEC Makeup Water Storage Pond, and the HEC Cooling Tower Outfall 001, during the period from February 2012 through September 2012.

$$X_f = \frac{X_0 \times CW + X_{mw} \times MW - X_0 \times BW - 0 \times EW - X_0 \times DL}{CW - EW - DL - BW + MW} \qquad \textbf{(Equation 1)}$$

$$Cycles\_of\_Concentration = \frac{X_f}{X_{mw}} \qquad \textbf{(Equation 2)}$$

Where,

X_f = Nitrate-N concentration at Outfall 001, mg/L

X_0 = Initial nitrate-N concentration in the system, mg/L

X_{mw} = Nitrate-N concentration in the makeup water, mg/L

CW = Cooling tower water circulation flow rate, mgd

MW = Makeup water flow rate, mgd

EW = Rate of evaporation, mgd

DL = Rate of Drift Loss, mgd

Table 11: Cooling Tower

Cooling Tower Water Circulation Flow Rate	242 mgd
Evaporation Rate	2.55 mgd
Rate of Drift Loss	0.0012 mgd
Average Daily Average Flow Rate at Outfall 001	0.61 mgd
Maximum Daily Maximum Flow Rate at Outfall 001	0.94 mgd

Also factored into the cycles of concentration calculations are the portion of makeup water provided to the cooling tower system from the blowdown out of the heat recovery steam generator, the evaporative coolers, and the wastewater from the boiler water treatment system (reverse osmosis generation reject stream and backwash from the ion exchange). Calpine typically adds a total of 4 gallons of 19% active aqueous ammonia-N solution to the steam cycle. A sample was obtained from this additional combined stream of makeup water on November 20, 2012. Ammonia-N was detected at a concentration of 1.71 mg/L and nitrate-N was not detected. The typical flow rate of this combined stream was recorded to be 0.12 mgd.

Table 12: Cycles of Concentration (February 2012 through September 2012)

	Cycles of Concentration		
Parameter	**Mean**	**Standard Deviation**	**Typical Range**
Ammonia-N	2.6	1.3	1.3 to 3.9
Nitrate-N	4.3	0.9	3.4 to 5.2

Based on the recent operational history of the cooling tower system from February 2012 through September 2012, the cycles of concentration based on ammonia-N would typically range between 1.3 and 3.9, compared to a range between 3.4 and 5.2 based on nitrate-N.

The lower cycle of concentration determined for the cooling tower system based on ammonia-N is due to the fact that some of the ammonia-N input to the system reacts with the chlorine fed to the cooling tower system as sodium hypochlorite. Thus, the ammonia-N concentration of the recirculated water after a number of cycles of concentration is lower than what it would have been if that reaction did not take place. Consequently, the determination of the number of cycles of concentration in the cooling tower system is best determined based on nitrate-N. The cumulative probability of the cycles of concentration for ammonia-N and nitrate-N, based on the historical operating data, are shown in **Figure 6**.

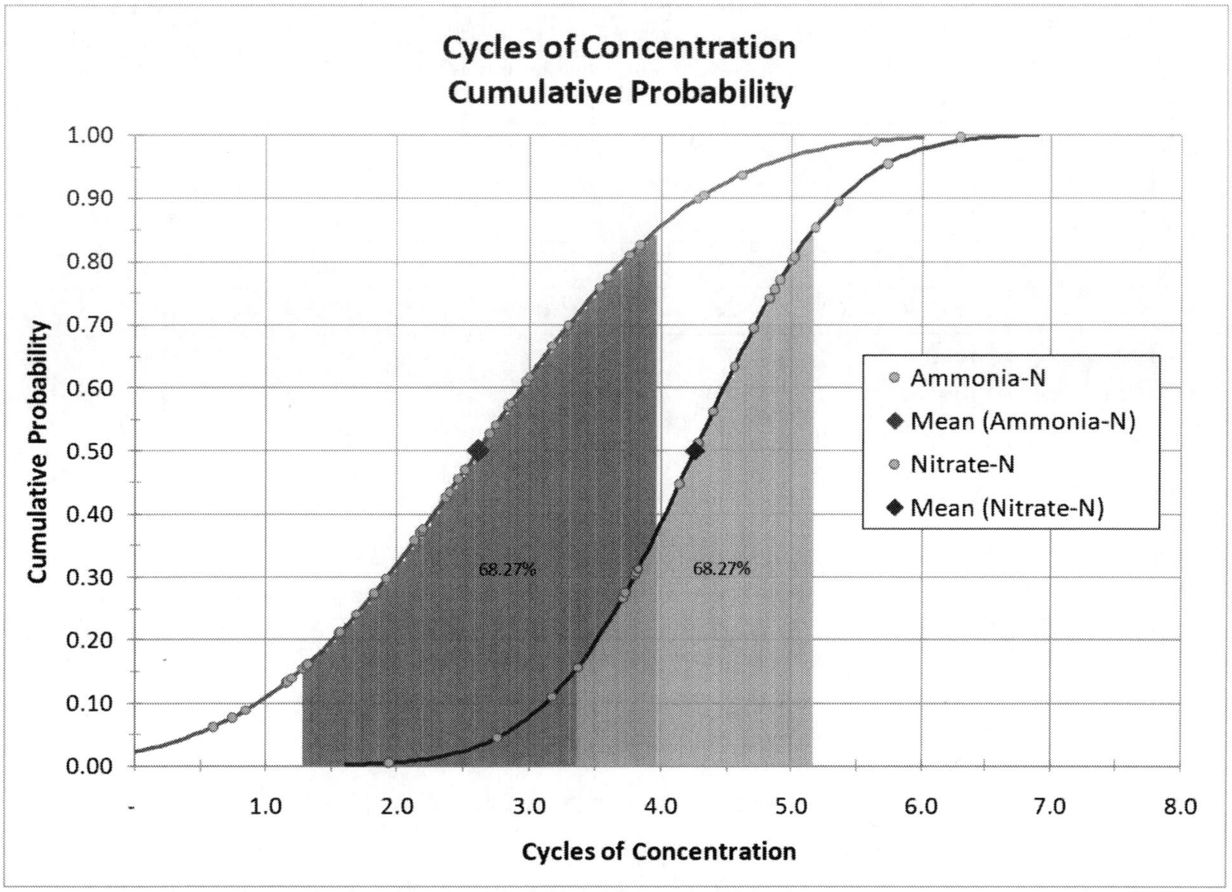

Figure 6: Cycles of Concentration Cumulative Probability

WATER QUALITY DATA ANALYSIS AND MASS BALANCE

Using the expected influent concentrations and the current cycles of concentration a mass balance around the cooling tower system was developed to determine the expected cooling tower effluent ammonia-N and nitrate-N concentrations.

Expected Effluent Concentration of Ammonia-N at Outfall 001

The expected ammonia-N concentrations in the cooling tower effluent at Outfall 001 are estimated in **Table 13** given the current cycles of concentration and the expected influent water quality.

Table 13: Expected Concentration of Ammonia-N at Outfall 001

		Cooling Tower Influent Ammonia-N Concentration (mg/L)	Current Cycles of Concentration	Outfall 001 Ammonia-N Concentration (mg/L)	Permit Discharge Limitations (mg/L)
Expected Daily Average	Mean	0.15	2.6 +/- 1.3	0.39 +/- 0.20	1.0
	Mean Plus 1 STDEV	0.20	2.6 +/- 1.3	0.52 +/- 0.26	1.0
	Maximum	0.24	2.6 +/- 1.3	0.62 +/- 0.31	1.0
Expected Daily Maximum	Mean	0.28	2.6 +/- 1.3	0.73 +/- 0.36	3.0
	Mean Plus 1 STDEV	0.39	2.6 +/- 1.3	1.01 +/- 0.51	3.0
	Maximum	0.46	2.6 +/- 1.3	1.20 +/- 0.60	3.0

Given that ammonia-N concentrations from the McAllen WWTP remain within the expected range, based on the data obtained since February 2012 and the assessment that the new, upgraded facility will be able to maintain this level of operation, the daily average effluent concentrations of ammonia-N at Outfall 001 are expected to remain below the permit discharge limitation of 1.0 mg/L. Likewise, the daily maximum effluent concentrations of ammonia-N at Outfall 001 are expected to remain below the permit discharge limitation of 3.0 mg/L.

Expected Effluent Concentration of Nitrate-N at Outfall 001

The expected nitrate-N concentrations in the cooling tower effluent at Outfall 001 are estimated in **Table 14**, given the current cycles of concentration and the expected influent water quality.

Given that nitrate-N concentrations from the McAllen WWTP remain within the expected range, based on the data obtained since February 2012 and the assessment that the new, upgraded facility will be able to maintain this level of operation, the daily average effluent concentrations of nitrate-N at Outfall 001 are expected to exceed the permit discharge limitation of 137 mg/L. However, the daily maximum effluent concentrations of nitrate-N at Outfall 001 are expected to remain below the permit discharge limitation of 289 mg/L.

Table 14: Expected Concentration of Nitrate-N at Outfall 001

		Cooling Tower Influent Nitrate-N Concentration (mg/L)	Current Cycles of Concentration	Outfall 001 Nitrate-N Concentration (mg/L)	Permit Discharge Limitations (mg/L)
Expected Daily Average	Mean	33.06	4.3 +/- 0.9	142 +/- 30	137
	Mean Plus 1 STDEV	35.66	4.3 +/- 0.9	153 +/- 32	137
	Maximum	35.84	4.3 +/- 0.9	154 +/- 32	137
Expected Daily Maximum	Mean	35.90	4.3 +/- 0.9	154 +/- 32	289
	Mean Plus 1 STDEV	39.44	4.3 +/- 0.9	170 +/- 35	289
	Maximum	41.30	4.3 +/- 0.9	178 +/- 37	289

Cycles of Concentration to Meet Permit Discharge Limitations

Given that approximately 84% of future daily average concentrations of nitrate-N are expected to be below 35.66 mg/L in the makeup water pond, based on the mean plus one standard deviation of the data set, in order for the nitrate-N daily average concentration at Outfall 001 to remain below the permit discharge limitation of 137 mg/L, the cooling tower cycles of concentration are required by the analysis to remain below 3.8 (See **Table 15**).

Table 15: Maximum Cycles of Concentration to Meet Discharge Limitations

	Mean Plus 1 STDEV Nitrate-N Conc. (mg/L)	Permit Discharge Limitations (mg/L)	Maximum Cycles of Concentration to Meet Discharge Limitation
Daily Average Concentration	35.66	137	3.8
Daily Maximum Concentration	39.44	289	7.3

Maximum Nitrate-N Concentration at Cooling Tower Influent

Looking at it another way, in order to maintain the nitrate-N concentrations in the cooling effluent at Outfall 001 below the permit discharge limitations of 137 mg/L for the daily average and 289 mg/L for the daily maximum, the maximum daily average and daily maximum nitrate-N concentrations in the cooling tower influent would need to remain below 31.9 +/- 6.7 mg/L and 67.2 +/- mg/L, respectively, given that the current cooling tower cycles of concentration remain at 4.3 +/- 0.9 (See **Table 16**).

Table 16: Recommended Maximum Nitrate-N Influent Concentration

	Permit Discharge Limitations (mg/L)	Current Cycles of Concentration	Recommended Maximum Nitrate-N Influent Concentration (mg/L)
Daily Average Concentration	137	4.3 +/- 0.9	31.9 +/- 6.7
Daily Maximum Concentration	289	4.3 +/- 0.9	67.2 +/- 14.1

Analysis of Cooling Tower Operating Range

As shown in **Figure 7** and **Figure 8**, in order to not exceed the daily average permit discharge limitation of 137 mg/L for the nitrate-N concentration, the cooling tower operating range is to remain below 3.8 cycles of concentration corresponding to the blowdown flow rate ranging from 0.90 mgd to 1.84 mgd. However, it should be noted that per the TCEQ permit that the Outfall 001 maximum daily average flow rate is to be less than 0.92 mgd and the maximum daily maximum flow rate is to be less than 1.84 mgd. The blowdown flow rate of 0.92 mgd corresponds to the cooling towers operating at 3.77 cycles of concentration. Therefore, the ideal operating point is at 3.8 cycles of concentration that corresponds to the blowdown flow

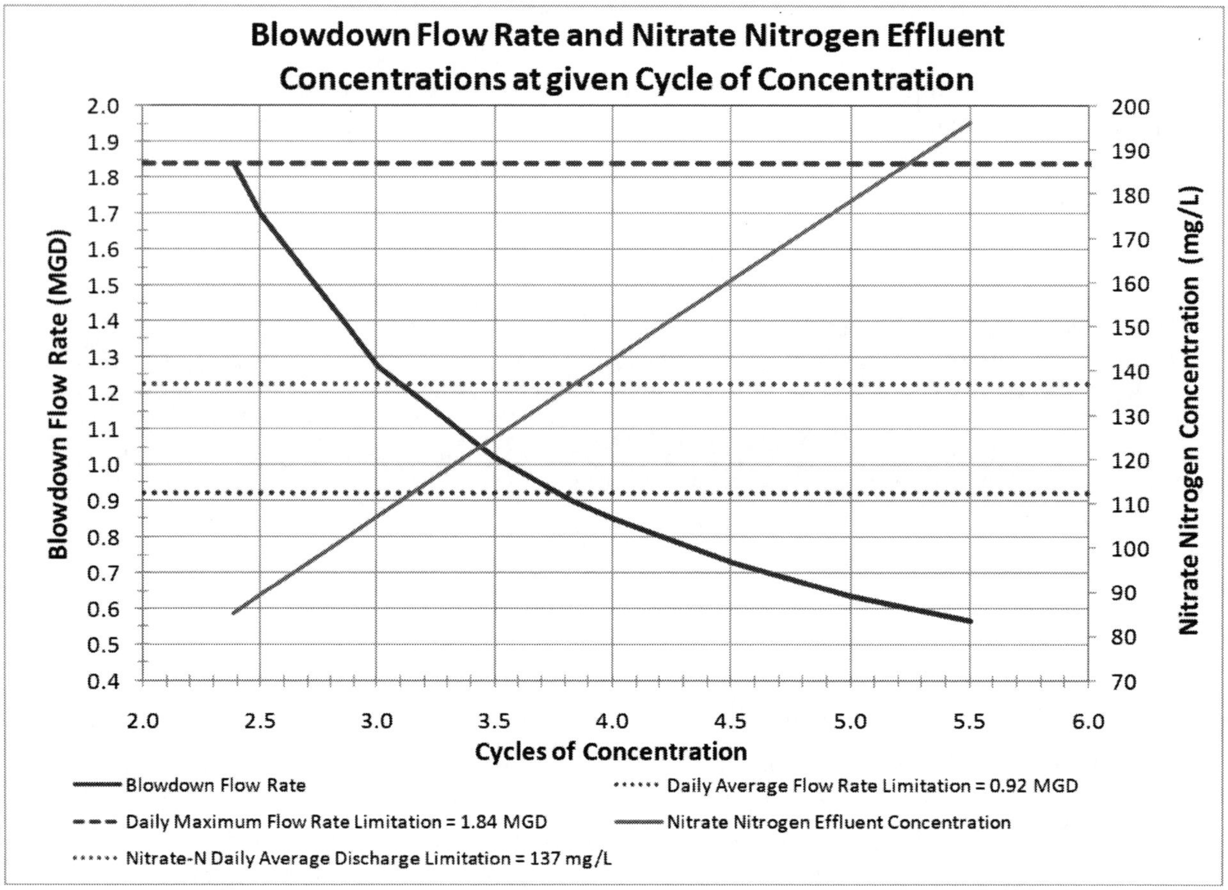

Figure 7: Blowdown Flow Rate and Nitrate-N versus Cycles of Concentration

rate of 0.90 mgd. The blowdown flow rate may be increased periodically to as high as 1.84 mgd (daily maximum flow rate permit limitation), if there are high influent concentrations of nitrate-N ranging from 36 mg/L to 121 mg/L (See **Figure 8** and **Table 17**).

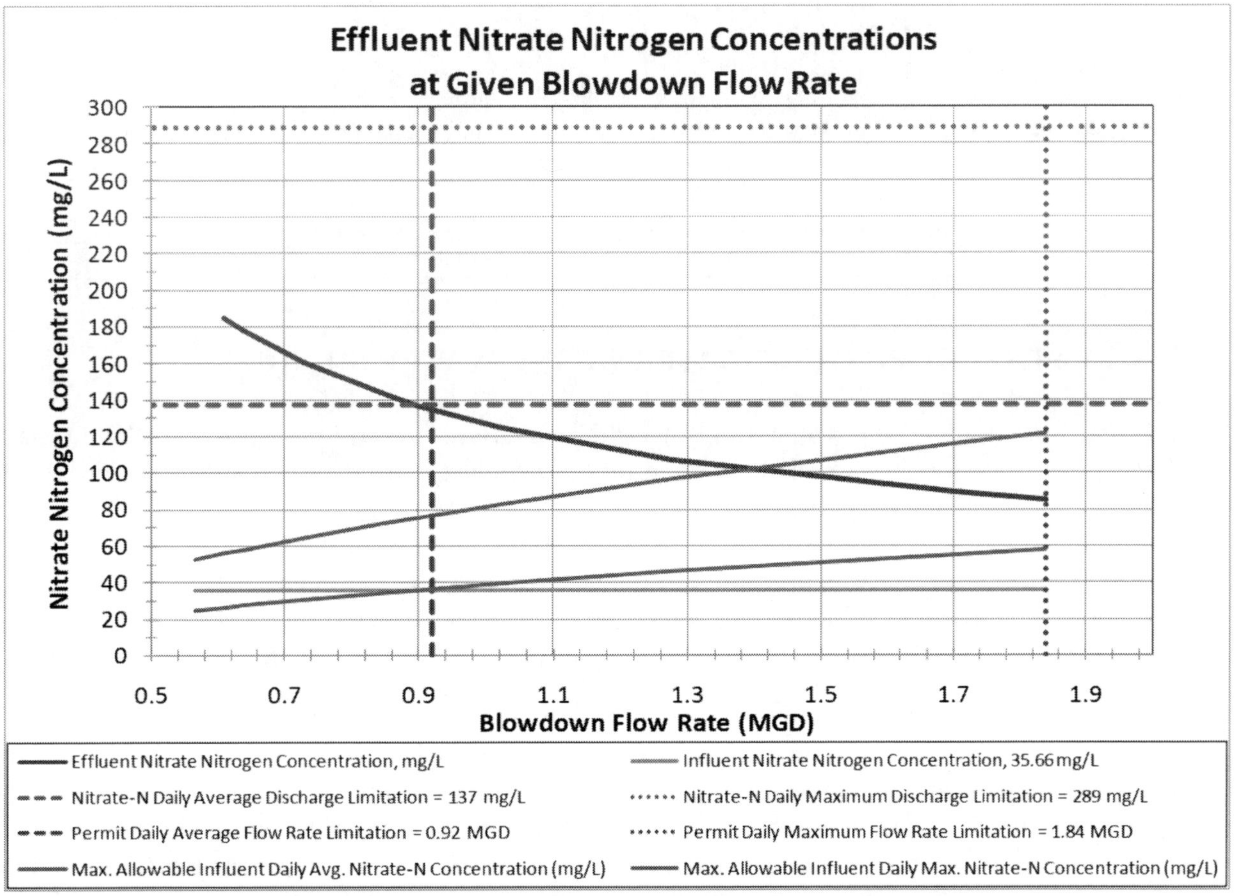

Figure 8 Effluent Nitrate-N Concentrations versus Blowdown Flow Rate

Table 17: Cooling Tower System Model Output

Mean Plus 1 STDEV Nitrate-N Conc. in Cooling Tower Influent (mg/L)	Blowdown Flow Rate (mgd)	Predicted Nitrate-N Conc. at Outfall 001 (mg/L)	Cycles of Conc.	Maximum Permitted Influent Daily Avg. Nitrate-N Conc. (mg/L)	Maximum Permitted Influent Daily Max. Nitrate-N Conc.(mg/L)
35.66	1.84	85	2.4	57	121
35.66	1.70	89	2.5	55	115
35.66	1.27	107	3.0	46	96
35.66	1.02	125	3.5	39	83
35.66	0.90	137	3.8	36	75
35.66	0.85	143	4.0	34	72
35.66	0.73	161	4.5	30	64

Table 17: Cooling Tower System Model Output

Mean Plus 1 STDEV Nitrate-N Conc. in Cooling Tower Influent (mg/L)	Blowdown Flow Rate (mgd)	Predicted Nitrate-N Conc. at Outfall 001 (mg/L)	Cycles of Conc.	Maximum Permitted Influent Daily Avg. Nitrate-N Conc. (mg/L)	Maximum Permitted Influent Daily Max. Nitrate-N Conc.(mg/L)
35.66	0.64	178	5.0	27	58
35.66	0.61	185	5.2	26	56
35.66	0.57	196	5.5	25	53

EVALUATION OF POSSIBLE ALTERNATIVE WATER SOURCES

To determine if a back-up water supply would be available should the McAllen WWTP be determined to be inadequate for long-term reliable water quality or to be able to avoid using the reclaimed water source during periods of process upsets at the McAllen WWTP, CDM Smith investigated potential alternative supplies in the vicinity of the Calpine Hidalgo Energy Center (HEC). The analysis presented in this section provides a general summary of alternative water sources that may be available to the plant based on the literature review and other desktop evaluations; and other potential sources of reclaimed water.

The State of Texas has established a state water planning program to develop water supplies and ensure that plans are in place to meet the State's water needs into the future. The State of Texas is comprised of 16 different planning groups that are responsible for developing a 50-year regional water plan every five years to evaluate the water supplies and water demands in the region and to identify potential water management strategies to meet projected water needs. HEC is located in Region M, a region that extends from Maverick County to Cameron County following the Rio Grande River and Texas/Mexico border. Within the most recent Region M water plan prepared in 2011, several regional alternative water supply options were identified. These include:

- Expand existing Groundwater wells (Gulf Coast Aquifer); and
- Purchase Additional Surface Water (Amistad-Falcon Reservoir System water right or contract),
- Reclaimed Water.

These options were reviewed to evaluate relevant alternative water supply options for Calpine during conditions where the water quality of the reclaimed water supply from the City of McAllen is not adequate or during potential drought conditions. Below is a discussion of each of these additional sources.

Expanded Use of Groundwater

There are three issues that directly affect the expanded use of groundwater, which include:

- Regulatory Framework;
- Groundwater Availability; and
- Groundwater Quality.

Regulatory Framework

For many parts of Texas, locally controlled groundwater conservation districts (GCDs) provide the regulatory framework which governs the use and availability of groundwater. The HEC is located in southern Hidalgo County, which is partially covered by three groundwater conservation districts (GCDs):

- Brush Country GCD (northwestern corner of county),
- Kennedy County GCD (remainder of northern Hidalgo County), and
- Red Sands GCD (small area in central portion of county).

However, as shown on **Figure 9**, the area surrounding the HEC is not within the jurisdictional boundary of any GCD with the authority to regulate the use of groundwater as authorized in Chapter 36 of the Texas Water Code. Therefore, the use of groundwater in this area is governed strictly by the rule of capture. This means that the owner of a parcel of land, regardless of size, may pump to the surface as much groundwater as physically possible, as long as the water is being put to a beneficial use (not wasted) and there is no malicious intent to harm neighboring land owners (see Texas Water Code §36.002 for additional description).

For non-public water supply wells (as would be the case for a water supply well drilled and completed for the HEC), the only regulatory/permitting requirement in this area is for the well driller to file drilling reports with the Texas Department of Licensing and Regulation. If desalination of brackish groundwater were to be considered, there would also be permitting issues regarding concentrate disposal that would need to be evaluated.

The HEC is underlain by the Gulf Coast Aquifer; both the Chicot and Evangeline aquifers make up the majority of what is designated as the Gulf Coast Aquifer as shown in **Figure 10**. Based on the 2011 Rio Grande Regional Water Plan and the 2012 State Water Plan, there is an abundance of groundwater available from the Gulf Coast Aquifer (specifically the Evangeline Aquifer) in Hidalgo County and groundwater availability on a sustainable basis from the Gulf Coast Aquifer in Hidalgo County is as presented in **Table 18**.

Calpine currently has access to groundwater at another facility that they operate nearby, called Magic Valley, located several miles to the southeast of the HEC. Based on discussions with the Magic Valley maintenance manager, these groundwater wells can produce up to 600 gpm of water and the water supply is treated with reverse osmosis to provide steam generation for the boilers. However, if large volumes of water are used from the wells, Magic Valley staff noted that it is possible to stress the wells. It is expected that similar productivity could be obtained at HEC should a groundwater source be considered.

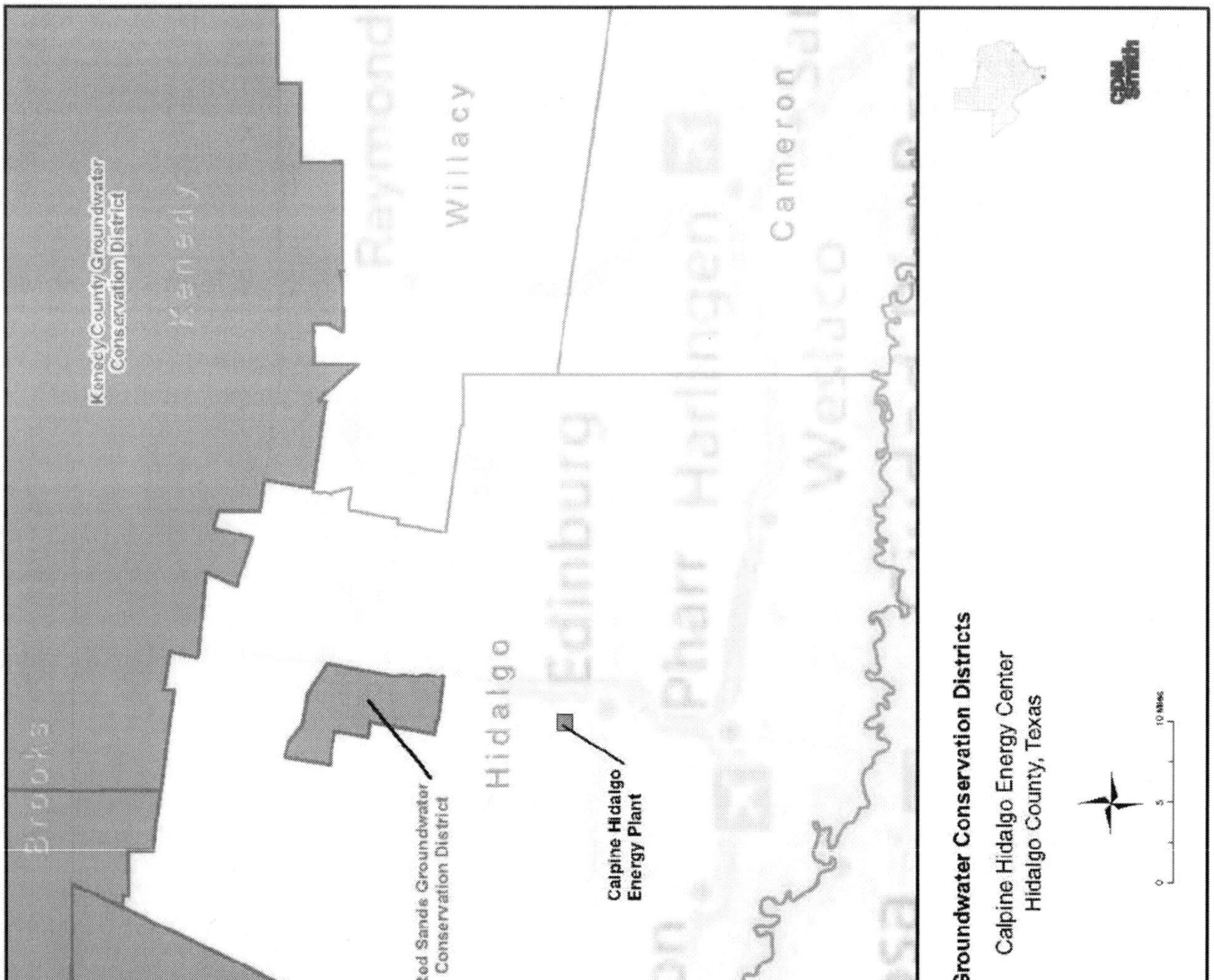

Figure 9 Groundwater Conservation Districts near HEC

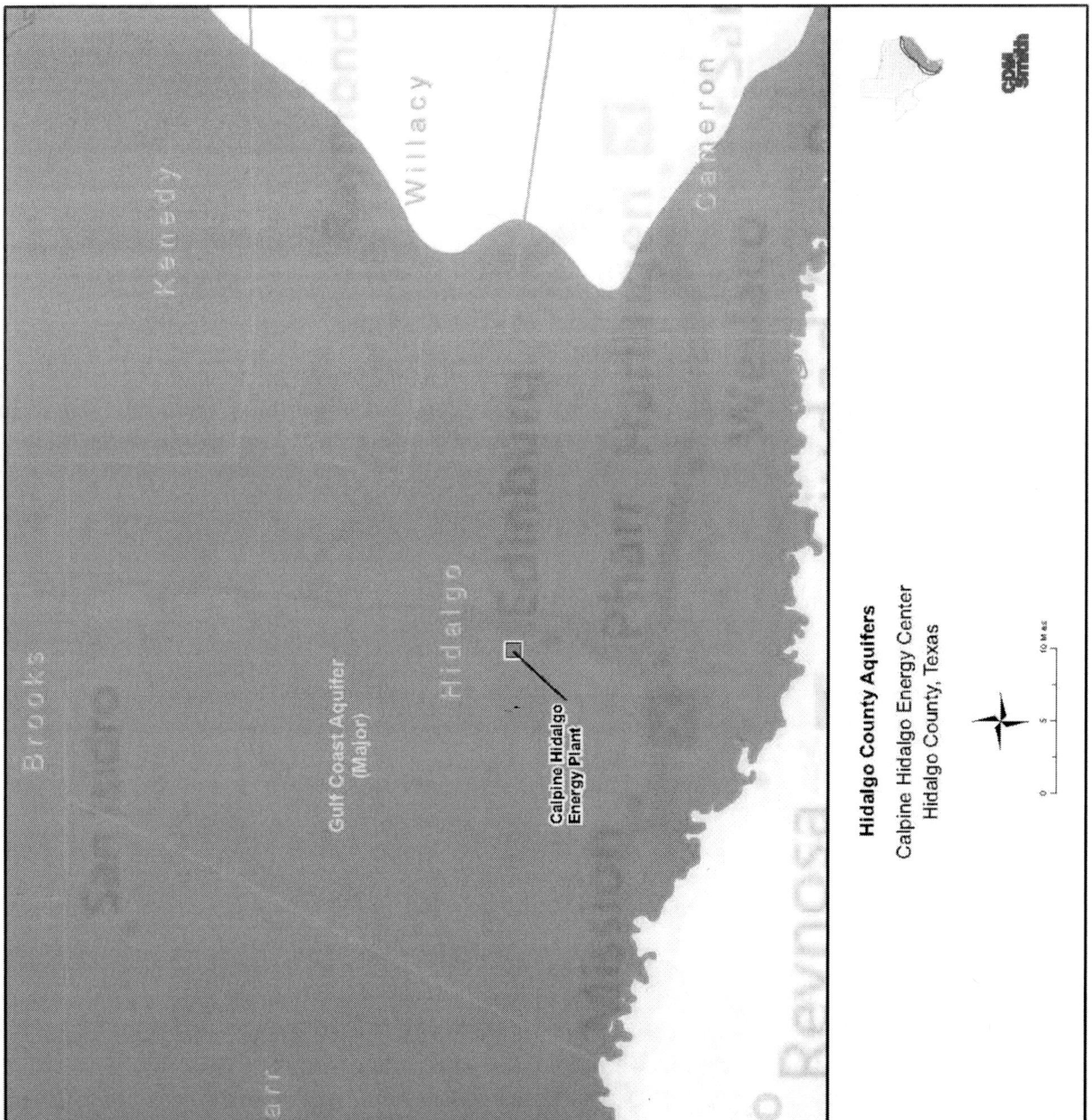

Figure 10 Hidalgo Energy Center Aquifers

Table 18:Groundwater availability from the Gulf Coast Aquifer in Hidalgo County by Basin (acre-ft/year)

Basin	2010	2020	2030	2040	2050	2060
Nueces-Rio Grande	47,277	47,277	47,277	47,277	47,277	47,277
Rio Grande	5,253	5,253	5,253	5,253	5,253	5,253
Total	52,530	52,530	52,530	52,530	52,530	52,530
Note: excerpted from Table 3.11 in 2011 Rio Grande Regional Water Plan						

Groundwater Quality

There have been several investigations regarding groundwater quantity and quality in the Gulf Coast Aquifer of the Lower Rio Grande over the last 50 years. In addition, the Magic Valley Groundwater wells also provide an indication as to expected groundwater quality. The following sections provide an overview of these topics.

The results from Chowdhury and Mace (2007) are the latest and most comprehensive investigation regarding groundwater quantity and quality in the Gulf Coast Aquifer of the Lower Rio Grande. Chowdhury and Mace (2007) conclude that although there are significant quantities of groundwater in the Gulf Coast Aquifer in areas where sands are dominant, much of this groundwater resource is not directly usable due to its moderate to high salinity. McCoy (1990) in a comparison of groundwater quality with surface water from the Rio Grande did note that there are isolated areas with alluvial and deltaic deposits and shallow sediments covering small areas in southeastern Hidalgo, southwestern Cameron, and north central Hidalgo counties that contain fresh groundwater, indicating effects of local recharge. However, based on an examination of McCoy (1990), this area does not extend to the area of the HEC.

According to Chowdhury and Mace (2007), groundwater in the Lower Rio Grande Valley varies widely in composition from fresh to brine (198 to 37,752 milligrams per liter). In this study, the Evangeline Aquifer has 27 percent fresh, 65 percent slightly saline, and 6 percent moderately saline waters, however, statistical comparisons specific to Hidalgo County were not presented in this study. A visual examination of spatially presented water quality data presented by Chowdhury and Mace (2007) in the area of the HEC does indicate that the vast majority of samples used in this study are for groundwater with greater than 1,000 milligrams per liter with only 3-4 wells observed in the general area measuring less than 1,000 milligrams per liter.

Because quality of the water is quite poor (with respect to TDS), expansion of groundwater use is not expected to be desirable for the facility. In fact, it is our understanding that the facility would like to limit groundwater use entirely.

Local Groundwater Options

As noted, Calpine currently has access to groundwater on the Magic Valley Site, located several miles to the southeast of the HEC. The quality of these wells is quite poor and requires treatment with reverse osmosis in order to use the groundwater for boiler make-up water.

Freshwater zones may be available at HEC, but field investigation would be required to identify these areas.

Surface Water Alternatives

Another potential source of water for Calpine is to purchase additional water by acquiring additional water rights or obtaining more water via contract with another supplier. This may be of particular interest for the facility because the groundwater that is currently available is quite poor, with high TDS and sulfate.

The vast majority of water supply available to water users in the Lower Rio Grande is derived from surface water in storage in the Amistad-Falcon Reservoirs located on the Lower Rio Grande. However, the management and release to water rights holders of surface water stored in the Amistad-Falcon Reservoirs is the responsibility of the Rio Grande Water Master. Because all available surface water in storage in Amistad-Flacon Reservoirs and all run-of-the river rights have been fully adjudicated (allocated), the only process available to acquire additional water from the Rio Grande is through the lease or purchase of water rights from an existing water rights holder, such as a local irrigation district or individual water rights holder.

Two primary local water supply options were evaluated for the HEC:

- Obtain water from local irrigation district
- Obtain water from Magic Valley Irrigation District

CDM Smith contacted Hidalgo Irrigation District 13 and Santa Cruz District 15, both in the vicinity of HEC, to determine their ability to serve the facility. Based on information available from the TCEQ Water Utilities Viewer, the HEC is within the service area of the Hidalgo Irrigation District 13 (HID13) and the district was responsible for providing water to the area in the past. However, although they have lines in the area, they no longer provide water because individuals in the area no longer wanted to pay for the district services. Based on conversations with HEC and Magic Valley facility staff, CDM Smith understands HID13 canals are in the vicinity of the HEC and it may be possible to conduct a more detailed evaluation to evaluate feasibility of connecting to the HID13 and nearby canals.

CDM Smith understands that the reclaimed water pipeline from the City of McAllen runs in very close proximity to the Calpine Magic Valley facility. Therefore, it may be possible to use the surface water supply and/or storage pond at Magic Valley to provide a backup water supply to HEC. This would provide flexibility to the HEC to obtain the multiple sources of water available at Magic Valley (groundwater, irrigation district and City of Edinburg potable water supplies). In addition, this option would provide access to a large pond on-site at Magic Valley with up to 4 days supply of water which could be used to mix the various sources of water to ensure adequate water quality for the HEC.

SUMMARY OF FINDINGS

Based on a review of historical data from the McAllen WWTP before and after the completion of its major upgrade project and an evaluation of the facility and its current operations, it appears that the effluent quality received at HEC since February of 2012 is reflective of the effluent quality that can be expected for the foreseeable future. The ammonia-N concentrations in the McAllen WWTP effluent since February of 2012 have been consistently below 0.24 mg/L. Based on these ammonia-N concentrations, HEC should be able to operate their cooling tower system with their current cycles of concentration, while still meeting and maintaining their permit limits. The nitrate-N concentrations are of greater concern, given the current operation of the cooling tower system. Limiting the cycles of concentration, particularly when the nitrate-N concentrations in the storage pond are high, would still allow HEC to meet and maintain their permit limits.

With regard to alternate water sources, it was determined that groundwater would not be a feasible alternative to supply the quantity and quality of water that would be required for the cooling tower operation. The Hidalgo Irrigation District 13 has water available for purchase, which may meet the needs of HEC, but further evaluation would be required to determine if this is the most economical solution. Lastly, the McAllen WWTP effluent line runs within a short distance of the Calpine Magic Valley Generating Station, which has existing access to multiple water sources. It may be possible to utilize those sources and receive water from Magic Valley, particularly as a supplement to the existing McAllen WWTP source.

RECOMMENDATIONS

Although the reclaimed water that HEC receives from the McAllen WWTP is forecasted to have very low concentrations of ammonia-N that would enable HEC to maintain compliance with permit limits under normal operating conditions, CDM Smith has prepared a series of recommendations for possible action that would increase the reliability of plant operations, provide additional assurance of meeting permit limits, and potentially allow operating the cooling tower at higher cycles of concentration thereby reducing water demand.

Monitor Daily Ammonia-N and Nitrate-N Concentrations

Calpine HEC has historically not monitored the influent nitrate-N concentrations in the reclaimed water that is provided by the McAllen WWTP. It is recommended that Calpine HEC maintain a monitoring program so that operations can be adjusted in the event of high influent nitrate-N concentrations. Since June of 2012, HEC has been monitoring ammonia and nitrate-N on a weekly basis and unless an alternate water source is provided for the cooling tower make-up water, a more frequent monitoring program would not be necessary. Continued routine monitoring of ammonia and nitrate-N, as well as regular communication with the McAllen WWTP, to ensure that the water received is of consistent quality, will allow HEC to adjust the cooling tower operations to avoid ammonia and nitrate-N exceedances when possible.

If HEC were to expand its monitoring practice, the most robust monitoring option would be to install a probe in the pond, near the discharge pipe from the McAllen WWTP, with a transducer that can be periodically checked or even report back to the HEC control room. There are commercially available probes that will specifically measure nitrate-N. A handheld probe would be less expensive and allow for instantaneous readings, which could be used for routine checking of nitrate-N concentrations in the pond. The least expensive options that would allow for regular monitoring would be a lab spectrophotometer or handheld colorimeter, which would require more labor, but would still give relatively instantaneous feedback. Routinely sampling the makeup water pond to monitor the nitrate-N and ammonia-N concentrations will also allow HEC to monitor the cooling tower performance (cycles of concentration) and identify any above average concentrations that may be coming from the McAllen WWTP.

Proposed Cooling Tower Operation

The recommended operating point for the cooling tower system based on the nitrate-N concentration is at 3.8 cycles of concentration, which corresponds to the blowdown flow rate of 0.90 mgd. The blowdown flow rate may be increased periodically to as high as 1.84 mgd (daily maximum flow rate permit limitation) if there are high influent concentrations of nitrate-N ranging from 36 mg/L to 121 mg/L. CDM Smith understands that since initiating this study, HEC has been operating at 3.5 cycles of concentration, which based on historical ammonia-N and nitrate-N concentrations will consistently maintain compliance with permit limits.

Increase Makeup Water Storage Capacity

The Calpine HEC currently has approximately 1.2 million gallons of stored makeup water available for use in the cooling tower system. This provides about a half a day of storage for cooling tower makeup water. Given the small storage pond, HEC is at risk to exceed the permit discharge limitations at Outfall 001 if there is an upset at McAllen WWTP that leads to higher concentrations in their effluent.

HEC currently receives on average 3 mgd of makeup water from the McAllen WWTP. It is recommended that HEC increase the makeup water storage volume to provide two or more days of storage, which is approximately 6 million gallons. Along with the increase in storage volume, a more continuous monitoring program should be implemented, such as an online probe or regular grab samples. This will allow HEC to make the necessary adjustments if there is an upset at the McAllen WWTP, or to shut off flow from the McAllen WWTP and utilize the stored water while the McAllen WWTP returns to normal performance and compliant effluent. The data analysis performed indicates that with the discharge effluent quality of the new McAllen plant, HEC will be able to meet the conditions of its permit if limiting the cooling tower cycles of concentration to a maximum of three. At higher cycles of concentration, there is a risk at times of exceeding the permit limitations for nitrate-N given the current conditions at both the McAllen WWTP and HEC.

Coordinate with McAllen WWTP to Reduce Nitrate-N Concentrations

Because McAllen does not currently have limitations in their TPDES permit for nitrate, there is limited current incentive for this facility to do anything to lower nitrate-N in their wastewater effluent. However, as discussed, there are operational modifications and possible additional treatment steps that could be put in place at the McAllen WWTP to reduce the nitrate-N in the water sent to HEC. Calpine may be able to negotiate either an increase in water rates or partnering in capital expenditures or both to provide the necessary incentive for McAllen to improve the quality of the effluent pumped to HEC. These alternatives are not particularly ideal, as they would likely require significant costs for Calpine and would continue to require HEC to be dependent on the McAllen WWTP operations for continued delivery of quality effluent. As it stands, the quality of effluent is currently adequate for HEC to operate its cooling tower without violating its permit, but these additional treatment steps would potentially lower the nitrate-N concentrations in the effluent received even more and in theory, provide consistently high quality water. Coordination with McAllen wastewater staff would be required to determine if this is a feasible alternative and what scope of modifications or improvements would be required. Alternatives would include operation of their existing plant with a pre-anoxic zone either with or without internal recycle, depending on the level of denitrification required, or installation of denitrification filters on the clarified effluent, which would only be required for the 3 mgd that would be pumped to HEC. Important factors to consider would be the quality of the effluent and the consistency of that quality relative to both ammonia-N and nitrate-N after denitrification. For sewage treatment effluent, there is a need to consider temporal variation in quality. It is important to monitor makeup water used. By linking a normal usage pattern, any exceedances in ammonia-N and/or nitrate-N can be identified easily.

Obtain Water from Alternative Source

CDM Smith understands that the reclaimed water pipeline from the City of McAllen runs in very close proximity to the Calpine Magic Valley facility. Therefore, it may be possible to use the surface water supply and/or storage pond at Magic Valley to provide a backup water supply to HEC. This would provide flexibility to the HEC to obtain the multiple sources of water available at Magic Valley (groundwater, irrigation district and City of Edinburg potable water supplies). In addition, this option would provide access to a large pond on-site at Magic Valley with up to 4 days supply of water which could be used to mix the various sources of water to ensure adequate water quality for the HEC. It may also be possible to conduct a more detailed evaluation to evaluate feasibility of connecting to the HID13 and nearby canals.

CONCLUSION

In conclusion, given the data received from the McAllen WWTP since the improvements were made in early 2012, HEC should be able to continue to operate its cooling tower while meeting the permit requirements for ammonia and nitrate-N, given the recommended cycles of concentration outlined in this report. Additional steps to increase the operational flexibility of the facility and provide opportunities for HEC to overcome potential upsets from the McAllen WWTP should be considered with one or a combination of the following recommended actions: increasing onsite storage capacity, working with McAllen to improve nitrate-N removal, or use of other sources of water with low concentrations of ammonia-N and nitrate-N. However, an engineering and economic evaluation is necessary to determine the feasibility and benefits associated with these actions.

REFERENCES

Chowdhury, A., and Mace, R. E. 2007, Groundwater resource evaluation and availability model of the Gulf Coast Aquifer in the Lower Rio Grande Valley of Texas: Texas Water Development Board, Report 368, 130 p.

McCoy, T. W., 1990, Evaluation of Ground-Water Resources in the Lower Rio Grande Valley, Texas: Texas Water Development Board, Report 316, 48 p.

Get It Done! Using Performance Contracting To Implement Energy-Related Improvements

Authors: Rich Atoulikian[1], Clark Wiedetz[2], Tom Kochaba[1], Kim Kennedy[1], Bob Huffmyer[1]

[1]HDR Incorporated
[2]Siemens Energy

I. ABSTRACT

Due to our current economic situation, there is an increased focus on fiscal responsibility and cost reductions for utilities, coupled with a higher and ever increasing level of environmental quality standards which they must meet. This puts a brighter spotlight on improving operational efficiencies, which is driving utilities to optimize the performance and operation of their facilities. Doing so not only saves energy and mitigates rate increases to customers, but also makes them more environmentally sustainable.

Utilities can reduce their power bill by implementing energy efficiency related improvements while also looking for opportunities to supplement their needs through the implementation of a renewable energy strategy. Energy audits are the first step in identifying potential savings. High level energy audits can identify and rank potential savings, and investment grade audits document and quantify those savings. These audits can also identify savings of other resources, such as process chemicals, solids handling and disposal, non-energy related utilities, or parasitic water loss.

The following are typically evaluated as part of the energy audit process:

- Current energy use, patterns of use, billing rate structures and opportunities to conserve. This also includes development of an energy baseline and benchmarking.
- Treatment process configuration and operation, including identifying operational modifications or capital upgrades that can be made to reduce energy or lower operating costs.
- Building envelope and support facilities, such as HVAC systems and lighting.
- Alternative energy related to biosolids, solar, wind, small hydro, and geothermal; including onsite production and procurement from third parties.

Unfortunately, most utilities have difficulty using traditional sources of financing and conventional project delivery models to make the initial investment required to implement these improvements. This is where performance contracting can be used. Performance contracting uses the associated savings from the improvements to pay off the initial capital investment required to effect the improvements needed to realize the savings. Therefore, a large initial capital investment on the part of the utility is not required; rather, guaranteed long-term operating savings are leveraged to secure financing provided by an energy service provider or other third party. Most times, the risk of achieving these savings resides with the energy service provider, although this requires a properly defined baseline and effective measurement and verification.

Once the potential energy savings opportunities are identified, they are prioritized, and the ones which are most attractive are included in the performance contract. Typically, those with the best return on investment or payback are the ones that are included. Similar to an annuity, savings from initial improvements can be used to fund the implementation of subsequent improvements. Using this model, the initial cost is annualized, and paid off with guaranteed savings generated by the improvements.

This paper will describe the procedure used to identify and develop the cost savings to fund a performance contract at a wastewater treatment plant. It will highlight some of the key areas of opportunity that are often explored, including the unique considerations that must be made to maximize the benefit of the project to the owner. Renewable generation and both process and also non-process energy efficiency and operational improvements will be discussed with experience from case studies interwoven throughout the discussion.

In summary, performance contracting is a viable alternative approach to funding and implementing improvements at municipal water and wastewater facilities. A successful program must go beyond the initial implementation of audit results, and include ongoing measurement and verification, as well as continuous improvement practices. Measurement and verification must be carefully planned and incorporated into the facility, as this is the basis for the guaranteed savings.

II. INTRODUCTION/CHANGE DRIVERS

According to the United States Environmental Protection Agency (US EPA) and market research, combined drinking water and wastewater systems account for approximately 3 to 4 percent of energy use in the United States, adding over 45 million tons of greenhouse gases annually. Further, drinking water and wastewater plants are typically the largest energy consumers of municipal governments, accounting for 30 to 40 percent of total energy consumed. Energy as a percent of operating costs for drinking water systems can also reach as high as 40 percent and is expected to increase 20 percent in the next 15 years due to population growth and tightening drinking water and wastewater discharge National Pollutant Discharge Elimination System (NPDES) regulations.

Wastewater aeration blowers, sludge processing and pumping the largest consumers of energy, followed by lighting and buildings. This is shown on the following Figure 1, which is taken from the Water Environment Federation "Energy Conservation in Water and Wastewater Facilities," Manual of Practice 32. However, each plant has its own unique energy consumption profile, which can be developed then used during the energy audit to focus on areas of greatest consumption. For potable water systems, pumping is the largest component (raw water pumping, intermediate in-plant pumping, and finished water pumping). Processes associated with water treatment generally consume much less energy than the pumping described above. Lighting and building energy consumption can also be significant on the water side.

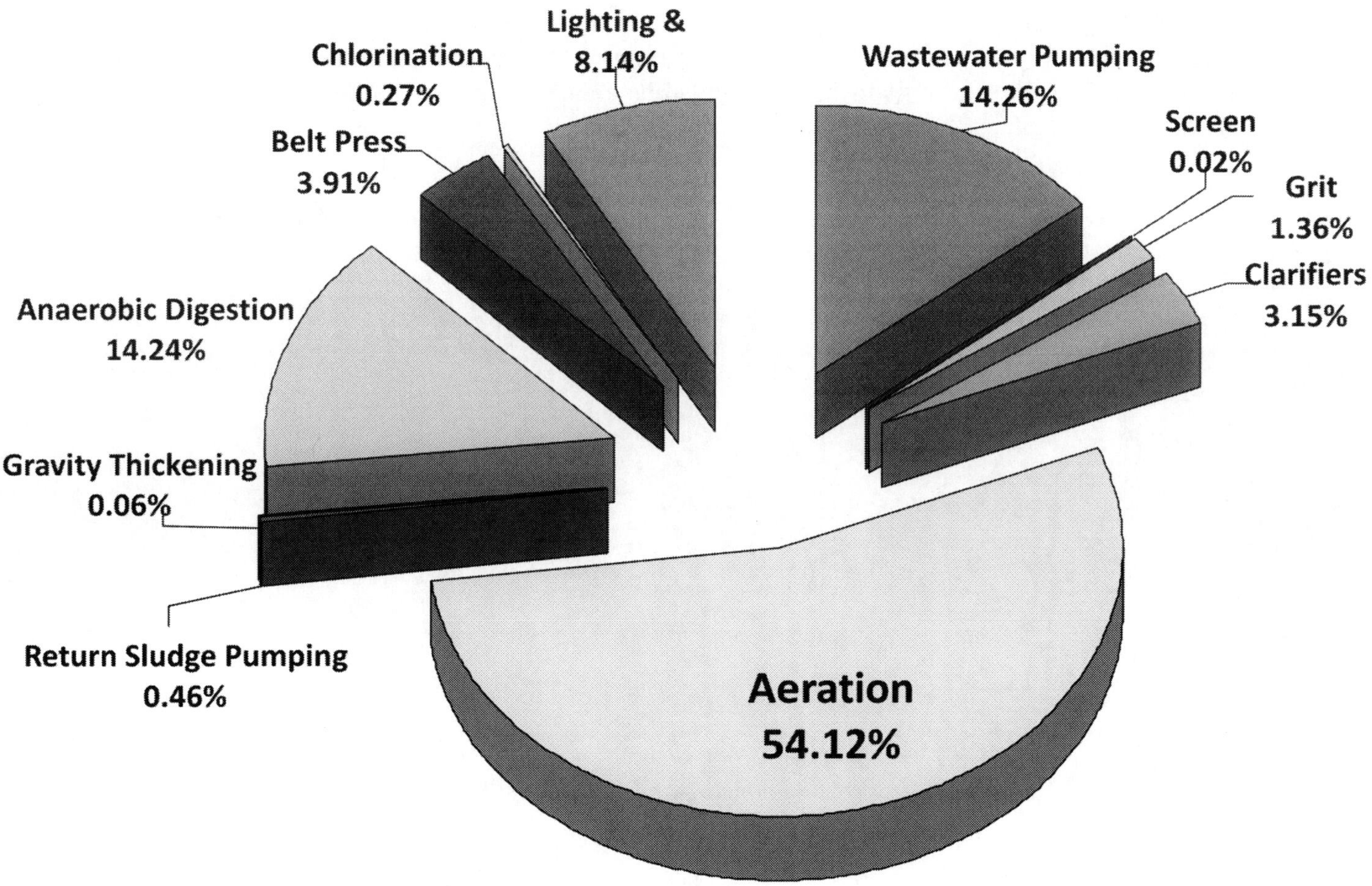

Figure 1 – Example Energy Consumption Delineation by Process/Function (WEF)

The recession has hit municipal revenues hard, resulting both from losses in income taxes as well as real estate taxes due to the housing market collapse and vacant properties. Vacant properties also do not consume water nor produce wastewater, further exacerbating the revenue problem. In addition, there has been a necessary focus in some locales on meeting US EPA Wastewater Consent Decrees, which many times necessitate annual rate increase in the 10 to 20 percent range (or more.) Combined, these market drivers have forced utilities to better optimize and prioritize expenditures and use an integrated, balanced planning approach.

These challenging economic times are also driving utilities to look at alternative means of energy generation through "renewable" sources. Renewable energy includes using digester gas resulting from anaerobic digestion for either heat, electricity generation or even to fuel vehicles. In addition, sludge pretreatment processes upstream of anaerobic digesters show promise in producing more biogas and reducing overall solids production. Utilities are also looking at the incorporation of solar, wind or hydropower energy generation into their facilities. To emphasize the desired change in philosophy, the Water Environment Federation is now moving towards re-branding what had formerly been called "wastewater treatment" plants to "resource recovery" plants since they can provide the ingredients for energy production.

As utilities reduce their energy footprint costs, alternative contracting strategies become more interesting, especially with the incorporation of renewable energy. Partnering with a private entity can increase the attractiveness of renewable energy, because the private side can take advantage of tax breaks not available to the public sector.

III. ENERGY "MODELS" / COMPARISON

Implementation of identified energy savings opportunities can follow different models, depending on what is most appropriate for the utility owner. As the utility evaluates which model is most appropriate for them, considerations such as financing, risk management, internal resource availability to manage consultants/contractors/energy service providers are all considerations. Of the models discussed, the energy service providers require the least amount of internal utility resources to manage.

The **first model** is traditional delivery, which consists of the utility performing or hiring a consultant to conduct an energy audit, followed by quantifying and prioritizing opportunities in an Energy Savings Plan. A manageable number of the highest priority savings would then be identified for implementation. Depending on size, generally a reasonable number of projects to move forward to implementation initially may range from three to six. These projects would then move forward to detailed design, followed by bidding to construction contractors, finally construction, commissioning and operation. Alternative delivery models, such as design-build, construction management at risk, etc. can also be used to deliver these projects.

The **second model** is energy performance contracting. Energy performance contracting is a turnkey model, in which one entity is hired to study, design, and implement the projects included in the performance contract. This starts with the initial energy audit and Energy Savings Plan, then implementation, and continues through long-term monitoring over the performance period. With respect to energy, this model often includes improvements related to energy efficiency, renewable energy and other power related improvements. Due to the nature of this work, where the Energy Service Company (ESCO) provides the up-front capital and guarantees, a formal contract between the ESCO and utility is entered into. This contract includes the necessary guarantees since the savings produced by the projects are expected to be sufficient to finance the full cost of the project. The ESCO also provides, through a third-party, long term financing of the project. There are a number of different financing vehicles available. The performance contract also defines the roles/responsibilities of each party, particularly related to the operation of the facility. In addition, being a private entity, the ESCO is eligible for tax credits related to renewable energy that municipal clients cannot receive.

IV. IDENTIFYING ENERGY REDUCTION OPPORTUNITIES

The identification and subsequent prioritization of energy reduction and renewable energy opportunities requires a systematic approach. This starts with gaining an understanding of how the utility operates and performs, both with respect to their physical facilities as well as their overall management strategy, and how it is executed across the organization. To gain this understanding, both data/information reviews and client interactions/interviews are equally important. Changing the way a utility functions and implements energy improvement recommendations must include both technical and cultural considerations. Recognizing the importance of each of these components will enable a utility to take the first step in its journey of reducing its energy costs.

If energy projects are delivered either traditionally or through an ESCO, then the next step is development of an Energy Audit followed by an Energy Savings Plan. The Energy Savings Plan would outline prioritized recommendations, and would be followed by implementation of the recommended conceptual designs to enable initial capital costs and operational/maintenance costs to be developed. This will enable the utility to quantify the cost benefits of the projects. Non-cost considerations, such as public perception may also be a factor. For those projects that the utility decides to move forward, detailed design is conducted followed by bidding and construction when traditional delivery is utilized.

If a performance contracting approach is utilized, then the findings from the Energy Audit and Energy Savings Plan could proceed to design and implementation. A performance contracting approach offers the utility a more expedited means of implementation, as well as an assignment of the risk associated with initial investment and realized savings to a contractor. Utility staff involvement in the design process is important in either case.

For either delivery mechanism, the next step after implementation of the recommendations is monitoring, verification and control processes. Monitoring, verification and control enables a utility to confirm whether or not the proposed savings are actually being achieved. It also can provide visibility into the process to determine if additional enhancements (i.e. continuous improvement) may be possible to enable the utility to sustain and further improve their energy savings program moving forward, looking for incremental changes to further reduce energy consumption.

Monitoring, verification and control are important whenever performing energy improvement work, whether using traditional delivery or performance contracting. These items are particularly important with a performance contracting approach since the savings are used to pay off the capital investment. If the projected savings are not being realized, then further investigation is required to determine if input parameters have changed, if operation/maintenance reasons are causing this, or if there are other root cause issues resulting in performance goals not being met. It is important to include all important input parameters. These are all elements that must be prescribed in the performance contract.

This overall process, from audit through continuous improvement is shown in Figure 2, which is prepared based on information included in the aforementioned WEF Manual of Practice 32.

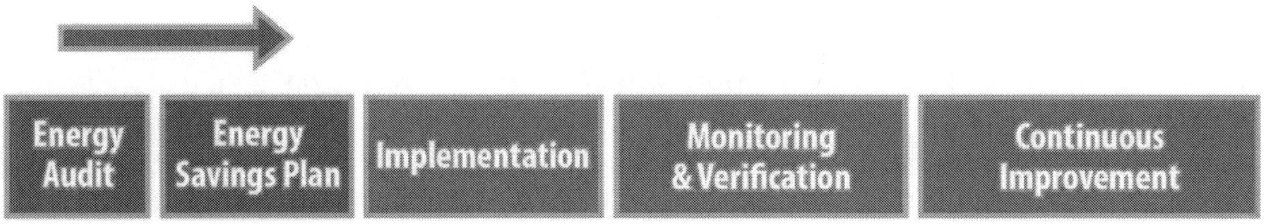

Figure 2 - Energy Reduction Process

A typical approach that is followed when performing and Energy Audit and preparing an Energy Savings Plan is shown in Figure 3. This starts with data collection and a review of existing information, followed by development of an energy baseline, opportunity identification and development, project prioritization, and implementation. Each step is discussed below:

Figure 3 – Energy Reduction Opportunity Evaluation Process

Data Gathering, Interviews and Site Visits

The existing information to be reviewed includes energy bills that show both unit consumption, demand, and other charges; electrical sub-metering data where it exists; monthly operating reports that show in influent and effluent (or raw and finished water) quality; pumping records; precipitation records; other monthly operating information from the SCADA system; existing record drawings; operations and maintenance manuals; service area characteristics, NPDES permit requirements, etc. Often there is a single individual in an organization that can provide access to this information – the challenge in any energy audit project is to identify and make contact with that individual.

Where available, process models using software such as BioWin™ and GPS-X® can provide enlightening information on existing performance. These tools can also to identify and potential capital and operational enhancements, and compare different options for optimizing performance from an energy and carbon standpoint.

Once this existing information is reviewed, site visits and interviews are scheduled. The site visits enable a visualization of what had been reviewed in the paper/electronic documentation, to see if anything has been changed in the field and to provide the opportunity to interact with operations staff. Each of the facilities included in the energy audit should be visited. While at the sites, operations and maintenance staff on location should be interviewed to gain a greater understanding of system operation, rules of thumb that are frequently used, and other operating guidelines and cultural considerations beyond what can be identified in the pure data collection phase.

In addition to interviews with the operations and maintenance personnel located at the various sites, interviews should occur with other individuals across the organization. As noted above, this should include those on the front lines, as well as personnel from various Departments such as: Planning, Engineering, Construction, Finance, Operations and Maintenance, and Information Technology. Depending on the utility, meetings with other Departments may also be beneficial. Also, as these interactions occur, possible options for renewable energy, as well as smart grid and smart meter options should also be explored.

Multiple utility interactions are important for gathering information and communications. They enable the energy team to communicate the importance and organization motivators driving the utility's emphasis on reducing their energy footprint. They also enable communication of management's energy reduction objectives and encourage staff to engage in the process. Ideally, these interactions reinforce the message being conveyed from senior utility management. These interactions will also enable the "energy" team to introduce the project and team to the utility's organization. In doing this, they will describe the project and their approach, and solicit ideas that could potentially reduce energy consumption. This approach will provide a vehicle for the organization to bring forward their ideas that can reduce energy consumption, that, for whatever reason, have not been brought to light previously. It will also enable the "energy" team to build credibility with the utility's staff.

If done properly, taking these steps will enable the organizational change management process to begin. In other words, these face-to-face meetings create a common understanding and enable the "energy" team to begin to know the organization from various perspectives, and (more importantly) vice versa. These communications begin the process of enabling the organization to "buy into" the Energy Audit and Energy Savings Plan process. Multiple interactions will further enhance understanding and credibility in developing a common understanding of the importance of focusing on reducing energy consumption.

To supplement the one-on-one and small group interactions, workshops should also be held with key staff to use the "power of the group" to first "truth" then prioritize the concepts developed. Workshops should ideally be conducted with representatives across utility Departments. In support of each of these workshops, an in-depth analysis can be conducted of the data and resulting recommendations of energy savings measures. These workshops also help focus the various client entities on the overarching goals of this project. These interactions, both one-on-one and those in small groups were all part of the overall organizational change management process.

Develop Energy Baseline

As a utility embarks upon an energy reduction program, it is important to establish an energy baseline to better understand energy use across the utility's facilities. The energy baseline approach can also be used to gauge the utility's greenhouse gas (GHG) emissions. Ideally, the baseline is performed on a "facility" or "process" level, to enable specific and detailed results to be obtained. The baseline is developed through the analysis of existing data, ideally quantified to the sub-metering level.

The results of the energy baseline preparation will provide a framework to identify areas of greatest energy use, which may be those areas where the greatest potential savings can be found. This energy baseline data can also be used to perform comparative analyses between various facilities to identify higher priority opportunities. The baseline provides a point of reference to measure the impacts of subsequent improvements against. This baseline can be readily modified to incorporate ongoing improvements and then reflect the energy consumption breakdown between the various processes.

Opportunity Formulation/Data and Opportunity Analysis

The next step is opportunity formulation, which includes both the identification and development of potential opportunities. In this step, the data obtained from the field visits, interviews and otherwise received is utilized as the basis for developing specific opportunities. This starts with identifying potential energy savings opportunities. As this process occurs, obvious, easy to implement energy savings are identified that can be implemented for very little capital cost. These are typically brought forward immediately for implementation. Often these improvements are related to lighting and/or HVAC improvements.

Projects and other opportunties that are identified to reduce energy consumption can generally be delineated into the following categories. Of these, the Capital projects category is the one which is most applicable to performance contracting delivery.

- Capital Projects, which require detailed design, followed by construction. These are generally of sufficient size that they are identified for the utility's Capital Improvements Program.

- Procedural/Programmatic Activities, such as development of Standard Operating Procedures, such as in the areas of use of high efficiency or premium motors, or modifying building thermostat set points.

- Maintenance Related Activities, such as installation of power monitoring to enable equipment operating in parallel to be compared, to see if any individual device may be operating out of standard performance, and consuming extra power (e.g. on a pump, due to a clogged line or broken impeller.)

The remaining opportunities are then analyzed at a high level to determine what would be required to implement them (from a capital and organizational standpoint) and well as what positive benefits they would achieve. This would then be followed by a prioritization process. The importance of prioritizing multiple project opportunities is necessary since typically resources are limited and only the most attractive opportunities should be brought forward initially for implementation. More about the prioritization process is described below.

On traditional delivery projects, development of an overall Energy Savings Plan that includes an implementation "path forward" is prepared. Moving through this process quickly enables energy savings opportunities to be brought forward and implemented sooner. On performance contracts, the same processes generally apply, however projects can move more quickly into design and implementation.

Electric or gas utility rebate programs exist which can offset the capital costs associated with implementing energy efficiency projects. These are identified early in the process, to enable utilities to optimize their use of such programs to fund some or all of their projects.

As part of this process, a further look at renewable energy is conducted, such as the use of wind turbines, solar power or small hydro by recovering excess energy from the hydraulic profile.

The amount of wind energy generation potential can be conducted by looking at previaling wind information. If there are any impediments, such as proximity to airport landing paths. Solar energy is typically worth considering and can be constructed on building roofs, tanks, etc. if they are structurally able to support them. However, solar's financial payback period is relatively long, even with rebates and other financial incentives. The shortcoming associated with solar and wind technologies is that the energy must be used as it is generated, unless complex storage facilities are provided. Another renewable energy source are micro turbines located in the liquid flow stream if 20 to 30 feet or more of head is available. However, these would need to be located downstream of major treatment processes where where the water quality is expected to be good.

In addition, some clients also welcome the positive public perception associated with the use of renewable energy. In this era where "green" has a positive conotation, demonstration of renewable energy use brings with it an acceptance from the public. One large eastern utility has installed small solar panels to power the entrance sign to their facilities, to demonstrate their commitment to "green."

Project Prioritization and Implementation

Since capital funds are limited, it is imporant to prioritize opportunities. The first step is to identify those projects which that meet the appropriate financial and implementability criteria for the utility. Each potential opportunity is developed to sufficient detail to enable the capital cost and operating/maintenace costs to be initially quantified. Often the priorization process entails rating the projects based on different agreed-upon criteria and seeing how their ranking changes, i.e. ease of implementation to determine what can be done most readily to demonstrate success, low capital cost if funds are limited, or payback period/return on investment if that metric is most important to the organization in terms of prioritizing project opportunities for implementation.

The means of prioritization is utility-specific, as to whether it should be solely based on financial conditions, public perception, or other considerations. Financial ranking metrics can be initial capital cost or simple payback period or return on investment (depending on client preference.) For utilities with a small capital budget will be limited in the size of the project(s) that they can undertake. In addition, many times ease of implementation is also an important consideration. The potential project opportunites which pass these tests are then developed further to better quantify these metrics.

A client workshop can be held to weigh the non-financial considerations. Typically, a decision criteria software is used to enable the weighting of various criteria to be conducted, objectively, and to enable various perspectives to be discussed and weighed.

Monitoring and Control (Energy Dashboard)

As part of this process, the provision of an energy dashboard should be considered for incorporation into the utility's overall control system. A dashboard is a visual display of the most important information needed, particularly by management, to achieve a limited number of higher level objectives. Energy dashboards collect data from instruments and systems across the utility, analyze and trend it, then present it in a meaningful fashion. The information is generally consolidated and presented on a single screen to enable the person reviewing the dashboard to quickly gauge performance at a single glance. The energy dashboard also allows for the utility's energy management strategy to be translated from their planning and business functions to those persons making operational and management decisions, including those individuals on the front lines.

Incorporating an energy dashboard approach would reinforce the cultural change that is part of any energy management initiative, and raise the awareness of the importance of energy management in utility employee's day-to-day activities. Such a dashboard would show the respective organizational goals and real-time progress toward those goals. This enables operations staff to make informed decisions that won't compromise reliability or the utility's ability to meet regulatory demands. Decisions such as whether or not to start equipment with large motors can be made, based on actual consumption of the facility, time of day demand charges which may be incurred, and what is occurring at the time in the utility company power grid.

The dashboard would visually present the utility's progress against key performance indicators (KPIs) related to energy consumption, and indicate trending. For example, KPIs could be kWh/MGD of flow, kWh per lb/pollutant removed (such as oxygen demand), or some combination thereof. Demand would be tracked on a real time basis, and trending would be provided, both at the sub-metering level as well as a rollup across each facility, group of facilities, and enterprise-wide.

Ideally, the dashboard would be made available for viewing by all levels in the organization, from the front lines up to the Executive Director. Dashboards can also be customized, depending on the viewer. In other words, the Executive Director may wish to see higher level information presented than others in the organization. Along the same lines, those on the front lines will likely want to see more specific and detailed information. In addition to the informational benefits, the dashboard can also be used by operations staff to make informed decisions, along the line of "what gets measured gets managed."

V. CHALLENGES – CULTURAL, TECHNICAL AND FINANCIAL

Adopting an energy savings mentality and the embracing of that culture across the organization starts with the technical findings which can produce the potential savings. As noted above, these findings will be prioritized. The Energy Savings Plan would also identify a path forward which will achieve the utility's goals and objectives, while providing discrete measures to move that Plan forward properly.

However, cultural change is just as important as the implementation of the technical changes. Changing an organization's behavior requires this cultural change occur, which starts with the goals and objectives being clearly articulated from the top of the organization, and then being embraced throughout the organization, particularly on the front lines where daily decisions can impact energy consumption. It also requires a passionate advocate for change, the individual that keeps the possibilities in front of the group and challenges them to improve. In the case of many utilities, this would be the Energy Manager.

Another consideration when rolling out such an energy programs is that, most progressive agencies have a number of projects underway, to meet more stringent regulations and position them for the future as well as to replace aging facilities. In order to move the utility forward on the change management process, the energy perspective must be considered for ongoing projects, noting that the greatest opportunity to impact a project is early in its project lifecycle, i.e. in the planning stage. The ability to impact a project is progressively less as a project moves from design, through procurement, implementation and closeout.

In the new world of energy management, the impacts of energy on cost, operations, etc. must be considered. It is important to understand the extent of projects that the utility already has programmed into their Capital Improvement Program, and how they will impact the energy baseline, when doing energy evaluations. If possible, the Energy Manager should interject herself to ensure that new projects are being done in conformance with the utility's energy standards. In addition, if performance contracting is being utilized, the impact of these projects on those projects that are part of the performance contract must be measured and known.

Finally, a challenge with any energy savings programs is finding the money available funds to implement the recommendations even when energy savings projects may have a relatively short payback period of less than 5 to 7 years. Utilities may also have other priorities related to their ability to reliably meet their permit requirements. This is where performance contracting as a delivery mechanism can be attractive because the performance contractor can provide the capital necessary to implement the selected energy savings project. The savings are then used to pay back that initial cost, over a prescribed time period, and within defined performance parameters.

VI CASE STUDY – PERFORMANCE CONTRACTING

The authors were recently involved in a performance contracting opportunity at two wastewater treatment plants in a northern climate. The utility had desired to improve their biosolids treatment, prior to disposal to a landfill, and well as reduce their energy consumption. As part of this evaluation, an energy audit and a process evaluation were conducted to meet their objectives. Improvements that were identified included:

- New anaerobic digesters with solids dewatering and energy recovery,
- HVAC and lighting improvements,
- Hydropower to recover energy from the effluent of the treatment plant, and
- Replacement of standard blowers with new high-efficiency blowers.

For each of these potential opportunities, preliminary design was conducted to the extent that both capital as well as operational costs could be identified. This information was used to prioritize the recommendations and conduct preliminary equipment selections. It was determined that more traditional items, such as lighting and HVAC replacement had a shorter payback period, and also required less initial capital investment. Alternatively, as expected, incorporation of the anaerobic digestion/sludge dewatering facilities had a much longer payback period and a higher capital cost. This is even when considering the use of the digester gas as an energy source both for heating the digesters and buildings, plus utilization of excess gas to produce electricity. Also considered was the near elimination of lime for sludge stabilization, and the resultant reduction in solids quantities as a result of the digestion process.

Once the technical opportunities were identified, the next step was to prioritize the opportunities using the metrics that the client would be most comfortable with, whether this be simple payback period, return on investment, visible progress on sustainability for public benefit, achieving longer term reliability for sludge disposal, etc. This is important for two reasons:

1. It is important to know your client well at various levels so that the opportunities can be ranked in accordance with their goals and objectives, expectations and priorities.

2. It is important to understand what challenges exist that would prevent the utility from achieving these currently, so that the performance contracting strategy can properly address them.

In other words, the performance contracting structure is very client specific, and must be customized to an individual client and not approached generically.

Also, on the technical side, the engagement of staff that are on the front lines is important, both in calibrating the nature and implementation of the recommended improvements as well as the perceived or real value that they would bring. Also, it is not generally practical to perform everything all at once, due to initial cost and resource constraints. It is better to prioritize and attack the higher priority opportunities first, and gain some wins, then tackle the next higher priority items, etc.

Prioritizing the opportunities based on the client's metrics, and developing conceptual designs first on those items best meets client goals and objectives in a strategic fashion. This approach also reduces the overall effort of developing detailed designs for all the items, which results in unnecessary design and construction estimating costs for items that will not be implemented first. It is not uncommon for client "wish list" projects to get identified for implementation.

As noted above, it is not practical nor desirable to perform all the projects in one package, but to package projects that support an "adaptive management" approach to implementation, that are matched to priorities and good stand-alone solutions, and which build towards the entire portfolio of wish list projects in a strategic, phased approach. It is also important for the performance contractor to take a step back and provide solid recommendations based on their experience, and not just defer to the utility's wishes.

VII. CONCLUSIONS

The reality of today's economic times is that many utilities have opportunities to reduce their energy consumption, but do not have the necessary capital funding to implement the projects which can deliver these savings. This is unfortunate, given that the payback period on the investment can be rather short and that other benefits, such as greenhouse gas reductions would also be realized. One of the project delivery tools available to utilities is performance contracting. Performance contracting can provide the up-front capital to implement these projects. However, it is important to take the proper precautions when developing a performance contract, such as provide adequate definition of roles/responsibilities between the utility and the ESCO, and a means of properly quantifying the savings.

Demand Management Strategies Resulting In "Zero and Low Cost" Energy Saving Opportunities for Water and Wastewater Treatment Facilities.

Bryan Lisk PE, CEM – Hazen and Sawyer P.C.

ABSTRACT

Reducing energy costs is rapidly becoming a major priority for water and wastewater utilities. As a result, many water and wastewater utilities are performing energy management audits and evaluations to identify energy saving opportunities that will provide energy benefits at the lowest possible capital cost. Many water and wastewater facilities can realize "zero or low cost" energy saving opportunities through managing their energy demand in coordination with the energy billing rates. This is especially true for facilities whose electric energy rate structures include demand ratcheting and time of use billing. In order for demand management strategies to be effective, the demand management strategy must not have any negative impacts to the treatment process and must be coordinated with the energy billing rate structure(s).

This paper will present multiple demand management strategies that were identified from multiple energy management audits performed for multiple wastewater treatment facilities. The results of multiple demand management case studies will be presented with an emphasis on demand management strategies that provided a payback period of **1 year or less**. Specifically this paper will include the following as a minimum:

- A description of common electric utility rate schedules (i.e. Time of Use, Demand Ratchets) and how different rate schedules impact the demand management strategies for water and wastewater treatment facilities.
- Typical demand management practices commonly used by water and wastewater utilities including flow equalization, peak shaving, and operations management.
- The role power monitoring capabilities has on the demand management capabilities.
- The results of multiple case studies where demand management opportunities were identified as a part of an energy management project.

KEYWORDS

Energy management, demand management, combined heat and power, biogas utilization, electric utility rates

INTRODUCTION

Purchased electrical energy is a major cost for most water and wastewater utilities. With the recent emphasis on energy conservation and greenhouse gas reduction, many water and wastewater utilities are seeking opportunities to reduce energy costs. The strategies water and wastewater generally employ to reduce energy costs fall into 3 basic categories:

- Increase process and equipment energy efficiency to reduce energy usage
- Recover resources such as biogas to offset purchased energy
- Manage demand to reduce purchased energy costs

Energy efficiency improvements and resource recovery projects are focused on reducing the energy usage to lower energy costs where as demand management focuses less on reducing energy usage and more on managing a plants energy demand to reduce electric utility demand and energy charges. Most demand management opportunities can be realized through plant operation changes and can be implemented at a very low or zero costs. In order to identify and implement beneficial demand management opportunities, the following information must be understood as a minimum:

- The purchased energy source billing rate structure.
- Diurnal flow fluctuations and the resulting plant demand profile
- The plant's ability to reduce demand during times when energy costs are high (on-peak periods) without negative impacts to the process.
- The plant's power monitoring and energy usage trending capabilities

Electric utility billing structures have the highest impact on demand management strategies and must be thoroughly understood in order to develop beneficial demand management strategies. Most demand management strategies are based on managing the plant load in coordination with the electric utility billing rate to minimize energy usage during the periods when energy costs are high (on peak periods) and/or to reduce the plant's peak demand to reduce electric demand charges. Other demand management opportunities exist by utilizing on-site power generation systems to participate in electric utility emergency response and load curtailment programs.

Overview of Electric Utility Billing

Electric utilities billing rates vary greatly among the utilities and the regions they serve; however, most electric utilities base their billing rate structures on energy usage charge (kilowatts-hours), demand charge (kilowatts), and a facility charge for the utility owned equipment. The following is a description of each of these billing components.

Energy Usage Charges. Energy usage is the product of the plant demand and time. For example, a 1 kilowatt (1000 watts) heater operating for one hour will use one kilowatt-hour of energy. Typically, the electric utility billing energy usage charge is either a flat rate (i.e. ¢4.25/kWh) or the energy usage charge can vary with the "time or use". Time of use rates vary the cost of energy with time of day the energy is used. Generally, utility companies change their time of use rate for the summer and winter seasons so that the higher energy rates coincide with the seasonal high demand periods.

Metered Demand Charges. Demand charges are used by the utilities to provide the capital cost for the facilities to provide the electric service to the customer. Demand charges are typically based on the peak 15 to 30 minute plant demand metered by the electric utility during a billing period. For example, the demand charge for the billing period shown on Figure 1 will be based on the metered peak 15 minute demand (6500kW) for the 30 day billing period even though the average demand is well below the peak demand.

Figure-1 – Single Billing Period Demand Profile

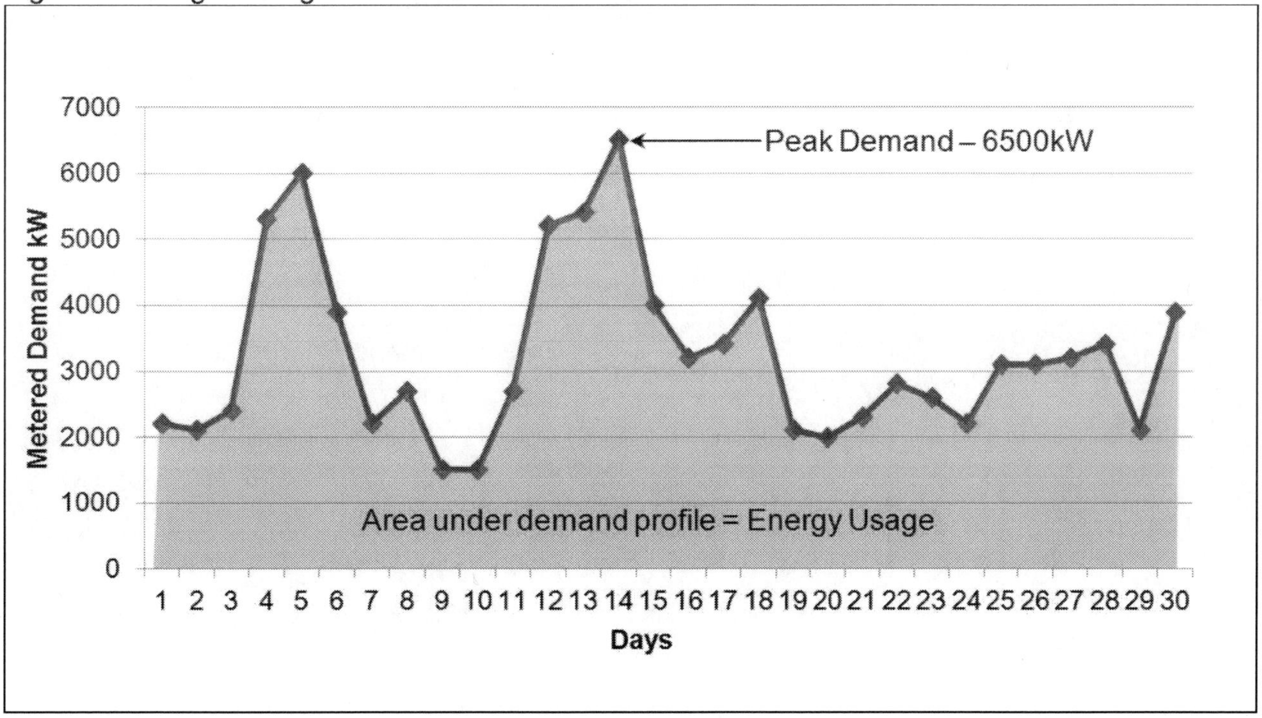

Minimum Billing Demand Charges. Many electric utility billing rates will include a minimum billing demand charge that will be charged to the customer if the metered demand charge is below the minimum billing demand charge threshold. In most cases, the minimum billing demand charge is defined in the electric utility service contract. Some electric utilities will establish the minimum billing demand charge as a percentage of the peak metered demand from the preceding 12 billing periods. Figure 2 shows a example of a 12 month minimum demand charge based on 80% of the peak demand. Demand management opportunities can only generate benefit from demand reduction if the plant demand is above the minimum billing demand level. This example illustrates the importance of understanding minimum billing demand charges when developing demand management strategies.

Figure-2 – 80% Minimum Demand Charge Example

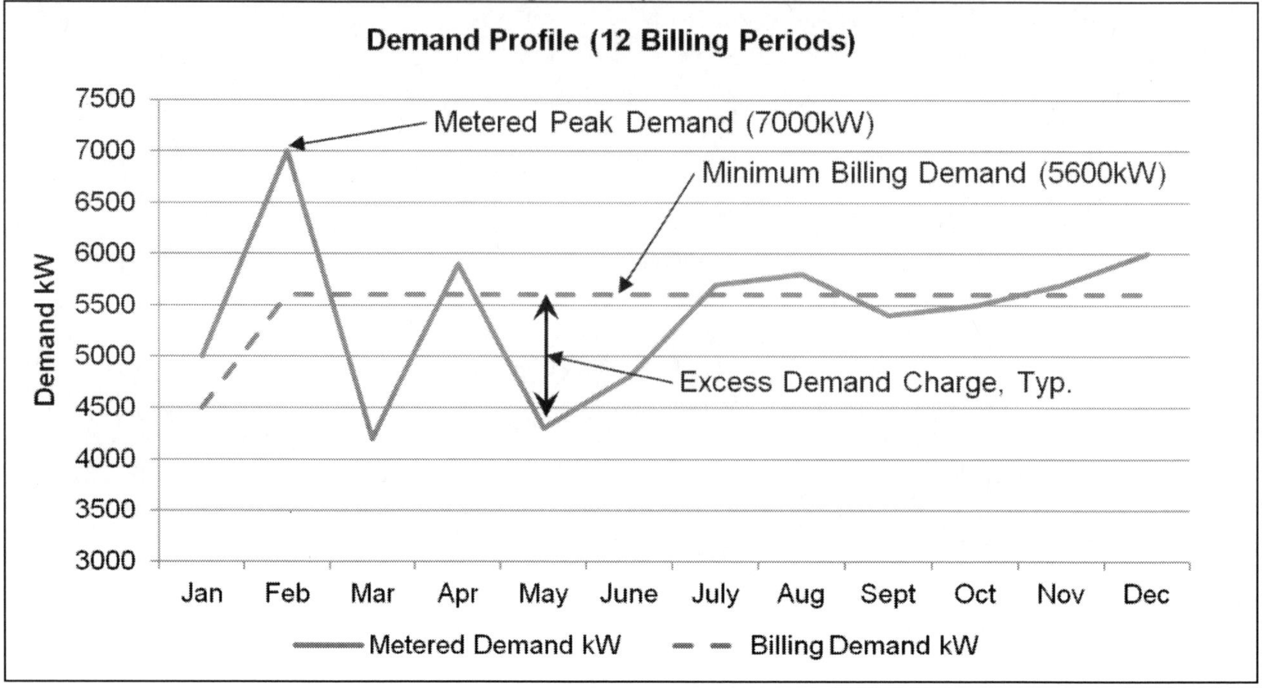

RESULTS AND DISCUSSION

Developing Demand Management Strategies

The goal of managing demand is to minimize cost by minimizing the electrical demand when energy and demand charges are at their highest per the electric utility billing rate. This is typically achieved by manage plant operations to reduce demand during on-peak hours and/or utilize on-site power generation capacity to manage plant demand. The demand management strategy will depend on the cost ratio between the energy and demand charges and the differences in cost between the on-peak and off-peak billing periods as defined in the facility's electric utility rate structure. The first step in developing demand management strategies is to evaluate the electric utility billing rate structure and the average plant demand profile to understand how the plant demand profile in conjunction with the electric utility billing rate is impacting the overall electric energy cost. In many cases, the plant peak demand charges will have a significant impact on the overall cost of electric energy. To illustrate this concept, Figure 3 shows two hypothetical plant demand profiles, one with a high peaking factor and the other with a lower peaking factor. For each profile, the energy usage for the billing period is the same (2330400 KWH each). Table 1 shows the resulting electric utility cost using a simple, but realistic electric energy charge of 4.0₵/KWH and a demand charge of $12.00/KW. As shown on Table 1, the electric utility costs for the high peak scenario are ~20% higher due to the elevated demand charges than the low peaking scenario even though the energy usage was the same in both cases. This illustrates the potential benefits from managing a plant's operations to minimize the peak demand charges.

Figure 3 – High and Low Peaking Plant Demand Profiles

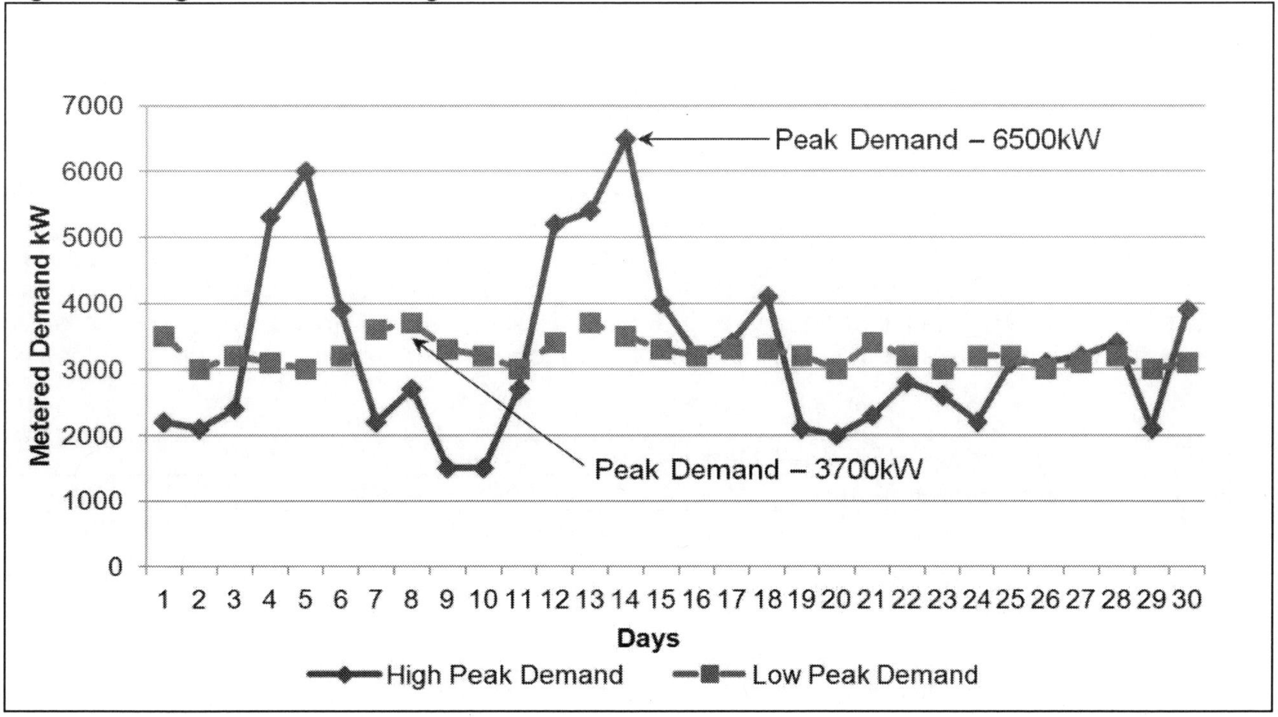

Table 1 – Energy Cost Comparisons Between High and Low Peaking Plant Demand Profiles

	Energy Usage	**Energy Charge @ 4.0¢/KWH**	**Metered Demand**	**Demand Charge @ $12.00/KW**	**Total Charges**	**Average Cost per/KWH**
High Peaking Scenario	2330400 KWH	$93,216	6500kW	$78,000	$171,216	7.3¢/KWH
Low Peaking Scenario	2330400 KWH	$93,216	3700kW	$44,000	$137,616	5.9¢/KWH

Common demand management strategies that can be used to control demand and energy costs include:

- Defer non-critical loads to off peak hours when energy and demand charges are lower. Non critical loads such as dewatering, filter backwashing, and some mixing loads are examples of loads that are commonly deferred to off-peak periods. Deferring dewatering operations has a secondary benefit by reducing the nitrogen loading from the filtrate return to the secondary process during the on peak periods which can increase the aeration demand.
- Utilize equalization storage to reduce diurnal fluctuations to minimize high pumping and aeration loads.
- Interlock intermittent loads such as filter backwash pumping, air scour blowers, and large air compressor loads so that they do not operate concurrently resulting in elevated demand charges. It is also possible to interlock intermittent loads with other continuous

loads (i.e. backwash pumps and digester mixing pumps) that can be stopped without negatively impacting the process to offset the demand.

Demand Management Case Studies

South Durham Water Reclamation Facility (NC) Demand Management Opportunities. An Energy Management Plan was developed for the South Durham Water Reclamation Facility (SDWRF) to identify long term energy saving opportunities. The South Durham Water Reclamation Facility is a 10MGD (Average Flow) facility located in Durham NC. The biogas generated from the SDWRF anaerobic sludge digestion process is used for digester heating purposes and is used to fuel two 300hp (approximate) digester gas fueled engine driven aeration blowers. The SDWRF electric utility service is supplied by Duke Energy via three separate services and is billed per Duke Energy's OPT-G (Optional Power Service, Time of Use General Service) rate schedule. The rate description for this service is included in Table 2 below.

Table 2 – Duke Energy OPT-G Rate Description		
Basic Facility Charge, per Month	$36.51	
On-Peak Demand Charges*	Summer Months June 1 – September 30	Winter Months October 1 – May 31
For the first 2000 kW of Billing Demand per month, per kW	$14.0080/KW	$8.2460/KW
For the next 3000 kW of Billing Demand per month, per kW	$12.8319/KW	$7.0591/KW
For all over 5000 kW of Billing Demand per month, per kW	$11.6449/KW	$5.8613/KW
Off Peak (Economy Demand) Charge	$1.1136/KW	$1.1136/KW
On Peak Energy Charge, per KWH	5.6821¢/KWH	5.6821¢/KWH
Off Peak Energy Charge, per KWH	3.3789¢/KWH	3.3789¢/KWH
On Peak Hours. All other hours are off-peak including holidays.	1:00 p.m. – 9:00 p.m. Monday – Friday	6:00 a.m. – 1:00 p.m. Monday – Friday
*The billing demand is the maximum integrated thirty-minute demand during the applicable summer or winter on-peak period during the month for which the bill is rendered or 50% of the Contract Demand whichever is greatest.		

As described in Table 2 above, the SDWRF utility demand charge is established during the seasonal on-peak hours. The SDWRF currently manages demand charges for the Blower Building service by stopping all motor driven blowers and operating the engine driven blowers to reduce the demand during the on-peak hours. The SDWRF was entered into Duke Energy's Energy Profiler program so that an evaluation of the plant's demand profile could be performed. The program records and makes available the 30 interval demand data for each of the three SDWRF utility services. The average demand profile for 10 billing periods was developed for the Blower Building electric utility service and is shown on Figure 4 below.

Figure 4 – SDWRF Blower Building Electric Utility Service Average Demand Profile

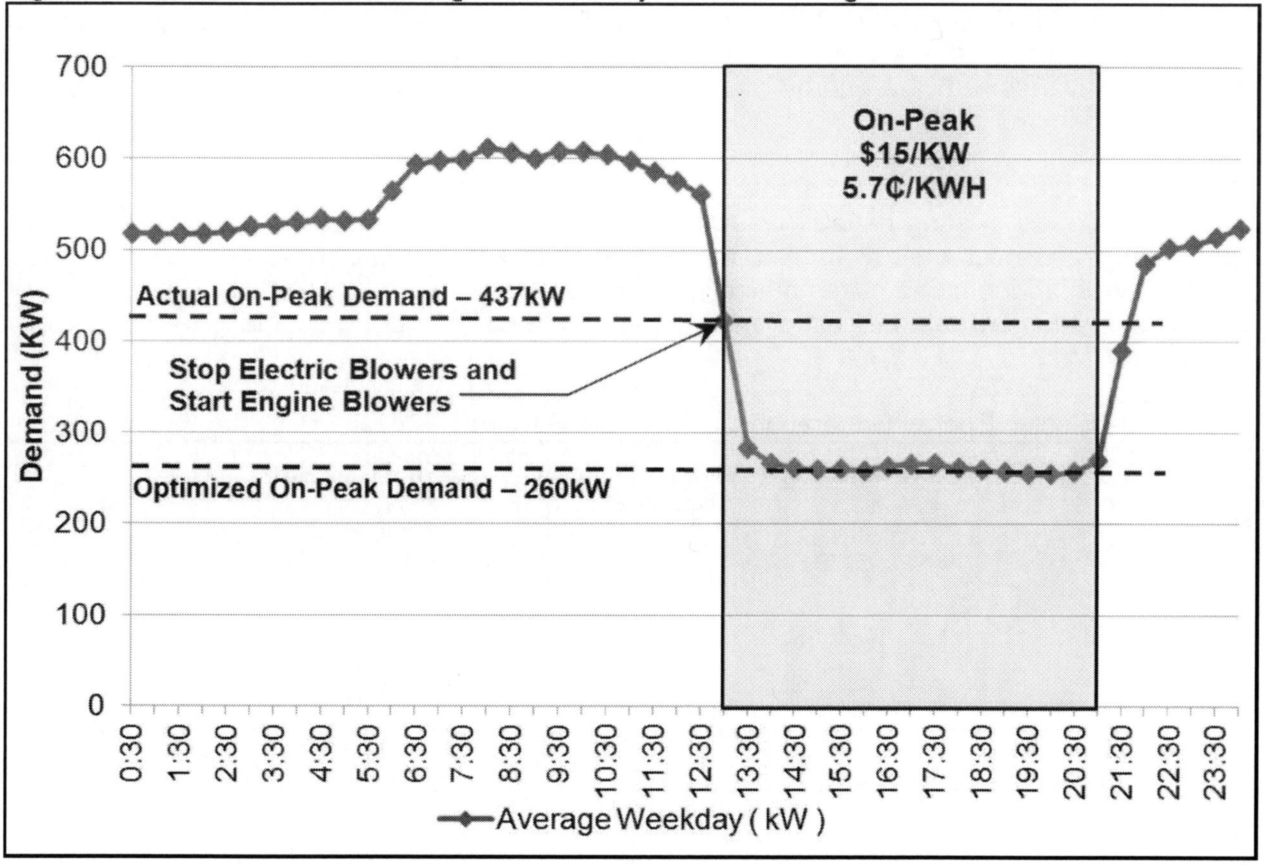

As shown on Figure 4, the actual metered average on peak demand (437kW) is significantly higher than the optimized on peak demand (260kW) resulting in ~$2,600 of additional demand charges per month. A possible explanation for this condition is the motor driven blower is stopped shortly after the start on-peak periods start. To resolve this issue, it was recommended that the SWDRF stop the motor driven blower 15 minutes before the on-peak period begins and start the motor driven blower 15 minutes after the on-peak period ends. Interval data collected after this recommendation showed no increase for the maximum peak demand indicating that this operating schedule provides billing demand commensurate with the optimized on-peak demand. This was a "zero cost" demand management opportunity that resulted in ~$2,600 per month reduction in demand charges. This opportunity also illustrates the importance of having the power monitoring capabilities to monitor, trend and analyze the plant demand profile.

Village Creek WWTP (AL) Demand Management with On-Site Power Generators. The Jefferson County Department of Environmental Services operates the Village Creek WWTP located in Birmingham AL. The plant operates two (2) separate treatment trains (Plant-001 and Plant-002). Plant-002 uses eight (8) 3200kW (26.5MW total), natural gas fueled generators to supply standby power to the plant and to supplement the utility source in the event of a peak flow event. The utilization of these generators to manage plant demand during the on peak billing periods was evaluated.

The Village Creek WWTP is served by two (2) electric service from Alabama Power. Plant-001 receives a primary voltage (13.2kV) service and is billed based on Alabama Power's Large Light and Power Service (Rate-LPL). Plant-002 receives a transmission voltage (115kV) service and is billed based on Alabama Power's Large Light and Power –Time of Use Service (Rate-LPTL).

This rate charges an energy charge of 13.13¢/KWH during on-peak periods (June 1 – Sept 30 12:00PM - 7:00PM Monday – Friday), 5.36¢/KWH during Intermediate-peak periods (10:00AM - 9:00PM Monday-Friday), and 2.70¢/KWH during off-peak periods (all remaining hours). In addition to these rates, Plant-002 is also charged a 2.5¢/KWH (average) for all KWH used. The fuel charge varies with the energy market priced during each billing period. Plant-002 is not charged a demand charge.

A cost comparison between the electrical energy costs and the cost to generate electrical energy using the existing natural gas fueled generators revealed that the plant can benefit by generating electrical energy in lieu of purchasing electrical energy during on-peak hours (15.6¢/KWH) and in some cases during intermediate-peak hours (7.9¢/KWH). Figure 5 shows the relationship between the cost of natural gas ($/MMBTU) and the cost to generate electrical energy ($/KWH).

Figure 5 – Electric Energy Generation Cost ("Spark Spread")

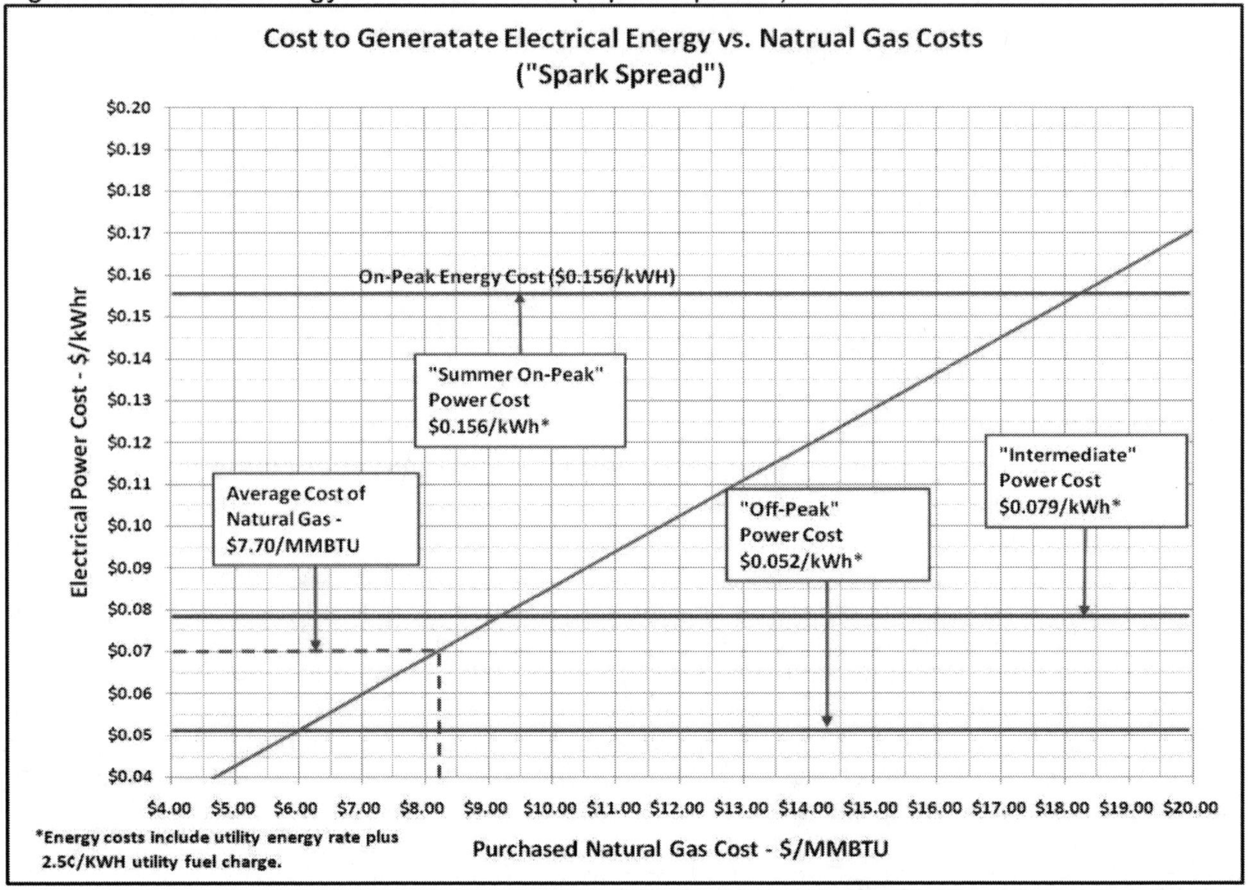

With an average demand during the on-peak periods of 2100kW and an average natural gas cost of $7.70/MMBTU, plant-002 can save approximately $22,000 per summer month by operating the natural gas fueled standby generators during the on peak billing periods. This demand management opportunity was recognized as a beneficial opportunity, however it could not be implemented due to air permitting limitations and restrictions set by Alabama Power's Standby Power Generation Capacity Credit Program. It was determined that entering the Plant-002 standby generators into Alabama Power's Standby Generation Capacity credit program would generate more benefit. The benefit from Alabama Power's Standby Generation Capacity credit program is described below.

Village Creek WWTP (AL) Generator Capacity Credit Program. Alabama Power offers a Standby Generator Capacity credit to customers with onsite power generators having a capacity of 250kW or more. In return, Alabama Power reserves the right to operate the generators in the event of a utility emergency up to 240 hours per year. The owner of the generator(s) is responsible for providing and paying for the fueling and maintenance of the generator(s). The credit offered by Alabama Power is $2.31 per kW per month of contracted standby power generation capacity minus a facility charge of $620/month. Many North Carolina utilities (Duke Energy & Progress Energy) offer standby power generation capacity credit programs similar to this program. An evaluation of the potential benefit with respect the contracted standby capacity and annual run time (hrs/year) is shown on Figure 6 below. The benefit is based on an average natural gas cost of $7.70/MMBTU.

Figure 6

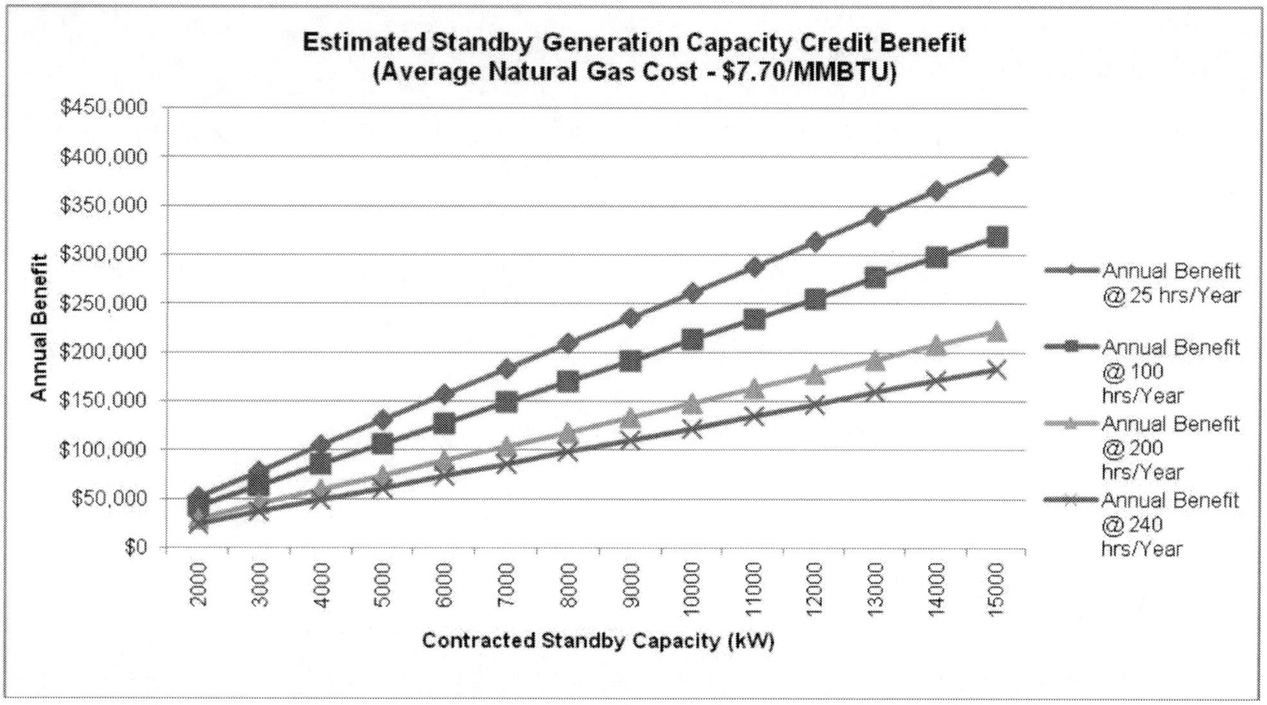

The Village Creek entered into Alabama Power generator capacity credit program with a contracted capacity of 15MW. The historical average annual run time under this program is less than 25 hours for the past 10 years which allow the plant to save nearly $400,000 per year. In order to participate in this program, approximately $200,000 in modifications the existing generator switchgear were required.

SUMMARY

As shown in the case studies above, demand management opportunities can result low or no-cost energy savings. The key to developing beneficial demand management strategies is to understand the plant's electric utility billing rate and identify process and operation changes that, in coordination with the electric utility rate schedule, reduce energy and demand charges. Specific electric utility billing elements such as minimum demand charges and other fixed charges that will not change for each billing period will also have a significant impact on the demand management strategies. Demand management strategies can include process/operations changes and the use of on-site standby power generation capacity. Significant cost savings have been realized using onsite standby power generators as a part of electric utility generator capacity credit programs. The use of onsite power generators to manage demand is considered non-emergency use by the EPA and therefore must comply the non-emergency use emission requirements of the local air permitting board, the EPA New Source Standards, and the EPA National Emission Standards for Hazardous Air pollutants.

REFERENCES

Bullard, C. M., Fishman, M. A., Lisk, B. R., and Hardy, S. A., Putting Digester Gas to Work: Economic and Environmental Sustainability Via on-Site Energy Production", 2010 NC AWWA-WEA 90th Annual Conference, Winston-Salem, NC, November 2010.

Capehart, B.L., Turner, W.C., and Kennedy, W.J., Guide to Energy Management, Sixth Edition. Fairmont Press, Inc. Liburn GA.

Aeration Efficiency in Oxidation Ditches

John Lindam, Xylem Inc.
Xylem Water Solutions
Gesällvägen 33
174 87, Sundbyberg, Sweden
+46 (0)73 236 6180

ABSTRACT

Oxygen transfer and efficiency in water reclamation facilities are studied. In particular, design aspects of the factors affecting aeration performance in biological reactors are emphasized.

Due to the growing importance of energy consumption, a comparison study has been conducted, evaluating the features and performance of mechanical surface aerators and diffused aeration systems. This has been done with special consideration to oxidation ditches which need to be assessed with regard to creating sufficient horizontal flow in addition to high oxygenation capacity and efficiency.

It was concluded, based on a number of case studies that diffused aeration systems offer significant advantages over mechanical surface aerators in terms of increased oxygen transfer capacity, enhanced levels of process control capability, and reduced energy requirements to produce high quality effluent.

The potential to improve energy efficiency even further with the use of high density, low flux fine bubble diffusers in combination with mid-size, low speed mixers is also discussed.

KEYWORDS

Aeration, mechanical surface aerators, fine bubble aeration, oxidation ditch, horizontal flow, aeration efficiency, energy, power, oxygen transfer, control system

INTRODUCTION

Research and experiences within biological treatment have led to greater ability among municipal water resource recovery facilities to review and upgrade equipment and subsystems to reduce energy consumption. In most cases, processes and equipment concerning aeration are the major consumers of energy. It has been estimated that up to 60% of a plant's power consumption can be attributed to aerobic reactors in the activated sludge process of secondary treatment.

Factors pertaining to power requirement and oxygen transfer of aeration systems in biological treatment are discussed, with emphasis on the difference and effects on energy efficiency between fine bubble aeration and mechanical surface aerators. In particular, oxidation ditches are discussed, which are common in the United States and worldwide, presenting a number of engineering challenges with respect to aeration and mixing.

GENERAL CONSIDERATIONS IN AERATION EFFICIENCY

Oxygen Transfer and Efficiency

In order to discuss aeration efficiency, the term efficiency must be clearly defined. It is generally acknowledged that oxygen transfer efficiency (OTE) plays an important role.

$$OTE = \frac{\text{oxygen transferred}}{\text{oxygen supplied}} \qquad \text{eq. 1}$$

Where OTE = oxygen transfer efficiency, %
Oxygen transferred = oxygen transferred from bubble to mixed liquor, kg (lb)
Oxygen supplied = oxygen supplied aeration equipment, kg (lb)

If OTE is measured under standardized conditions, as described by American Society of Civil Engineers (2007), the term Standard OTE (SOTE) is used. When comparing performance of different equipment for aeration, it must be done under standardized conditions to validate and normalize performance data.

With regard to energy consumption, the common expression for efficiency is the standard aeration efficiency (SAE) which factors in pressure and efficiency of the mechanical equipment required to achieve a de facto oxygen transfer. The standard aeration efficiency is the amount of oxygen transferred per unit of energy consumed:

$$SAE = \frac{SOTR}{P} \qquad \text{eq. 2}$$

Where SAE = aeration efficiency, kg O_2/kWh (lb O_2/hp.h)
SOTR = oxygen transfer rate, kg/h (lb/hr)
P = power required to transfer one mass unit of oxygen per unit time, kW (hp)

It should be noted that the definition of power must always be stated clearly. In this paper, power is defined as the total input power of any mechanical equipment required to achieve the rate of oxygen transfer for the biological process as stipulated by the European Committee of Standardization (2009).

Technologies

There is a wide range of equipment suitable for delivering oxygen to aerobic reactors in biological treatment. Aeration devices such as mechanical aerators, and submerged coarse and fine bubble diffusers cover a large portion of aeration equipment installed at water resource recovery facilities globally. The aeration equipment selection process should consider system capacity, operating range, and equipment reliability. Assessment of the aeration system performance should consider total system efficiency, evaluating individual aeration system component efficiencies along with the aeration control algorithm.

Table 1 shows a brief summary of estimated SAE values for common types of aeration equipment used for biological treatment. Aerator types include high and low speed surface aerators, submersed jet aerators (such has nozzles which distribute air below the liquid surface) fine bubble disc diffusers and high density low flux (HDLF) fine bubble diffusers.

Table 1. Summary of efficiency ranges for various types of aeration equipment (Tschobanoglous *et al*, 2003).

Type	SAE, kg O_2/kWh	SAE, lb O_2/hp. h
Low speed surface aerators	1.5-2.1	2.5-3.4
High speed surface aerators	1.1-1.4	1.8-2.3
Submersed jet aerators	0.9-1.4	1.6-2.3
Fine bubble diffusers, discs	2-7	3-10
HDLF fine bubble diffusers	3-8	5-13

Note that the SAE values in Table 1 should be used as reference, exact figures are equipment and application specific.

Fine bubble diffusers in particular present major advantages in terms of both capacity and power consumption, however in order to achieve maximum oxygen transfer to energy consumed (SAE), a number of factors relating to the floor density and air flux of the diffusers must be evaluated.

When full floor diffuser coverage systems are considered, diffuser floor density refers to the ratio of area covered by diffuser membranes to the total floor area. Diffuser flux, expressed as flow per membrane area is directly related to the diffuser density if a fixed oxygen transfer rate requirement is considered. In order to model oxygen transfer accurately, Computational Fluid Dynamics (CFD) may provide great insight into the mass transfer mechanisms present in any biological reactor on a case by case basis. However, with extensive oxygen transfer test data at hand, oxygen transfer can be accurately predicted utilizing modern multivariate statistical methods by taking into account the variables associated with the configuration of any full floor coverage diffuser installation.

In summary, the driving forces behind oxygen transfer in fine bubble systems in clean water can be attributed to the following:

- Diffuser flux
- Diffuser density
- Depth

Obtaining a high transfer efficiency requires optimizing the bubble retention time, thereby increasing the total amount of oxygen transferred per volume of air supplied to the system.

Bubble retention time correlates to spiral flow effects, which are more or less present in any bottom diffuser installation and depend on the density and membrane flux of the installed system. The phenomenon of spiral flows is described in Figure 1 below. A densely installed diffuser system with low membrane flux mitigates secondary flow effects, allowing slower bubble rise within the reactor.

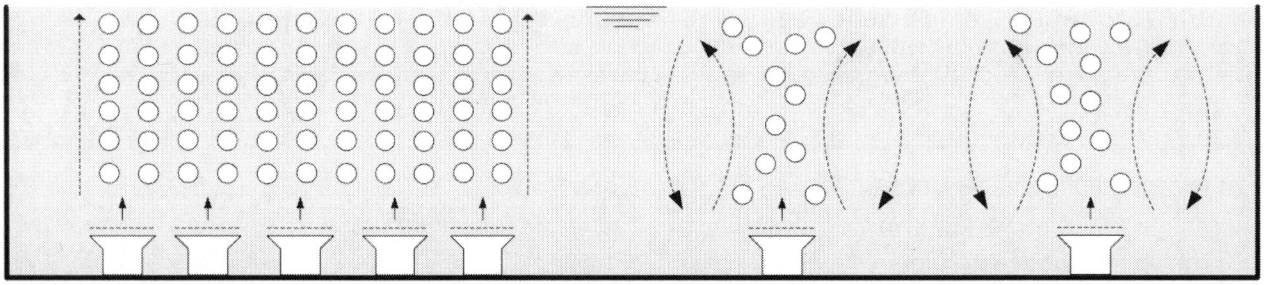

Figure 1. Principle of spiral flows. Left: high membrane density fine bubble diffuser system. Right: low density-high flux installation, generating secondary spiral flows causing reduced bubble detention time and oxygen transfer rate.

The response to power requirement of a fine bubble full floor coverage diffuser system can be described conceptually in Figure 2 below. The system as a whole, including air blowers, responds positively by an increase in diffuser density, leading to lower flux and higher SOTE. In what can be described as a positive feed-back response, high SOTE generates lower air flow, pressure and power requirement.

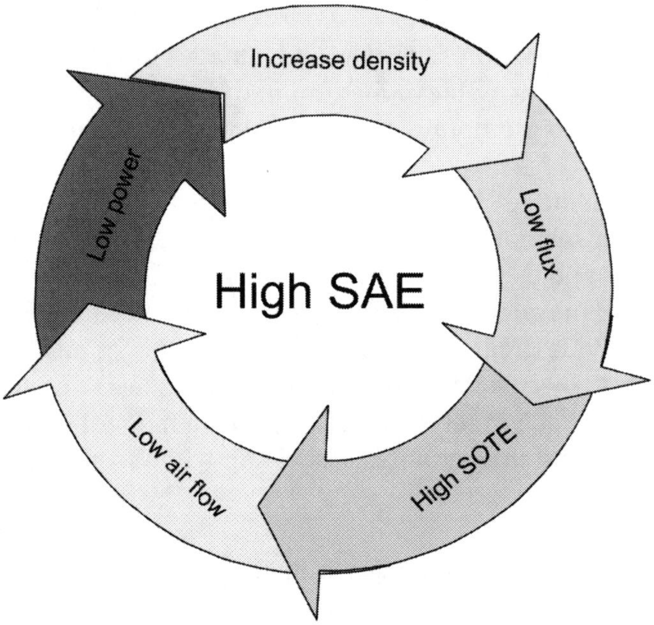

Figure 2. Dynamics of the design of a fine bubble diffuser system and the effect on aeration efficiency.

The depth of the aerobic reactor is a vital factor in assessing the oxygen transfer capacity of the aeration system as well as the OTE and the overall SAE. While increasing reactor depth increases the oxygen transfer rate or transfer efficiency SOTE of a bottom diffuser system, its relationship is sub-linear, i.e. SOTE increases slower than linearly versus diffuser submergence. At the same time, system pressure has a nearly linear relationship with diffuser submergence. For a given reactor footprint and oxygen transfer requirement, there exist a power minimum and an SAE optimum. This is portrayed in Figure 3.

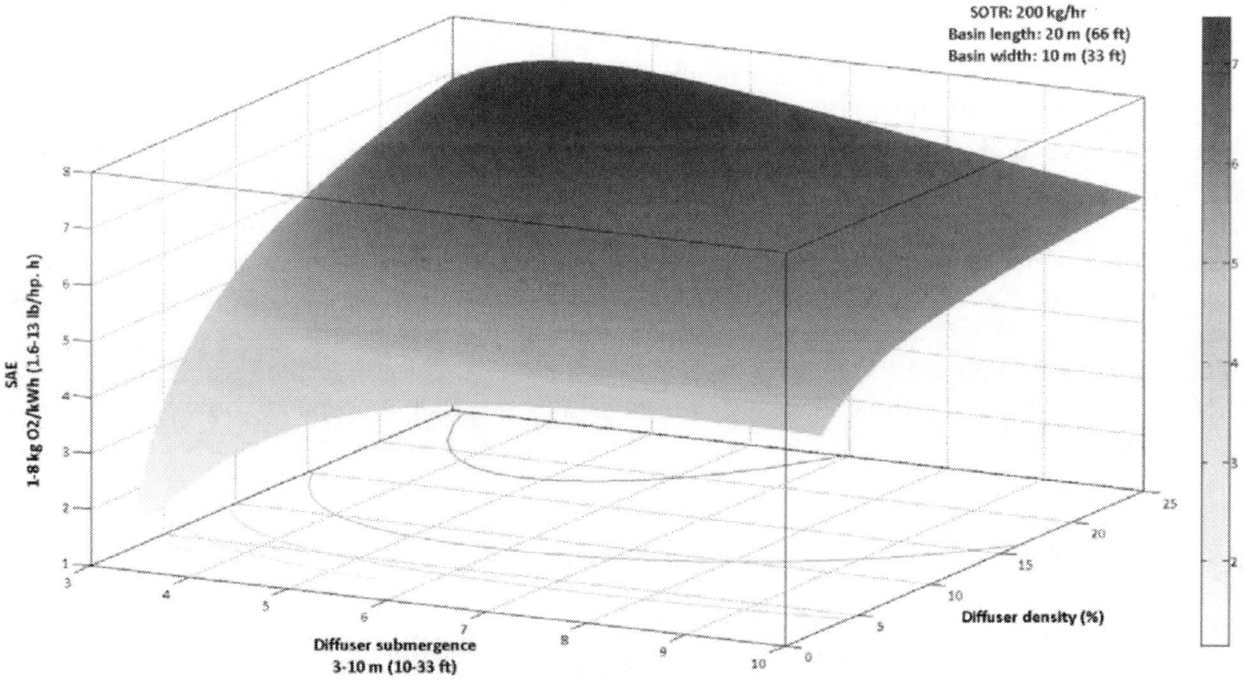

Figure 3. Optimal diffuser submergence of an aerobic reactor with fine bubble diffused aeration. Example plot under standard conditions according to ASCE (2007) based on Xylem diffuser performance, assuming an overall blower efficiency of 80%.

The positive effects of diffuser density is clearly visible, where the optimal operating point from an energy perspective is a diffuser submergence of 4-5 m (13-16 ft) assuming a constant blower efficiency of 80%. Calculations are based on oxygen transfer performance data of Sanitaire Silver Series II fine bubble diffusers.

The dynamics of the three major factors in aeration system design should be taken into consideration during the engineering and design phase of an activated sludge facility. The optimal design point from an aeration performance perspective varies with the unique characteristics and design of the fine bubble diffuser.

OXYGEN TRANSFER IN OXIDATION DITCHES

Above discussions pertain mainly to conventional activated sludge processes, however the dynamics of air flux, diffuser density and depth affecting performance are relevant to all variations of the activated sludge process. Oxidation ditches traditionally operate on the basis of supplying oxygen to the biological process in dedicated aeration zones, utilizing simultaneous mechanical generation of horizontal flow to ensure mixing of liquor and transfer of dissolved oxygen to non-aerated zones. Depending on the design of the ditch, non-aerated zones may also provide anoxic conditions for denitrifying bacteria.

Surface Aerators Versus Fine Bubble Diffused Aeration in Oxidation Ditches

Similar to conventional activated sludge reactors, the oxygenating capabilities of mechanical surface aerators and their SAE should be considered when assessing the power requirements for an oxidation ditch application. Surface aerators rely on producing plumes of droplets above the liquid surface which creates a large water to air surface area for oxygen transfer from the air to the liquid droplets. Oxygen transfer also occurs as air is drawn into the bulk liquid with the mechanical device. The mixing flow pattern in the reactor transfers oxygenated liquid from the surface. Inevitably, a gradient of dissolved oxygen levels forms, where the highest level of dissolved oxygen exists close to the surface and in proximity of the aerating device. In the first generation oxidation ditches, the mixing requirements within an oxidation ditch were satisfied by the aeration device, as vertical or horizontal shaft aerators create a horizontal flow pattern. Installation of equipment above the liquid surface was seen as an attractive feature of surface aerators.

There may be a number of limitations in such installations, including limited reach of oxygenated liquid in deep tanks, insufficient mixing of liquor, limited turndown capacity, low SAE, and generation of aerosol in the vicinity of the reactor as well as frequent maintenance of rotating parts. The mixing capabilities of such equipment may also limit the horizontal transfer of dissolved oxygen along the lanes of the ditch. Table 2 summarizes features and advantages of surface aerators and diffused aeration. See Figure 4 for the working principles of vertical shaft mechanical aerators and fine bubble diffusers with low speed horizontal mixers.

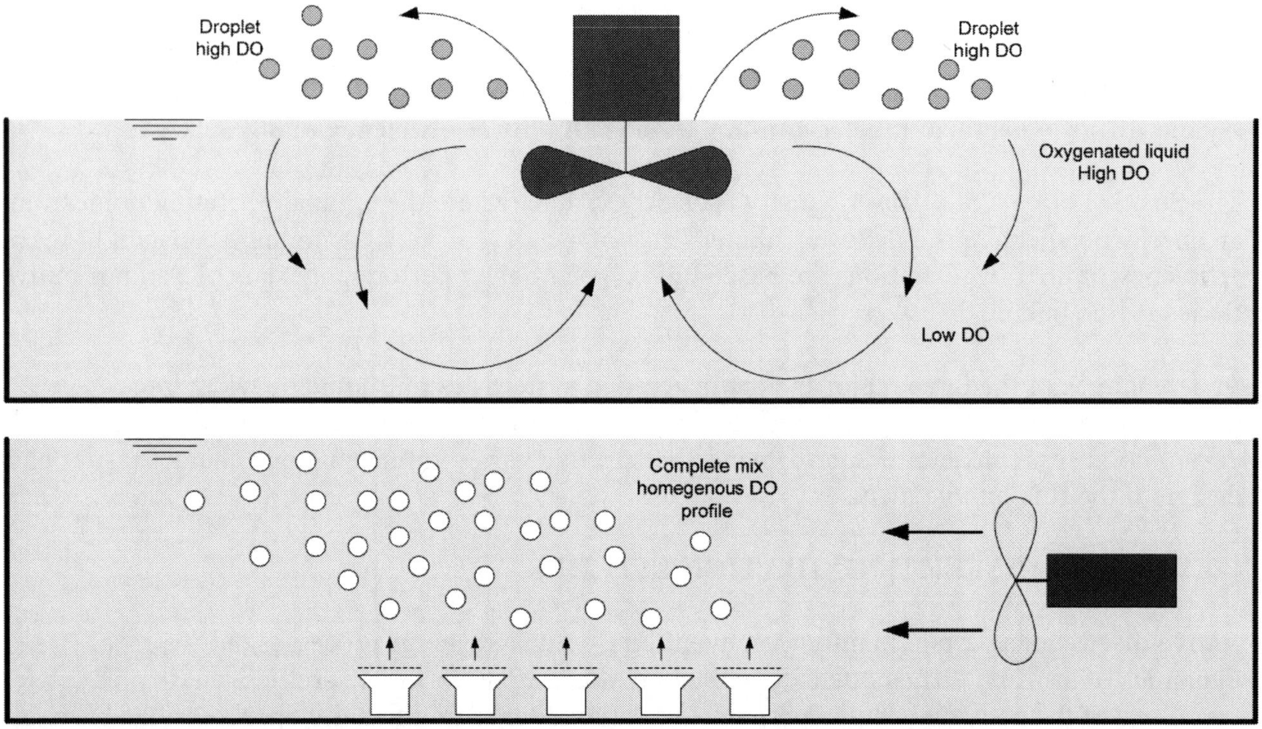

Figure 4 . Upper image: vertical shaft surface aerator which oxygenates liquid by generation of plumes of droplets above the surface. Lower image: diffused aeration in combination with submersible low speed mixers, enabling an effective horizontal flow and enhancing the oxygenation of mixed liquor.

Fine bubble diffused aeration offers many advantages to mechanical surface aeration—as discussed previously—in oxidation ditches in particular. General benefits of diffused aeration in combination with submersible mixers include increased SAE (including mixer power), high oxygen transfer rates, independent mixing and aeration devices, completely submersed system with low maintenance needs, and elimination of aerosol formation from rotating parts at the reactor liquid surface.

With independent aeration and mixing devices, fine bubble diffused aeration system not only provide the highest level of OTE, but also provide the capacity to efficiently address a wide range of operating conditions. Independent mixing and aeration offers another level of process control to minimize air flow requirement, adapting blower air flow or control valves against required and observed process parameters such as ammonia-nitrogen and dissolved oxygen concentrations. This level of operational control is restricted with traditional single device serving both aeration and mixing demands.

Table 2. Features of diffused aeration and surface aeration in oxidation ditches.

Feature	Diffused aeration	Mechanical surface aerators
SAE	Potentially as high as 7-8 kg O_2/kWh	Limited to a approximately 2 kg O_2/kWh
Installation	Diffuser system with piping and blowers. Submersible mixers to generate horizontal flow.	Single unit installation. May require additional stand-alone mixers.
Flexibility in engineering design and capacity	Stand-alone units enable tailor-made solutions which meet oxygen transfer demands.	Limited oxygen transfer variability. Number and size of units is the main flexibility factor.
Installation cost	Depends to a large extent on diffuser density and number of blowers and mixers required. Cost highly related to efficiency.	Relatively low-cost with few additional components. May require additional concrete reinforcements.
Aerosols	Very limited to none.	Common issue due to droplet generation required for oxygen transfer.
Process control capacity	Highly flexible with independent blower air flow and/or valve control.	Very limited.
Maintenance	Few mechanical parts which need maintenance.	Frequent maintenance of rotating aerating parts.

Combining fine bubble aeration with horizontal flow

In order to mix liquor in a biological reactor, sufficient thrust must be generated to keep particles in suspension. The use of thrust as a standardized parameter to measure and compare mixer performance is described by ISO (2007). Practically, a certain flow velocity should be maintained throughout an oxidation ditch. The relationship between the required thrust and horizontal flow velocity can be described by equation 3 (Uby, 2012):

$$F \sim k \cdot u^2 \qquad \text{eq. 3}$$

Where F = thrust, N
k = reactor momentum loss factor, -
u = horizontal flow velocity, m/s (ft/s)

For unaerated ditches, a horizontal flow velocity of 0.3 m/s (1 ft/s) most often satisfies particle suspension and mixing of liquor. The loss factor k is a function of tank geometry and obstacles present in the reactor. For oxidation ditches with dedicated aeration zones, the values of both k and u in equation 3 need to be determined as functions of the aeration/mixing horizontal flow effects expanded on below.

Using fine bubble diffused aeration and low speed mixers to generate a horizontal flow around an oxidation ditch presents additional consideration of the momentum losses generated by the vertical rise of bubbles across aerated zones. Proper placement of aeration and mixing equipment is required to ensure optimum performance with respect to mixing and aeration, otherwise spiral flow effects may develop, reducing transfer capacity and efficiency.

Assessment of the ability to produce an even horizontal flow can be done by considering the modified Froude number for such systems. The modified Froude number (Uby, 2012) can be correlated to the total aeration loss factor, where a low loss factor for optimal performance is desired.

$$Fr = \frac{u^2}{gS} \cdot \frac{u}{u_q} \qquad \text{eq. 4}$$

Where Fr = Froude number, -
u = average cross-section horizontal flow velocity, m/s (ft/s)
g = acceleration by gravity, 9.8 m/s^2 (32 ft/s^2)
S = submergence of diffuser, m (ft)
u_q = air flow per aerated total aerated area, m^3/m^2/s (ft^3/ft^2/s)

Figure 5 describes in qualitative terms the correlation between the Froude number and the aeration loss factor. For oxidation ditches, a Froude number value estimated using equation 3 should exceed 0.3 (an approximate critical Froude number) to eliminate spiral flows, ensure high oxygen transfer rates and reliable mixing of wastewater (Uby, 2012).

Designing a robust oxidation ditch with stable performance requires a critical Froude number which strikes a proper balance between design depth (relating to S), diffuser flux (relating to u_q), and mixer thrust (relating to u). The need to establish a critical Froude number relates to the point at which the horizontal thrust produced by submersible mixers breaks through the bubble curtain generated by the diffusers.

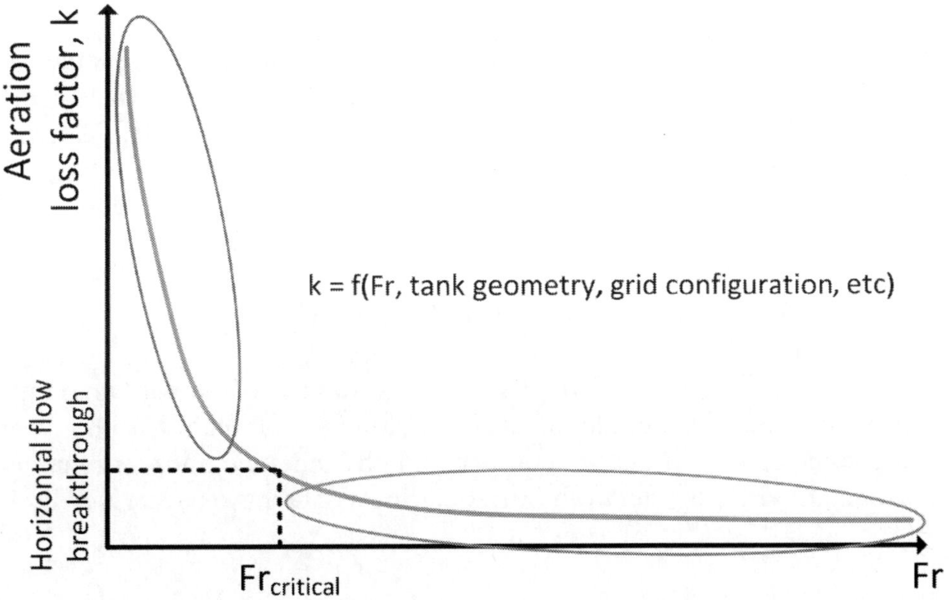

Figure 5. Depiction of the correlation between the Froude number and the aeration loss factor in horizontal flow-aeration dynamics.

While the aeration loss factor can be implicitly quantified in part by calculation of the Froude number; tank geometry, liquid viscosity, aeration grid layout and mixer positioning also constitute important parameters, as they influence the shape of the aeration loss curve in Figure 5.

The need to appreciate an oxidation ditch as a system with dynamic behavior—whose individual components operate in unison—must be emphasized. Parallel to the pathways of achieving high efficiency aeration systems in general (outlined previously in Figure 2), the aeration-horizontal flow dynamics in oxidation ditches can be described in a similar manner by Figure 6.

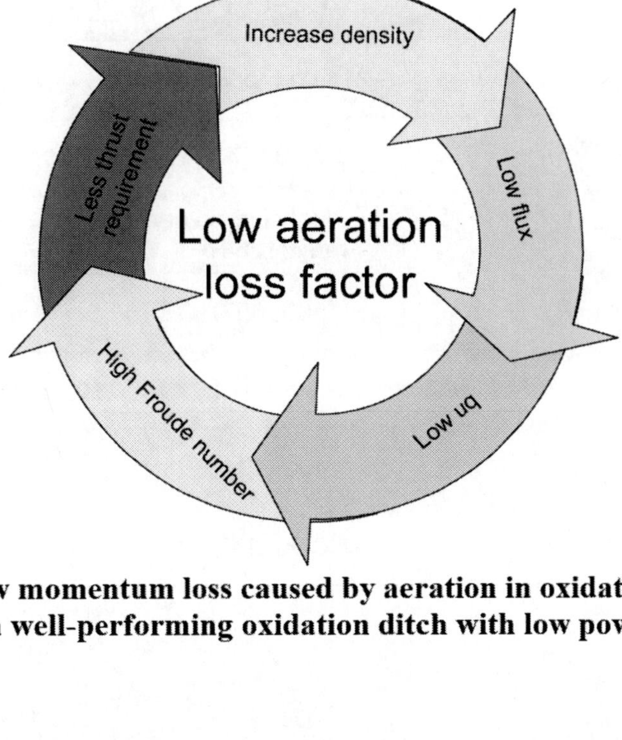

Figure 6. Achieving low momentum loss caused by aeration in oxidation ditches. A low loss factor is desirable for a well-performing oxidation ditch with low power requirement.

LESSONS LEARNED

In recent years, significant improvements have been observed at treatment plants which have operated oxidation ditch biological reactors with the use of mechanical surface aerators. Such plants consist of Eunice, LA, South Water Reclamation Facility, FL and Big Gulch, WA. These plants have all replaced mechanical aeration with diffused aeration in combination with submersible low speed mixers, experiencing improved performance and reduced maintenance needs.

Comparison study – South Water Reclamation Facility, FL

In 2002, South Water Reclamation Facility (SWRF), FL, upgraded one of two identically sized oxidation ditches, installing fine bubble diffused aeration in combination with low speed mixers. The upgrade facilitated an opportunity to compare side-by-side two nutrient removal processes with two fundamentally different aeration-horizontal flow systems as described by Reardon *et al* (2003).

The two ditches were subject to data collection during the same six months. Xylem Sanitaire fine bubble diffusers and Xylem Flygt mixers were installed in one ditch, designed to achieve both nitrification and denitrification. The other ditch maintained the original vertical/horizontal shaft surface aerators employed to provide both mixing and aeration in a single device. A summary of the performance after the six-month evaluation period is shown in Table 3. The fine bubble ditch received a higher hydraulic load, roughly three times higher, compared to the companion mechanically aerated oxidation ditch. Both lines produced very similar effluent quality. However, the upgraded fine bubble-low speed mixer ditch did exhibit significantly lower power costs per volume of treated wastewater, which was the factor used for normalization of energy comparison.

Table 3. Summary of performance comparison of the two ditches at South Water Reclamation Facility after six months of testing from 2002-2003 (Reardon *et al*, 2003).

Design feature/result	South East reactor	South West reactor
Tank volume, m^3 (ft^3)	21,405 (755,295)	20,837 (735,243)
Design flow capacity, m^3/day (MGD)	28,400 (7.5)	77,700 (20.5)
Number of air supply units	5 mechanical surface aerators rated 125 HP	2 air blowers rated at 600 HP
Average power draw, kW (HP)	505 (667)	664 (864)**
Power cost, $/kWh	$0.07	
Cost per treated volume of wastewater, $/km3 ($/MG)	432 (114)	204 (54)
aSAE, kg O_2/kWh (lb O_2/hp- h)	0.95 (1.6)	1.8 (2.9)**

In Table 3 above aSAE estimation is based on design AOR, measured power uptake and measured alpha factors at both process lines. Alpha at the SE reactor was 0.67 and 0.54 at the SW reactor. Power draw did not include mixer power for the SW reactor.

Note that if an AOR to SOR conversion was to be made using alpha as conversion factors, the South West process would operate at an SAE of approximately 3.3 kg $O2/kWh$ (5.4 lb $O2/hp.$ h) and the South East process at 1.4 kg $O2/kWh$ (2.3 lb $O2/hp.$ h), consistent with theoretical tabulated values shown previously in Table 1.

Even though mixer power should be included in the overall energy calculations, the SWRF facility clearly demonstrates the operational advantages that can be made from an energy perspective with a well-working fine bubble system with high oxygenating capacity, without any compromise of the biological process. In fact, a significantly larger biological load could be treated while meeting required effluent targets, with roughly equal operating power.

Creating a low maintenance environment with energy cost saving at Big Gulch, WA

At Big Gulch WWTP, WA, until 2009, two oxidation ditches were in operation using mechanical brush aerators. The plant experienced maintenance issues concerning the mechanics of the surface aerators in addition to aerosols around the bioprocess basins. To cope with increased influent TSS and BOD loadings, the two ditches were upgraded to operate with a diffused aeration system with automatically dissolved oxygen controlled turbo blowers. Low speed mixers were installed to generate the required horizontal flow (US Environmental Protection Agency, 2010).

Although the intent with the upgrade was primarily to improve operation, reduce maintenance and noise levels, the plant also enjoyed lower power consumption following the upgrades. The diffused aeration upgrade provided an annual energy savings of $10,000 compared to the four years before the upgrade when mechanical aeration was used.

From a process perspective, the new diffused aeration system, improved performance by reducing the impact of filamentous organisms on sludge settling. Chlorine usage, previously used to control filamentous outbreaks has been reduced with the aeration system upgrade. Fine bubble diffused aeration has provided a much wider control range, enabling the operators to control DO levels during the aerobic phases of the process, along with Oxidation Reduction Potential (ORP) monitoring during the anoxic phases.

The facility has also benefited from the elimination of aerosol previously produced by the brush aerators providing a safer area around the basins.

As a complement study to the South Water Reclamation Facility, Big Gulch showcases operational advantages apart from energy savings alone.

Solving performance problems at Eunice Wastewater Treatment Plant

As discussed previously, oxidation ditch design can be complex with regard to the dynamics of the aeration-horizontal flow. Plants using oxidation ditch technologies should consider installing equipment from suppliers with experience combining low speed mixers and diffused aeration to ensure proper equipment application to provide process and hydraulic performance. The Eunice, LA facility contacted Xylem to investigate the supply of Sanitaire diffused aeration and Flygt low speed submersible mixers to replace the existing aeration and eliminate effluent problems at the site. The previous ceramic diffusers and mixers did not generate the desired process performance over time.

Previously, the facility could not deliver adequate oxygen to maintain the dissolved oxygen levels required for the reduction of ammonia. The highest level of dissolved oxygen was 1.75 mg/L. The system now supplies an average of 5,851 kg standard O_2/day (12,900 lb/day) to achieve a minimum of 2 mg/L dissolved oxygen. The mixer upgrade provided a healthy 0.3 m/s (1 ft/s) horizontal flow velocity, producing a clear effluent with BOD, ammonia and suspended solids concentrations well below plant permit limits (London, 2010).

TAKING PROCESS AND AERATION PERFORMANCE TO THE NEXT LEVEL—ADVANTAGES OF HIGH DENSITY, LOW FLUX DIFFUSER INSTALLATIONS AND IMPROVED MIXER HYDRAULICS

Taking into account the positive effects of low flux diffuser systems for suspended growth aerobic biological reactors in general, fine bubble high density diffusers offer another choice along with disc diffusers to satisfy oxidation ditch aeration needs. Increasing diffuser floor density not only positively influences SOTE and SAE, but also contributes to high Froude number (and low aeration loss factor due to a low u_q) as can be derived from equation 4.

Installations with HDLF diffusers may favor the development of a homogeneous horizontal flow throughout the ditch due to significant reduction of the aeration loss factor. Figure 7 shows HDLF diffusers installed at a membrane density of 40%, making possible a low air flux, effectively reducing mixer power as well as enhancing OTE and SAE when compared to an equivalent installation with lower diffuser density.

Figure 7. Left: example of an installation of Sanitaire (a Xylem brand) Gold Series HDLF diffusers on a liftable grid with a 40% diffuser density. Right: single Gold Series diffuser.

For ditches with limited possibilities—for example due to available footprint—of achieving desired basin depth or diffuser density to reach a critical Froude number for proper horizontal flow, HDLF diffusers are viable alternatives.

Steady mixer hydraulics product development has yielded new low speed mid-size mixers with high efficiency with lower propeller diameter compared to large low-speed submersible mixers. For example shallow or narrow oxidation ditch channels can still take advantage of diffused aeration in combination with submersible low speed mixers with maintained high SAE and capacity, without resorting to a mechanical surface aeration installation. Figure 8 displays a comparison of 1.2 m (3.9 ft) mid-size and 1.4-2.5 m (4.6-8.2 ft) large low speed submersible mixers, suitable for oxidation ditch designs.

Figure 8. Left: mid-size low speed mixer suitable for oxidation ditches where reactor geometry does not allow for large, low speed mixers (Xylem Inc., 2012a). Right: large, low speed mixer highly suitable for oxidation ditches, capable of high thrust at low power input (Xylem Inc., 2012b).

Thus, the development of mid-size, high thrust and efficiency propellers provide design flexibility even for ditches less than 3 m (10 ft) deep.

SUMMARY AND CONCLUSIONS

Two viable alternatives for aerating biological reactors in municipal water resource recovery facilities often stand side by side: mechanical surface aerators and fine bubble diffused aeration. Mechanical surface aerators are often found in oxidation ditches but lack many of the features required to provide low operating cost with optimized aeration efficiency while maximizing oxygen transfer capacity without compromising mixing. Diffused aeration is an engineered solution which requires insight into the factors affecting the degree of oxygen transfer for a given reactor geometry and diffuser configuration. However, with experience, such installations offer tangible advantages to mechanical surface aeration including high aeration efficiency, low maintenance, a safer working environment, and increased process control flexibility.

Oxidation ditches present additional challenges in terms of achieving an adequate horizontal flow throughout the reactor. Highlighted in this study are examples where Xylem has successfully incorporated diffused aeration systems with low speed horizontal flow mixers to lower operational costs and at the same time improve effluent quality along with reduced equipment maintenance. It has been demonstrated with the proper engineering experience, diffused aeration and submersible low speed mixers can provide increased treatment capacity, improved energy efficiency, and operational flexibility when compared to mechanical surface aerators.

Due to the nature of the dynamics of the aeration-horizontal flow phenomenon, the use of low flux, HDLF diffusers has the ability to further improve oxygen transfer efficiency and mixer thrust requirement without compromising effluent quality. Further, as an alternative to large, low speed mixers, the next generation of mid-size mixers may offer extended design flexibility.

REFERENCES

American Society of Civil Engineers (2007). *Measurement of Oxygen Transfer in Clean Water;* ASCE/EWRI 2-06. Reston, VA.

European Committee for Standardization (2003) *Wastewater Treatment plants – Part 15: Measurement of the Oxygen Transfer in Clean Water in Aeration Tanks of Activated Sludge;* EN 12255-15:2003 E. Brussels, Belgium.

Tschobanoglous, G.; Burton, F. L.; Stensel, H. D. (2003) Wastewater Engineering: Treatment and Reuse. McGraw-Hill Education. London, UK.

ISO (2007) *Pumps – Testing – Submersible Mixers for Wastewater and Similar Applications*; ISO 21630:2007.

Uby, L. (2012) Handbook of Mixing for Wastewater and Similar Applications. Xylem Water Solutions AB. Sundbyberg, Sweden.

Reardon, R. D.; Karmasin, B. M.; Prieto, L. M.; Hurley, W. J.; VonMutius, J. (2003) A Side-by-Side Comparison of Step-Feed BNR and Simultaneous NDN. CDM: Maitland, FL.

US Environmental Protection Agency (2010) *Evaluation of Energy Conservation Methods for Wastewater Treatment Facilities;* EPA 832-R-10-005, 1-14.

London, S. (2010) Elected Changes Restore Compliances at Eunice, LA, Wastewater Treatment Plant. Water Online The Magazine: Erie, PA.

Xylem Inc. (2012a) *Flygt mid-size Mixers: Optimal Efficiency in a New Dimension*. 1411. Flygt mid-size mixers.1. Master.1. 20120419.

Xylem Inc. (2012b) *Flygt low-speed mixers: outstanding efficiency.*

Guide to Bioaugmentation for Systems Experiencing Chemical Inhibition using an Operating Envelope

Robin Schroeder[1], Martin Maxwell[1], and Dong Hyun[1]
[1]Environmental Business Specialists, LLC, Mandeville, LA 70470

Abstract

Improved treatment and enhanced biogas production is contingent on the activity and overall health of the biomass. Chemical inhibition and toxicity in wastewater treatment plants has notoriously been difficult to diagnose much less prevent. This paper applies the operating envelope developed by Torrens and Salerno (1996) to inhibition in a treatment plant and thereby provides guidelines for monitoring and controlling the system when chemical inhibition is present. The operating envelope can be used to determine the optimum MLVSS required to treat an inhibited system, but many times increasing the MLVSS alone is not enough. Bioaugmentation can be used to alter the effective biological kinetics both by increasing degradation of the chemical components and increasing the number of active bacteria in the system. Field trials shows that bioaugmentation improved the kinetics in an inhibited sytem increasing the K value from an inhibited 2.1 and increased the K level above normal operating values of 12.2. Bioaugmentation will also secondarily increase the %MLVSS in a system improving the operating envelope. Operator guidelines are established using a bioaugmentaiton-modified kinetic line to determine when to increase the MLVSS and at what point to begin wasting. In practice, spiked DOUR compared to ATP generation and MLVSS values can be used in the field to determine where the system is in the envelope and thereby what response is required.

1. Introduction

Wastewater treatment system optimization and treatment efficiency depends on the health of the biological secondary treatment system in both aerobic and anaerobic systems. Improved treatment and gas production therefore also relies on system stability and biological health. For example, there have been many studies detailing the optimization of biological nitrogen removal (BNR) from high strength digester side streams and increasing anaerobic gas production by various methods. Many of these articles indicate system physical-chemical conditions that must exist to promote the process including: heat balance (Tatersall et al., 2011), consistent influent flow and loading (Sarioglu, 2011), alkalinity (Steidel et al., 2011), and pH (Khanal et al., 2004). Most growth pressures are adjusted chemically or through engineering controls and there are monitoring strategies to account for them. However, bacterial inhibition and toxicity from chemical components is more difficult to identify and monitor because most of these chemicals cannot be detected by typical laboratory testing.

Chemically driven toxicity and inhibition in a secondary wastewater system has several pathways of action including physically coating the surface of the bacteria, solubilizing the cell membrane, disrupting enzymatic activity, and generally creating uncoupling (Liu, 2003) leading eventually to cell death. The observed result of these actions can include higher effluent TSS and poor settling due to deflocculation. The loss of bacteria to the effluent will lead to trending lower MLSS which will eventually lead to a loss of treatment as the Food:Mass ratio shifts. BOD breakthrough can occur at any time if the chemical involved has a higher BOD associated with it or if the concentration is high enough to cause rapid cell death in the system.

Chemical inhibition and toxicity can have acute (occurring rapidly after the chemical is introduced) and/or chronic (periodically causing an upset and generally lowering system health) impacts. Acute chemical impact occurs when the chemical is present in the system above its threshold of inhibition, or at a concentration high enough to cause a significant impact on the system. Every chemical has a unique threshold or concentration above which that chemical is inhibitory. Chronic impacts are generally associated with chemicals that preferentially adsorb onto the floc or bioaccumulate in the system. The symptoms of chemical inhibition include uncoupling (dissipation of the proton gradient of the cell membrane before it can be used to provide the energy for oxidative phosphorylation) and deflocculation (Nedergaard et al.,2005). Over time bioaccumulation can cause the chemical to concentrate above its threshold for inhibition. There are many scenarios by which chemicals can negatively impact the system, however once the negative symptoms are observed the response is fairly generic. Most standard operator manuals recommend decreased wasting as the first response and most consultants will then recommend bioaugmentation to supplement. Unfortunately, there is little guidance on when one or the other option is appropriate. If the chemical is bioaccumulating, decreasing wasting could further degrade the system while bioaugmentation is expensive and cannot be used as the sole option for response.

Torrens and Salerno (1998) developed a comprehensive method of looking at system capacity and design using aeration capacity of the reactor, solids capacity of the secondary clarifiers, and bacterial kinetics for substrate removal. They call this model an operating envelope. This process can be enhanced and slightly modified to also represent defined responses to chemical upset conditions and theoretically can be applied to both aerobic and anaerobic conditions.

2. Materials and Methods

An industrial activated sludge plant volunteered data for this analysis. Their BOD_5 data was used to determine the normal condition kinetics for the system using the equation developed by Eckenfelder et al. (1974) and modified by as shown by Torrens and Salerno (1996):

$$X_v = \frac{S_o\ (S_o * S_e)}{S_e * t * K}$$

X_v = MLSS; mg/L
S_o = Influent BOD_5; mg/L
S_e = Effluent BOD_5; mg/L
t = hydraulic retention time; days

K = Organic removal rate/day; mg/L*day

A full year of historical lab data was provided, which included two months where the system experienced an upset involving chemical inhibition where bioaugmentation was used to help the system recover. This data included BOD from the influent to the effluent, MLSS, and MLVSS. The K value was calculated for each season during normal operation, during inhibited conditions, and during bioaugmentation activities.

MLSS for this particular system is generally maintained at or below 4500 mg/L to stay within the performance limitations of the secondary clarifiers with 5500 mg/L as a maximum operating standard. Sludge age is maintained and monitored, however an expansion of the operating envelope by Boero and Eckenfelder (2003) shows that higher sludge ages can also change the operating envelope and this consideration may need to be included for other applications.

Aeration capacity limits BOD_5 to 54,000 lbs O_2/day. For this analysis, oxygen is not a limiting factor so the aeration limitation line is removed from the graphs, however it should be noted that if both inhibitory conditions and aeration limitations existed, this additional line would reduce the operating envelope as described by Torrens and Salerno (1996).

Chemical analysis of system influent and effluent was completed during the upset and each month afterwards. Dr. Clifford Lange at Auburn University completed the chemical analysis using a combination of high-performance liquid chromatography (HPLC), a gas chromatograph installed with a FID, and a gas chromatograph. The resulting chemical fingerprint provided a detailed list of chemicals in the waste stream as well as their relative concentrations.

Biological analysis was completed by both Dr. Lange and Environmental Business Specialists, LLC using staining techniques to quantify the number of total bacteria compared to the actively respiring fraction (Bartosch et al., 2003). Viability plates were used to quantify colony forming units and nutrient content of the biomass was monitored to ensure the bacteria were not limited with respect to N and P. Nutrient data is not shown but supplemental nutrients were applied to the system as needed.

Under normal operating conditions, a sample was analyzed using a Surcis BM-Advance open-system batch respirometer to compare treatment efficiency of mixed liquor from the wastewater system with and without bioaugmentation supplement (Surcis, S.L., Barcelona, Spain). The COD fractions were analyzed along with the resulting chemical fingerprint for each scenario.

3. Results and Discussion

An analysis of the organic removal rate (K value) of the wastewater system is shown in Table 1. From this analysis it is obvious that upset conditions lead to decreased bacterial activity, in this instance due to the presence of inhibitory compounds causing uncoupling (analysis not shown). Dr. Lange observed poor ATP production at a high DOUR and low floc particulate charge leading to deflocculation as well as the inability to form floc in samples taken during the upset earlier in the year. A chemical analysis confirmed the presence of high concentrations of inhibitory chemicals (data not shown).

Table 1. K value analysis between normal and upset conditions			
Upset Conditions	Heavy Bioaugmentation	Conservative Bioaugmentation	Normal Conditions
2.1	77*	17	12.2

*calculation biased due to high dispersed bacteria population

Figure 1 shows the system MLVSS in relation to K values and known system operations. This figure details two upset scenarios. The K values during the first upset are artificially elevated due to the high number of dispersed bacteria in the samples. Dispersed bacteria are not captured on a typical TSS filter pad and therefore the MLVSS is reported lower than the actual biomass in the system. The lower reported MLVSS results in a falsely elevated K value. Despite the bias, K values were much higher than under normal conditions. Bioaugmentation was added conservatively at the beginning of the first upset; more aggressive bioaugmentation was implemented to try to increase MLSS/MLVSS concentrations. Once the system began to recover, bioaugmentation was reduced back to conservative concentrations and then stopped when the system returned to normal operation. The successful implementation of bioaugmentation led operators to begin feeding bacteria earlier during the second observed upset leading to a shorter recovery time and less overall application of bioaugmentation.

Figure 2 shows the operating envelope of normal operations and inhibited conditions compared to enhancements with bioaugmentation. In bench scale respirometry testing under normal conditions, bioaugmentation increased the K value by a factor of 1.7 under normal (conservative) dosing of bacteria.

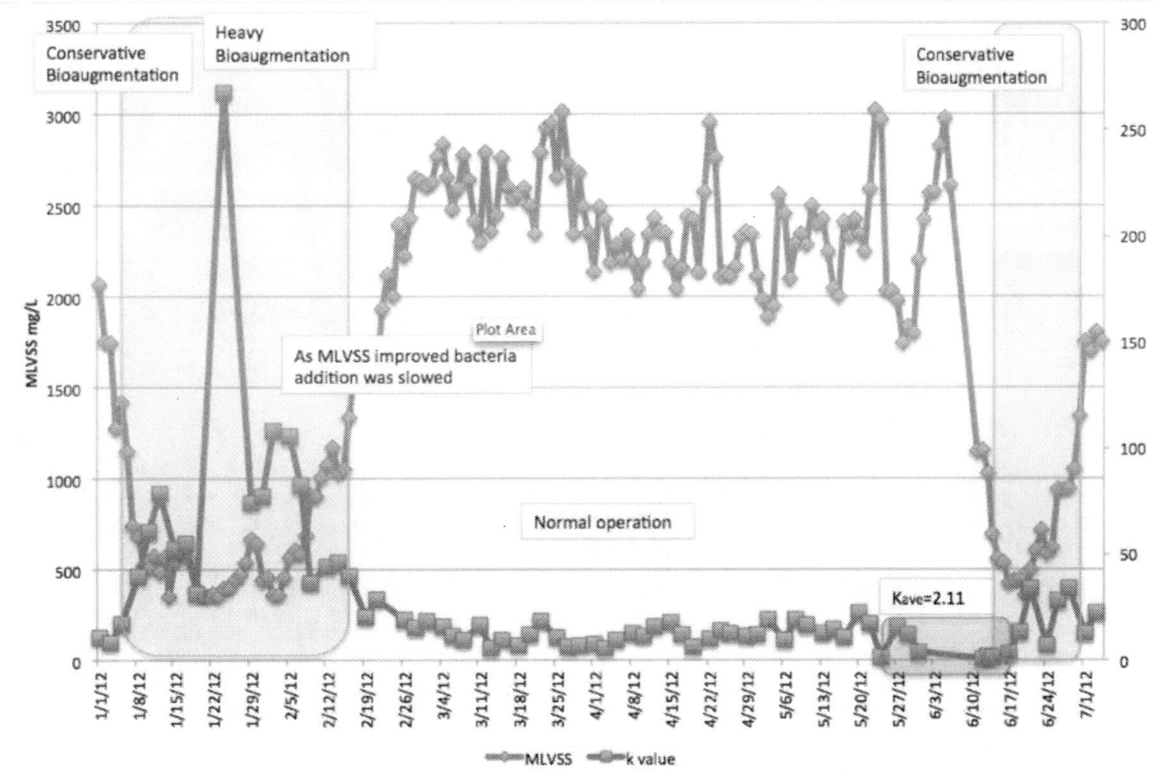

Figure 1. Analysis of K value changes with respect to MLVSS and known operational changes.

The inhibited K value shows that bacterial activity is heavily suppressed even when the MLVSS is increased to its maximum value; however, with bioaugmentation the K value can be brought back to normal operating conditions (based on field values). This represents an 8-fold increase over measured inhibited K values.

The dramatic increase in K values when applying bioaugmentation is a combination of two factors:

1. Bacterial supplements increase the concentration of bacteria in a given sample above what increases in MLVSS alone can accomplish, and
2. Bacterial supplements have additional enzymatic pathways that can degrade a more diverse food source.

Figure 3 shows that a process stream with high concentrations of inhibitory compounds can be degraded to a larger extent when bacterial supplements (EBS Biostar© P) are added to the sample. An analysis of COD fractionation in the same study showed the readily degradable fraction was increased from 35% to 50% and the slowly degradable fraction was increased from 5% to 17% with use of bioaugmentation products (Unpublished Data).

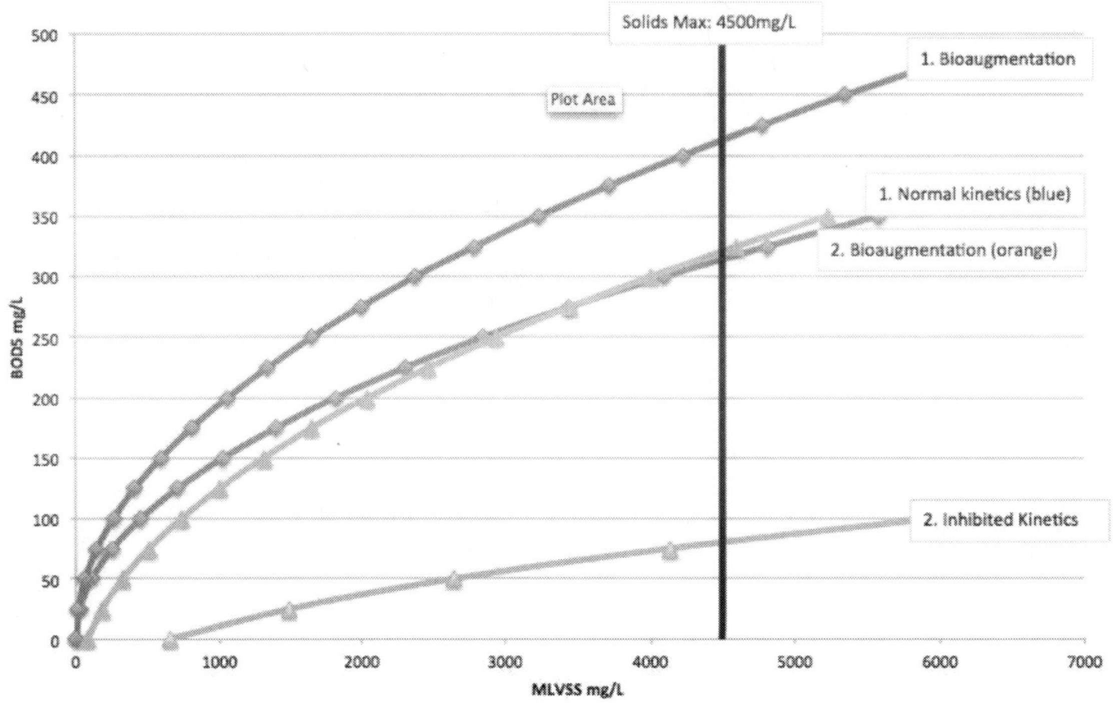

Figure 2. Operating envelope for paper mill under normal and upset conditions.

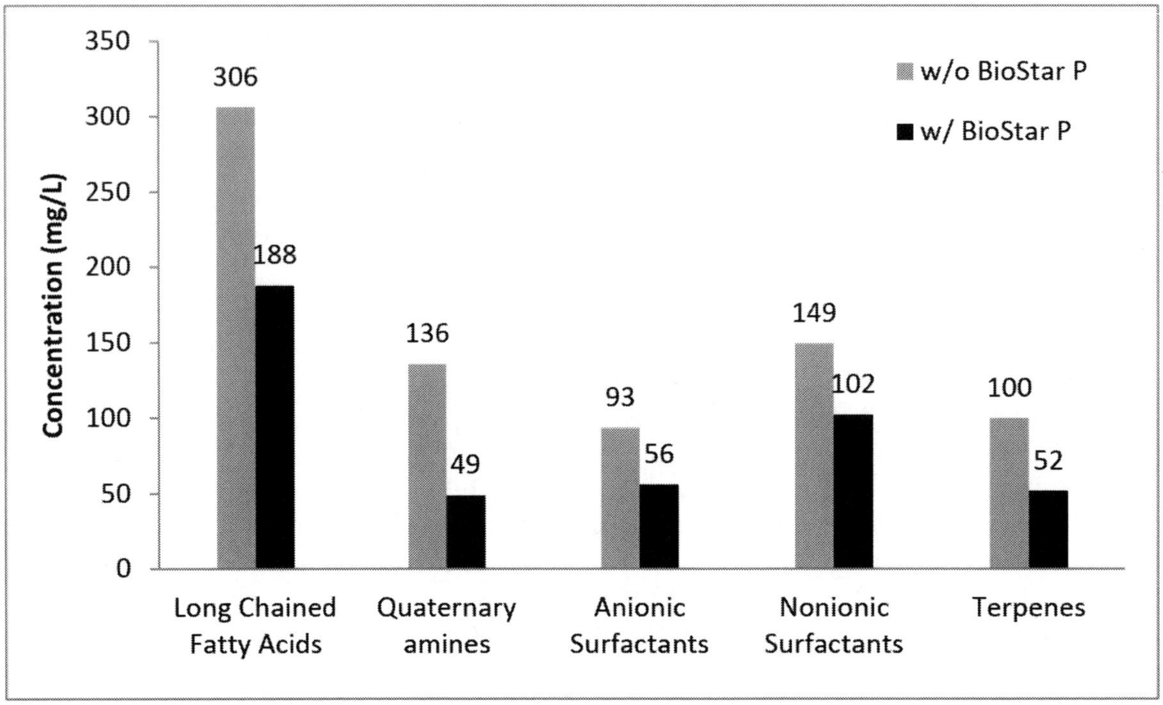

Figure 3. Analysis of chemical constituents after a respirometry study with and without bioaugmentation.

4. Conclusions

The data presented can be used to apply the operating envelope to suggest a guideline for operation under inhibited conditions as shown in Figure 4. The threshold for inhibition is the maximum K value that the biomass can operate at without beginning to fail (deflocculate) when inhibitory chemicals are introduced to the system. The percent activity in the MLVSS at the time the system is impacted will determine if the K value is higher or lower than the established line. Bioaugmentation can increase the percent activity of any given system and thereby boost the health and activity of the bacteria by adding more active bacteria and increasing the degradation of inhibitory chemicals, which will expand the operating envelope. The MLVSS can also be increased by decreasing solids wasting to better protect the system from failure up to a critical point. Once the maximum solids are entrained in the system, the solids should be wasted to prevent excessive bioaccumulation and supplemented with bioaugmentation to enhance treatment. At this critical point, wasting should be controlled by intensive monitoring of bed levels in the secondary clarifiers to keep the maximum number of solids in the system without overloading the equipment. Proper bioaugmentation and wasting can also secondarily increase percent MLVSS by increasing the number of bacteria per floc that will also lead to a larger envelope.

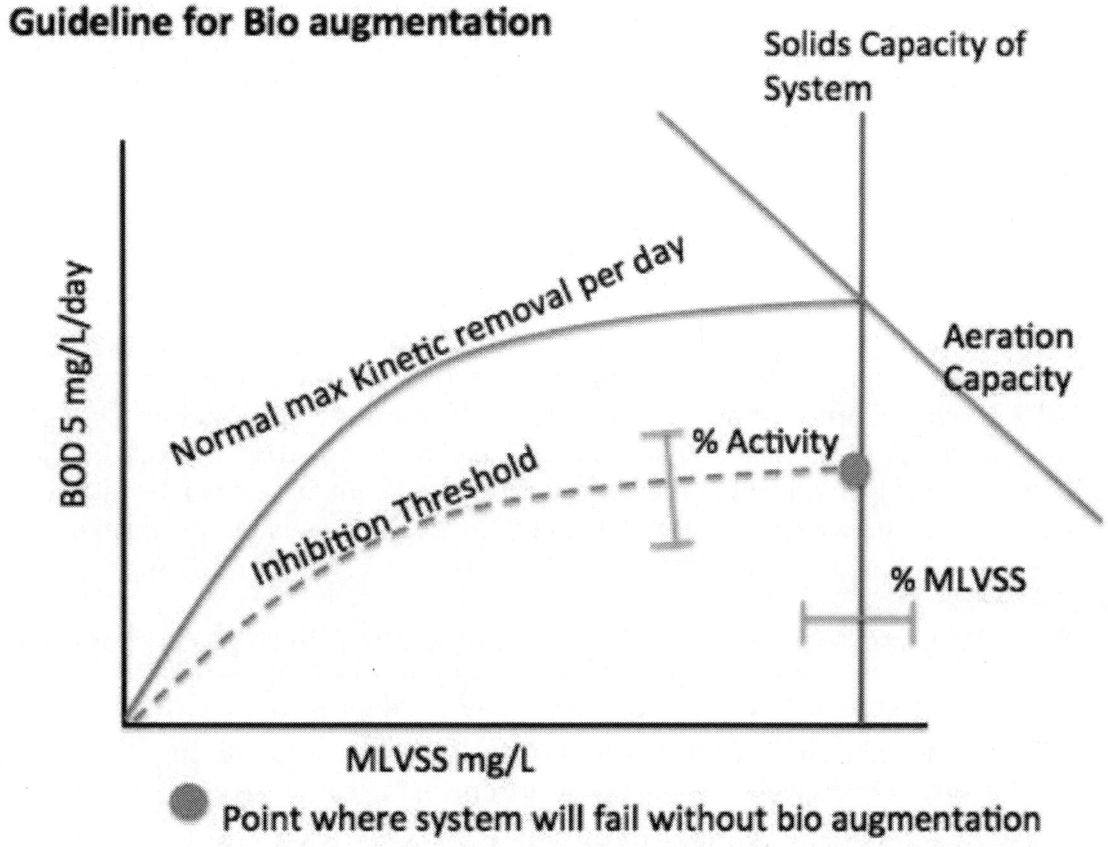

Figure 4. Application of operator envelope for inhibited conditions.

From an operator's standpoint, the symptoms to look for in an inhibited system are as follows:

- High effluent TSS
- Low fed dissolved oxygen uptake rate (DOUR) relative to optimal
- Poor settling
- No drop in DO residual (assuming the residual is normally above 1 mg/L)

The following operator guidelines can be used to respond to symptoms of inhibition:

- Failure of a fed DOUR test is the most immediate way of checking for possible inhibition. Stop wasting and increase MLVSS.
- High effluent TSS with no higher loading (immediately signaled with DO residual; later with BOD testing) can be the first stage of an inhibited system. At this point stop wasting and start bio augmenting until the symptoms stop appearing.
- If the maximum solids are reached under this scenario, begin wasting again and more aggressively feed bio augmentation products. Monitor sludge bed levels in the clarifier to determine when to waste rather than the MLSS concentration in the reactor. Maintain as high solids as possible until the symptoms of inhibition cease.

The application of this model to inhibited systems is intended to provide operators with guidelines that will prevent system failure and answer the age old question: When is bioaugmentation necessary? There are several field methods to address inhibition as stated above, but also note that advanced biological testing and chemical analysis can also be of great use. Monitoring the ATP production and activity of the biomass can be used to determine when bioaugmentation is no longer required and chemical analysis can be used to track the source of the chemical that is causing inhibition in the system.

Theoretically, this model can also be applied to anaerobic reactors. Anaerobic bacteria are susceptible to many of the same inhibitory chemicals that attack aerobic systems and bioaugmentation of anaerobic systems can be used to boost MLVSS. However, under anaerobic conditions the monitoring guidelines will need to be revised. Regardless of the metabolic regime of a system, this application of the operating envelope will lead to better operation and control of a wastewater treatment system, which in turn will lead to more reliable system performance.

References:

Boero, V.J., and W. W. Eckenfelder. (2003) The operating envelope concept and the Eckenfelder activated sludge model used for COD removal design. Industrial Wastes Conference in San Antonio, TX, 2003. Pp. 892-921.

Bartosch, S., R. Mansch, K. Knötzsch, and E. Bock. (2003). CTC staining and counting of actively respiring bacteria in natural stone using confocal laser scanning microscopy. Journal of Microbiological Methds. 52:1. Pp. 75-84.

Liu, Y. (2003). Chemically reduced excess sludge production in the activated sludge process. Chemosphere, 50:1. Pp 1-7.

Nedergaard, J, D. Ricquier, L.P. Kozak (2005) "Uncoupling proteins: current status and therapeutic prospects" EMBO reports 6, pp 917-921..

Khanal, S.K., WH Chen, L. Li, and S. Sung. (2004). Biological hydrogen production: effects of pH and intermediate products. International Journal of Hydrogen Energy. 29:11. Pp 1123-1131.

Sarioglu, M. (2011). Startup and operation of a large wastewater treatment plant in warm climate and under modified conditions- a case study. Water Environment Federation Technical Exhibition and Conference 2011.

Steidel, R., C. Watson, E. Whitehurst, E. Cronin, A. Umble, L. Zuravnsky, Y. Fang, and A. Warakomski. (2011). Design of sidestream treatment for bioaugmentation at the City of Richmond, VA WWTP. Water Environment Federation Technical Exhibition and Conference 20ll.

Tattersall, JM, G. Knight, and B. Ning. (2011). 10 years of integrating thermal hydrolysis into biosolids treatment- What are the lessons learned. Water Environment FederationTechnical Exhibition and Conference 2011.

Torrens, K.D., and L.P. Salerno. (1996). Calculated Effort, Industrial Wastewater. Pp 39-44.

CHP System Implementation and Operations Optimization at F. Wayne Hill Water Resources Center

Richard Schoeck, P.E., PMP[1], Srinivas Jalla, P.E., LEED AP[2], and Richard Porter[1]

[1]*Gwinnett County Department of Water Resources* [2]*CH2M HILL*, srinivas.jalla@ch2m.com,

ABSTRACT

The F. Wayne Hill Water Resources Center (FWHWRC) is Gwinnett County's largest and most advanced wastewater treatment facility, with a maximum monthly average daily flow (MMADF) treatment capacity of 227,000 cubic meters per day (m^3/d) or 60 million gallons per day (mgd). The effluent from the facility is discharged to Lake Lanier, the County's drinking water source.

In 2008, to address the rising cost of operations, the Gwinnett County Department of Water Resources (DWR) initiated the Gwinnett Resource Recovery Program (GRRP). This program was designed to identify opportunities for more effectively managing the resources used and available within its operations, including reducing energy costs (Jalla and Richards, 2011). A business-case evaluation (BCE) concluded that combined heat and power (CHP) generation, utilizing biogas from FWHWRC's anaerobic digestion process, could be an economically viable alternative. The selected CHP system included a 2-megawatt (MW) internal combustion engine (ICE) and a fats, oil, and grease (FOG) and high-strength wastes (HSW) Receiving Station for co-digestion of external wastes. Bench-scale testing performed to verify the feasibility of co-digestion indicated enhanced biogas production and increased solids destruction. However, these co-digestion benefits decreased as the percent of co-digestates exceeded 20% of the overall volatile solids loading (Tandukar and Pavlostathis, 2011).

DWR implemented the CHP and FOG/HSW Receiving Station projects, called Gwinnett POWER (Processing Organic Waste for Energy Recovery), using a design/build (DB) procurement approach. The CHP system began operation in May 2011 and the FOG/HSW Receiving Station was put into operation in November 2011. The facility started accepting FOG wastes at the end of 2011. In parallel, DWR initiated primary clarification and thickening improvements that further optimized biogas production and CHP system operation. Based on the first full year of operations experience, the project has been a success and the ownership costs have been in-line with the initial planning estimates.

Key Words: CHP, co-digestion, energy recovery, operations optimization, ownership costs

INTRODUCTION

The F. Wayne Hill Water Resources Center (FWHWRC) is the Gwinnett County Department of Water Resources' (DWR) largest and most advanced wastewater treatment facility with a maximum monthly average daily flow (MMADF) treatment capacity of 227,000 cubic meters per day (m^3/d) or 60 million gallons per day (mgd) . The effluent from the facility is discharged to Lake Lanier, the County's drinking water source. The facility currently treats about 129,000 m^3/d (34 mgd) and receives primary and biological sludge via collection system from the Yellow River Water Reclamation Facility, an 83,000-m^3/d (22-mgd) MMADF Membrane Bio-Reactor (MBR) facility.

The FWHWRC liquid treatment train includes preliminary and primary treatment, secondary treatment with biological nutrient removal, tertiary chemical clarification, filtration (granular activated carbon or hollow-fiber membranes), and ozone disinfection. The solids treatment train includes thickening, anaerobic digestion (primary sludge and waste activated sludge), and dewatering (anaerobically digested and chemical sludges). The thickening system is currently being upgraded to enable primary sludge thickening to improve concentration feed solids to the anaerobic digesters. Biogas from five 1-million-gallon (MG) egg-shaped anaerobic digesters is stored in four steel storage tanks after moisture removal using a refrigerant dryer. The storage tanks have a working volume of approximately 98,000 cubic feet (ft^3) (0.73 MG) at an operating pressure of 55 pounds per square inch (psi) and a minimum required tank outlet pressure of 10 psi. The biogas from the storage tanks is routed through hydrogen sulfide (H_2S) and siloxane removal systems before use in the new CHP system. Excess biogas that cannot be stored is sent directly to the flare.

BUSINESS CASE FOR CHP

In 2008, the rising energy costs (Table 1) and declining revenues resulting from extreme drought conditions (Level IV) and the economic downturn forced DWR to initiate the Gwinnett Resource Recovery Program (GRRP), which included measures to review and optimize use of all the resources under DWR's control (Jalla and Richards, 2011). These factors created the perfect storm for these hardships to be turned into an opportunity.

TABLE 1

Historical Power Costs at FWHWRC

Year	Annual Power Consumption (kiloWatt-hours [kWh])	Annual Average Power Cost (/kWh)	Peak Monthly Average Cost (/kWh)	Peak Demand (kW)	Annual Power Cost
2006	42,898,407	5.26¢	7.80¢	6,840	**$2,255,970**
2007	41,155,895	5.65¢	9.39¢	7,229	**$2,324,790**
2008	47,280,512	7.37¢	10.21¢	7,783	**$3,485,254**

The GRRP resulted in a series of workshops during the first quarter of 2009 culminating in the identification of a broad range of energy recovery alternatives that included biogas (produced from anaerobic digestion at FWHWRC), solar energy, wind energy, and hydropower (in reuse line from FWHWRC). These alternatives were evaluated using DWR's business case evaluation (BCE) protocols. The beneficial use of biogas from anaerobic digesters at FWHWRC presented the most economically viable alternative (Forbes et al., 2010).

The BCE results showed that a single biogas generator would achieve a shorter payback period than two biogas generators in all the cases evaluated because of the reduced operations and maintenance (O&M) costs from economy of scale. Payback periods for various-sized generators were found to be similar at the current biogas production rate of 250 standard cubic feet per minute (scfm). However, as the biogas production increases, the annual O&M savings become much greater for the larger generators. The BCE determined that the payback period for the CHP system were highly dependent on power costs and biogas production (Figure 1) (CH2M HILL, 2009).

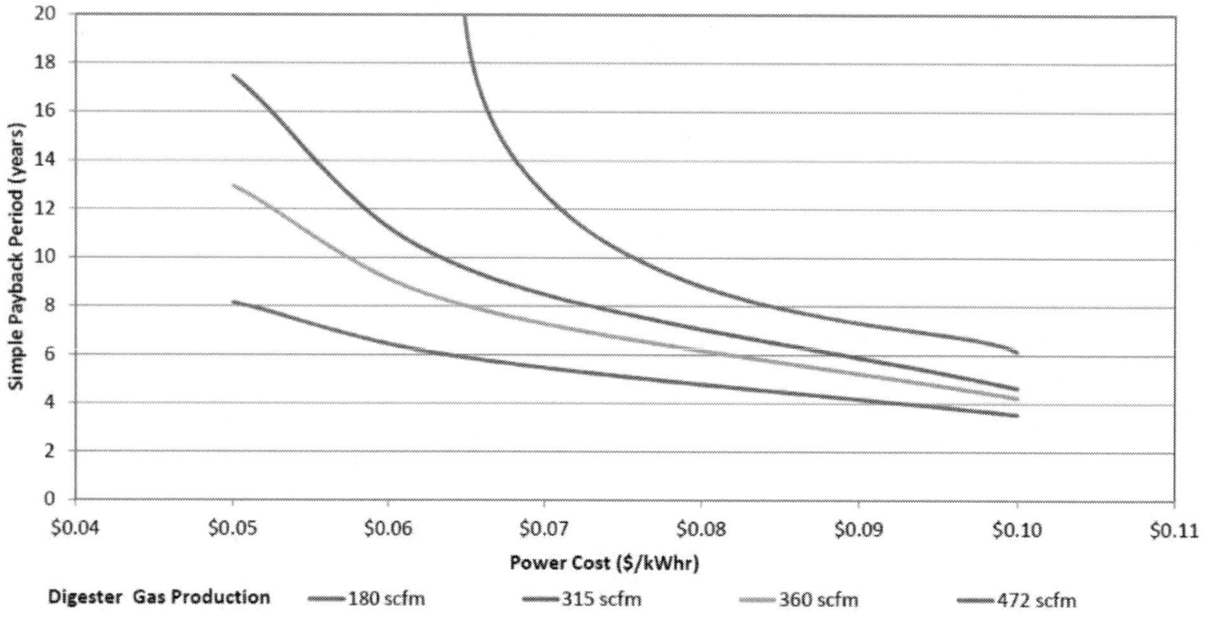

Figure 1: Impact of Biogas Production and Power Cost on Payback Period

The BCE also determined that the addition of a FOG Receiving Station would increase biogas production, provide a new revenue stream, and reduce the payback period of the combined project to approximately 9 years. The estimated capital cost of the combined FOG Receiving Station and CHP system project was $10.13 million. The annual O&M savings, Table 2, were estimated assuming revenues (tipping fees) from users of the FOG Receiving Facility at $0.10 per gallon and purchased power at $0.07/kWh.

TABLE 2

Estimated Annual O&M Savings with CHP System at $0.07/kWH

Year	Estimated Plant Influent Flow (mgd)	2-MW CHP System without Co-digestion	2-MW CHP System with Co-digestion
2014	34.7	$241,157	$1,054,595
2020	43.7	$328,682	$1,264,709
2030	58.6	$362,204	$1,458,524

The recommended project, Gwinnett POWER (Processing Organic Waste for Energy Recovery), Figure 2, included the following elements:

- Biogas treatment (hydrogen sulfide [H_2S], moisture, particulate, and siloxane removal)
- One 2-MW biogas-fueled internal combustion engine (ICE) generator
- Engine jacket and exhaust waste heat recovery systems
- Switchgear and related devices for power transfer and synchronization
- Fats, oil, and grease (FOG) / high-strength wastes (HSW) Receiving Station

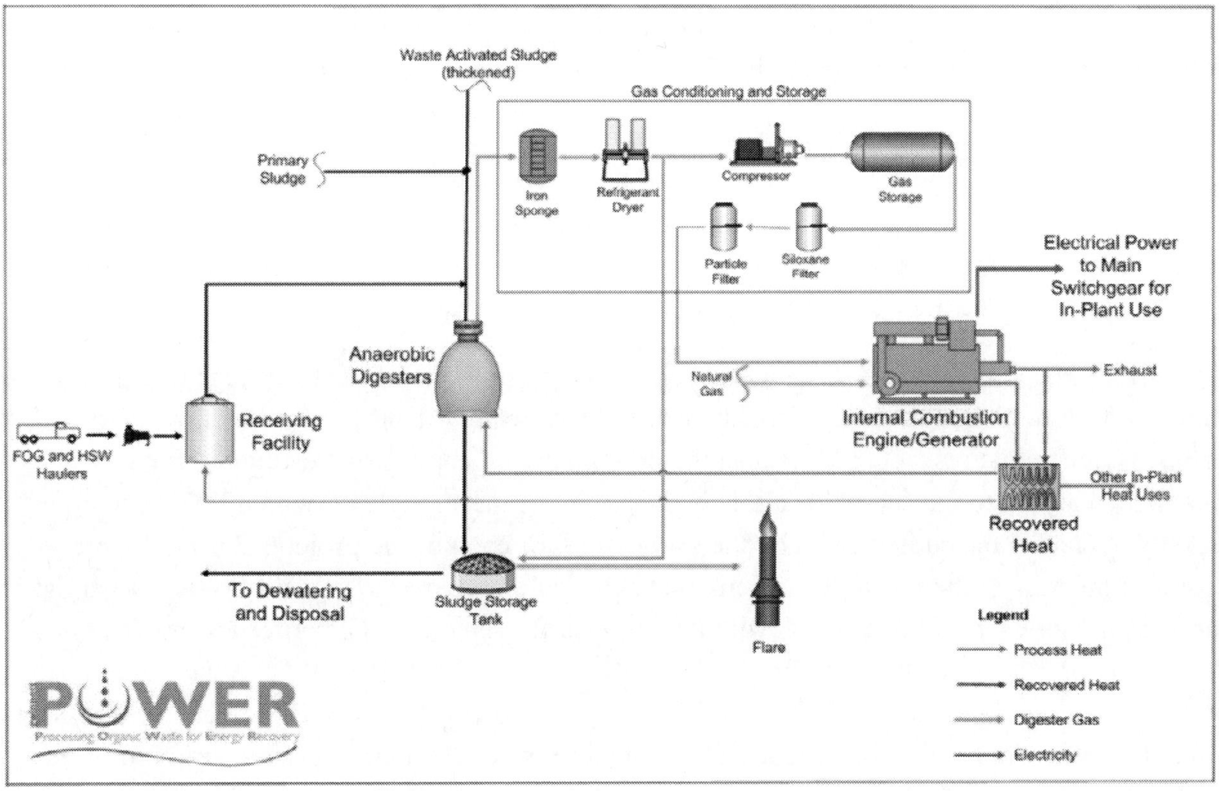

Figure 2: Gwinnett POWER Project at FWHWRC

WHEN PLANNING MEETS REALITY AND OPPORTUNITY

It is not unusual for owners to be conservative in both estimating capital costs and predicting the operational efficiency of a project, especially when it is a new process being evaluated. Likewise, when estimating operating costs and the costs of energy, owners use a conservative approach. As a result, DWR was reluctant to move forward with this project at the time the BCE was completed. However, when the option of grant funding from the American Reinvestment and Recovery Act (ARRA) became available, the initial borderline payback projections of the project were significantly improved and some of the risks associated with implementation were not only mitigated, but removed. DWR applied for and received the ARRA funds allocated for shovel-ready green projects through the Clean Water State Revolving Fund (CWSRF) and the Department of Energy's Energy Efficiency and Conservation Block Grant (EECBG). To meet the objectives of this project, it was still important that the project implementation costs not exceed those developed at the planning level and that significant energy benefits be realized.

Funding Influence

To take advantage of the ARRA grant funds, the Gwinnett POWER project was divided into two separate procurement packages. The $5 million ($3 million grant + $2 million loan at 2% interest rate to be paid over 20 years) from the CWSRF was used to procure the CHP system, which included a 2.1-MW engine generator, associated biogas conditioning, and waste heat recovery systems. The $4 million grant from the EECBG was used for implementing the FOG/HSW Receiving Station.

In order to meet the schedule requirements for the shovel-ready green projects, DWR utilized an alternate delivery, design-build (DB) method for both projects. The long lead times required for engine generators (approximately 8 months) led to quick approval of the engine generator design and fabrication while the CHP system design was in progress.

The DB approach is often used by owners on projects when one of the three legs of the "iron triangle" (Scope/Schedule/Budget) must be restricted, while the other two are allowed to compensate for that restriction. Because the planning level capital cost estimates were established as not-to-exceed values, this is the aspect of the triangle on which DWR remained firm by declaring the construction budgets available for each of the projects during the pre-proposal meetings. The DB procurement packages were prepared to present DWR's wish list of scope items for each of the projects with key criteria for selection. The criteria were designed to select the DB contractor that offered the innovation necessary to deliver the most scope for the budget available. Interestingly, in each of these two instances, the overall project schedule was not a driver except for the construction start date to meet the shovel-ready requirement of the ARRA funding. This flexibility within the Schedule leg of the iron triangle allowed more effective overall DB management during the implementation and start-up of each project, once the contracts were awarded. As a result, the emphasis on innovation was truly anchored in getting the most scope for the limited budget. The following benefits were observed by DWR from this procurement approach:

Opportunity for Innovation

The selected DB contractors for both the engine generator and FOG/HSW Receiving Station projects offered innovative alternatives in their proposals to maximize scope delivered without exceeding the DWR-announced budget. This approach allowed for the DB contractor to price the design options in real time to ensure that the final design is implementable within the contracted budget. An additional benefit of the DB approach was that the approach offered in the initial proposal remained consistently under review for improvement throughout the design, permitting, construction, and start-up phases of each project.

In the engine generator project, the DB proposal offered a larger engine generator (2.185-MW instead of the 2-MW specified) at a construction cost of \$5.2 million, \$0.54 million below the BCE's planning estimate. The reduced cost proposal required moving the engine generator closer to the digester complex to reduce the length of pipe runs (digester gas and hot water loop) and to take advantage of the switchgear previously installed for standby generators during Phase 1 of FWHWRC construction. The selected DB proposal offered the exhaust heat recovery system as an additional scope item at a cost of \$147,000 and used the existing liquid ring compressors, which consume more power to operate, instead of new centrifugal blowers specified in the request for proposals.

Figure 3: Containerized 2.1-MW Engine Generator being Delivered to FWHWRC

In the FOG / HSW Receiving Station project, the selected DB proposal relocated the station closer to the existing anaerobic digester complex to reduce the earthwork and avoid building modifications associated with the location proposed in the procurement package. Other innovative ideas included removing the decommissioned Regenerative Thermal Oxidizer (RTO) units at no additional cost to the owner to locate the receiving station. The DB contractor delivered the majority of the scope included in the procurement package at $3.16 million, with 80,000 gallons of storage tanks. The DB contractor scope also included $150,000 for bench-scale testing at the Georgia Institute of Technology (Georgia Tech) to determine the optimal loading rates and quantify the benefits of various FOG and HSW streams identified by DWR. The bench-scale testing not only identified the largest FOG hauler currently contracted by DWR but also showed a reduced benefit from FOG addition when volatile solids (VS) loading from FOG exceed 20 percent of the total VS load to the digesters (Tandukar and Pavlostathis, 2011).

The DB proposal also offered DWR the engineering support for development of a comprehensive FOG program, utilizing a benchmarking approach with other entities across the United States that already had similar FOG receiving programs. The DB contractor also investigated other Capital Improvement Program (CIP) projects currently underway and planned at the FWHWRC to look for mutually beneficial components. As a result, DWR was able to receive additional tankage in lieu of a dedicated odor control unit by increasing the scope of another project with an odor control system, thus reducing the overall combined cost.

Single Point of Responsibility to Reduce Risk

CHP systems are new to DWR and recognizing the complexity of these systems and the potential for start-up issues, DWR preferred a single point of responsibility offered by DB procurement for ensuring delivery of a functioning system that meets the specifications. The importance of a single point of responsibility was demonstrated during start-up, when problems with the exhaust heat recovery system (fire tube/water shell heat exchanger), Figure 4, resulted in multiple engine shut-downs. The exhaust heat recovery system experienced excessive pressures from formation and expansion of gases, resulting in the cracking of glycol into hydrogen and carbon dioxide. Initial efforts to solve the issues concentrated on the expansion tank sizing.

Several weeks later, when the tubes in the exhaust heat recovery system cracked, causing water leaks, the team realized that the gases were accumulating in the heat recovery system, displacing the water and resulting in elevated temperatures. Solving the issues associated with the exhaust heat recovery system involved computational fluid dynamics (CFD) modeling of the heat exchanger, a review by independent evaluators, and an inventory of glycol cracking related issues at other projects around the world. The solution included installing baffles inside the exhaust heat recovery system to eliminate short circuiting, increasing the heat exchanger operating pressure, and decreasing the glycol concentration (from 33% to 20%) to eliminate cracking issues (Hardy et al., 2012)

Figure 4: Inside of Exhaust Heat Recovery System Heat Exchanger (water shell/fire tube)

The waste heat recovered in excess of process heating demands was dissipated in a radiator. The radiator was initially located near the air inlet louvers of the engine generator. During start-up, the engine generator experienced frequent stops, resulting from high inlet air temperatures produced from the radiator-dissipated heat entering the engine. The DB contractor relocated the radiator to resolve the issue.

Trading Scope to Maximize Project Benefit

The DB approach allowed for implementation of insights gained during the various stages of project development (vendor procurement, permitting, construction layout, start-up, etc.) to adjust and confirm the scope so that the end product maximized the benefits DWR received. To provide a snapshot of the value that was offered, a couple of simple examples from the CHP project are summarized.

As discussed above, the DB contractor proposal only included the engine jacket heat recovery. To meet the BCE-estimated waste heat recovery goals, DWR worked with the contractor during the design phase to review various scope items and evaluate cost impacts resulting in multiple negative scope changes to accommodate the exhaust waste heat recovery system ($147,000).

These changes included elimination of the maintenance agreement for the H_2S treatment system ($50,000) and elimination of oil changes from the engine generator maintenance contract ($120,000).

The initial biogas sampling showed the H_2S concentrations to be significantly lower, approximately 100 to 115 parts per million by volume (ppmV), than inlet concentrations used for the design basis. Based on the lower inlet concentrations, the media is estimated to last longer than the project-required 2-year maintenance agreement. Considering the low estimated media replacement cost of approximately $15,000 the maintenance agreement was removed to receive a $50,000 credit.

Regenerative and non-regenerative siloxane treatment systems were also evaluated during the design phase. Based on the low siloxane concentrations, approximately 11 to 22 ppmV (as silicon [Si]) observed in the samples collected during the design phase, the media replacement frequency ranged from 300 to 600 days at a replacement cost of $12,000. DWR selected a non-regenerative siloxane treatment system based on the life-cycle cost analysis, resulting in a capital cost savings of $100,000 and maintenance agreement savings of $68,000. These savings were used to install additional natural gas and biogas flow meters required for emissions permit reporting.

Constructability and O&M issues were also effectively addressed. For example, DWR was able to question the DB contractor and ask for supporting documentation on the lack of a coating system being applied to the Calgon VaporPac stainless steel tank installations. As a result, Calgon re-evaluated and included bead blasting on these units based on the operating environment of these units. Also, the FOG/HSW project's DB proposal offered a credit to DWR for using existing hose pumps from FWHWRC. However, when the existing spare hoses for these transfer pumps started experiencing ruptures and clogging, the DB team committed the time and resources to investigating the issue--leading to the conclusion that a different hose material was needed.

An additional benefit offered in this overall approach was the opportunity for DWR to allocate costs to hire a Program Manager (PM) for both projects. This PM helped DWR in developing the procurement documents, reporting to state and federal agencies that provided grant funds to the project, and managing the DB team through design, permitting, construction, start-up, and operations.

CHP SYSTEM OPERATIONS OPTIMIZATION

Grant funding reduced DWR's net project cost to approximately $2 million at an annual interest rate of 2% for 20 years. Despite the low capital costs to DWR, the success of the project still depended on realizing significant energy benefits. The CHP system came online in May 2011 and the FOG/HSW Receiving Station was available for use in November 2011.

CHP System Operation Aligned with Power Tariff

The FWHWRC changed its power purchase tariff during implementation of the CHP system from Real Time Pricing (RTP) - Day Ahead (RTP-DA) to RTP - Hour Ahead (RTP_HA). The new tariff offered a lower average cost of power by increasing the variability in the unit cost of power during a day. The RTP-HA is ideally suited for facilities like FWHWRC that can reduce their energy load during the peak hours when energy costs are at their highs. In this tariff, FWHWRC received e-mail notification from the power utility on the day before with estimated unit power costs for every hour of the following day. The power cost profile from the notification allowed DWR to plan CHP system operations to maximize power cost savings.

In the example presented in Figure 5, the power cost ranged from 3 ¢/kWh to 40 ¢/kWh and the CHP system operation was coordinated to maximize power production during the hours having highest rates to reduce power purchased. This strategy resulted in average savings of 30¢/kWh for the power produced during the peak hours.

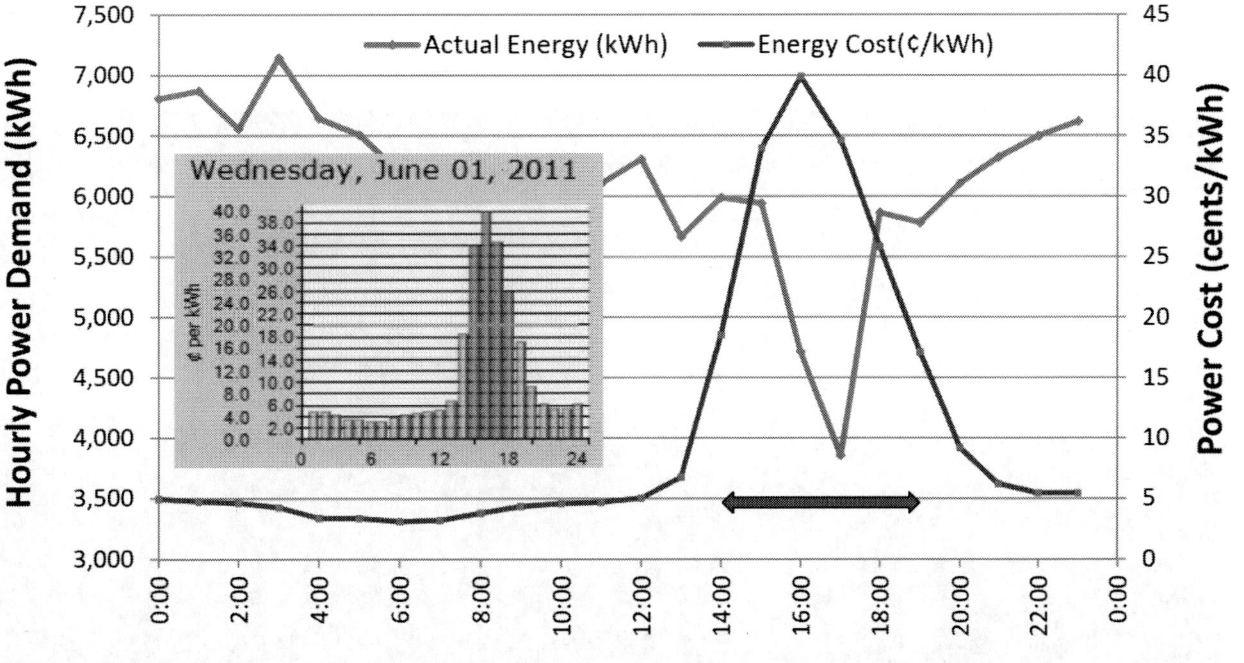

Figure 5: CHP System Operated to Maximize Savings per kWh Generated on June 1, 2011

The above strategy allowed for maximizing the cost savings per kWh generated during 2011 when the biogas production was low and power costs were highly variable. However, in 2012 the low utility power costs, averaged at 3.6 ¢/kWh, and stable hourly costs resulted in significantly fewer opportunities to maximize savings.

Non-Hazardous Trucked Waste Receiving Program

Georgia Tech conducted bench-scale testing for the purpose of establishing loading rates, biogas production, and recycle stream characteristics, for various potential waste streams considered for co-digestion with primary and waste activated sludges. Waste streams tested included dissolved air flotation (DAF) skimmings from a food processing facility, DAF skimmings from a poultry processing facility, HSW from chewing gum manufacturing, and commercial restaurant grease trap waste (raw and concentrated) (Tandukar and Pavlostathis, 2011). Based on the study findings, DWR negotiated an agreement with a grease trap waste (FOG) supplier for a portion of the receiving station capacity. Figure 6 shows the completed FOG/HSW Receiving Station.

Under the agreement, DWR received a monthly advance payment of $19,800 for a volume of up to 30,000 gallons per weekday. The agreement allowed for additional revenue at 3¢/gallon if the flows exceeded the calendar month's contracted volume. DWR signed agreements with other suppliers for up to 20,000 gallons per weekday in June 2012, also at 3¢/gallon received. During the first year of operation, DWR collected revenues of $288,000 from the FOG tipping fees and reached an average day FOG flow of approximately 50,000 gallons per weekday. DWR is currently reviewing and re-negotiating the agreements and anticipates the revenues to exceed $400,000 during 2013.

Figure 6: FOG / HSW Receiving Station and Egg-Shaped Anaerobic Digesters

The FOG program has significantly exceeded the planning goal of 30,000 gallons/day by 2014. However, the tipping fee supported by the market conditions has been significantly lower than the estimated 10¢/gallon. Continuous reliable operation of the receiving station over the last year has resulted in requests for capacity purchases from additional waste suppliers. DWR hopes the renewed interest from suppliers and the reputation gained will allow for increased leverage in setting the tipping fee and selecting higher quality waste streams.

Process Improvements Boost Biogas Production

In addition to the co-digestion of HSW like FOG, other facility improvements implemented under separate construction contracts have contributed to increasing biogas production. Figure 7 shows the increases in primary sludge and biogas production observed at FWHWRC during full-scale testing of the operational and hydraulic improvements currently under construction.

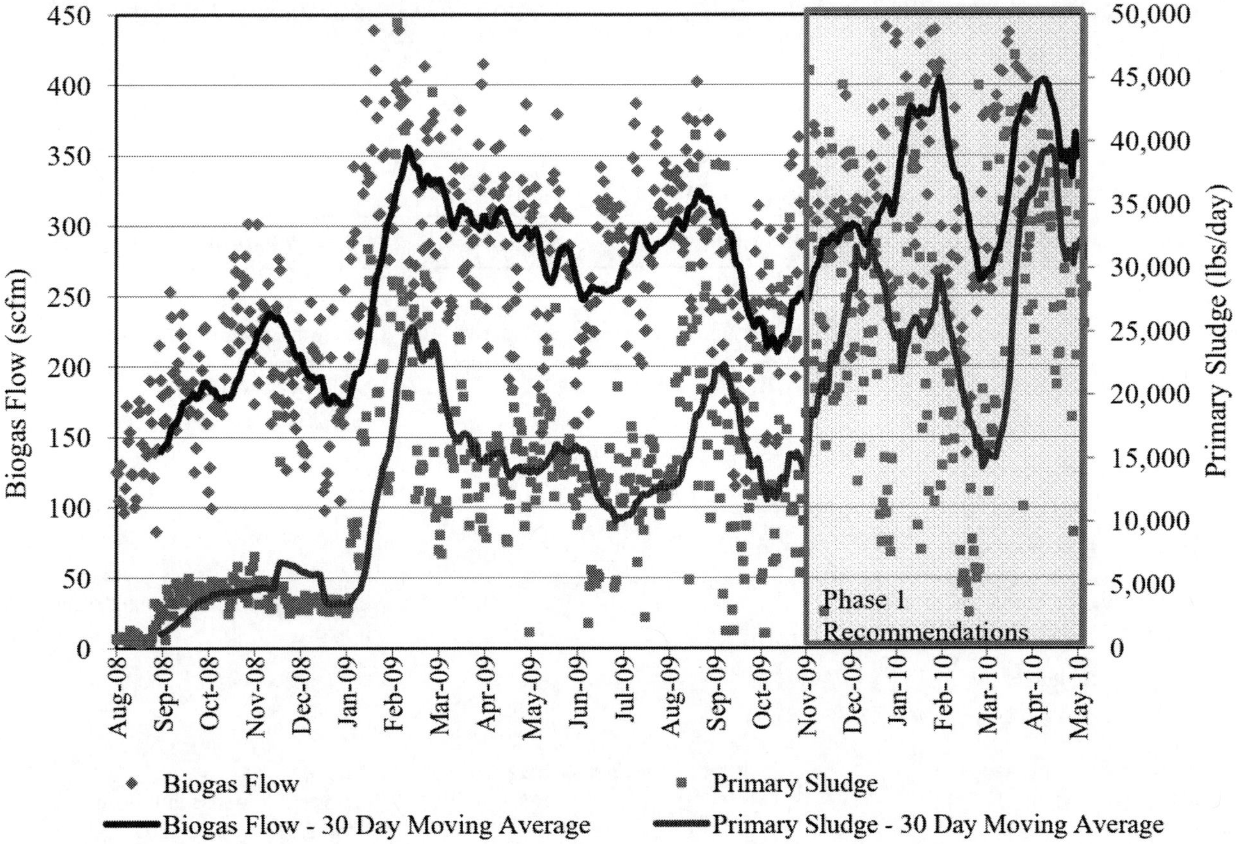

Figure 7: Primary Sludge and Biogas Production Data Before and During Phase 1

Additionally, DWR is currently replacing the thickening centrifuges with rotary drum thickeners (RDTs) to co-thicken primary and secondary sludges to improve overall digester feed solids concentration to 5.5%. The resulting increase in the anaerobic digester hydraulic retention time

(HRT) coupled with FOG waste streams will result in higher VS destruction and further increases in biogas production.

The cumulative effects of co-digestion and ongoing facility improvements discussed above were an increase in biogas production and a corresponding increase in power production. During 2012, the CHP system produced on average over 593,000 kWh of electricity each month, a 25% increase compared to the average monthly production in 2011. The onsite power accounted for approximately 14% of the overall power use at FWHWRC and resulted in avoided power costs of over $260,500. The increased power production has provided more control over energy costs, though reduced cost savings have been slightly offset by a reduction in the cost of utility power. The lower savings is a result of record low annual average utility power cost of approximately 3.59¢/kWh during 2012 compared to annual average cost of 7.37¢/kWh in 2008.

First Full Year Ownership Costs Consistent with Expectations

Costs incurred during 2012, the first full year of operation for Gwinnett POWER, are summarized in Table 3.

TABLE 3

Gwinnett POWER Program Costs during 2012

Labor Costs	CHP System	FOG/HSW Receiving Station
Labor Costs:	$54,600 (preventive maintenance, contractor coordination, and response to system alerts)	$72,800 (preventive maintenance, waste supplier tracking & coordination, contractor coordination, and responses to system alerts)
Permit Compliance Costs[a]		Not applicable
Permit Fee	$3,300	
Emission Testing	$4,500	
Preventive Maintenance Costs		No significant costs during 2012
Engine Maintenance	$30,470	
Oil Purchases	$13,350	
Oil Testing	$ 6,000	
Corrective Maintenance	No Significant Costs during 2012	$ 35,160 (Replaced 32 feed pump hoses and 12 transfer pump hoses)

[a]Addition of the CHP system resulted in modification of FWHWRC's air permit to a "synthetic minor" permit with annual testing and reporting requirements.

The rebuild and replacement costs cannot be estimated at this time because of the limited operational experience. But, the ownership costs incurred during 2012 were in-line with the planning estimates.

CONCLUSIONS

The GRRP helped turn the hardships resulting from the economic downturn and the statewide water restrictions into an opportunity. The BCE process was useful in developing an economically viable biogas utilization project and establishing the financial expectations for construction and operation. DWR considered the Gwinnett POWER projects a success if the right facilities could be procured within the limited funding available and operated to realize the projected benefits. The alternate delivery approach not only helped DWR meet the ARRA's shovel-ready requirements but also manage the scope components to maximize project benefits. The strategy resulted in lower capital cost, Table 4, and facilities with larger generation capacity than planned.

TABLE 4

Gwinnett POWER Program Cost Summary

O&M Cost Components	CHP System	FOG/HSW Receiving Station	TOTALS
Actual Capital Costs	$5,500,000	$3,500,000	$9,000,000
2012 Ownership Costs	$112,220	$107,960	$220,180
2012 Revenues/Savings	$260,500	$288,000	$548,500

The CHP system provides DWR greater control over the FWHWRC's energy costs and protects rate payers from the impact of future increases in utility power costs. Significantly lower than estimated utility power costs observed during 2012 resulted in lower avoided energy costs than anticipated. The non-hazardous trucked waste receiving program has been successful in attracting waste streams and generating new revenues for DWR. The revenues from this program are anticipated to increase significantly in future years with minimal increases in ownership costs. The ownership costs of Gwinnett POWER projects incurred during the first year of operation have been consistent with the planning estimates. Based on the actual costs and benefits, the projects have proved to be economically viable.

In the future, as biogas production is improved from the ongoing primary clarifier and thickening process improvements, increased influent flows to FWHWRC, and improved quality of co-digested wastes, the power generated will increase, resulting in greater savings and more control over DWR's future energy costs.

REFERENCES

CH2M HILL (2009). Technical Memorandum No. 3 – Alternatives Evaluation, Energy Recovery Business Case Evaluation prepared for Gwinnett County Department of Water Resources. July 2009.

Forbes, B., S. Jalla, and J. Horton (2010). Fast Track to New Digester Gas-to-Energy Facilities at the F. Wayne Hill Water Resources Center, Gwinnett County, Georgia. Residuals and Biosolids 2010, Savannah, Georgia. Water Environment Federation.

Hardy, S., M. Reisinger, R. Schoeck, A. Minchey, R. Porter, and R. Bush (2012). Lessons Learned in Designing and Commissioning of Digester Gas CHP Energy Recovery Projects in A Design/Build Project Delivery Environment. WEFTEC 2012, New Orleans, Louisiana. Water Environment Federation

Jalla, S. and T. Richards (2011). Advancing Sustainable Energy Management Wastewater Utilities. The Georgia Engineer, June – July, 2011

Tandukar, M., and S.G. Pavlostathis (2011). Evaluation of High Strength Waste Streams for Biogas Production from the Anaerobic Co-digestion of Municipal Sludge. Report to Hazen & Sawyer, PC for Gwinnett County, Department of Water Resources, Lawrenceville, Georgia.

LIFE CYCLE WATER USE FOR ELECTRICITY GENERATION: IMPLICATIONS OF THE DISTRIBUTION OF COLLECTED ESTIMATES

James Meldrum, University of Colorado
Jordan Macknick, National Renewable Energy Laboratory
Garvin Heath, National Renewable Energy Laboratory
Syndi Nettles-Anderson, National Renewable Energy Laboratory
University of Colorado
Cooperative Institute for Research in Environmental Sciences (CIRES)
Box 216 UCB
Boulder, CO 80309
(303) 492-1143

ABSTRACT

This article provides consolidated estimates of water withdrawal and water consumption for the full life cycle of selected electricity generating technologies, which includes component manufacturing, fuel acquisition, processing, and transport, and power plant operation and decommissioning. Estimates were gathered through a broad search of publicly available sources, screened for quality and relevance, and harmonized for methodological differences. Published estimates vary substantially, due in part to differences in production pathways, in defined boundaries, and in performance parameters. Despite limitations to available data, we find that: water used for cooling of thermoelectric power plants dominates the life cycle water use in most cases; the coal, natural gas, and nuclear fuel cycles require substantial water per megawatt-hour in most cases; and, a substantial proportion of life cycle water use per megawatt-hour is required for the manufacturing and construction of concentrating solar, geothermal, photovoltaic, and wind power facilities. On the basis of the best available evidence for the evaluated technologies, total life cycle water use appears lowest for electricity generated by photovoltaics and wind, and highest for thermoelectric generation technologies.

KEYWORDS

Energy-water nexus, electricity, life cycle assessment, harmonization, renewable

INTRODUCTION

Water requirements throughout the electricity generation life cycle have important implications for the electricity sector. Thermoelectric power plant operations are estimated as responsible for around 36% (Averyt et al., submitted) to 41% (Kenny et al., 2009) of total freshwater withdrawals in the United States and 3% of total freshwater consumption (Solley et al., 1998; Averyt et al., submitted). However, the life cycle of electricity generation consists of many stages besides power plant operation, including component manufacturing, fuel acquisition, processing, and transport, and power plant decommissioning. The water requirements associated with choices along this life cycle, such as the selection of fuel type or cooling technology, are not well understood.

Recent research (EPRI, 2012; Meldrum et al., 2013) consolidates and harmonizes estimates of water withdrawal and water consumption for the full life cycle of selected electricity generating technologies: coal, natural gas, nuclear, concentrating solar power (CSP), geothermal, photovoltaics (PV), and wind. Harmonization reduces analytical variability by adjusting published estimates to ones based on a more consistent set of methods and assumptions (Heath and Mann, 2012).

This study presents the results of this recent analysis and expands upon previous results by investigating the implications that the distributions of collected estimates have upon estimated life cycle water requirements of different electricity generation choices. This study thereby offers additional insight for understanding the water use implications of electricity generation.

METHODS

As described in more detail elsewhere (Meldrum et al., 2013), water use estimates were gathered through a broad search of publicly available sources, including peer-reviewed scientific literature, government reports and statistics, and corporate sustainability reports. Inclusion of gathered estimates in reported results depended on those estimates passing a series of three screens that focused on methods quality, completeness of reporting, and current technological relevance. Only quantified water use, supported by sufficient information and adhering to basic engineering principles, was included. Out of the more than 1,000 references consulted, 138 passed all three screens and contributed unique, usable data. The majority of these references provided primary data on only one or two technologies each.

Usable data on water use were classified as referring to either water withdrawals, in the case of water diverted from a water source for use, or water consumption, in the case of the portion of withdrawn water not returned to the immediate water environment after use. Where possible, estimates were harmonized to common boundaries and performance parameters, including thermal efficiency, fuel heat content, and plant lifetime, that match accepted values used elsewhere in the literature and previously reported.

Water use factors were developed for each of the three main life cycle stages of seven different electricity generation technologies. As Figure 1 depicts, these three main stages are the fuel cycle, power plant life cycle, and power plant operation; the fuel cycle includes fuel extraction, fuel processing, and fuel transportation, and the power plant life cycle includes component manufacturing, power plant construction, and power plant decommissioning. Where available data substantially differentiate water use among sub-categories of generation technologies, cooling technologies, or fuel cycle characteristics, separate estimates were developed for each sub-category; life cycle water use factors were constructed as weighted sums of the water use factors for each of the three major life cycle stages.

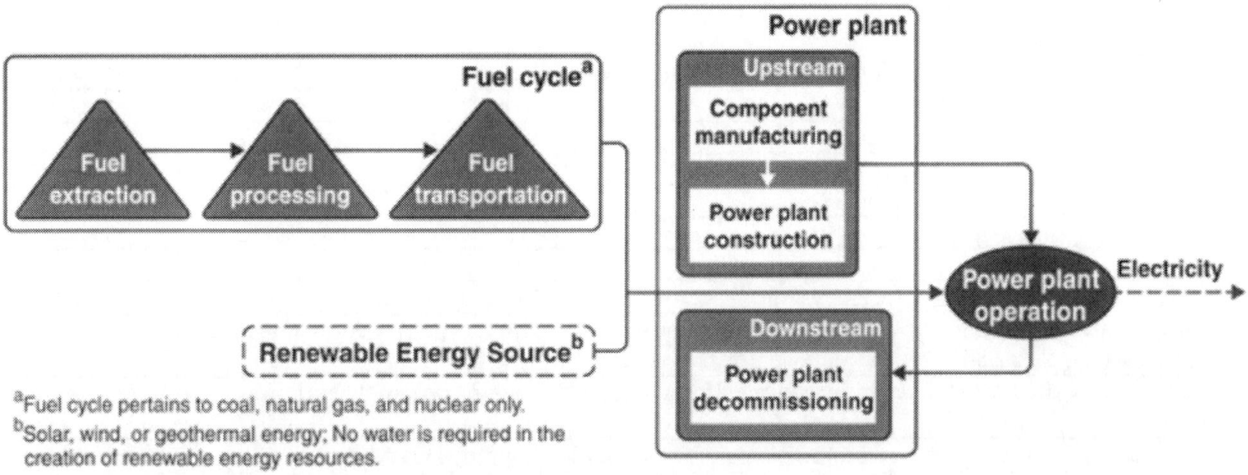

Figure 1. A schematic of the significant life cycle stages for each electricity generation technology.

For the many cases in which the water use type was not clearly specified, it was either inferred from context if possible, and if not, the related data were omitted. Limitations in relevant data availability made harmonization impossible for many of the collected water use estimates. Analytical boundaries around life cycle stages, such as whether indirect water was included in reported data, vary by study but were not consistently reported and were therefore often either inferred or assumed. Finally, this analysis focused only on externally sourced water and did not explicitly address water quality or the use of degraded or recycled water as a mechanism to reduce water use.

RESULTS AND DISCUSSION

Despite extensive collection, screening, and harmonization efforts, gathered estimates for most generation technologies and life cycle stages remained few in number and wide in range. For example, representative estimates or descriptive statistics of consumption in some cases exceed the corresponding median estimate for withdrawal. Estimates for nearly all processes and life cycle stages vary significantly, reflecting a combination of issues including methodological inconsistency, subcategory heterogeneity, and the effect of local conditions on water use.

Tables 1 through 3 present summary statistics (minimum, median, and maximum) for the water consumption and water withdrawals associated with each of these three main life cycle stages for a selection of electricity generation technologies and important sub-categorizations for those. Summary statistics pertain to screened estimates of the volume of water consumed or withdrawn (m3) per unit of electricity produced (kWh) under harmonizing assumptions for consistent performance parameters throughout the electricity generation life cycle. The number of collected estimates contributing to each reported range (n) is presented as an indicator of the depth of literature estimates available for each presented category; numbers range from 1 to 26. Details e.g., on reported estimates and further-disaggregated life cycle stages, are reported elsewhere (Meldrum et al., 2013).

Table 1. Range of fuel cycle water consumption and withdrawal estimates for select generation technologies and sub-categories. Assembled from previously published data (Meldrum et al., 2013).

Generation Technology	Sub-category	Consumption (m^3/kWh)				Withdrawal (m^3/kWh)			
		Min.	Median	Max.	n[a]	Min.	Median	Max.	n[a]
Coal	Surface Mining[b,c,d]	23	83	220	8	23	83	230	8
Coal	Underground Mining[b,c,d]	64	210	870	7	64	220	870	7
Natural Gas	Conventional Gas[b,c,d,e]	4	15	98	9	15	19	130	8
Natural Gas	Shale Gas[b,c,d]	11	61	790	20	19	64	830	18
Nuclear	Centrifugal Enrichment[b,f]	49	210	1,100	5	49	210	1,100	5
Nuclear	Diffusion Enrichment[b,f]	160	330	1,200	5	230	530	1,600	5

[a]For estimates constructed from multiple disaggregated stages or processes, 'n' reports the average number of estimates over each of the stages.

[b]Includes estimates constructed from estimates for individual stages in addition to estimates only for aggregated fuel cycle water use.

[c]Assuming train transportation (coal) or pipeline transportation (natural gas). Mine-mouth coal conversion decreases estimates negligibly; slurry pipeline coal transport increases estimates substantially. Estimated water use for liquid natural gas (LNG) ranges widely and is not reported here.

[d]Harmonized to the thermal efficiency of a sub-critical pulverized coal power plant or a combined cycle natural gas plant, as relevant.

[e]Defined as that natural gas not requiring fracture stimulation.

[f]Represents a combined estimate for different extraction types and storage and disposal for spent fuel; reprocessing and recycling of spent fuel is not currently practiced in the United States but would increase estimates substantially.

Ranges in reported estimates reflect not only variability in the analytical reliability of collected estimates but also the aggregation of many potential sub-categories within technologies and life cycle stages. Despite the breadth of many of these ranges, the minimum and maximum in the available literature may not represent the true minimum or maximum considering all deployment conditions, technological permutations, etc. Reflecting both the variability and uncertainty in the estimates for broad technology categories, reported results are limited to two significant digits.

As reflected in Table 1 and Figure 1, the fuel cycle only pertains to the subset of generation technologies not dependent on renewable solar, wind, or geothermal energy, namely coal, natural gas, and nuclear technologies. The fuel cycle encompasses extraction, processing, and transportation. Focusing on median water use estimates, natural gas requires the lowest water withdrawal and water consumption per unit of electricity, regardless of whether hydraulic fracturing is used (for shale gas) or not, and nuclear technology with diffusion enrichment uses the most water per unit of electricity. However, the full ranges of harmonized estimates overlap each other for nearly all presented sub-categories, tempering conclusions about relative ranking. In addition, nearly all ranges span at least an order of magnitude, demonstrating wide variability. To the extent that that variability reflects variation in technology sub-categories and operating characteristics, rather than analytical variability such as differing life cycle stage boundaries or measurement techniques, the evidence suggests there is no absolute ordering of the technologies in terms of water use; different circumstances and sub-category combinations could lead to different relative rankings of generation technologies.

Table 2. Range of power plant equipment water consumption and withdrawal estimates for select generation technologies and sub-categories. Assembled from previously published data (Meldrum et al., 2013).

Generation Technology	Consumption (m^3/kWh)				Withdrawal (m^3/kWh)			
	Min.	Median	Max	n[a]	Min.	Median	Max	n[a]
Coal[b,c]	<2	4	95	8	<2	4	45	8
Natural Gas[b,c]	<2	4	4	2	<2	<2	4	3
Nuclear[d]	<2	<2	<2	2	<2	<2	<2	2
CSP[c,e]	300	610	640	3	370	610	640	3
Geothermal[d]	8	8	8	1	<2	11	38	11
PV (C-Si)[c,e]	38	310	790	3	4	360	6,100	24
PV (other)[c,e]	19	23	26	3	<2	68	5,300	19
Wind[c]	<2	4	34	12	49	98	310	19

[a]For estimates constructed from multiple disaggregated stages or processes (reported in table 3), 'n' reports the average number of estimates over each of the stages, plus any included estimates that are not disaggregated. For categories with exactly 2 estimates, the median is defined as the arithmetic mean.

[b]Harmonized to the thermal efficiency of a sub-critical pulverized coal power plant or a combined cycle natural gas plant, as relevant.

[c]Includes both upstream water use estimates and downstream water use estimates (for decommissioning power plants). The latter contributes negligibly to the total for this life cycle stage.

[d]Due to limited information, upstream water use includes estimates only for manufacturing power plant raw materials. In addition, although the stage is expected to be negligible, no estimates for power plant downstream water use were found.

[e]CSP = concentrating solar power; PV = photovoltaics; C-Si = crystalline silicone; "Other" is primarily thin-film

Table 2 presents summary statistics for the ranges of harmonized water use estimates for the power plant, including upstream processes such as component manufacturing and power plant construction and downstream processes of power plant decommissioning (although the latter is negligible in most cases). Although many of the ranges still overlap at their extremes, the generation technologies based on non-renewable energy sources of coal, natural gas, and nuclear power tend to be estimated as requiring substantially less water withdrawals and water consumption than the technologies based on renewable energy sources. The former also are less frequently addressed in the literature, particularly compared to withdrawal estimates for the latter. Of renewable energy sources, water use estimates for solar-powered technologies are higher than those for geothermal- or wind-powered technologies and span ranges generally similar to the fuel cycle estimates reported in Table 1. However, these relative ranking results are tempered by the fact that withdrawal estimates for photovoltaics (PV) range particularly widely, across more than 3 orders of magnitude, in reflection not only of widely varying but also of rapidly advancing production processes (Meldrum et al., 2013).

Table 3 presents summary statistics for the ranges of harmonized water use estimates for power plant operations for a selection of cooling technology and prime mover combinations. A more complete set of operations water use estimates is available elsewhere, both harmonized to consistent parameters as shown here (Meldrum et al., 2013) and without harmonization (Macknick et al., 2012) (with the latter based on a similar but not identical dataset).

Although some of the reported ranges span roughly an order of magnitude, as before, many categories are more tightly estimated compared to previous tables despite similar numbers of collected estimates as before, reflecting perhaps greater scrutiny on operations water withdrawals and consumption. In contrast to previous tables, two generation technologies (PV and wind) are clearly differentiated with substantially less water use per unit of electricity. These two technologies only require limited operations water, primarily for occasional cleaning purposes, in contrast to the large water demands typically required for the steam processes of most thermal electric generation technologies.

Table 3. Range of power plant operations water consumption and withdrawal estimates for select generation technologies and sub-categories. Assembled from previously published data (Meldrum et al., 2013).

		Consumption (m^3/kWh)				Withdrawal (m^3/kWh)			
Generation Technology	**Sub-category**	Min.	Median	Max.	n	Min.	Median	Max.	n[a]
Coal: Pulverized Coal (sub-critical)	Cooling Tower	760	2,000	4,900	20	1,700	2,500	4,500	21
Coal: Integrated Gasification Combined Cycle	Cooling Tower	130	1,200	1,700	14	610	1,500	25,000	16
Natural Gas: Combined Cycle	Cooling Tower	180	790	1,100	19	570	950	2,900	16
Natural Gas: Combustion Turbine	No Cooling	190	190	1,300	3	1,600	1,600	1,600	1
Nuclear	Cooling Tower	2,200	2,700	3,400	9	3,000	4,200	9,800	7
CSP: Trough[b]	Cooling Tower	2,100	3,400	7,200	26	3,300	3,600	4,200	2
CSP: Power Tower[b]	Cooling Tower	2,800	3,100	3,300	5	2,800	2,800	2,800	1
Geothermal	Binary, Dry Cooling[c]	1,000	1,100	2,400	3	1,000	1,100	2,400	3
PV[b]	Flat Panel[c]	4	23	98	9	4	23	98	9
Wind: On-shore	n/a	4	4	8	10	4	4	4	2

[a]For categories with exactly 2 estimates, the median is defined as the arithmetic mean.

[b]CSP = concentrating solar power; PV = photovoltaics

[c]Reflecting data limitations and the nature of water use, we assume withdrawal and consumption are equal for all estimates in this category.

In addition to separate water use estimates for the three main life cycle stages, estimates for water consumption and withdrawal across the entire life cycle are generated as the sum of harmonized water use estimates for each of the three main life cycle stages. However, it is critical to remember that estimates aggregated across multiple life cycle stages gloss over the importance of spatial and temporal details of water use for understanding the impacts and implications of that water use. Due to the many dimensions of variability in the underlying data, previous research presents life cycle water use estimates aggregated from the median of the gathered, screened, and harmonized estimates. Acknowledging that the combination of factors that contribute to the variability in these data pertains to important considerations such as operating conditions and the specifics of implemented technologies, this study expands that analysis to consider the sensitivity of the life cycle water withdrawal and water consumption estimates to the wider range of estimates assembled previously. A sample of these results, covering a subset of the assessed technology sub-category permutations, is presented in Table 4.

Table 4. Range of life cycle water consumption and withdrawal estimates aggregated from minimum, median, and maximum estimates for each of the three main life cycle stages, for select generation technologies and sub-categories. Assembled from previously published data (Meldrum et al., 2013).

		Consumption (m^3/kWh)			Withdrawal (m^3/kWh)		
Generation Technology	**Sub-category**	Min.	Median	Max.	Min.	Median	Max.
Coal: Pulverized Coal (sub-critical)	Surface Mining, Cooling Tower	780	2,100	5,200	1,800	2,600	4,800

Coal: Integrated Gasification Combined Cycle	Surface Mining, Cooling Tower	160	1,300	2,000	630	1,600	26,000
Natural Gas: Combined Cycle	Conventional Gas, Cooling Tower	190	810	1,200	590	970	3,000
Natural Gas: Combustion Turbine	Conventional Gas, No Cooling	200	210	1,400	1,600	1,700	1,800
Nuclear	Centrifugal Enrichment, Cooling Tower	2,200	2,900	4,500	3,100	4,400	11,000
CSP: Trough[a]	Cooling Tower	2,400	4,000	7,800	3,700	4,200	4,800
CSP: Power Tower[a]	Cooling Tower	3,100	3,700	3,900	3,200	3,400	3,400
Geothermal	Binary, Dry Cooling	1,000	1,100	2,400	1,000	1,100	2,400
PV[a]	Crystalline-Silicon, Flat Panel	42	330	890	8	380	6,200
Wind: On-shore	n/a	8	8	42	53	100	320

[a]CSP = concentrating solar power; PV = photovoltaics

Reflecting the substantial proportional contribution of operations to life cycle water use for most generation technologies, comparisons of life cycle water use across different generation technologies shown in Table 4 is similar to that in Table 3. However, this similarity can be deceptive as it glosses over the different relative contributions of different main life cycle stages. In particular, the fuel cycle is estimated as contributing the majority of additional (non-operations) water use for technologies based on non-renewable energy, compared to low contributions for the power plant life cycle. In contrast, for most those technologies based on renewable technologies and not requiring a fuel cycle, the power plant life cycle is estimated as requiring much higher water use per unit of electricity compared to the power plant life cycles of non-renewable technologies requiring fuel.

In addition, focusing only on median estimates can lead to different conclusions versus considering the entire range of estimates for each category. For example, wind is fairly consistently ranked as the lowest life cycle water user and is more sharply differentiated from PV by full life cycle water use, but upper estimates of water withdrawals associated with wind greatly exceed lower estimates of those for PV. Focusing only on median estimates, PV is also a relatively low life cycle water user, as is natural gas, geothermal, and the more efficient, integrated gasification combined cycle coal generation technology. However, the wide range of estimates for PV's power plant life cycle places its maximum life cycle water withdrawal estimate as the third highest of the categories considered.

CONCLUSIONS

This study builds on the recent, concise presentation of life cycle water use factors gathered from a thorough review of the available literature, screened for relevance and quality, and harmonized for consistency (Meldrum et al., 2013). As found previously, although operations in many cases dominate life cycle water use in absolute magnitude, the water implications of choices about electricity generation technologies clearly do not end at those due to the cooling water demands for thermoelectric generation. The fuel cycles of coal, natural gas, and nuclear power all require significant water volumes, and renewable generation technologies require significant water for manufacturing and construction.

In general, based on review and harmonization of the available evidence, total life cycle water use across the generation technologies considered here appears lowest for electricity generated by PV and wind, and highest for nuclear and conventional coal technologies, with the relative position of natural gas and CSP technologies varying substantially. However, the wide ranges of estimates gathered from the literature leads to correspondingly wide variability in the estimated life cycle water withdrawal and water consumption associated with each generation technology. In some cases, this variability leads to different ordering of technologies in terms of life cycle water use. Therefore, although this body of research establishes a foundation for estimating water requirements for the three main life cycle stages of different electricity generation choices, this sensitivity analysis emphasizes the importance of using the most specific data possible for estimation of water use for actual projects.

In addition, although the influence of non-operations water use on total life cycle water use estimates of different types technology may to some extent be balanced by the opposing effects of no fuel requirements for technologies based on renewable energy but relatively higher power plant water use for many of those same technologies, the different spatial and temporal requirements associated with fuel cycle versus power plant life cycle water requirements suggest different implications for the two. For example, the majority of fuel cycle water requirements are associated with the extraction of fuels, meaning that this water use is constrained to places where the fuel is available, regardless of local water resources. In contrast, the majority of power plant water requirements are associated with the manufacturing of power plant materials. Although many considerations go into the siting of manufacturing facilities, the availability of water resources can be considered and even, if necessary, siting could be optimized accordingly. Such considerations demonstrate the value of considering life cycle water use, not just as an aggregated volume but also as disaggregated across different life cycle stages, when considering the water use implications of different electricity generation technologies.

REFERENCES

Averyt, K., Meldrum, J., Caldwell, P., Sun, G., McNulty, S., Huber-Lee, A., and Madden, N., 2013 (submitted), "Sectoral Vulnerabilities to Changing Water Resources in the Conterminous U.S.," Environmental Research Letters.

Electric Power Research Institute (EPRI), 2012, "Water Use in Electricity Generation Technologies: A Life Cycle Perspective," 1023820, EPRI, Palo Alto, CA.

Heath, G., and Mann, M., 2012, "Background and Reflections on the Life Cycle Assessment Harmonization Project," Journal of Industrial Ecology, 16, pp. S8-S11.

Kenny, J., Barber, N., Hutson, S., Linsey, K., Lovelace, J., and Maupin, M. 2009. "Estimated Use of Water in the United States in 2005." US Geological Survey Circular 1344, USGS, Reston, VA.

Macknick, J., Newmark, R., Heath, G. and Hallett, K., 2012, "Operational Water Consumption and Withdrawal Factors for Electricity Generating Technologies: A Review of Existing Literature," Environmental Research Letters, 7, 10pp.

Meldrum, J., Nettles-Anderson, S., Heath, G., and Macknick, J., 2013, "Life Cycle Water Use for Electricity Generation: A Review and Harmonization of Literature Estimates," Environmental Research Letters, 8, 18pp.

Solley, W., Pierce, R., and Perlman, H., 1998, "Estimated Use of Water in the United States in 1995," US Geological Survey Circular 1200, USGS, Reston, VA.

Water and Energy Efficiency through Steam Systems

Angela Ganley, City West Water
Lesley Smith, City West Water
Audra Liubinas, City West Water
Sophea Eng, City West Water
City West Water
247-251 St Albans Road
Sunshine, Melbourne VIC 3020
Australia
(+61) 3 9313 8300

ABSTRACT

The Steam System Efficiency Program was undertaken by City West Water (CWW) to improve the efficiency of its business customer's steam systems by reducing water, energy and chemical use. This was achieved by providing free steam system audits and training to selected industrial and institutional customers to identify and quantify saving opportunities. This paper outlines the outcomes of the program and key ways to improve water and energy efficiency in steam systems.

KEY WORDS

Resource efficiency, steam systems, demand management, greenhouse gas reduction, water energy nexus, industry, water efficiency, energy efficiency, trade waste reduction.

INTRODUCTION

Energy and water use are embedded throughout the water cycle, but particularly in the end use of water. The total energy use by Australian water utilities is less than 15 per cent of the energy used for residential water heating (Kenway et al, 2008) and heating water is some four to five times more energy intensive than desalination using membranes (PMSEIC, 2010). Steam systems are particularly energy intensive end uses of water: an average steam system will use around 2500MJ/kL. This is roughly one thousand times more than the energy used in City West Water's supply and treatment networks, illustrated in figure 1.

CWW has strong established relationships with over 100 steam system operators through its Business Resource Efficiency Program. The program works with business customers to improve their water and energy efficiency and reduce contaminants in trade waste discharge. The Business Resource Efficiency Program has developed strong releationships with customers, and as such, CWW was well placed to assist its customers to reduce the water and energy use of their steam system. Generally these customers lacked knowledge on how to make steam systems more efficient, and how to quantify savings and costs of opportunities. Through provision of expertise CWW was able to assist customers to overcome these barriers.

Many water and trade waste efficiency initiatives important to CWW, such as salt reductions in trade waste to enable wastewater reuse, do not present attractive financial returns so are sometimes difficult for business to justify. Conversely, steam system operation can account for a large proportion of a business' total operating cost and generally has much higher financial returns than salt reduction projects. Through assisting business with knowledge on steam system upgrades and associated energy saving CWW was able to leverage trade waste and water efficiency savings.

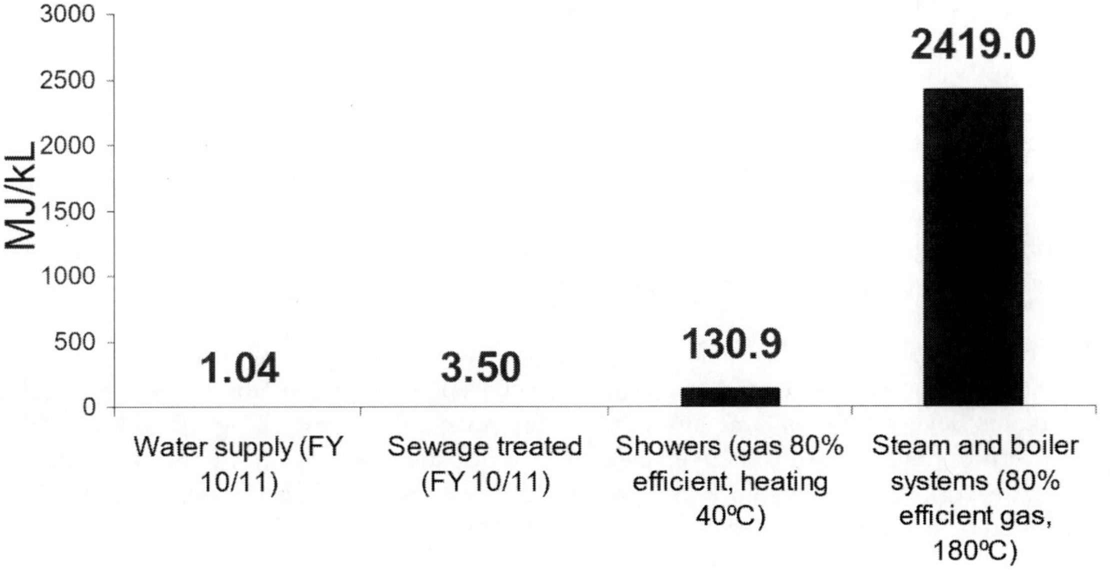

Figure 1 - Energy intensity of water end uses compared to supply and treatment

BACKGROUND

City West Water Overview

CWW is one of three Victorian Government owned retail water businesses in metropolitan Melbourne. Corporate activities are outlined in our Statement of Obligations and include delivery of drinking water, sewerage, trade waste and recycled water services to customers in Melbourne's central business district, inner and western suburbs. Our statement of obligations also requires us to deliver programs that improve the water efficiency of our customers. This requirement is in part fulfilled by the Business Resource Efficiency Program.

As well as assisting customers with energy efficiency initiatives, the management of energy within its own operations is a priority for CWW. This is guided by an Energy Strategy which prioritises the following energy management actions based on the Victorian Environmental Protection Authority's Carbon Management Hierarchy: measure, set objectives, avoid, reduce, switch, sequester, assess, and offset.

CWW's corporate vision is to be "a truly sustainable water business" and its mission to "guarantee affordable and safe water for today and tomorrow". The company operates on a triple bottom line, with a number of social, environmental and economic goals. These include environmental goals to "balance the needs of our customers in the environment in managing supply and demand for water" and a social goal to assist customers to be "efficient water users and clean and efficient producers". The steam system program contributes to CWW's corporate vision, mission and the fulfilment of these goals.

Business Resource Efficiency Program

CWW's Business Resource Efficiency Program works with business customers to reduce water use and contaminants of concern in trade waste. The program formally began targeting reductions in trade waste volumes and pollutants with non-residential customers in 2003. Later the program was expanded to include water conservation, as the longest running drought in Australian history led to the development of a suite of programs to ensure the security of future water supplies.

Business' participate in the Business Resource Efficiency Program by completing a resource management action plan (resourceMAP). The resourceMAP outlines how water is used in customers business' and documents water efficiency and trade waste improvements that have or could be implemented. The development and annual review of resourceMAPs is facilitated by CWW consultants who work one on one with customers to identify and implement opportunities. CWW consultants possess specialist expertise across a range of industries and are principally trained in chemical engineering and environmental science.

The Business Resource Efficiency Program has successfully delivered reductions in water use and trade waste contaminants, principally based on strong relationships with business customers developed over time and through the practical support being offered. CWW uses these strong relationships to deliver messages of holistic resource efficiency. To this end the program works in partnership with government and industry stakeholders to deliver a co-ordinated approach to resource efficiency, including energy efficiency. Through such a partnership with the Environmental Protection Authority Victoria, CWW received $50,000 contribution to the Steam System Efficiency program.

METHODOLOGY

CWW engaged a consultant to provide free steam system audits to selected customers. The intent was to show how water, chemical and trade waste costs could contribute significantly to steam system operating costs. Other steam system programs, as well as approaches by customers, have assessed opportunities based on energy savings only. This program aimed to achieve a more holistic approach to steam systems and hence resource efficiency in general.

Customers from various industries were targeted to demonstrate the value of steam system auditing, maintenance, and retrofitting within a particular industry sector. The target industries were food and beverage, meat processing, chemical/petrochemical, laundry, tanker washers, textile, and hospital and medical.

The program also included a training course aimed at facility managers, site engineers and maintenance personnel. The purpose of the training was to improve knowledge and promote behavioural change with respect to ongoing maintenance of steam systems as well as investigating and implementing improvements.

The program was conducted in two rounds. The first round was conducted in 2010/11 and the second in 2011/12.

Phase One Audits

Customers from various industries were targeted to demonstrate the value of steam system auditing, maintenance, and retrofitting. A survey conducted in 2009 on behalf of CWW found that 34% of customers had a boiler/steam related action item in their resourceMAP. The survey data was used by the project team to identify target industries and consolidate a list of customer sites with the greatest potential for steam system savings.

General audits conducted in the first round included the following deliverables:

- Identify steam system equipment at each site
- Describe inefficiencies
- Provide recommendations on improving
    - Water and energy losses
    - Steam generation and distribution
    - Engineering practices and correct applications
    - Return of condensate and heat recovery
- Potential water, energy, chemical and trade waste savings
- Cost of implementation
- Return on investment or payback period.

Phase Two Audits

Phase two audits were open to all resourceMAP customers on a first-come-first-served basis. This change was integrated into the selection process as an improvement action identified in the phase one review. It was seen that this approach would improve customer buy in and commitment to implementing opportunities that were identified.

The audits in the second round were expanded from the general steam system audits offered in the first round to include focus area audits for:

- Condensate return
- Boiler room equipment
- Steam trap audit
- Pipe insulation survey.

Training

The program also included a training course aimed at facility managers, site engineers and maintenance personnel. The purpose of the training was to improve knowledge and promote behavioural change with respect to ongoing maintenance of steam systems as well as investigating and implementing improvements. Training was funded by City West Water and delivered by steam system providers Spirax Sparco. Spirax Sarco are experienced in delivering such training through demonstrations using a live steam system.

Communications

To disseminate learning's to a broader audience CWW has developed communications material to promote steam system savings opportunities. This includes a short video case study filmed at Victoria Wool Processors to highlight boiler start-up and shut-down optimisation and a factsheet detailing the top tips for steam system optimisation that is available on the CWW website.

The knowledge gained by the Business Resource Efficiency team participating in the audits is used to help other customers who have not received an audit to identify and implement resource efficiency in their steam systems. This is disseminated at one-on-one resourceMAP reviews.

Program Review

Once audit recommendations were identified customers were expected to drive the implementation of opportunities identified. Assistance and resources to achieve this were provided by CWW through its Business Resource Efficiency program. This included the provision of hot water and cold water checkmeters to encourage ongoing monitoring of steam system savings, and providing co-funding to encourage the implementation of capital projects that have water savings.

CWW continues to work with the customers to implement the identified actions. Energy, water, trade waste and chemical treatment costs continue to rise each financial year. Therefore, opportunities need to be revisited annually to ensure that they can be implemented when the necessary payback period is achieved. The Business Resource Efficiency team re-visit these opportunities in annual resourceMAP reviews with customers so that businesses can implement opportunities once they become financially feasible.

RESULTS

Developing Holistic Resource Efficiency Considerations with Business

The Steam System Efficiency Program encourages CWW's business customers to take a holistic approach to improvement projects that account for water and wastewater savings as well as energy savings. This approach encourages business customers to think about resource efficiency projects beyond the Steam System Efficiency Program. For example, the opportunities for heat recovery around the steam system encouraged a textile manufacturer to investigate other areas where they could also implement heat recovery. As a result of the steam audit, the company now views waste heat as a resource that can be utilised for other processes, including wastewater treatment. CWW has built on this by linking this business with emerging technologies and processes to utilise the waste heat.

Simultaneously addressing water and energy issues is more cost effective than addressing only water or only energy issues. This program has helped industry see the benefits of addressing all of the factors and inputs into their production processes. This will develop ongoing behavioural changes in steam system maintenance practices and the approach to identifying and improving retrofitting and operating practices. The result will be improved economic performance for businesses through sustainable business practices. The program's success was recognised by winning both State and National Australian Water Association 2012 Awards for Program Innovation.

Audit Outcomes

Thirty audits were completed. In total, 152 actions were identified in the steam audits. If all projects were implemented, annual savings of 294 megalitres of water, 185 terajoules of gas, and greenhouse gas reductions of 9,960 tonnes of CO_2-e would be achieved. To date, 23 actions have been implemented, achieving annual savings of 10 megalitres of water, 14,000 gigajoules of natural gas, and greenhouse gas reductions of almost 720 tonnes of CO2-e. Savings associated with projects that have been implemented and those that were identified are shown in the figure 4.

Opportunities identified included condensate return, boiler and steam system insulation, automatic blowdown control on boilers, boilers economisers, and steam trap and leak repair. All actions implemented have had low associated capital costs, with only $25,500 spent by customers to achieve these savings. CWW is currently working with customers to assist implementation of the remaining action items by providing resources through the Business Resource Efficiency Program.

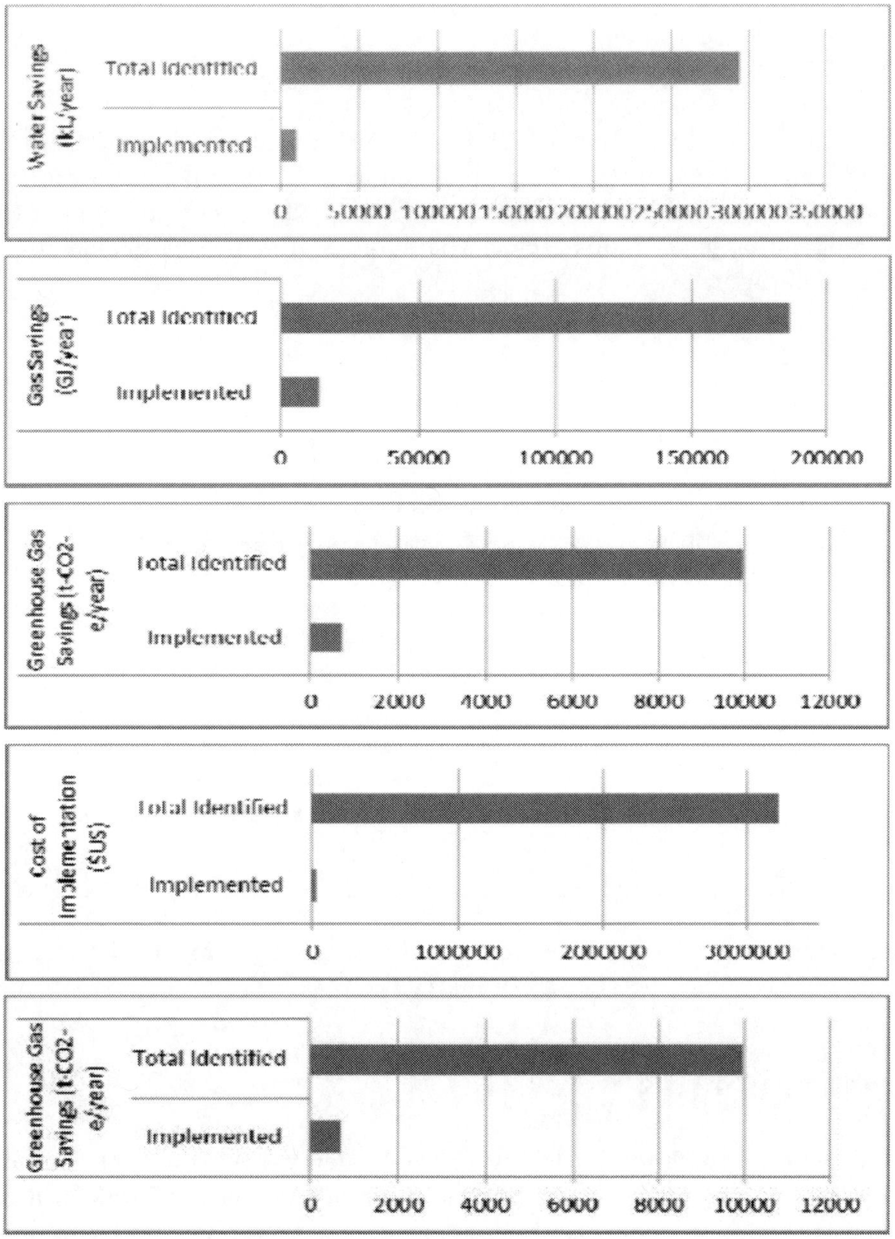

Figure 2 - Quantum of Savings Implemented and Identified in the Program

Industry Trends

It was found that hospitals were generally very well maintained with few leaks detected and good steam trap maintenance. Most condensate was returned to the boiler, limiting the opportunities to improve condensate return. The main recommendations were insulation for flanges and pipework and boiler economisers.

Audit reports for other industrial sites identified more leaks and steam trap maintenance requirements. Condensate return opportunities were common as were boiler economisers, monitoring and targeting, and insulation. Less common opportunities were boiler blowdown and boiler efficiency tune-ups.

The uptake of audits was particularly high from the food and beverage industry. Eight of the thirty audits were for this sector. This could indicate a general need from the food and beverage sector in auditing their steam systems and should be a consideration in any resource efficiency program targeted to this sector. The number of audits completed at each industry type are shown in Figure 3.

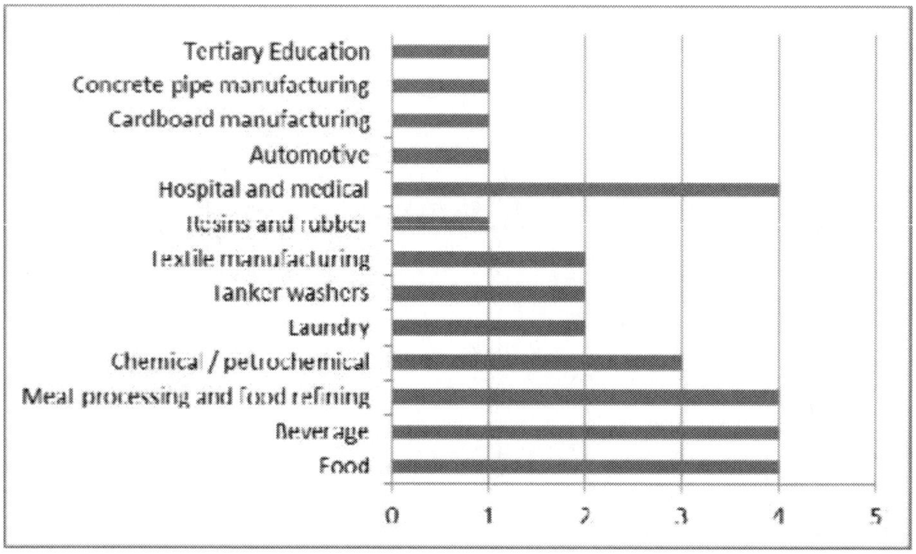

Figure 3 - Number of customer targeted in the Steam System Efficiency Program by Industry type

OPPORTUNITY DETAILS

Figure 4 shows a breakdown of the water and energy savings identified for the main opportunity types. The main water and energy savings were found with steam trap maintenance, condensate return and monitoring and targeting. Boiler economiser also presented significant potential for energy savings. Insulation, boiler blowdown and boiler efficiency opportunities did not produce a significant amount of water and energy savings for the audited sites. Each of these opportunities is discussed in more details below.

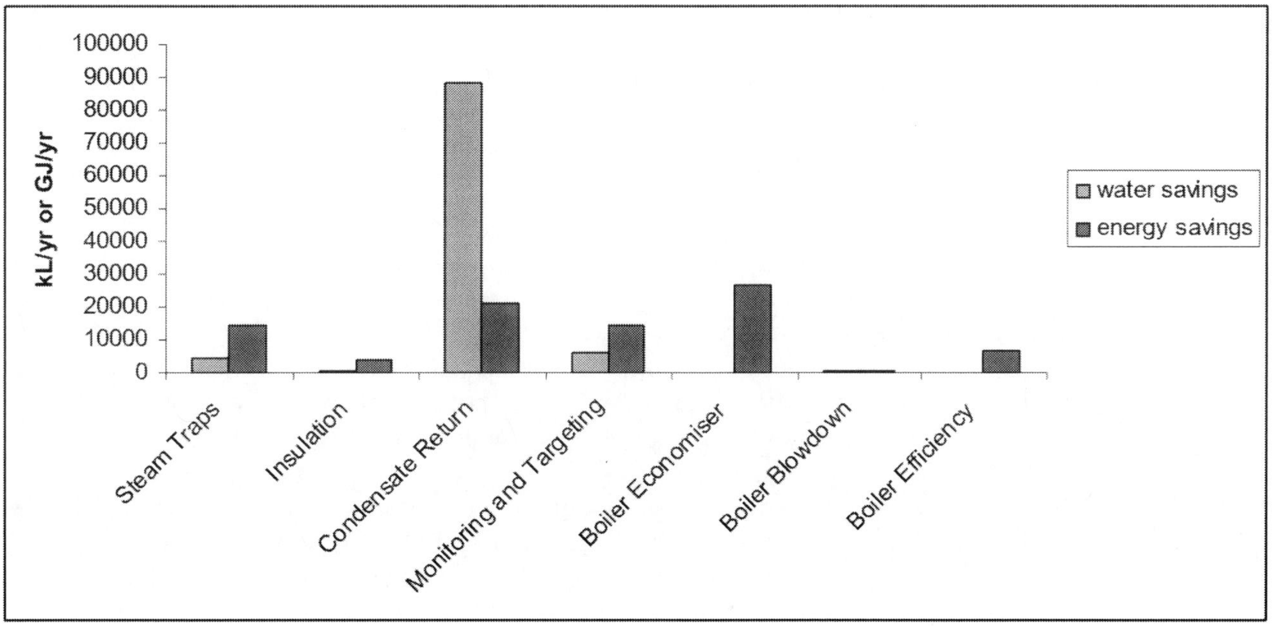

Figure 4 - Average water and energy savings for each opportunity type

Steam Trap Maintenance

Steam traps are devices that collect and discharge condensate and air that collects in the steam pipework with minimal loss of live steam from the system (Risko, 2011). Most steam traps are essentially automatic valves that automatically open, close or modulate.

If a steam trap fails fully open, there will be loss of steam in the system and more energy and water will be required to maintain the required steam load for the process. If a steam trap fails fully closed, there will be condensate build-up in the steam system, which will cause operational problems. The most common problem is water hammer on start-up of the system. This is where the pressure of the steam pushes the condensate along the line, where it contacts the bend in the pipework at high velocity. As well as being noisy, water hammer stresses the pipe, which can eventually lead to pipe failure (Risko, 2011).

The audit reports recommended that all steam traps are tested at least annually, but preferably every six months. Twelve of the sites audited did not have a regular steam trap maintenance program. It was estimated that 30-40% of steam traps have failed in systems that have never been tested. Each site inspection would cost approximately $1,200/yr, depending on the number of steam traps on site.

Monitoring and Targeting

The audits found that there was a general lack of monitoring and targeting of steam systems. This was particularly the case for the small to medium sized customers, which do not have advanced metering systems. Without measuring the gas and water use in the steam system, it is difficult to identify and quantify problems and opportunities. It is estimated that 5% savings in greenhouse gas emissions can be achieved by implementing advanced metering systems (The Carbon Trust, 2007).

Metering and monitoring gas consumption and boiler feedwater can help identify problems in the steam system. If there are losses in the system, make-up water will increase and more energy will be required to produce the same amount of steam. The audit reports showed that a leak will cost a business anywhere between $0.10/hr and $62.20/hr depending on the steam pressure and the size of the leak, so timely identification and maintenance can have significant financial benefits.

It is also important that the target boiler energy use is appropriate for site processes. Some businesses have boilers idling for short periods as it is not worthwhile to shut the boilers down and turn them on again when required. Steam systems generally take around two hours to ramp up before they are ready to provide the required steam load. If a simple GJ/Tonne is used as a target, the efficiency will improve when production increases and therefore system improvements and inefficiencies are not observed. In this case it is better to have a base load measurement that is separate from target energy use per tonne of production.

When building a business case it is preferable to have precise data to estimate savings and payback periods. Equally important is being able to prove that the savings have occurred to justify further improvement opportunities. The identification of resource efficiency improvements is a key driver for the Business Resource Efficiency program. As such, CWW has provided hot and cold water check meters to customers to help build the business case for improvements to steam systems.

Condensate Return

Condensate return opportunities are site specific and few generalisations can be made as to installation costs and payback periods. However, some notes can be made in the approach of identifying condensate return opportunities.

Many sites do not consider these opportunities in detail because there is a general perception that condensate streams that are far from the boiler will be financially and technically unfeasible to return. Some sites had never considered water, trade waste and chemical treatment costs in their business case. Once these are included with energy costs, the business case becomes attractive. This was the case even for a stream that was 50m from the boiler house. Another business saved 5.7ML/yr, 211GJ/yr, and 11 tonnes of CO_2-e/yr and over $32,500/yr by repairing their condensate pump and recommissioning a condensate return line. When the pump first failed it was not found to be financially feasible to repair it. However, this was based on energy savings only. Once water, chemical treatment and trade waste costs were included, payback period was calculated to be 2 months.

Insulation

Most sites had extensive insulation on the steam pipework, with only two sites noted as having large sections of pipework uninsulated. Energy losses from uninsulated pipework vary according to the size of the pipe and steam pressure/temperature. An average figure of 23GJ/m/yr of pipework was found during the audits and is a good rule-of-thumb for businesses wanting a back of envelope calculation for insulation savings.

Flanges and valves were an area with very little insulation at all sites. The main reason for this was maintenance issues associated with removing the insulation. The insulation would also make it more difficult to observe leaks. Depending of the size of the flange or valve, the energy losses could be between 35GJ/yr and 52GJ/yr which equates to between 1.8 and 2.7 tonnes of CO_2-e / yr. There is a clear trade-off between energy savings and maintenance issues, with most businesses deciding that the maintenance issues were significant enough not to install the insulation.

Other areas to consider for insulation are the boiler water feed tank and condensate return lines, although the payback periods on these tend to be longer because the heat losses are lower.

Boiler Economiser

An economiser uses waste heat from the boiler flue gases to pre-heat the feed water to the boiler. In doing so there can be significant energy savings in the operation of the steam system. The audit reports found that economisers typically reduce gas consumption by 4%. Payback periods tend to be less than 5 years if the boiler is used all the time. Many sites had multiple boilers which would share the load. In this situation the payback periods were not attractive.

Boiler Blowdown

Boiler blowdown refers to the bleeding of the water to remove excess salts. Salts build-up as the water is circulated through the steam system and can cause scaling and corrosion. Blowdown can have automatic or manual operation. Manual systems usually involve instructions for an operator to open the blowdown valve for a certain period of time and at certain intervals. There is potential here to have excess blowdown, which means that more water, energy and treatment chemicals are required to replace what is lost from the system. If the blowdown is less than required, there is a risk of scaling or corrosion in the boiler, which would be very costly to rectify.

Automatic blowdown can be controlled by either a timer or electrical conductivity sensor. The time controlled blowdown essentially removes the risk of operator error when opening and closing the blowdown valve. The electrical conductivity control means that blowdown only occurs at a certain total dissolved solids (TDS) concentration. This is the ideal way to control blowdown because it eliminates the risk of removing too much or too little water from the system. However, the steam audits generally found that the manual and time based controls did not adversely affect the chemical treatment program and therefore that the chemical treatment supplier and operators did a good job in controlling the blowdown. It was generally not recommended to install electrical conductivity controlled blowdown because the savings were negligible.

Boiler Efficiency

Boiler efficiencies were generally found to be within acceptable ranges. A flue gas analyser provides concentrations of CO, CO_2, O_2 and flue gas temperature. The audit reports stated that the oxygen concentration in the flue is a indication of the excess oxygen used in the boiler and should typically be between 4.5% to 5.5%. There were few instances where this was outside the range. This was probably due to the fact that boiler efficiency is checked once per year at the annual boiler inspection and that this is satisfactory for good boiler operation throughout the year.

Other Lessons Learned

In the first round, the contractor went to site to gather the information and complete the audit. This was not always successful as not every customer had the information as requested on the day of the audit. This lead to delays in the preparation of the audit report as the contractor had to keep requesting the information from the customer.

In the second round, once the customer had registered for a steam audit they had to complete a pre-audit checklist before proceeding to the site audit. This provided the consultant with information before going on site. This reduced time spent on site and produced better quality audit reports. Having customers spend time on data gathering also helped to ensure customer buy in.

After the first round of audits, some businesses reported that they knew the general issues with their steam system but required more in depth information in specific areas. As a result the second round of the program offered focus area audits, which meant that customers did not have to dedicate time to areas that they felt were not beneficial.

Ultimately, the programs success was a direct result of the strong relationships with customers. Strong customer relationships enabled CWW to selectively identify who was likely to get most benefit from the project. It also allowed for ongoing communication during implementation and follow up at the post project review stage. These strong customer relationships where crucial for the projects success.

CONCLUSIONS

The Steam System Efficiency Program is an effective way to identify water, trade waste and energy savings in businesses concurrently. Examining the savings opportunities of resource efficiency holistically (rather than addressing a single saving such as water or energy) helps develop strong business cases for implementing resource efficiency projects.

The program has also demonstrated that steam system improvements represent an extremely cost effective way to achieve greenhouse gas reductions: the projects implemented have saved over 700 tonnes of carbon dioxide equivalent greenhouse gases, with implementation costing less than US$27,000. For water authorities seeking to reduce the greenhouse gases associated with their operations such opportunities represent a significant opportunity: if all the actions identified by the program were implemented, greenhouse gases would be reduced by 10,000 tonnes of carbon dioxide equivalent gases a year – over half of the 14,000 tonnes associated with CWW's entire business.

Main opportunities for improvement were found to be condensate return, steam trap maintenance, and monitoring and targeting. Through the development of strong customer relationships water authorities can help provide customers with the knowledge and resources that will benefit both business and the environment.

ACKNOWLEDGMENTS

I would like to acknowledge my colleague Angela Ganley for her instrumental role in initiating and delivering the program. Thanks also to the members of the Business Resource Efficiency team who provided customer liaison and supported the program's delivery -

especially to the Business Resource Efficiency Program Managers, Audra Liubinas and John Fawcett for the original vision of integrating energy efficiency into CWW's water demand management programs. For conducting the steam system audits thanks are owed to consultants Worley Parsons and KBR - it was their expertise that made the audits so sucessful. Thanks are also owed to the EPA Victoria for their financial contribution. Finally, I would also like to acknowledge those customers that participated in the program without whose participation the program could never have happened at all.

REFERENCES

Kenway, S.J. et al. 2008. Water-energy futures for Melbourne: the effect of water strategies, water use and urban form, CSIRO: Water for a Healthy Country National Research Flagship, Australia.

PMSEIC, 2010, Challenges at Energy-Water-Carbon Intersections, Canberra, ACT, Australia.

Risko, J.R. 2011. Understanding Steam Traps, American Institute of Chemical Engineers, New York, NY, USA

The Carbon Trust 2007. Advanced Metering for SMEs Carbon and Cost Savings, The Carbon Trust, London, UK

CREATING NEW REVENUE WITH EXISTING PROCESSES: EVALUATING GRID BALANCE™ WITH DEMAND-SIDE LOADS AT WASTEWATER TREATMENT PLANTS USING DYNAMIC MODELING

Daryl Letto[1*], Spencer Snowling[2], Raoul George[1], Christopher Morrissey[1], Hank Andres[2], and Conrad Fox[1]

[1]ENBALA Power Networks, Toronto, Ontario, M5H 2V6, CANADA
[2]Hydromantis Environmental Software Solutions, Inc., Hamilton, Ontario, L8P 4R5, CANADA
*email: dletto@enbala.com

ABSTRACT

The electricity system requires real-time balancing through Grid Balance® traditionally provided by generation, but recently provided by demand-side loads. ENBALA Power Networks' platform enables loads at wastewater treatment plants to provide this service, specifically aeration basin blowers due to the large duty load, process flexibility, and available process storage.

Grid Balance® was demonstrated using dynamic modeling and simulation software GPS-X (Hydromantis ESS Inc., Hamilton, Ontario, Canada, 2012). Models of wastewater treatment processes are used to demonstrate how demand-side loads can be connected to the grid to generate revenue, help the electricity systems run more efficiently, provide reliability, and lengthen the lifespan of existing infrastructure while reducing greenhouse gases.

The models demonstrated that Grid Balance® did not affect the process and maintained fully treated wastewater for discharge. Mechanics of blower operation were determined to be typical in terms of the frequency and the size of the power change request.

KEYWORDS: Smart Grid, energy storage, aeration, demand-side loads, wastewater, Grid Balance®

INTRODUCTION

The Electricity System

Across North America, many electric power utilities have been restructured to create open and competitive energy markets. The adoption of competitive markets required that the regional electricity generation (supply), transmission, and distribution (demand) be separated.

After this restructuring, a central authority in each region was required to manage the electricity system and ensure market quality from generation through to consumption. To achieve this, Independent System Operators (ISOs) and Regional Transmission Organizations (RTOs) were created.

Regional electricity system operators must maintain a continuous balance between market demand and generation supply to ensure all consumers have a reliable supply of electricity. Grid Balance® (also known as Regulation Service by electricity system operators) is an ancillary service where system operators traditionally rely on generators to increase and decrease their power generation to match the real-time demand (consumption).

To help deal with pressure of continuous electricity balance, large electrical loads (demand-side loads) such as those found at water treatment and wastewater treatment plants, can be used to connect to the grid to participate in the management of the electricity system by responding to fluctuations in supply and demand. Using demand-side loads to balance the system is an innovative method to provide Grid Balance® and it is fundamental to the operation of a modern electric power system.

With an increased amount of power from renewable sources (solar and wind) being generated, the electricity system operator has a greater and greater challenge in integrating these intermittent sources to the grid in real-time. The ENBALA Power Network provides balance to the system on the demand-side, allowing the electricity system operator to make increased use of renewable energy sources.

By integrating demand-side loads into the electric power system, electricity systems run more efficiently, provide reliability, and lengthen the lifespan of existing infrastructure while reducing greenhouse gases.

The ENBALA Power Network

The ENBALA Power Network is the network effect of coordinating many loads in a robust, reliable, and resilient way. The electric power system requires near real-time balancing, but there are few, if any, assets/processes that can tolerate unpredictable fluctuations every few seconds. The ENBALA Power Network optimizes which loads to adjust, how much to adjust them, and when to adjust them, based on their owner-defined operating constraints.

Until the advent of the Smart Grid —a bi-directional, real-time communications system that monitors and optimizes the operation of the power system — participation by demand-side loads was restricted. As process control equipment and the data services telemetry speed has improved, the Smart Grid has evolved to allow demand-side load participation. The service procured by the ISOs/RTOs is a required ancillary service, Grid Balance®, and network participants, such as water and wastewater treatment plants, receive a new revenue stream to help offset existing electricity charges. This unique platform connects participants to a Smart Grid network and allows them to specify their own customized constraints on flexibility so that modulation of their power use has no impact on the effective operations of their equipment or processes.

Using a secure, private communications platform, ENBALA connects large electricity users to the ENBALA Power Network. This sophisticated Smart Grid platform understands and monitors the real-time health, status and flexibility of participants.

The ENBALA Power Network pays companies to help keep the electric power system balanced in their regions by increasing its efficiency and reliability and helps reduce greenhouse gases.

The water-energy nexus is a term that describes water used to produce energy and the energy used to provide water-related services. There is emerging interest worldwide in identifying opportunities for water conservation and energy reduction. There are significant opportunities within major water-using sectors, specifically water and wastewater treatment plants, to manage the way electrical equipment uses power to smooth out the energy demand curve thereby improving energy use and energy infrastructure.

At wastewater treatment systems, aeration basin blowers, pumping stations, and recycled activated sludge (RAS) pumps are typical loads well suited for Grid Balance™. Specifically, aeration of wastewater is the largest electricity consuming process in a wastewater treatment plant. Minor adjustments can be made to aeration flow rates which shifts the time when the power demand occurs. The overall large retention time of an activated sludge plant, combined with the relatively slow growth rate of biomass, minimizes the effect of minor temporal fluctuations in aeration on effluent quality. This means that, on average, the dissolved oxygen concentrations and effluent quality is maintained within the limits set forth by the operators. For example, ENBALA defines influence as a request by the ENBALA Power Network for a blower to speed up or slow down when the dissolved oxygen (DO) is within the acceptable range of 2.0 mg/L ± 1.0 mg/L set within the supervisory control and data acquisition system (SCADA).

If the ISO/RTO requests a net increase in power consumed, the ENBALA Power Network will request the process to speed up the blower variable frequency drive (VFD) per the dissolved oxygen and blower control loop sequence. If the constraints in the plant control systems are not satisfied, then Grid Balance® will not happen and the process will not deviate from the current operation. The opposite is true for when the ISO/RTO requests a net decrease in power consumed. The ENBALA Power Network will request the process to slow down the blower VFD per the dissolved oxygen and blower control loop sequence. As measured over hours or days, the DO concentration and the energy consumed will be the same under normal operating conditions or with regulation enabled. The difference will be in the detailed timing of events.

The critical quality parameters of a wastewater treatment plant are measured by the quality of the plant effluent and steady operation of the treatment processes. Specific treatment plants can be modeled to show the effect of process adjustments due to external Smart Grid inputs using the industry-recognized wastewater treatment plant dynamic modeling and simulation software GPS-X (by Hydromantis ESS, Inc.). Models of wastewater facilities typically found in Canada and the United States are used to demonstrate how these demand-side assets can be connected to the grid, providing significant opportunities to balance the real-time supply and demand for electricity and generating revenue without affecting the effluent quality or upsetting plant processes.

METHODOLOGY

Models of typical wastewater treatment plants are set up in the Hydromantis GPS-X software. Activated sludge plants using traditional fine-bubble diffuser aeration blowers were investigated. A nitrifying plant and a denitrifying plant were modelled using a variety of scenarios including average flow, waste strength, diurnal, and steady state influent flow conditions. Each plant was simulated with and without ENBALA's Grid Balance® enabled to compare the effluent quality and to assess the potential for revenue generation from Smart Grid. Modelled influent wastewater conditions and effluent requirements are described in Table 1. The denitrify effluent conditions were selected to describe a large majority of US facilities that have low total nitrogen effluent requirements. The nitrify effluent conditions were used to describe typical Canadian facilities. Further descriptions of the models are provided in Table 1.

Table 1: Influent and effluent conditions

Parameter	Wastewater Influent	Facility Effluent	
		Denitrifying (mg/L)	Nitrifying (mg/L)
Total N (mg/L)	N/A	< 10	N/A
TKN (mg/L)	40	N/A	N/A
Ammonia (mg/L)	N/A	N/A	< 10
$cBOD_5$ (mg/L)	175	< 10	< 25
TSS (mg/L)	175	< 10	< 25
Temperature (°C)	20	N/A	N/A

Dynamic Wastewater Treatment Plant Modeling Software: GPS-X

GPS-X Version 6.1.1 (Hydromantis Environmental Software Solutions Inc., Hamilton, Ontario, Canada, 2012) can be used to assess the changes in plant performance arising from changes in the application of controller movements in aeration and pumped flow controllers. The model predicts and compares performance (oxygen transfer, biological treatment, effluent quality, etc.) with ENBALA's Grid Balance® solution when enabled and disabled.

GPS-X uses a set of dynamic mechanistic models for biological conversion, oxygen transfer (via diffused aeration system) and secondary clarification, to predict the transformation of BOD, COD, nitrogen and phosphorus. The models predict effluent concentrations, aeration demand, oxygen transfer, and energy usage.

The conventional GPS-X models were modified for this project by changing the input to the aeration model from airflow (m^3/d) to power (kW) for the diffuser system. The airflow is then calculated from energy, and subsequently used as an input for calculating oxygen mass transfer coefficient (k_{La}) for the aeration model.

Two models were prepared to assess the changes in performance when Grid Balance® is applied to typical plant operation. Models for typical North American activated sludge plants were constructed and sized for 50 mgd (190,000 m^3/d) influent flow. The nitrifying plant model and denitrifying plant model consists of an influent stream, an activated sludge basin, and a circular secondary clarifier. The influent stream is characterized for a typical primary effluent, by setting the TSS, COD, TKN, ammonia concentrations and stoichiometry ratios to those typically found in nitrifying plants in North America. The modeled parameters are found in Table 1.

Nitrifying Model

The nitrifying plant model consists of a plug-flow activated sludge tank comprised of four reactors in series of equal volume, with a fine-bubble diffused aeration system. The airflow is tapered through the length of the plug flow tank to maximize oxygen uptake in the head of the tank, where demand is highest. Recycled activated sludge (RAS) is pumped at a recycle ratio equal to the influent wastewater flow rate (1Q) back to the head of the aeration tank, and waste activated sludge (WAS) is pumped from the bottom of the clarifier. The wastage rate is set appropriately to maintain a mixed liquor suspended solids (MLSS) of ~2000 mg/L and a sludge retention time (SRT) of approximately 8 days.

Denitrifying Model

The denitrifying plant model consists of a plug-flow activated sludge tank comprised of four reactors in series, with two smaller anoxic zones up front, followed by two larger aerated zones that make up 70% of the total volume. The aerated zones use a fine-bubble diffused aeration system. There is an internal mixed-liquor recycle from the end of the aerated zones to the head of the anoxic zones, at a rate equal to twice the influent wastewater flow rate (2Q), to optimize denitrification. RAS is pumped at a recycle ratio of 1Q, back to the head of the aeration tank, and waste activated sludge (WAS) is pumped from the bottom of the clarifier. The wastage rate is set appropriately to maintain a MLSS of ~2000 mg/L and a SRT of approximately 6-7 days.

Process Characterization using GPS-X

For the modelled nitrifying only and denitrifying plants, Grid Balance® participation is simulated for a 24 hour period and a 7 day period. This process involves characterizing the available flexible power range of the plants' aeration blowers, simulating their participation within a network, and then analyzing the effects of the participation on plant effluent and dissolved oxygen (DO) levels.

The limiting constraint in determining the blower power range was determined to be dissolved oxygen (DO) since the wastewater leaving the aeration basins were required to be within the aerobic zone DO levels. Aerobic zones in the aeration basins were required to maintain a DO of 1.0-3.0 mg/L. GPS-X was used to do a power translation to determine the blower power for each modelled plant that results in a DO of 1.0 - 3.0 mg/L in the aeration basin's final tank. This is done using a steady-state analysis where the GPS-X software determines, assuming all other factors including influent flow remain constant, the DO level that would eventually be settled at for a given blower power set point. The influent characteristics used in this analysis are described in Table 1. This process characterization is described as a power translation for the limiting constraint and blower power. Anoxic zones in the denitrifying model were not adjusted and DO was maintained at 0 mg/L.

Once the power translation was completed, a script was generated to feed the information into the Grid Balance® simulation, where they are simulated within a network of nine other participating assets.

Table 2 lists the network assets and characteristics.

Table 2: Network assets

NETWORK DEVICE #	Industry Type	Asset Type	Range (kW)	Process Storage (h)	Energy Storage (kWh)	Ramp (kW/s)
1	Wastewater Treatment Plant #1	aeration blower	1,000 – 1,400	10	4000	40
2	Water Treatment Plant #1	pump	1,000 – 1,200	0.5	100	10
3	Commercial Building	electric boiler	0 - 500	30.3	15275	83
4	Wastewater Treatment Plant #2	aeration blower	812 – 1,000	10	1884	19
5	Water Treatment Plant #2	pump	1,000 – 1,200	0.7	134	10
6	Wastewater Treatment Plant #3	aeration blower	544 – 622	10	785	8
7	Wastewater Treatment Plant #4	aeration blower	505 – 555	10	510	5
8	Wastewater Treatment Plant #5	aeration blower	833 – 1,000	10	1671	17
9*	NITRIFYING MODEL	aeration blower	1,319 – 1,752	10	4330	43
10*	DENITRIFYING MODEL	aeration blower	579 – 730	10	1510	15
11	Water Treatment Plant #3	pump	1,280 – 1,520	3	720	24

* The modelled wastewater treatment plant network devices

ENBALA Power Network Simulation

ENBALA's simulation software (ENBALA Power Networks Inc., Toronto, Ontario, Canada, 2012) enables ENBALA to create a complex network of demand-side assets and have them participate in a simulated Grid Balance® market. During simulated operations, the real-time control optimization engine (R3OE) divides the ISO/RTO signal between all participating assets. The ISO/RTO regulation signal used in simulations was generated from a historical real-time Regulation Service signal from PJM. Figure 1 shows the signal used in the simulations to determine the R3OE Grid Balance® request for the simulation network. The ISO/RTO signal is normalized and represented in terms of relative movement required within the ISO/RTO. Thus the y-axis of the chart is represented as a range between -1 and +1. A normalized ISO/RTO signal of -1 corresponds to over generation, which requires a reduction of generation and an increase in consumption. A normalized ISO/RTO signal of +1 corresponds to under generation, which requires an increase in generation and a decrease in consumption. The R3OE ensures that the set points sent to each asset is within their operating constraints and that our network is respecting market performance requirements. The regulation market requires that our network of assets is able to:

- accept new ISO/RTO set points in intervals up to 2 second
- bid into market an hour in advance
- maintain a bid for an entire hour
- satisfy market performance metrics

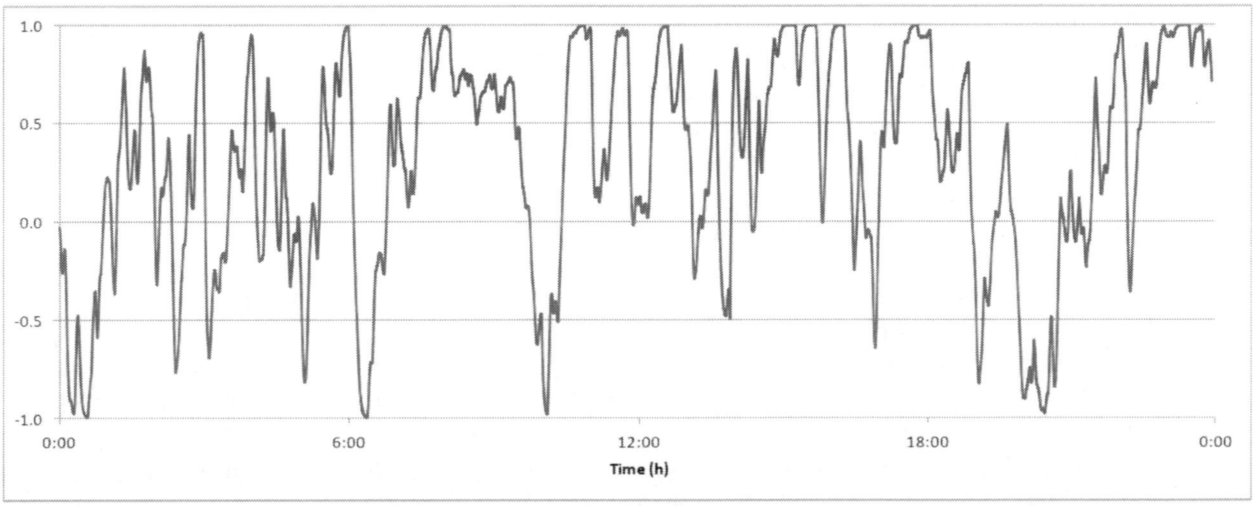

Figure 1: Normalized Regulation Service Signal[1]

The R3OE is a mixed integer programming solver which optimizes the allocation of the ISO/RTO signal to each asset based on their cost functions. Cost functions are a real-time analysis of a device's attractiveness to be bid into the market. The R3OE optimization engine attempts to determine the optimal collection of control set points for each asset. It does this while trying to maximize the regulation capability. The cost function and characteristics of each asset are set in a simulation configuration file. Common simulation configuration inputs include:

- power constraints
- power to asset rate conversion
- storage operating levels limits
- costs associated with movement/storage level/work demand
- power and process ramp rates
- storage dead band
- process rate trends

A script then generates a network which can be controlled by the R3OE. If an asset is forecasted to be unable to provide regulation for an entire hour, the R3OE takes the asset out of the network until it is again able to satisfy the R3OE's operating requirements.

In parallel with the control optimization, is the market bidding. It is through bidding that ENBALA is able to optimize long term network performance by maintaining stable operating conditions. Bidding a network involves determining:

- which assets are going to participate,
- bid range,
- base point

Bid range is the amount of power an individual asset is capable of sustaining for an entire hour given the composition of the network. The sum of the individual asset capabilities then determines the network bid. Base point is total operating base-load plus half of the asset range.

[1] http://www.pjm.com/markets-and-operations/ancillary-services/mkt-based-regulation.aspx

The results from a simulation can then be exported into a CSV file. For this simulation, the timestamp and specific device power information in response to the ISO signal were exported.

Constraint Validation with GPS-X: Modelling Effluent and Process Effects
The results from the Grid Balance® simulation are then passed back through GPS-X to observe the effects of participation of the wastewater plants' blowers and the resulting effluent. The ENBALA Power Network provides data with each blower's simulated response to Grid Balance® requests over 24 hours and 7 days of participation. This data shows the power set points that each blower is sent throughout the simulated time period.

The set points are then scripted into GPS-X to model the reaction of each plant model to the simulated changes in blower power. The model is set up such that the plant is assumed to have been steadily maintaining a DO level of 2.0 mg/L in its last aeration tank before regulation begins. Modelling at this stage is done in both, a steady flow scenario with influent flow constant at 50 mgd, and in a more dynamic and realistic scenario with influent following a typical diurnal flow pattern. The steady flow scenario isolates the effects of Grid Balance® participation on effluent and DO, while the diurnal flow scenario shows the effects that can be experienced in live implementation where there are daily fluctuations in flow rate and plant loading. The models will also be used to show if Grid Balance® changes plant performance by presenting data with Grid Balance® enabled and disabled. Effluent data, specifically for TKN, $cBOD_5$, TSS, and ammonia, is observed throughout the process to see the effects that Grid Balance® participation and if they violate the effluent limits described in Table 1.

RESULTS AND DISCUSSIONS

Minor adjustments to facility processes demonstrated that participation in the Smart Grid with wastewater treatment plant assets, such as aeration blowers, can generate Grid Balance® revenue without affecting plant operations or effluent water quality. The operator defined constraints accurately defined acceptable process ranges to allow the demand-side loads access to the Smart Grid.

Process Characterization and Power Translations

The wastewater models from GPS-X first needed to characterize the aeration process to determine the power translations for dissolved oxygen. The process was characterized using steady state conditions rather than with diurnal flow. The diurnal flow variations are used to simulate real world operation and determine the impact on effluent quality. The steady state flow can be characterized by the average flow of the diurnally modelled cases – 50 mgd. The simulation was conducted over a practical operating range as typical found in wastewater treatment plants. For aerobic zones, the dissolved oxygen was defined to be 1.0 – 3.0 mg/L. For anoxic zones, the dissolved oxygen was defined to be 0 mg/L. The final tank was determined to be the limiting process constraint for both models – nitrifying and denitrifying. Thus the power translations were generated from the final tank power translations of dissolved oxygen.

Figure 2 shows the power translation of the four aeration tanks in the aeration basin for the nitrifying model. All four tanks have a similar profile, but tanks 1-3 have a lower specific energy associated with them since the waste strength associated with the wastewater in the preceding tanks is higher and oxygen diffusion is stronger. The power translation was modeled for a dissolved oxygen range of 0.0 mg/L – 7.0 mg/L. The power translation generated a blower power range of 1319-1752 kW for the 50 mgd wastewater treatment plant operating the nitrifying aeration process between 1.0 – 3.0 mg/L of dissolved oxygen.

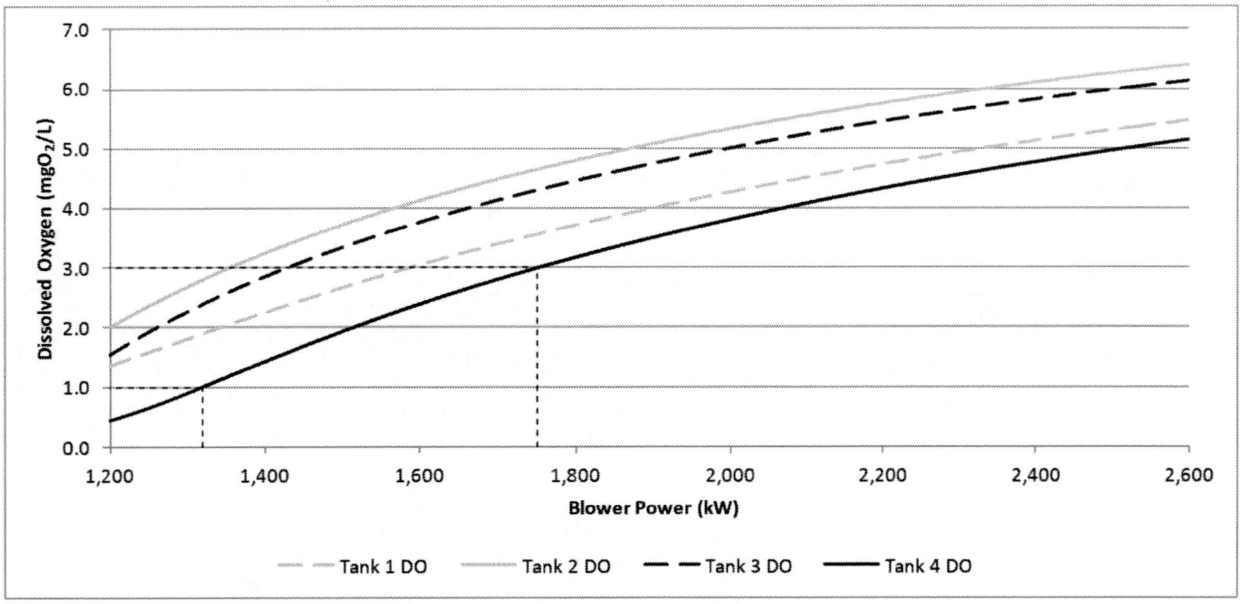

Figure 2: Nitrifying Power Translation

Figure 3 characterizes the aeration process for the nitrifying plant with about 6.5-hours of total storage. Storage is a term that is used to describe the length of time between the upper and lower limits of a limiting process constraint in response to a defined power input. In this case, the power translation for a DO of 1.0 mg/L – 3.0 mg/L was used. The step input to the power shows a nearly instantaneous response in power but a 6.5-hour lag in the DO before it reaches the lower DO limit of 1 mg/L. This is an advantageous characteristic for the aeration process as this long storage parameter provides stability when responding to smaller inputs from the R3OE. To practically evaluate the amount of storage, the storage associated with the time constant value was calculated. The time constant value represents the time it takes to reach 63.2% of its final value of 1 mg/L of DO when given a step power input. The storage time calculated from the time constant is about 26-minutes where the DO value at this time is 1.74 mg/L.

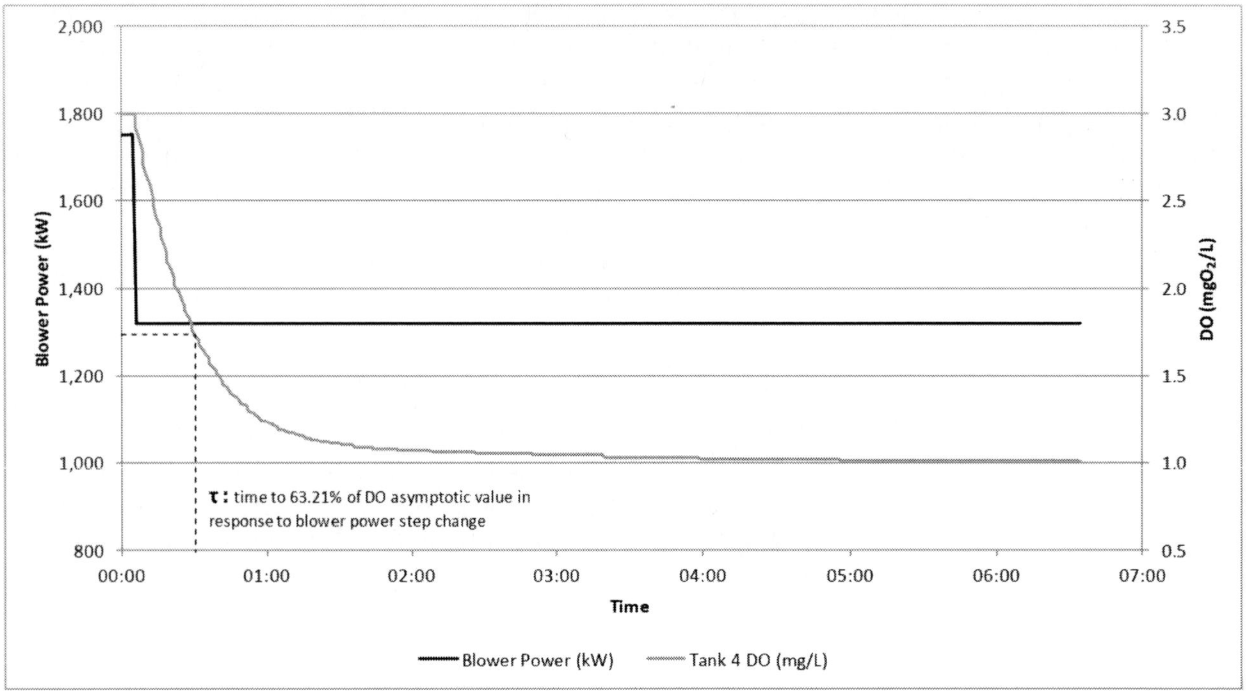

Figure 3: Nitrifying Process Storage Characterization

Figure 4 shows the power translation of the four aeration tanks in the aeration basin for the denitrifying model. Tanks 1 and 2 show that there is no power associated with the anoxic aeration tanks since the DO is maintained at 0 mg/L. Tank 3 has a power profile associated with it that shows a higher specific energy since it is receiving anoxic waste from tanks 1 and 2. Final tank shows the power translation for the full range and the chart identifies the aerobic zone range. The power translation was modeled for a dissolved oxygen range of 0.0 mg/L – 7.0 mg/L. The power translation generated a blower power range of 579-730 kW for the 50 mgd wastewater treatment plant operating the denitrifying aeration process between 1.0 – 3.0 mg/L of dissolved oxygen. This has a lower overall power associated with it since the process flexibility is only associated with two of the four tanks.

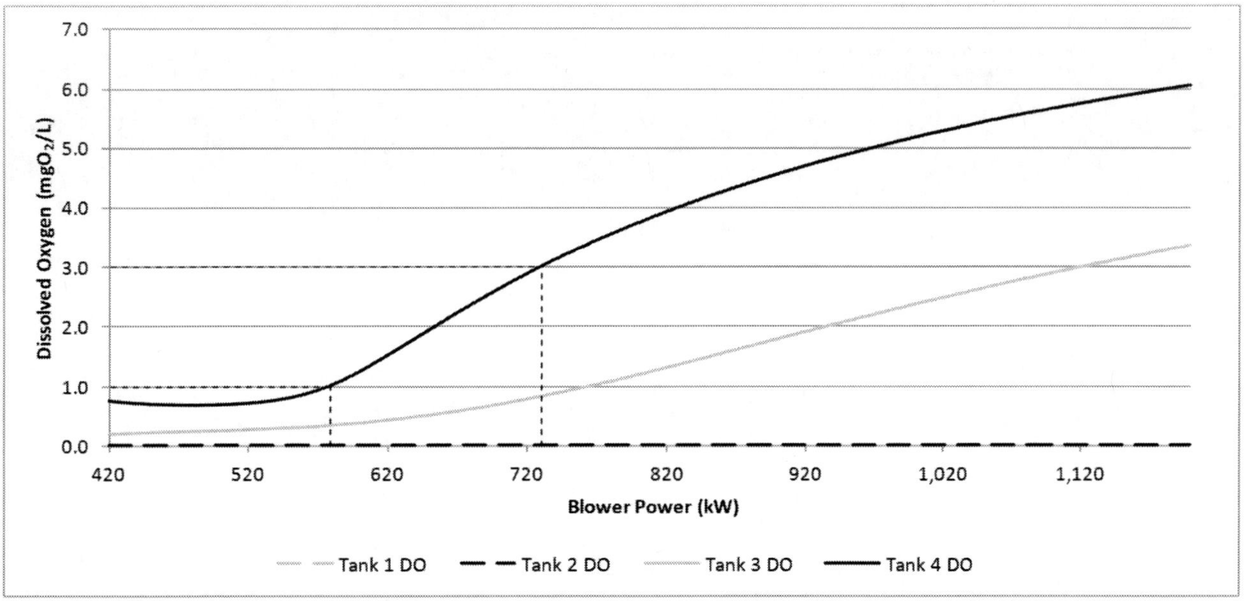

Figure 4: Denitrifying Power Translation

Figure 5 characterizes the aeration process for the denitrifying plant with about 6.5-hours of storage. Storage is a term that is used to describe the length of time between the upper and lower limits of a limiting process constraint in response to a defined power input. In this case, the power translation for a DO of 1.0 – 3.0 mg/L was used. The step input to the power shows a nearly instantaneous response in power but a 6.5-hour lag in the DO before it reaches the lower DO limit of 1.0 mg/L. This is an advantageous characteristic for the aeration process as this long storage parameter provides stability when responding to smaller inputs from the R3OE. To practically evaluate the amount of storage, the storage associated with the time constant value was calculated. The time constant value represents the time it takes to reach 63.2% of its final value of 1 mg/L of DO when given a step power input. The storage time calculated from the time constant is about 35-minutes where the DO value at this time is 1.74 mg/L.

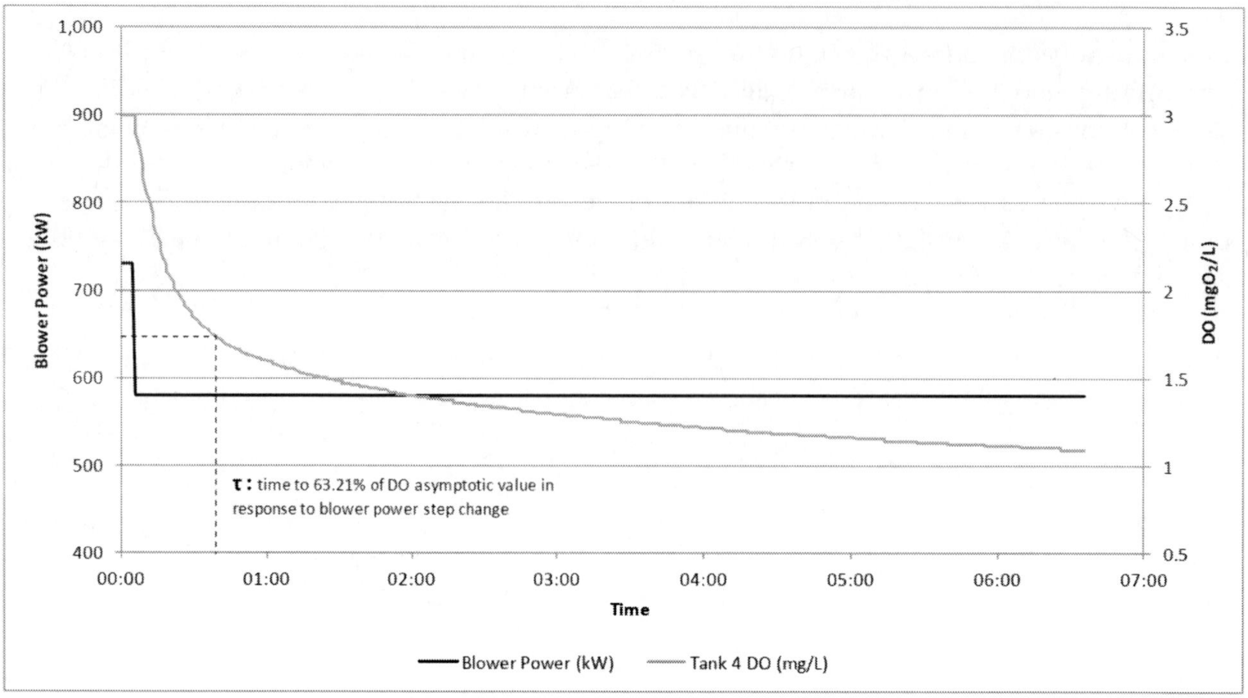

Figure 5: Denitrifying Process Storage Characterization

Nitrifying Model Results

Figure 6 below shows the aeration blower response when simulated in an eleven (11) asset network when responding to a signal from the ISO/RTO over a 24-hour period with the wastewater treatment plant receiving wastewater in a typical diurnal pattern. Blower power is given a power set-point as directed by the R3OE characterized from the power translations. The dissolved oxygen responds to changing basin conditions. The GPS-X model and the R3OE simulations are not connected in the real-time simulations and therefore is not a closed loop control scenario. For this reason, the dissolved oxygen in few occasions migrates outside of the range of 1.0 – 3.0 mg/L. In live implemented systems, there is a closed loop control system. This means that if the dissolved oxygen migrates outside of the specified range, the system will disable Grid Balance® and the control system will push the system back into the required operating range.

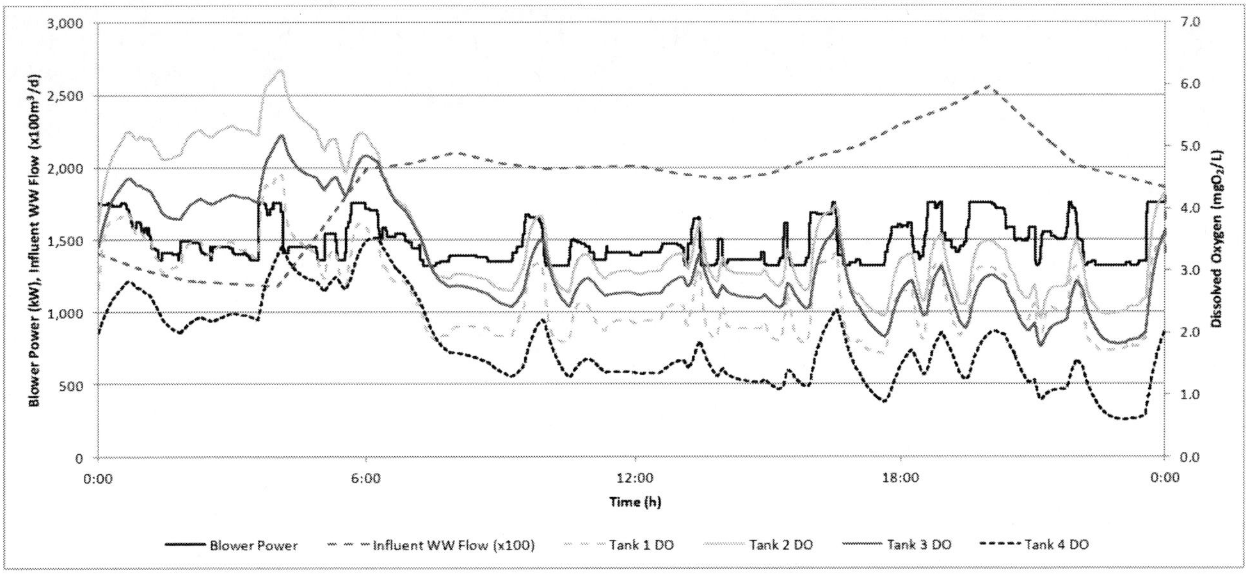

Figure 6: Nitrifying Aeration Tank and Blower Response Under Diurnal Influent Flow Over a 24-hour Period

The system was shown to have hours of dissolved oxygen storage in Figure 3. However, the critical operating parameter is the wastewater effluent parameters. It is known that the microbes responsible for treating the wastewater are much more resilient to change and that a change in dissolved oxygen does not change the treatment capacity over short periods of time (1-2 hours). Figure 7 shows the resulting effluent quality parameters total suspended solids (TSS), cBOD5, ammonia, and total Kjeldahl nitrogen (TKN). The defined discharge limits were described in Table 1 and are also noted on the charts below. Figure 7 shows that there are minor variations in the effluent quality despite end-to-end changes in the aeration basin dissolved oxygen. In the instances where DO limits were exceeded, the effluent quality parameters were still within the specified limits.

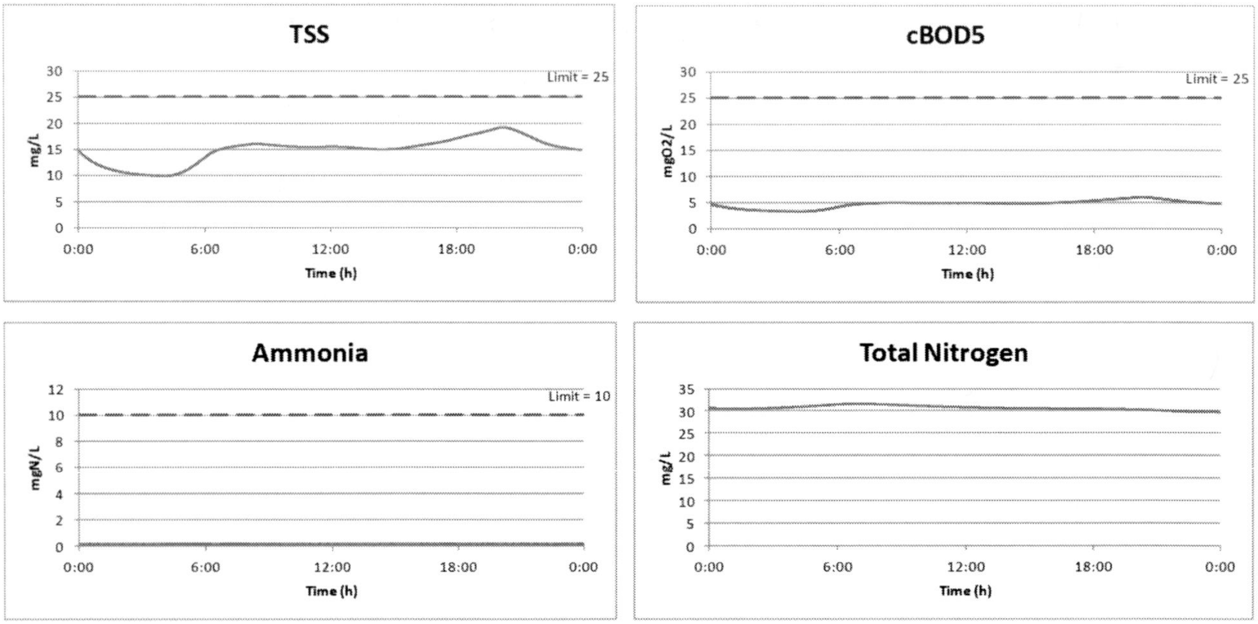

Figure 7: Nitrifying Wastewater Effluent Parameters and Limits

Figure 8 was generated for comparative purposes to show the aeration blower response under steady state flow conditions. The effluent graphs were not shown here but all parameters (TSS, $cBOD_5$, and ammonia) was below the specified limits.

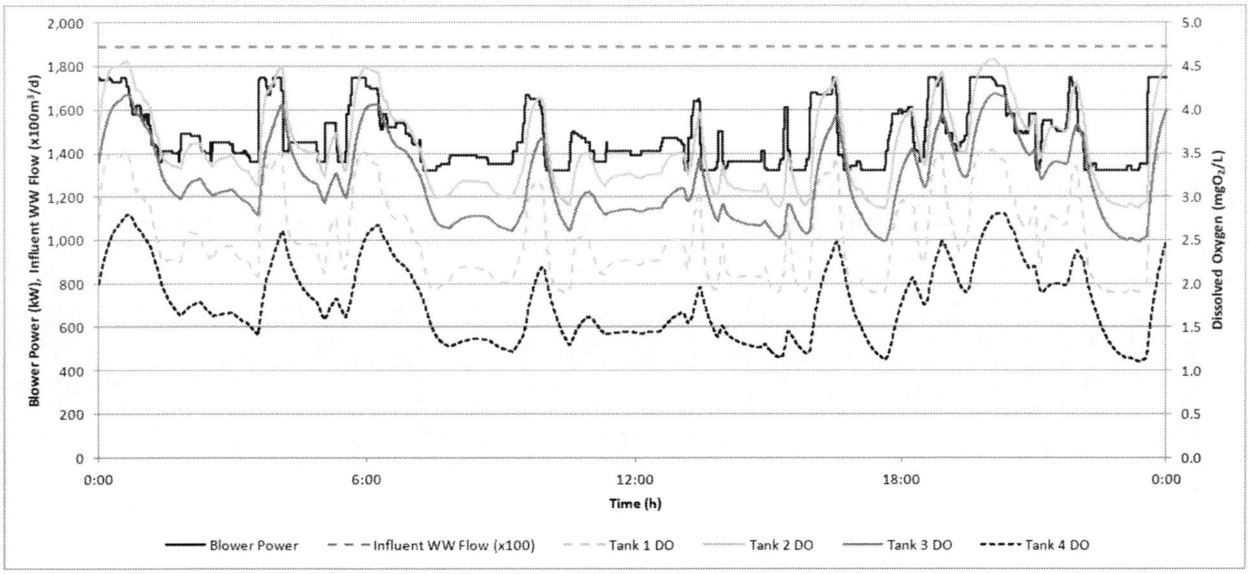

Figure 8: Nitrifying Aeration Blower Response under Steady State Influent Flow Over a 24-hour Period

Figure 9 shows the blower response over a 7-day simulation. Since influent conditions remained constant, but blower response was varied, a similar repeating pattern for dissolved oxygen is observed. There are slight variations in the reactor dissolved oxygen due to a changing blower power in response to an ever changing ISO/RTO Grid Balance® request signal. The effluent graphs were not shown here but all below the specified limits for TSS, $cBOD_5$, and ammonia.

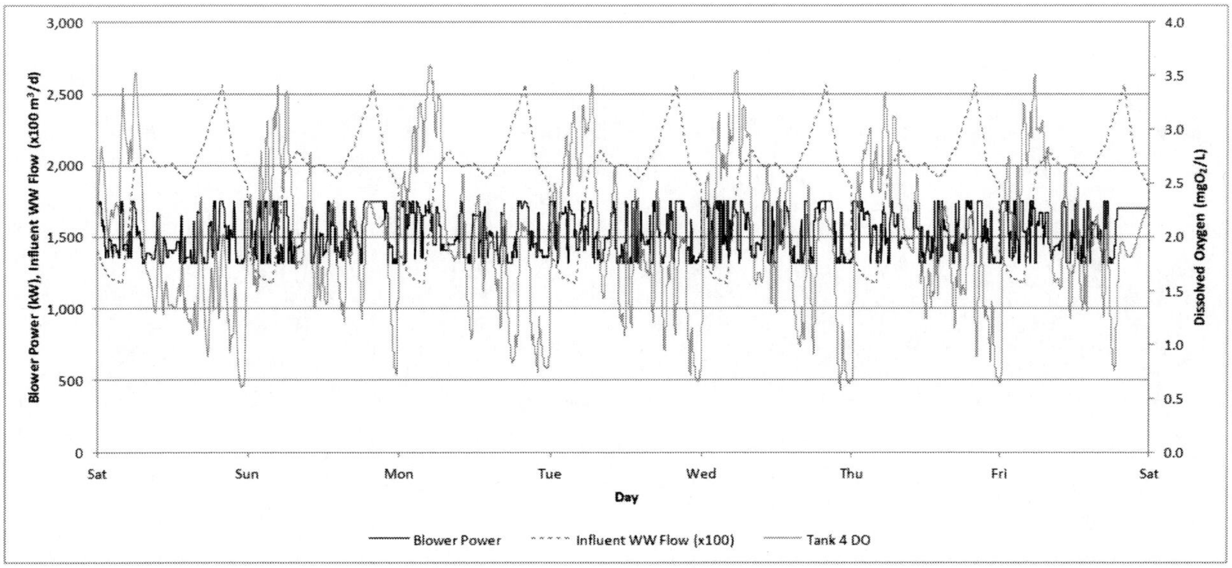

Figure 9: Nitrifying Aeration Blower Response under Diurnal Influent Flow Over a 7-day Period

Denitrifying Model Results

Figure 10 shows the aeration blower response when simulated in an eleven (11) asset network when responding to a signal from the ISO/RTO over a 24-hour period with the wastewater treatment plant receiving wastewater in a typical diurnal pattern. Blower power is given a power set-point as directed by the R3OE characterized from the power translations. The dissolved oxygen responds to changing basin conditions. The GPS-X model and the R3OE simulations are not connected in the real-time simulations and therefore is not a closed loop control scenario. For this reason, the dissolved oxygen in final tank migrates outside of the range of 1-3 mg/L in a few instances. In live implemented systems, there is closed loop control system. This means that if the dissolved oxygen migrates outside of the specified range, the system will disable Grid Balance® and the control system will push the system back into the required operating range.

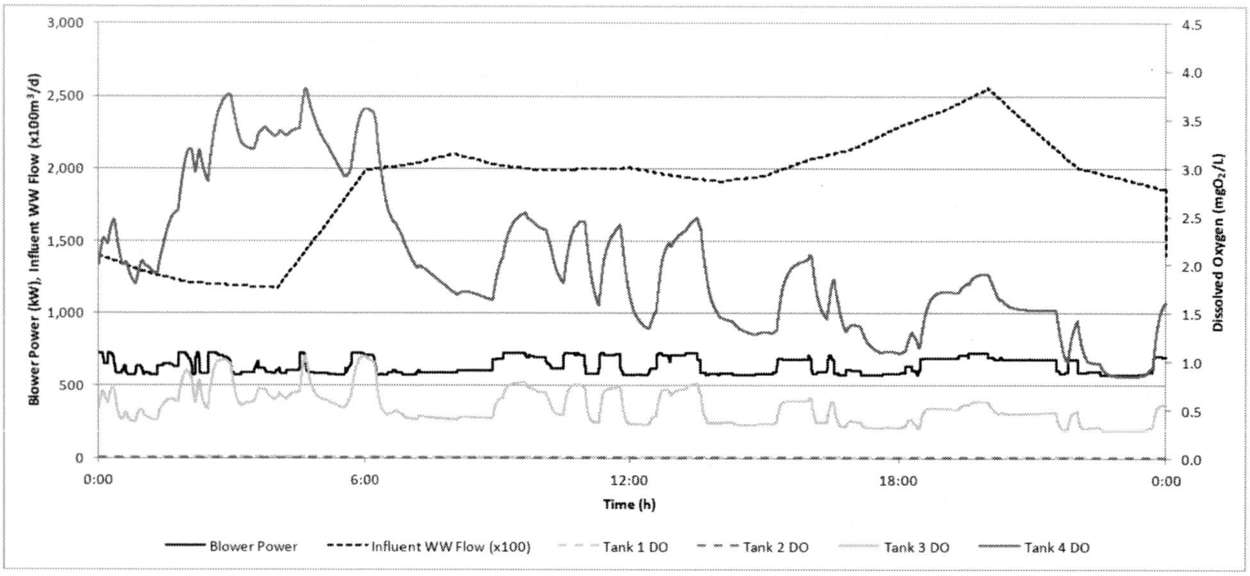

Figure 10: Denitrifying Aeration Tank and Blower Response Under Diurnal Influent Flow Over a 24-hour Period

With hours of dissolved oxygen storage (Figure 5) and the associated microbial resiliency, slight changes in the aeration basin dissolved oxygen does not change the treatment capacity. The resulting effluent parameters are shown in Figure 11 and show that only the TSS does not meet the defined limits defined in Table 1. This can be corrected in the model by changing the secondary clarifier design, but was not the focus of this paper.

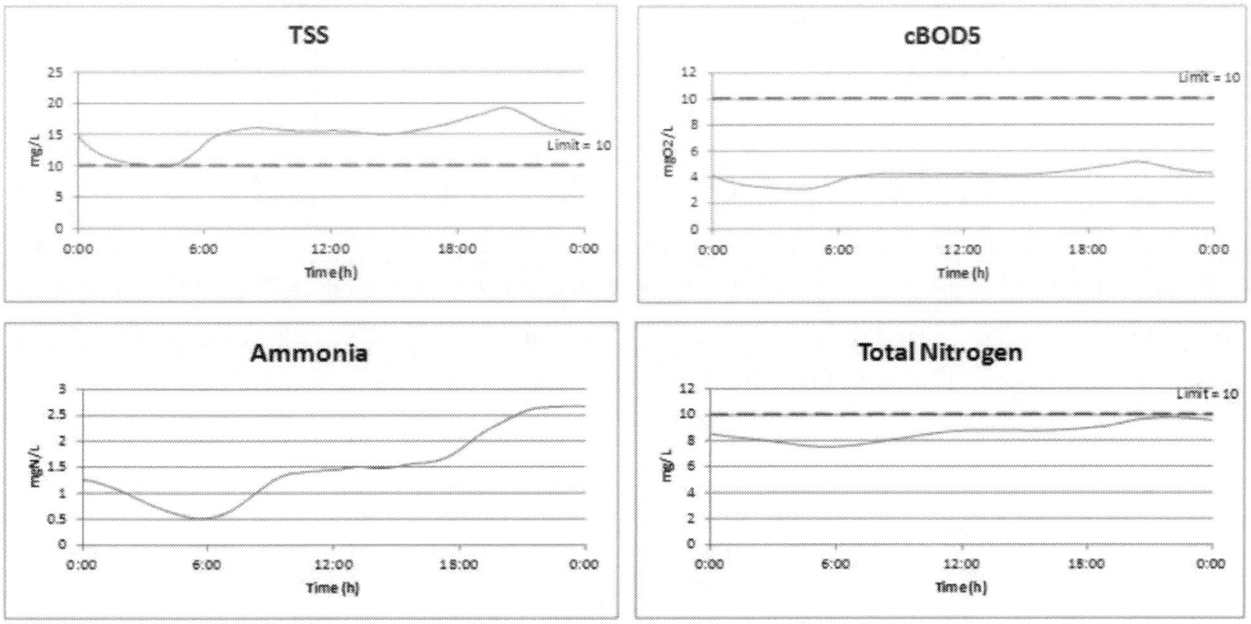

Figure 11: Denitrifying Wastewater Effluent Parameters and Limits

Figure 12 shows the effluent limits of total nitrogen for the denitrifying model when Grid Balance® is enabled and disabled. It appears that the total nitrogen approaches the defined limit of 10 mg/L. However, when compared to the model when Grid Balance® is disabled, there is very little difference and thus treatment capacity is considered to be unaffected. Total nitrogen in effluent from facilities with the most stringent effluent requirements in the Northeastern US often requires total nitrogen in the effluent of less than 3 mg/L. The scope of this paper is to explore the effects on wastewater operations due to Grid Balance® and not to optimize the wastewater treatment models. The models were constructed to provide typical results, but have not been optimized for specific operating requirement. Furthermore, to validate the differences in the models, the total nitrogen in the nitrifying model is three times higher than that of the denitrifying model. The nitrifying model total nitrogen chart is show in Figure 7.

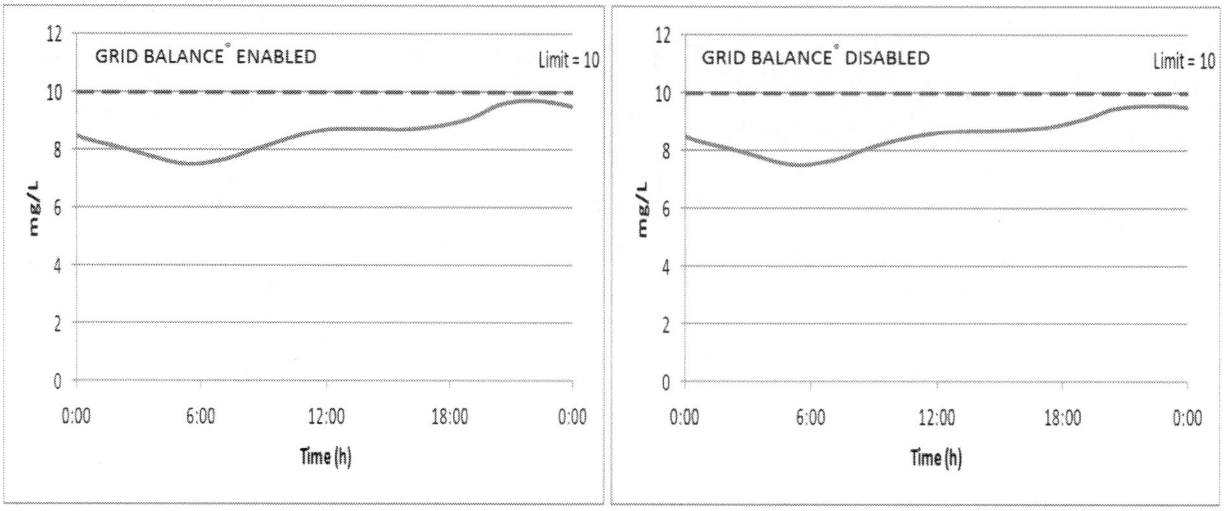

Figure 12: Effluent Total Nitrogen for the Denitrifying Model under Diurnal Flow with Grid Balance® Enabled and Disabled

Figure 13 was generated for comparative purposes to show the aeration blower response under steady state flow conditions. The effluent graphs were not shown here but all below the specified limits.

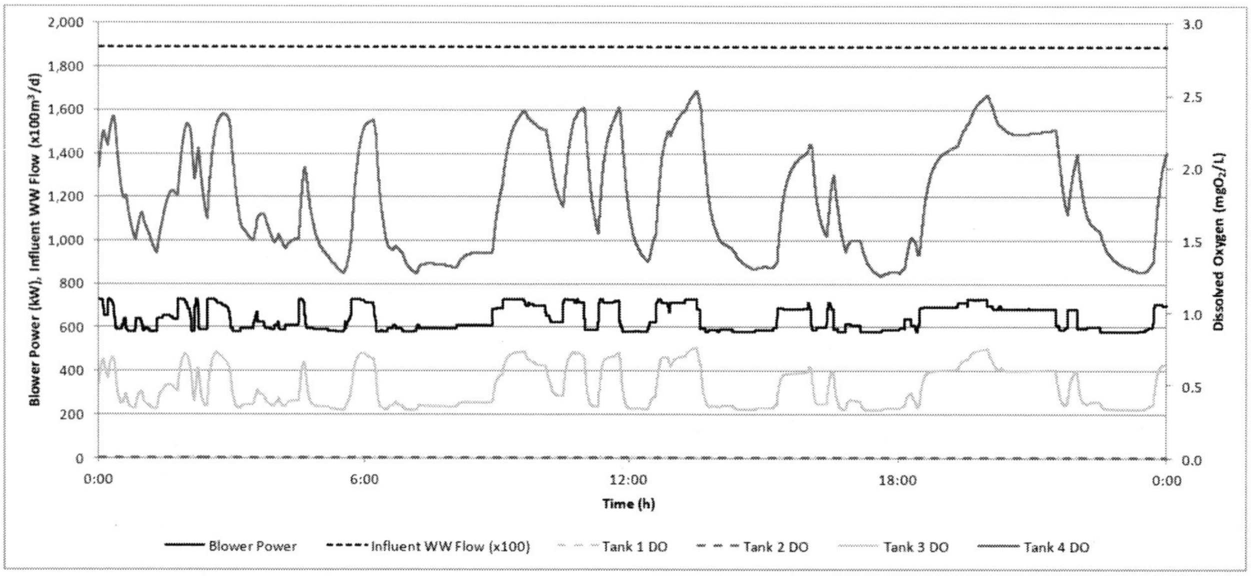

Figure 13: Denitrifying Aeration Blower Response Under Steady State Influent Flow Over a 24-hour Period

Figure 14 shows the blower response over a 7-day simulation. Since influent conditions remained constant, but blower response was varied, a similar repeating pattern for dissolved oxygen is observed. There are slight variations in the reactor dissolved oxygen due to a changing blower power in response to an ever changing ISO/RTO Grid Balance® request signal. The effluent graphs were not shown here but all below the specified limits.

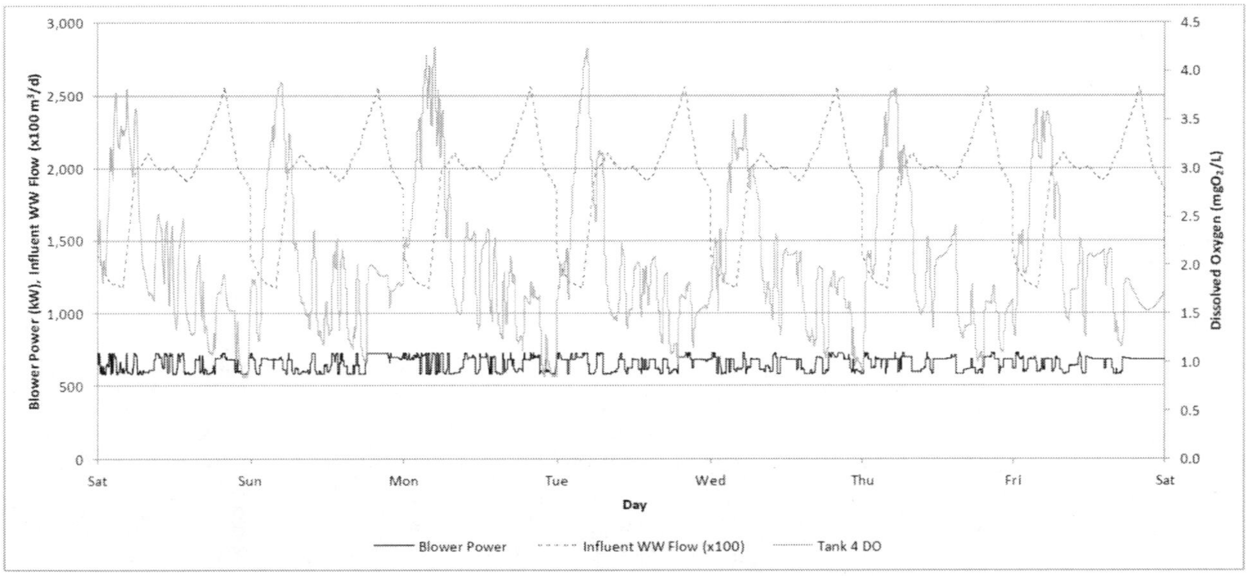

Figure 14: Denitrifying Aeration Blower Response under Diurnal Influent Flow over a 7-day Period

Network Response

The network response is shown in Figure 15. The network power response is the complete network response of the eleven (11) assets responding to the R3OE in response to the ISO/RTO signal. The chart below it shows the normalized range of network device over the 24 hour period. Normalized range is the sum of the absolute power movements over the given period of time. The modeled nitrifying WWTP contributed a normalized 503 kW of Grid Balance® over the 24 hour period. To calculate annualized revenue, this level of participation would result in approximately $25,000 of annual recurring revenue. Likewise for the modeled denitrifying WWTP, it contributed a normalized 220 kW of Grid Balance® over the 24 hour period. This level of participation would result in approximately $11,000 of annual recurring revenue.

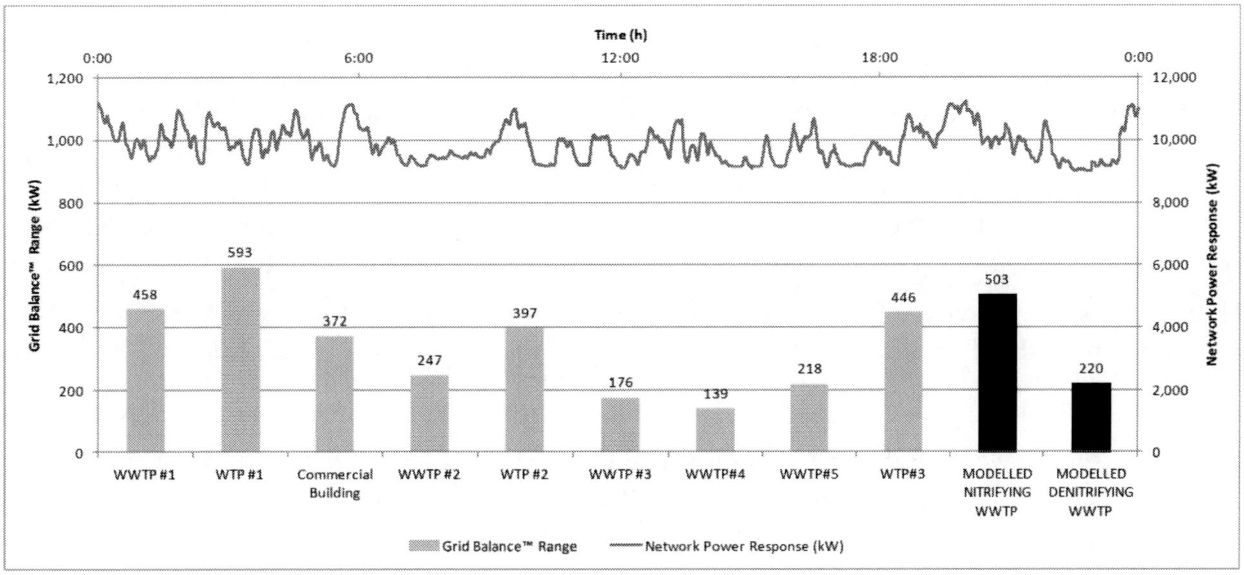

Figure 15: 24-Hour Network Response to Grid Balance® and Normalized Range for Each Network Device

The models were also run for a 7-day period to understand network performance as shown in Figure 16. The nitrifying model showed an increase in range over the course of the simulation. The denitrifying model showed a net decrease in range.

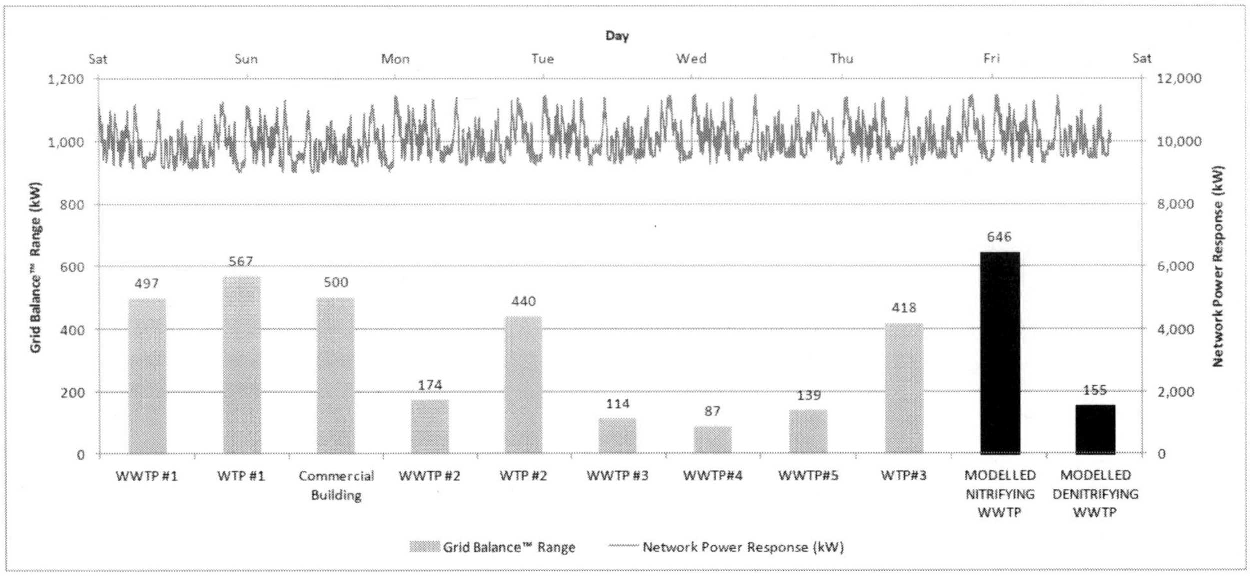

Figure 16: 7-Day Network Response to Grid Balance® and Normalized Range for Each Network Device

In addition to the water permit compliance, operations and maintenance personnel are concerned with device performance and operation. From a Grid Balance® participation perspective, this means the frequency and magnitude of power set point requests. Generally speaking, large and very frequent changes in power are not desirable. Both modelled WWTPs were analyzed using the distribution of movements (relative size of change in power) and the frequency of the movement (number of requests to change power). These distributions have been captured in the histograms shown in Figure 17 and Figure 18.

The minimum threshold for a change in power to be considered a move was an absolute value of 5 kW. A Matlab script was written to look at a 30-second time intervals and determine if all the moves are in the same direction. If during that 30-seconds there is a move in one direction and then in the other, it will separate those two into separate moves. For example, if the blower power was asked to increase power (positive value) and then the power was asked to decrease (negative power), this would be the indication of the start of another movement or request to change power. In the histogram, a movement at +100% of asset range means a full swing from bottom limit to top limit. A movement of -100% means a full swing from top to bottom.

Figure 17 is the histogram for the nitrifying WWTP aeration blower for the 24-hour and 7-day modeling periods. The histograms show that most of the movements are within +/- 20% of the range. The range for the nitrification blower is 503 kW. This means that most of the movements are +/- 100 kW. This is not to be confused with the operating power range of 1319 – 1752 kW. The operating power range is the total power of device operating as would be measured by a device specific power meter. This means that the device typically is only asked to move 5-10% of its operating power. There are some larger power movements, but these are infrequent events.

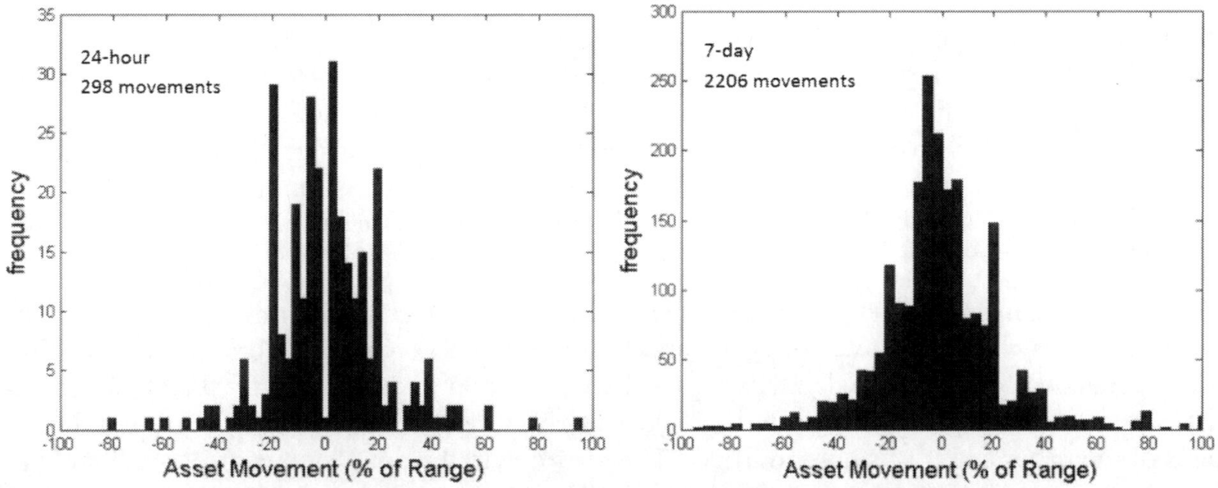

Figure 17: Histogram of Nitrifying WWTP Aeration Blower for 24-hour and 7-day Period

Figure 18 is the histogram for the denitrifying WWTP aeration blower for the 24-hour and 7-day modeling periods. For this model, the movements are concentrated inside +/-20% of the range. The range for the denitrifying blower is 220 kW. This means that most of the movements are +/- 44 kW. This is not to be confused with the operating power range of 579 – 730 kW. The operating power range is the total power of device operating as would be measured by a device specific power meter. Again, this means that the device typically is only asked to move 5-10% of its operating power. There are more power movements outside +/-20% of the range and there are less total movements than the nitrifying model likely due to the smaller operating range and therefore required to have larger power movements to keep responding to the Grid Balance® request.

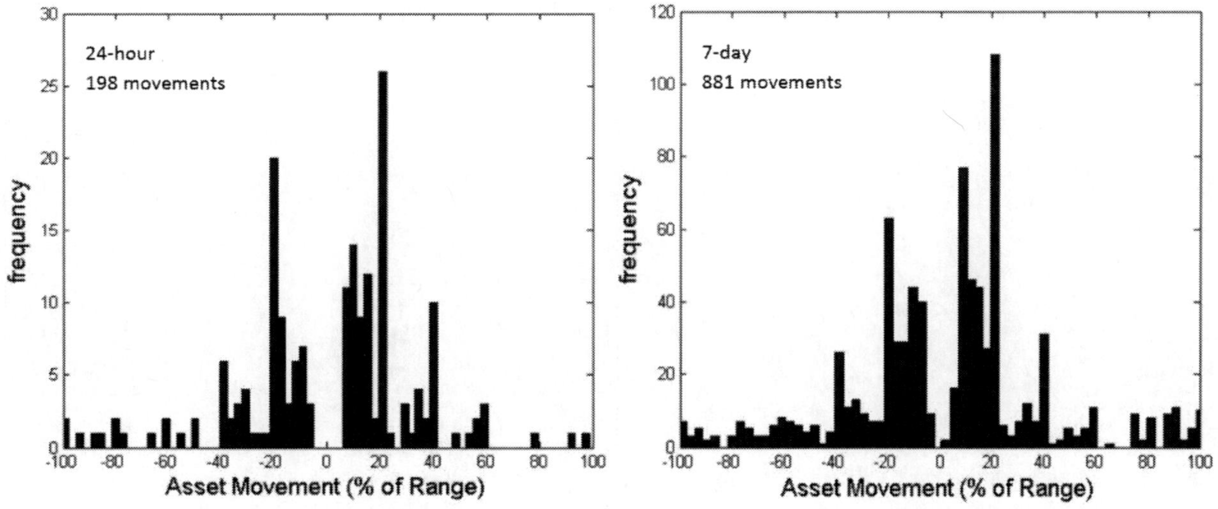

Figure 18: Histogram of Denitrifying WWTP Aeration Blower for 24-hour and 7-day Period

CONCLUSIONS

This paper has shown that aeration blowers can provide Grid Balance® in wastewater treatment plants with both nitrifying and denitrifying activated sludge processes while maintaining effluent quality. Both models provided predictable power translations and demonstrated that there is sufficient storage. These characteristics of aeration blowers present a significant opportunity for wastewater plants to generate revenue and help the electricity systems run more efficiently, provide reliability, and lengthen the lifespan of existing infrastructure while reducing greenhouse gases.

The results demonstrated that the processes stayed within the acceptable limits, with exception of a few occasions where dissolved oxygen concentrations were exceeded for brief periods of time. Total suspended solids for the denitrifying model were found to exceed the set threshold, a result that would require making changes to secondary clarifier design. However, this was not the focus of the paper and it was not modified. This was due to the open loop control methodology used. Future work to integrate GPS-X and the ENBALA Power Network R3OE to have closed-loop control is being investigated. The 24-hour and 7-day models provided similar effluent results. The 7-day nitrifying results demonstrated that there was more Grid Balance® range available than was available when modelled for only 24-hours. The denitrifying model showed that less Grid Balance® range was available for the 7-day simulation.

Wastewater effluent was also carefully analyzed and the results were all below the defined limits as defined in typical wastewater operations. The purpose of the model was not to refine the wastewater process and optimize the effluent. More importantly, the purpose was to show that the effluent quality was not affected by Grid Balance®, which was effectively demonstrated.

Finally, blower operation was analyzed to determine how frequent and how big the changes in power were. The relative changes in power are small compared to duty power and the majority of the frequent requests are small changes in duty power. This means that blower operation is within the typical mechanical operating conditions of aeration blowers.

Future work to validate a facility model with live facility Grid Balance® participation is work reserved for future investigation.

Looking at the Future: How Metropolitan Sewer District of Greater Cincinnati (MSDGC) is Turning Waste into Resources and Energy

Ting Lu[1*], Biju George[2], Joo-Youp Lee[3] and Mingming Lu[4]
[1] Black and Veatch Inc., U.S.A.
[2] Metropolitan Sewer District of Greater Cincinnati, Cincinnati, OH, USA
[3] Chemical Engineering Program, School of Energy, Environmental, Biological, and Medical Engineering, University of Cincinnati, Cincinnati, OH, U.S.A.
[4] Environmental Engineering Program, School of Energy, Environmental, Biological, and Medical Engineering, University of Cincinnati, Cincinnati, OH, U.S.A.

*Corresponding authors. Mailing address: 1600 Gest Street, Cincinnati, OH 45204. Phone: (513) 244-5137. Fax: (513) 557-7175. E-mail: Ting.Lu@cincinnati-oh.gov

Abstract

The Metropolitan Sewer District of Greater Cincinnati (MSDGC) is responsible for conveying and treating an average of 192 million gallons per day (MGD) of wastewater and it operates seven major treatment plants in Hamilton County. Common to many utilities in North America, MSDGC is facing a lot of challenges including reducing combined sewer overflows (CSO), aging infrastructure and declined water usage etc. One important strategy that helps the utility grow for the future is to turn wastes into resources and energy, as energy is the biggest contributor to our current operation budgets. **This study presents two case studies that MSDGC is collaborating with University of Cincinnati for sustainable future growth and community benefit. This is a paradigm shift in our industry to utilize innovative ideas to solve today's environmental problems, and also rethink wastes as resources for the future.**

Keywords: *Chlorella vulgaris*, Wastewater Effluent, Nitrogen and Phosphorus Removal, FOG, Biodiesel

Introduction

The first case study investigates the synergistic relationship between nutrient removal and algae growth in the municipal wastewater treatment plants. Traditional nutrient removal involves using selective nitrifying bacteria and excess oxygen, which consumes a significant amount of energy. Microalgae-derived biodiesel production has received great attention due to their ability to produce triacylglycerols for biodiesel while using solar energy and carbon dioxide (CO_2). The use of wastewater for algae growth will reduce not only the cost associated with nutrients but also reduce the residual nutrient concentrations present in the wastewater effluent stream. Thus there is a potential synergy between algae growth (e.g. for biodiesel production) and wastewater treatment.

The second case study investigates on how to convert fat, oil and grease into biodiesel. As indicated by the U.S. Environmental Protection Agency (EPA), fat, oil and grease (FOGs) is the number one cause of sewer line blockages in the US. An analogy of the cause was drawn of bad cholesterol to the blood vessels as to FOG to the sewer lines. Consequently, sewer blockage leads to combined sewer overflow (CSO) and sanitary sewer overflow (SSO). This research is important to not only reduce the cause of the CSO and SSO, but also utilize it as resources to produce biodiesel.

1. Investigation of the Synergistic Relationship between Nutrient Removal and Algae Growth in Municipal Wastewater Treatment Plants

The interest of microalgae has been increased as a source of biodiesel due to their substantial amount (e.g., 15-50% dry cell weight) of lipids (Hu et al., 2008). In addition, they grow fast, use a high degree of solar energy (3-10%), and use $CO_2(g)$ as a carbon source which is a major contributor of global warming (Larsen et al., 2009; Lu et al., 2011; Zou et al., 2010). However, nutrient supply is one of the major costs for the microalgae cultivation according to the life cycle assessment study (Clarens et al., 2010; Lardon et al., 2009). In this sense, the use of wastewater was suggested to reduce the cost of cultivation because wastewater includes nitrogen and phosphorus (Clarens et al., 2010; Lardon et al., 2009). This method can reduce not only the cost of nutrient but also remove nitrogen or phosphorus from the wastewater which is a major contributor of ecological eutrophication.

The removal of nitrogen and phosphorus in wastewater was studied by using *Chlorella* sp., *Scenedesmus obliquus and Phaeodactylum tricornutum* (Craggs et al., 1995; González et al., 1997; Tam et al., 1994; Voltolina et al., 2005). However, the availability of wastewater for the growth of microalgae as a nutrient source and the growth kinetics of microalgae are little studied. The Mill Creek plant located in Cincinnati, Ohio, U.S.A. is a major wastewater treatment facility where ~70% is collected from industrial source and the other is collected from domestic source. The wastewater effluent from the plant is clear and contains ammonia/ammonium ion with a low concentration of orthophosphate ion. In this sense, the wastewater was selected as a nutrient source for the growth of microalgae. In this study, the growth kinetic of *Chlorella vulgaris (C. vulgaris)* was investigated by using the wastewater effluent because the algae are one of the fastest microalgae.

Methods

Cultivation Medium and Conditions

The wastewater collected from the Mill Creek plant located in Cincinnati, Ohio, U.S.A. was used for the growth of *C. vulgaris*. Two initial algal cell densities (44.37±0.60 mg/L and 354.48±2.14 mg/L) were used in this study. During the culture, light intensity and light/dark cycle were set to 100.8 $\mu mol \cdot m^{-2} \cdot s^{-1}$ and 16-h/8-h, respectively, and NH_3/NH_4^+, PO_4^{3-}, and total inorganic carbon (TIC), cell density, and the pH of the medium were measured once daily. In addition, the pH of the mediums was controlled to 7 by using $CO_2(g)$ for a high initial algal cell density because microalgae grew well in the narrow pH range of 7-8 (Kim et al., 2012) and the mediums were mixed by using a magnetic stirrer with a speed of 350 rpm.

Determination of Cell Density of *C. vulgaris*

The cell density of *C. vulgaris* was determined by measuring the optical density of a 15-mL sample at 682 nm (Widjaja et al., 2009) for every 24 h by using UV-vis spectrophotometer (UV-1800, Shimadzu Scientific Instruments). The calibration curve was prepared by comparing the absorbance of UV spectrophotometer at 682 nm and the dry weight of *C. vulgaris.*

Determination of Total Inorganic Carbon (TIC) Concentration

An acid-base titration method (Fischer & Peters, 1968) was used to determine the concentrations of TIC (i.e. = $[HCO_3^-]+[CO_3^{2-}]+[H_2CO_3]+[CO_2(aq)]$) in a 15-mL sample using 0.1 N and 0.01 N HCl solutions.

Determination of Orthophosphate (PO_4^{3-}) and Ammonia (NH_3)/Ammonium Ion (NH_4^+) Concentrations

PO_4^{3-} is measured using the Phosver 3 phosphate reagent (HACH Model PO-19 reactive phosphorus measurement kit). A 5-mL sample was taken from a culture medium and a Phosver 3 phosphate reagent was added to it. After 2 min, blue color appeared due to the presence of phosphorus, and the color intensity was proportional to the amount of an PO_4^{3-} concentration in the solution. The color intensity was measured using a UV-Visible spectrophotometer (UV-1800, Shimadzu Scientific Instruments) at 890 nm.

The nitrogen concentrations in NH_3 and NH_4^+ were measured using an ammonia probe (Model: 9512HPBNWP Orion Thermo Scientific) (APHA, 2005). All ammonium ions were converted into ammonia by raising the pH of the sample solution of the culture medium above 12, and the resultant ammonia concentration was determined by the ammonia probe. A 15-mL sample was filtered out using a syringe filter (0.45 μm nominal pore with 24 mm diameter, Whatman filter) in order to avoid potential blockage of the membrane of the ammonia probe.

Results and Discussion

Growth and Nutrients Removal with Low Initial Cell Density

The wastewater contains NH_4^+/NH_3 with a low concentration of PO_4^{3-} shown in Table 1.

Table 1. TIC, NH_4^+/NH_3, PO_4^{3-}, and pH of wastewater samples collected from Mill Creek plant

Parameter	Wastewater 1 (W1, mg/L)	Wastewater 2 (W2, mg/L)
TIC	50.18 ± 0.83	72.24 ± 1.81
NH_4^+/NH_3	8.05 ± 0.16	18.31 ± 0.53
PO_4^{3-}	1.85 ±0.10	1.37 ±0.01
pH	7.34±0.05	7.88±0.07

The growth of *C. vulgaris*, TIC consumption, and pH change are shown in Figure 1. The lag-phase period was short (24 h) and the TIC was not decreased during the phase. Then, the cell density of *C. vulgaris* significantly increased with the decrease of TIC until 96 h.. However, the growth was stopped after 96 h due to the high pH (higher than 11) of mediums because *Chlorella* sp. grew well in the pH range of 7-8.

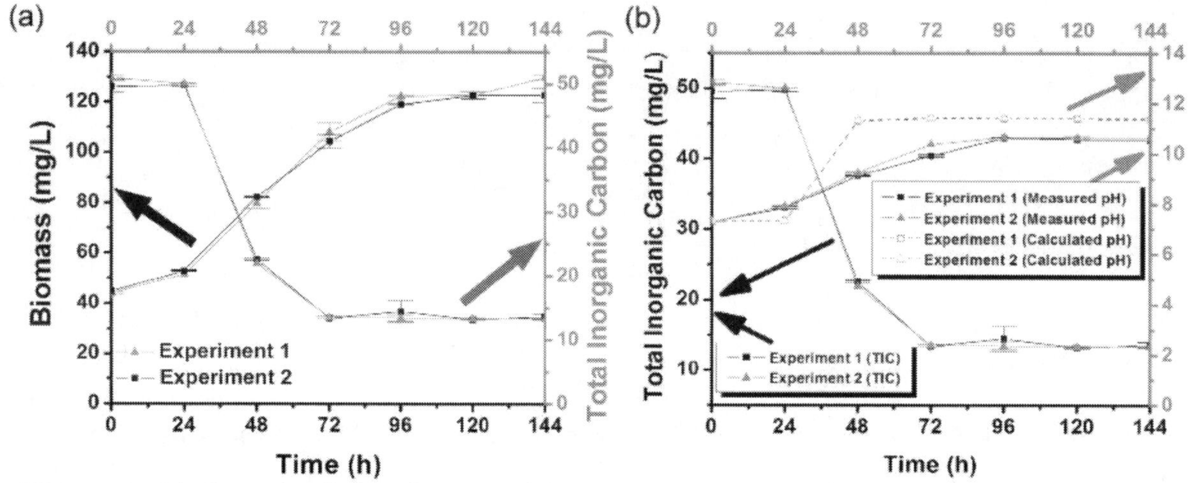

Figure 1. (a) Growth of *C. vulgaris* and TIC consumption and (b) TIC consumption and pH change with low initial algal cell density.

The removal of NH_3/NH_4^+ and PO_4^{3-} are shown in Figure 2. Similar with the TIC consumption, a small amount of nitrogen (1.11±0.31 mg/L) and phosphorus (0.08±0.02 mg/L) were decreased during the lag phase, but the nitrogen and phosphorus was decreased from 6.94±0.18 mg/L to 0.56±0.04 mg/L and 0.47±0.03 mg/L to 0.03±0.01 mg/L, respectively until 96 h.

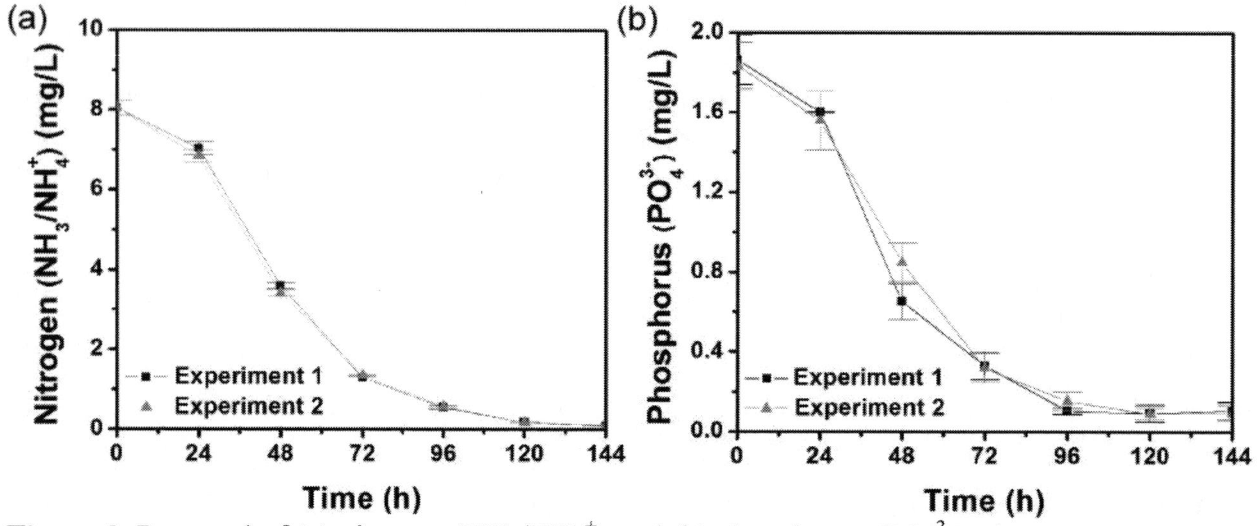

Figure 2. Removal of (a) nitrogen (NH_3/NH_4^+) and (b) phosphorus (PO_4^{3-}).

Growth and Nutrients Removal with High Initial Cell Density

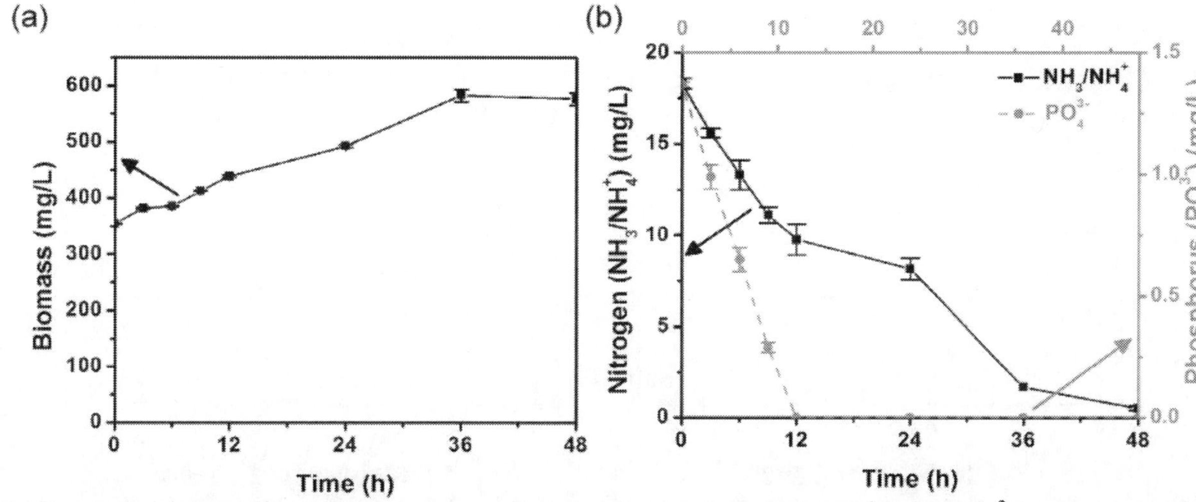

Figure 3. (a) Growth of *C. vulgaris* and (b) removal of NH_3/NH_4^+ and PO_4^{3-} with high initial algal cell density (354.48±2.14 mg/L).

The growth of *C. vulgaris* and the removal of nutrients with pH control to 7 are shown in Figure 3. PO_4^{3-} was removed within 12 h and NH_3/NH_4^+ were also removed within 48 h without a lag-phase due to the high initial cell density and pH control to 7 which is the suitable for the growth of microalgae.

Growth Kinetics of *C. vulgaris*

The Monod equation is used to describe the growth of *C. vulgaris* as shown in equation (1):

$$\mu = \frac{\mu_{\max} S}{K_s + S} \tag{1}$$

where μ_{max} is a maximum growth rate coefficient (h^{-1}), μ is a specific growth rate (h^{-1}), K_s is the Monod coefficient (mg/L), and S is the concentration of a limiting nutrient (mg/L). For the low initial cell density, nitrogen was determined to be a limiting substrate for the model because average R^2 value (0.97) for nitrogen is higher than the value (0.64) for phosphorus. Then, μ_{max} (h^{-1}) and K_s (mg/L) were determined to be 0.01245 (h^{-1}) and 0.0696 (mg/L) for Experiment 1, and 0.01241 (h^{-1}) and 0.0864 (mg/L) for Experiment 2, respectively. The re-plot by using the determined Monod equations were well matched with the experimental data, shown in Figure 4.

For the high initial cell density, phosphorus was determined to be a limiting substrate for the model because R^2 value (0.99) for phosphorus is higher than the value for nitrogen (Figure 5). Then, μ_{max} and K_s were determined to be 32.85 (h^{-1}) and 0.99 (mg/L). The μ value for the culture with a high initial cell density and pH control was ~40 times higher than the value for the culture with a low initial cell density and no pH control. Thus, it is noted that a high initial cell density and pH control could promote the growth of *C. vulgaris* by using nitrogen and phosphorus in wastewater. In this study, the C. vulgaris was successfully by using wastewater containing with nitrogen and phosphorus. In addition, the determined Monod equations can be incorporated into a reactor model such as batch, CSTC, and PFR.

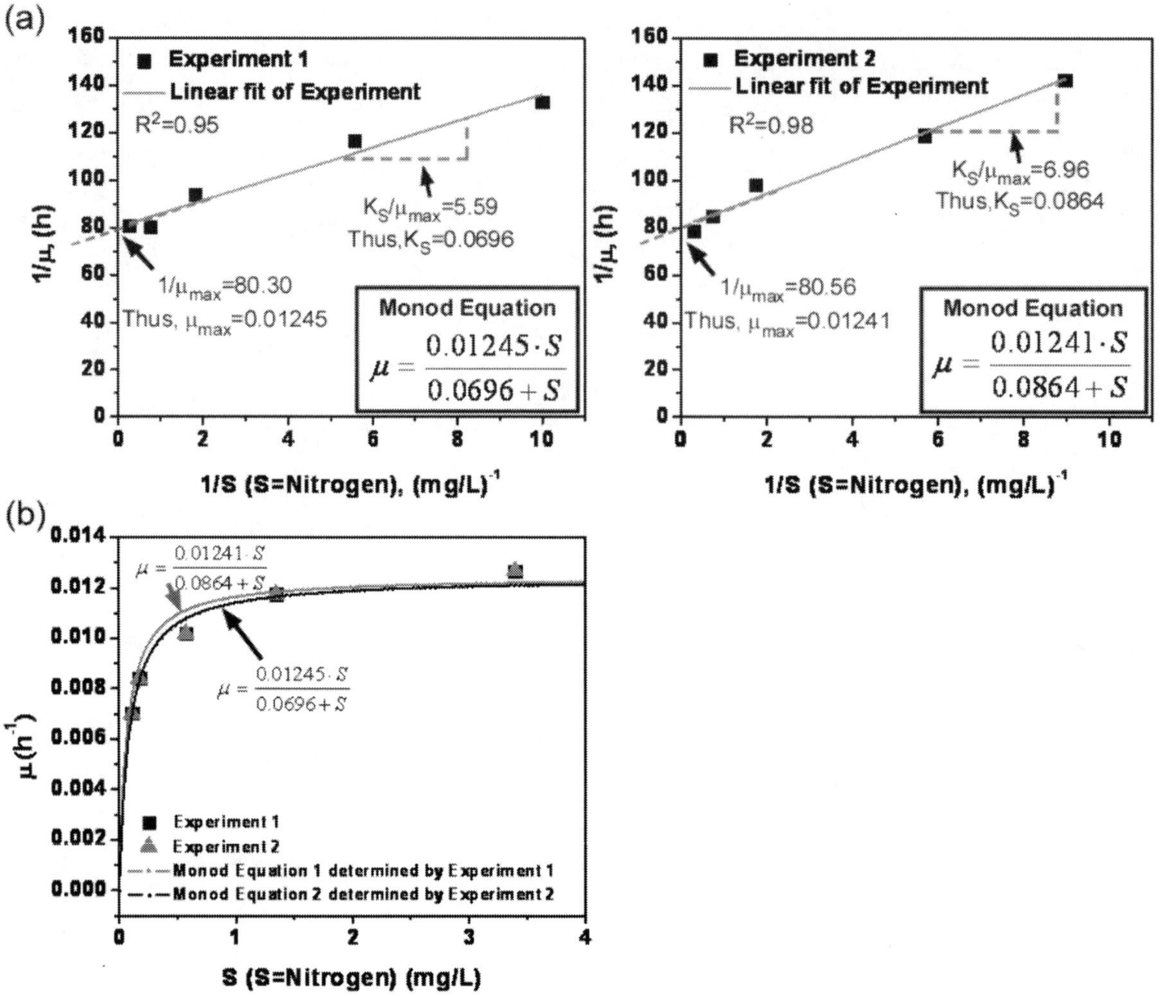

Figure 4. (a) Determination of μ_{max} (h^{-1}) and K_s (mg/L) of the Monod equation, and (b) Monod plot (μ vs. S) for low initial algal cell density.

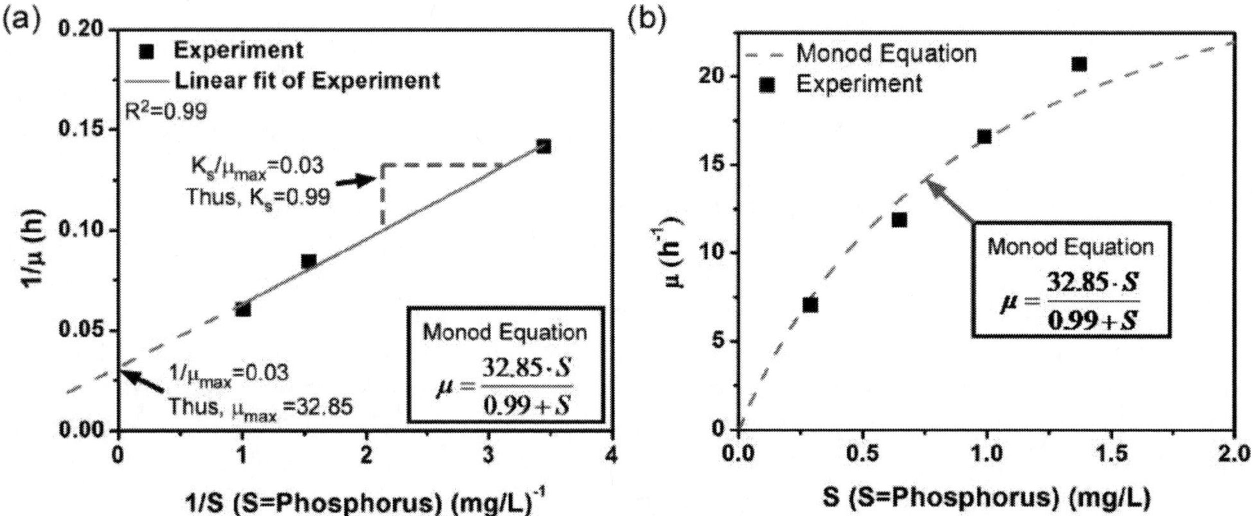

Figure 5. (a) Determination of μ_{max} (h^{-1}) and K_s (mg/L) of the Monod, and (b) Monod plot (μ vs.

S) for high initial algal cell density.

Summary

The secondary wastewater collected from a local wastewater treatment plant was successfully used for the growth of *C. vulgaris* as a nutrients source. The removal rate of nitrogen and phosphorus in wastewater can be enhanced by using a high initial cell density and pH control to 7. In addition, the Monod equation can be used to describe the algal growth and apply to the design of photobioreactor.

2. Converting fat, oil and grease (FOGs) into Biodiesel

FOGs are generated from restaurants and food processors as well as individual households when the waste cooking oil, oil containing wash liquids etc are discharged. Commercial food services usually are required to install grease traps, which need to be pumped regularly by grease haulers, while the FOG from individuals tend to end up in the pipelines.

In the Metropolitan Sewer District (MSD) of Greater Cincinnati, trap grease comes from both the restaurants (sent in by grease haulers) and the public sewer lines. Both streams were then mixed, and the oils are skimmed off the primary settling tanks. This grease, called trap grease by the MSD, is processed through screening and primary tanks, being skimmed off, and hauled for the landfill eventually (Tu et al, 2011). At MSDGC, an estimated 10,000,000 gallons of grease trap are discharge discharged into the collection system annually. Treatment and disposal cost for trap grease is much higher compared with the charge for those grease haulers.

Trap grease inventory and properties:

UC trap grease to biodiesel team estimated the inventory based on the data collected (email and phone) from registered haulers. Out of the 15 grease haulers serving the Greater Cincinnati area, 12 responded. The total amount of grease trap waste collected by the 12 out of the 15 grease haulers is 11.16 million gallons per year. So the average trap grease in Cincinnati is estimated at 5.58-11.16 lb/person/year from the haulers, while the national averaged data of 13.37 lb/person/year (Wiltsee 1998).

The results indicated that the per person trap grease generated in Cincinnati is much lower than the national average, and therefore it is essential to conduct the inventory study as the basis for trap grease reuse.

Compositional analysis shows that the MSD trap grease consists of water (57% on average), free fatty acids (FFAs) (21%) and the rest non-extractable part/solids. According to experts in the brown grease area, the oil fraction of the trap grease is mostly all FFAs.

Preliminary practice of making biodiesel

A non-traditional biodiesel feedstock due to its lower quality, challenges remain in making trap grease into biodiesel: much higher impurities, extremely high moisture content, and the free fatty acid contents much higher than the "traditional" waste cooking oil from kitchens, or the pure vegetable oils.

Our goal is to extract the oil content out of the trap grease low cost, and with a method compatible to the biodiesel industry. Although chemical solvents, such as hexane, are commonly used in laboratories, it will be rejected by the biodiesel industry, due to the increased cost and extra product handling requirements, etc.

UC's **patent pending** process creatively uses waste cooking oil to extract the oil fraction from the trap grease. The mixture can then be processed via a two-step process: first d to convert the FFA into esters and then the regular alkali transesterification. This will open the doors for the trap grease to biodiesel process for communities that are plagued by the FOG issues.

The optimum extraction ratio is found to be 3.5 to 4:1 of waste cooking oil to trap grease. The effect on extraction temperature was found at 60°C (Figure 2), and the optimum extraction duration is 90 minutes (Wang, 2012). Other possible technologies to convert trap grease into biodiesel will also be evaluated along with the cost.

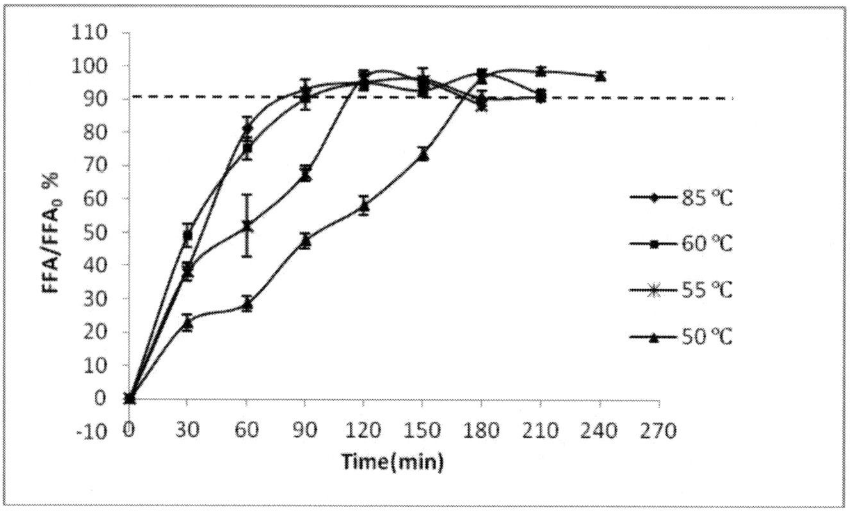

Figure 2. Effect of temperature on extraction time (error bars represent standard error)

In addition to the technology, a detailed trap grease inventory, grease collection and handling, and cost evaluation will follow, to make this a viable business. It is expected that in the future, biodiesel produced can be used not only by the diesel equipment in the MSD facilities but also by the City's fleet.

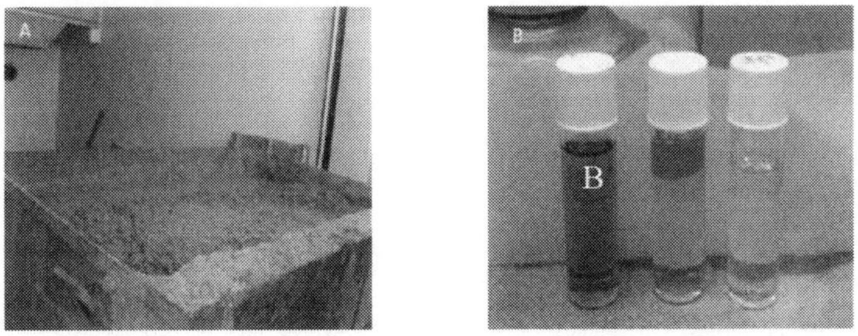

Figure 3. A) Trap grease that currently stored in the holding tank for landfill at MSDGC; and B) biodiesel product converted from regular vegetable oil, waste cooking oil and trap grease (right to left)

References

APHA. 2005. *Standard Methods for the Examination of Water and Wastewater. Twenty first Edition ed.* American Water Works Association and Water Pollution Control Federation, New York.

Clarens, A.F., Resurreccion, E.P., White, M.A., Colosi, L.M. 2010. Environmental Life Cycle Comparison of Algae to Other Bioenergy Feedstocks. *Environ. Sci. Technol.*, **44**(5), 1813-1819.

Craggs, R.J., Smith, V.J., McAuley, P.J. 1995. Waste-water nutrient removal by marine microalgae cultured under ambient conditions in mini-ponds. *Water Sci. Technol.*, **31**(12), 151-160.

Fischer, R.B., Peters, D.G. 1968. *Basic theory and practice of quantitative chemical analysis. Third edition ed.* W. B. Saunders Company., Philadelphia, London, Toronto.

González, L.E., Cañizares, R.O., Baena, S. 1997. Efficiency of ammonia and phosphorus removal from a colombian agroindustrial wastewater by the microalgae Chlorella vulgaris and Scenedesmus dimorphus. *Bioresource Technol.*, **60**(3), 259-262.

Hu, Q., Sommerfeld, M., Jarvis, E., Ghirardi, M., Posewitz, M., Seibert, M., Darzins, A. 2008. Microalgal triacylglycerols as feedstocks for biofuel production: perspectives and advances. *Plant J.*, **54**(4), 621-639.

Kim, W., Park, J.M., Gim, G.H., Jeong, S.-H., Kang, C.M., Kim, D.-J., Kim, S.W. 2012. Optimization of culture conditions and comparison of biomass productivity of three green algae. *Bioprocess and biosystems engineering*, **35**, 19-27.

Lardon, L., Helias, A., Sialve, B., Stayer, J.P., Bernard, O. 2009. Life-Cycle Assessment of Biodiesel Production from Microalgae. *Environ. Sci. Technol.*, **43**(17), 6475-6481.

Larsen, T.A., Alder, A.C., Eggen, R.I.L., Maurer, M., Lienert, J. 2009. Source Separation: Will We See a Paradigm Shift in Wastewater Handling? *Environmental Science & Technology*, **43**(16), 6121-6125.

Lu, J., Sheahan, C., Fu, P. 2011. Metabolic engineering of algae for fourth generation biofuels production. *Energy & Environmental Science*, **4**(7), 2451-2466.

Tam, N.F.Y., Lau, P.S., Wong, Y.S. 1994. Waste-water inorganic N-removal and P-removal by immobilized *Chlorella-vulgaris*. *Water Sci. Technol.*, **30**(6), 369-374.

Tu, Q., Wang, J., Lu, M, Chai, M and Lu, T. Feasibility and Practices of Making biodiesel Out of Low Quality Greases. 2012. Environmental Manager. January Issue, pp 26-29.

Voltolina, D., Gómez-Villa, H., Correa, G. 2005. Nitrogen removal and recycling by Scenedesmus obliquus in semicontinuous cultures using artificial wastewater and a

simulated light and temperature cycle. *Bioresource Technol.*, **96**(3), 359-362.
Wang, JJ, "Feasibility study for a community scale trap grease to biodiesel process", MS thesis, University of Cincinnati, 2012.
Widjaja, A., Chien, C.-C., Ju, Y.-H. 2009. Study of increasing lipid production from fresh water microalgae Chlorella vulgaris. *Journal of the Taiwan Institute of Chemical Engineers*, **40**(1), 13-20.
Wiltsee, G. 1998. Urban Waste Grease Resource Assessment. Report No. NREL/SR-570-26141; National Renewable Energy Laboratory: Golden, CO.
Zou, S., Wu, Y., Yang, M., Li, C., Tong, J. 2010. Bio-oil production from sub- and supercritical water liquefaction of microalgae Dunaliella tertiolecta and related properties. *Energy & Environmental Science*, **3**(8), 1073-1078.

Energy Consumption of MBR for Municipal Wastewater Treatment: Current Situation and Potential

P. Côté, COTE Membrane Separation Ltd
T. Young,GHD
S. Smoot, GHD
J. Peeters, GE Water & Process Technologies

ABSTRACT

Energy consumption factors were developed for membrane bioreactor (MBR) and conventional activated sludge (CAS) flowsheets using consistent assumptions based on a Metcalf & Eddy average-strength domestic wastewater, for a plant with an average daily flow (ADF) rate of 18,925 m^3/d (5 MGD) and a peak hourly flow factor of 2.0.

For a flowsheet without primary clarification, the energy consumption factors for MBR and CAS plants were 476 Wh/m^3 and 406 Wh/m^3, respectively; so the energy consumption for the MBR plant was 17% higher than for the CAS plant. The fraction of the energy consumption factor for the MBR plant attributed to the membrane system was 92 Wh/m^3, or about 20% of the total plant energy consumption.

The energy consumption factors estimated for MBR plants in this paper are lower by about 50% when compared to most values reported in the literature. One reason is the rapid evolution of MBR technology which has led to a reduction in the energy required for membrane air scouring by a factor of 15 over the past 15 years. Another reason is the plant utilization factor which penalizes MBR when the flow is less than design because membrane scouring and RAS flow rate are typically constant and independent of the feed flow rate. It is suggested that the mode of operation of the membranes could be modified to make energy consumption proportional to the feed flow rate. With this method it is shown that the energy penalty for MBR could be reduced to as little as 10% over CAS when the plant utilization factor is 60%.

KEYWORDS

MBR, membrane bioreactor, energy consumption, sewage treatment, municipal wastewater, activated sludge.

INTRODUCTION

Membrane bioreactor (MBR) technology has developed rapidly in the past decade in applications where a filtered effluent was valued, such as water reuse or discharge to sensitive surface water. A number of studies have shown that a MBR has a lower capital cost than a conventional activated sludge (CAS) plant followed by tertiary filtration (Brauer 2005; Brepols et al, 2010; Cote et al, 2004; Verrecht 2010; Young et al, 2012).However, a MBR carries additional operating costs associated with membrane replacement and the energy for operating the membranes. In a recent analysis of the ZeeWeed membrane (GE Water) for MBR, it was determined that membrane life for the current generation (ZW500D) should be 10 years or longer (Cote et al 2012); this makes membrane replacement a very small component (less than 5%) of the O&M costs (Young et al 2012).

The focus of this paper is on energy consumption. Published numbers for MBR energy consumption vary over a broad range. In recent literature surveys, energy consumption factors (ECF) ranging between 400 and1,800 Wh/m^3 were reported for large-scale municipal MBR (Wallis-Lage and Levesque, 2009; Fenu et al, 2010; Owerdieck et al, 2011; Palmowski et al, 2011;Tao et al, 2010; Barillon et al, 2012). In general, membrane air scouring was identified as the largest contributor, greater than the energy required for biological aeration.

The energy consumption for CAS also varies within a wide range, depending on flow variability, level of nitrogen removal and the method for oxygen transfer (coarse or fine bubble aeration). Based on a literature survey and modeling, Monteith et al (2007) reported energy consumption factors between 350 and 650 Wh/m^3.

Based on a review of the literature, MBR would consume approximately two times more energy as compared to CAS. The wide range and relatively high energy values for MBR as compared to CAS are attributed to the following causes: i) the type of membrane is often not specified (membranes are not all equal); ii) recent improvements in membrane technology (e.g. higher fluxes or different modes of intermittent scouring aeration) are not reflected in the literature numbers; iii) the definition of what is included in the quoted numbers is not always clear and does not account for significant site-specific factors related to topography (pumping), water temperature and flow peaking; and iv) the plant utilization factor is not specified; MBR plants are often over-designed, with membranes operated continuously under peak hydraulic design conditions.

METHODOLOGY

Flowsheets

There is no publication that compares the energy consumption of CAS and MBR using consistent assumptions. In this paper we therefore compare side-by-side the energy required for MBR and CAS for all major energy components based on the complete process flow diagrams shown in Figure 1 (without primary clarification) and Figure 2 (with primary clarification). In both cases, a conventional activated sludge (CAS) and a membrane bioreactor (MBR) version were developed.

The comparison was made using a medium strength domestic wastewater as described in Metcalf & Eddy (2003). The CAS and MBR processes were designed to provide an effluent quality of 10/10/10/0.2 mg/L (BOD/TSS/TN/TP). The activated sludge processes were designed using Biowin (EnviroSim Associates Ltd) for an ADF of 18,925 m^3/d (5 MGD) at a temperature of 12°C and with a peak hourly flow (PHF) rate of 2.0 times the ADF (Young et al, 2012).

The flowsheets are similar except for the following elements. The CAS flowsheets included tertiary sand filters to meet the TSS effluent requirements of 10 mg/L consistently; they also included an intermediate pumping station because the total head loss through the plant exceeded 5m. The MBR flowsheets have smaller activated sludge reactors because they are designed to operate at a higher average mixed-liquor suspended solids (MLSS) concentration (6,000 versus 3,000 mg/L) and include fine screening to protect the membranes.

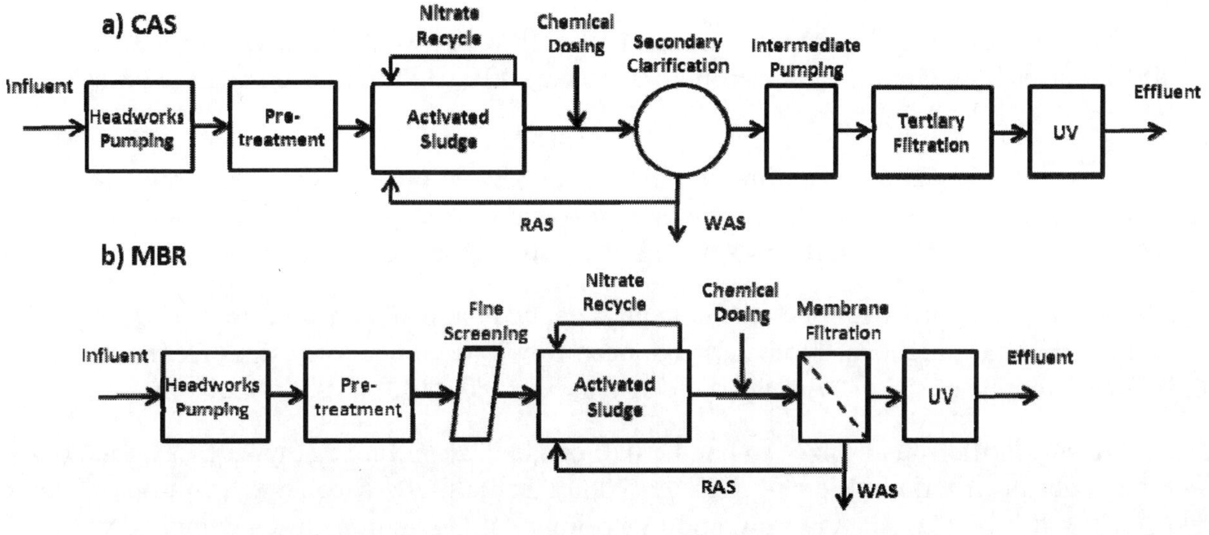

Figure 1 Flowsheet without primary clarification

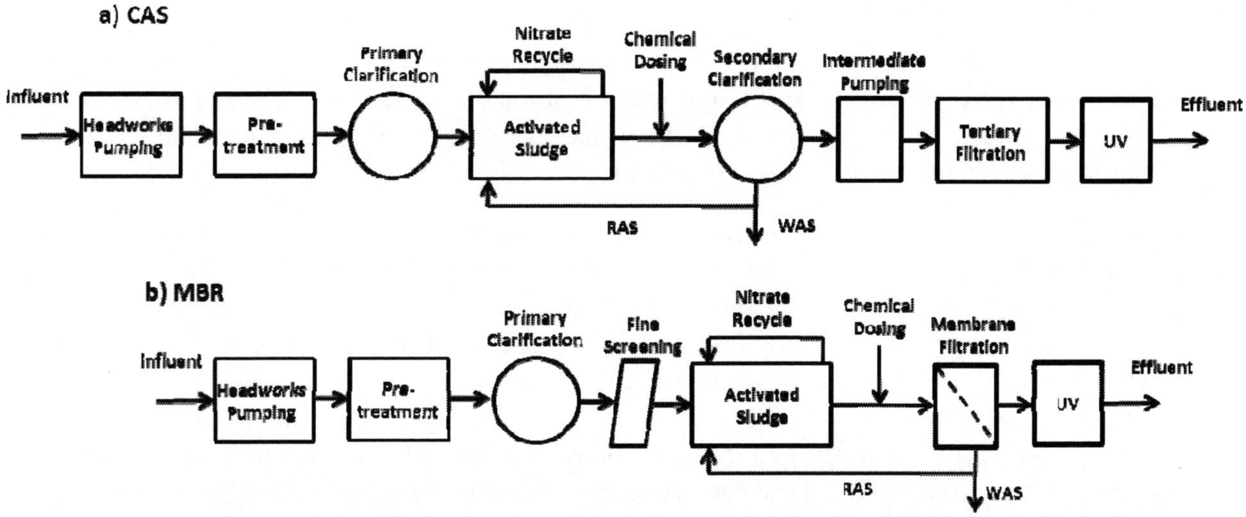

Figure 2 Flowsheet with primary clarification

Energy Estimation

The estimated power draw for each unit process is described in this section. Unit processes were designed with adequate hydraulic capacity to handle a peak flow rate of 2.0 times ADF, and meet effluent requirements with a maximum monthly load of 1.35 times the average load. In general, motor sizes are quoted; to calculate the resulting power draw, a look-up function was used in the model to determine the motor efficiency as a function of its size.

Headworks Pumping

The installed power for the lift pumps was based on a vertical lift of 8m with pressure losses of 4m for a total dynamic head of 12m. For the average flow, two pumps are running, each with a motor size of 18.75 kW (25 HP).

Preliminary Treatment

Two 6-mm coarse screens are running in parallel, each with a washer and a compactor sized to handle the peak flow rate. The washer motors are 0.56 kW (0.75 HP) each and the compactor motors are 2.25 kW (3.0 HP) each.

Two vortex grit removal units are running in parallel, sized to handle the peak flow rate. The motor sizes for the grit pumps, the grit chamber drive and the grit washer/classifier are 3.75 kW (5 HP), 0.75 kW (1.0 HP) and 0.75 kW (1.0 HP), respectively.

The MBRs are equipped with two mechanically-cleaned band screens (2 mm openings) with washer / compactor, each sized to handle the peak flow rate. The motor sizes for the fine screen and washer / compactor are 0.56 kW (0.75 HP) and 2.25 kW (3.0 HP), respectively.

Two primary clarifiers, each sized to handle half of the peak influent flowwere designed based on typical design criteria in Metcalf & Eddy (2003), Table 5-20. Mechanical equipment includes two clarifier drives, two scum pumps, and two primary sludge pumps, each equipped with motors of 0.38 kW (0.5HP) and 2.25 kW (3.0 HP), and 3.75 kW (5.0 HP), respectively.

Activated Sludge Reactors

The biological reactors were designed as four parallel trains in a modified Ludzack-Ettinger (MLE) process, with two anoxic zones and four aerobic zones in series. Reactor zones and equipment sizes were determined based on maximum month loading conditions while energy was calculated based on average loading conditions.

Key process parameters obtained from Biowin simulations for each of the four flowsheets are summarized in Table 1. The average COD mass load to the biological process was 8,138 kg/d and 5,127 kg/d for the cases without and with primary clarification (PC), respectively (it was assumed that PC can remove 60% of TSS and 37% of COD). The HRT ranged from 23.6 h for the CAS without PC to 7.0 h for the MBR with PC. The SRT was 15 days for all systems. The CAS reactors were designed for an average MLSS of 3,000 mg/L while the MBR reactors were designed at 6,000 mg/L. The oxygen transfer correction factor for wastewater (alpha) was calculated as a function of MLSS concentration according to the relationship proposed by Krampe and Krauth (2003), where MLSS is expressed in g/L).

$$\alpha = e^{-0.08766*MLSS} \quad \text{Equation 1}$$

Nitrate recycle pumps were designed to pump from the last aerobic zone to upstream of the first anoxic zone. For the two cases without primary clarification, there was sufficient COD in the wastewater to achieve adequate denitrification and meet TN effluent requirements; for the cases with primary clarification, methanol addition was required.

The activated sludge reactors consume energy for fine bubble diffusion, anoxic zone mixing, and nitrate recycling. The estimated sizes for blowers and motors are presented in Table 1. The energy consumption of the reactor blowers was calculated based on the methodology presented in USEPA's Design Manual for Fine Pore Aeration Systems (USEPA, 1989). The energy consumption of the anoxic zone mixers was based on manufacturer proposals and interpolated to the reactor zone sizing.

RAS Pumping

For CAS flowsheets, the RAS pumping was sized to provide 1.0 times the ADF with two pumps of 5.2 kW (7 HP) each.

For MBR flowsheets, the RAS pumping was sized to provide 4.0 times the ADF with four pumps of 11.2 kW (15 HP) each. The rationale for selecting a higher RAS pumping flow rate for MBR is discussed below.

Table 1 Selected biological treatment process parameters and power requirements

	Without Primary Clarification		With Primary Clarification	
	CAS	MBR	CAS	MBR
Process parameters				
COD load, kg/d	8,138	8,138	5,127	5,127
MLSS, average month, mg/L	3,000	6,000	3,000	6,000
MLSS, maximum month, mg/L	3,500	8,000	3,500	8,000
Alpha factor	0.77	0.59	0.77	0.59
HRT, h	23.6	12.2	12.8	7.0
SRT, d	15	15	15	15
Anoxic volume, fraction	0.35	0.40	0.40	0.45
Methanol addition	no	no	yes	yes
Oxygen demand, kg/d	5,793	5,875	4,422	4,478
Power estimation				
Reactor blowers, kW	92.9	105.0	70.9	74.3
Anoxic zone mixing, kW	52.2	29.8	37.3	22.4
Nitrate recycle, kW	14.9	14.9	14.9	14.9
Total power for reactors, kW	**160**	**150**	**123**	**112**

Intermediate Pumping Station (IPS) and Filters

The plant designs were based on lifting the wastewater 4m above ground level at the head of the plant, and discharging the treated effluent 1m below ground level, i.e. a total allowable head difference of 5m. The head-losses through each process and between processes were estimated based on best practice. For the CAS flowsheets, the presence of the tertiary sand filters, made the total head-loss larger than the maximum head difference available (i.e. 7.5m and 9.0m without and with PC, respectively). Therefore, intermediate pumping was required to lift the wastewater, as shown in Figure 1. For the MBR flowsheets, the permeate pumps were designed to provide a lift of 1m and a separate IPS could be avoided.

The IPS was designed to lift 2.5m and 4.0m without and with PC, respectively. It involved four pumps, each with a capacity of half of the average flow. The pump power was 10 kW (13.4 HP) and 12.3 kW (16.5) for the flowsheets without and with PC.

Three gravity-fed deep bed sand filters were designed with a capacity to pass peak day flow with one filter offline and peak hour flow with all filters in operation as per Metcalf & Eddy 4th Edition, Table 11-8. Motors for the backwash pump, backwash blower and mud-well pump (one per filter) were sized at 26 kW (35 HP), 30 kW (40 HP) and 5.6 kW (7.5 HP), respectively. For energy calculation, it was assumed that backwash pumps and blowers would run a total of 3 hours per day and that the mud-well pumps would run continuously.

Secondary Separation

Secondary separation consisted of secondary clarification for CAS flowsheets and membrane filtration for MBR flowsheets.

For the CAS flowsheets, two secondary clarifiers, each capable of handling half of the peak influent flow rate were designed based on typical design criteria in Metcalf & Eddy (2003), Table 8-7. Mechanical equipment included two clarifier drives and two scum pumps, each with motors of 0.75 kW (1.0 HP) and 2.25 kW (3.0 HP), respectively.

The membrane system was designed based on the LEAPmbr product supplied by GE Water & Process Technologies. The ZeeWeed 500D modules were deployed in 4 parallel tanks. Each tank was equipped with a permeate pump and a blower for membrane air scouring, each with a VFD and sized to process a flow rate up to the peak flow of 2 times the average flow rate. Each permeate pump has a motor of 9.0 kW (12.0 HP); each blower has a motor of 22 kW (30.0 HP). The permeate pump and blower run at half capacity to process the average flow. The LEAPmbr aeration system involves switching the aeration rate between LEAP-Lo and LEAP-Hi as a function of the membrane flux. For energy calculation, it was assumed that the membranes were operating at the LEAP-Lo level for 11 months of the year and at the LEAP-Hi level for 1 month of the year.

UV System

The UV disinfection system consisted of two channels in parallel, each designed to pass the peak flow and achieve 200 MPN/100 mL fecal coliform based on a transmissivity of 65% for CAS and 70% for MBR. The power for the CAS and MBR systems was 10 kW and 6.0 kW, respectively.

Others

This category, which included WAS pumping, chemical feeding systems, compressors for valve activation, process control, building heating/lighting, etc., represented a power draw of 30 kW all together for both the CAS and MBR scenarios.

RESULTS

The estimated energy consumption factors (ECF) for CAS and MBR plants are shown in Figure 3 for the flowsheets without and with primary clarification (PC), expressed on a per m^3 treated basis. The ECF for CAS flowsheets are 406 Wh/m^3 and 377 Wh/m^3 without and with PC, respectively. Primary clarification thus reduced the energy consumption by 7% as compared to the baseline.

The ECF for MBR flowsheets are 476 Wh/m^3 without PC and 436 Wh/m^3 with PC. The additional energy requirement for MBR flowsheets is 70 Wh/m^3 (without PC) and 59 Wh/m^3 (with PC), or on average about 65 Wh/m^3 or 17% higher than CAS flowsheets. Undeniably, if the ECF associated with IPS and tertiary filtration is not considered for the CAS worksheet, the energy consumption for the CAS reference facility would be significantly reduced. However, tertiary filtration is required in the CAS worksheet to result in an effluent quality comparable to an MBR process and to achieve the effluent requirements defined in the methodology. In other words, exclusion of the IPS and tertiary filtration would result in an unfair comparison between the CAS and MBR worksheets.

The ECF for preliminary treatment are slightly higher for MBRs because of the fine screens. However, the ECF for the MBR reactors are lower than for the CAS reactors; the details in Table 1 show that the blower power is higher for MBRs (because of the lower alpha factor), but this is more than compensated by a lower energy input for the anoxic zone mixing (because the tanks are half the size). The ECF for RAS pumping is higher for the MBR processes due to the increased RAS flow rate (4x ADF for MBR versus 1x ADF for CAS).

The major differences in ECFs for CAS and MBR is associated with secondary separation: the MBR system consumes 92 Wh/m^3 while the secondary clarifier consumes 6 Wh/m^3. However, when one compares the sum of the membrane and the UV systems for MBRs (92 + 8 = 100 Wh/m^3) to the equivalent secondary and tertiary processes for CAS (secondary clarification, IPS, tertiary filtration and UV = 69 Wh/m^3), the difference is significantly reduced.

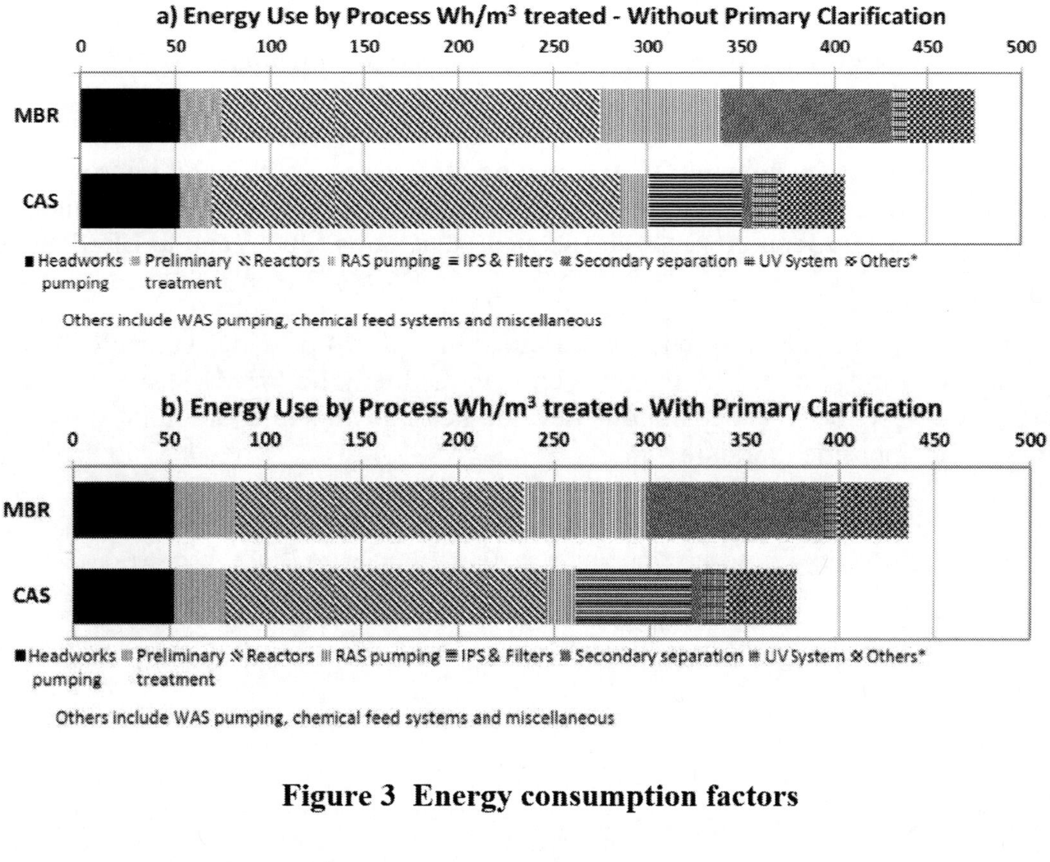

Figure 3 Energy consumption factors

DISCUSSION

Three factors that affect the energy consumption of MBR and represent potential explanations for the large difference observed in the literature between MBR and CAS will be discussed in this section:

1. Membrane technology
2. Biological design
3. Plant utilization factor

Membrane Technology

The two major energy consumers for an immersed membrane system are permeate pumping and air scouring. For the MBRs studied in this analysis, they together represent on average 92 Wh/m^3, or about 20% of the total plant energy consumption. However, since the introduction of immersed MBR technology in the mid-1990s, there has been a rapid and significant evolution of energy consumption for MBR technology, which was described by Cote et al (2012); key elements that relate to energy consumption are summarized below.

The amount of energy required for permeate pumping has changed little over the years. The reason is that the maximum trans-membrane pressure (TMP) that can be applied to immersed membranes is practically limited to 70-80 kPa; this limit can be reached when the membranes are fouled and processing the peak flow rate, just before a cleaning event. In practice, the design suction pressure for average conditions is much smaller (30-40 kPa). While technological advancements have resulted in improved membranes over the years, these improvements have largely been used to increase the design flux as opposed to being used to decrease the maximum TMP and the associated pumping energy.

The evolution of air scouring requirement for the ZeeWeed 500 membrane is illustrated in Figure 4, and shows a decrease by a factor of 15 over a 15 year period, to reach the values used in the current analysis. This evolution is due to several factors, including increases in flux, increases in membrane tank intensity (membrane surface area per unit tank volume), and improvements to the aeration method. The latter is the most significant, with an evolution from continuous aeration (1995), to cyclic aeration (2000), to sequential aeration (2006) and to LEAP aeration (2011).

The dramatic reduction in air scour energy presented in Figure 4 provides valuable insight in explaining the wide range of energy numbers cited in the literature for MBR systems. There is often a lag of 4-5 years between technology development, its implementation in full-scale plants, and reports in the scientific literature.

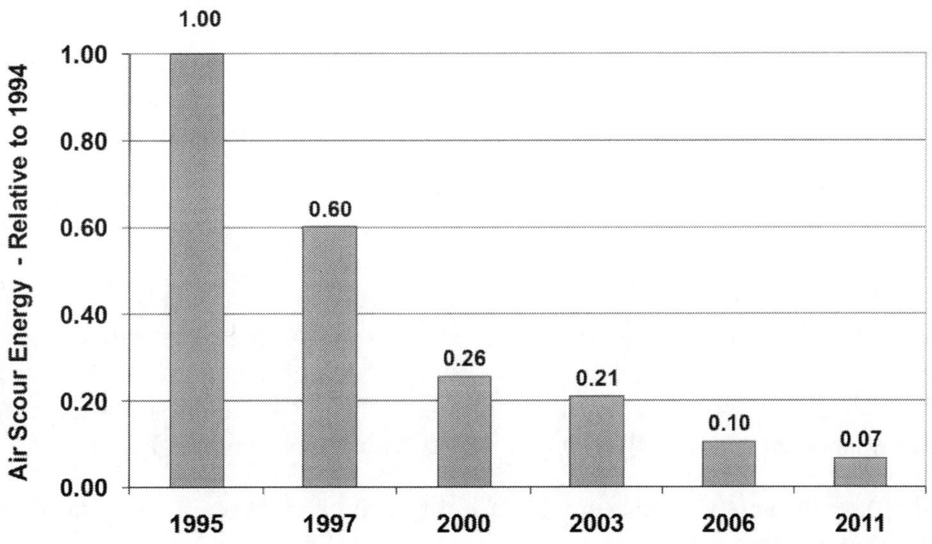

Figure 4 Evolution of air scouring requirements for the ZeeWeed 500 membrane

Biological Design

Early MBR plants in the 1990s were designed with membranes immersed directly into the aeration tank and were operated at very high MLSS concentrations of 12,000 to 18,000 mg/L. The driving force was to build very compact treatment plants. However, design MLSS concentrations for MBRs have decreased over the years to the current level of 6,000-8,000 mg/L to optimize the performance of the membranes and allow for their installation in separate tanks.

As discussed above, the alpha factor exponentially decreases as MLSS concentration is increased. Equation 1 demonstrates that the alpha factor is almost doubled by reducing the MLSS from 15,000 mg/L to 6,000 mg/L, resulting in nearly half as much energy required for aeration.

When the membranes are installed in a separate tank, there is a simple relationship derived from a mass balance that links the MLSS in the aeration tank (C_A) to the maximum concentration in the membrane tank (C_M), and the RAS ratio ($R = Q_{RAS}/Q_{feed}$):

$$C_A = C_M \frac{R}{(R+1)} \qquad \text{Equation 2}$$

$C_M$is set by the membrane manufacturer and varies with flux and temperature; a value of 10,000 mg/L was used in this study. With C_M=10,000, an R value of 4 results in a C_A of 8,000 mg/L. In this study, the maximum month loadings were assumed to be 35% higher than average; therefore, the MLSS for the average month would stabilize at approximately 6,000 mg/L if the plant is run at constant SRT. An alternative to this design would be to reduce the RAS ratio to 1.0, which for the same value of C_M, would reduce the design MLSS in the aeration tanks to 5,000 mg/L for max month, and 3,500 mg/L for average month. These conditions are identical to the CAS worksheet and would reduce the energy consumption factors for the reactors and RAS pumping to the level of CAS, albeit at the cost of larger reactor tanks.

The current MBR design practice strikes a balance between capital costs (smaller reactors) and the incremental energy cost to operate at high MLSS. If energy were the overwhelming factor, there are no process constraints that would prevent the design of an MBR process that operates at the same MLSS concentrations used for CAS plants.

Plant Utilization Factor

The plant utilization factor (PUF) is the ratio of actual flow rate over design flow rate. Very few plants operate at a PUF approaching 100%; it is not uncommon for new plants to start operation with an initial PUF of 50% and for plant expansion to be triggered when the PUF reaches 80-85%. The analysis in this paper has been based on a PUF of 100%. The purpose of this section is to explore the impact of a reduced PUF on energy consumption factors (ECF).

Wastewater treatment plants are designed for a future projected flow rate and tanks and equipment are sized for that future flow rate. The electrical power draw can be classified as "base power" if it is constant and independent of flow rate, and "variable power" if it can be modulated based on flow rate treated. Examples of base power include clarifier drive mechanisms, tank mixing, preliminary equipment and UV disinfection. Examples of variable power include pumping for headworks, nitrate recycle, RAS, intermediate pumping and filtration, and reactor aeration (assuming that the blowers are equipped with VFDs and controlled based on DO).

The ECF for all the motors and blowers included in the flowsheets without primary clarification presented in Figure 3a were classified as base and variable power as shown in Table 2. Two versions of the MBR plant are presented labeled "typical" to represent past and current practice, and "optimized" to represent an energy-optimized design approach. For the CAS plant, all ECFs were classified as "base" except for pumping (headworks, nitrate recycle, intermediate and RAS), reactor blowers and filters. For the "typical" MBR plant, all ECFs were classified as "base" except for pumping (headworks, nitrate recycle, permeate) and reactor blowers; for the "optimized" MBR plant, it was additionally assumed that RAS pumping and membrane aeration were part of the variable fraction. Design options to make these variables will be explored below.

Table 2 Base and Variable energy fraction for treatment flowsheets without primary clarification

	$ECF_{100\%}$ (Wh/m^3)	Base Fraction (F_b)	Variable Fraction (F_v)
CAS	406	37%	63%
MBR – typical	476	52%	48%
MBR - optimized	476	24%	76%

The impact of the base fraction (F_b) and variable fraction (F_v) on the actual energy consumption factor (ECF_a) can be estimated with the following equation as a function of the plant utilization factor (PUF).

$$ECF_a = \frac{ECF_{100}F_b + ECF_{100}F_v\ PUF}{PUF} \qquad \text{Equation 3}$$

The actual ECF is plotted versus the PUF in Figure 5 for the CAS flowsheet and the two MBR flowsheets. The ECFs at 100% PUF are identical to those shown in Figure 3. However, as the PUF is reduced, the actual ECF for the MBR-Typical flowsheet increases faster than that of the CAS flowsheet because its Base Fraction is higher (52% versus 37%). As a result, for a PUF of 60%, the actual ECF for the MBR-Typical flowsheet would 639 Wh/m^3, 27% higher than for the CAS flowsheet (505 Wh/m^3). The gap is larger than at a PUF of 100% because Equation 3 is not linear.

Figure 5 also shows a MBR-Optimized curve where the Base Fraction for the MBR flowsheet has been reduced to 24% (versus 37% for the CAS flowsheet). This changes the rate of ECF increase for the MBR flowsheet as the PUF is reduced and actually bridges the gap with the CAS flowsheet. As a result, for a PUF of 60%, the actual ECF for the MBR-Optimized flowsheet would be 550 Wh/m^3, only 9% higher than for the CAS flowsheet (505 Wh/m^3).

Reducing the energy consumption of an MBR flowsheet to within 10% of a CAS flowsheet (at 60% plant utilization factor) is very desirable, but how can it be achieved? Let us consider how membrane aeration and RAS flow rate can be managed differently.

The past (and often current) practice for the operation of an MBR plant has been to modulate the permeate flow rate and leave membrane aeration constant, at the level necessary to process the design flow rate (initially the peak flow rate). The LEAPmbr aeration method introduces LEAP-Lo and LEAP-Hi aeration levels for average and peak flow conditions. In order to make membrane aeration part of the Variable Fraction, it would be necessary to modulate the aeration level to track the permeate flow rate; this is not practical because scouring aeration devices do not have an infinite turn down ratio. However, it is possible to design a multi-train plant so that membrane trains are always running at either average or peak flow rate and trains are turned off when not needed. Intermittent, short periods of aeration could be used to keep the tanks mixed and aerobic.

A similar situation exists for RAS pumping; RAS pumps are often continuously to maximize de-concentration of the membrane tanks. An alternate method approach consists in installing RAS pumps with VFDs, or to run the RAS pumps only when a membrane train is running.

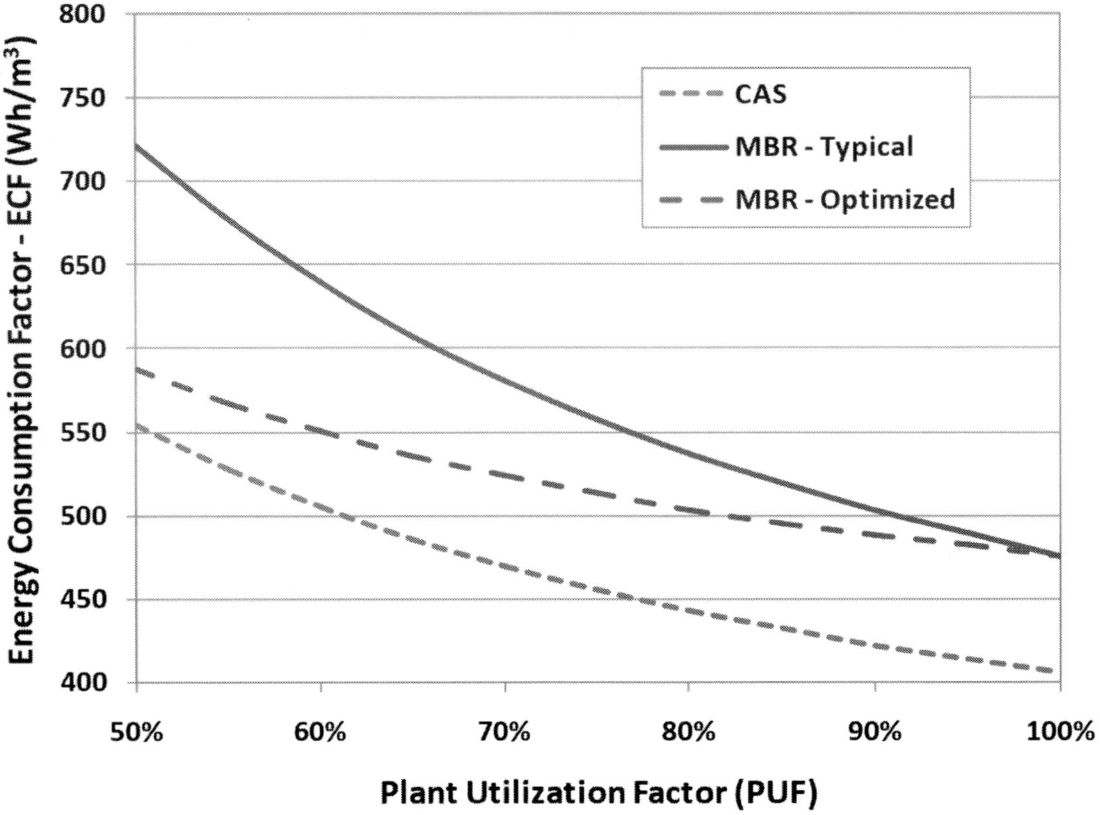

Figure 5 Impact of the plant utilization factor on energy consumption

Life Cycle Cost Impact

In a companion paper (Young et al, 2012), the capital cost difference for the flowsheets without PC described in this paper (CAS - MBR) was estimated to be -3.7 M$ (MBR less expensive). The energy cost was also estimated (at 0.10 $/kWh) and its present worth was calculated (20 years, nominal discount rate 3.5%, inflation rate 4.9%); the energy present worth difference was +0.5 M$ (CAS less expensive). So, although the MBR consumes more energy, this penalty is only a small fraction of the capital cost savings; this would remain so, even at a low plant utilization factor.

SUMMARY AND CONCLUSIONS

Energy consumption factors were developed for membrane bioreactor (MBR) and conventional activated sludge (CAS) flowsheets using consistent assumptions based on a Metcalf & Eddy average strength domestic wastewater, for a plant with an average daily flow rate of 18,925 m^3/d (5 MGD) and a peak hourly flow factor of 2.0.

For a flowsheet without primary clarification, the energy consumption factors for MBR and CAS plants were 476 Wh/m^3 and 406 Wh/m^3, respectively; so the energy consumption for the MBR plant was 17% higher than for the CAS plant.

The fraction of the energy consumption factor for the MBR plant attributed to the membrane filtration system was 92 Wh/m^3,or about 20% of the total plant energy consumption.

The energy consumption factors estimated for MBR in this paper are lower by about 50% compared to values reported in the literature. One reason is the rapid evolution of MBR technology over the past 15 years, which has led to a reduction in energy for membrane air scouring by a factor of 15. For early MBR plants the energy consumption for membrane air scouring was greater than that for biological aeration. With the current technology, membrane air scouring energy has been reduced to a fraction of biological aeration energy.

Another reason for the higher reported values for MBR is the plant utilization factor (PUF). Numbers estimated in this paper are for plants running at full capacity (i.e. PUF = 100%). In practice, new plants are sized for future flows and initial PUF can be as low as 50-60%. This increases the energy consumption factors depending on whether the energy consumption is classified in the "Base Fraction" (independent of flow) or "Variable Fraction" (proportional to flow). The analysis in this paper shows that current MBR plants have a typical Base Fraction of 52%, higher than that of CAS (37%); as a consequence, the energy gap between MBR and CAS increases with decreasing PUF.

The energy consumption factors for MBR at low PUF could be decreased by decreasing their Base Fraction. One way to do this would involve changing the mode of operation of the membranes so that their energy consumption becomes proportional to the feed flow rate. With this method it was shown that the energy penalty for MBR could be reduced to as little as 10% over CAS when the plant utilization factor is 60%.

REFERENCES

Barillon, B., Martin-Ruel, S., Langlais, C. and Lazarova, V., « Energy Optimization in Membrane Bioreactors », IWA Conference, 2011

Brauer, R., Okazaki, J., Pelissier, D., Morine, M., Klein, A., and Crawford, G., "Henderson's New Water Reclamation Facility: Choosing Between Membrane and non-Membrane Solutions", WEFTEC, 2005.

Brepols, C., Schafer, H., and Engelhardt, N., "Considerations on Design and Financial Feasibility of Large-Scale Membrane Bioreactors", Water Science & \technology, 6(10): 2461-2468, 2010.

Cote, P., Masini, M., and Mourato, D., "Comparison of Membrane Options for Water Reuse and Reclamation", Desalination, 167:1-11, 2004.

Cote, P., Alam, Z., and Penny, J., "Hollow Fibre Membrane Life in Membrane Bioreactors", Desalination 288:145-151, 2012.

Fenu, A., Roels, J., Wambecq, T., De Gussem, K., Thoeye, C., De Gueldre, G. and Van De Steene, B., "Energy Audit of a Full Scale MBR System", Desalination, 262: 121-128, 2010.

Krampe J, and KrauthK.,"Oxygen transfer into activated sludge with high MLSS concentrations", Water Sci Technol. 2003;47(11):297-303, 2003.

Metcalf & Eddy Inc., "Wastewater Engineering: Treatment and Reuse", Fourth Edition, McGraw Hill, 2003.

Monteith, H., Kalogo, Y., and Louzeiro, N., "Achieving Effluent Limits Takes a Lot of Energy", Proceeding of WEFTEC 2007.

Owerdieck, C., Penny, J., Peeters, J., Baumgarten, S., Di Pofi, M., and Kicsi, G., "Trends in MBR Energy Consumption: Past, Present and Future",6thIWA Specialist Conference on Membrane Technology for Water and Wastewater Treatment, Aachen, Germany, October 4-7, 2011 .

Palmowski, L.M., Veltmann, K. and Pinnekamp, J., "Optimization Measures of Large Scale MBR Plants to Lower Energy Consumption",6th IWA Specialist Conference on Membrane Technology for Water and Wastewater Treatment, Aachen, Germany, October 4-7, 2011.

Tao, G., Kekre, K., HtunOo, M., Lew, D. and Seah, H., "Energy Reduction and Optimization in Membrane Bioreactor Systems", Water Convention 2010, Singapore, 28 June – 2 July, 2010.

USEPA, "Design Manual: Fine Pore Aeration Systems", EPA/625/1-89-023.US Environmental Protection Agency Office of Research and Development.September 1989.

Verreth, B., Maere, T., Nopens, I., Brepols, C., and Judd, S., "The Cost of Large-Scale Hollow Fibre MBR", Water Research, 44(18):5274-5283, 2010.

Wallis-Lage, C.L. and Levesque, S.D., "Cost-Effective & Energy Efficient MBR Systems",Singapore International Water Week, Singapore, June 22-26, 2009.

Young, T., Muftugil, M., Smoot, S. and J. Peeters, "MBR vs. CAS: Capital and Operating Cost Evaluation", Water Practice & Technology, Vol 7(4), 2012

Environment Enhancing Energy

--Innovative adsorptive-biological process to recovery/recycle carbon and nitrogen from aqueous product of hydrothermal liquefaction to maximize bioenergy production from wastewater

Yan Zhou*, Lance Schideman, Yuanhui Zhang, Xin Zhao

University of Illinois at Urbana-Champaign, Illinois, U.S.

*yanzhou1@illinois.edu, 217-552-5672

1. Introduction

A novel wastewater treatment system, Environment Enhancing Energy (E^2-Energy, as shown in Figure 1.) was previously proposed, which synergistically combines large scale algal biofuel production and wastewater treatment. This system maximizes biofuel potential from wastewater inputs by internally capturing and recycling carbon and nutrients--a powerful concept referred to as ***multi-cycle nutrient reuse*** that amplifies waste nutrients into multiple cycles of algal biomass and bioenergy production. Our previous research (Zhou WEFTEC 2012, Zhou WEFTEC 2011) has confirmed the basic steps of the E^2-Energy process and projected that this approach has the potential to ultimately replace all U.S. petroleum imports using only current wastewater feedstocks and carbon dioxide from the atmosphere or point sources.

As shown in Figure 1, Biosolids in the incoming waste stream would be separated and then pumped into a hydrothermal liquefaction (HTL) unit to convert the volatile solids into self-separating biocrude oil. The wastewater liquid flow and any supernatant from the biosolids with high nitrogen and phosphorus will be pumped to an algal bioreactor, where algae and bacteria grow symbiotically to remove both organic carbon and nutrients. The algal-bacteria biomass will then be harvested and fed back into the HTL reactor for bio-fuel production. The post-HTL process wastewater (PHWW) and carbon dioxide will be fed to the algae bioreactor to recycle nutrients for multiple cycles of algae growth. It can potentially bring paradigm change to current wastewater treatment system because 1) it is turning the carbon-emitting wastewater treatments plant into a carbon sequestration stations by cultivating photosynthetic algae; 2) through nutrient recycling, the carbon capture potential can be maximized; 3) the energy consumptive wastewater treatment process is turned into a bioenergy generating process.

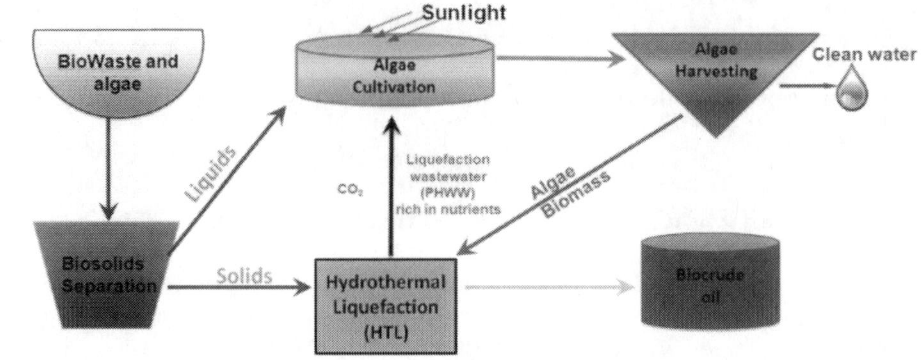

Figure 1 Process Configuration of Environment-Enhancing Energy (E^2-Energy).

However, it was also found that when cultivating bacterial-algal biomass in post-hydrothermal liquefaction wastewater (PHWW) for both biomass production and wastewater treatment, there was 15-40% slow-degrading or recalcitrant organic/nutrient residue, which limits the potential to increase the nutrient/carbon recycling efficiency and the overall biofuel production capacity of this novel system. This study addresses this issue in pursuit of an optimized system integrating algal wastewater treatment and bioenergy production including original experimental data of a hybrid adsorbent-membrane photobioreactor for enhanced nutrient/organic recovery, and a process modeling that quantifies the benefit of increased nutrient recycling efficiency from the perspective of energy production and environmental impact.

2. Material and Methods

2.1. Continuous flow photobioreactor

Algal biomass was also produced in an 11 L continuous flow photobioreactor (PBR). A PVDF hollow fiber ultrafiltration membrane module (ZW-1000 Jr, GE-Zenon) was placed in the center of the reactor, which has a nominal pore size of 0.04 μm and membrane area of approximately 1 m^2 as shown in Figure 2a. Two peristaltic pumps were used to deliver influent and withdraw effluent through the membrane module with a hydraulic retention time of 2 days. Light was provided by eight 22 W fluorescent lights around the bioreactor. The light intensity at the center of empty reactor is between 200-250μmol photon m^{-2} s^{-1}. Air enriched with 1% CO_2 (v/v basis) was provided through a diffuser at the bottom of the reactor at a flow rate of 4 LPM. The influent was a mixture of 99%

filtered municipal WW and 1% PHWW-Spirulina. At a certain stage, 200g Calgon 400 granular activated carbon was added into the reactor.

(a) (b)

Figure 2. Continuous-flow hollow fiber membrane photobioreactor: (a) without algae; (b) with algae.

2.2. Adsorption batch test

Two different adsorbents were tested in this study: nonionic porous resin SEPABEADS™ 700 (SP 700) and granular activated carbon Calgon 400. Different doses of these two adsorbent was put into a 250 ml flasks containing 100ml from the continuous operating PBR (HRT=5 days, SRT=infinite). The flasks were put on a shaker table rotating at 100 rpm and samples were obtained at different time point for water quality analysis. GC-MS analysis was also performed to identify the recalcitrant compounds and how the adsorbent could help remove those compounds.

3. Results and Discussion

3.1. Adsorption batch test

Batch test adsorption result showed that both SP 700 resin and Calgon 400 GAC could remove the slow-degrading organics from the algal-bacteria treated PHWW as shown in Figure 3. However, activated carbon showed better performance compared to resin. GAC achieved solube chemical oxygen demand (SCOD) removal over 90% at dosage of 15g/L while only 68% SCOD removal was achieved at a resin dosge of 22g/L. The better removal effciency of carbon was probably because the wider pore ranges of activated carbon compred to resin, which resulted in wider range of target compounds size. Thus, activated carbon was chosen to be used in the photobioreactor.

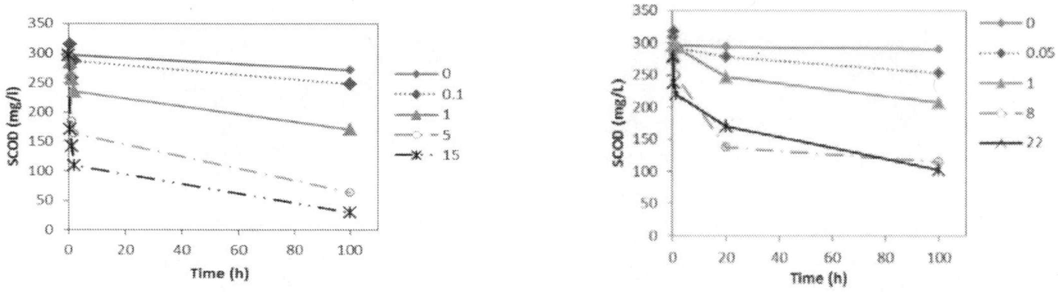

(a) (b)

Figure 3 Soluble chemical oxygen demand (SCOD) removal from effluent of photobioreactor using different adsorbent; (a) activated carbon; (b) nonionic resin. The legend is showing the adsorbent dose (g/L).

3.2. Continuous photobioreactor operation

The effect of hydraulic retention time (HRT)

The hydraulic retention time affected nutrient removal but not organic removal. Soluble chemical oxygen demand (SCOD) got consistent removal around 60% when HRT ranged from 2 days to 10 days. However, nutrient removal was greatly affected by the HRT. When the HRT was 2 days and solid retention time was infinite, total soluble nitrogen got around 30% removals and almost no ammonia nitrogen was removed. When the HRT increased to 10 and 5 days (stage I, II and III), the ammonia removal increased, ranging between 20% to 84% as shown in Figure 4b. The biomass concentration kept steady at around 2000mg/L when the SRT is 20 days and HRT is 2 days.

The effect of solid retention time (SRT)

The solid retention time (SRT) didn't show any effect in organic removal, but did show a significant effect on the nutrient removal. When the SRT is infinite, the nutrient removal rates depended on the biomass growth rates, since nutrient removal was mainly achieved by nutrient being taken up by microbes and being converted into biomass. The more biomass increase there is every day, the better nutrient removal there will be. As we can see in Figure 4, during stage I, since algae was rapidly growing indicated by total suspended solid (Figure 4d), large amount of ammonia was being taken up. As the biomass concentration increased, light attenuation started to occur resulting from light absorption and scattering by suspending cells. Limited light delivery caused decreased algal growth rate, and the subsequent decreased ammonia removal rate at the end of stage I and with entire stage II. However, as SRT was turned from infinite to 20 days, the trend of increasing ammonia in effluent stopped, and ammonia removal quickly stabilized between 20%-30%. Therefore the nutrient removal could be controlled by the biomass growth, and biomass growth could be controlled by adjusting SRT.

The effect of granular activated carbon (GAC)

The addition of GAC had an obvious positive impact on both organic and nutrient removal. Soluble COD in PBR effluent dropped to around 100 mg/L compared with the ~300mg/L without GAC (Figure 4a). The ammonia even got 100% removal after the addition of GAC as shown in Figure 4b. TN removal increased from around 10% without GAC to around 70% as shown in Figure 4c.

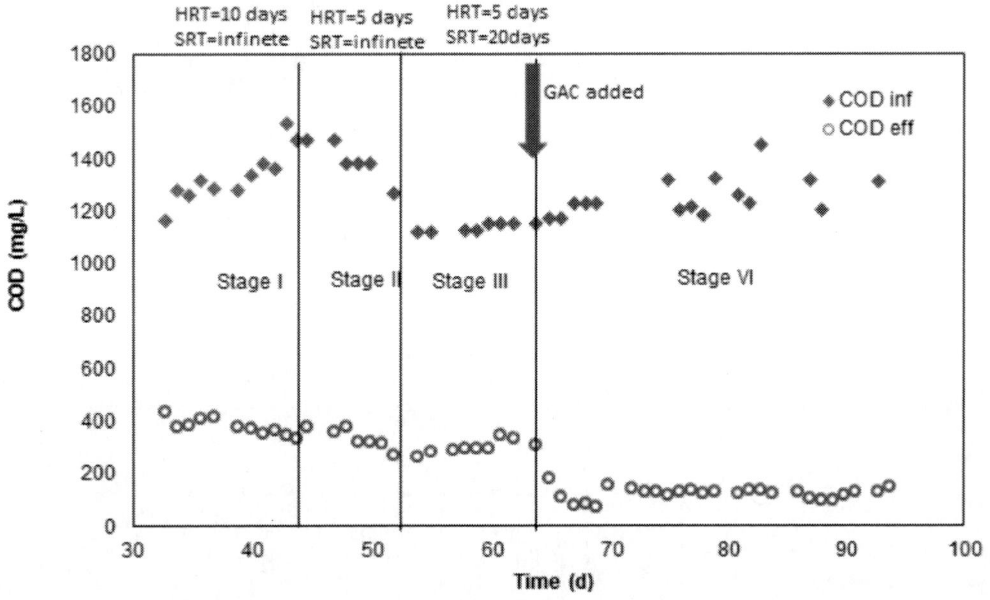

(a)

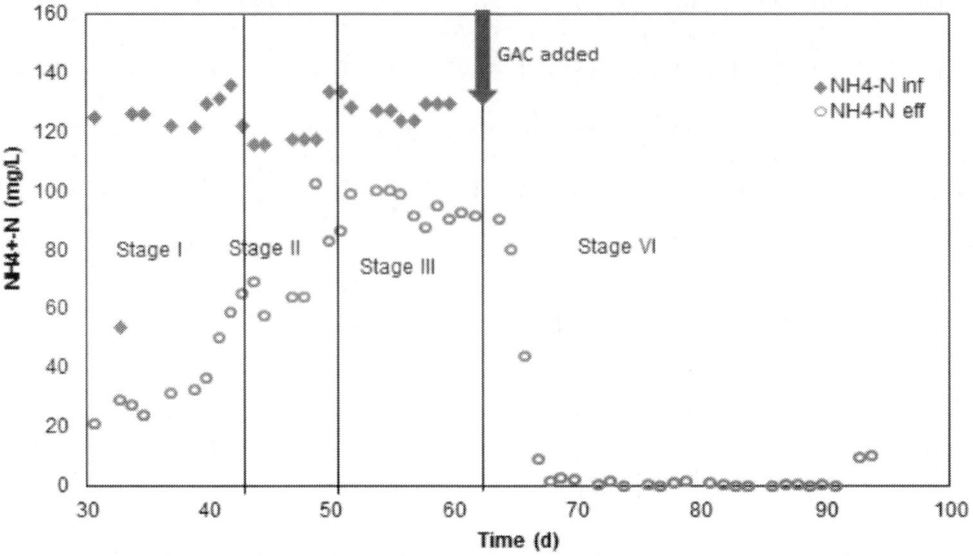
NH4+-N (mg/L)
Time (d)
GAC added
NH4-N inf
NH4-N eff
Stage I
Stage II
Stage III
Stage VI

(b)

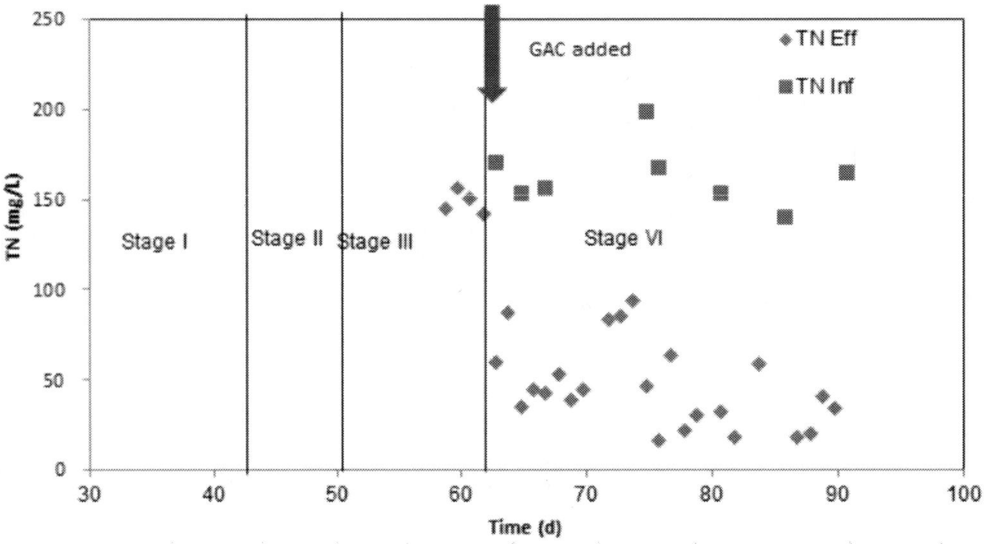
TN (mg/L)
Time (d)
GAC added
TN Eff
TN Inf
Stage I
Stage II
Stage III
Stage VI

(c)

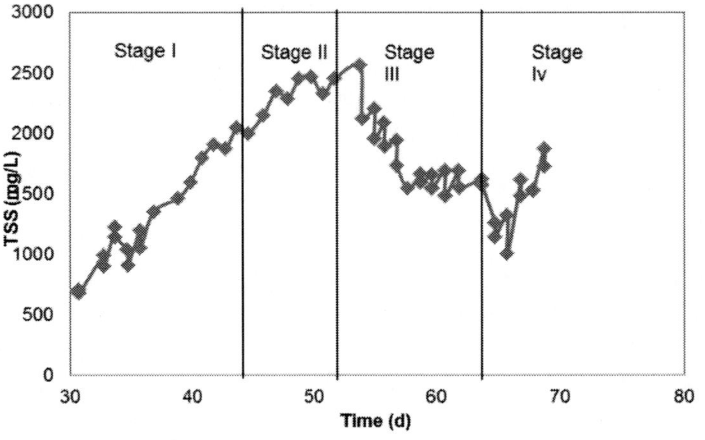
TSS (mg/L)
Time (d)
Stage I
Stage II
Stage III
Stage Iv

(d)

Figure 4 Continuous operation photobioreactor performance: (a) soluble chemical oxygen demand; (b) ammonia nitrogen; (c) total dissolved nitrogen; (d) total biomass indicated by total suspended solid. There were four stages of reactor operation based on hydraulic retention time (HRT), solid retention time (SRT) and the existence of adsorbent (granular activated carbon, GAC).

3.3. Increased nutrient recycling efficiency

This enhancement in organic and nutrient removal could increase the overall algal biomass production capacity of the E^2-Energy system. We previously constructed a mathematic model (Zhou and Schideman, L.C., Yu, G., Zhang,Y. Under review) to simulate the mass flow and energy production of E^2-Eenrgy system, and sensitivity analysis revealed that one of the most sensitive parameter was the nutrient removal efficiency during algal-bacteria cultivation in PHWW. As discussed above, this innovative hybrid wastewater treatment system could improve organic removal from 60% to 90%, and nutrient removal from 30% to even 100% (only ammonia data is available now, we are analyzing more nutrient data). Using the input from the current experiment, model simulation result showed that the overall energy production of algal biofuel production system could be increased by 47% because more nutrient/organics would be recycled to support multiple cycles of algal-bacterial biomass production.

4. Conclusion

A hybrid process which incorporates adsorbents into membrane photobioreactor was developed in order to increase the carbon and nutrient recovery efficiency from aqueous product of hydrothermal liquefaction (post-HTL wastewater, or PHWW) in order to maximize carbon utilization efficiency and bioenergy production. Batch tests showed that both activated carbon and porous resin could efficiently remove slow-degrading or recalcitrant compounds from PHWW. The continuous membrane PBR showed over 90% COD and 70% nutrient removal with addition of activated carbon compared to around 70% COD and 10% removal without activated carbon. During the past three months of operation, the reactor showed consistent removal of organics and nitrogen, the amount of which far exceeded the adsorption capacity of added activated carbon (obtained in batch tests), highlighting the self bioregeneration of the adsorbents. The adsorption process could capture the slow-degrading compounds in a much faster rate (in seconds or minutes) than regular biological degradation process. In addition, it can serves as a rapid physico-chemical sink of various dissolved organic substrates (including toxic compounds) and therefore adsorption process is much less sensitive to upset and changes in operating conditions. The bioregneration also significantly increased the economic feasibility of the proposed novel process

References

Zhou Y, Schideman, L.C., Yu, G., Zhang,Y. (Under review) Environment-Enhancing Energy: A synergistic combination of algal wastewater treatment and biofuel production maximized by nutrient and carbon recycling. Energy Environ Sci

Zhou Y, Schideman LC, Zhang Y, Yu G (2011) Environment-Enhancing Energy: A Novel Wastewater Treatment System that Maximizes Algal Biofuel Production and Minimizes Greenhouse Gas Emissions

Zhou, Y., Schideman, L.C. (2012) The analysis of carbon capture, nutrient reuse and energy production of a novel wastewater treatment system that incorporates large scale algal biofuel productions

Sustainable Solutions for the 21st Century – Integration of Water Treatment Systems with Energy Derived from Municipal Waste

Authors: Paul L. Hauck, P.E., CDM Smith, Tampa, Florida USA
Diane C. Kemp, CDM Smith, Tampa, Florida, USA
Valerie P. Going, P.E., CDM Smith, Tampa, Florida, USA

Abstract

Hillsborough County is a Florida municipality that has seen rapid growth and utility demands over the past 50 years. In response, it has implemented an innovative project to reduce the cost of wastewater and water treatment processes by using renewable electrical energy produced from municipal solid waste.

The county developed a modern energy-from-waste (EfW) facility 27 years ago, recognizing that solid waste was capable of producing significant amounts of renewable energy in the form of electricity, and to limit landfill disposal. The electricity was sold to a local utility to help offset the costs of solid waste disposal to local rate payers. The EfW facility was also located adjacent to one of the county's advanced wastewater treatment plants (AWTP), allowing the county to take advantage of reclaimed water at the EfW facility for cooling, irrigation, and other process uses.

Six years ago, faced with continuing growth, the county expanded their EfW facility by approximately 50 percent capacity. At this time, they also decided to disconnect the adjacent AWTP from the local electric grid and use renewable electricity supplied from their adjacent EfW facility to reduce the cost of wastewater treatment at the AWTP. This synergistic arrangement resulted in savings to both the Solid Waste and Water Resources Departments. Following the success of this initial project, the county expanded their internal use of their electrical power by adding a new electrical power feeder circuit to a nearby potable water booster station for additional shared savings.

The Hillsborough County Florida success may serve as the catalyst for other municipal entities to develop integrated utility projects. The opportunities will vary greatly from one municipal utility to another based on local variables and markets. Ideally, all of the energy demand for a municipal utility's water and wastewater treatment processes would be met by electrical output from the EfW facility. The percent of internal use should be maximized in order to optimize the shared benefits.

To evaluate whether this ideal synergistic arrangement could be feasible, an analysis was performed for three communities with population sizes ranging from small (176,000), medium (700,000) to large (2 million). The potential to match the electrical output from the EfW facility to the energy requirements of the water and wastewater processes was examined.

Keywords: renewable energy from wastes, internal use of electricity for municipal water resources, integrated resource management, sustainability

Introduction

There are numerous paradigms emerging within the disciplines of what was once the purview of Public Works. These changes are being driven by a growing awareness of the need to address climate change; reduce greenhouse gasses; improve municipal energy efficiency; and implement renewable energy projects, including renewable energy derived from municipal wastes. Faced with an aging but rapidly expanding infrastructure, and the mandate to efficiently manage their assets, municipal governments are caught in a perplexing situation of becoming more efficient while working under the pressure of smaller budgets, fewer resources, and less staff.

Methodology

Many communities are faced with the need to develop alternate water supplies to accommodate future growth. With decreasing high-quality water supply sources, greater quantities of energy are generally required to process lower quality water sources using advanced water treatment technologies. This paradox presents an opportunity for Public Works to explore the synergies of using renewable energy from wastes to power advanced water treatment facilities for the benefit of the local environment and economy.

This paper explores the synergies that can be realized through the integration of renewable energy systems derived from municipal solid waste with several municipal water treatment processes. The evolution of the modern energy-from-wastes (EfW) industry has not been without major hurdles. Often referred to as incineration, the first generation of plants typically did not recover energy, nor did they have efficient and modern air pollution controls. In the 1980s, many of the current generation EfW facilities were constructed and equipped with efficient energy recovery systems and modern air pollution controls. Today, EfW facilities in the United States convert approximately 13 percent of the nation's waste at 86 operating EfW facilities, processing approximately 25 million tons of solid waste with a generation capacity in excess of 2,700 megawatts (MW).

Recent benchmarking studies have shown that the U.S. EfW industry continues to advance in the direction of sustainability. Modern EfW facilities have improved with reduced air and wastewater emissions, greater processing and energy recovery efficiencies, advanced metal recovery, and other benefits for their respective communities. Recent trends in the EfW industry are noted below.

Increasing EfW trends include:

- Number of EfW facility expansions
- More stringent emission limits and green-house gas (GHG) reporting
- Higher heating value (HHV) of waste
- Higher boiler and turbine-generator availability
- Greater use of reclaimed water for cooling
- Higher gross and net electric generation per ton of solid waste
- Recognition and classification of electricity from EfW as renewable energy
- Marketable value of the environmental attributes of EfW electricity
- Addition of non-ferrous metal recovery systems, including fine materials < 12mm

- Integrated solid waste management systems on an eco-campus

Decreasing EfW trends include:

- Lower air emissions
- Lower reagent consumption
- Lower water consumption
- Lower wastewater discharges
- Lower electric payments

As noted above, the trend for net electrical energy production from modern EfW facilities continues to increase as the industry matures. The initial generation of EfW plants built in the early 1980s were designed around low superheat steaming conditions in the range of 600 psig / 750°F, which resulted in a net electrical generation in the range of 450 - 550 kWh per ton of combusted waste. Second generation EfW facilities were designed for medium pressure and temperature superheat steaming conditions, in the range of 865 psig / 830°F, which increased the net electrical generation rates to approximately 550 - 600 kWh. Looking to the future, waste combustion boilers will likely be designed for higher superheat steaming conditions in the range of 1350 – 1450 psig and 900 - 930°F which will result in net electrical generation rates in the range of 700 – 800 kWh per ton of waste combusted. Experience garnered over the past 25 years of commercial operation of EfW facilities has provided reliable data and confidence to facility operators to mitigate higher wear rates on boiler pressure boundary components at these higher superheat steaming conditions. Advanced metallurgy and regularly scheduled maintenance cycles will allow future EfW facilities to generate 50% more electricity than the first generation plants.

The U.S. EfW market appears to be on the verge of renaissance, with several plants recently expanding their capacity, one new facility currently under construction in Florida, and several new projects in development. One of the key drivers in the cost-effectiveness of EfW is the price paid for renewable electricity. As noted above, one of the recent trends in the U.S. electric power industry has been a decline in retail sales of electricity for the first time in modern history. This reduced electric demand, in addition to the current abundance and low cost of natural gas, has led to a low demand for new electric power plant capacity, and consequently, a lower avoided cost to produce electricity. The combination of these conditions has resulted in lower payments to independent power producers, which includes EfW facilities.

Unless there is a turnabout in the methodology for determining how municipal EfW facilities are compensated for their renewable energy, the opportunity to use all or some of the electric power “behind the meter” (internally) within Public Works may present the highest and best use of this renewable electricity. One such opportunity for internal use of electricity will be for the treatment of municipal water resources. This broad class of vital municipal services includes treatment of wastewater, reclaimed water, stormwater and potable water.

Matching a modern EfW facility’s electrical output with water treatment demand varies significantly based on the source of the raw water, quality, treatment and pumping system compatibility with the community’s existing water distribution system. Energy input for water treatment plants varies widely with the raw water quality and treatment process type required to meet potable water standards. The typical range of energy input for various water treatment processes—including raw water withdrawal and transfer, treatment, disinfection and distribution—is summarized below in Table 1.

Table 1 – Summary of Electrical Energy Intensity for Water Treatment

Water Source	**Treatment Technology**	**Electrical Energy Intensity** kWh/M^3 (kWh/MG)
Groundwater	Conventional softening, filter, and disinfection	0.07 – 0.4 (250 – 1500)
Surface Water	Conventional softening, filter, and disinfection	0.07 – 0.4 (250 – 1500)
Brackish Water	Reverse osmosis / membrane	1.06 – 2.64 (4,000 – 10,000)
Seawater	Reverse osmosis / membrane	2.64 – 5.28 (10 – 20,000)
Seawater	Multi-stage flash evaporation (MSF) Multiple effect distillation (MED)	5.28 – 26.4 (20,000 – 100,000)
Reclaimed Water	Reverse osmosis / membrane	2.64 – 3.96 (10,000 – 15,000)
Reclaimed Water	MSF/MED	3.96 – 5.28 (15,000–20,000)
Wastewater	Biological treatment / disinfection / membrane bio-nutrient removal	0.26 – 1.32 (1,000 – 5,000)

Figure 1 below illustrates the relationship between the size of an EfW facility processing 907 metric tons per day (mtpd) of municipal solid waste and a number of potential water treatment processes, assuming that all of the net electricity produced by the EfW was used for water treatment. This medium size EfW plant is capable of serving a community of approximately 500,000 people.

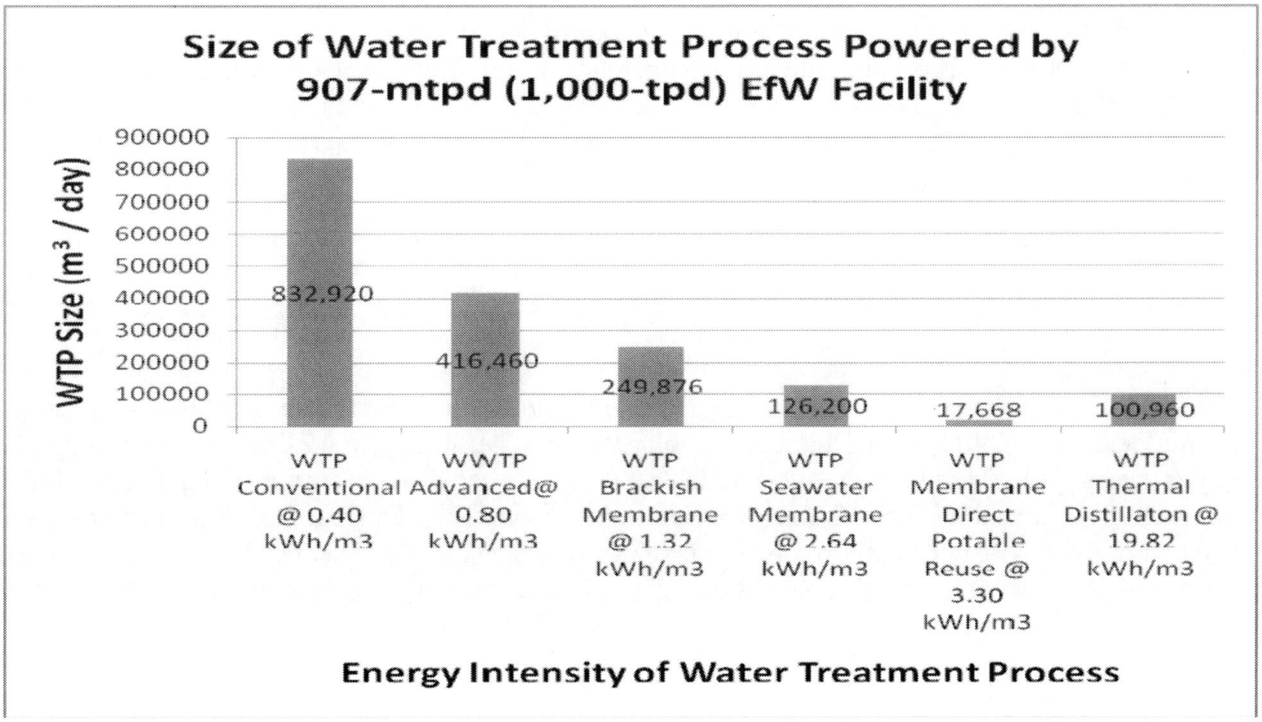

Figure 1 – Size of Water Treatment Process Powered by 907- mtpd (1,000-tpd) EfW

As shown above, the net electrical output of a medium sized EfW facility may provide more electrical energy than needed to meet the community's demand for convention potable water and wastewater treatment services. For communities in need of securing additional water supplies from alternate water sources, including lower quality surface water, brackish water, reclaimed water, or seawater, the compatibility of EfW and advanced water treatment processes improves as the demand for energy increases. An ever-growing percentage of the global population curently resides along coastal states. Much of this population resides in large- and medium-sized coastal communities, where the demand for additional water supply may be provided by seawater treatment technologies. In these communities, the use of reverse osmosis (R/O), multi-stage flash (MSF) evaporation and multiple-effect distillation (MED) processes may be ideally suited to use 100 percent of the electricity from an EfW facility that is sized to treat the municipal solid waste generated in the community. Likewise, inland communities may also be located in watersheds where clean surface and ground water sources have been depleted, and alternate supplies may have to be provided by brackish or reclaimed water sources or regulations for wastewater discharge require significant decreases in nutrient content.

Results

The Hillsborough County Florida municipal EfW facility was originally constructed between 1985 and 1987 to process 1,089-mtpd (1,200 tpd) of municipal solid waste (MSW). The facility, as originally designed, generates 29 MW of renewable electricity which was sold to a local utility. Located adjacent to a 30,283 $m^{3/}$day (12 million-gallon-per-day (mgd)) advanced wastewater treatment plant (AWTP), the EfW facility receives up to 4,164 $m^{3/}$day (1.1 mgd) of high quality reclaimed water for multiple purposes noted below:

- The largest user, at approximately 3,785 $m^{3/}$day (1 mgd), is makeup water to the EfW's cooling system which employs a wet evaporative cooling tower,
- The second largest user, at approximately 212 $m^{3/}$day (56,000 gpd), is dilution water, which is used in the spray dryer scrubber air pollution control unit,
- Approximately 41.6 $m^{3/}$day (11,000 gpd) is used for plant wash down water,
- Another 20.1 $m^{3/}$day (5,300 gpd) is used for ancillary cooling of equipment, and
- Landscape irrigation water supply is also provided by the AWTP reclaimed water system.

The AWTP facility also accepts sanitary waste along with process wastewater from the EfW facility.

The EfW facility was expanded between 2007 and 2009 to accommodate the continuing growth of the local community and its increasing solid waste generation. An additional 544 mt/day (600 tpd) of waste processing capacity was added along with a separate steam turbine generator for the production of an additional 17 MW of renewable electricity. During the addition of the new expansion unit, Hillsborough County Water Resource Services staff worked in conjunction with the Solid Waste staff to investigate the technical and economic feasibility of internally using some of the EfW renewable electricity to power the AWTP facility.

After review and analysis, it was decided that the AWTP facility would disconnect from the local utility grid, and be powered entirely by renewable electricity from the EfW facility. Figure 2 is a recent aerial photograph of the two adjacent municipally owned facilities.

Figure 2 - Hillsborough County, Florida 1,633 mt/day (1,800-tpd) EfW Located Adjacent to 30,283 $m^{3/}$day (12-mgd) AWTP.

Backup diesel generator power was already available at the AWTP plant to provide full operating capacity for periods when the EfW facility is not available (for example, during periodic turbine generator outages that are performed approximately every 5-6 years for maintenance). A total of four diesel-powered emergency generators located at the AWTP site can provide up to 2,500 kW of electricity for periods when the EfW facility is at reduced load due to planned and unplanned maintenance outages. Figure 3 is a photograph of one of the four diesel-powered backup generators located at the AWTP.

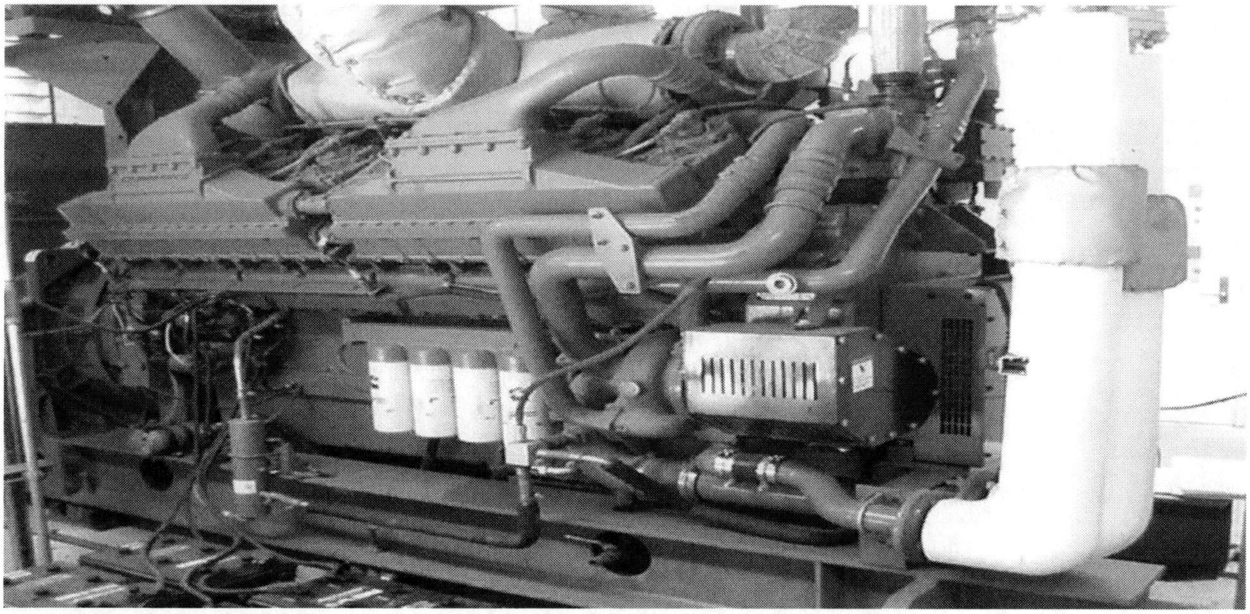

Figure 3 - One of Four Diesel Powered Backup Generators Located at the AWTP.

Commencing in the summer of 2009, the AWTP facility was placed entirely on the EfW electrical system and has been consuming approximately 1.7 MW of electricity. This synergistic relationship allows the new Public Utilities Department (combined solid waste and water resource services groups) to share resources. The solid waste group sells the remaining power to the local electric utility for approximately 6 cents per kilowatt-hour (kWh), whereas the AWTP was formerly purchasing electricity from the local electric utility for approximately 9 to 11 cents per kWh under several commercial tariff rates.

The current internal sharing rate for the EfW electricity utilization benefits the Public Utilities Department in two ways: the Solid Waste Enterprise system receives additional revenues because the self-service rate paid by the Water Service Enterprise system is slightly higher than current electric market rates; and, the Water Service Enterprise system has significantly lowered its energy cost by approximately 50 percent because the self-served facilities no longer purchase commercial energy from a local electric utility. The estimated overall annual benefit to the Public Utilities Department is approximately $600,000 per year assuming an average use of 1.7 MW of electricity valued at 4 cents/kWh differential (cost of electricity purchased by the AWTP minus the current EfW power sales price to the local electric utility).

Other services recently powered include a water treatment facility, a customer service center, and a warehouse complex utilizing approximately 0.3 MW of electricity. The estimated overall annual benefit to the Public Utilities Department is an additional \$100,000 per year assuming an average use of 0.3 MW of electricity valued at a differential of 4 cents/kWh.

In response to this initial successful synergistic project, Hillsborough County Public Utilities Department is evaluating the potential to continue expanding the internal use of its renewable electricity by up to an additional 5 MW of electric power for other essential municipal services that are managed on its municipal campus.

Modifications to the existing electrical switchgear and transmission system required for the new electrical supply to the AWTP located south of the EfW facility were completed during the expansion project in 2009 by the EfW operator, Covanta Energy Inc. However, a new electrical switchyard and transmission line was required for the additional electrical supply to the county facilities located north of the EfW facility.

Figure 4 shows the new 13.2 kV electrical substation in the foreground, with the existing switchyard in the background.

Figure 4 - New 13.2 kV Substation (foreground) for Internal Use of Electricity by Hillsborough County, Florida.

Discussion - Future Opportunities
For communities considering similar synergistic arrangements, the potential cost savings can be significant, depending upon the percentage of EfW net electrical output that can be used "internally," or "behind the meter" for treatment of water and wastewater resources.

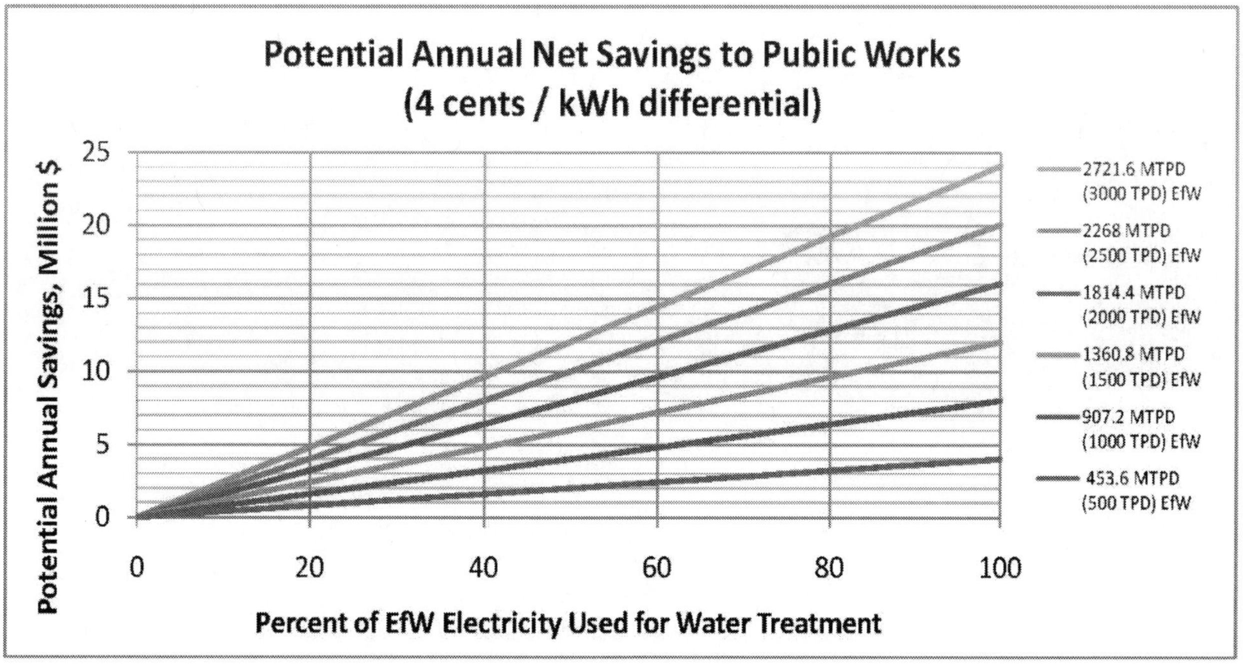

Figure 5 - Potential Benefits for use of Renewable Electrical Energy from EfW Facilities.

The above figure illustrates the potential savings to Public Works for a variety of EfW facility sizes based upon the percentage of electricity that is used internally for the treatment of water resources. As shown, the savings can be significant, potentially tens of millions of dollars per year. As a point of reference, the Hillsborough County Florida EfW internal use of electricity is currently only 5.1% of the total net generation. The county continues to explore additional opportunities for ways to use a greater percentage of their renewable energy for other municipal services and public work projects.

A synergistic approach to managing several municipal processes on a single campus is compatible with the goals of sustainability, waste reduction, and development of alternate water supplies, while answering the challenge of the water-energy nexus. In addition to this successful Florida project, there are numerous opportunities to integrate energy from waste with water treatment processes for various alternate water sources, including surface water, wastewater, reclaimed water, stormwater, brackish water, and seawater as viable options for the future era of sustainable public works.

Advanced treatment requiring greater demands of energy for water resource treatments is becoming common when processing lower quality raw water, while meeting higher environmental standards for a growing family of water chemistry parameters and chemicals of concerns. Ultraviolet light, ozone, ultrasonics, membrane technology and other electrically derived disinfection and filtration technologies continue to evolve, with the net effect of increasing the overall energy intensity of future municipal water treatment systems.

Figure 6 illustrates how a future integrated municipal water utility may be configured and operated with renewable electricity from EfW.

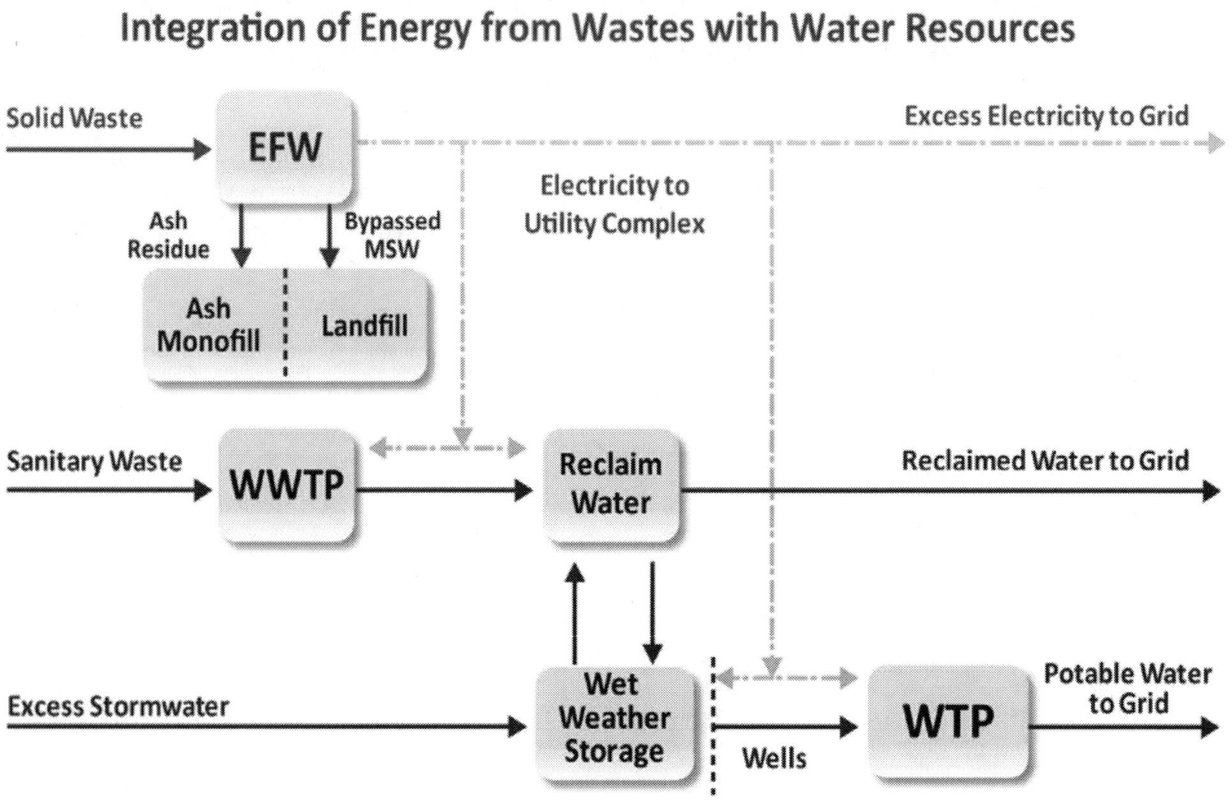

Figure 6 - Future Integrated Campus for Municipal Utilities.

Often referred to as the "water-energy nexus," water and energy are inextricably linked in the goal of public works to provide a clean and affordable municipal water supply. High quality solid waste management and water treatment delivery are also two key processes critical to the successful development of sustainable economic development on local, regional, state and national levels.

In addition to the uses of EfW electricity for water treatment and associated distribution systems, other uses for energy derived from municipal wastes on an integrated campus may include:

- EfW electricity or steam, along with biomethane from landfill gas or anaerobic digesters for drying WWTP biosolids.
- Operation of recycling facilities (construction and demolition waste, household hazardous waste, electronic waste, and material recovery facilities).
- Operation of municipal buildings via combined heat and power (CHP) using steam, hot and chilled water.

Planning for a Future Integrated System

When considering the implementation of a system integrating energy from waste with water treatment processes, a municipal utility will need to assess and evaluate the potential beneficial synergies that may be shared between the utility groups (potable water, wastewater and solid waste) and identify possible ways to maximize those synergies. The opportunities will vary greatly from one municipal utility to the other based on such things as current infrastructure and future infrastructure needs, existing and future available water supply sources, existing and future wastewater treatment requirements, commercial electric rates, and the size of the EfW facility, among others. Ideally, all of the energy demand for a municipal utility's water and wastewater treatment processes would be met by electrical output from the EfW facility. The percent of internal use should be maximized in order to optimize the shared benefits.

To evaluate whether this ideal synergistic arrangement could be feasible, an analysis was performed for three communities with population sizes ranging from small (176,000), medium (700,000) to large (2 million). The potential to match the electrical output from the EfW facility to the energy requirements of the water and wastewater processes sized to meet the community's needs was examined. In addition, the potential cost savings to the water utility groups and the revenues to the solid waste utility group were quantified for the various-sized communities. Finally, several optimization alternatives were evaluated for the EfW facility and process to maximize revenues with the results indicating the potential to essentially drive the cost to zero.

Base Alternative

An examination of various planning elements is presented below, with various-sized EfW facilities combined with typical demand factors for water, wastewater and solid waste. The demand assumptions used were as follows:

- Potable Water – 0.38 m^3/day/capita (100 gallons/day/capita)
- Wastewater – 0.30 m^3/day/capita (80 gallons/day/capita)
- Solid Waste – 1.59 kg/day/capita (3.5 pounds/day/capita)

The solid waste demand is based on normal usage patterns, with a reduction for a portion of the waste being recycled. The normally reported municipal solid waste generation rate published by EPA is typically 9.9 kg/day/capita (4.5 pounds/day/capita). Curbside and drop-off recycling programs across the U.S. continue to advance with more materials collected, ranging from the popular basics (metal, plastic, and glass containers; newspaper, and yard waste) to an ever growing menu of items (e-waste, junk mail, food waste, etc.). The assumed value of 7.7 kg/day/capita (3.5 pounds/day/capita) reflects the future contribution of recycling to reduce the amount of combustible waste in a typical urban community.

The population ranged from 175,000 to 700,000 to 2 million, in order to present small, medium and large EfW facility results.

Using the above criteria, the total demands for each utility element is as follows:

	Population 175,000	Population 700,000	Population 2 million
▪ Potable Water –	66,240 m^3/Day 17.5 MGD	265,000 m^3/Day 70.0 MGD	757,100 m^3/Day 200 MGD
▪ Wastewater -	53,000 m^3/Day 14.0 MGD	212,000 m^3/Day 56.0 MGD	605,700 m^3/Day 160 MGD
▪ Solid Waste -	272 MTPD 300 Tons/Day	1,089 MTPD 1,200 Tons/Day	2,721 MTPD 3,000 Tons/Day

The most critical need for reduced energy costs occurs when the planned raw water source is high in salinity or the wastewater pollutant reduction need is high. When this is combined in an area of the country with high costs for electricity, an EfW should at least be considered as a source of power, as an option for integrated resource management and opportunity to minimize costs to the rate payers.

An example of some actual planned projects for desalination water plants in the U.S. provides the following prospective:

	Population 175,000	Population 700,000	Population 2 million
Desalination Planned Size	28,390 m^3/Day 7.5 MGD	34,070 m^3/Day 9.0 MGD	189,300 m^3/Day 50 MGD

Given that there are planned desalination water plants in the feasibility study stage, the average energy intensity and water usage parameters for these projects were used in our analysis:

	Energy Intensity
▪ Potable Water – desalination	4.0 kWh/m^3 15,000 kWh/MG
▪ Potable Water – surface/groundwater	0.20 kWh/m^3 750 kWh/MG
▪ Wastewater	0.80 kWh/m^3 3,000 kWh/MG
▪ Solid Waste – net supply	606 kWh/Metric Ton 550 kWh/Ton

The combination of the energy demands per service element and the electrical output of the EfW facility is provided for each population size below. For the case of the small population, the electrical output of the EfW facility is less than the demand for water treatment processes. However in the cases of the medium and large population, the EfW facility produces more electricity than needed by the water treatment processes. In these cases, the excess electrical energy would need to be sold to the local utility grid. Final sizing of the EfW facility could result in a closer match of electrical output to the demand, if this were found to be the best scenario.

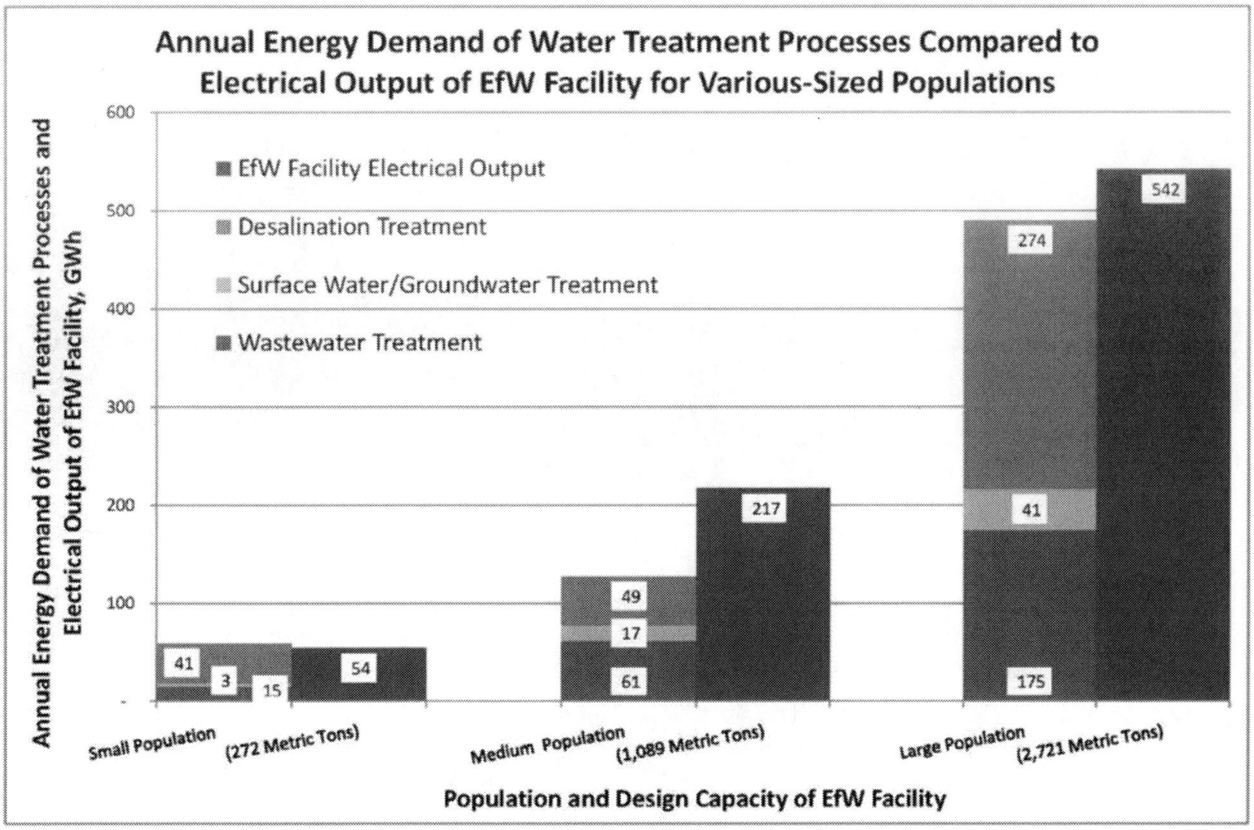

Figure 7 – Match of Water Treatment Processes Energy Demand to EfW Facility Output.

Other elements to take into consideration are the commercial cost of power and the estimated sale price of power back to the grid. It is the differential between these two numbers that create the ability to share any savings between the utility services. There is a wide range of prices being charged per kWh, depending on location. Examples of this range are identified below.

	Continental U.S.		Island Communities	
	Normal	High	Hawaii	Islands
Cost of commercial power per kWh	\$0.09	\$0.12	\$0.20	\$0.47
Estimated sale price to grid	\$0.06	\$0.09	\$0.15	\$0.30

The figure below presents the total energy costs to the utilities with a sales price of \$0.06/kWh for each size of EfW facility along with the value of the savings at \$0.03/kWh, also shown is the existing energy costs for the utilities at \$0.09/kWh.

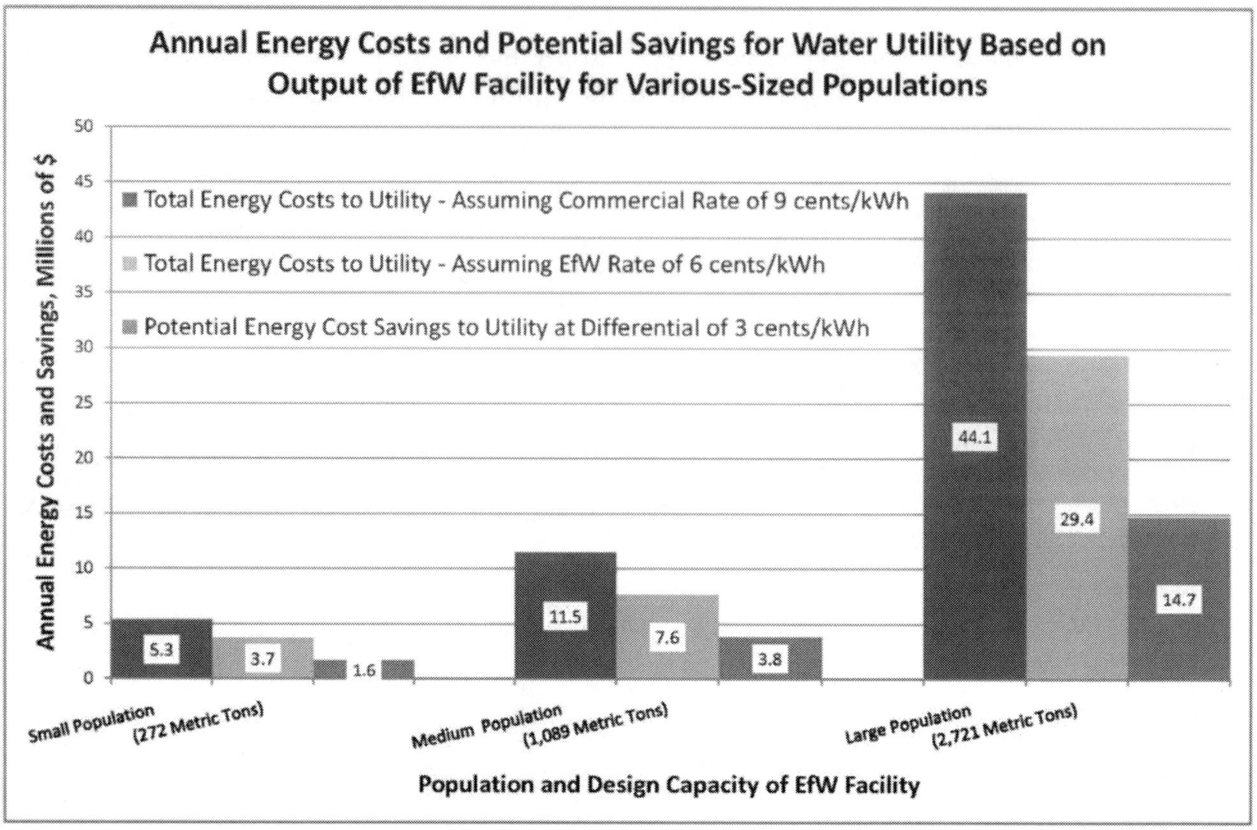

Figure 8 – Energy Costs and Potential Savings to Water and Wastewater Utility Groups.

The balance of the energy revenue at $0.06/kWh (charges to the water and wastewater utilities and income from sales to the grid) has been assigned to the solid waste utility in our analysis, although any range of sharing could be negotiated amongst the Public Works / Utility Departments. The total revenues for the EfW facilities are presented below, with both the sales to the utilities and to the grid shown independently and as a total.

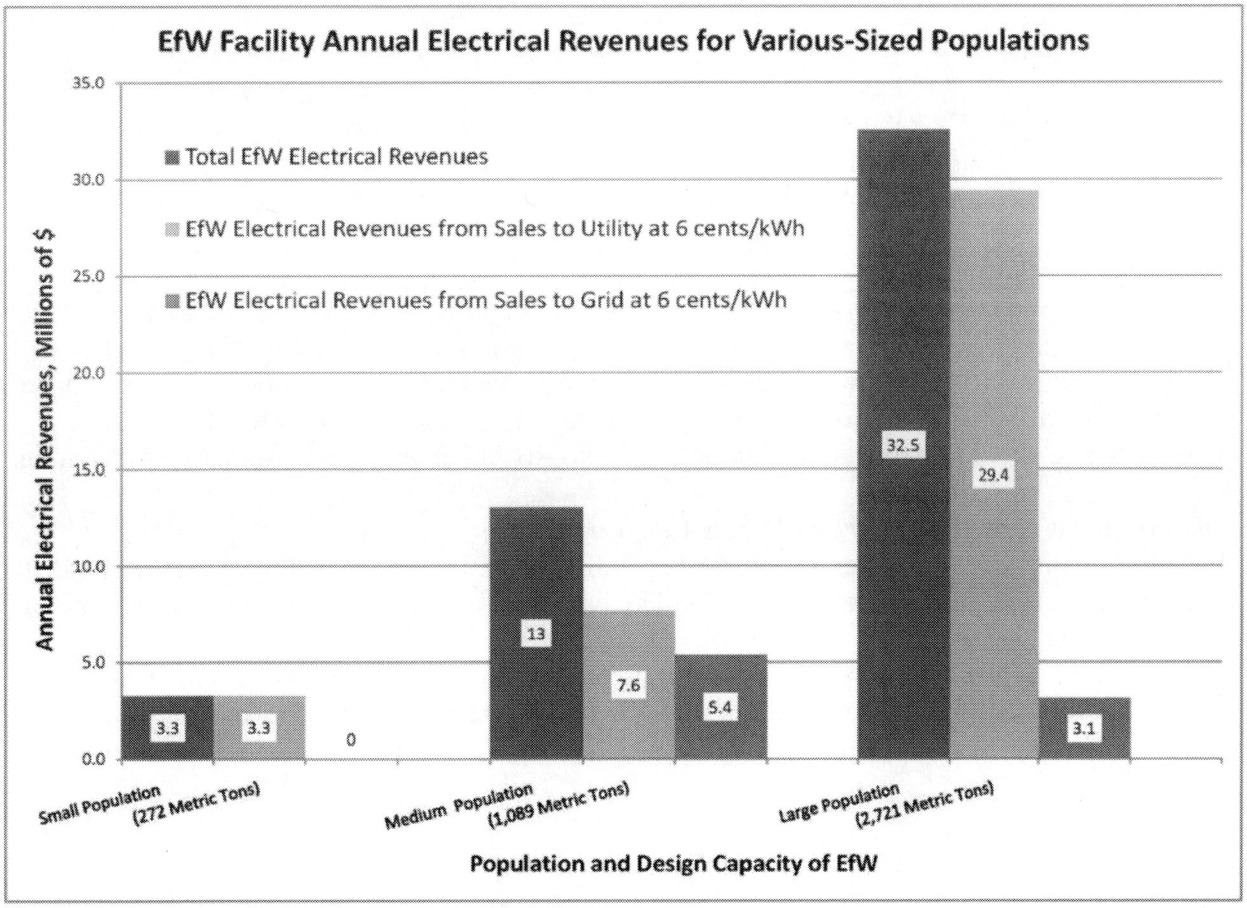

Figure 9 – Potential Revenues to Solid Waste Utility Group.

Optimization Alternatives

Although the above discussion relates to the normal operating outcomes from existing EfW facilities, there are opportunities to dramatically improve the results. By optimizing certain aspects of the EfW facility and process, costs can be decreased and revenues increased. Lowering the costs of EfW results in a viable alternative for solid waste disposal that historically has not been able to compete with other lower cost options, such as landfilling. EfW also provides the added benefit of providing electricity for internal use, reducing the energy costs for the municipal utility. The various optimization alternatives that were considered are discussed below.

Design for Lowered Capital Costs

There are opportunities to provide cost savings for the EfW facility with attention to design which can significantly reduce the capital cost. Some of these cost decreases include providing a properly sized and positioned tip floor and refuse pit (using common wall construction and no over-sizing), appropriately sizing and location of the ash storage area, and eliminating unnecessary redundancy and expenses. The estimate is that a ten-percent reduction in capital costs can be realized.

High Pressure EfW
There are high pressure EfW facilities currently being constructed or in operation that provide much higher units of gross and net kWh per ton of waste processed. Existing normal pressure of the current generation of EfW boilers is 900 psi. Higher pressure facilities can be constructed in the 1350 to 1450 psi range, thereby improving the boiler efficiency and electrical energy production. The initial cost of the EfW facility is estimated to increase by ten percent. However, the net increase in electrical generation will increase by approximately 36 percent.

Managing the Waste Stream to Increase HHV
In addition to the above analysis, the heating value of municipal waste may be further increased by removing food and yard wastes for processing in compost or anaerobic digestion systems.. The goal for this alternative would be to reduce the percentage of food and yard waste entering the EfW by fifty percent. In this scenario, the heating value is estimated to increase by 15 percent. There should not be any additional capital costs in achieving this higher heating value.

Additional Ferrous/Non-Ferrous Metals Recovery
An additional source of revenue would be available to the EfW facility by maximizing the recovery of ferrous and non-ferrous metals. The base case analysis assumed a recovery rate of 3.2 percent of the waste stream for ferrous metals and 0.3 percent for non-ferrous metals. With proper attention to design, operation and maintenance of the post-combustion metal recovery system, the optimal metal recovery rates of 5.0 percent ferrous and 0.5% non-ferrous could be achieved. The net result of the improved metal recovery system is greater revenues to the EfW project.

Renewable Energy Credits
Depending on the location of the EfW facility, there may be Renewable Energy Credits (RECs) available. The current U.S. market value for RECs is quite low; however, they may increase in value in response to expected future clean energy legislation and federal and state incentives The current assumption used was a market value of $2.00/REC or $2.00 per MW.

Recycled Bottom Ash
Depending on the quality of the bottom ash, there may be potential to use this residue as construction aggregates, such as road base material. Should this be the case, the cost for landfilling the ash would be reduced, resulting in a lower estimated cost per ton and savings to the EfW project.

Combination of All Alternatives
By combining the elements identified above, the waste processing cost per ton can be dramatically reduced, thereby maximizing future benefits to the solid waste and water resources rate payers.

Combination of All Alternatives Including Processing of Special Waste
Additional revenues could be available to the EfW facility from the contracting and processing of special wastes (out-of-date and off-spec pharmaceuticals, medical wastes, industrial waste, and other wastes requiring assured destruction.. These wastes command a higher tipping fee, and thereby generate additional revenues for the EfW project.

A summary of the various optimization alternatives and the percent decrease in the tip fee from the base case is presented below. This graph illustrates the significant reductions in tipping fees to the EfW project. As noted previously, the overall cost savings from an integrated resource management project can be shared internally between the solid waste and water resource department, with the ultimate benefit to the system rate payers.

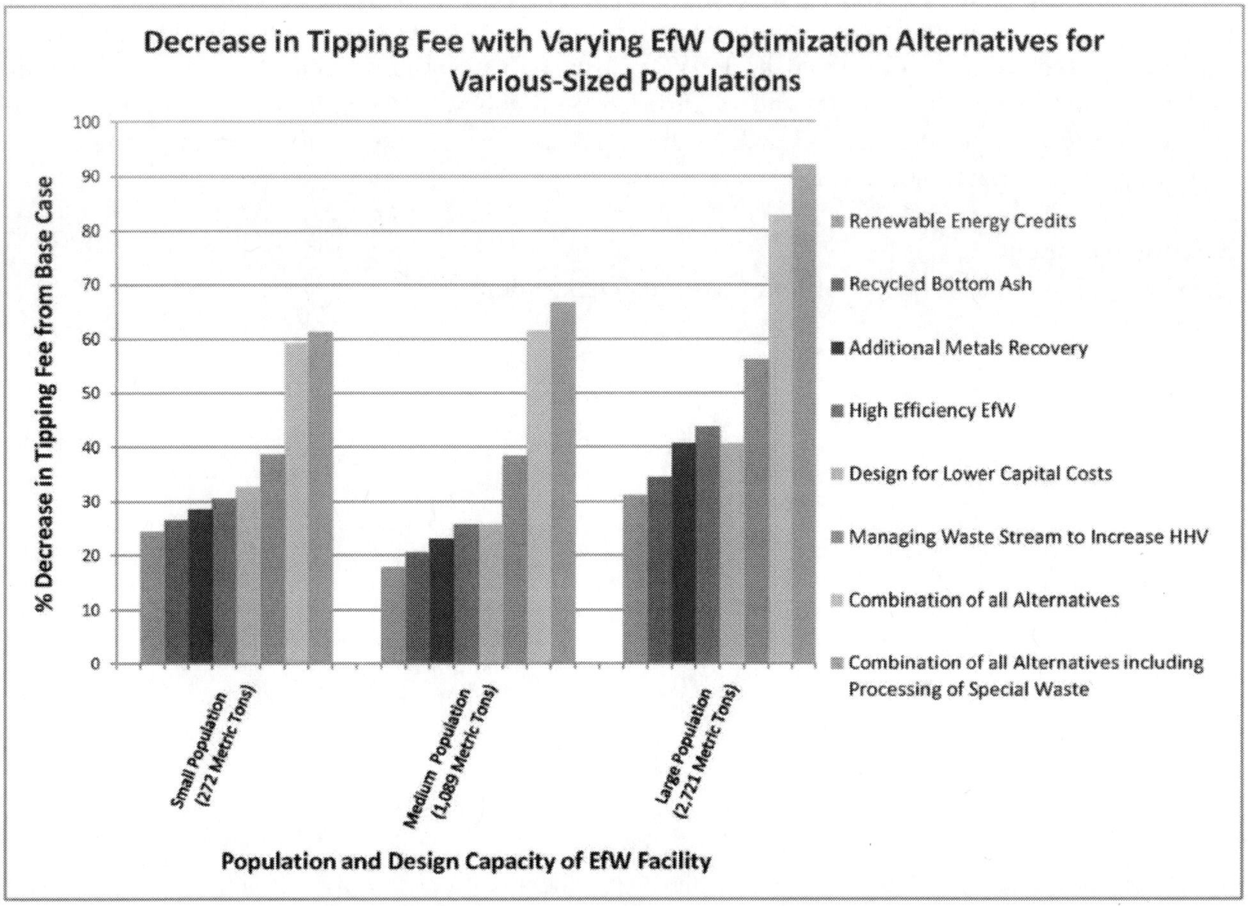

Figure 10 – Effect on Tipping Fee for Various Optimization Alternatives.

Conclusions

Overall, a strong argument can be made in favor of EfW facilities integrated with water resource facilities. In addition to cost savings for the treatment of water resources, greatly reduced solid waste disposal costs can be realized (potentially by as much as 90 percent, approaching zero cost).

While there may appear to be institutional barriers to the integration of EfW with water resource, those perceived barriers are soon to be shattered by the grim reality of the current economic climate in the U.S. The economic and political forces working behind the scenes are related to the following:

- Increasing cost/value of water and energy

- Economic need to do more with less
- Need for upgrades to aging facilities
- Increasingly stringent regulations
- Desire for beneficial use of resources
- Commitment to environmental stewardship

The recent success story in Hillsborough County Florida may serve as the catalyst for other municipal entities to develop integrated utility projects. The water resources staff of the Hillsborough County Public Utilities Department took the first step to explore the feasibility to use the renewable electricity from its EfW facility, and ultimately made the decision to boldly move forward. Both the solid waste department and the water resources department are now happy to jointly share the savings, while the local rate payers indirectly reap the benefit in the form of reduced cost of services!

Acknowledgments
Hillsborough County, Florida – Public Utilities Department

Integrating Major Components of a New Landfill Gas System – Coordination of Multiple Contracts for Gas Treatment, Pipeline Transmission and Turbine Energy Generation

By
Ralph B. "Rusty" Schroedel, Jr., P.E., BCEE, Brown and Caldwell
Michael J. Martin, P.E., BCEE Milwaukee Metropolitan Sewerage District
Cary J. Solberg, P.E., Milwaukee Metropolitan Sewerage District
William Krill, P.E., Milwaukee Metropolitan Sewerage District

ABSTRACT

The Milwaukee Metropolitan Sewerage District (MMSD) is one of the leading wastewater agencies addressing sustainability. One opportunity for renewable energy production which MMSD has undertaken is a project to utilize landfill gas from a private landfill to the 100 mgd (average flow) Jones Island Water Reclamation Facility (JIWRF) in order to produce electricity and heat.

In order to help the MMSD coordinate and manage startup of the project, Brown and Caldwell was retained to provide assistance in several areas. This paper will describe the landfill gas system; discuss the Coordination and Management Services effort; and how the challenges of coordination with multiple engineers, owners, contractors, and operators were addressed.

KEYWORDS

Landfill gas, energy, turbines, sludge drying, gas pipeline, project management

DESCRIPTION OF LANDFILL GAS SYSTEM

The landfill gas system collects and processes LFG at Emerald Park Landfill (EPL), conveys it JIWRF, and combusts it in a new turbine generator facility to produce electricity and heat. The use of the LFG in this manner is often referred to as combined heat and power (CHP). The new Gas Turbine Facility at JIWRF will use LFG to replace natural gas (NG) as the primary source of energy for treating wastewater and producing Milorganite® at JIWRF. Use of LFG in place of NG will result in operational cost savings, reduced emissions, and an overall reduction in greenhouse gases.

JIWRF is owned by MMSD and operated under contract by Veolia Water Milwaukee (VWM). EPL is owned and operated Advanced Disposal. MMSD has a long term contract with Advanced Disposal for the purchase of LFG.

Organic material degradation in a landfill produces landfill gas. The gas is comprised primarily of carbon dioxide and methane. The LFG also contains small amounts of other constituents including water vapor, hydrogen sulfide (H_2S), oxygen, and siloxanes. Landfill gas commonly contains about 45-50% methane. For comparison, wastewater treatment plant anaerobic digester gas typically contains 55-65% methane. In order to prevent the gas from permeating off-site and to minimize air pollution, a system of gas extraction wells and piping collects the gas from the landfill. The extracted gas at the EPL is currently flared to control emissions.
A graphic and schematic of the overall landfill gas system are presented in Figures 1 and 2 respectively.

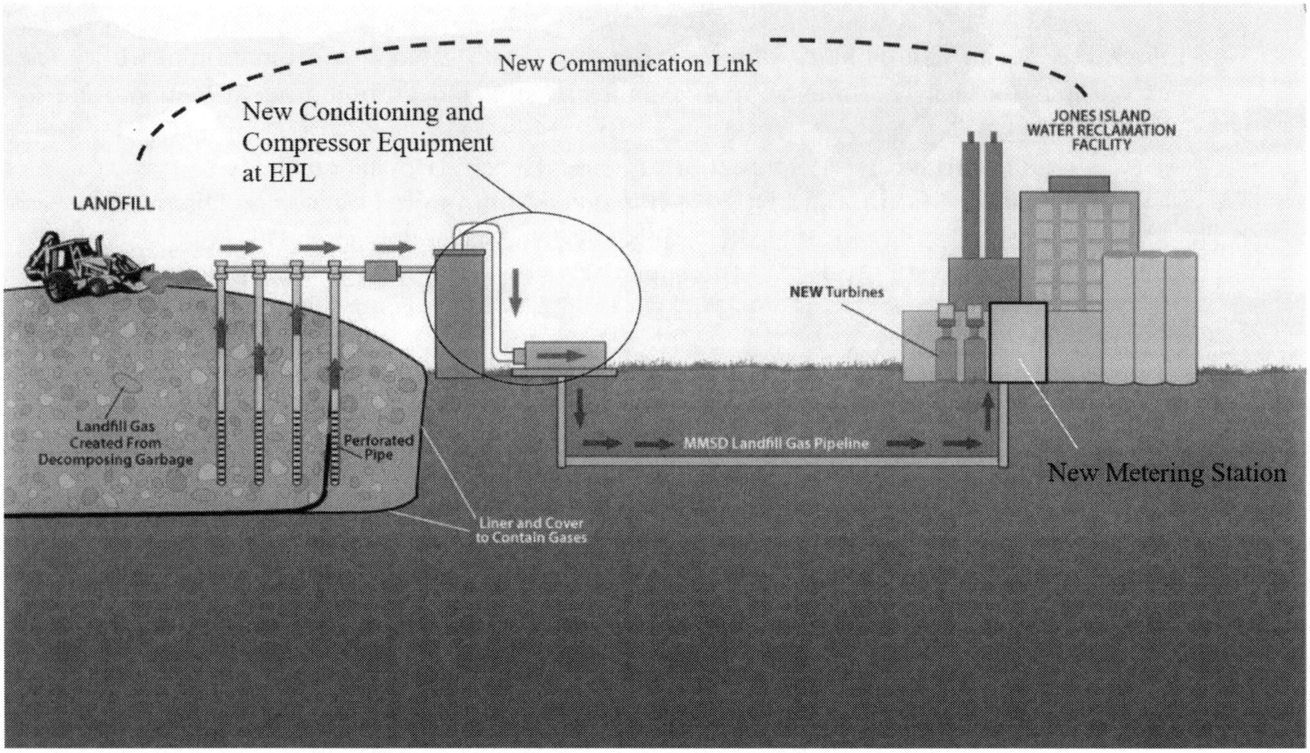

Figure 1 – Overall Landfill Gas System Graphic

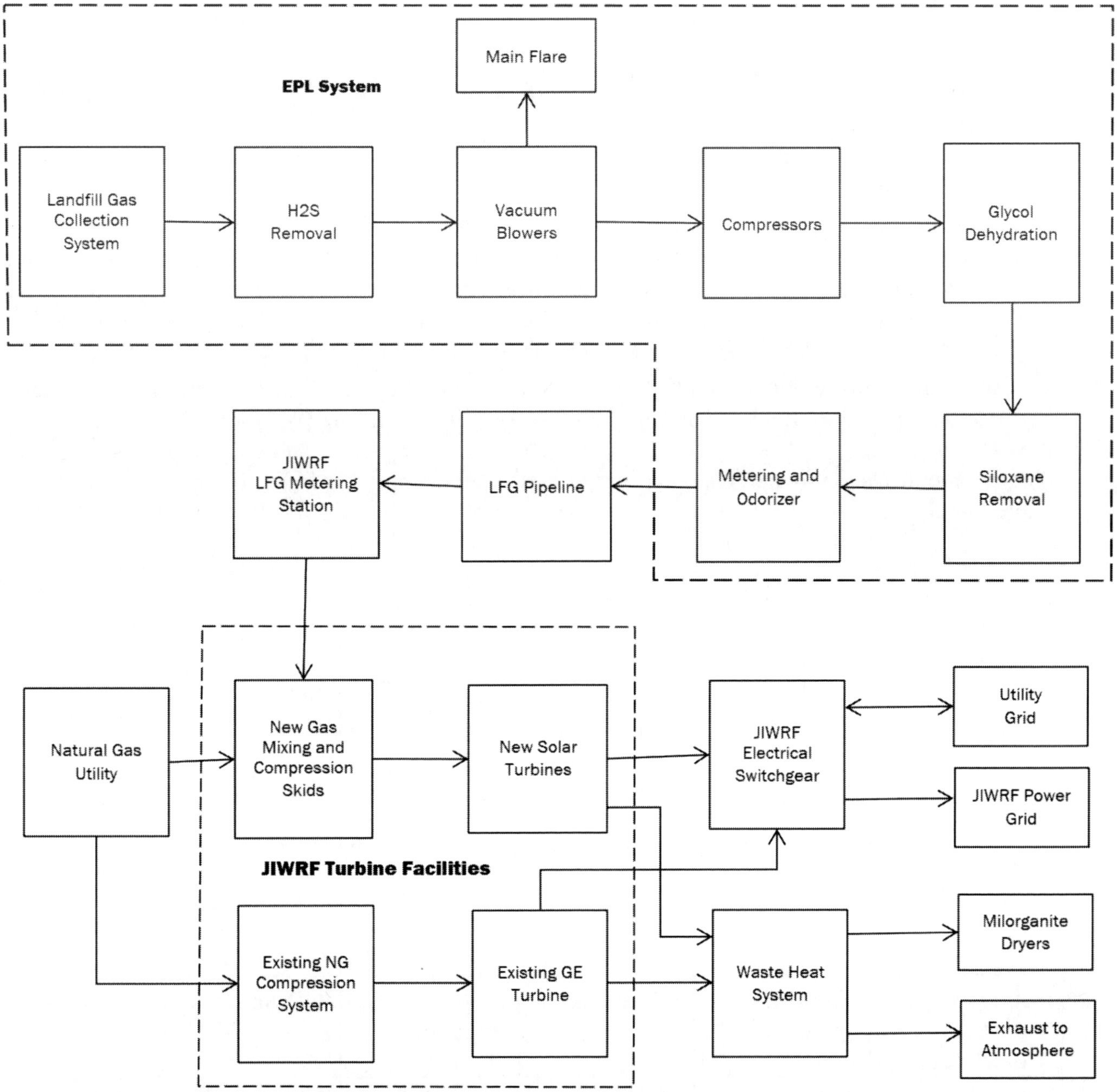

Figure 2 – Landfill Gas System Schematic

The main components of the landfill gas system are:

- LFG Collection, Compression and Conditioning System at EPL - This system consists of new vacuum blowers, H_2S removal system, compressors, glycol dehydration, siloxane removal, metering, and odorization equipment.
- LFG Pipeline - The pipeline is owned by MMSD and is operated and maintained by Douglas Pipeline Company (DPC) to meet federal and state gas pipeline safety regulations. The pipeline consists of:
  - 13.5 miles of new 16 inch diameter high density polyethylene (HDPE) pipe with appurtenances.

- Approximate 5.5 mile segment of existing 16 inch diameter steel pipe with appurtenances.

- Metering Station at JIWRF - The metering station contains: flow, pressure, temperature, oxygen, methane, and hydrogen sulfide (H_2S) measurement; and appurtenances. The metering station is owned by MMSD and will be operated by VWM. A small portion of the metering station piping falls under the jurisdiction of federal and state gas pipeline safety regulations. DPC will ensure that the regulated portions of the metering station piping are operated to meet these regulations.
- New Turbine Generator Facility at JIWRF - Three new turbine generators will use landfill gas as the primary fuel. These turbines and supporting equipment are installed in a new building. One existing turbine in the JIWRF Powerhouse remains in service and uses natural gas. VWM will operate and maintain the turbine generator facilities. Electricity produced by the turbines powers JIWRF. Waste heat from the turbines is used to produce Milorganite.
- Communication Link between EPL and JIWRF - Note that while the link does communicate information, it does not provide capability to control EPL from JIWRF or vice versa.

Table 1 identifies the owner, designer, major equipment supplier, contractor, and operator for each component.

Table 1 - LFG System Component Project Information

Owner	MMSD				EPL
Component Project	Turbine Facility and Equipment Installation	Pipeline		Metering Station	Conditioning and Compression Equipment
		New HDPE Pipe	Steel Pipe Rehab		
Design Consultant	CH2M Hill	CDM Smith	CDM Smith	CDM Smith	Cornerstone
Owner-Furnished Equipment Supplier(s)	Solar Turbines	NA	NA	NA	Unison, MV Tech, CVIC, Venture
Prime Construction Contractor	Findorff & Sons	Meade	Michels Pipeline	CD Smith	VES
Operator	VWM	DPC		VWM / DPC	VES

The LFG system incorporates a flexible design which allows for additional energy to be provided at JIWRF. Specifically, there is additional pipeline capacity to allow for other future sources of biofuel. An additional pipe connection is available for the future addition of LFG from another landfill near EPL. Other unknown sources of biogas could also become available in the future. The current turbine building and systems have been designed to enable the future addition of two turbine systems.

LANDFILL GAS PURCHASE CONTRACT

A "Landfill Gas Purchase and Sale Agreement" between Advance Disposal and MMSD defines the terms and conditions by which Advance Disposal (EPL) shall deliver and sell Conforming LFG and that MMSD (JIWRF) shall accept and purchase Conforming LFG. For 2013, the annual quantity of conforming LFG is 684,982 MMBtu, which is equivalent to 78.2 MMBtu per hour, assuming equal distribution over the entire year. The annual amount of conforming LFG increases between 1-2 percent per year, peaking out at 1,031,180 MMBtu in 2031.

Conforming LFG requires the fuel content to be:

- 450 Btu per cubic foot heating value minimum
- 2% oxygen (O_2) content maximum
- 100 ppmv hydrogen sulfide (H_2S) concentration maximum
- 5 mg Si/nm3 CH_4 siloxane concentration maximum
- 105 ppmv moisture content maximum

Nomination Process. The nomination process requires that the MMSD will prepare and deliver monthly to Advance Disposal a written rolling twelve (12) month schedule (broken down by month) setting forth (in MMBtu and decatherm) the reasonably projected volumes of conforming LFG that MMSD intends to accept from EPL over the ensuing year (referred to as "Yearly LFG Nominations"). In addition, each month MMSD will prepare and deliver to Advance Disposal a written schedule setting forth (in MMBtu and decatherm) the reasonably projected volumes of conforming LFG that MMSD intends to accept from EPL during the immediately forthcoming month (referred to as the "Monthly LFG Nominations"), broken down on a daily basis (referred to as the "Daily LFG Nominations"). Advance Disposal will acknowledge receipt of and notify MMSD in writing of any anticipated problem in delivering the amount of conforming LFG set forth in such nominations.

It is anticipated that this will be a joint, cooperative effort to optimize both landfill gas system operation at EPL and energy production at JIWRF. For example, should landfill system conditions warrant extraction of less LFG, EPL will need to communicate this to JIWRF in a timely fashion so that MMSD and VWM can plan for the power needs at JIWRF.

The agreement also addresses issues including:

- A "Nomination Process" by which the MMSD provides an annual and monthly estimate of how much LFG it will accept and Advance Disposal notifies of any anticipated problem delivering that amount of LFG.
- Construction, testing and startup obligations
- Excess landfill gas quantities
- Interruptions in delivery or acceptance of landfill gas
- Scheduled outages
- Non-Conforming LFG
- Construction of additional Conforming LFG acceptance capacity.

Table 2 provides the Conforming LFG Specification and Table 3 provides the Conforming Pressures.

Table 2 - Conforming LFG Specification					
Gas Parameter	Unit of Measurement	Maximum	Minimum	Test Method	Monitoring Frequency
Water Vapor	PPMV	<105	-	ARI Meter	Continuous
Pressure (based on Flow SCFM)					
0-1543 (SCFM)	Psig	95	60*	Gauge	Continuous
1543-4630 (SCFM)	Psig	95	65*	Gauge	Continuous
4630-6173(SCFM)	Psig	95	75*	Gauge	Continuous
6173-7716 (SCFM)	Psig	95	80*	Gauge	Continuous
7716-9259 (SCFM)	Psig	95	90*	Gauge	Continuous
9259 and above (SCFM)	Psig	95	95*	Gauge	Continuous
Hydrogen Sulfide	PPMV	100	-	ARI Meter	Per Manufacturer
Oxygen	%	2	-	ARI Meter	Per Manufacturer
Heating Value (LHV)	BTU/SCFM	-	450	ARI Meter & Thermal Mass Flow Meter	Continuous
Siloxanes	Mg Si/nm3 CH4	5	-	Jet-Care Si Test	Annual
Temperature	Degrees Fahrenheit	-	-	Thermometer	Per Manufacturer
Carbon Dioxide	%	-	-	ARI Meter	Per Manufacturer

*Note: *- At Delivery Point*

Table 3 - Conforming LFG Pressures		
Flow Rate Range* (MMBTU/Day)	Flow Rate Range* (SCFM)	Minimum Delivery Pressure (PSI)
0-1000	0-1543	60
1000-3000	1543-4630	65
3000-4000	4630-6173	75
4000-5000	6173-7716	80
5000-6000	7716-9259	90
6000 and above	9259 and above	95

*Note: *- At Delivery Point*

COMPONENT FACILITIES DESCRIPTION

EPL Systems

EPL is a municipal solid waste landfill located in Muskego, Wisconsin. Organic material degradation in a landfill produces LFG. The gas is comprised primarily of carbon dioxide and methane. The LFG also contains small amounts of other constituents including water vapor, hydrogen sulfide (H_2S), oxygen, and siloxanes. LFG commonly contains about 45-50% methane. For comparison, water reclamation facility anaerobic digester gas typically contains 55-65% methane. Withdrawal of LFG is controlled to maximize the methane content and minimize the oxygen content.

In order to prevent the gas from permeating off-site, a system of gas extraction wells and piping collects the gas from the landfill. Prior to completion of this project, the extracted gas at the EPL was flared at the site to minimize air pollution and greenhouse gas emissions.

EPL constructed a LFG processing facility consisting of blowers, compressors, filters, absorption systems, dehydration systems, gas metering and odorization system. The processing facility is currently sized to process 6,000 standard cubic feet per minute (scfm). These systems consist of the following components in order of treatment/processing:

1. H_2S Treatment System – Reduces the concentration of H_2S within the LFG to less than 100 parts per million.
2. Two multistage blowers capable of applying approximately 100 inches w.c. of vacuum to the LFG collection system have been installed.
3. Aftercoolers positioned downstream of the blowers reduce the gas temperature prior to compression.
4. Three 2,000 scfm compressors pressurize LFG to approximately 100 psi.

5. Aftercoolers downstream of the compressor skids reduce temperature prior to dehydration.
6. A glycol dehydration system removes moisture from the LFG in order to allow conveyance through MMSD's carbon steel pipeline segment.
7. A thermal oxidizer controls off gases generated from the glycol dehydration system as well as additional piping to supplement destruction of these off gases above and beyond what can be used to operate the glycol system's boiler.
8. A siloxane treatment system reduces siloxane concentrations in the LFG.
9. A metering system incorporating a laser gas analyzer and gas thermal mass flow meters measure the heat value, flow and other properties of the LFG prior to delivery to the MMSD pipeline. The metering system at EPL is the "billing meter".
10. The LFG is odorized to smell like NG prior to delivery to the MMSD pipeline.
11. A candlestick flare, which formerly combusted all of the collected LFG, functions as the backup system if LFG cannot be delivered to or accepted by MMSD.

LFG Pipeline System

The LFG Pipeline consists of:

- 13.5 miles of new 16 inch diameter high density polyethylene (HDPE) pipe with appurtenances including a pig launching station at EPL, three concrete buried vaults with 16-inch diameter isolation valves, two concrete buried vaults with condensate/moisture traps and pig receiving station at the JIWRF Metering Station.
- Approximate 5.5 mile segment of existing 16 inch diameter steel pipe with appurtenances. cathodic protection systems, pig launching/receiving stations at each end of the steel pipeline, corrosion coupon monitoring stations at each end of the steel pipeline, three concrete buried vaults with 16-inch diameter isolation valves, and an aerial crossing over a Milwaukee city street.

Figure 3 shows the general route of the LFG pipeline and location of the major appurtenances along the route. Isolation valves are provided at key locations and at maximum 5 mile intervals. Condensate blow down structures are provided in the event that any moisture condenses in the pipeline after LFG leaves EPL. Because the gas is dehydrated, high volumes of condensation are not expected.

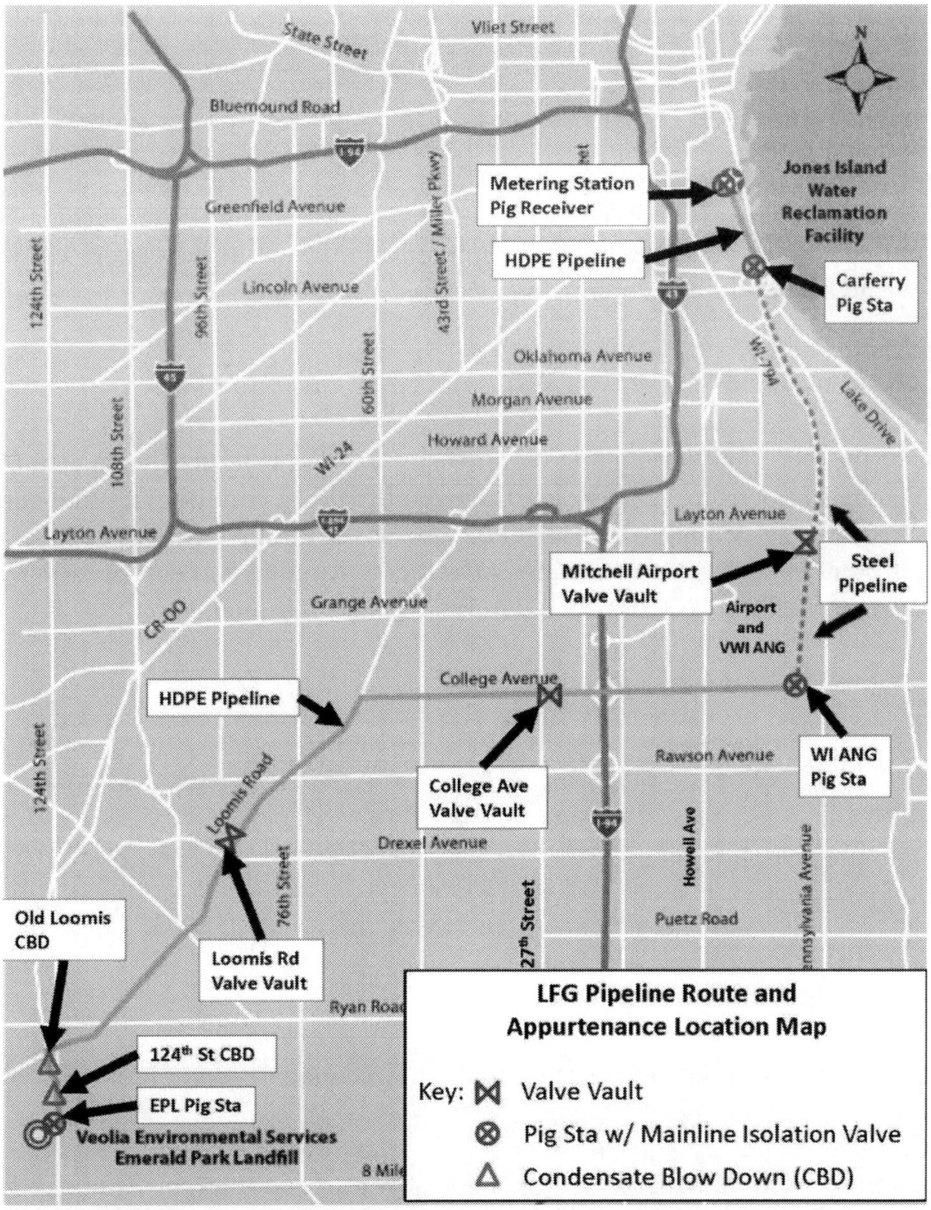

Figure 3 – General Route of LFG Pipeline and Location of Major Appurtenances

Pipeline construction, operation and maintenance is regulated by the U.S. DOT Pipeline and Hazardous Materials Safety Administration (PHMSA) and the Wisconsin Public Service Commission (PSC).

LFG Metering Station

The JIWRF Metering Station at the downstream end of the Pipeline provides monitoring of condition, flow and pressure of the LFG. The landfill gas is metered and analyzed for methane (CH_4), oxygen (O_2), carbon dioxide (CO_2), moisture, and hydrogen sulfide (H_2S). Gas temperature and pressure are also measured. These parameters are critical to proper turbine operation. The Metering Station measurements are used to verify "billing" measurements obtained at the EPL metering system.

Turbine Facility

The new Turbines Facility produces electricity and heat for use at JIWRF. The heat from will be used to assist in the drying process for manufacturing Milorganite®. The Turbine Facility is primary comprised of gas conditioning and compression, LFG turbine generators, electrical switchgear, waste heat and other supporting systems.

Fuel Gas Conditioning and Compression

The purpose of the gas conditioning and compression system is to produce a clean, high pressure fuel source for the LFG turbines. There are three new fuel gas conditioning and compression skid units, each sized to provide fuel to one turbine. Crossover piping with manual isolation valves does allow different operating combinations. Each conditioning and compression system skid can receive LFG from EPL, NG, or mix a combination of both fuels.

Each skid contains the following components:

- Control valves
- Fuel blending vessel
- Scrubbers
- Gas compressor and motor
- Lube oil system,
- Heat exchangers, and
- Instrumentation and control panel.

The gas compressors boost the gas pressure to about 255 psig as required for the operation of the turbines.

Instruments are provided to measure the gas inlet flow rate and inlet pressure. Each compressor discharge has instruments to measure the pressure and temperature of the gas stream before it enters the turbines. Solar turbine operation will normally rely on LFG quality data measured at the Metering Station. The gas compressors boost the gas pressure to approximately 240 psig as required for the operation of the turbines. There is room in the new Turbine Facility to install two additional conditioning and compression skids when future conditions warrant.

LFG Turbine Generators

There are three LFG gas turbine generators, each rated for 4.4 megawatts (MW) at 50 degree F ambient air. As ambient air temperatures increase, the rated output from a turbine decreases and as ambient temperatures decrease, the turbine output increases. There is room in the new Turbine Facility to install two additional turbine units when future conditions warrant.

The main fuel source for the turbines is LFG, with NG as a supplemental fuel. LFG is a medium BTU fuel when compared to NG. Each turbine however, can use LFG, NG, or a combination of the two fuels, with the blending of the two gasses occurring at the fuel gas conditioning and compression skid.

The 2013 equalized quantity of conforming LFG will allow 2 turbines to run at 90 percent load (78.2 / 2x43), or 3 turbines at 60 percent load (78.2 / 3x43). The turbines are designed to run mainly on LFG, with NG as supplemental fuel. NG, when blended with the LFG, can run the turbines at higher electrical output to meet plant demand when all the available LFG is consumed. Additionally, NG can be used for short periods of time if there is a disruption in the LFG supply.

Existing GE turbine No. 1 will be retained for continued use. The GE turbine is rated 18 MW for peak use and operates primarily on NG with fuel oil as a backup. The other GE turbine will be decommissioned during the startup of the new turbines, in accordance with the requirements of MMSD's construction air permit.

Electrical Switchgear

The turbine generators generate 3-phase electrical energy at 13,200 volt (V), which matches the main power distribution into JIWRF. The electrical output from each generator is connected to the switchgear. Each generator breaker within the switchgear is equipped with electronic controls that monitor the quality of the power from the operating generator and will automatically open the breaker if bad quality power is detected. This monitoring system is also used in the synchronization process and will prevent the breaker from closing into a live bus unless the frequency and voltage are within a narrow range of the frequency and voltage on the bus. The switchgear provides power to building electrical loads and to the existing Power House switchgear for distribution to the rest of the treatment plant.

Waste Heat System

Approximately 30% of the energy input into a turbine generator comes out in the form of electricity. The remaining energy output is in the form of heat in the air mass used by the turbine (waste heat). The waste heat is typically used to dry solids into Milorganite but can be exhausted to the atmosphere. A large duct system conveys waste heat from the Turbine Facility and Powerhouse to the Milorganite dryers in the Dewatering & Drying Facility. Turbine exhaust dampers are used to maintain the waste heat duct pressure.

Waste heat temperatures from the turbines vary between 650 and 750 degrees F depending on ambient air temperature, fuel source, and turbine output. Each turbine operating at full load will have about 30,000 scfm of exhaust. Each dryer typically uses 18,000 to 20,000 scfm air flow and requires a temperature around 850°F.

Consequently, the turbine exhaust is neither sufficient in volume nor temperature to meet the needs of Milorganite drying, which typically has 6 to 10 dryers in service. Additional air volume and heat can be provided by the existing dryer burner system at each dryer. The dryer burner system uses NG only. The GE turbine can produce waste heat temperatures and exhaust flows to support operation of all dryers in the Dewatering & Drying Facility. Waste heat can be directed to the dryers from the LFG turbines or the GE Turbine, but not both at the same time. Future use of this turbine is expected to be only for short periods of time to supply peak power demands for wet weather events.

Other Support Systems

Other support systems include cooling water, fire protection, instrument air and "black start" generator. The black start diesel engine generator is provided in case service from WE Energies is lost, the turbines are off and the GE turbine is not available to operate. Output from the generator is sufficient to start one LFG turbine. Once a turbine is started, the output can be used to start other turbine generators.

COORDINATION AND MANAGEMENT CONTRACT

The District determined that the management and coordination during construction and startup of the multiple owners, engineers, contractors, and operators of the system required outside consultant assistance. A contract was developed to engage this assistance. The Contract Scope of Services includes the tasks listed below:

- Review Design Information - Review for information relevant to startup and commissioning
- Site Meetings – Attend to become familiar with the management team, contractors, engineers, and current activities for the major system components
- Obtain and Review Schedules for Each Component – Perform a logic review for issues relating to startup and commissioning.
- Prepare Overall Startup Schedule – Prepare using information from individual contract schedules and merges the activities of the individual contract schedules
- Schedule Analysis – Evaluate to identify critical elements required for on-schedule startup and commissioning
- Identify and Track Startup Items – Identify conflicts and constraints that could impact or prevent efficient startup
- Schedule Updates – Update schedule and identify issues that may impact milestones.
- Track Recommendations from Process Hazard Analysis (PHA) to Completion – Assist the District in tracking completion of the items identified in the PHA.
- Communication Protocols for Startup and Continued Operation – Provide written communication protocol.
- Overall Startup Sequence and Operation Plan – Develop plan for system startup and operation.
- Track Milestones – Track progress towards achieving system operation milestones

- LFG System Functional Testing – Review reports for individual component systems and identify deficiencies
- Startup Coordination Meetings – Coordinate, prepare for, and attend regular startup coordination meetings. Prepare and distribute meeting agendas and minutes.
- Distribute Documentation- Provide copies of documents to all system parties
- Consultant Team Coordination – Provide coordination of subconsultants.
- Coordination with LFG Stakeholders – Coordinate with respect to startup and commissioning.

There is an allowance for post-startup assistance during initial operation.

Benefits and Challenges

With many stakeholders with varying priorities and interests, ongoing communication was critical. Multiple, frequent meetings, both regularly scheduled and to address specific issues, were necessary to assure proper project coordination. Scheduling these meetings could be challenging at times. However, considering the project involvement by two owners, three operating entities, several owner supplied equipment items, three consulting engineers (not counting Brown and Caldwell's coordination contract), and five construction contractors, coordination and cooperation was excellent.

All of the stakeholders participated in meetings at each site (EPL and JIWRF) to gain knowledge of each others systems. This assisted in a better understanding of each others needs and enabled cooperation. It is believed this was an important task for facilitating system startup.

A significant effort was spent on regular coordination meetings, agendas and minutes. Initially, all involved team members met bi-weekly to share information and to coordinate design related and construction activities. As construction neared completion, the meetings were adjusted to every four weeks. Meeting minutes and agendas were prepared by Brown and Caldwell, reviewed by the MMSD Project Manager, and issued in advance of the next meeting. The agenda was updated based on the previous meeting notes, new information, and also provided in advance of the next meeting. As startup neared, the engineer for the turbine facility organized weekly meetings to coordinate specific turbine startup related activities. Although a substantial effort, these meetings have resulted in outstanding and open communication which has resulted in positive relationships and a unified team approach towards successful project completion.

An early task of the management contract was to compile schedules for each component of the system into an overall startup system startup sequence schedule. This required obtaining good schedule information from several involved parties. The effort to check for dependencies, constraints and conflicts that could affect sequence and schedule provided confidence to the MMSD and the entire team that system startup would be successful.

Individual operation and maintenance manuals were developed for each system component, but there was not a single document for the entire system. Brown and Caldwell's contract included development of a combined, overall, system plan, based on the information in all of the other manuals. The purpose of the Overall Startup Sequence and Operation (OSSO) Plan for the Landfill Gas (LFG) system is to define the system, the procedure to startup and operation of the overall system. The OSSO plan provides LFG system owners and operations with procedures to follow during start up and ongoing operation of the overall LFG System including changes in operating modes, parameters for system shut downs, abnormal conditions and emergency operations.

A critical issues list was developed and maintained to help the team focus on those issues that were most likely to result in startup coordination problems or delays. The spreadsheet format used colors to reflect the status of each item: Red for unresolved and requiring attention, yellow for in progress by not completed, and green for addressed and completed. The list was organized by system designated as EPL for issues at the landfill, PIPE for the pipeline, MS for the Meter Station, TF for the turbine facility, and OSSO for the Overall Startup Sequence and Operations Plan. The list contains 30 items at the time of this writing. The critical issues list provides several examples of issues which were tracked and resolved through coordination and management effort.

Could discuss need for overall system facility description and operation manual.

Several challenges were addressed throughout the project. Changes needed to be made to the landfill gas moisture removal due to the use of the existing steel pipeline. This change created a delay in the completion of the EPL gas processing facility and impacted the date on which LFG would first be available at JIWRF. Coordination of the required training of the VWM staff created a need to carefully develop a training schedule that minimized the impact on the VWM staff while remaining within the budgets and specification training requirements for the equipment vendors.

Some initial startup on NG and LFG was performed in January and February of 2013. Several systems were able to be functionally tested. As is typical for major, complex systems, several issues have arisen:

- EPL
    - The delivery of the glycol system was delayed by the manufacturer
    - The glycol system experienced excessive outlet LFG temperatures during initial operation requiring evaluation and adjustment by the vendor
    - Excessive amounts of condensate collected at the inlet to the vacuum blowers requiring blower bearings to be replaced. One unit was off site for repair and the second unit was repaired on-site.
- Turbine Facility
    - A compressor on one of the fuel gas conditioning and compression skids was damaged during initial startup due to issues with the gas mixing valves, requiring replacement.
    - The LFG and NG control valves on the compressor skids needed to be replaced.
    - A regulator on the NG line was identified as needing to be added.

Ownership of a gas pipeline is new for the MMSD requiring continued communication and coordination with MMSD's contract pipeline operator as the project moves through start up and initial operation. Completion of Operational Readiness Testing (ORT) and scheduling of Functional Acceptance Testing (FAT) has provided confidence that all system components will work as intended

Project Status

The project is currently near completion of all construction elements. The first fire of the turbines on NG occurred on January 3, 2013 which initiates a timeline for air permit requirements and decommissioning of one of the two existing GE Turbines. Full scale operation is expected in the spring of 2013.

Conclusion

The MMSD has embarked on a significant project to sustainably make electricity using a renewable resource at the Jones Island Water Reclamation Plant. Outstanding communication, cooperation, and diligence by all parties has resulted in a major, complex facility being on track to result in a system achieving a high level of sustainable energy. With ongoing oversight and management by MMSD, and Brown and Caldwell's assistance, successful completion of startup and full scale operation is anticipated.

Integrated Net Zero for Energy, Water and Waste at Army Installations: Establishing Priorities in a Fiscally Constrained Context

By Elizabeth Keysar, PhD – Concurrent Technologies Corporation, Johnstown, PA

Session No. 13, 9:30 am

Aggressive goals for achieving "Net Zero" installations are being pursued by pilot site installations within the United States Army as part of an innovative initiative to go beyond regulations and executive orders in support of long-term sustainability. Only two of the 17 pilot sites are attempting to achieve net zero for all three – energy, water and waste – simultaneously. Under the Army's Net Zero Initiative, Army installations seek to: (1) produce as much energy on-site as they use annually; (2) limit the consumption of freshwater resources and return water back to the same watershed so as not to deplete the groundwater and surface water resources of that region in quantity or quality; and/or (3) reduce, reuse, and recover solid waste streams by converting them to resource values with zero solid waste to landfill. This presentation will summarize the challenges faced in developing an integrated Net Zero Roadmap for a large military installation, and how these challenges were addressed.

Multiple Courses of Action (COAs) have been identified to assist installations in achieving their Net Zero goals. In a constrained fiscal context, it is not possible to execute them all and those that have a positive Return on Investment (ROI) typically get first priority as they are easy to defend. Logically, planning for COAs at the integrated sites should be based on achieving overall installation Net Zero objectives, but balancing water, energy and waste is often a challenge. Multiple planning processes with many funding streams and numerous stakeholders complicate the prioritization process. Once the COAs have been selected, it is critical to effectively communicate these to a broad-base of stakeholders across the installation as a unified vision of sustainability. Decision support tools, therefore, must not only assist in prioritizing but must also easily communicate those decisions once made, as Net Zero COAs require support and behavior change throughout the organization for success.

Creating an integrated perspective was the first challenge. In any organization the size of an active Army training installation with over 1 million acres of training areas, 31 million square feet of building space, 34,000 soldiers and 4,000 families, siloes and functional fragmentation are unavoidable. A framework based on major activities was used, with each COA assigned to applicable Activity Groups. These groupings are based on real property codes, similar to land use categories, and easily recognizable across the installation. This framework should allow for improved communication as well as improved tracking of metrics, as it is also linked to the physical infrastructure.

The next challenge was to develop a simple, yet robust, evaluation framework. The evaluation criteria must be acceptable to the leadership and major decision-makers and also capture negative and positive impacts of COAs across the three domains. The major evaluation criteria tested were: Net Zero Impact, Cost Effectiveness, Probability of Success, and Mission Impact. Scoring along these dimensions was addressed using quantitative and qualitative techniques, as much of the needed data is not available.

Cost effectiveness was particularly challenging to score and include in the evaluation matrix. There are a wide range of potential metrics that can be used for evaluating costs and condensing these into one or two was necessary. The COAs themselves vary widely in first cost, life cycle cost and ROI - by orders of magnitude. For instance, large-scale renewable energy projects are difficult to compare to small-scale plumbing fixture replacements on a cost dimension.

The low cost of delivered water and wastewater treatment makes it difficult to justify projects based on ROI. The value of water thus becomes central to planning and decision-making in a context where water is cheap and there is minimal budget for conservation and reclaimed water projects. Energy and water security criteria were therefore important discriminators and evaluation criteria. Finally, timeline considerations were included in the analysis as a measure of Critical Path Dependency. The extent to which a given COA is a predecessor to others helps determine its priority. Some COAs may not at first appear to be relevant to Net Zero, but are critical to implementing others that are, such as ensuring the purple pipe infrastructure is in place to deliver reclaimed wastewater.

Conclusions will address next steps in the development of this approach for both the case study installation and the Army as it seeks to institutionalize Net Zero across the enterprise.

SANTA ROSA CASE STUDY – WATER, ELECTRIC AND GAS UTILITY COLLABORATION

Adam Ross, P.E., Principal Engineer, Brown and Caldwell

David Guhin, P.E., Utilities Director, City of Santa Rosa

Mike Prinz, P.E., Deputy Director – Subregional Operations, City of Santa Rosa

Brown and Caldwell

201 N Civic Dr, Suite 115

Walnut Creek, CA 94596

(925) 210-2535

ABSTRACT

The City of Santa Rosa's Laguna Subregional Water Reclamation Facility is a fascinating case study for water, electric and gas utility collaboration. Three projects are presented as examples of integrated planning and the water/energy nexus. At a time when wastewater treatment facilities are aiming for a net-zero energy footprint, Santa Rosa's effluent is enabling the production of 100 megawatts of power with the Geysers Recharge Project. It is a unique example of integrated planning and highlights the "other" side of the water/energy nexus: the water demand for power generation. Within the Water Reclamation Facility, a recent integrated energy audit provides mutual benefit for the electric utility and the City. Finally, where most studies only address the best on-site use of biogas, the Combined Heat and Power Project asks the bigger question: "Who – the facility or the utility – can better use biogas and natural gas?"

KEYWORDS

Combined heat and power, energy audit, energy efficiency, power generation, utility collaboration, renewable energy, geothermal

BACKGROUND

The City of Santa Rosa's role as a leader in advancing innovations in water recycling, residuals management and environmental stewardship dates back to its initial cogeneration efforts (1976), the beginnings of its now extensive reclamation system (1979), and the start of beneficial land application of biosolids (1991). The Geysers Recharge Project (2003), Integrated Energy Audit (2012) and Combined Heat and Power Project (2012) are tremendous contemporary examples of how the City has continued implementing its philosophy by partnering with water and energy utilities, providing water for power generation, assuring efficient wastewater treatment, and upgrading to an advanced self-generation facility for biogas utilization.

The Laguna Treatment Plant is a tertiary-level treatment facility that has an average daily dry weather flow of 17.5 million gallons per day (mgd). Nearly all of the water is reclaimed – in addition to the Geysers, the reclamation system comprises 6,130 acres and uses 46 pump stations to deliver reclaimed water to the buried pipe and above-ground irrigation systems. Nearly all of the biosolids are beneficially reused via land application or Class A compost.

GEYSERS RECHARGE PROJECT

To date, Santa Rosa has successfully delivered 35 billion gallons of Title 22 tertiary recycled water to the Geysers steamfields. The Geysers steamfields are a rare geothermal occurrence – and site of the largest group of geothermal power plants in the world – in which natural steam is produced when underground water comes into contact with rock formations that have been heated by underlying magma (or molten rock). Because the magma in The Geysers area is relatively close to the Earth's surface, the steam escapes from the ground in the form of hot springs and fumaroles. When the steam reaches the surface in production wells that have been drilled by energy companies, it travels through insulated pipelines to a generator unit where it spins turbines to create electricity. Only 20 percent of the expended steam can be cost-effectively condensed and returned to the subsurface; therefore, the resource is slowly being depleted. Water from Santa Rosa's Geysers Recharge Project not only recharges this reservoir, but enables the production of even more steam, thus extending the life of the steamfields and allowing the generation of electricity to remain at its current level for the foreseeable future. The 12.6 mgd of recycled water sent to the steamfields through the Geysers Recharge Project boosts the operation's electrical output by 100 megawatts. Few, if any, wastewater treatment plants can claim that much energy production.

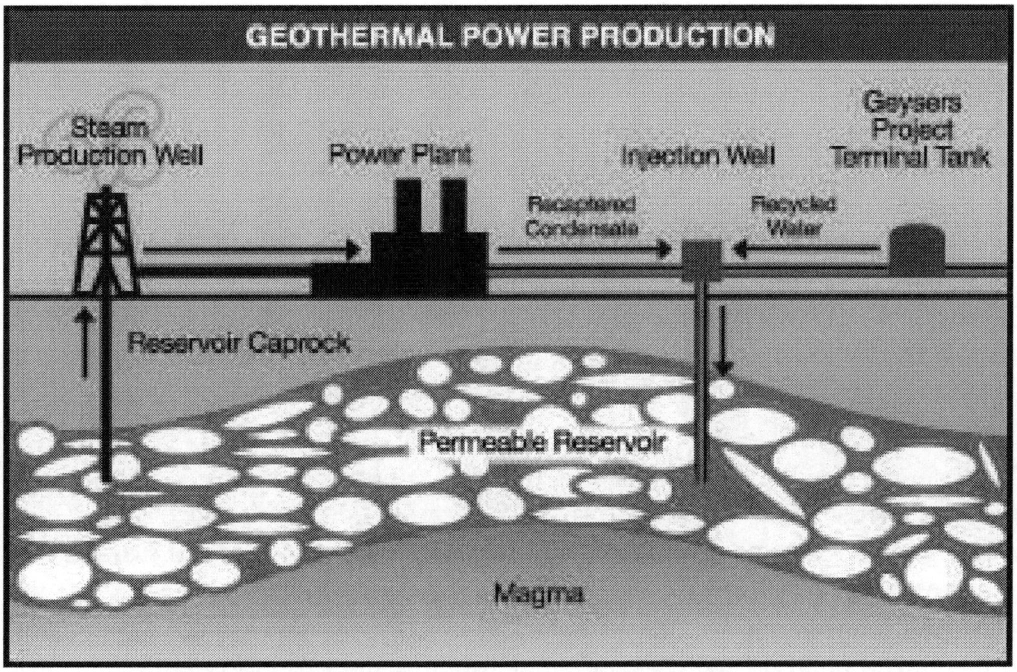

Figure 1. Geothermal Power Production (© City of Santa Rosa)

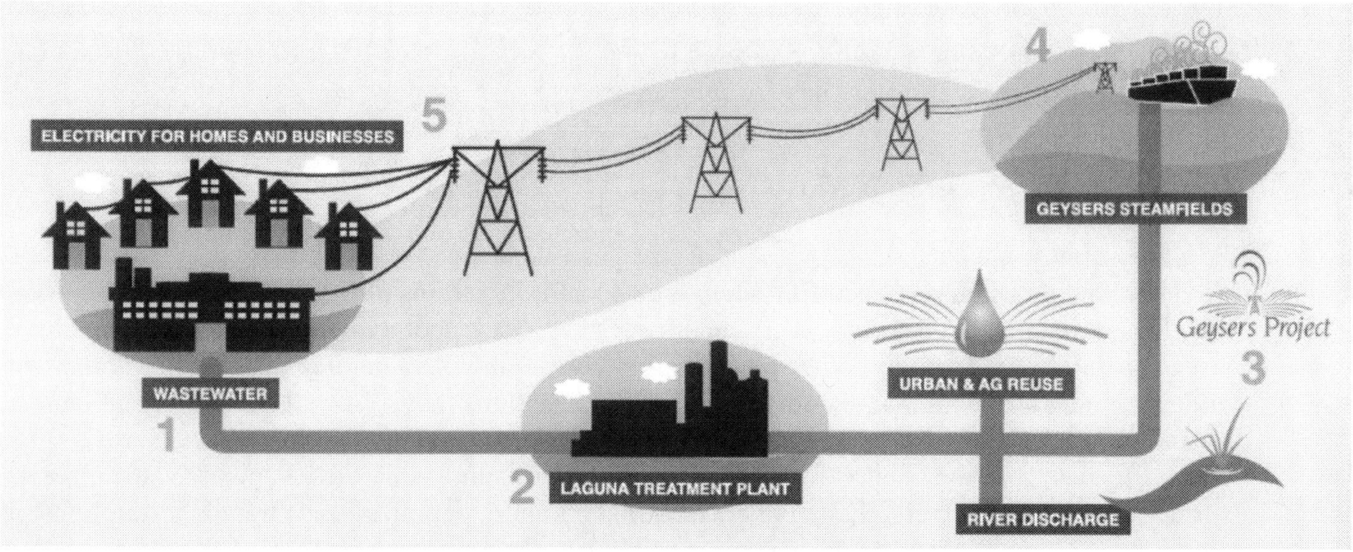

Figure 2. Water-Energy Cycle (© City of Santa Rosa)

The award-winning Geysers Recharge Project is a benefit to the Santa Rosa Laguna Subregional Water Reclamation Facility as well as Calpine, the steamfields' power plant owners. The Geysers Project provides a weather-independent effluent reuse alternative and keeps discharge out of the Laguna de Santa Rosa. The recycled water satisfies Calpine's water demand for power generation. The project is a high point on the water/energy nexus.

INTEGRATED ENERGY AUDIT

The Integrated Energy Audit was the result of a service of the Pacific Gas and Electric Company (PG&E) to its large commercial and industrial customers under its Large Integrated Audit Program (LIA). PG&E offers the LIA to its large customers to educate them about Integrated Demand Side Management (IDSM) as described in the California State Strategic Energy Plan. The California State Strategic Energy Plan espouses implementation of a mix of energy conservation, energy efficiency, demand response and self-generation (including renewables and cogeneration) to reduce energy consumption in California. The State of California believes IDSM will reduce per capita energy consumption, save California residents and businesses money, and reduce greenhouse gas (GHG) emissions within the state.

The Audit summarized the annual electrical and gas energy consumption at the Laguna Treatment Plant and recommended cost-saving measures. Six Energy Efficiency Opportunities (EEOs) were presented in the report. The City is moving forward with implementation of these EEOs, which will result in the following savings:

- Electrical energy savings of 363,118 kWh per year, which represents 1.1% of the facility's electrical energy consumption (1.7% of electrical energy procured from PG&E)
- Peak demand savings of 38.0 kW
- Potential cost saving of $32,277 per year, which represents 1.4% of the facility's total annual energy costs

- Total potential incentives and rebates of $27,688
- Total installed cost with incentives and rebates of $140,379
- Overall simple payback period with incentives and rebates of 4.3 years

Table 1 shows the energy-efficiency opportunities and associated benefits and costs.

Table 1. Energy Efficiency Opportunities (from BASE Energy, Inc., Integrated Audit)

Measure Number	Measure Description	Energy, Cost and GHG* Savings					Project Costs, Incentives, and Payback			
		Peak Savings (kW) **	Electricity (kWh/yr)	Natural Gas (Therms /yr)	Annual Cost Savings ($/yr)	CO2 Saved (Tons /yr)	Estimated Installed Cost ($)	Potential PG&E Incentive ($)	Net Measure Cost ($)	Pay-back Period (Yrs)
EEO-1	Optimize Control of Filter Influent Pumps to Increase Pumping System Efficiency	0.0	48,810	0	$4,218	14.0	$3,600	$4,393	-$793	0.0
EEO-2	Widen Deadband Between Cooling and Heating Setpoint Temperatures and Setback Zone Temperatures During Unoccupied Hours for Compost Facility Offices	0.0	12,558	0	$943	3.6	$0	$0	$0	0.0
EEO-3	Install Automatic Lighting Controls	0.0	59,112	0	$5,149	17.0	$10,903	$2,931	$7,972	1.5
EEO-4	Install a More Efficient VFD Air Compressor	5.2	45,534	0	$3,933	13.1	$22,031	$4,618	$17,413	4.4
EEO-5	Install High Efficiency Fluorescent Lighting in Various Areas	31.7	187,336	0	$17,191	53.9	$124,711	$14,757	$109,954	6.4
EEO-6	Install a More Efficient Air Dryer	1.1	9,768	0	$843	2.8	$6,822	$989	$5,833	6.9
Recommended Energy Efficiency Measure Totals		**38.0**	**363,118**	**0**	**$32,277**	**104.4**	**$168,067**	**$27,688**	**$140,379**	**4.3**

As the potential energy savings from the EEOs are small (only 1% of overall consumption), the primary conclusion of the study was confirmation that the Laguna Treatment Plant is operating extremely efficiently, which paved the way for future energy projects. The audit is an example of utility collaboration benefitting the facility (cost savings) as well as PG&E, the electric utility. The audit was paid for by PG&E.

COMBINED HEAT AND POWER PROJECT

The Combined Heat and Power (CHP) Project replaced the City's existing cogeneration facility with an advanced, efficient biogas utilization facility. Project predesign included evaluation of technologies, including engines, fuel cells, turbines and biomethane pipeline injection. The final project included 4.4 megawatts (MW) of generation capacity – four 1.1 MW engines – as well as a digester gas treatment system (hydrogen sulfide, moisture and siloxane removal), gas blending system (digester gas and natural gas), and a new building. The CHP facility will use all of the Laguna Treatment Plant's digester gas (approximately 950 kW of electricity, or one engine at nearly full load). The other engines provide redundancy and standby generation capacity. The CHP facility is dually permitted for normal operation using digester gas and for emergency operation using natural gas. This functionality saves the City the expense of renting seasonal standby power capacity.

The CHP project qualified for the California Public Utilities Commission's Self-Generation Incentive Program (SGIP). The SGIP provides financial incentives for installing new, qualifying self-generation equipment installed to meet all or a portion of the electric energy needs of a facility. SGIP is administered by the utility providers. The project meets the efficiency and heat recovery requirements for "conventional CHP" and receives additional incentive for using biogas – considered an emerging technology. In all, the CHP Project will receive $2.375 million.

Figure 3. Combined Heat and Power Internal Combustion Engine, 1100 kW

Figure 4. Engine installation mid-construction

Figure 5. Digester Gas Conditioning Equipment

Figure 6. CHP Building (shown during construction)

CHP PREDESIGN RESULTS AND CONCLUSIONS

The predesign process yielded interesting results. In addition to evaluating on-site generation options, the team also looked at "directed-use" options (e.g., biomethane pipeline injection). Although fuel cells and engines have high electrical efficiency – in the 40% range – this pales in comparison to the efficiency of a utility-scale natural gas power plant. A combined-cycle power plant can achieve greater than 60% electrical efficiency. In this context, the most efficient use of biogas is to deliver it to the utility (after satisfying on-site heat needs with a boiler). This isn't necessarily the most economical option for the facility, nor is the practice accepted by all utilities. The gas utility, PG&E, will accept biomethane pipeline injection from manure digesters but not municipal digesters.

The overall ranking of alternatives is shown below. The rankings are based on quantitative and qualitative criteria, and City-assigned scores and weighting factors. Criteria included:

- Initial cost
- 20-year present worth
- Reliability and maintainability of equipment
- Proven technology, availability of equipment
- Ability to meet plant's heat needs
- Minimal gas treatment requirements
- Operational flexibility (turndown)
- Greenhouse gas emissions / global carbon footprint
- Air quality impact, ease of permitting
- Constructability
- Versatility and capacity for standby power

Rankings are as follows (with greatest benefits in parenthesis):

1. New, advanced internal combustion engines in a new building. (Constructability, Versatility)
2. Existing engines in a new building. (Proven technology)
3. Biomethane pipeline injection – scrub, compress and sell digester gas methane to gas utility. (GHG emissions)
4. Fuel cells. (Air quality, Initial cost)

NEXT STEPS

Santa Rosa's next steps include a Strategic Biosolids Management Plan to optimize biosolids management, increase energy production, expand digestion capacity, reduce cost, reduce GHGs and enhance flexibility of future disposal options. The City is interested in exploiting the energy potential contained in the biosolids and assuring a diversified portfolio of solid residual management options in the future.

Additionally, the City is evaluating the economic and environmental feasibility of operating one or two of the CHP facility's four engines on natural gas by adding a selective catalyst reduction system. The current low price of natural gas, compared to the value of electricity, makes the project financially attractive.

REFERENCES

City of Santa Rosa, Geysers Project Webpage (2013)
http://ci.santa-rosa.ca.us/departments/utilities/irwp/geysers/Pages/default.aspx

Calpine, Geysers Project Webpage (2012)
http://www.geysers.com/

Integrated Energy Audit for City of Santa Rosa, BASE Energy, Inc., March 2012.

Integrating Liquid and Solid Waste Planning to Make Green Infrastructure Pay

Stephen J. Salter, PEng, LEED AP®, Farallon Consultants Limited
Victoria, British Columbia, Canada
Email: *farallon@telus.net*

ABSTRACT

This paper describes the results of an Integrated Resource Recovery study in the Canadian community of Comox Valley, British Columbia. Separate departments of the regional government had planned to double the size of the existing biosolids composting facility, to build two new landfills, and also to build a new composting facility for source-separated organic waste. The study found that if the community's needs for wastewater and solid waste infrastructure are reviewed in an integrated manner, the results for the environment and for taxpayers can be better than the business-as-usual approach. The study predicted significant benefits from the construction of a new anaerobic digestion facility for the digestion of biosolids and source-separated organic waste, and co-treatment and upgrading of captured landfill gas and biogas from anaerobic digestion into pipeline-quality biomethane for use in regional transit buses and for sale to a utility. This integrated approach would result in an estimated $2.9 million per year of increased value to the community versus the business-as-usual approach. The integrated option would also reduce greenhouse gas emissions by an estimated 23,000 tonnes per year, and result in eight fulltime jobs.

KEY WORDS

Integrated resource recovery, zero net energy facilities, industrial ecology, greenhouse gas reduction, waste-to-energy, solid waste diversion, anaerobic digestion, biomethane, biosolids, wastewater treatment.

INTRODUCTION

The Comox Valley Regional District (CVRD) in British Columbia (BC), Canada manages waste infrastructure for a population of approximately 100,000 people, including residents of the City of Courtenay and the Town of Comox. The CVRD is interested in building infrastructure that supports the sustainability objectives in its *Regional Growth Strategy* and *Comox Valley Sustainability Strategy*. To that end, in 2012 the CVRD commissioned Farallon Consultants Limited to complete an Integrated Resource Recovery study to examine the region's sources of waste, potential markets for recovered resources, and the costs and benefits of resource recovery options. This paper presents a summary of the results of the study, the goal of which was to find practical ways to reduce greenhouse gas emissions, reduce the lifecycle cost of water and waste management infrastructure, provide sustainable employment, and provide non-tax revenues for local governments.

METHODOLOGY

Community Interests and Initiatives

A local researcher, Jennifer Manuel, was hired to collect information from local governments, industry, agriculture, aquaculture, and individual building owners to develop an inventory of waste resources and of resources consumed, and raw materials. Options for resource recovery were identified, and the most promising options were analyzed in detail to estimate their capital costs, annual costs, and annual revenues.

Figure 1. Area of Study

The following community interests and opportunities were uncovered early in the study:

1. The CVRD plans to increase its diversion of solid waste from the current rate of 48% to a target rate of 90%.
2. The CVRD completed a diligent public consultation exercise in 2012, as part of the process of developing the Solid Waste Management Plan (SWMP), and the results showed that the public supports waste diversion efforts.
3. BC Ministry of Agriculture personnel reported that soils in the Comox Valley will benefit from the addition of carbon and nutrients in the form of compost or digestate.
4. The Comox Valley agricultural community has expressed interest in using fertilizer produced by natural means, to displace fertilizer from mineral and manufactured sources.
5. Abattoir and dairy waste is available in the region.
6. Approximately 1,200 Bone Dry Tonnes (BDT) or 5,000 wet tonnes/year of biosolids from the Comox Valley Water Pollution Control Centre (CVWPCC) are currently composted with approximately 5,000 green tonnes/year of wood residues to produce a compost marketed under the name SkyRocket.
7. The CVRD buys wood residues for the biosolids composting operation, but finds it challenging to obtain enough residues to meet their annual requirement.
8. The budget for the biosolids composting operation is approximately $500,000 per year, and sales of SkyRocket compost generate revenues of $60,000 per year.
9. The CVRD plans to expand the biosolids composting operation at an estimated capital cost of $3.5 million.
10. The region generates 123,746 tonnes per year of solid waste, and while one regional landfill has reached capacity, the CVRD anticipates that the second landfill (the Comox Valley Waste Management Centre or CVWMC) will reach capacity by 2018.

11. Planning is underway to replace the first landfill at an estimated capital cost of $19.8 million, and to expand the Comox Valley Waste Management Centre at an estimated capital cost of $15.5 million[1].

12. The biosolids composting operation is adjacent to the existing landfill.

13. The CVRD estimates that leachate treatment for one landfill expansion will cost $7.5 million in capital, and that operating costs for the leachate treatment facility will be approximately $200,000 per year.

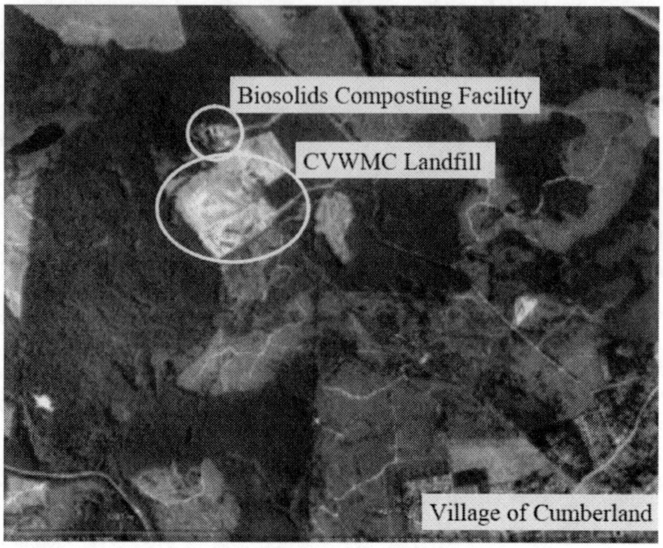

Figure 2. CVWMC Landfill and Biosolids Composting Sites

14. The British Columbia *Landfill Gas Management Regulation* requires that larger landfills be provided with a Landfill Gas (LFG) capture system, which the CVRD estimates will cost approximately $2.5 million per landfill.

15. The efficiency of Landfill Gas Capture systems to date has fallen short of the Provincial target of 75%. For example, a 2009 study estimated the efficiency of the Capital Regional District's Hartland Landfill (located near Victoria, BC) LFG capture system at 38.8%.[2]

16. The Comox Valley Transit buses burn approximately 400,000 litres of diesel fuel per year. Transit buses equipped to burn Compressed Natural Gas (CNG) or biomethane are in use in other British Columbia communities, including Vancouver.

17. The CVRD is considering relocating its bus depot: a new depot would include new fuel storage infrastructure.

The experience of other local governments illustrates that achieving high solid waste diversion rates requires transformative rather than incremental changes.

The table below compares the current and planned rates of diversion of solid waste by local governments in Canada and the United States as of 2006.[3]

Table 1 Solid Waste Diversion Rates in Canada and the USA

Organization	Current Rate	Target Rate
San Francisco, California	72%	100%
Regional District of Nanaimo, BC	68%	75%
Oakland, California	65%	90%
Halifax, Nova Scotia	60%	88%
Comox Valley Regional District, BC	48%	90%

Three scenarios to address the CVRD's interests and opportunities were compared in the study:

1. Waste-to-energy in which mixed municipal solid waste would be burned;
2. Composting of source-separated organics; and
3. Anaerobic digestion of both biosolids and source-separated organics to produce compost.

RESULTS

Scenario One: Waste-to-Energy

In one option, the CVRD would build a waste-to-energy (WTE) facility to burn 52,454 tonnes of mixed municipal solid waste per year by 2017, with an additional 5,563 tonnes per year deposited in a landfill (AECOM, 2011). Solid waste would not be imported from other communities such as Vancouver, since it is expected that these communities will develop their own capacity to recover resources from waste. The biosolids composting facility would undergo its planned expansion independently.

Advantages of WTE for the CVRD

1. Source separation would not be required.
2. The expansion of the existing landfill and construction of a new landfill in Campbell River could be deferred.
3. Emissions of methane from decomposing waste in the new landfills would be reduced (although this gain would be offset by higher greenhouse gas emissions from incineration).

Challenges of WTE for the CVRD

1. The estimated cost of $143 per tonne over forty years for an incineration in the Comox Valley (AECOM, 2011) is higher than the CVRD's current receiving fees of $80 per tonne for general refuse, and $65 per tonne for wood waste.
2. Although the efficiency of energy recovery is helped by the combustion of plastics in garbage, efficiency is reduced by the high water content of food waste.
3. Burning of plastic, which is derived from oil, produces significant greenhouse gas emissions.
4. Dioxins are formed in the WTE combustion processes, because of the presence of chlorine (from PVC plastic for example) in mixed municipal solid waste.[4]
5. NOX emissions from WTE are significant: NOX is formed when plastics and food waste are burned. N_2O is a strong greenhouse gas with a Global Warming Potential of 310, meaning that one tonne of N_2O makes the same contribution to global warming as 310 tonnes of CO_2.
6. Nutrients embodied in organic waste would not be returned to soil.
7. Ash from municipal solid waste incinerators can be contaminated, and would require additional processing before landfilling.
8. Incineration produces energy in less flexible forms than other alternatives: incineration can produce electricity and waste heat, but not biofuels for example.
9. Concerning planning for WTE facilities, the BC Ministry of Environment requires the following of local governments:

    The Ministry expects local governments to have a minimum target of 70% reduction of waste before utilizing a WTE facility as a waste management option. The 70% target is calculated only from Reduce, Reuse, and Recycle initiatives.

    The Ministry expects that resource recovery facilities (4th R) will obtain at least 60% of the potential energy from the MSW used as fuel.

    If a WTE facility does not achieve 60% energy efficiency, the Ministry will consider that WTE facility as a residual management facility (5th R).[5]

Scenario Two: Composting of Food Waste

In the second scenario, the CVRD would invest in source separation of organic waste as a means of achieving its diversion targets. 10,962 tonnes per year of food waste would go to an aerobic composting facility that would operate separately from the existing biosolids composting facility; the biosolids facility would undergo its planned expansion. The configuration of an aerobic composting facility is described in the *Regional Compost Facility Study* received by the CVRD in 2011.[6] The estimated capital cost of a regional composting facility was $19,831,000. The table below summarizes the estimated annual revenues and costs for this facility.

Table 2. Summary of Estimated Annual Revenues and Costs, Composting Facility

	Quantity	Units	Rate	Units	Value	Notes
Annual Revenues						
Receiving Fees for Food Waste Diverted from Landfill	10,962	tonnes/yr	$65.00	$/tonne	$713,000	1.
Sales of Compost, Dry Basis	4,778	BDT/year	$66.67	$/BDT	$319,000	2.
Total Estimated Annual Revenues					**$1,032,000**	
Estimated Annual Costs: Composting Facility					$750,800	3.
Estimated Annual Costs: Financing					$947,000	4.
Estimated Additional Collection Costs					$338,000	5.
Total Estimated Annual Costs					**$2,035,800**	
Estimated Net Annual Revenues					**($1,003,800)**	

Notes

1. The estimated receiving fees of $65 per tonne only includes an estimated 10,962 tonnes per year to be disposed of in landfills (Walker, 2011).
2. Based on the equivalent price of $30/green tonne for compost at 45% moisture.
3. Per Walker, 2011.
4. Based on a capital cost of $19,831,000 and an MFA interest rate of 2.5% over 30 years.[7]
5. Per Walker, 2011.

Advantages of Composting for the CVRD

1. Composting supports the CVRD's SWMP objectives to divert organic waste away from landfills and toward beneficial uses.
2. Compost would support agriculture in the Comox Valley. Nutrients embodied in organic waste would be returned to soil, and the residual carbonaceous materials in the finished compost would provide structure to soil and a food source for beneficial soil micro organisms. The nutrients in compost are also released more slowly than mineral and artificial fertilizers, which reduces pollution of surface waters. Finally, compost can be a less expensive form of fertilizer for farms.
3. Emissions of methane from decomposing waste in the new or expanded landfills would be reduced.
4. Organic material is stabilized, so that finished material from composting is less likely to attract vectors.
5. Because of somewhat lower volumes of waste received, the ongoing operating costs of new or expanded landfills could be reduced.
6. Composting is familiar to Canadian local governments.

Challenges of Composting for the CVRD

1. Organic materials (food waste, yard and garden waste, and so on) will need to be separated at the source by businesses and homeowners. Successful source separation programs have been developed elsewhere however, especially in the Regional District of Nanaimo, Halifax, San Francisco, Scandinavia, and Germany.
2. Since air does not reach every particle of organic material during composting, the process produces modest amounts of methane. In addition, composting operations give rise to greenhouse gas emissions from fuel combustion in mobile equipment. The greenhouse gas emissions from composting are estimated later in this paper.
3. Composting generates odours, that can be managed if an in-vessel process is used; air leaving the closed vessels can be passed through a biofilter. In other options such as open windrow composting, containing odours is not practical, which presents a risk that an investment of public funds in a composting operation could result in additional costs. The City of Gothenburg, Sweden, for example, discontinued its composting operations in 2005 in response to public odour complaints, and now diverts its food waste to anaerobic digestion.[8]
4. Composting does not provide an opportunity to recover the energy embodied in organic waste in the form of biofuels.

Scenario Three: Integrated Anaerobic Digestion

Anaerobic digestion is a form of composting that preserves the inorganic nutrients (compounds of nitrogen, phosphorous, sulphur as well as micronutrients) while converting the majority of carbon in organic waste to methane and carbon dioxide. Like aerobic composting, the residual material is stable and provides an excellent form of slow-release fertilizer. Unlike aerobic composting, anaerobic digestion provides biogas (approximately 2/3 methane, and 1/3 carbon dioxide) that can be upgraded to replace fossil fuels in vehicles, replace natural gas for heating, burned to produce electricity and heat, or converted to liquid fuels.

During anaerobic digestion, bacteria decompose organic matter in the absence of oxygen, typically over a period of twenty days. A large number of digester configurations are available, but digesters are grouped according to their operating temperature: mesophilic digesters operate at 37°C, while thermophilic digesters operate at 55°C.

Anaerobic digestion is a natural process that was described in 1808 by Sir Humphry Davy, who discovered that "marsh gas" contained methane. The first anaerobic digester was built in 1895, and anaerobic digestion is used in North America to generate biogas from manure on farms and from biosolids in wastewater treatment plants. In addition, anaerobic digestion is commonly used to generate biogas and compost from food waste in Europe and elsewhere: over three thousand anaerobic digesters operate in Europe while an estimated two million digesters operate in India.

Figure 3. Community Biogas Plant, Sweden[9]

Interest in anaerobic digestion of food waste in Europe has been supported by the shortage of land for landfills, the high cost of building new landfills, bans on organic waste deposits to landfills, carbon taxes, and objectives for greenhouse gas reductions at the national, regional, and municipal levels. Many of these conditions now exist in British Columbia: Provincial regulations governing the design and construction of new landfills include environmental safeguards that have increased their cost, a carbon tax has increased the economic value of biofuels, and objectives for reducing greenhouse gas emissions are in place at the Provincial, regional, and municipal levels. Local governments in North America are also becoming more interested in anaerobic digestion as a means of deriving greater value from food waste. For example, the East Bay Municipal District integrates anaerobic digestion of food waste and biosolids in its Oakland wastewater treatment facility.

The section that follows gives an indication of the costs and benefits of an anaerobic digestion facility based on the CVRD's circumstances.

1. Food waste from homes and businesses could be collected together with yard and garden waste with existing collection methods, and with the addition of kitchen pails and green carts of 120 to 360 litre capacity. The incremental cost of collecting food waste with yard and garden waste has been estimated at $338,000 per year (Walker, 2011) and this figure has been added to the estimated annual costs of the composting option and of the integrated anaerobic digestion option.

2. Diversion of food waste, leaf and yard waste, and soiled paper (e.g. food wrappers, paper plates, and so on) could be accelerated in the CVRD, effectively limiting the amount of organic waste deposited in landfills. If the regulatory trigger threshold of 1,000 tonnes of methane per year is avoided in this way, then new LFG capture systems would not be required, and the CVRD could save the $2.5 million budgeted for each new LFG system.

3. Source-separation of solid waste could be encouraged by charging individual homeowners and businesses according to the weight of waste disposed, a practice used in Germany. Diligent homeowners and businesses would pay less by separating recyclables from waste more carefully. This approach is analogous to charging users of municipal water systems for consumption, which is a basic tool to steer behaviour toward conservation.

4. Diverted organic waste would be processed in an Integrated Anaerobic Digestion Facility (IADF) which could be built on the site of the existing biosolids composting facility, and which use existing material transportation and handling infrastructure. By building the facility on this site, the CVRD could avoid the challenges of developing and permitting a Greenfield site. In addition, the CVRD could save the $3.5 million budgeted for expansion of the biosolids composting facility.

5. The IADF would incorporate:

    a. A receiving hall operating under negative pressure, with air vented to a biofilter to remove odours associated with deliveries of organic waste;

    b. Pulping and maceration processes to finely divide organic waste;

    c. Mixing of macerated waste with water to form a slurry;

    d. Pre-processing to remove contaminants such as plastic, metal, and grit;

    e. Heating of the slurry to digester temperatures with heat provided from a spiral heat exchanger recovering heat from digestate exiting the digester, and with energy recovered from biogas upgrading compressors;

    f. Mesophilic anaerobic digestion in two separate digestion processes: one for food and garden waste, and a separate process for biosolids;

    g. Digestate dewatering, drying, and storage of dry digestate/compost;

    h. Collection and upgrading of biogas to biomethane; and

    i. On-site wastewater treatment.

6. Raw biogas from both digesters, and also from the LFG capture system to be installed in the existing landfill, would be upgraded to pipeline quality biomethane. Waste energy from the upgrading plant's compressors would be used to provide heat to the digesters.

7. The new transit depot could be built within the industrial-zoned lands near the existing landfill and biosolids composting facility, and the new IADF. Biomethane produced by the upgrading plant would contain approximately 97% methane, and could therefore be burned in vehicles equipped to burn natural gas. Upgraded biomethane would be piped to the new transit bus depot, where buses could be refilled overnight. If the transit depot is not built near the existing Comox Valley landfill and biosolids composting facility, biomethane could still be delivered to transit buses through the natural gas utility distribution system.

8. The balance of the upgraded biomethane could be sold to the natural gas utility, at a premium over the wholesale price of fossil natural gas. Demand exists for this greenhouse gas-neutral fuel, which is marketed to consumers as Renewable Natural Gas.

Figure 4. Biomethane-Powered Bus, Stockholm

9. Organic material that has passed through an anaerobic digester is known as digestate. As with compost, nutrients from food waste are concentrated in the digestate. Unlike compost, digestate is richer in nitrogen compounds that are retained in the liquid fraction. Digestate typically contains 85% to 95% water; if it is deposited directly on farmland, the soil receives the benefit of this higher nitrogen concentration. Since it is not currently known how many farms could receive whole digestate, estimates of capital and operating costs in this study are based on the assumption that digestate would be dried and stored for sale in the Spring.

10. Feedstocks for the food digester would include the following:

    a. Food waste from the region, estimated on the basis of the amount of waste *generated,* as opposed to the amount of waste currently *disposed*;

    b. Soiled paper associated with food waste;

    c. Dairy whey;

    d. Grease trap materials currently received by the CVWPCC; and

    e. Yard and garden waste (leaves and grass).

11. Since biosolids have a low carbon to nitrogen (C:N) ratio, approximately half of the soiled paper stream would be mixed with biosolids to bring the ratio to between 25 and 30 in the biosolids digester.

12. Wood residues that are currently used to balance the C:N ratio in the biosolids compost operation could be diverted to a biomass boiler for district energy.

13. The CVRD could investigate potential capital and operating savings from using a single facility to treat wastewater from the IADF and also landfill leachate.

14. During the conceptual design phase of the IADF, CVRD wastewater treatment personnel could study the economics of:

   a. Transporting primary and secondary biosolids from the CVWPCC to the digestion facility separately, to allow secondary Waste Activated Sludge to undergo pre-treatment by cell lysation (for example through electrolysis or pressure gradients) in order to increase biogas production from this source.

   b. Including additional fertilizer recovery from the biosolids digester supernatant in the form of struvite (magnesium ammonium phosphate or $MgNH_3PO_4 \cdot 6H_2O$).

The table below lists the proposed input materials to the food waste digester. Composition values were chosen from a 2006 Metro Vancouver study, since this region's waste diversion rate was then comparable to the current rate of the CVRD.[10] These rates have been applied to the quantity of waste *generated* in the region for food waste and soiled paper as of 2012. This approach assumes that food waste would be diverted from material that is currently disposed of (estimated at approximately 11,000 tonnes per year in Walker, 2011), and that an additional 8,000 tonnes/year of food waste would be diverted from other fates, for a total of approximately 19,000 tonnes per year.

Table 3. Substrates for the Food Waste Digester (A)

Substrate	**% of Generated Waste**	**% Diverted to AD**	**Wet Mass** (tonnes/yr)	**Moisture Content**	**Dry Mass** (BDT/year)
Food Waste	21.83%	70%	18,910	70%	5,673
Food Factories		100%	325	60%	130
Dairy Whey		100%	400	93%	28
Soiled Paper	14.94%	42%	7,765	30%	5,435
Grease Trap Material		100%	290	65%	102
Totals			**27,690**		**11,368**

Note that the available supply of soiled paper would be divided between the food digester and the biosolids digester, in order to balance the C:N ratios of the two processes. The table below shows the proposed input materials to the biosolids digester.

Table 4. Substrates for the Biosolids and Paper Digester (B)

Substrate	**% of Generated Waste**	**% Diverted to AD**	**Wet Mass** (tonnes/yr)	**Moisture Content**	**Dry Mass** (BDT/year)
Biosolids	n/a	100%	5,114	75%	1,279
Soiled Paper	14.94%	42%	7,765	30%	5,435
			12,879		6,714

The IADF would include a biogas scrubber and upgrading process to remove sulphur compounds (trace H_2S and SO_2) from the digester gases, and in addition siloxanes from landfill gas. The table below estimates the flow rate of raw landfill gas from the existing landfill, as a means of finding the required capacity for the upgrading equipment. The CVRD estimates that 1,941 tonnes of methane are generated per year from the existing CVWMC landfill, and if 50% of this gas is captured, then 1,482,000 m^3/year would be available. The table below summarizes the estimated quantities of biomethane from the three sources.

Table 5. Summary of Biomethane Sources

Source	**Input Quantity** (tonnes/yr)	**Residuals Quantity** (BDT/year)	**Upgraded Biomethane** (m^3/year)	**Energy Content** (GJ/year)	**Energy Contribution**
Food Digester A	27,689	6,739	2,301,000	87,700	45%
Biosolids Digester B	12,879	3,316	1,302,000	49,600	26%
Landfill Gas Capture	n/a	n/a	1,482,000	56,300	29%
Totals	**40,568**	**10,055**	**5,085,000**	**193,600**	**100%**

The capital cost of the IADF is estimated to be $24,997,000. Cost estimates in this study are based on manufacturers' quotations prepared for installations of comparable size on Vancouver Island, and include a 25% contingency.

The CVRD could also consider that in the integrated anaerobic digestion scenario, two future expenditures would be avoided: $2.5 million per LFG capture system for new landfills, and $3.5 million for the expansion of the biosolids composting facility. The net effect of these savings would be to reduce the cost of the IADF to approximately $19 million. The table below summarizes the estimated annual revenues and costs for the IADF.

Table 6. Summary of Estimated Annual Revenues and Costs, IADF

	Quantity	Units	Rate	Units	Value	Notes
Annual Revenues						
Receiving Fees:						
Food Waste Diverted from Landfills	10,962	tonnes/yr	$65.00	$/tonne	$713,000	1.
Soiled Paper Diverted from Landfills	15,530	tonnes/yr	$80.00	$/tonne	$1,242,000	
Food Waste Diverted from Other Fates	9,015	tonnes/yr	$-	$/tonne	$-	
Sales of Biomethane to Comox Valley Transit	15,500	GJ/year	$25.81	$/GJ	$400,000	2.
Sales of Balance of Biomethane	178,100	GJ/year	$7.00	$/GJ	$1,247,000	3.
Sales of Compost, Dry Basis	10,050	BDT/year	$66.67	$/BDT	$670,000	4.
Total Estimated Annual Revenues					**$4,272,000**	
Estimated Annual Costs: Operations and Maintenance					$1,257,000	5.
Estimated Annual Costs: Financing					$1,194,000	6.
Estimated Additional Collection Costs					$338,000	7.
Estimated Annual Savings					($400,000)	8.
Total Estimated Net Annual Costs					**$2,389,000**	
Estimated Net Annual Revenues					**$1,883,000**	9.

Notes

1. The CVRD's receiving fee of $65 per tonne has been applied to food waste of 10,962 tonnes per year that is currently disposed of in landfills. The CVRD's receiving fee of $80 per tonne has been applied to soiled paper. Modelling in this study assumes that receiving fees would not be charged for the additional quantity of approximately 9,000 tonnes per year of organic material that is currently generated but diverted from the landfill. It is also assumed that receiving fees would not be charged for biosolids.

2. It is recognized that Comox Valley Transit buses would need to be replaced with buses capable of burning Compressed Natural Gas in order to realize this economic benefit. The estimate here is based on the consumption of 400,000 litres/year in 2011, a value of $1.00 per litre, and an energy content of diesel fuel of 38.68 MJ/litre to calculate equivalency to biomethane.
3. The balance of the upgraded biomethane would be sold to the gas utility.
4. Based on the equivalent price of $30/green tonne for compost at 45% moisture, which is the average price of SkyRocket biosolids compost sold by the CVRD. This price is conservative, since the nutrient value of compost from anaerobic digestion of food waste and biosolids could justify a higher price.
5. Includes consumables, collection, utilities, operations, and maintenance.
6. Based on the MFA interest rate of 2.5% over 30 years. The cost of a wastewater treatment process at $1.8 million has been included in the capital cost estimate for the IADF.
7. Per Walker, 2011.
8. These savings would result from incorporating the functions currently provided by the biosolids composting facility into the IADF.
9. Cost estimates do not account for the potential to sell carbon offsets for greenhouse gas reductions over and above those required by Provincial regulations. Cost estimates do not account for the economic value of diverting waste from landfills, which could defer the capital expenditure required to develop landfill expansions.

Advantages of Anaerobic Digestion for the CVRD

1. Integrated anaerobic digestion supports the CVRD's draft *Solid Waste Management Plan* objectives to divert organic waste away from landfills, and toward beneficial uses.
2. The anaerobic digestion scenario modelled in this study has the potential to provide net revenues to the CVRD of an estimated $1.9 million per year, versus the estimated net loss of an estimated $1 million per year for conventional composting. The estimated difference in economic value between these two options is an estimated $2.9 million per year.
3. Anaerobic digestion would avoid the estimated $3.5 million cost of expanding the CVRD's existing biosolids composting facility.
4. If diversion of organic waste from new landfills is implemented adequately, methane emissions from these landfills could fall below the Provincial trigger of 1,000 tonnes of methane emitted per year, so that the CVRD could avoid the estimated $2.5 million cost of landfill gas capture in each landfill.
5. Emissions of methane from decomposing waste in the new landfills would be avoided, and in total an anaerobic digestion would reduce regional greenhouse gas emissions by an estimated 23,000 tonnes per year.
6. Compost from an anaerobic digestion facility would support agriculture in the Comox Valley. Nutrients embodied in organic waste would be returned to soil, and the residual

carbonaceous materials in the finished compost would provide structure to soil and a food source for beneficial soil micro organisms. The nutrients in compost are also released more slowly than mineral and artificial fertilizers, which reduces pollution of surface waters. Finally, compost can be a less expensive form of fertilizer for farms.

7. Organic material would be stabilized, so that finished material is less likely to attract vectors.
8. Anaerobic digestion would provide carbon-neutral biofuel in the form of biomethane that can be used to replace fossil fuels in transit busses initially, and which could replace fossil fuel in other local government vehicles as they are replaced.
9. Anaerobic digestion would provide biomethane that could be sold to the natural gas utility, which markets the fuel as Renewable Natural Gas. Renewable Natural Gas could in turn be bought by publicly-owned buildings to achieve carbon-neutrality. For scale, the amount of biomethane/Renewable Natural Gas that could be sold, net of the biomethane used in transit buses, would equal the natural gas consumption of approximately 2,200 homes.
10. The biomethane upgrading facility could process both landfill gas (approximately 55% methane) from the existing Comox Valley landfill, as well as biogas from the food digester and biosolids digester (approximately 65% methane). The costs of a single upgrading facility to handle these three sources has been included in this study.
11. Greenhouse gas emissions and conventional air pollution from fossil fuel combustion would be reduced when fossil fuels are displaced with biomethane from digestion.
12. Anaerobic digestion can use soiled paper and cardboard as the source of carbon to achieve target C:N ratios. This provides the double benefit of providing a beneficial reuse for this fraction of solid waste, and also making wood waste available for a higher-value purpose, such as providing heat in a district energy system.
13. After organic waste is diverted to anaerobic digestion, the volumes of waste requiring disposal in landfills would be reduced. As a result, construction of new or expanded landfills could potentially be delayed, resulting in a savings in financing costs. Potentially, construction could also be completed in smaller increments. Lower volumes of waste would also result in operating savings.
14. Operation of an anaerobic digestion facility falls within the skill sets of operators of the CVRD's existing wastewater treatment and biosolids composting facilities. These operators could be provided with equipment-specific training by the firm chosen by the CVRD to design and build the facility.
15. Separation of contaminants such as plastic packaging from organic waste is a challenge for both traditional composting and anaerobic digestion. Since organic waste would be pulped and strained before digestion however, contaminants will be more easily removed from an anaerobic digestion process than from a traditional composting process.

Challenges of Anaerobic Digestion for the CVRD

1. Anaerobic digestion is new to CVRD staff, so that expertise would need to be sourced from firms with extensive experience building facilities for food waste.
2. Although the anaerobic digestion process is contained in sealed vessels, odours can arise from unloading raw material and from processing digestate. Mitigation measures can include operating the receiving hall under negative pressure, and passing air from the hall and from digestate drying through a biofilter to remove the compounds responsible for odours (a biofilter is used successfully at the CVRD's existing biosolids composting facility). In addition, the anaerobic digestion facility could be located at the site of the existing biosolids composting facility, which is zoned for this type of operation.
3. Organic materials (food waste, yard and garden waste, and so on) will need to be separated at the source by businesses and homeowners. Successful source separation programs have been developed elsewhere however, especially in the Regional District of Nanaimo, Halifax, San Francisco, Scandinavia, and Germany.

It is also recognized that collection of source-separated organic waste will require a long-term commitment from the CVRD's member municipalities, as well as long-term arrangements to ensure the supply of organic waste is maintained in the future. Given the economic, social, and environmental benefits of integrated anaerobic digestion, member municipalities are likely to recognize that collection of source-separated organic waste aligns with their own interests.

Greenhouse Gas Implications

The greenhouse gas (GHG) emissions and reductions associated with each of the three options for organic solid waste have been estimated by asking the following questions:

1. What greenhouse gas emissions would be directly generated by the chosen option?
2. What greenhouse gas emissions would be indirectly avoided by the chosen option?

Greenhouse gas emission estimates associated with the combustion of fossil fuels have been prepared using the methods and emission factors provided by the *BC Reporting Regulation Methodology Manual*. Emission factors for landfill gases and composting have been prepared following BC Ministry of Environment (MOE) and Intergovernmental Panel on Climate Change (IPCC) guidelines. Deposits of waste in a landfill decay to produce methane over time, and modelling of landfill gas emissions takes this into account through a time-series calculation. Emissions are compared in this study on a steady-state basis to provide a clearer comparison among options.

Consistent with the guidance in the *BC Reporting Regulation Methodology Manual,* carbon dioxide emissions from biogenic material are considered current and therefore carbon-neutral. Emissions of N_2O and CH_4 from combustion of biogenic material are not considered to be carbon-neutral however, and a number of climate scientists are pointing out that efforts to reduce methane emissions will have a significant benefit in the near term.[11] In addition, the *BC Reporting Regulation Methodology Manual* Global Warming Potential (GWP) of 21 for methane is based on effects over 100 years: the GWP of methane over 20 years is 72. One recent study found that because of leaks from natural production via hydraulic fracturing, the global warming effect of natural gas can be twice as high as coal over a twenty-year period.[12]

The estimate of landfill gas emissions that would be avoided by diverting organic waste from the regional landfill were completed on First Order Decay (FOD) method, per IPCC and Conestoga-Rovers & Associates Ltd. 2009:[13]

Methane Emissions $=L_0*(EXP(-k*c)-EXP(-k*T))*(1-Ox) = 1.0875$ *tonnes* CO_2e*/tonne of waste*

Based on:

1. A landfill age of 40 years.
2. A k value of 0.11 based on average annual precipitation of 1,179 mm for the Comox Valley and an L_0 value of 160 for organic waste in on Vancouver Island.[14]
3. 10% of methane generated in the landfill would be oxidized in soil.
4. A Global Warming Potential value of 21 for methane, per the *BC Reporting Regulation Methodology Manual* (this value was revised by the IPCC in 2007 to 25, to reflect the current understating of the capacity of the atmosphere and biosphere to degrade methane into carbon dioxide).[15]
5. A capture rate of methane from the existing regional landfill of 50% after the LFG capture system is installed.

The value of 50% for the estimated efficiency of landfill gas capture is based on the following information:

1. Studies show that a significant proportion of the total amount of methane generated by mixed municipal waste in a landfill is lost between the time the waste is deposited and the time that LFG capture piping can be installed. One study from the United States surveyed 327 landfills in 36 states, and found that on average only 57% of the active area of these landfills is devoted to LFG recovery. The same study surveyed twenty four landfills in California with LFG capture systems and found that on average only 35% of generated methane is captured.[16] In addition, the rate of generation of methane from organic waste is strongly influenced by moisture and temperature conditions within the landfill. The Comox Valley receives an average of 1,179 mm of precipitation per year, making this environment wetter than California's. The higher methane generation rate before LFG capture piping can be installed would likely result in even greater losses of methane to the atmosphere.
2. Temperatures in landfills receiving organic waste rise above ambient temperatures, which accelerates the rate of decomposition and generation of methane. This temperature rise can be observed in landfill leachate, and temperatures above 40°C have been reported in the scientific literature.[17] These temperatures are higher than the operating range of mesophilic anaerobic digesters (the most common class), which convert the majority of available volatile solids in organic materials to methane and carbon dioxide within twenty days. This period is

shorter than the time that normally elapses between the point where waste is deposited in a landfill and the time when LFG collection piping is installed to recover methane.

3. The Capital Regional District has evaluated the LFG capture rate of its Hartland Landfill at 38.8%.[18]
4. After diligently investing time and capital in improving the LFG capture rate of its landfill, the Regional District of Nanaimo estimates its capture efficiency at approximately 40%.[19]

The estimate of greenhouse gas emissions from incineration is based on the *BC Reporting Regulation Methodology Manual* Table 20-7: Other Emission Factors for Municipal Solid Waste, and on an energy content for MSW of 10.5 GJ/tonne (AECOM, 2011) and are shown in the table below.

CO_2 Emission Factor (kg/GJ)	**CH_4 Emission Factor** (g/GJ)	**N_20 Emission Factor** (g/GJ)	**Combined Emission Factor** (tonnes CO_2e / tonne of MSW)
91.7	30.0	4.0	1.075

The table below shows a summary of the greenhouse gas emissions for the WTE scenario.

Table 7. Estimate of GHG Emissions and Reductions, WTE Scenario

Source	**Activity Data** (tonnes/yr)	**Emission Factor** (tonnes CO_2e/ tonne of waste)	**GHG Emissions** (tonnes of CO_2e per year)	**Notes**
Emissions from WTE	52,454	0.9825	51,535	1.
Emissions from Composting Biosolids and Wood Waste	10,000	0.193	1,930	2.
Avoided Emissions from Composting of Food Waste	10,962	0.193	(2,116)	3.
		Total	**51,349**	

Notes

1. Per the *BC Reporting Regulation Methodology Manual* Table 20-7.
2. The CVRD's existing biosolids composting would be separate from food waste composting in this scenario, but its emissions are included here since biosolids composting would continue.

3. Per 2006 IPCC Guidelines for National Greenhouse Gas Inventories, Table 4.1, wet tonnes basis, and based on the component of organic waste diverted from composting, not from landfilling.

The table below shows a summary of the greenhouse gas emissions for the composting scenario.

Table 8. Estimate of GHG Emissions and Reductions, Composting Scenario

Source	**Activity Data** (tonnes/yr)	**Emission Factor** (tonnes CO_2e per tonne of waste)	**GHG Emissions** (tonnes of CO_2e per yr)	**Notes**
Emissions from Composting Food Waste	10,962	0.193	2,116	1.
Emissions from Composting Biosolids with Wood Waste	10,000	0.193	1,930	2.
Avoided Emissions, Landfill Methane	10,962	1.075	(11,787)	3.
		Total	**(7,741)**	

<u>Notes</u>

1. Per 2006 IPCC Guidelines for National Greenhouse Gas Inventories, Table 4.1, wet tonnes basis.[20]
2. Per 2006 IPCC Guidelines for National Greenhouse Gas Inventories, Table 4.1, wet tonnes basis. Biosolids composting would proceed separately from food waste composting in this scenario, but its emissions are included here since biosolids composting would continue in this scenario.
3. Based on the component of organic waste diverted from landfilling. Methodology based on IPCC, 2006 and Conestoga-Rovers & Associates Ltd., 2009.

Table 9. Estimate of GHG Emissions and Reductions, Integrated Anaerobic Digestion

Source	Activity Data	Units	Emission Factor (EF)	EF Units	GHG Emissions (tonnes of CO_2e per year)	Notes
Methane Slip, Upgrading Process	18	tonnes CH_4 per year	21	tonnes CO_2e per tonne of CH_4	383	1.
Methane from Anaerobic Digestion	40,568	tonnes/yr	0.500	g CH_4 per tonne of substrate	426	2.
Avoided Emissions, Transit Buses	400,000	litres/yr	2.790	kg CO_2e per litre	(1,116)	3.
Avoided Emissions, Displaced Natural Gas	178,100	GJ/yr	50.312	kg CO_2e/GJ	(8,960)	4.
Avoided Emissions, Landfill Methane	10,962	tonnes/yr	1.075	tonnes CO_2e per tonne of waste	(11,787)	5.
Avoided Emissions from Composting, Biosolids and Wood Waste	10,000	tonnes/yr	0.193	tonnes CO_2e per tonne of compost	(1,930)	6.
				Total	**(22,984)**	

Notes

1. Limited to 0.5% by passing off-gas from the biomethane upgrader through a flameless oxidizer or catalytic converter.
2. IPCC provides a range of 0 to 8 grams of CH_4 per wet kg of substrate, with a default value of 1 gram. The value of 0.5 grams has been used here since the specified digester configuration includes a secondary digestion and storage stage with gas capture.
3. Based on CO_2: 2.663 kg/litre, CH_4: 0.133 g/litre, N_2O: 0.23 g/litre, per the *BC Reporting Regulation Methodology Manual.*
4. Based on CO_2: 50 kg/GJ, CH_4: 0.966 g/GJ, N_2O: 0.913 g/GJ, per the *BC Reporting Regulation Methodology Manual.*
5. Per 2006 *IPCC Guidelines for National Greenhouse Gas Inventories*, Table 4.1, wet tonnes

basis. Based on the component of organic waste diverted from landfilling.

6. Per 2006 *IPCC Guidelines for National Greenhouse Gas Inventories*, Table 4.1, wet tonnes basis.

DISCUSSION

Anaerobic digestion to produce biomethane is a natural process that has been exploited since the 1800s, and anaerobic digestion of food waste is well-established in Europe, China, India and elsewhere, but relatively new to North America. Local governments could implement anaerobic digestion in stages, as described earlier in this paper.

An anaerobic digestion facility could be developed in stages, with each successive stage designed to reduce real and perceived risks to the CVRD. Typical stages are as follows;

1. The CVRD could request proposals from qualified firms with a proven track record of designing and building anaerobic digesters for food waste. This process would be inexpensive for the CVRD, but would yield a high quantity and quality of practical, current information.
2. One or more proponents could be invited to collect samples of food waste from the region, to test its biogas yield when blended with soiled paper (or even biosolids) under laboratory conditions.
3. Following laboratory testing, one or more proponents could be invited to conduct pilot-scale tests with the proposed feedstocks for the anaerobic digestion facility. The outcome of these tests would be used to confirm biogas yields, evaluate the quality of digestate/compost, propose the optimum configuration for the digestion process (retention time, vessel capacity, digester configuration, mixing processes, and so on), and as the basis for refined capital and operating cost estimates.
4. On the basis of the estimated biogas yields, refined capital and operating cost estimates, and information about the quantity and quality of digestate/compost, the CVRD could prepare its business case for anaerobic digestion.

This staged implementation process would also help the CVRD staff and political leadership to develop a better understanding of anaerobic digestion. Staff could then also be in a position to find further synergies between anaerobic digestion and other regional initiatives, beyond those suggested in this study.

The risk of failing to obtain regulatory approval is low, given the number of Provincial strategic objectives that an IADF would support.

The CVRD's public consultation work has found that the public supports diversion of waste from landfills, and resource recovery. In addition, if the facility is built at the site of the existing biosolids composting facility, additional impacts of traffic and odours would be limited.

The costs and benefits of the three scenarios are summarized in the table below. To ensure the comparison is made on an apples-to-apples basis, elements that would be common to all options (e.g. solid waste collection and transportation) have not been included.

Table 10. Summary of Options for Resource Recovery From Solid Waste

Scenario	**GHG Emissions** (tonnes/yr)	**Estimated Capital Cost**	**Estimated Annual Cost** ($/year)	**Estimated Annual Revenues** ($/year)	**Estimated Annual Net Revenues** ($/year)	**Notes**
WTE	+ 51,300				($3,300,000)	1.
Composting	- 7,700	$19,831,000	$2,035,800	$1,032,000	($1,003,800)	2.
Integrated AD	- 23,000	$24,997,000	$2,389,000	$4,272,000	$1,883,000	3.

Notes

1. The estimated annual net revenues for the WTE scenario are based on receiving fees for 52,454 tonnes per year at $80 per tonne, and costs of $143 per tonne over forty years (AECOM, 2011).
2. The estimate of annual costs for the composting option includes operations and maintenance costs of $750,000 (Walker, 2011). The estimate also includes the financing cost for the facility at the MFA interest rate of 2.5% over thirty years, equaling $947,000 per year. The estimate of net revenues is based on receiving fees for 10,962 tonnes per year at $65 per tonne, and sales of 4,778 BDT per year of compost at $66.67 per BDT (based on an average selling price of SkyRocket compost of $30 per green tonne at 45% moisture). The estimate of 4,778 BDT per year of compost is based on an output of 540 m^3/week of finished compost, an average bulk density of 410 kg/m^3, and an average moisture content of 59% (Walker, 2011).
3. Note that the capital cost for the IADF does not take into account the avoided cost of $2.5 million for LFG capture systems for each the new landfills, the economic value of deferring landfill expansions, or the avoided cost of up to $3.5 million for an expansion of the biosolids composting facility. The estimate of annual costs does include a credit of $400,000 per year from incorporating the functions currently provided by the biosolids composting facility into the IADF. The estimate of annual costs also includes financing for the facility at the MFA interest rate of 2.5% over thirty years. The estimate of revenues includes receiving fees only for materials that are currently disposed of in the landfill.

CONCLUSIONS

1. The preliminary technical and economic modelling described in this paper appears to indicate that an IADF can be economically self-supporting, with the potential to support the economic, social, and environmental goals in the CVRD's *Regional Growth Strategy,* the *Comox Valley Sustainability Strategy*, and the *Solid Waste Management Plan*.

   An IADF would support the following objectives in the *CVRD Regional Growth Strategy*:

   a. *Provide affordable, effective and efficient services and infrastructure that conserves land, water and energy resources (p. 21).*
   An IADF would produce an estimated $1.9 million per year for the CVRD in non-tax revenues and savings, which is an estimated $2.9 million per year more value than could be provided by the next best option.

   b. *Minimize regional greenhouse gas emissions and plan for adaptation (p. 21)* and *Reduce GHG emissions in the solid waste sector (p. 74).*
   An IADF would reduce GHG emissions from landfills, transit buses, and consumers of natural gas by an estimated 23,000 tonnes per year.

   c. *Encourage renewable energy generation and related jobs (p. 45).*
   An IADF would produce biomethane to displace an estimated 193,600 GJ/year of fossil fuel. The facility would also involve an estimated eight full-time jobs, and would support the local economy by keeping energy expenditures in the community.

   d. *In anticipation of the Comox Valley Waste Management Centre reaching capacity, the Regional Solid Waste Management Plan will consider a new landfill site, waste-to-energy and other long-term alternatives (p. 52).*
   An IADF could divert an estimated 26,500 of organic waste from regional landfills. The IADF would also substitute a waste material, soiled paper, for the 5,000 tonnes of wood waste per year currently used to produce compost from biosolids.

   An IADF would support the following objectives in the *Comox Valley Sustainability Strategy:*

   a. *The* Comox *Valley will achieve a 90% diversion of waste from landfills (p. 9).* and *Develop a regional Zero Waste strategy, including promotion of recycling that is aimed at waste reduction and diversion from landfill disposal (p. 52).*
   An IADF could accept soiled paper and food waste, diverting an estimated 26,500 of organic waste diverted from regional landfills.

   b. *Reduce energy consumption and GHG emissions in municipal infrastructure (p. 45)* and *The Comox Valley will reduce overall greenhouse gas emissions by 80% from 2007 levels by 2050.*

An IADF would reduce greenhouse gas emissions by an estimated 23,000 tonnes per year.

c. *Establish a diverse network of clean and renewable local energy supply systems (p. 47)* and *Energy is harnessed from waste sources in the community (p. 50).*
An IADF would produce an estimated 193,600 GJ/year of renewable energy from waste.

d. *Promote eco-industrial development that turns wastes into resources and builds cooperative relationships amongst businesses and industrial and other land uses (p. 66).*
An IADF would be the first example of an eco-industrial development in the region.

e. *Advance the integrated management of infrastructure systems and resources (p. 67).*
An IADF would be a clear example of integrated management among the CVRD departments responsible for Solid Waste, the Wastewater Services, Transit Department, and the CVRD's member municipalities.

f. *Bolster the local farm and food economy (p. 100)* and *Improve the ecological sustainability of local food systems through increasing energy efficiencies, improving land stewardship practices, and limiting or recycling waste (p. 108).*
An IADF could provide an estimated 20,000 tonnes of compost/digestate per year to farms.

g. *Encourage potential opportunities in new green business sectors, particularly those focused on reducing climate emissions (p 128).*
Operation of IADF will create local, sustainable employment in the region.

2. The CVRD could also consider that in the IADF scenario, two future expenditures could be avoided: $2.5 million per LFG capture system for new landfills, and $3.5 million for the expansion of the biosolids composting facility.

3. An IADF has the potential to simplify the CVRD's management of solid waste and biosolids, by consolidating facilities, infrastructure, and operations.

4. Local governments can find infrastructure options which are simultaneously economically and environmental sustainable if:

    a. Infrastructure decision-making is integrated across the conventional "silos" of liquid waste, solid waste, transportation, sustainability, and land use planning.

    b. Infrastructure decisions are based on lifecycle revenues and lifecycle costs, rather than initial cost alone.

    c. Staff are empowered to make infrastructure decisions that result in transformative rather than incremental improvements.

ACKNOWLEDGEMENTS

The author would like to acknowledge Jennifer Manuel for her research work, Michael Zbarsky, CVRD Manager of Transit and Sustainability for his assistance during this study and for kindly agreeing to allow the results of the study to be published, and Marc Rutten, PEng, CVRD Senior Manager of Engineering Services, Tom Boatman, PE, CVRD Manager of Solid Waste, Jim Elliott, CVRD Manager of Wastewater Services, and Lisa Levesque, BC Ministry of Agriculture Renewable Energy Specialist for their generous assistance during this study.

END NOTES

---

[1] AECOM (2011) Waste-to-Energy Options Assessment. 58pp.

[2] Envirochem Services Inc. for the Capital Regional District (2009) Review of Hartland Landfill gas Monitoring Program and the Data Collected from April 2008 to March 2009. 252pp.

[3] Source: Statistics Canada, Environment Accounts and Statistics Division. Waste diversion data are derived from a survey administered by RECYC-QUÉBEC. Last modified: 2008-07-10.

[4] Watanabe, N. et al (2004) Combustible and Incombustible Speciation of Cl and S in Various Components of Municipal Solid Waste. Waste Management 24 (2004) 623-632.

[5] Province of British Columbia (2010) Information Sheet, Waste-to-Energy and Solid Waste Management Plans. 2pp.

[6] Maura Walker and Associates Environmental Consultants (2011) Regional Compost Facility Study. 84pp.

[7] Per the British Columbia Municipal Finance Authority (MFA) net actuarial value of financing.

[8] Personal communication between Mr. Kaj Andersson, Manager, Sustainable Water and Waste Management, City of Gothenburg and Stephen Salter, P.Eng. in October, 2006.

[9] Source: Municipality of Kristianstad

[10] Salter, S., Jackson, D., Corps, C., Barraclough, C., Lucey, Wm.P. (2011) *Integrated Resource Recovery study, Metro Vancouver North Shore Communities Technical Appendix*. Metro Vancouver. 302pp.

[11] Cox, P. et al. (2010) Methane radiative forcing controls the allowable CO_2 emissions for climate stabilization. Current Opinion in Environmental Sustainability. Volume 2, Issue 5, Pages 404-408.

[12] Howarth, R. et al. (2011) Methane and the greenhouse-gas footprint of natural gas from shale

formations. Climatic Change. Volume 106, Number 4, Pages 679-690.

[13] Intergovernmental Panel on Climate Change (2006) 2006 IPCC Guidelines for National Greenhouse Gas Inventories, Chapter 2: Waste Generation, Composition and Management Data. 26pp. and Intergovernmental Panel on Climate Change (2006) 2006 IPCC Guidelines for National Greenhouse Gas Inventories. Volume 5: Waste. Chapter 4: Biological Treatment of Solid Waste. 8pp.

[14] Province of British Columbia Ministry of Environment and Conestoga-Rovers & Associates Ltd. (2009) Landfill Gas Generation Assessment Procedure Guidance Report. Conestoga-Rovers & Associates Ltd., Richmond B.C. 81pp.

[15] Intergovernmental Panel on Climate Change (2007) Climate Change 2007: The Physical Science Basis. Chapter 2: Changes in Atmospheric Constituents and in Radiative Forcing, Table 2.14. 106pp.

[16] Themelis, N. et al. (2007) Methane Generation in Landfills. Renewable Energy 32 (2007) 1243–1257

[17] Themelis, N. et al. (2007) Methane Generation in Landfills. Renewable Energy 32 (2007) 1243–1257

[18] Envirochem Services Inc. for the Capital Regional District. 2009. Review of Hartland Landfill gas Monitoring Program and the Data Collected from April 2008 to March 2009. 252pp.

[19] Personal communication between Stephen Salter, PEng and Helmut Blanken, PEng, Superintendent of Engineering and Disposal Operations, Regional District of Nanaimo.

[20] Intergovernmental Panel on Climate Change. 2006. 2006 IPCC Guidelines for National Greenhouse Gas Inventories. Volume 5: Waste. Chapter 4: Biological Treatment of Solid Waste. 8pp.

REFERENCES

AECOM (2011) *Waste-to-Energy Options Assessment.* 58pp.

Brown, S. 2010 (2010) *Putting the Landfill Energy Myth to Rest.* BioCycle; May 2010; 51, 5. pg 23-35.

Comox Valley Regional District (2010) *Comox Valley Sustainability Strategy.* 158pp.

Comox Valley Regional District (2010) *Comox Valley Regional Growth Strategy.* 147pp.

Comox Strathcona Waste Management (2010) *Solid Waste Management Plan.* 108pp.

Comox Valley Regional District (2009) *Corporate Operations Energy and Greenhouse Gas Emissions Inventory*. 15pp.

Comox Valley Regional District (2012) *Draft 2010 Community Energy and Emissions Inventory*. 15pp.

Cox, P. et al. (2010) *Methane radiative forcing controls the allowable CO_2 emissions for climate stabilization*. Current Opinion in Environmental Sustainability. Volume 2. Issue 5, Pages 404-408.

Cuéllar, A.D.; Webber, M.E. *Wasted Food, Wasted Energy: The Embedded Energy in Food Waste in the United States* (2010) Environmental Science & Technology (August 2010), 44 (16), pg. 6464-6469

Davidsson, A. et al. (2007) *Anaerobic digestion potential of urban organic waste: a case study in Malmö*. Waste Management & Research. 2007. 25: 162–169

Davidsson, A. et al. (2006) *Methane yield in source-sorted organic fraction of municipal solid waste*. Waste Management. 27 (3): 406-414.

Edelmann, W., H. Engeli, and M. Gradenecker (2000) *Co-digestion of organic solid waste and sludge from sewage treatment*. Water Science and Technology. 41 (3): 213-221.

Envirochem Services Inc. for the Capital Regional District (2009) *Review of Hartland Landfill gas Monitoring Program and the Data Collected from April 2008 to March 2009*. 252pp.

Ghafoori, E. and Flynn, P. (2006) *Optimum Sizing for Anaerobic Digestion. BIOCAP Research Project*. Department of Mechanical Engineering, University of Alberta. 31pp.

Gnansounou, E. et al. (2008) *Accounting for indirect land-use changes in GHG balances of biofuels*. Ecole Polytechnique Fédérale de Lausanne, Switzerland. 22pp.

Hong, J. et al. (2009). *Environmental and economic life cycle assessment for sewage sludge treatment processes in Japan*. Waste Management 29 (2009) 696–703 8pp.

Howarth, R. et al. (2011) *Methane and the greenhouse-gas footprint of natural gas from shale formations*. Climatic Change. 2011, Volume 106, Number 4, Pages 679-690.

Intergovernmental Panel on Climate Change (2006) *2006 IPCC Guidelines for National Greenhouse Gas Inventories. Volume 5: Waste. Chapter 4: Biological Treatment of Solid Waste*. 8pp.

Intergovernmental Panel on Climate Change (2006) *2006 IPCC Guidelines for National Greenhouse Gas Inventories, Chapter 2: Waste Generation, Composition and Management Data*. 26pp.

Intergovernmental Panel on Climate Change (2007) Climate Change 2007: *The Physical Science Basis. Chapter 2: Changes in Atmospheric Constituents and in Radiative Forcing*, Table

2.14. 106pp.

Intergovernmental Panel on Climate Change (2010) *Emissions Factor Database.*

International Energy Agency (2009) *Biogas upgrading technologies – developments and innovations*. IEA Task 37. 20 pp.

Juniper Consultancy Services Ltd. (2007) *Commercial Assessment: Anaerobic Digestion Technology for Biomass Projects*. (Prepared for Renewable East.) Juniper Consultancy Services Ltd, England. 174pp.

Kongshaug, G. (1998) *Energy Consumption and Greenhouse Gas Emissions in Fertilizer Production*. IFA Technical Conference, Marrakech, Morocco, 28 September-1 October, 1998. 18pp.

Lefebvre, X. et al. (2000) *The Role of Aerobic Activity on Refuse Temperature Rise*. Waste Management & Research. 2000 18: 444

Maura Walker and Associates Environmental Consultants (2011) *Regional Compost Facility study*. 84pp.

National Energy Board (2012) *Short-term Canadian Natural Gas Deliverability 2012-2014 - Energy Market Assessment*. 24pp.

Nova Scotia Environment. 2009. *Renewal of Nova Scotia's Solid Waste Resource Management Strategy: Consultation Summary Report 2009*. 18 pp.

Province of British Columbia and Connestoga-Rovers and Associates (2010 *Landfill Gas Management Facilities Design Guidelines*. 230 pp.

Province of British Columbia and Conestoga-Rovers & Associates Ltd. (2009) *Landfill Gas Generation Assessment Procedure Guidance Report*. Conestoga-Rovers & Associates Ltd., Richmond B.C. 81pp.

Province of British Columbia (2007) Bill 44-2007 *Greenhouse Gas Reduction Targets Act*

Province of British Columbia (2008) Bill 37-2008 *Carbon Tax Act*

Province of British Columbia (2008) *Landfill Gas Management Regulation.* 12pp.

Province of British Columbia (2009) *BC Reporting Regulation Methodology Manual.* 118pp.

Province of British Columbia (2009) *BC Reporting Regulation.* 61pp.

Province of British Columbia (2009) *Resources From Waste: A Guide to Integrated Resource Recovery*. 64pp.

Province of British Columbia (2010) *Information Sheet, Waste-to-Energy and Solid Waste*

Management Plans. 2pp.

Ralevic, P., David B. Layzell (2006) *An Inventory of the Bioenergy Potential of British Columbia*. BIOCAP Canada, Kingston, Ontario. 8pp.

Ralevic, P., Karau, J., Smith, T. and Richardson, J. (2008) *IEA Bioenergy Task 31 Country Report: Canada*.

Salter, S., Jackson, D., Barraclough, C., Lucey, Wm.P., Corps, C. (2011) *Integrated Resource Recovery study, Metro Vancouver North Shore Communities Technical Appendix*. Metro Vancouver. 302pp.

Salter, S. (2012) *Integrated Resource Recovery Options for the Comox Valley Regional District. CVRD*. 127pp.

Stroot, P. et al. 1999. *Anaerobic codigestion of municipal solid waste and biosolids under various mixing conditions - digester performance*. Water Resources Vol. 35, No. 7, pp. 1804–1816

Tchobanoglous, G. (1991) *Wastewater Engineering: Treatment, Disposal and Reuse. Metcalf and Eddy, Inc - 3rd Edition*. 1,848 pp.

Themelis, N. et al. (2007) *Methane Generation in Landfills*. Renewable Energy 32 (2007) 1243–1257

Trivedi, B. (2008) *Dinner's dirty secret*. New Scientist, 13 September 2008, 4pp.

US EPA (1995) *Compilation of Air Pollutant Emission Factors*. 2,038pp.

Walker, S. et al. (2004) *Nova Scotia GPI Solid Waste-Resource Accounts. Executive Summary. Genuine Progress Index for Atlantic Canada. Measuring Sustainable Development: Application of the Genuine Progress Index to Nova Scotia*. 10 pp.

Watanabe, N. et al. (2004) *Combustible and Incombustible Speciation of Cl and S in Various Components of Municipal Solid Waste*. Waste Management 24 (2004) 623–632.

Wood, S. et al. (2004) *A Review of Greenhouse Gas Emission Factors for Fertiliser Production*. (Prepared for IEA Bioenergy). New South Wales, AUS: Research and Development Division, State Forests of New South Whales: Cooperative Research Centre for Greenhouse Accounting. 20pp.

Zábranská, J., M. Dohányos, P. Jenícek, P. Zaplatílková, and J. Kutil (2002) *The contribution of thermophilic anaerobic digestion to the stable operation of wastewater sludge treatment*. Water Science and Technology. 46 (4-5): 447–453.

Zah, Rainer et al. (2007) *Life Cycle Assessment of Energy Products: Environmental Assessment of Biofuels*. 20pp.

Embodied Energy of Lost Water: Evaluating the Energy Efficiency of Infrastructure Investments

Craig Aubuchon, Analysis Group, Inc.
111 Huntington Avenue, 10th Floor
Boston, MA 02199
(617) 425-8000

J. Alan Roberson, American Water Works Association
West 1300 Eye Street NW
Washington, DC 20005
(202) 628-8303

ABSTRACT

In recent years, the water and wastewater industries have seen an increased interest in research, funding, and application of energy efficiency investments related to the production and distribution of clean water. However, during this same period, funding needs for traditional distribution and infrastructure issues have also continued to increase. Since a primary concern with infrastructure investments is to reduce the amount of non-revenue or lost water, the current research combines data from three novel sources to estimate the embodied energy of non-revenue water. We find that the mean embodied energy of lost water for 270 utilities in 2010 was 1,849 MWh of electricity and 1,360 short tons of CO_2. These findings compare favorably to other energy efficiency programs implemented for water and wastewater treatment plants under the 2009 American Recovery and Reinvestment Act. Therefore, the magnitude of these findings suggests that for some utilities, investment in infrastructure replacement may provide significant energy efficiency benefits.

KEYWORDS

Non-Revenue Water; Embodied Energy; Carbon Emissions

INTRODUCTION

The Environmental Protection Agency ("EPA") estimates that drinking water and wastewater systems account for almost 4% of total U.S. energy use. [1] In part to address the significance and magnitude of this number, the 2009 American Recovery and Re-investment Act ("ARRA") initiated several new investments in energy efficiency projects for water and wastewater utilities.[2] These investments came at a critical time and demonstrated a renewed interest in water utility infrastructure, particularly as it relates to energy use . However, as the groundbreaking AWWA report, 'Buried No Longer' highlighted, the necessary investments to replace underground infrastructure could approach over $1 trillion over the next 25 years.[3] This fact alone explains why 'infrastructure' is consistently identified as one of the top industry concerns over the next five years (Murphy, 2011).

However, the same State of the Industry Report also identified 'infrastructure' as the most inadequately addressed issue. To help address the current gap between funding available for energy efficiency energy projects and the growing financial needs for underground infrastructure investment, this analysis estimates the embodied energy non-revenue/lost water for a subset of utilities from the 2010 American Water Works Association/Raftelis Financial Consulting (AWWA/RFC) rates survey. In particular, we find that by combining simple engineering estimates of unit electricity consumption with regional emissions rates, the mean embodied energy of lost water for 270 utilities in 2010 was 1,849 MWh of electricity and 1,360 short tons of CO_2. The magnitude of these findings suggests that for some utilities, investment in infrastructure replacement may provide significant energy efficiency benefits.

Where does Efficiency fit into the Water-Energy Nexus?

Two statistics are often used in the context of a discussion of the water-energy nexus; the first has already been mentioned: EPA estimates that drinking water and wastewater systems account for approximately 4% of total U.S. energy use. The second refers to water consumption by energy utilities: thermoelectric power generation water withdrawals account for roughly 50% of total U.S. withdrawals (USGS, 2009). Others have offered refinements on these general estimates, but the end goal is the same – to estimate the resource consumption of one sector from another sector.[4]

Typical energy efficiency investments in the water sector have focused on efforts to reduce energy use for water treatment and distribution. These include strategies to optimize pumping systems, treatment systems and to generate energy resources on-site (EPA, 2008, 2010; ACEEE, 2013). These investments are an important strategy to reduce overall energy use and lower carbon emissions since they target one of the largest sectors of electricity nation-wide. The point of the present research is not to suggest that energy efficiency investments are poor allocations of public funds or revenue; rather, the question is how to best maximize scarce resources among often times competing and important priorities. Thus, consideration of the water-energy nexus helps to show that reductions in water demand can also reduce energy consumption, since less raw water must be treated, regardless of the pump efficiency. Equally important, reductions in water demand at the distribution/retail stage also benefit electricity generation units to the extent that there is less demand on raw water sources.

Non-revenue water, defined as the difference between the water that leaves the treatment plant and the water that is metered at the customer tap requires energy to produce and at the same time, represents raw water that is not available to other uses such as power generation. As the name suggests, it costs the utility money to produce the water, but because the water is lost during distribution, it is not available to generate revenue. Thus, non-revenue water represents an important win-win-win opportunity by reducing energy use through an optimization of treatment volumes, increasing raw water supplies for other uses, and increasing revenues (or decreasing production costs) for water utilities. By presenting the energy-efficiency benefits from investments in distribution infrastructure, it may be possible to align funding needs more closely with spending priorities.

The growing gap between funding priorities and needs is illustrated by the 2007 Drinking Water Infrastructure Needs Survey and Assessment (DWSRF, 2009), which estimated that infrastructure spending needs for the 2007-2026 period was $334.8 billion, with 60% of that need for transmission and distribution and 22% for treatment. In contrast, during the 2009 program cycle, the Drinking Water State Revolving Fund ("DWSRF") allocated 37% of assistance to transmission and distribution and 43% to treatment needs. The 2009 DWSRF annual report notes that funding for treatment projects better meet the public health protection objective of the DWSRF, while distribution projects are more often financed by cash reserves or on a pay-as-you-go basis. However, the 2009 ARRA funding demonstrated a higher level concern with energy efficiency: the Department of Energy received $28.5 billion in awards for contracts/grants/loans, while the EPA and the Clean Water and Safe Drinking Water Revolving Funds received $4 and $2 billion, respectively.

METHODOLOGY

To calculate the embodied energy in lost water, we combine several data sources in a novel manner. In particular, we use:

- The 2010 AWWA/RFC rates survey, which sampled over 300 water and wastewater utilities. This survey includes data on Utility size, region, rates, capacity, demand, and lost water.
- Unit Energy Consumption estimates from the 2002 Electric Power Research Institute (EPRI) report that underlies the 4% energy consumption number routinely cited from the EPA.
- CO_2 emissions rate models from the EPA Emissions & Generation Resource Integrated Database (eGRID) for the year 2009 power generation sector.

The 2010 AWWA/RFC rate survey is among the largest and most detailed rate surveys of its kind. Figure 1 shows the distribution of surveyed utilities across the U.S. An important caveat to using the AWWA/RFC data is that the survey is not a random sample of utilities; in particular, it is known to oversample larger utilities, particularly in the west and southwest. Therefore, it is important to consider that the average energy use values or average carbon emissions presented here may not be representative of a true national average. Instead, what is important is the general magnitude of the findings and the method by which individual utilities can easily replicate these results using their own utility specific data in a benefit cost analysis test for future infrastructure projects. The rate survey classifies utilities into three groups (group A utilities have average daily sales greater than 75 MGD; group B utilities serve greater than 20 MGD per day, and group C utilities serve less than 20 MGD) and four regions (Northeast, South, West, Midwest). Within the survey, not all utilities reported values for daily treatment capacity, lost water as a percentage of total sales, and annual rates. After dropping utilities without relevant data, we were able to estimate the embodied energy of 270 utilities; 39 group A utilities, 78 group B, and 153 group C utilities. Of these 270, 82 were located in the West, 112 in the South, 52 in the Midwest, and 24 in the Northeast.

Figure 1. Water and Wastewater Utilities in the 2010 American Water Works Association/ Raftelis Financial Consulting Survey

For each utility, the embodied energy of lost water was calculated as:

Equation 1. $Embodied\ Energy\ (MWh)\ of\ Lost\ Water = Total\ Gallons\ Sold\ (MMGal) * Lost\ Water\ (\%) * Unit\ Electricity\ Consumption\ \left(\frac{kWh}{Mgal}\right) * \left(\frac{1MWh}{1000kWh}\right)$

Table 1 provides the EPRI (2002) estimates for the total unit electricity consumption by treatment plant size and source water.

Table 1: Total Unit Electricity Consumption by Plant Size

Treatment Plant Size (MGD)	Unit Electricity Consumption (kWh/Mgal)
1	1,483
5	1,418
10	1,406
20	1,409
50	1,408
100	1,407

Note: GW is 30% higher, independent of the size of pump

From the AWWA/RFC data, we were able to estimate unit electricity consumption for each utility based on its source water and size. First, the utility was matched to the EPRI data based on its daily treatment capacity (for example, a utility with a daily treatment capacity of 17 Mgal was assumed to have a unit electricity consumption of 1,409 kWh/Mgal). Second, we calculated the weighted average of electricity consumption from the proportion of source water from surface and ground water, with ground water requiring 30% higher energy consumption. We assumed that all wholesale purchases of water came from surface water plants; this is a conservative estimate for the embodied energy consumption, particularly since it doesn't account for the energy required for transmission between wholesale provider and the distribution utility.

Table 2 provides descriptive statistics for the lost water (in pct and Mgal) and embodied electricity consumption for the utilities in our sample.[5] The AWWA/RFC survey includes the question:

"Please provide the estimated percentage of water loss for your utility in your 2009 operating period"

Unfortunately, this question does not differentiate between lost water and non-revenue water. AWWA (1996) defines 'lost water' as: unidentified leakage, meter inaccuracies, theft, under-estimated accounts, improperly typed and sized meters, meter reading errors, and accounting errors. It is unclear how utilities interpreted this question; in contrast to non-revenue water, lost water could include known leaks and losses. The water loss percentage reported on the survey form could either overstate non-revenue water, if it includes known leaks, or it could understate non-revenue water if it does not fully account for unknown losses. In the present analysis, we do not differentiate among the reasons for lost water. Although our focus is on recapturing lost water due to leakage, increases in meter accuracy and accounting errors could also reduce demand, particularly if they require the customer to pay for water they had previously been unbilled.

Most utilities self-reported lost water values less than 10%, the recommended measure from AWWA (1996). However, the utilities with the greatest water loss all reported values greater than 20%, irrespective of size or region. Within the current panel, water loss is highest, both in maximum percentages and in mean loss, for the Northeast and Midwest. In both regions, nearly 60% of the current infrastructure was built prior to 1940 (AWWA, 2012) and contains a high percentage of cast iron pipes. The same is true for the total amount of water lost (in Mgal).

In contrast, the Northeast has the lowest unit electricity cost, possibly reflecting a higher proportion of surface water systems in the sample. The mean unit electric rate is 1,480 kWh/Mgal in the Northeast, 1,577 in the Midwest, 1,522 in the South and 1,530 in the West. In terms of gross embodied energy, the Northeast still has the highest average, while larger systems (with a larger volume of total water lost) also have higher rates of lost embodied energy.

Table 2: Reported Values for Lost Water (% and Mgal/year) and Estimated Unit Energy Consumption of Lost Water

	Utility Sub-Sample	Min	Max	Mean	Median	Std	Count
Lost Water (pct)	Total	0.01	0.38	0.11	0.09	0.07	275
	A	0.01	0.38	0.11	0.10	0.08	41
	B	0.01	0.38	0.10	0.08	0.07	79
	C	0.01	0.32	0.11	0.09	0.07	155
	West	0.01	0.23	0.08	0.07	0.05	84
	South	0.01	0.31	0.11	0.10	0.06	114
	Midwest	0.01	0.38	0.13	0.12	0.08	53
	Northeast	0.03	0.38	0.17	0.16	0.09	24
Lost Water (Mgal)	Total	0.11	35,837	1,591	522	3,474	269
	A	613	35,837	6,576	4,977	7,052	39
	B	86	5,181	1,317	1,071	955	78
	C	0.11	7,923	452	274	824	152
	West	0.11	6,768	1,040	455	1,473	81
	South	14.28	25,316	1,516	523	2,883	112
	Midwest	23.11	19,357	1,487	435	3,174	52
	Northeast	65.42	35,837	4,027	1,628	7,957	24
Embodied Energy Consumption of Lost Water (MWh)	Total	0.18	50,423	2,369	787	5,059	269
	A	863	50,423	9,762	7,468	10,175	39
	B	134	7,290	1,955	1,561	1,353	78
	C	0.2	11,147	685	423	1,178	152
	West	0.18	13,048	1,586	703	2,304	81
	South	20	38,826	2,288	791	4,439	112
	Midwest	42	27,235	2,231	613	4,547	52
	Northeast	94	50,423	5,695	2,292	11,185	24

Next, we use emissions rate data from the EPA eGRID database[6] to calculate the embedded CO_2 emissions in lost water as a function of the embodied energy used during production. Carbon emission rates vary regionally based on the electricity generating units within each power region. The eGRID database calculates the average annual emission rate based on 2009 power plant data for actual CO_2 emissions and generation rates. Regions with more coal units have higher emission rates, while regions with more natural gas or hydropower generation units have lower CO_2 rates. Aggregated up from the EPA subregions to the AWWA/RFC regions, the mean emission rate for energy consumed by water utilities was 817 lbs CO_2 per MWh for the Northeast, 934 lbs CO_2/MWh in the West, 1200 lbs CO_2/MWh in the South, and 1617 lbs CO_2/MWh in the Midwest. Figure 3, taken from the EPA eGRID database, shows the 22 sub-regions of the North American Electric Reliability Corporation (NERC).

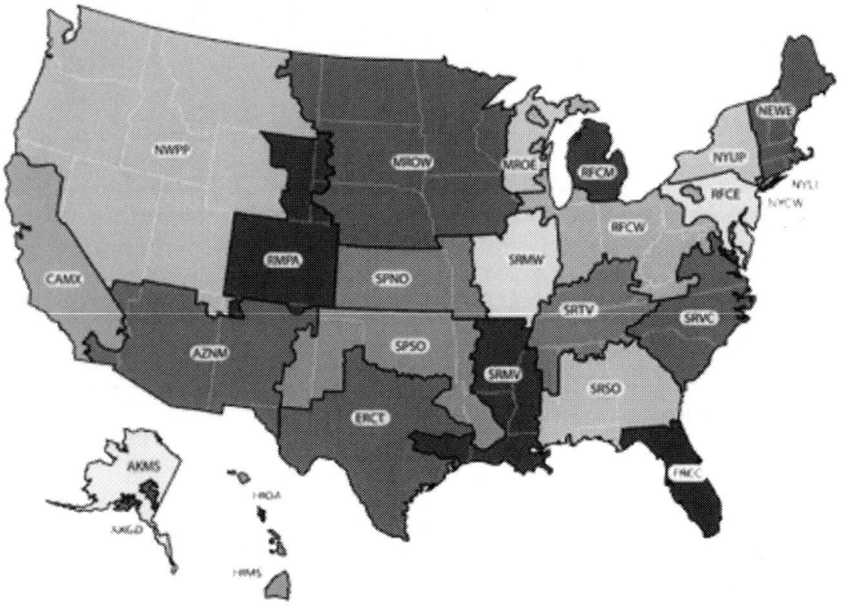

Figure 2.EPA eGRID sub-region representational map

The current version of the eGRID power profiler zip code tool allows data users to input up to 2000 zip codes, and it then automatically determines the eGRID sub-region and displays total output emission rates for CO_2, $N_2 0$, and CH_4. Previous iterations of the eGRID database have also provided data for non-base load emission rates. The total CO_2 emissions embodied was determined as:

Equation 2.

$$Embodied\ Emissions\ (short\ tons\ CO2)\ of\ Lost\ Water = Embodied\ Energy\ of\ Lost\ Water\ (MWh) * eGRID\ Total\ Emission\ Factor\ \left(\frac{lbs\ CO2}{MWh}\right)$$

RESULTS

We were able to calculate embodied CO_2 emissions for 270 utilities in the AWWA/RFC rate survey. The wide range of values depends on three main factors: the volume of lost water, the amount of energy required to treat and produce the water, and the underlying generation technology to produce the electricity. Table 3 provides additional details on the embedded CO_2 emissions. Values range from a low of 0.06 short tons of CO_2 in the 2009 year for a small utility in the West to a high of 22,941 short tons of CO_2 for a large utility in the South. The average embodied emissions for the full sample is 1,333 short tons of CO_2, but due primarily to the volume of lost water, larger systems tend to have higher total embodied CO_2 emissions. Utilities in the Northeast have the highest mean level of embodied CO_2 at 2,094 short tons of CO_2, while the West has almost a third as much, at 780 short tons of CO_2 for the average utility.

Table 3: Estimated Embodied CO_2 Emissions in Lost Water

	Utility Sub-Sample	Min	Max	Mean	Median	Std	Count
Embodied Emissions (Short tons CO_2)	Total	0.06	22,941	1,333	452	2,707	270
	A	788	22,941	5,601	4,413	5,192	39
	B	80	3,053	1,039	917	681	78
	C	0.06	5,281	395	218	623	153
	West	0.06	10,395	780	264	1,403	82
	South	12	22,941	1,376	468	2,652	112
	Midwest	32	20,706	1,764	501	3,493	52
	Northeast	34	15,396	2,094	814	3,953	24

SUMMARY AND CONCLUSIONS

These findings compare favorably to other energy efficiency programs implemented for water and wastewater treatment plants. For example, in 2007, the Massachusetts Department of Environmental Protection (MassDEP) initiated a pilot program to reduce greenhouse gas emissions and energy use at water treatment facilities by 20%. This program later received significant funding under the 2009 ARRA and highlighted as an EPA pilot project.[7] The project included 14 utilities (7 each from the water and wastewater sectors) and included a wide variety of size and technologies. The program included a wide range of stakeholders, including academic institutions, energy efficiency non-profits, regional government divisions, and major electric and gas providers, who offered energy audits to the water participants. The MassDEP estimates that these green investments, at a total of 21 completed project sites, will ultimately save 22,000 tons of CO_2 and 29 million kWh of electricity. Within this total, the majority of water and wastewater plants identified annual CO_2 reductions of approximately 400 tons. The largest exceptions were for wastewater treatment plants that identified Digester and HVAC improvements, resulting in large decreases in natural gas usage and a larger GHG reduction.[8] Indeed, for just the Northeast region alone, the estimated embodied CO_2 emissions in lost water was slightly more than 50,000 tons of CO_2 and 136 million kWh. This suggests that lost water may be a critically important area for energy efficiency measures.

The 2009 American Recovery and Reinvestment Act provided much needed funding to the Clean Water and Safe Drinking Water State Revolving Funds. The majority of this money was spent on traditional projects, while 29% was allocated to Green Project Funding in terms of both water efficiency and energy efficiency. However, as this research demonstrates, investments in water efficiency, and even more narrowly, investments in distribution infrastructure, are not exclusive from energy efficiency investments. The estimates presented here suggest that the embodied energy of lost water is comparable to, and in many instances greater than, the reported energy efficiency savings from other ARRA EE investments in the water sector. The mean embodied energy of lost water in the AWWA/RFC sample was 2,300 kWh and 1,333 tons of CO_2, with findings varying significantly by region and utility size. This research presents an important first step in addressing the stated need for infrastructure funding and the actual allocation of investment dollars. Future research should consider how water utilities can apply and present generalized versions of rate payer funded energy efficiency cost-effectiveness tests to their more traditional resource and infrastructure spending needs. Under these cost-effectiveness tests, utilities should consider the non-resource values embedded in electricity costs and external social pollution, such as carbon dioxide. By presenting these benefits to other stakeholders, including energy providers, state regulators, and energy-efficiency non-profits, the water utility sector can broaden its base of supporters for infrastructure funding and continue to address the significant infrastructure challenges on the horizon.

. . . .

---

[1] For example, see: http://water.epa.gov/infrastructure/sustain/energyefficiency.cfm. This 4% number appears to be based on the initial work presented in EPRI (2002) and accounts for the production, conveyance, and treatment of water and wastewater. More broadly, estimates indicate that energy for water services such as end-use heating may be as high as 12% of total U.S. energy use (Sanders and Webber, 2012).

[2] For example, see: http://water.epa.gov/infrastructure/sustain/energyefficiency.cfm and http://www.mass.gov/dep/water/priorities/eerewwu.htm.

[3] http://www.awwa.org/portals/0/files/legreg/documents/buriednolonger.pdf

[4] For example, Macknick et al. (2011) have refined the USGS estimate of water needs by the power sector and developed technology specific estimates of water withdrawal and consumption rates by generation technology.

[5] We have excluded utilities that reported zero lost water or zero annual total sales.

[6] The EPA eGRID database is available for public download at: http://www.epa.gov/cleanenergy/energy-resources/egrid/index.html

The 2012 release presents emissions rate data from 2009 power plant reported values. The power profiler tool calculates the average CO2 emissions at a zip code level.

[7] See:
http://water.epa.gov/aboutow/eparecovery/upload/2010_01_26_eparecovery_ARRA_Mass_EnergyCasyStudy_low-res_10-28-09.pdf

[8] The primary constituent of natural gas is methane (CH_4), which is 21 to 25x more potent than CO_2 as a GHG over a 100 year time period (IPCC, 2007). Totals by facility: the Barnstable Wastewater Treatment Facility (1,931 tons of CO2), Falmouth Wastewater Treatment Facility (3,200 tons of CO2), Greater Lawrence Sanitary District (5,887 tons of CO_2), and Pittsfield Wastewater Treatment Facility (3,252 tons of CO_2). See: http://www.mass.gov/dep/water/priorities/empilot.htm for specific details of plant programs.

REFERENCES

American Council for an Energy Efficient Economy. Water and Wastewater Treatment. Accessed February 2013. Available at: http://aceee.org/sector/local-policy/toolkit/water.

Aubuchon, C. 2013. From What Perspective? Distributional Accounting within Cost Benefit Analysis. Under Review, *Journal American Water Works Association.*

Aubuchon, C.P. and P. Hibbard. 2013. Summary of Quantifiable Benefits and Costs Related to Select Targeted Infrastructure Replacement Programs. Prepared on behalf of the Barr Foundation. Available at: http://www.analysisgroup.com/uploadedFiles/Publishing/Articles/Benefits_Costs_TIRF_Jan2013.pdf.

AWWA Leak Detection and Water Accountability Committee. 1996. Committee Report: Water Accountability. p. 108-111.

AWWA. 2012. Buried No Longer: Confronting America's Water Infrastructure Challenge. Available at: http://www.awwa.org/Portals/0/files/legreg/documents/BuriedNoLonger.pdf

Environmental Protection Agency. 2008. Ensuring a Sustainable Future: An Energy Management Guidebook for Wastewater and Water Utilities. EPA 832-R08-002. Available at: http://water.epa.gov/infrastructure/sustain/upload/Final-Energy-Management-Guidebook.pdf.

Environmental Protection Agency. 2010. Evaluation of Energy Conservation Measures for Wastewater Treatment Facilities. EPA 832-R-10-005. Available at: http://water.epa.gov/scitech/wastetech/upload/Evaluation-of-Energy-Conservation-Measures-for-Wastewater-Treatment-Facilities.pdf.

Forster, P., V. Ramaswamy, P. Artaxo, T. Berntsen, R. Betts, D.W. Fahey, J. Haywood, J. Lean, D.C. Lowe, G. Myhre, J. Nganga, R. Prinn, G. Raga, M. Schulz and R. Van Dorland, 2007: Changes in Atmospheric Constituents and in Radiative Forcing. In: Climate Change 2007: The Physical Science Basis. Contribution of Working Group I to the Fourth Assessment Report of the Intergovernmental Panel on Climate Change [Solomon, S., D. Qin, M. Manning, Z. Chen, M. Marquis, K.B. Averyt, M.Tignor and H.L. Miller (eds.)]. Cambridge University Press, Cambridge, United Kingdom and New York, NY, USA. Cited as (IPCC, 2007).

Kenny, J.F., Barber, N.L., Hutson, S.S., Linsey, K.S., Lovelace, J.K., and Maupin, M.A., 2009, Estimated use of water in the United States in 2005: U.S. Geological Survey Circular 1344, 52 p. Cited as USGS (2009).

Kushler, M., Nowak, S., and P. Witte. 2012. A National Survey of State Policies and Practices for the Evaluation of Ratepayer-Funded Energy Efficiency Programs. ACEEE Report Number U122.

Macknick, J., Newmark, R., Heath, G., and KC Hallett. 2011. A Review of Operational Water Consumption and Withdrawal Factors for Electricity Generating Technologies. National Renewable Energy Laboratory Technical Report 6A20-50900.

Murphy, M. 2011. State of the Industry Report 2011: Economic Stagnation Dampens Industry Outlook. *Journal American Water Works Association*, 103:10:42.

Overcast, H.E. 2012 Is it time for water utilities to rethink pricing? *Journal American Water Works Association*, 104:6:59.

Sanders, K.T. and M.E. Webber. 2012. Evaluating the energy consumed for water use in the United States. *Environmental Research Letters*, (7), 11 p [Insert page number].

THE MOST ENERGY EFFICIENT PUMP IS THE ONE THAT'S NOT RUNNING: A CASE STUDY OF ENERGY SAVINGS DUE TO AN I&I REDUCTION PROGRAM

Kevin Colvett[1], Brent Freeman[2] , Todd Spangler[3]

[1] CH2M Hill
[2]Metro Water Services
[3] City of Brentwood, TN

KEYWORDS

Collection System Rehabilitation, Energy Efficiency, I&I Reduction

INTRODUCTION AND BACKGROUND

There are many justifiable reasons to undertake an Inflow and Infiltration (I&I) reduction program – regulatory, environmental, and operational just to name a few. Correspondingly, there are many methods to judge the relative success of the program – order compliance, reduced overflows, additional capacity, and so on. One component that is often overlooked is the simple truth that reduced flow in a system means reduced pump run times and energy consumption in the collection system. Cost savings attributable to reductions in O&M efforts, such as fewer overflow responses and costly and inefficient reactive maintenance, are real but can be difficult to objectively quantify. However, whenever extraneous flow has been removed from a sewer collection system upstream of a pumping facility, there is a quantifiable savings in both dollars and in carbon footprint.

In 2007, the City of Brentwood, TN, began a system-wide sewer rehabilitation program. Situated just south of Nashville, Brentwood's collection system connects via a master meter to Metro Water Services' system at what is referred to as "The Brentwood Pumping Station" by Metro Water Services and as "The Metro Pumping Station" by Brentwood. Because approximately 10% of the flow at the facility originates from Metro Water Services customers, both systems share in the operational costs, and subsequently both systems bear the burden of increased energy costs and share the benefits of reduced pumping. This paper documents the relationship between the reduced flows at the pumping station and the reduced electrical costs realized by both systems and generally provides a narrative of discovering an unanticipated benefit of rehabilitating a sewer collection system. Further, a method for estimating savings for other programs' efforts is proposed, not as a strict calculation but as a reasonable approximation that could be utilized when considering what type of rehabilitation to utilize and to what extent rehabilitation of the collection system is practical.

THE BRENTWOOD NARRATIVE

Situated just south of Nashville, Brentwood's collection system is fairly typical of most suburban systems:

- Approximately 200 miles of gravity pipe, ranging from 8" to 30"
- 10 small sewer lift stations and one comparatively large one with a capacity of 9.5 MGD
- A master metered connection to Metro Water Services, the regional water provider

When initiated, the goals of the$30,000,000 sewer rehabilitation program were straightforward and simple: comply with the regulatory order and the accepted Corrective Action Plan and accomplish this as quickly and economically as possible. As the years 2007-2010 progressed, measurable progress was made on many "success factors", but the primary metric that regulatory agencies were concerned with was the wet-weather overflow frequency, which was reduced by ¾.

A side benefit to reduced wet-weather flows is that in many cases dry weather flows are reduced as well, since the lining of pipes, coating of manholes, and replacement of sewer laterals reduced the insidious ongoing groundwater leaks in the collection system. The result of this was that the entire collection system was collecting less stormwater, and over time the multitude of 1 gpm sorts of leaks being eliminated began to be noticed at the master meter connection to Metro Water Services, and by 2010, Brentwood was paying approximately $1,000,000 less per year in metered costs to Metro than it would have been with the same flows from 2007.

Anecdotally, there was a realization that the main pumping station was operating less frequently, and certainly operating at a higher speed less frequently. In fact, in drier times, the station began to commonly shut down for extended periods of time during the daily low-flow times. Because Brentwood, like many other utilities, was implementing procedures for CMOM programs and similar efforts, there was increased scrutiny of things like pump run times and pump preventive maintenance. Coupled with the reduced flows at the main pumping station, the assumption that the electric usage and subsequent charges were reducing at the station was tested.

GENERAL APPROACH AND RESULTS TO DATE

There are only two pieces of data necessary to quantify the cost savings associated with an I&I program's reduced energy consumption: 1) measured flows, and 2) electric bills. One would need to be careful about drawing conclusions for a short period of time because some years are drier than others and could skew results, so there are numerous methods available to normalize this quantification. However, for this example, actual flows are utilized because the data set includes over three years, with the first year (the end of 2007) being a historic drought for middle Tennessee and the successive years experiencing more normal precipitation.

The graph below shows the flow reduction trend resulting from the rehabilitation of approximately 25 miles of sewer lines, 500 manholes and numerous service repairs. Approximately 1,000,000 gallons of flow per day were removed from the system for a total cost of approximately $10,000,000 (additional work is ongoing that could include equalization and additional rehabilitation). This constitutes approximately a 20% reduction in overall flow to the pumping station. Solely from reduced flow through the master metered connection to Metro Water Services, the rehabilitation program will pay for itself in approximately 10 years. By any measure, the program has been a success.

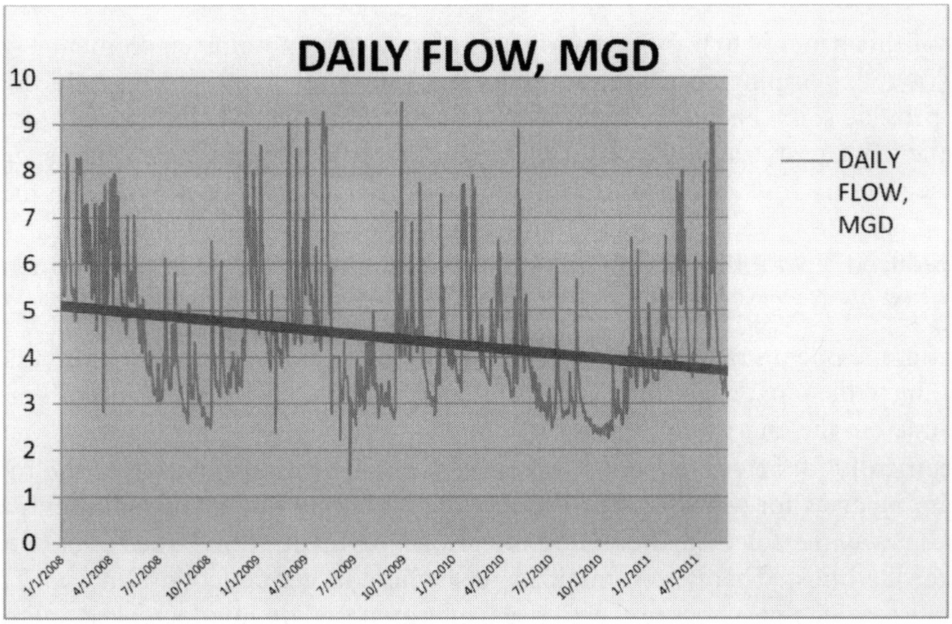

About halfway through the program, it was noticed that the monthly electrical costs at the pumping station were being reduced. Brentwood is no different than most utilities in that the personnel most able to control and track O&M costs are not privy to electric bills – they are received and paid by the Finance Department. When the bills were requested from the Finance Department and plotted onto a simple spreadsheet, an interesting trend was evident: the electrical costs were lowering at the station at approximately the same rate as the flows. While this might seem like an elementary observation, it became evident that the cost savings associated with flow reduction dwarfed previous "optimization" efforts to increase pump efficiency at the station. More savings was realized by not having the pumps run than solely by having them run more efficiently.

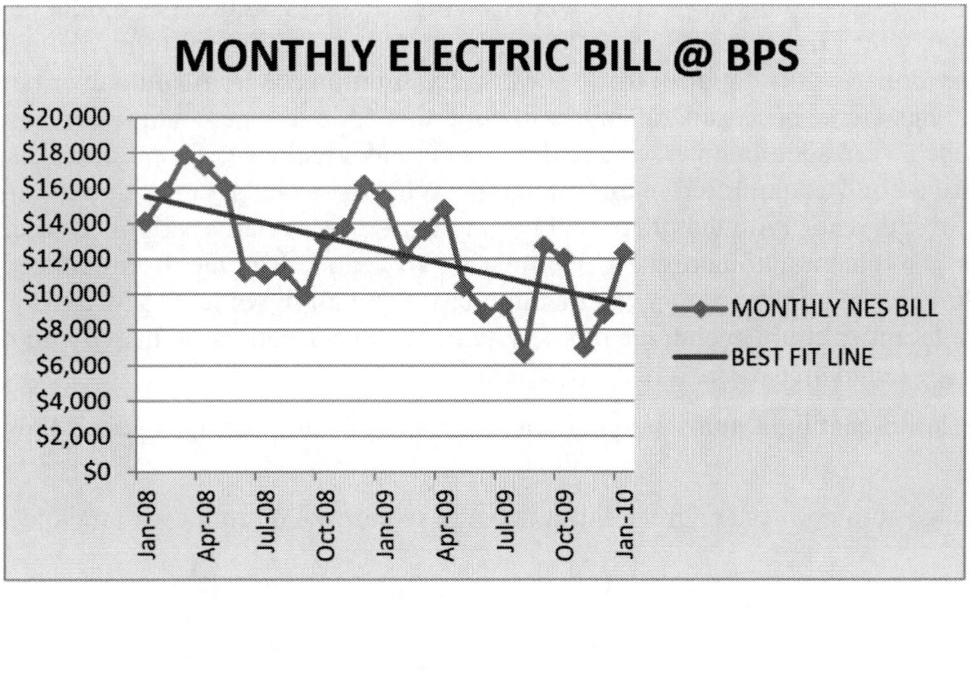

RECOMMENDATIONS FOR OUR INDUSTRY

This intent of this paper is to provide tangible evidence of something so elementary that it can be easily overlooked: pumping costs are reduced when pumping quantities are reduced. Further, the intent is to provide some encouragement to both water and energy professionals to, whenever possible, quantify energy savings even if the goal of a project or program is not specifically for energy conservation. It is a story worth telling, and often a narrative provided in the form of dollars saved is more effective than the same narrative communicated in the form of sewer overflows reduced, operability improved or orders complied with. Other recommendations include:

- Encourage operators to review energy consumption bills. They are ultimately the ones who have the most capability to control energy costs, and too infrequently are they included in the energy conservation conversation.
- Specifically for sewer rehabilitation projects, include an estimation of the reduced pumping costs for estimated flow reductions. Making this inclusion might, in some cases, sway the selected alternative toward fixing infiltration issues over constructing larger facilities. There are some instances where larger collection lines and equalization basins are necessary, but in these cases, energy costs are not decreased and often they are increased.
- When negotiating regulatory compliance actions such as Corrective Action Plans / Engineering Reports include energy in the discussion.
- As an academic exercise, if you, as a consultant, owner or contractor, have been associated with a sewer rehabilitation project in the last 5 years that appears to have been successful from a flow reduction standpoint, take the (relatively small amount of) time to quantify energy reduction costs corresponding to that work. This might be quantifiable at both a pumping station or the treatment facility, or both.

CONCLUSIONS

On this average-sized pumping station, a cost savings of about $5,000 per month has been documented, which translates into $60,000 per year, or, over a 20-year term, about $1,200,000 in present-day dollars. It is doubtful that a sewer rehabilitation program could ever be justified by energy savings alone, but a general approximation that for every percentage decrease in flow there will be a corresponding percentage decrease in electrical costs for pumping facilities is not only intuitive but is exemplified in this situation. What the exact savings amount might be is secondary to the realization that there will be savings – a fact often overlooked when weighing the merits of a rehabilitation program. Comparing 20 years of savings from reduced pumping costs to 20 years of pumping to an equalization basin and all the other O&M costs associated with those facilities could encourage more systems to concentrate more heavily on rehabilitation, at least in the beginning stages of their programs.

Compact fluorescent light bulbs are great, but it is better to turn off lights when they aren't needed.

High-mileage automobiles are great, but it is better to carpool or cut down on unnecessary driving.

Efficient pumps are great, but it is better to have the pumps sitting idle and not pumping rainwater out of a sewer collection system.

Small Wastewater Treatment Facilities Can Be Energy Producers

Joseph Cantwell, P.E.

Science Applications International Corporation

ABSTRACT

Wastewater treatment facilities (WWTFs) are implementing energy efficiency and renewable energy modifications to reduce energy consumption and demand from the electric power grid, as a result they are moving toward energy neutrality.

To meet this objective, WWTFs first need to become energy efficient to reduce their energy consumption, which will allow them to minimize the size of the renewable energy system they will need to become energy neutral. They can then modify their solids processing strategy to increase the production and beneficial utilization of biogas. Many WWTFs that have anaerobic digesters are not operating the digesters at design (full load capacity) conditions and maximizing their biogas production. However, WWTFs are learning they can generate energy (electric generation and waste heat capture) sufficiently to offset major energy consumption at their facilities and, in some locations, be exporters of energy.

KEYWORDS: Energy efficiency, renewable energy, anaerobic treatment, biogas, beneficial utilization

INTRODUCTION

Municipal wastewater treatment facilities (WWTFs) are recognizing the need to become proactive toward energy efficiency and utilizing renewable energy sources. They are having their wastewater utilities audited to identify how they can improve their treatment processes and operations to become energy efficient. In the process of becoming energy efficient, WWTFs are experiencing 20 percent to 50 percent and greater reduction in energy consumption. After the wastewater utilities have become energy efficient, they are assessing the potential application of renewable energy resources, in particular the potential and value of utilizing biogas.

The following is a municipal example of a community that became proactive and proceeded along the path to become energy efficient and to utilize renewable energy resources. Their journey will be presented from identification, conception, and implementation to operation, resulting in the reduction of electric grid consumption by producing a major portion of all of their own energy consumption requirements through utilization of a renewable energy source: biogas.

Figure 1 – Stevens Point Wastewater Treatment Facility

The City of Stevens Point WWTF, with a design flow of 4.6 MGD but an existing flow of 2.9 MGD, recognized there was value in becoming energy efficient. The city government became proactive by having an energy survey completed on its WWTF. The survey identified a number of energy efficiency opportunities. The community implemented many of the identified energy efficiency opportunities:

1. Educated their operators on how they are billed, including a review of demand charges
2. Monitoring their electrical and demand usage daily
3. Installed a new plant air compressor for providing air to their air diaphragm sludge pumps. Reduced from 50 Hp down to 10 Hp
4. Changed out outside building lighting to LED lighting
5. Changed aeration basin dissolved oxygen (DO) set points from 2 ppm to 0.7 ppm
6. Installed heat pumps in the WWTF's lab/office and large garage that run off the plant effluent
7. Installed a heat loop from biogas-fired boilers to multiple buildings for heating purposes in the winter months
8. Optimized waste-activated sludge (WAS) thickening to remove "water" from entering the anaerobic digester, which reduced biogas being consumed to heat the WAS (less volume to heat up)
9. Purchased a more efficient aeration blower (rotary screw technology)
10. Replaced 9-inch ceramic diffusers with 9-inch flexible membrane diffusers.
11. Installed a new ultraviolet (UV) light system with automatic cleaning.
12. Installed revised supervisory control and data acquisition (SCADA) programming, which does not allow more than one screw pump (raw sewage pumps) to run at a time (100 Hp); installed programming that does not allow more than one aeration blower to run at a time(150 Hp); added programming that rotates the sludge storage mixers to allow only one mixer to run at a time

13. Replaced three anaerobic digester submersible mixers with lower Hp mixers (going from 8.3 Hp to 2.5 Hp)
14. Turned off un-needed outside night lighting (parking lot areas and driveways)

The WWTF also modified their approach when purchasing equipment or making treatment process modifications:

- Considering plant equipment purchases with energy usage in mind by incorporating life-cycle-cost analyses into the decision-making process
- Evaluating WWTF process changes with energy consumption in mind

The Stevens Point WWTF implemented the projects on a budget-available basis. So they realized a steady reduction in yearly energy consumption (Fig. 2) as they continued to implement energy efficiency projects. By realizing the steady reduction, they also recognized the monetary value it brought to the WWTF.

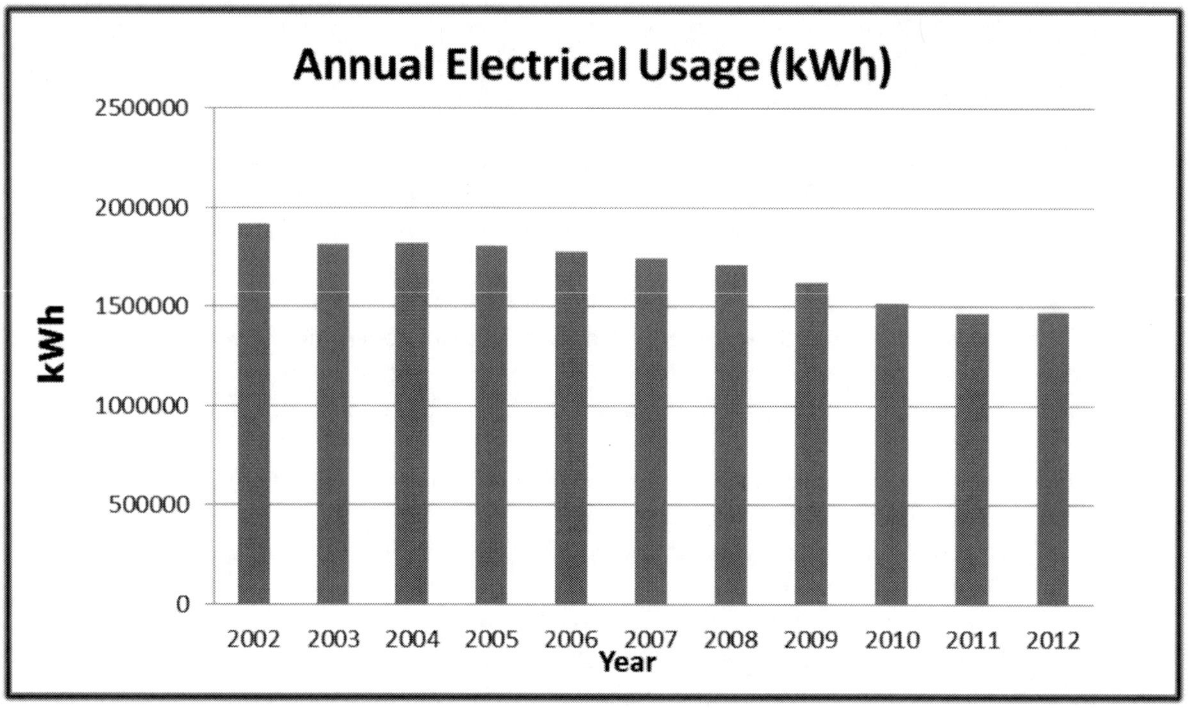

Figure 2

During the steady reduction in energy consumption, the organic loading to the WWTF continually increased (Fig. 3) while the average flow (Fig. 4) remained relatively constant. So, the impact of energy efficiency became a more valuable asset for the WWTF because they were treating more organic waste while still reducing their energy consumption.

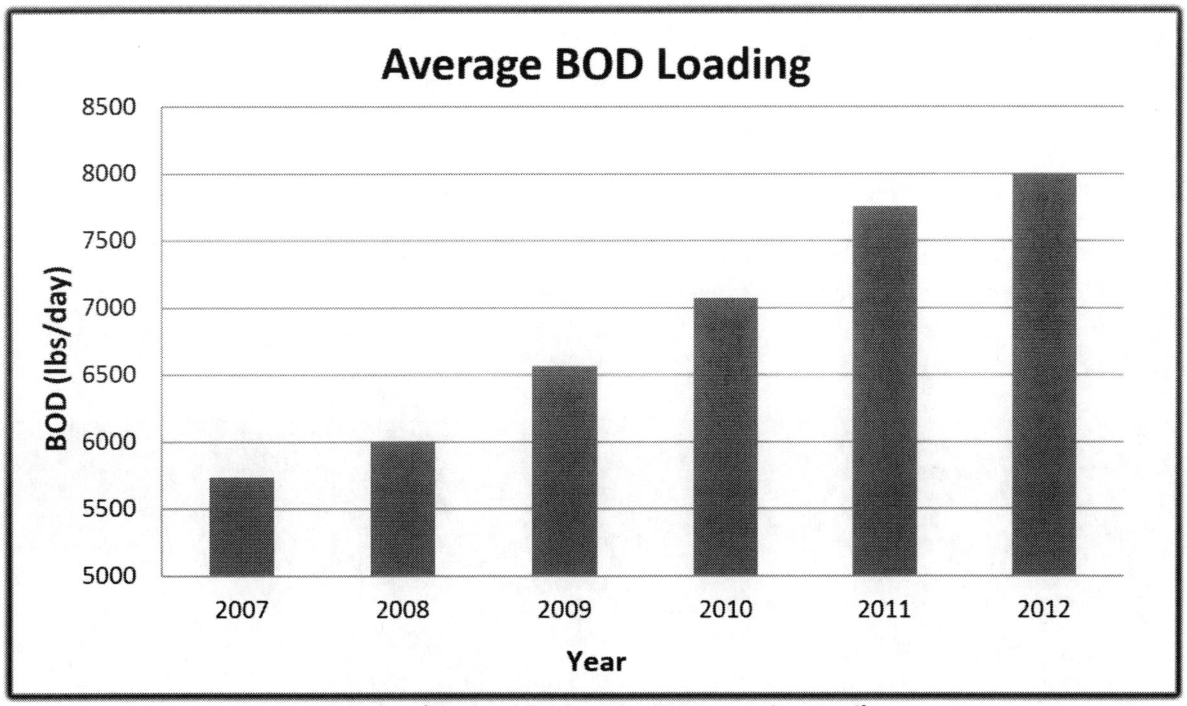

Figure 3 (BOD – biochemical oxygen demand)

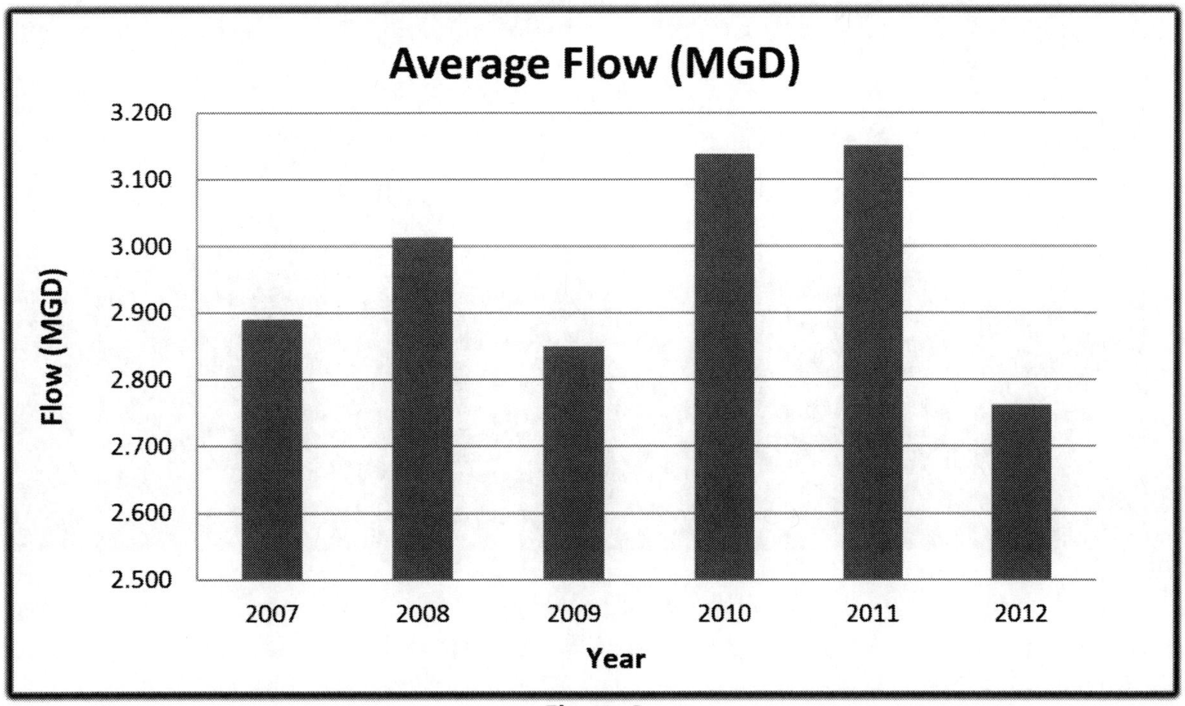

Figure 4

Having completed energy efficiency projects and the community realizing energy consumption reductions, they moved forward and addressed what they could do with anaerobic digesters. They acknowledged they had a sunken capital investment, yet the digesters were only being utilized to a fraction of their treatment capability and capacity. The community moved forward with an analysis of their anaerobic digestion system to identify what modifications were necessary to utilize it at its functional capacity. The study identified modifications that could be implemented to improve its volatile solids reduction. This resulted in a reduced volume of biosolids to dispose, but more important, the increased volatile solids reduction increased its biogas production, providing a viable renewable energy fuel source for the WWTF. The modifications identified were the following:

1. Install a new anaerobic digester mixer (linear motion mixer) on one digester, consuming low Hp and providing improved mixing resulting in increased biogas production (Fig. 5).
2. Modify operation to divert as much high-strength waste directly into the anaerobic digesters to increase organic loading to the digester and decrease loading to the aeration system.
3. Install a combined heat and power (CHP) unit (180 kW) with auxiliary biogas conditioning system to ensure clean biogas being provided to the engine (Fig. 6 and 7).

Figure 5

Figure 6

Figure 7

With modifications to its anaerobic digestion system installed and operating, the system could produce more biogas and beneficially utilize it. The installed modifications resulted in the digestion system having the capability to produce sufficient biogas to offset the WWTF's entire electric consumption on a periodic basis. Usually the installed modifications generate sufficient biogas and electricity to provide 80 percent to 90 percent of the WWTF's electric consumption. Further, the modifications incorporated heat recovery from the engine driving the generator, resulting in the system also offsetting the process heat requirements of the anaerobic treatment system. The installed system is presently offsetting approximately 75 percent to 80 percent of daily electric consumption and, during low flow periods, can offset the entire electrical needs of the WWTF. Fig. 8 presents the amount of energy purchased in 2011 as compared to 2012, representing the first year of operation for the generation system. As presented, it can be observed the generation system is offsetting a major portion of the electric consumption at the WWTF.

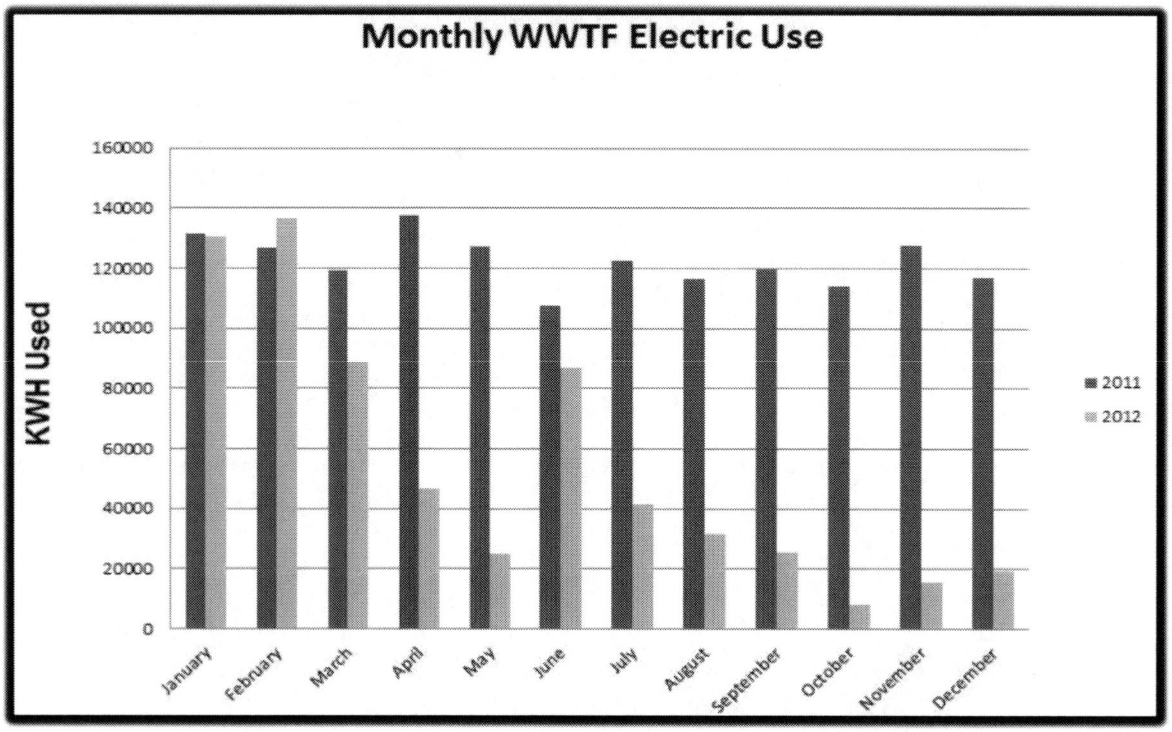

Figure 8

However, what has also been noticed is a change in the composition of the electric bill (Fig. 9). The electric bills received prior to the generation of onsite power were approximately 24 percent demand and 76 percent electric consumption. However, since the generation system has been in operation, the electric bill now consists of 40 percent demand charge and 60 percent electric consumption. Therefore, the generation system is assisting in reducing consumption, but now the WWTF is seeking to identify opportunities to reduce their demand charge as well.

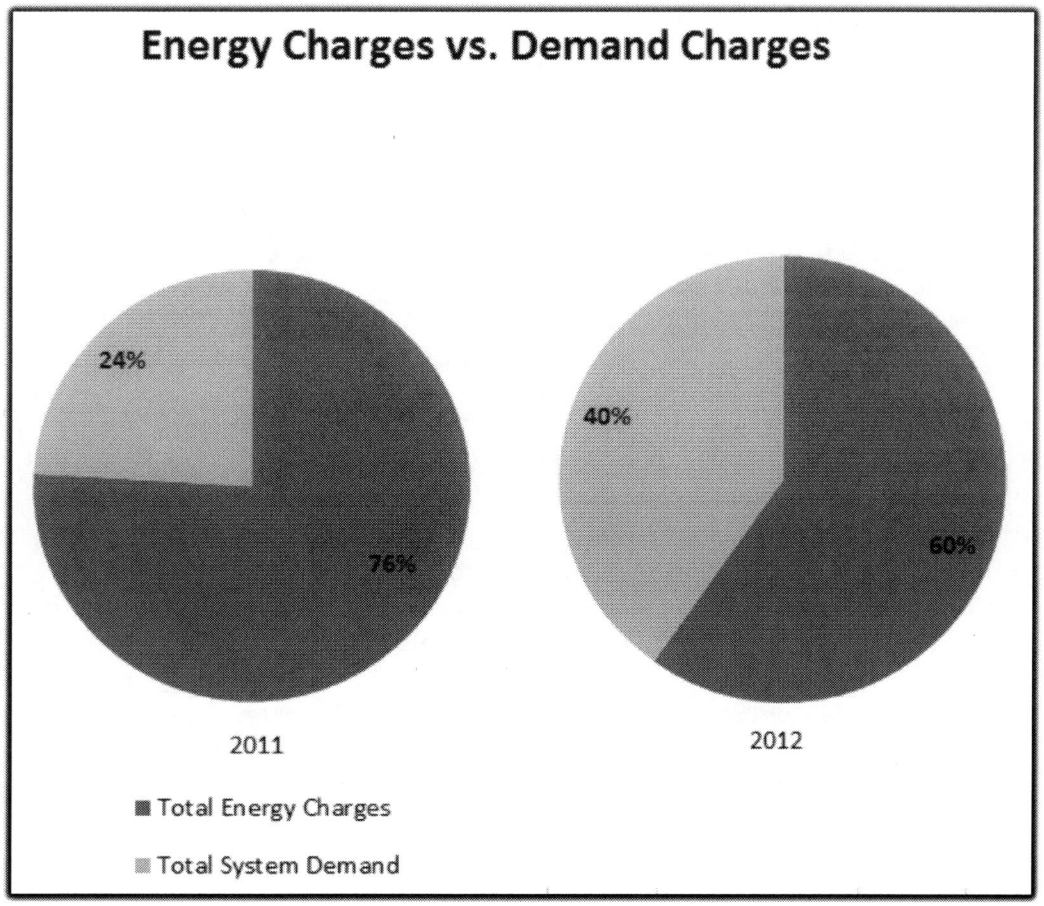

Figure 9

The WWTF also became active in pursuing auxiliary feedstock materials, because they understood to make the system most beneficial to the community, the anaerobic digestion system needed to operate near its treatment capacity. To achieve design-loading conditions, the WWTF sought and procured outside auxiliary feedstock for the anaerobic digestion system. The auxiliary feedstock consists of

- Dairy waste
- Marinade waste from a chicken facility (food waste high in chemical oxygen demand, COD)
- Concentrated brewery waste
- Miscellaneous food waste (small particle size)
- Fats, oils, and grease (FOG)

SUMMARY

The WWTF has moved forward on becoming energy efficient and utilize their biogas for electric generation and heat capture. However, they have learned, and they recognize this effort needs to continue to move toward more energy consumption reduction and increase their beneficial utilization of the biogas they produce. The following is a list they are targeting as efforts or goals to continue to achieve additional movement toward energy neutrality as well as efforts once achieved to maintain energy neutrality.

- Install additional biogas storage capacity.
- Construct an auxiliary feed stock receiving station.
- Assess cell lyses' technology and its potential installation.
- Construct additional biogas conditioning capacity.
- Install an additional generator.
- Eliminate gas flaring.
- Identify and procure additional auxiliary feed stock sources.
- Install new technology mixing equipment into the other existing digesters.
- Assess the potential to interconnect a local industry directly to the digester complex so its high strength waste can be utilized as a direct auxiliary feedstock for the digesters (Also, by directly connecting it to the WWTF, it will lower the organic load to the aeration system so it is a potential dual advantage to the WWTF.)

CONCLUSION

The City of Stevens Point WWTF revitalized the beneficial utilization of a well-documented wastewater treatment process anaerobic digestion. Many wastewater treatment facilities have anaerobic treatment processes at their facility but are under-utilizing their investment, and many are wasting the biogas (energy) they are producing by simply flaring off the gas. This project summary of a recently completed project at a smaller-size WWTF (less than 5 MGD) presents the value of becoming energy efficient, beneficially utilizing the biogas produced, managing ever-increasing energy consumption and costs, and pursuing energy neutrality. All can be achieved irrespective of the size of the WWTF. Further, all of the achievements were realized while maintaining biological phosphorus removal and earning revenue, approximately $60,000, for treating the high strength auxiliary feedstock.

THE TREATMENT PLANT OF THE FUTURE
INTEGRATED CO_2 CAPTURE FOR ENHANCED PLANT GROWTH

Jim Schettler, Brown and Caldwell
Nancy Andrews, Brown and Caldwell
Brown and Caldwell,
201 North Civic Drive,
Walnut Creek, CA 94596
(925) 2100-2343

ABSTRACT

At over 800 commercial greenhouse facilities in western Europe, and now one facility in the USA, carbon dioxide (CO_2) gas is now captured and beneficially used to very substantially enhance plant growth in advanced and sophisticated commercial hydroponic greenhouses. The CO_2 at these sites is obtained directly from the carefully purified exhaust of large internal combustion engines in combined heat and power (CHP) systems.

Engine exhaust gases from a typical low emissions lean-burn engine fueled by natural gas contain about 6% CO_2, and in this manner the CO_2 is captured and productively reused. With CO_2 comprising 58 to 62% of anaerobic digester gas by weight, integrating this atmospheric CO_2 enhancement with a CHP system fueled by digester gas can achieve multiple benefits. Ultimately, combining the municipal wastewater treatment and greenhouse process, including anaerobic digestion for biosolids stabilization, might be a natural fit for progressive communities.

KEYWORDS

Biogas, CHP, cogeneration, chlorophyll, anaerobic digester, carbon capture.

INTRODUCTION

Modern large-scale sophisticated commercial greenhouses have become carefully managed agribusinesses. The recent growth in the greenhouse industry is focused on high-value crops such as tomatoes, specialty herbs, flowers, and cucumbers. For example, in Canada greenhouse-grown tomatoes dominate the market with 89% of the fruits produced in commercial greenhouses (USDA, 2004). Atmospheric CO_2 enrichment is a common feature of these greenhouses, boosting their commercial viability.

The atmosphere that you and I are now breathing generally contains about 350 to 390 parts per million (ppm) CO_2, but this exact amounts depends on the specific location and season. The current US average atmospheric CO_2 concentration is over 380 ppm.

This relatively small amount of CO_2 is used by the chlorophyll in green plants for growth per the following well known photosynthesis reaction:

$$6CO_2 + 6H_2O + \text{radiant energy (light)} \rightarrow C_6H_{12}O_6 + 6O_2 \ (1)$$

$$\text{Carbon dioxide + water} \rightarrow \text{carbohydrates + oxygen}$$

In an enclosed greenhouse, CO_2 levels often drop below normal atmospheric levels as a result of photosynthesis. Plant growth is halted in atmospheric levels below 200 ppm CO_2. Conversely, plant growth can be stimulated in the presence of light by adding supplemental CO_2 gas to the growing atmosphere within the greenhouse. As illustrated in Figure 2, optimal CO_2 levels of 500 to 1450 ppm can greatly increase plant growth. However, greenhouses generally limit their CO_2 levels to about 650 to 700 ppm to be well below the human toxic CO_2 concentration level (approximately 1900 ppm). CO_2 enrichment is often suspended at night, as needed for the normal diurnal photosynthesis patterns of the specific crops.

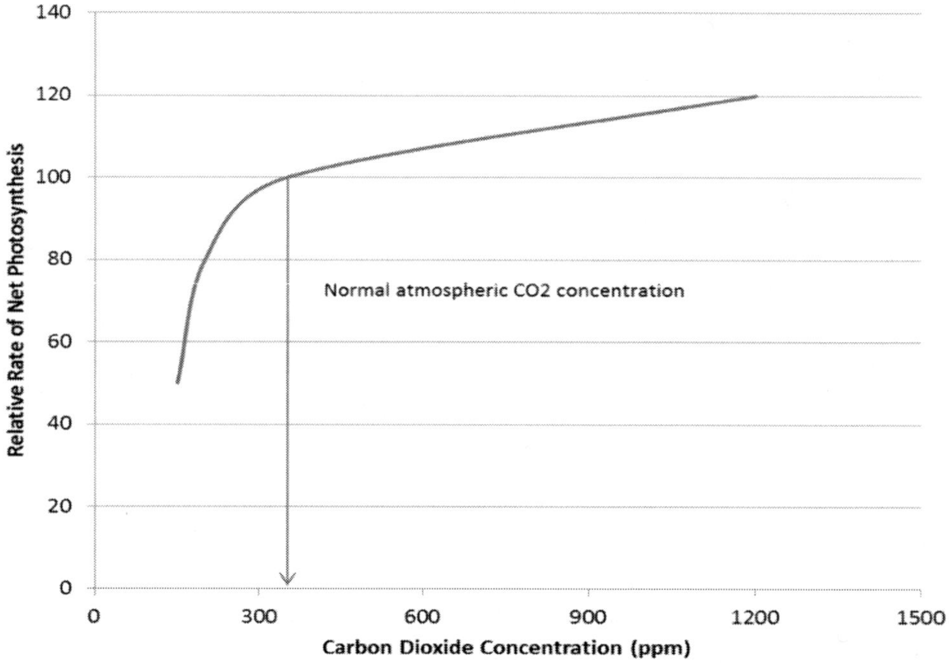

Figure 2. Relative rate of photosynthesis increases in enriched CO_2 environments (Blom, 2002)

Plant yield at some of these CO_2 enrichment greenhouses is reported to be ten times or more higher than in traditional open farmlands (Kren, 2004) due to the combination of CO_2 fertilization, optimized lighting, temperature and nutrient addition, hydroponic culture, and plant density. These enriched atmospheric greenhouses are also pesticide free with air locks to prevent entry of insects.

Conventional Greenhouse CO_2 Enrichment Methods

Typically, CO_2 for atmospheric enrichment is either generated on site by combusting fossil fuels or purchased as bulk liquid from an oil refinery. On-site CO_2 generators are often simple propane-fueled carbon dioxide generators which are installed throughout the greenhouse. These units directly combust the fuel and release the CO_2 into the greenhouse space. With these units, the heat and moisture generated by combustion are also released into the greenhouse and must be managed. In some cases CO_2 is recovered from low-NO_x (nitric oxide plus nitrogen dioxide) boiler exhaust (Blom, 2002, Buijk, 2013).

CO_2 Enrichment Methods via Internal Combustion Engines

In lieu of liquid CO_2 or fossil-fuel combustion systems, exhaust from internal combustion (IC) engine-generators is increasingly being used as a source of CO_2 for fertilization In over 800 commercial greenhouse facilities in Western Europe, and now one facility in California, CO_2 gas is captured and used beneficially, converting greenhouse gas emissions into plant material. The IC engine exhaust is carefully purified and the combined heat and power (CHP) systems also provide an economical source of heat to the greenhouse.

To date, these IC engines have been fueled by natural gas, but CO_2 recovery from biogas engines has the potential to produce CO_2 at an even greater rate. Engine exhaust gas from a typical low emissions lean-burn engine fueled by natural gas contains about 6% CO_2, and in this manner the CO_2 is captured and productively reused. With CO_2 comprising 58 to 62% of anaerobic digester gas by weight, integrating this atmospheric CO_2 enhancement with a CHP system fueled by digester gas can achieve multiple benefits. Thus, combining the municipal wastewater treatment and greenhouse process, including anaerobic digestion for biosolids stabilization, might be a natural fit for progressive communities (Figure 1).

Figure 1. Conceptual wastewater treatment plant of the future with beneficial greenhouse CO_2 recovery

EXAMPLE GREENHOUSE FACILITIES

The two following example facilities are described to illustrate existing facilities that utilize IC engines for CO_2 enrichment or biogas-fueled IC engines in partnership with greenhouse facilities.

The 125-acre Houweling Tomatoes greenhouse facility in Camarillo, California recovers CO_2 and heat from an 8.7 MW natural gas fueled CHP system, the first of its kind in the United States when it was commissioned in August 2012 (Slater, 2012) (Figure 3). The NO_x from the CHP engine exhaust was already required to be reduced to 5 ppm NO_x by the Ventura County Air Quality Management District (AQMD). The exhaust gas is further purified to meet greenhouse requirements.

A condensing system on the CHP engine exhaust recovers additional energy by capturing the latent heat of the water vapor in the exhaust. The condensing system produces 9,500 gallons of water per day for use in Houweling's greenhouse irrigation. The plant also has a hot water heat storage system on site since the engines do not operate at night when electricity for illumination and the CO_2 fertilization is not required.

Figure 3. Intensive hydroponic tomato production at the Houweling greenhouse in Camarillo, California

One parallel to the wastewater industry is the integration of a municipal solid waste landfill with a greenhouse via a CHP system. For example, the four acre Rock Solid Produce greenhouse in Taylorville, Illinois (Figure 4) was constructed in 2008 and utilizes biogas from the Five Oaks Landfill to fuel a 3.2 MW CHP system. The cost of the combined CHP/greenhouse system was $10 million. This facility utilizes heat and power from the CHP system, but does not recover CO_2 from the exhaust gas. This CHP system is owned and operated by Waste Management of Illinois, which sells the heat and a portion of the electric power to Rock Solid Produce.

Figure 4. Rock Solid Produce greenhouse in Taylorville, IL with biogas-fired CHP

DIGESTER GAS CO_2

Digester gas from a municipal WWTP could be a substantial source of beneficial CO_2. Anaerobic digester gas is composed mainly of methane (CH_4) and carbon dioxide (CO_2) Molecular CH_4 gas, with a specific gravity of 0.55, is much lighter than CO_2 (specific gravity of 1.53) so that on a mass flow basis digester gas is mainly CO_2.

In theory, CO_2 for greenhouse enrichment could be recovered either from the raw digester gas or from the exhaust gas of a CHP system. Table 1 summarizes the mass of CO_2 in various-sized biogas streams. Table 1 also tabulates the CO_2 in the exhaust gas of a CHP fueled by the same biogas stream. The CO_2 mass in the CHP exhaust gas is more than twice the CO_2 mass from the biogas due to the additional CO_2formed by methane combustion.

Currently, common greenhouse CO_2 application rates range from 45 to 90 kg/ha/hr or higher (40 to 80 lb/acre/hr) (Blom, 2002, Caterpillar, 2009). Flower growing greenhouses use less CO_2 than vegetable greenhouses (Clary, 1998). Considering the smallest example wastewater plant in Table 1, the CO_2 in 5,000 m^3/day of uncombusted digester gas could provide the CO_2 needs of a 3 hectare (7 acre) greenhouse at an application rate of 50 kg/ha/hr. The exhaust from a 500 kW engine fueled by this same biogas flow could support a much larger 7.5 hectare (19 acre) greenhouse at the same CO_2 application rate.

Table 1. Biogas flow summary for typical WWTPs

Digester gas flow, total, cubic meters per day	Digester gas mass flow, total, kilograms per day	Digester gas CH_4 mass flow, kilograms per day	Digester gas CO_2 mass flow kilograms per day	CO_2 made from complete CH_4 combustion, kilograms per day	Combustion exhaust CO_2 total, kilograms per day
5,000	5,600	2,000	3,400	5,500	9,000
10,000	11,200	4,000	6,900	11,000	18,000
15,000	16,800	6,000	10,400	16,600	26,900

POTENTIAL BIOGAS UTILIZATION SCENARIOS

There are at least three ways a wastewater treatment plant could to supply CO_2 enrichment to a future greenhouse partner:

Alternative 1: Combusting the digester gas on site in a CHP engine and piping the purified engine exhaust gases to a nearby greenhouse (Figure 5).

The electricity and heat from the CHP system are used within the treatment plant, with any excess exported to the greenhouse facility. The practical limitation on transferring CO_2 from engines to a greenhouse is the size and expense of the conveyance duct for the hot exhaust.

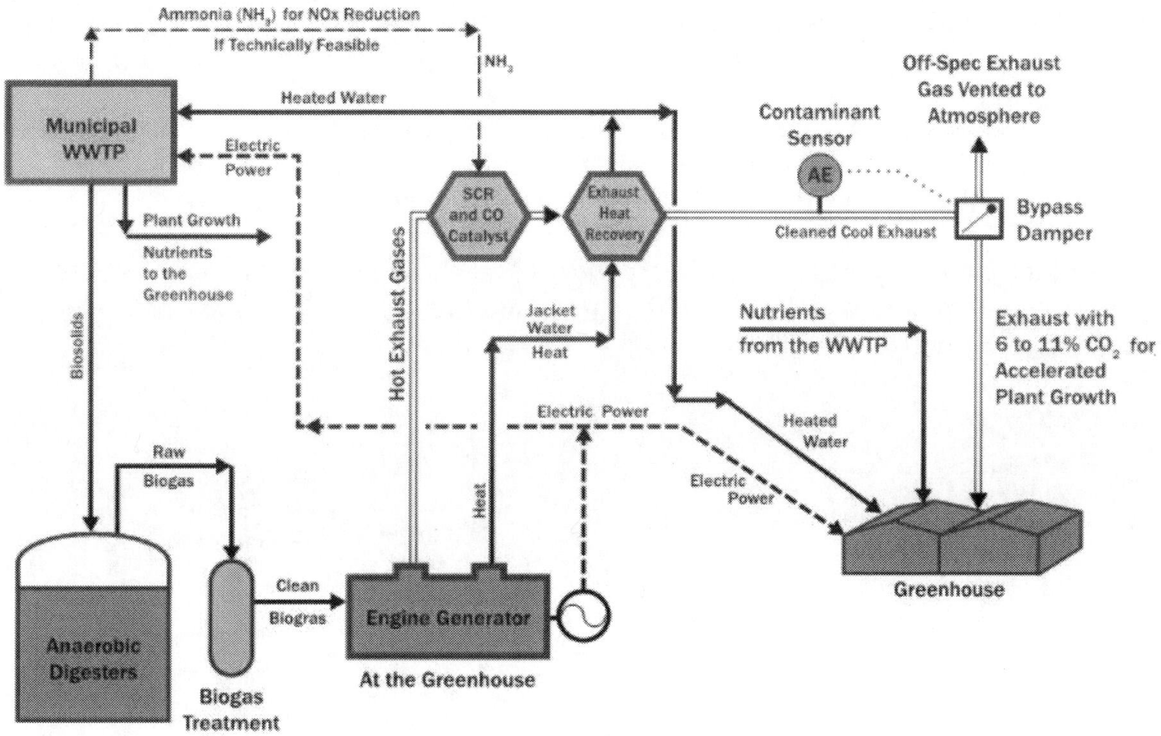

Figure 5. Combusting the digester gas in a CHP engine at a greenhouse located close to the WWTP

Alternative 2. Separating the anaerobic digester gas CH_4 and CO_2 components and conveying compressed and purified CO_2 to the greenhouse, with the CH_4 used to generate power at the WWTP or sold and direct injected into an existing compressed natural gas utility pipeline (Figure 6).

The separation technology used for this alternative has been proven at several WWTP facilities, as discussed in later sections of this paper. The CO_2 stream piping would be somewhat less expensive to specialized ductwork required to convey the hot exhaust gas stream.

The CO_2 for this alternative is generated by the biogas separation process without combustion in an IC engine. As a result, this alternative eliminates the capital cost and maintenance associated with the IC engine generator.

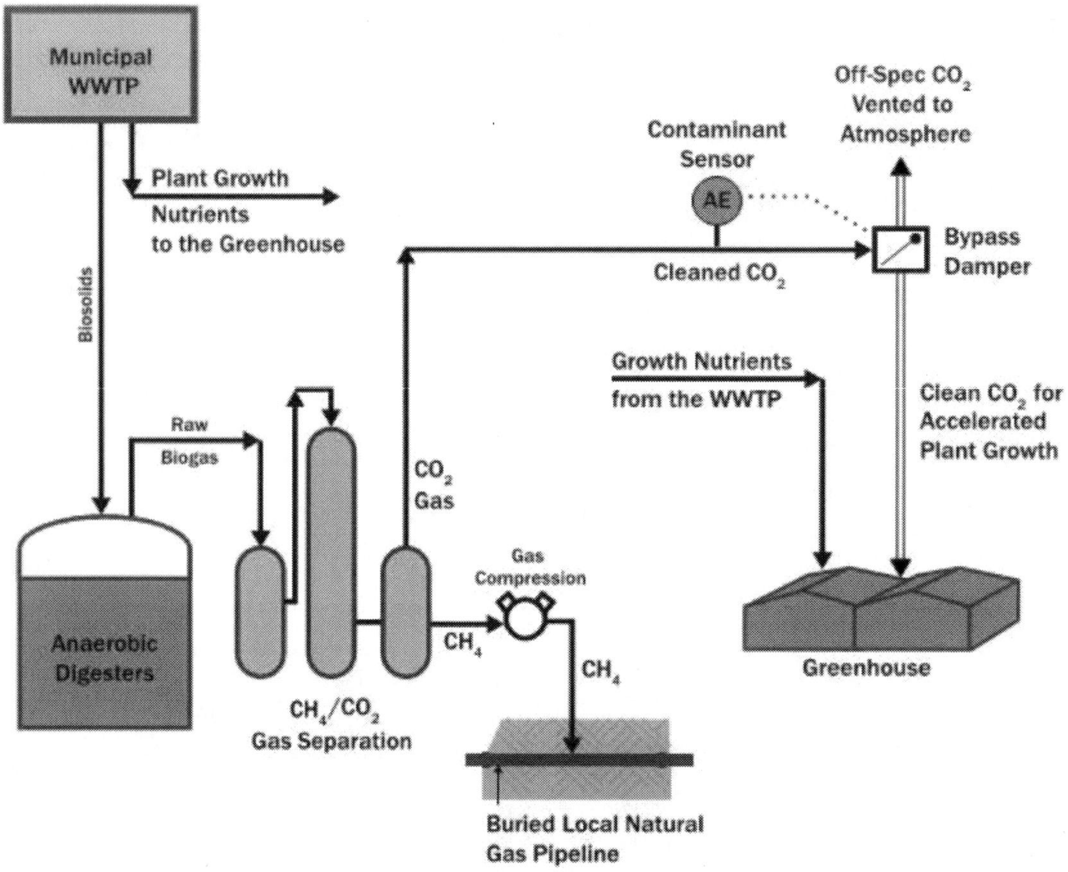

Figure 6. Greenhouse with CO_2 enrichment from an adjacent WWTP

Alternative 3. Separating the CH_4 and CO_2 and piping the CH_4 to the greenhouse via the natural gas pipeline distribution system where the CH_4 fuels the engine in a CHP system and the CO_2 for greenhouse enrichment is recovered from the engine exhaust (Figure 7).

The biomethane derived from the biogas separation process would be sold or transferred to the greenhouse via paper ownership transaction. The greenhouse would burn pipeline quality natural gas, so the exhaust gas could be purified with equipment that has been proven to work with pipeline quality natural gas. The separated CO_2 extracted from the digester gas is not recovered and used in this alternative.

This arrangement would be suitable for greenhouse locations located at some distance from the WWTP. However, the electricity and heat from the CHP system would be generated at the remote greenhouse, so heat would be provided by effluent heat recovery and heat pumps possibly augmented with solar energy.

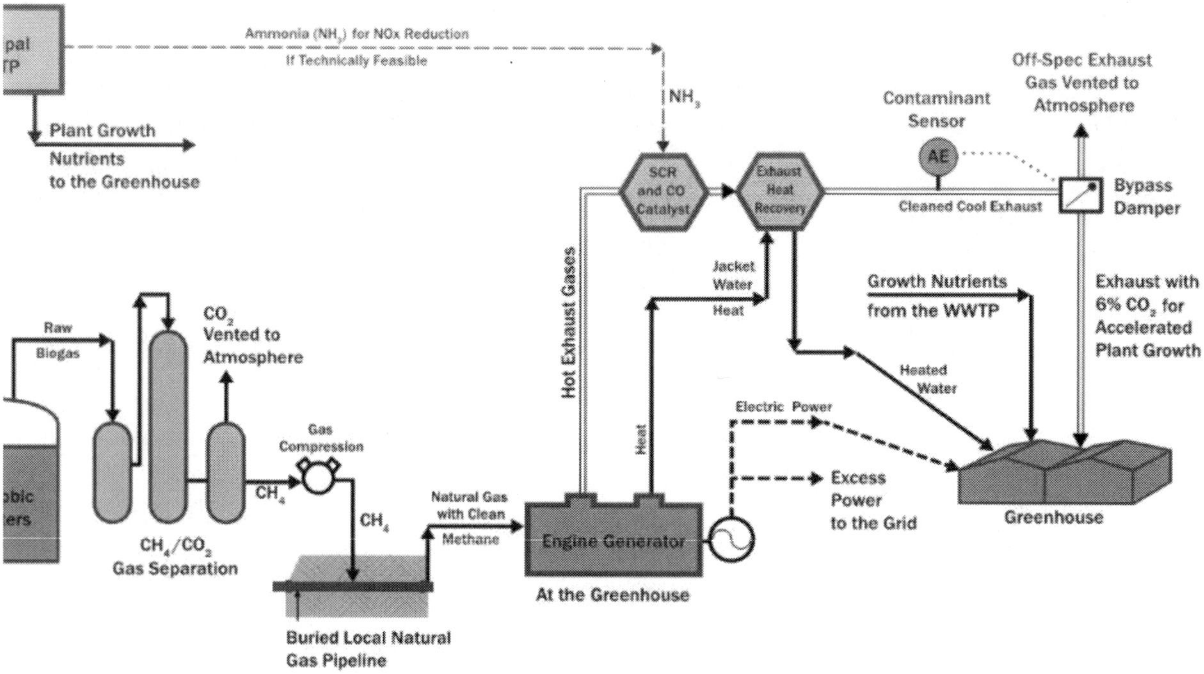

Figure 7. Greenhouse with CO_2 enrichment using CH_4 from a remote WWTP

The following sections summarize the technology that could be used for CO_2 fertilization from either engine exhaust or separated biogas systems.

ENGINE EXHAUST PURIFICATION TECHNOLOGY

Engine exhaust purification technologies have been developed to remove gases that could damage plants, such as ethylene (C_2H_4), sulfur dioxide, carbon monoxide and NOx from the CO_2 fertilization flow. Ethylene is especially problematic because it acts as a hormone that stimulates plants to ripen, mature, age, and rot. Because of its detrimental effect on fruit ripening, the ethylene gas concentration within the greenhouse atmosphere is a primary concern, even in small amounts down to the parts per billion range.

One well-known large stationary engine manufacturer's greenhouse gas enrichment applications exhaust gas quality specs are for:

- C_2H_4 not to exceed 0.46 ppm
- CO not to excess 10 ppm
- NO_X below 10 ppm

The first North American greenhouse to utilize engine exhaust for CO_2 enrichment was the Rosa Flora Limited in Dunnville, Ontario, Canada (Clary, 1998) in 1998 and featured a 2-stage Selective Catalytic Reduction (SCR) system. More recently, the Huewellings Tomato greenhouse engine-exhaust CO_2 fertilization system was installed with an exhaust purification system consisting of three stages (Slater, 2012) (Figures 8 and 9):

- **Oxidation Catalyst.** The catalytic oxidation of VOCs, CO and ethylene with excess oxygen in the exhaust produces CO_2 and water.
- **SCR Catalyst.** Following urea injection and pyrolyzing to ammonia, the SCR catalyst converts NO_X and ammonia to nitrogen gas and H_2O under carefully controlled temperature conditions.
- **Final Oxidation Catalyst.** The final oxidation catalyst stage removes residual ethylene and ammonia slip.

The quality of the treated engine exhaust is continuously monitored for personnel safety and phytotoxicity. If contaminant concentrations exceed alarm values for any reason, the exhaust is vented to atmosphere until the quality is restored. The treated exhaust monitoring system also modulates the urea feed rate to maintain exhaust quality.

Wastewater plants in southern California are on notice to lower their digester-gas fueled NO_X emissions to 11 ppm by January, 2016. NO_X has historically has been controlled via addition of SCR systems to their CHP engine exhausts. The exhaust purification technology noted here is a natural extension of the exhaust gas air pollution control system.

Figure 8. Exhaust purification system installation

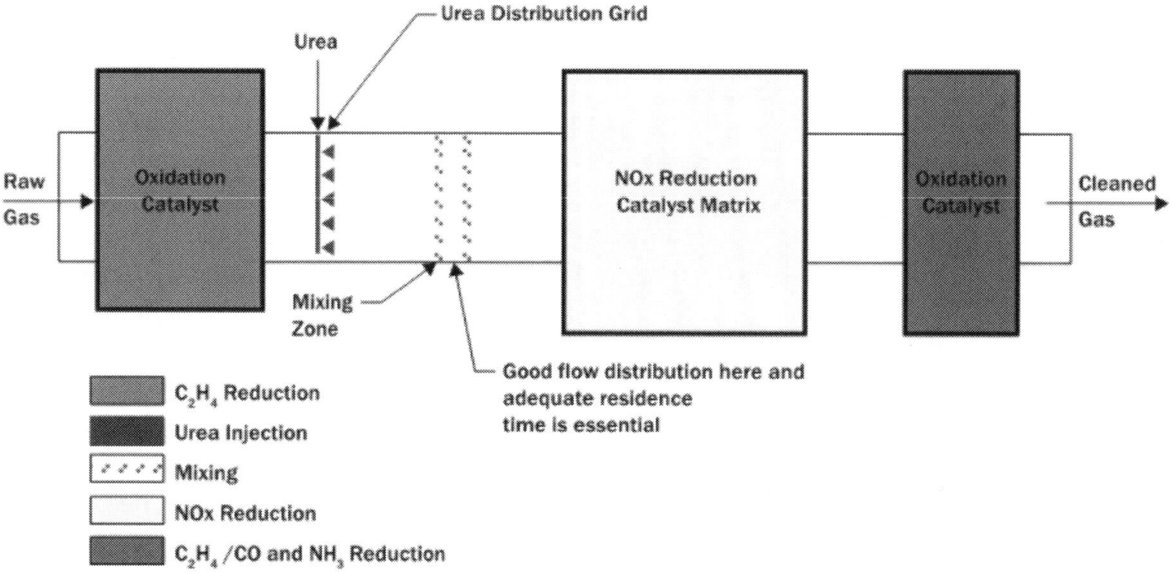

Figure 9. Engine Exhaust Purification Process Flow Diagram

Trace fuel contaminants and catalytic exhaust treatment systems

Historically, natural gas fuels large internal combustion (IC) engines used in greenhouse atmosphere CO_2 enrichment applications. Natural gas is primarily methane but it also contains up to 11 percent ethane (C_2H_6), propane (C_3H_8), and butane (C_4H_{10}) in addition to smaller amounts of C4+ hydrocarbon. While natural gas is a well known and widely used clean fuel, few detailed gas quality standards are in place in North America for limiting natural gas trace contaminants.

To our knowledge, the exhaust purification technology used for greenhouse applications has only been applied on a trial basis with biogas-fired engines. During the trials performance of the biogas exhaust purification systems was found to be highly dependent on removal of biogas contaminants prior to combustion to minimize fouling of the oxidation and SCR catalysts. Of particular concern to wastewater treatment plants, sulfur and inorganic silica must be removed from the biogas stream. Table 2 summarizes one manufacturer's guidelines for these contaminants.

Considerable information is available on wastewater treatment plant anaerobic digester gas, and this biogas has been combusted in large stationary internal combustion engines for over 70 years. Much less is known about some trace contaminants unless these contaminants are recorded in standards laboratory analysis tests such as EPA TO-15, ASTM D-1945, or ASTM D-5504. Little ethylene data is currently found in anaerobic digester gas analyses.

Ironically because of the abundance of CO_2 in the digester gas, often 33 to 43 percent CO_2 by volume, and due to the absence of ethane and higher paraffinic hydrocarbons, digester gas produces fewer NOx emissions than does natural gas. The only hydrocarbon in digester gas in quantities greater than the parts per billion range, is methane (CH_4). Anaerobic digester gas does need to be purified to remove H2S, water, and siloxanes, but these fuel contaminants are commonly removed very successfully at many treatment plants. Digester gas must be consistently and carefully conditioned or purified to be clean enough for use by the selective catalytic reduction system (SCR) and this purified digester gas should be a good candidate to fuel a large stationary internal combustion engine used for atmospheric greenhouse CO_2 enrichment.

Moreover, since urea or ammonia is continuously required by the SCR system for NOx reduction, the amount of urea consumed by an SCR system fueled by digester gas could easily be less than consumed when burning natural gas, which inherently burns with a hotter flame temperature and thus produces more NOx emissions in the combustion zone.

Table 2. Manufacturer's recommended biogas contaminant concentrations example (HUG)

Compound	Recommended Gaseous Fuel Concentration, mg/kg elemental
Sulfur	20
Inorganic Silica	15

BIOGAS SEPARATION TECHNOLOGY

Digester Gas CO_2 Separation: Operational Examples

At the wastewater treatment plants in Dayton, Ohio, San Antonio, Texas and Escondido California, and other locations, methane gas CH_4 is separated from the digester gas CO_2. The King County South treatment plant near Seattle, Washington operated a similar system for over 26 years. This treated methane gas is sold to the local natural gas utility via direct injection to the underground natural gas distribution system. The separated $CO_2$is not further purified and is released to the atmosphere. This gas is separated as summarized in Table 3.

Table 3. Digester Gas Separation Facilities

Treatment Plant	Treatment plant flow, average, mgd	Digester gas flow, average, cubic feet per day	Gas Separation process used	Gas separation system operational period
King County South plant, Seattle, Washington	75	1,200,000	Water Scrubber	1985 to 2011
SAWS, San Antonio, Texas	120	900,000	Molecular Sieve	2010
Dayton, Ohio	72	1,000,000 (capacity)	Molecular Sieve	2010
Escondido, California, Hale Av. Plant	16	360,000	Pressure Swing Adsorption	2011
Newark, Ohio	7.5	100,000	Molecular Sieve	2011
Janesville, Wisconsin	14	102,000	Membrane	2012

Gas Separation Component Stream Quality and Methods for Improved Quality

Ideally a digester gas separation process would split the gas flow into one stream of 100% pure methane and another gas stream of 100% pure carbon dioxide. This is not practical and the three common gas separation processes have different separation efficiencies. Table 4 compares the characteristics of the three prevalent separation technologies.

The waste gas (or tail gas) from the molecular sieve and membrane processes have a significant concentration of CH_4, so the CO_2 stream may require thermal oxidation, or other means of eliminating the trace methane prior to greenhouse use. The CH_4 concentration in the waste gas from the water scrubber separation process is much lower, but it is wasted from an air scrubber, so it is mixed with a significant quantity of nitrogen.

While recovering this $CO_2$is not practical today, this CO_2 byproduct could be simply too valuable to discard in the future.

Table 4. Digester Gas Processes: Comparison of Separation Efficiencies

	Example Methane Stream Quality			Example Waste Gas Stream Quality		
	CH_4 percent	CO_2 percent	Other	CO_2 percent	CH_4 percent	Other
Water scrubber	97 to 98	2 to 3	If N_2, O_2 present	25 to 35	<1	Balance Air
Molecular sieve	98	2	N/A	75 to 90	10 to 25	N_2
Membrane	97 to 98	2 to 3	N/A	80 to 85	10 to 15	N/A

WASTEWATER FACILITIES AND GREENHOUSES: ADDITIONAL SYNERGIES

Many WWTPs would like to install biogas CHP units, but have difficulty financially justifying these installations due to low local electrical rates or small plant size. Export of exhaust CO_2 to a greenhouse facility provides a potential revenue source for the WWTP, improving the economic feasibility of the CHP installation. The exhaust CO_2 has significant value since it would replace either liquid CO_2, which can cost as much as $100 per ton (Caterpillar, 2009), or propane/natural gas.

High CHP system efficiencies are achieved when virtually all of the CHP system heat and power are beneficially utilized. WWTP CHP systems often generate surplus heat when digester and seasonal plant building heat demands are less than the engine heat output. It is likely that a greenhouse facility could make use of this surplus heat, especially during cool nighttime conditions, increasing the effective CHP system efficiency. This synergy is similar to the benefit currently realized by the landfill-gas fueled CHP-greenhouse systems in Taylorville, Illinois and Model City, New York. These biogas CHP systems have been economically viable even without CO_2 enrichment systems and can serve as a model for the wastewater industry.

In addition to CO_2, electric power and heat derived from biogas-fueled engines, commercial greenhouses and WWTPs have several other potential synergies. Byproducts from the wastewater treatment process that could be utilized by the greenhouse include:

- Nutrient capture, either via residual nutrients in the effluent or recovered nutrient products such as struvite pellets.
- Biosolids as soil amendment (subject to strict biosolids regulations regarding food crop or other applications)
  Irrigation water availability (subject to water re-use standards)
- Ammonia from the wastewater to replace the urea in the SCR system if practical

Similarly, byproducts from the greenhouse could benefit the WWTP. For example, some organic greenhouse wastes such as off-spec fruits may be suitable as supplemental digester feedstocks to increase biogas production.

Many WWTPs are sited with significant setbacks to other properties in order to provide a land buffer around the WWTP for noise and odor mitigation. A greenhouse complex could make added beneficial use of this land by providing economic development and locally grown produce or flowers for the community. Overall, the public relations value of such projects could help amplify the overall theme of viewing a WWTP as a sustainable resource.

CURRENT BARRIER TO THE TECHNOLOGY IN 2013

Despite the numerous synergies enumerated above, several barriers to digester gas-fueled CHP - greenhouse partnerships.

- **Biogas treatment and exhaust purification systems.** Experience with SCR exhaust treatment systems has shown that meticulous biogas treatment is critical to reliable, cost effective exhaust purification operations. Although effective biogas treatment systems are available that would likely meet these biogas quality required for exhaust purification, no full-scale digester-gas fueled engine exhaust purification systems have been implemented to prove the effectiveness.
- **Gas separation.** Efficient CH_4 – CO_2 separation is required to facilitate CO_2 or CH_4 transmission to remote greenhouse locations. Existing separation technologies provide high quality biomethane product streams, but the waste gas CO_2 streams contain either significant quantities of methane that require treatment or excess air volume.
- **Safety.** The safety risks from conventional contaminants such as CO have been found to be manageable with proper monitoring at natural-gas fueled CHP-greenhouse installations. Greenhouse employees work in the growing areas with no special respiratory protection equipment. Further research would be required to ensure that there would be no trace compounds in the biogas that would create an unsafe or phytotoxic atmosphere in the greenhouse.
- **Economic risk** Products grown in sophisticated greenhouses have extremely high market value. For example, failure of the exhaust purification system (and the downstream exhaust bypass for diverting high contaminant concentrations) could increase ethylene concentration in the greenhouse, causing spoilage and significant product loss even in small amounts.
- **Public perception.** As biosolids professionals have long known, the public can be very averse to event the remote potential for pathogen contamination of food products by wastewater pathogens. Careful communication of the pathogen control methods and overall project benefits and will be required to address these concerns.

CONCLUSIONS

By combining the plant growth enhancement concept, future treatment plants could use just one more valuable component derived from wastewater and use it in a beneficial, economically attractive and environmentally positive manner.

REFERENCES

Blom, T.J; Straver, W.A.; Ingratta, F.J.; Khosla, S; Brown,W. (2002), *Carbon Dioxide In Greenhouses*; 12/02, Ontario Ministry of Agriculture, Food, and Rural Affairs

Buijk, J, owner, European Power Systems, personal conversation, February 22, 2013

Caterpillar (2009), *Cogeneration in Greenhouses*; Electric Power & Cogeneration Application Sheet,

Clary, W, Stringary (1998), J-P, *SCR Catalysts for Greenhouse Plants Provide a Microcosm of Earth*; Diesel Progress, March 1998

Hug Engineering, *Concentrations of Poisonous Substances for Hug Catalysts*

Kren, L (2004) *Power Plant" Takes on New Meaning*; Machine Design September 16, 2004 (the mentions a plan to do a 100-MW 100 acre greenhouse by Greenhouse Corporation of America, Newton, Massachusetts

Slater, DJ (2012) *You Say Tomato, I say Energy Efficiency, GE Jenbacher Engines Power CHP Project That Uses CO_2 for Plant Fertilization*; Diesel and Gas Turbine Worldwide, October 2012

Thayer, R. (no date) *Carbon Dioxide Enrichment Methods;* Hydrofarm

USDA Economic Research Service (2004 est.), *Greenhouse Tomatoes Change the Dynamics of the North American Fresh Tomato Industry* ERR-2

Water Environment Federation (WEF)
601 Wythe Street
Alexandria, Virginia 22314
USA

ISBN 978-1-5108-7148-9
9 781510 871489